Optimization for Control, Observation and Safety

Optimization for Control, Observation and Safety

Special Issue Editors

Guillermo Valencia-Palomo
Francisco-Ronay López-Estrada
Damiano Rotondo

MDPI • Basel • Beijing • Wuhan • Barcelona • Belgrade • Manchester • Tokyo • Cluj • Tianjin

Special Issue Editors

Guillermo Valencia-Palomo
Tecnológico Nacional de
México/IT Hermosillo
Mexico

Francisco-Ronay López-Estrada
Tecnológico Nacional de
México/IT Tuxtla Gutiérrez
Mexico

Damiano Rotondo
Universitetet i Stavanger
Norway

Editorial Office
MDPI
St. Alban-Anlage 66
4052 Basel, Switzerland

This is a reprint of articles from the Special Issue published online in the open access journal *Processes* (ISSN 2227-9717) (available at: https://www.mdpi.com/journal/processes/special_issues/optimization_observation_safety).

For citation purposes, cite each article independently as indicated on the article page online and as indicated below:

LastName, A.A.; LastName, B.B.; LastName, C.C. Article Title. *Journal Name* **Year**, *Article Number*, Page Range.

ISBN 978-3-03928-440-5 (Pbk)
ISBN 978-3-03928-441-2 (PDF)

Cover image courtesy of Francisco-Ronay López-Estrada.
Hydraulic network at the Hydroinformatic Lab of the TURIX-Dynamics Diagnosis and Control Group, Tecnológico Nacional de México/IT Tuxtla Gutiérrez, Mexico.

© 2020 by the authors. Articles in this book are Open Access and distributed under the Creative Commons Attribution (CC BY) license, which allows users to download, copy and build upon published articles, as long as the author and publisher are properly credited, which ensures maximum dissemination and a wider impact of our publications.

The book as a whole is distributed by MDPI under the terms and conditions of the Creative Commons license CC BY-NC-ND.

Contents

About the Special Issue Editors ix

Guillermo Valencia-Palomo, Francisco-Ronay López-Estrada and Damiano Rotondo
Recent Advances on Optimization for Control, Observation, and Safety
Reprinted from: *Processes* 2020, 8, 201, doi:10.3390/pr8020201 1

Francisco-Ronay López-Estrada, Damiano Rotondo and Guillermo Valencia-Palomo
A Review of *Convex* Approaches for Control, Observation and Safety of Linear Parameter Varying and Takagi-Sugeno Systems
Reprinted from: *Processes* 2019, 7, 814, doi:10.3390/pr7110814 7

Shiquan Zhao, Anca Maxim, Sheng Liu, Robin De Keyser and Clara Ionescu
Effect of Control Horizon in Model Predictive Control for Steam/Water Loop in Large-Scale Ships
Reprinted from: *Processes* 2018, 6, 265, doi:10.3390/pr6120265 47

Aydın Mühürcü
FFANN Optimization by ABC for Controlling a 2nd Order SISO System's Output with a Desired Settling Time
Reprinted from: *Processes* 2019, 7, 4, doi:10.3390/pr7010004 61

Irandi Gutierrez-Carmona, Jaime A. Moreno and H.F. Abundis-Fong
On the Boundary Conditions in a Non-Linear Dissipative Observer for Tubular Reactors
Reprinted from: *Processes* 2019, 7, 8, doi:10.3390/pr7010008 83

Jeongeun Son and Yuncheng Du
Model-Based Stochastic Fault Detection and Diagnosis of Lithium-Ion Batteries
Reprinted from: *Processes* 2019, 7, 38, doi:10.3390/pr7010038 99

Shu-Kai S. Fan, Chih-Hung Jen and Jai-Xhing Lee
Profile Monitoring for Autocorrelated Reflow Processes with Small Samples
Reprinted from: *Processes* 2019, 7, 104, doi:10.3390/pr7020104 119

Haiyong Dong, Qingfan Gu, Guoqing Wang, Zhengjun Zhai, Yanhong Lu and Miao Wang
Availability Assessment of IMA System Based on Model-Based Safety Analysis Using AltaRica 3.0
Reprinted from: *Processes* 2019, 7, 117, doi:10.3390/pr7020117 141

Li Zeng, Wei Long and Yanyan Li
A Novel Method for Gas Turbine Condition Monitoring Based on KPCA and Analysis of Statistics T^2 and SPE
Reprinted from: *Processes* 2019, 7, 124, doi:10.3390/pr7030124 155

Patrick Piprek, Sébastien Gros and Florian Holzapfel
Rare Event Chance-Constrained Optimal Control Using Polynomial Chaos and Subset Simulation
Reprinted from: *Processes* 2019, 7, 185, doi:10.3390/pr7040185 167

Rashida Adeeb Khanum, Muhammad Asif Jan, Nasser Tairan, Wali Khan Mashwani, Muhammad Sulaiman, Hidayat Ullah Khan and Habib Shah
Global Evolution Commended by Localized Search for Unconstrained Single Objective Optimization
Reprinted from: *Processes* **2019**, *7*, 362, doi:10.3390/pr7060362 191

Andrés Morán-Durán, Albino Martínez-Sibaja, José Pastor Rodríguez-Jarquin, Rubén Posada-Gómez and Oscar Sandoval González
PEM Fuel Cell Voltage Neural Control Based on Hydrogen Pressure Regulation
Reprinted from: *Processes* **2019**, *7*, 434, doi:10.3390/pr7070434 205

Huu-Quyen Nguyen, Anh-Dung Tran and Trong-Thang Nguyen
The Bilinear Model Predictive Method-Based Motion Control System of an Underactuated Ship with an Uncertain Model in the Disturbance
Reprinted from: *Processes* **2019**, *7*, 445, doi:10.3390/pr7070445 221

Ziyad T. Allawi, Ibraheem Kasim Ibraheem and Amjad J. Humaidi
Fine-Tuning Meta-Heuristic Algorithm for Global Optimization
Reprinted from: *Processes* **2019**, *7*, 657, doi:10.3390/pr7100657 235

Seung-Jun Shin, Young-Min Kim and Prita Meilanitasari
A Holonic-Based Self-Learning Mechanism for Energy-Predictive Planning in Machining Processes
Reprinted from: *Processes* **2019**, *7*, 739, doi:10.3390/pr7100739 249

Clara M. Ionescu, Constantin F. Caruntu, Ricardo Cajo, Mihaela Ghita, Guillaume Crevecoeur and Cosmin Copot
Multi-Objective Predictive Control Optimization with Varying Term Objectives: A Wind Farm Case Study
Reprinted from: *Processes* **2019**, *7*, 778, doi:10.3390/pr7110778 277

Li Zeng, Shaojiang Dong and Wei Long
The Rotating Components Performance Diagnosis of Gas Turbine Based on the Hybrid Filter
Reprinted from: *Processes* **2019**, *7*, 819, doi:10.3390/pr7110819 293

J. Hernández, D. F. Galaviz, L. Torres, A. Palacio-Pérez, A. Rodríguez-Valdés and J. E. V. Guzmán
Evolution of High-Viscosity Gas–Liquid Flows as Viewed Through a Detrended Fluctuation Characterization
Reprinted from: *Processes* **2019**, *7*, 822, doi:10.3390/pr7110822 307

Hani Albalawi and Sherif A. Zaid
Performance Improvement of a Grid-Tied Neutral-Point-Clamped 3-φ Transformerless Inverter Using Model Predictive Control
Reprinted from: *Processes* **2019**, *7*, 856, doi:10.3390/pr7110856 323

Feng Zheng, Jiahao Lin, Jie Huang and Yanzhen Lin
Generalized Proportional Model of Relay Protection Based on Adaptive Homotopy Algorithm Transient Stability
Reprinted from: *Processes* **2019**, *7*, 899, doi:10.3390/pr7120899 341

A. Navarro, J. A. Delgado-Aguiñaga, J. D. Sánchez-Torres, O. Begovich and G. Besançon
Evolutionary Observer Ensemble for Leak Diagnosis in Water Pipelines
Reprinted from: *Processes* **2019**, *7*, 913, doi:10.3390/pr7120913 357

Shuang-xi Liu and Ming Lü
Fault Diagnosis of the Blocking Diesel Particulate Filter Based on Spectral Analysis
Reprinted from: *Processes* 2019, 7, 943, doi:10.3390/pr7120943 375

Husam Kaid, Abdulrahman Al-Ahmari, Zhiwu Li and Reggie Davidrajuh
Single Controller-Based Colored Petri Nets for Deadlock Control in Automated
Manufacturing Systems
Reprinted from: *Processes* 2020, 8, 21, doi:10.3390/pr8010021 387

Shijie Cui, Peng Zeng, Chunhe Song and Zhongfeng Wang
Robust Fault Protection Technique for Low-Voltage Active Distribution Networks Containing
High Penetration of Converter-Interfaced Renewable Energy Resources
Reprinted from: *Processes* 2020, 8, 34, doi:10.3390/pr8010034 407

Fatemeh Karimi Pour, Vicenç Puig and Gabriela Cembrano
Economic Reliability-Aware MPC-LPV for Operational Management of Flow-Based Water
Networks Including Chance-Constraints Programming
Reprinted from: *Processes* 2020, 8, 60, doi:10.3390/pr8010060 423

Citlaly Martínez-García, Vicenç Puig, Carlos-M. Astorga-Zaragoza,
Guadalupe Madrigal-Espinosa and Juan Reyes-Reyes
Estimation of Actuator and System Faults Via an Unknown Input Interval Observer for
Takagi–Sugeno Systems
Reprinted from: *Processes* 2020, 8, 61, doi:10.3390/pr8010061 449

Huu Khoa Tran, Hoang Hai Son, Phan Van Duc, Tran Thanh Trang and Hoang-Nam Nguyen
Improved Genetic Algorithm Tuning Controller Design for Autonomous Hovercraft
Reprinted from: *Processes* 2020, 8, 66, doi:10.3390/pr8010066 465

Zhong Lu, Lu Zhuang, Li Dong and Xihui Liang
Model-Based Safety Analysis for the Fly-by-Wire System by Using Monte Carlo Simulation
Reprinted from: *Processes* 2020, 8, 90, doi:10.3390/pr8010090 475

About the Special Issue Editors

Guillermo Valencia-Palomo was born in Merida, Yucatan, Mexico, in 1980. He received an Engineering degree in Electronics from the Instituto Tecnológico de Mérida, Mexico, in 2003; an M.Sc. in Automatic Control from the National Center of Research and Technological Development (CENIDET), Mexico, in 2006; and a Ph.D. degree in Automatic Control and Systems Engineering from The University of Sheffield, UK, in 2010. Since 2010, Dr. Guillermo Valencia-Palomo is a full-time professor at Tecnológico Nacional de México/Instituto Tecnológico de Hermosillo (Mexico). He is author/co-author of more than 80 research papers published in ISI-Journals and international conferences. As a product of his research, he has 1 patent in commercial exploitation. He has led a number of funded research projects and the funding for these projects is comprised of a mixture of sole investigator funding, collaborative grants, and funding from industry. His research interests include predictive control, descriptor systems, linear parameter varying systems, fault detection, fault-tolerant control systems, and its applications to different physical systems.

Francisco-Ronay López-Estrada received his Ph.D. in automatic control from the University of Lorraine, France, in 2014. He has been with Tecnológico Nacional de México/Instituto Tecnológico de Tuxtla Gutiérrez, Mexico, as a lecturer since 2008. He received his MSc degree in electronic engineering in 2008 from the National Center of Research and Technological Development (CENIDET), Mexico. He has lead several funded research projects. His research interests are in descriptor systems, TS systems, fault detection, fault-tolerant control, and its applications to unmanned vehicles, and pipeline leak detection systems.

Damiano Rotondo received his BS degree (with honors) from the Second University of Naples, Italy, M.Sc. degree (with honors) from the University of Pisa, Italy, and Ph.D. degree (with honors) from the Universitat Politècnica de Catalunya, Spain in 2008, 2011, and 2016, respectively. From May 2016 until April 2017, he was an ERCIM postdoctoral researcher at AMOS/NTNU (Norway). From May 2017 until January 2018, he was a postdoctoral researcher at the Research Centre for Supervision, Safety and Automatic Control (CS2AC) of the Universitat Politècnica de Catalunya (Spain). From February 2018 until January 2020, he was a Juan de la Cierva fellow at the Institut de Robòtica i Informàtica Industrial (IRI) at the Spanish Council for Scientific Research (CSIC). Since February 2020, he is an Associate Professor at the Department of Electrical Engineering and Computer Science (IDE) of the Universitetet i Stavanger (UiS), in Norway. His main research interests include gain-scheduled control systems, fault detection and isolation (FDI) and fault-tolerant control (FTC) of dynamic systems. He has published several papers in international conference proceedings and scientific journals.

Editorial

Recent Advances on Optimization for Control, Observation, and Safety

Guillermo Valencia-Palomo [1,*], **Francisco-Ronay López-Estrada** [2,*] **and Damiano Rotondo** [3,4,*]

1. Tecnológico Nacional de México/Instituto Tecnológico de Hermosillo, Av. Tecnológico y Periférico Poniente, S/N, Hermosillo 83170, Sonora, Mexico
2. Tecnológico Nacional de México/Instituto Tecnológico de Tuxtla Gutiérrez. TURIX-Dynamics Diagnosis and Control Group, Carretera Panamericana km 1080, Tuxtla Gutierrez C.P. 29050, Mexico
3. Institut de Robòtica i Informàtica Industrial, CSIC-UPC, Llorens i Artigas 4-6, 08028 Barcelona, Spain
4. Research Center for Supervision, Safety and Automatic Control, Universitat Politècnica de Catalunya (UPC), Rambla Sant Nebridi, 22, 08022 Terrassa, Spain
* Correspondence: gvalencia@ith.mx (G.V.-P.); frlopez@ittg.edu.mx (F.-R.L.-E.); damiano.rotondo@upc.edu (D.R.)

Received: 28 January 2020; Accepted: 31 January 2020; Published: 6 February 2020

1. Introduction

Mathematical optimization is the selection of the best element in a set with respect to a given criterion. Optimization has become one of the most commonly used tools in modern control theory to compute control laws, adjust the controller parameters (tuning), estimate unmeasured states, find suitable conditions to fulfill a given closed-loop property, carry out model fitting, among others. Optimization is also used in the design of fault detection and isolation systems, due to the complexity of automated installations and to prevent safety hazards and huge production losses that require the detection and identification of any kind of fault, as early as possible, as well as the minimization of their impacts by implementing real-time fault detection and fault-tolerant operations systems where optimization algorithms play an important role. Recently, it has been proved that many optimization problems with convex objective functions and linear matrix inequality (LMI) constraints can be solved efficiently using existing software, which increases the flexibility and applicability of the control algorithms. Therefore, real-world control systems need to comply with several conditions and constraints that have to be taken into account in the problem formulation, which represents a challenge in the application of the optimization algorithms.

This special issue aims at offering an overview of the state-of-the-art of the most advanced (online and offline) optimization techniques and their applications in control engineering.

2. Papers Presented in the Special Issue

The first paper, presented by López-Estrada et al. [1], offers an extensive review of the three main topics covered in this special issue. This literature survey presents different methodologies for analysis and control, observer synthesis, and fault-related strategies for convex systems under different representations: Takagi–Sugeno fuzzy models, linear parameter varying (LPV), and quasi-LPV systems.

Zhao et al. [2] perform an analysis on the selection of the length of the control horizon for a linear model predictive control, with application to steam/water loops in large-scale watercraft/ships, with an emphasis on the performance and computational complexity of the algorithm.

Aydın Mühürcü [3] considers a combination of a feed-forward artificial neural network (FFANN) and an artificial bee colony (ABC) optimization algorithm to ensure the settling time of a second-order system. The FFANN is the nonlinear control structure adopted for a buck converter and its parameters are optimized using the ABC algorithm.

Gutierrez-Carmona et al. [4] analyze the performance of a nonlinear dissipative observer for a tubular reactor. They show that, by simple considerations in the boundary conditions, the observer's convergence is improved regardless of the presence of perturbations. The sensor locations acquire physical meaning, and by simple numerical manipulations, the inflow perturbations can be estimated numerically.

Son and Du [5] develop a reliable thermal management system to predict and monitor precisely the thermal behavior of lithium-ion batteries. First, an iterative optimization algorithm corrects the model by incorporating the errors between the measured quantities and the model predictions. Then, an optimization-based fault detection and diagnosis algorithm provide a probabilistic description of the occurrence of possible faults, while taking into account the uncertainties.

Fan et al. [6] present a profile monitoring methodology that includes model fitting and statistical process (SP) control. In this paper, the authors consider non-linear profiles with correlated within-the-profile observations. Three profile models were studied: a traditional one (polynomial regression) but with added autoregression structure, and two known from the theory of non-linear regression, but relatively unknown for SPC practitioners.

Dong et al. [7] present a methodology to assess a specific critical avionic system: the integrated modular avionics (IMA) system. This methodology is derived from a model-based safety analysis performed using the AltaRica 3.0 modeling language. Moreover, the authors present a design optimization of the IMA system.

Zeng et al. [8] present a fault diagnosis and isolation method for gas turbines. First, the measured aerodynamic parameters are decomposed using the kernel principal component analysis. Then, they construct the Hotelling-T^2 (T^2) statistic, which is the application of the T-statistic in multivariate analysis in the principal space and squared prediction error (SPE) statistics in the residual space. Finally, they calculate the parameters' sensitivity to the T^2 and SPE statistics to locate the fault.

Piprek et al. [9] provide a sampling approach to approximate the chance constraints in the formulation of optimal control problems for stochastic dynamical systems to capture rare events. The applicability of the proposed approach is demonstrated in a battery charging-discharging problem.

Khanum et al. [10] describe an interesting algorithm approach for improving global search minimum optimizations and compare multiple existing algorithms to assess their ability to find optimal parameters for various functions.

Morán-Durán et al. [11] propose the use of a trained neural network to predict and control the voltage of a proton-exchange membrane (PEM) fuel cell. The approach uses principal component analysis (PCA) to reduce the dimensionality, aiming to eliminate non-significant variables with respect to the control objective.

Nguyen et al. [12] present the design of a bilinear model-based predictive control for the three-degrees-of-freedom model of an underactuated ship affected by uncertain disturbances. The bilinear model of the ship is obtained by linearizing each nonlinear model section and the uncertain components and random disturbances of the model are compensated with a state estimator.

Allawi et al. [13] report a novel fine-tuning meta-heuristic algorithm to solve global optimization problems. Also, the proposed algorithm has been validated by comparing it with some featured meta-heuristic optimization algorithms over different benchmark test functions.

Shin et al. [14] discuss a holonic-based mechanism for self-learning factories based on a hybrid-learning approach which is designed to obtain predictive modeling ability in both data-existent and data-absent environments via accommodating machine learning and transfer learning.

Ionescu et al. [15] study the case of an optimization method that considers short-term and long-term cost objectives. The problem of cost-effective optimization of the system's output is studied in a multi-objective predictive control formulation and applied to a windmill park case study.

Zeng et al. [16] provide a method that uses a hybrid filter for fault diagnosis in a gas turbine. The hybrid filter is based on the unscented Kalman filter and a particle filter with optimized weight.

It estimates the health parameters of the rotor components and builds a model in order to give a prediction for fault diagnosis.

Hernández et al. [17] characterize the high viscosity gas-liquid intermittent flows by detrended fluctuation analysis. Specifically, the authors investigated the long-term evolution of highly viscous two-phase pipe flows of glycerin/air blends. Then they apply a detrended fluctuation analysis of pressure measurements at various positions along the flow line to extract long-range correlations.

Albalawi and Zaid [18] introduce the application of a model-based predictive control algorithm to control and improve the performance of a grid-tied neutral-point-clamped 3-φ transformerless inverter powered by a photo-voltaic panel. The controller considers the filter elements, as well as the internal impedance of the grid.

Zheng et al. [19] establish a generalized proportional hazard model to exploit the monitoring condition information of a relay protection equipment to ensure the safe and stable operation of a power system.

Navarro et al. [20] propose a method to detect, locate, and estimate the magnitude of leaks in a pipeline using only flow rate and pressure head measurements at both ends of the pipe. The method develops a mathematical model that builds an observer ensemble using genetic algorithms.

Liu and Lü [21] focus on an approach for fault diagnosis of the blocking diesel particulate filter based on spectral analysis of the instantaneous exhaust pressure. The method is validated experimentally.

Kaid et al. [22] develop a two-step robust deadlock control approach based on Petri nets for automated manufacturing systems where the structural complexity of the Petri net supervisors is minimized.

Cui et al. [23] provide the infrastructure and mathematical tools necessary to face the detection of active distribution networks faults with a wide range of converter interfaces and, therefore, their protection.

Pour et al. [24] present an economic reliability-aware model predictive control based on a finite horizon stochastic optimization problem with joint probabilistic constraints for the management of drinking water transport networks.

Martínez-García et al. [25] propose a discrete-time interval observer for a class of discrete-time parametric uncertain systems modeled in the Takagi–Sugeno form, where the perturbation vector is considered to be unknown but bounded, to estimate state variables and actuator faults.

Tran et al. [26] provide a tuning method for a fuzzy proportional-integral-derivative controller based on a modified genetic algorithm that can speed up convergence and save operation time by neglecting the chromosome decoding step.

Lu et al. [27] study the safety performance of the fly-by-wire system of an aircraft. The safety analysis is based on stochastic simulations of a Simulink model. The Simulink model represents the nominal operation of the system, extended with failure mode. The safety requirements of the system are defined by presenting the thresholds of system performance metrics.

3. Conclusions

We believe that the papers in this special issue reveal an exciting area that can be expected to continue to grow in the very near future, namely, the use of advanced optimization strategies in engineering applications. The pursuit of work in this area requires expertise in control engineering as well as in systems design and numerical analysis. We hope that this issue helps to bring these communities into closer contact with each other, as the fruitfulness of collaboration across these areas becomes clear.

Finally, we would like to acknowledge the enthusiastic effort of all the authors, reviewers and editorial staff who have participated in this special issue.

Author Contributions: All authors contributed equally to this work. All authors have read and agreed to the published version of the manuscript.

Funding: This work was funded by Tecnológico Nacional de México under the program "Investigación Científica Básica y Aplicada". This work has been also supported by the AEI through the Maria de Maeztu Seal of Excellence to IRI (MDM-2016-0656) and the grant Juan de la Cierva-Formacion (FJCI-2016-29019).

Conflicts of Interest: The authors declare no conflict of interest.

References

1. López-Estrada, F.R.; Rotondo, D.; Valencia-Palomo, G. A Review of Convex Approaches for Control, Observation and Safety of Linear Parameter Varying and Takagi-Sugeno Systems. *Processes* 2019, 7, 814. [CrossRef]
2. Zhao, S.; Maxim, A.; Liu, S.; De Keyser, R.; Ionescu, C. Effect of Control Horizon in Model Predictive Control for Steam/Water Loop in Large-Scale Ships. *Processes* 2018, 6, 265. [CrossRef]
3. Mühürcü, A. FFANN Optimization by ABC for Controlling a 2nd Order SISO System's Output with a Desired Settling Time. *Processes* 2018, 7, 4. [CrossRef]
4. Gutierrez-Carmona, I.; Moreno, J.A.; Abundis-Fong, H. On the Boundary Conditions in a Non-Linear Dissipative Observer for Tubular Reactors. *Processes* 2018, 7, 8. [CrossRef]
5. Son, J.; Du, Y. Model-Based Stochastic Fault Detection and Diagnosis of Lithium-Ion Batteries. *Processes* 2019, 7, 38. [CrossRef]
6. Fan, S.K.S.; Jen, C.H.; Lee, J.X. Profile Monitoring for Autocorrelated Reflow Processes with Small Samples. *Processes* 2019, 7, 104. [CrossRef]
7. Dong, H.; Gu, Q.; Wang, G.; Zhai, Z.; Lu, Y.; Wang, M. Availability Assessment of IMA System Based on Model-Based Safety Analysis Using AltaRica 3.0. *Processes* 2019, 7, 117. [CrossRef]
8. Zeng, L.; Long, W.; Li, Y. A Novel Method for Gas Turbine Condition Monitoring Based on KPCA and Analysis of Statistics T2 and SPE. *Processes* 2019, 7, 124. [CrossRef]
9. Piprek, P.; Gros, S.; Holzapfel, F. Rare Event Chance-Constrained Optimal Control Using Polynomial Chaos and Subset Simulation. *Processes* 2019, 7, 185. [CrossRef]
10. Khanum, R.A.; Jan, M.A.; Tairan, N.; Mashwani, W.K.; Sulaiman, M.; Khan, H.U.; Shah, H. Global Evolution Commended by Localized Search for Unconstrained Single Objective Optimization. *Processes* 2019, 7, 362. [CrossRef]
11. Morán-Durán, A.; Martínez-Sibaja, A.; Rodríguez-Jarquin, J.P.; Posada-Gómez, R.; González, O.S. PEM Fuel Cell Voltage Neural Control Based on Hydrogen Pressure Regulation. *Processes* 2019, 7, 434. [CrossRef]
12. Nguyen, H.Q.; Tran, A.D.; Nguyen, T.T. The Bilinear Model Predictive Method-Based Motion Control System of an Underactuated Ship with an Uncertain Model in the Disturbance. *Processes* 2019, 7, 445. [CrossRef]
13. Allawi, Z.T.; Ibraheem, I.K.; Humaidi, A.J. Fine-Tuning Meta-Heuristic Algorithm for Global Optimization. *Processes* 2019, 7, 657. [CrossRef]
14. Shin, S.J.; Kim, Y.M.; Meilanitasari, P. A Holonic-Based Self-Learning Mechanism for Energy-Predictive Planning in Machining Processes. *Processes* 2019, 7, 739. [CrossRef]
15. Ionescu, C.M.; Caruntu, C.F.; Cajo, R.; Ghita, M.; Crevecoeur, G.; Copot, C. Multi-Objective Predictive Control Optimization with Varying Term Objectives: A Wind Farm Case Study. *Processes* 2019, 7, 778. [CrossRef]
16. Zeng, L.; Dong, S.; Long, W. The Rotating Components Performance Diagnosis of Gas Turbine Based on the Hybrid Filter. *Processes* 2019, 7, 819. [CrossRef]
17. Hernández, J.; Galaviz, D.F.; Torres, L.; Palacio-Pérez, A.; Rodríguez-Valdés, A.; Guzmán, J.E.V. Evolution of High-Viscosity Gas–Liquid Flows as Viewed Through a Detrended Fluctuation Characterization. *Processes* 2019, 7, 822. [CrossRef]
18. Albalawi, H.; Zaid, S.A. Performance Improvement of a Grid-Tied Neutral-Point-Clamped 3-φ Transformerless Inverter Using Model Predictive Control. *Processes* 2019, 7, 856. [CrossRef]
19. Zheng, F.; Lin, J.; Huang, J.; Lin, Y. Generalized Proportional Model of Relay Protection Based on Adaptive Homotopy Algorithm Transient Stability. *Processes* 2019, 7, 899. [CrossRef]
20. Navarro, A.; Delgado-Aguiñaga, J.A.; Sánchez-Torres, J.D.; Begovich, O.; Besançon, G. Evolutionary Observer Ensemble for Leak Diagnosis in Water Pipelines. *Processes* 2019, 7, 913. [CrossRef]
21. Liu, S.X.; Lü, M. Fault Diagnosis of the Blocking Diesel Particulate Filter Based on Spectral Analysis. *Processes* 2019, 7, 943. [CrossRef]

22. Kaid, H.; Al-Ahmari, A.; Li, Z.; Davidrajuh, R. Single Controller-Based Colored Petri Nets for Deadlock Control in Automated Manufacturing Systems. *Processes* **2019**, *8*, 21. [CrossRef]
23. Cui, S.; Zeng, P.; Song, C.; Wang, Z. Robust Fault Protection Technique for Low-Voltage Active Distribution Networks Containing High Penetration of Converter-Interfaced Renewable Energy Resources. *Processes* **2019**, *8*, 34. [CrossRef]
24. Pour, F.K.; Puig, V.; Cembrano, G. Economic Reliability-Aware MPC-LPV for Operational Management of Flow-Based Water Networks Including Chance-Constraints Programming. *Processes* **2020**, *8*, 60. [CrossRef]
25. Martínez-García, C.; Puig, V.; Astorga-Zaragoza, C.M.; Madrigal-Espinosa, G.; Reyes-Reyes, J. Estimation of Actuator and System Faults Via an Unknown Input Interval Observer for Takagi–Sugeno Systems. *Processes* **2020**, *8*, 61. [CrossRef]
26. Tran, H.K.; Son, H.H.; Duc, P.V.; Trang, T.T.; Nguyen, H.N. Improved Genetic Algorithm Tuning Controller Design for Autonomous Hovercraft. *Processes* **2020**, *8*, 66. [CrossRef]
27. Lu, Z.; Zhuang, L.; Dong, L.; Liang, X. Model-Based Safety Analysis for the Fly-by-Wire System by Using Monte Carlo Simulation. *Processes* **2020**, *8*, 90. [CrossRef]

© 2020 by the authors. Licensee MDPI, Basel, Switzerland. This article is an open access article distributed under the terms and conditions of the Creative Commons Attribution (CC BY) license (http://creativecommons.org/licenses/by/4.0/).

Review

A Review of *Convex* Approaches for Control, Observation and Safety of Linear Parameter Varying and Takagi-Sugeno Systems

Francisco-Ronay López-Estrada [1,*,†], **Damiano Rotondo** [2,3,*,†] **and Guillermo Valencia-Palomo** [4,*,†]

1. TURIX-Dynamics Diagnosis and Control Group, Tecnológico Nacional de México/Instituto Tecnológico de Tuxtla Gutiérrez, Carretera Panamericana km 1080, C.P. 29050 Tuxtla Gutierrez, Mexico
2. Institut de Robòtica i Informàtica Industrial, CSIC-UPC, Llorens i Artigas 4-6, 08028 Barcelona, Spain
3. Research Center for Supervision, Safety and Automatic Control, Universitat Politècnica de Catalunya (UPC), Rambla Sant Nebridi, 22, 08022 Terrassa, Spain
4. Tecnológico Nacional de México/Instituto Tecnológico de Hermosillo, Av. Tecnológico y Periférico Poniente, S/N, 83170 Hermosillo, Sonora, Mexico
* Correspondence: frlopez@ittg.edu.mx (F.-R.L.-E.); damiano.rotondo@upc.edu (D.R.); gvalencia@ith.mx (G.V.-P.)
† These authors contributed equally to this work.

Received: 25 September 2019; Accepted: 30 October 2019; Published: 4 November 2019

Abstract: This paper provides a review about the concept of convex systems based on Takagi-Sugeno, linear parameter varying (LPV) and quasi-LPV modeling. These paradigms are capable of hiding the nonlinearities by means of an equivalent description which uses a set of linear models interpolated by appropriately defined weighing functions. Convex systems have become very popular since they allow applying extended linear techniques based on linear matrix inequalities (LMIs) to complex nonlinear systems. This survey aims at providing the reader with a significant overview of the existing LMI-based techniques for convex systems in the fields of control, observation and safety. Firstly, a detailed review of stability, feedback, tracking and model predictive control (MPC) convex controllers is considered. Secondly, the problem of state estimation is addressed through the design of proportional, proportional-integral, unknown input and descriptor observers. Finally, safety of convex systems is discussed by describing popular techniques for fault diagnosis and fault tolerant control (FTC).

Keywords: linear parameter varying (LPV) systems; Takagi-Sugeno systems; convex systems; linear matrix inequalities (LMIs); fault diagnosis; fault tolerant control (FTC)

---

1. Introduction

Confucius once said *"the beginning of wisdom is to call things by their proper name"*. In this regard, it can be noticed that within the control community there is a big disagreement to call a certain class of multiple model systems by its proper name, in other words, to call a spade a spade. Multiple models were proposed in order to reduce the complexity of controller design for nonlinear systems by describing the latter as a combination of local linear models. To this end, several approaches have been proposed in the literature to deal with this problem, such as the linear parameter varying (LPV), the quasi-LPV (qLPV) and the Takagi-Sugeno (TS).

LPV systems were introduced by Refs. [1,2] as models used to design controllers that guarantee a suitable closed-loop performance for nonlinear plants working under time-varying operating conditions. This was achieved by embedding the plant's nonlinearities inside the so-called *scheduling parameters*. The term LPV was coined to differentiate the resulting class of systems from both linear time invariant (LTI) and linear time varying (LTV) systems. The difference with respect to LTI systems is clear because LPV systems are non-stationary. On the other hand, LPV systems are distinguished from LTV systems in the perspective taken on both analysis and synthesis. LPV systems can be seen as a generalization of a group of LTV systems, each one obtained by means of a predetermined trajectory of the weighing functions. Therefore, properties such as stability, disturbance rejection and tracking, among others, hold for a family of LTV systems, rather than for a single LTV system [3]. A typical LPV system is described by:

$$\dot{x}(t) = A(\theta(t))x(t) + B(\theta(t))u(t) \quad (1)$$
$$y(t) = C(\theta(t))x(t) \quad (2)$$

where $x(t) \in \mathbb{R}^{n_x}$ denotes the state vector, $u(t) \in \mathbb{R}^{n_u}$ is the input vector, $y(t) \in \mathbb{R}^{n_y}$ is the output vector and $\theta(t) \in \mathbb{R}^{n_\theta}$ is the vector of varying parameters, which can be a function of exogenous or endogenous variables (in the latter case, the system is referred to as *quasi-LPV*) and that takes values within a region Θ, that is, $\theta(t) \in \Theta \subset \mathbb{R}^{n_\theta}$.

TS models are similar to LPV systems, since they are obtained by considering some collection of linear models, although their overall blending is obtained by means of a set of fuzzy IF ... THEN rules [4]. At first, they were obtained by performing linearization of the nonlinear plant about different operating points [5]. Nevertheless, this conception was changed in the work by Ohtake et al. [6], who proposed a convex modeling technique via the so-called sector nonlinearity approach. In this case, the main idea is to obtain a convex system such that the global model matches the nonlinear system exactly in a compact subset of the state space. The number of sub-models is directly related to the number of nonlinear terms. For each nonlinear term, two sub-models are obtained such that for k nonlinear terms, the global model is composed of $h = 2^k$ sub-models. Therefore, the bigger is the number of nonlinear terms, the bigger becomes the conservatism of the global convex system and the computational burden of both analysis and synthesis.

The TS approach was adopted rapidly by the control community and was applied to state estimation [7], control [8], fault detection [7], descriptor systems [9], state observers [10], waste-water treatment plants [11], bioreactors [12], process industry [13,14], mechatronics [15,16], aeronautics [17,18] and automotive [19,20], among others. Comprehensive material about the topic can be found in Refs. [8,21–23] and the references therein. On the other hand, another school of thought named these approaches as quasi-LPV (qLPV) in order to differentiate fuzzy approaches from model-based approaches. Nonetheless, models obtained by means of the sector nonlinearity approach are not fuzzy, since the weighting functions are completely deterministic, as detailed in Ref. [24]. Literature on qLPV systems can be found in Refs. [25–33], just to mention a few.

It is clear that LPV and TS systems have been developed independently but recently some works have started discussing about the analogies between these paradigms [23,34,35]. For this reason, we find it appropriate to consider a terminology that includes both schools of thought and in this review we propose to denote both LPV and TS systems as *convex* systems. The idea of unifying these two paradigms under a single name is not new, as it was originally proposed in Ref. [36] and retaken in Refs. [37–40]. Nonetheless, in spite of the success of these paradigms, there is no literature review that allows tasting all the flavors offered by the vastness of convex approaches. Therefore, in this paper, three main aspects of convex systems are reviewed: control, observation and safety. The objective is to help the reader to locate themselves in the area of convex systems by learning about the main used techniques. It is worth highlighting that, although real-life applications of the reviewed methods are discussed whenever appropriate, the level of detail is kept low, since the main focus of this review is theoretical. The reader interested in a more extensive survey of experimental applications and validations based on high-fidelity simulations is referred to the excellent work in Ref. [41] and the references therein.

The overall structure of this review is provided in Figure 1.

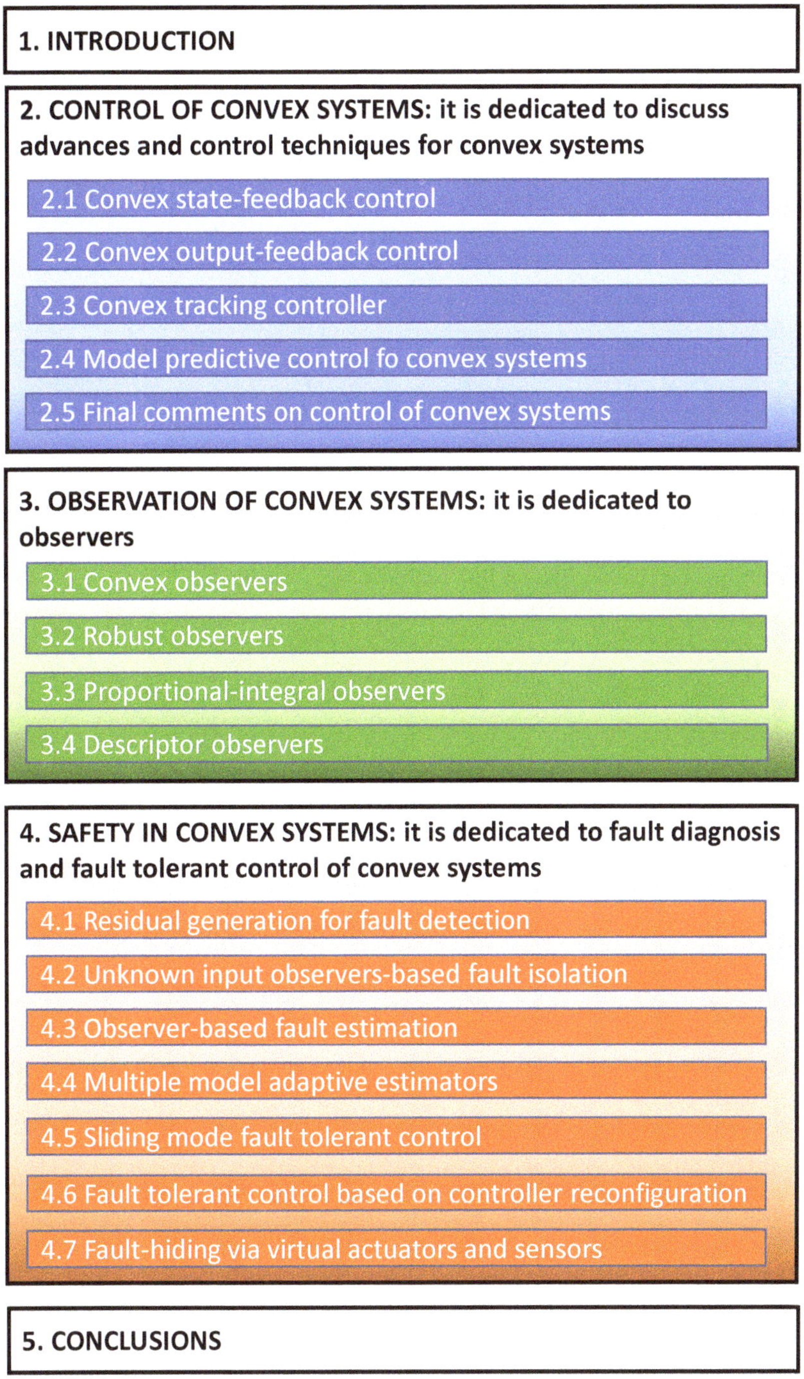

1. INTRODUCTION

2. CONTROL OF CONVEX SYSTEMS: it is dedicated to discuss advances and control techniques for convex systems
- 2.1 Convex state-feedback control
- 2.2 Convex output-feedback control
- 2.3 Convex tracking controller
- 2.4 Model predictive control fo convex systems
- 2.5 Final comments on control of convex systems

3. OBSERVATION OF CONVEX SYSTEMS: it is dedicated to observers
- 3.1 Convex observers
- 3.2 Robust observers
- 3.3 Proportional-integral observers
- 3.4 Descriptor observers

4. SAFETY IN CONVEX SYSTEMS: it is dedicated to fault diagnosis and fault tolerant control of convex systems
- 4.1 Residual generation for fault detection
- 4.2 Unknown input observers-based fault isolation
- 4.3 Observer-based fault estimation
- 4.4 Multiple model adaptive estimators
- 4.5 Sliding mode fault tolerant control
- 4.6 Fault tolerant control based on controller reconfiguration
- 4.7 Fault-hiding via virtual actuators and sensors

5. CONCLUSIONS

Figure 1. Structure of the review.

Notation: The notation used in this article is quite standard. $\mathbb{R}^{m \times n}$ denotes the set of all matrices with m rows and n columns. If a square matrix $A \in \mathbb{R}^{n \times n}$ is symmetric, this fact will be denoted by $A \in \mathbb{S}^n$. Given a matrix $A \in \mathbb{S}^n$, $A \succ 0$ ($A \prec 0$) denotes positive (negative) definiteness, that is, that all its eigenvalues are positive (negative). Similarly, $A \succeq 0$ ($A \preceq 0$) denotes positive (negative) semi-definiteness. For a matrix $A \in \mathbb{R}^{m \times n}$, A^T and $A^\dagger$ denote its transpose and pseudo-inverse, respectively. If $A \in \mathbb{R}^{n \times n}$ is non-singular, A^{-1} will denote its inverse. The symbol $*$ denotes the transposed element in a symmetric position of a matrix. Finally, $\text{He}\{A\}$ is used as a shorthand notation for $A + A^T$.

2. Control of Convex Systems

Convex systems can arise from three possible interpretations [42]: (i) they can be seen as linear systems subject to uncertainties for which the synthesis of a controller must be approached from a robust control perspective; (ii) they can be seen as a family of parameter varying systems, for which the instantaneous value of the varying parameters can be injected directly in the control structure, leading to a gain scheduled control [43,44]; and (iii) the two previous situations can be combined, as suggested by Ref. [45], where a double-layer polytopic framework was considered to this end.

In the last years, significant progress has been made in the control of convex systems. For example, in the presence of uncertainties or disturbances, LPV robust control techniques have shown to provide better performance than robust LTI controllers [46,47]. Indeed, many linear matrix inequality (LMI)-based solutions for LTI systems have been extended to LPV systems, for example, an LPV stabilizing controller was proposed for an arm-driven inverted pendulum in Ref. [48] and was shown to outperform classical robust control techniques, such as H_∞ and μ-synthesis. However, the method in Ref. [48] does not guarantee that the closed-loop system exhibits a robust performance. To handle this problem, a parametrized LPV H_∞ control was presented in Ref. [49], which showed good performance when applied to a turbofan jet engine. Other H_∞ controllers for systems affected by time-varying parametric uncertainties can be consulted in Refs. [50,51]. In order to improve the performance of H_∞ controllers, a switching controller designed with multiple Lyapunov functions was proposed by Ref. [52]. Similarly, an LPV control for switched systems with slow-varying parameters was proposed for an F-16 aircraft model in Ref. [53] by adopting the blending method developed by Ref. [54], which achieves the separation of the entire parameter set into overlapped subsets, such that the overall LPV controller can be blended over the entire region by means of regional controllers. In spite of the good achieved performance, the method is applicable only under the assumption that the scheduling parameters can be measured on-line, which is often difficult to satisfy in practice. To solve this problem, a robust compensator which considers prior and non-real-time knowledge of the varying parameters was proposed by Ref. [55] for stable polytopic LPV plants. Robust convex controllers have been also proposed in the context of networked nonlinear systems [56,57] where the communication channel is affected by package dropouts intermittently.

2.1. Convex State-Feedback Control

The first convex developments were proposed in Ref. [58] and subsequent papers [2,44,59]. The main difference with respect to the robust control theory is that the varying parameters are assumed to be known and they can be used to schedule the time-varying controller gain. The most widely applied control approach is the state-feedback, for which a conceptual scheme is given in Figure 2. This approach computes the control law as follows:

$$u(t) = K(\theta(t))x(t) \quad (3)$$

where $K(\theta(t)) \in \mathbb{R}^{n_u \times n_x}$ denotes the controller gain. It is the simplest control law that can be considered but its implementation requires knowing the full state of the system. Combining (1) and (3), the closed-loop system is described by the following autonomous convex system:

$$\dot{x}(t) = \Big[A(\theta(t)) + B(\theta(t))K(\theta(t))\Big]x(t) \tag{4}$$

Hence, by using the Lyapunov candidate function $V(x(t)) = x^T(t)Px(t) > 0$, with $P \in \mathbb{S}^{n_x}$ and requiring $\dot{V}(x(t)) < 0$, the so-called quadratic stability condition is obtained, as follows:

$$\text{He}\{PA(\theta) + PB(\theta)K(\theta)\} \prec 0 \qquad \forall \theta \in \Theta \tag{5}$$

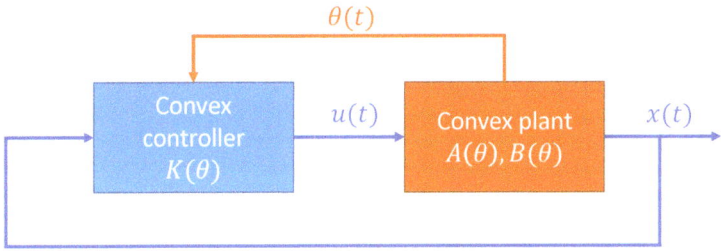

Figure 2. Conceptual scheme of convex control.

Equation (5) is a bilinear matrix inequality (BMI) as the unknown variables $K(\theta)$ and P appear in the same product of matrices $PB(\theta)K(\theta)$. However, it is possible to transform (5) into an LMI by pre- and post-multipliying (5) by $Q = P^{-1}$, thus obtaining [60] (similarity transformations do not change the eigenvalues of a matrix, hence its positive/negative definiteness):

$$\text{He}\{QPA(\theta)Q + QPB(\theta)K(\theta)Q\} \prec 0 \qquad \forall \theta \in \Theta \tag{6}$$

Note that in this case $PQ = (PQ)^T = I$ and therefore the following is obtained:

$$\text{He}\{A(\theta)Q + B(\theta)K(\theta)Q\} \prec 0 \qquad \forall \theta \in \Theta \tag{7}$$

Finally, the quadratic term is eliminated by using the change of variables $\Gamma(\theta) = K(\theta)Q$, so that (7) becomes:

$$\text{He}\{A(\theta)Q + B(\theta)\Gamma(\theta)\} \prec 0 \qquad \forall \theta \in \Theta \tag{8}$$

which is in an LMI form. It is important to mention that, in the case that multiple specifications are desired, the above change of variables introduces some conservatism, since it forces to use the same Lyapunov matrix Q for all specifications, whereas using different matrices for different specifications would lead to better performance. However, using LMIs instead of BMIs is convenient due to the computational efficiency of available LMI solvers, whereas BMIs are non-convex, so that there is no guarantee of obtaining a global minimum. Equation (8) represents an infinite number of constraints, therefore it presents a computational problem. Unfortunately, the direct application of a polytopic approach is not straightforward. One could rewrite (8) as:

$$\mathcal{M}_{ij} := \text{He}\{A_i Q + B_j \Gamma_i\} \prec 0 \qquad \forall i,j = 1,\ldots,h \tag{9}$$

and achieve stabilization by using $u(t) = K(\theta(t))x(t)$, with the feedback controller gain obtained as $K(\theta(t)) = \Gamma(\theta(t))Q^{-1}$, where $\Gamma(\theta(t)) = \sum_{i=1}^{h} \rho_i(\theta(t))\Gamma_i$ and $\rho_i(\theta(t))$ denotes the coefficients of the following polytopic decomposition:

$$\begin{bmatrix} A(\theta(t)) \\ B(\theta(t)) \end{bmatrix} = \sum_{i=1}^{h} \rho_i(\theta(t)) \begin{bmatrix} A_i \\ B_i \end{bmatrix}, \quad \sum_{i=1}^{h} \rho_i(\theta(t)) = 1, \quad \rho_i(\theta(t)) \geq 0 \quad \forall \theta \in \Theta \tag{10}$$

However, this solution has the drawback that a vertex gain K_i must be robust with respect to all possible values of $B(\theta(t))$, which corresponds to a high degree of conservatism. For this reason, alternative solutions can be found in the literature, such as that proposed by Ref. [61] which consists in pre-filtering the control input $u(t)$. The combination of the filter and the system (1) leads to a convex system with constant input matrix since $B(\theta(t))$ appears embedded into the state matrix of the augmented system. However, it must be mentioned that some recent work has questioned the advantages of the pre-filter against using directly the LMIs (9) for the controller design [62,63].

Other alternative solutions aim at relaxing (9), although usually the requirement of low conservatism is associated with an increase in the computational load. Among these solutions, it is worth mentioning the conditions proposed by Ref. [64], who presented a fuzzy control application of the Polya's theorems on positive forms in the standard simplex. The result is a set of sufficient conditions to prove the positiveness of double sums, which are progressively less conservative as a complexity parameter n increases. These conditions are asymptotically exact, that is, necessary and sufficient when n tends to infinity. Other conditions are those obtained by generating partitions of the polytope through the triangulation method [65], which allows to obtain a family of sufficient conditions for positivity/negativity of double sums and, in parallel, another family of necessary conditions, which become asymptotically exact by decreasing the size of the partitions. In addition, one can recall the conditions proposed by Ref. [66], that allow to relax the conditions of double polytopic sum to take into account, for example, the existence of gaps in the set Θ. Nonetheless, the most popular relaxation is the one proposed by Tuan et al. [67], which considers that an LMI in the form of (9) is equivalent to:

$$\mathcal{M}_{ij} \prec 0 \qquad i \in [1, 2, ..., h] \tag{11}$$

$$\frac{2}{h-1}\mathcal{M}_{ii} + \mathcal{M}_{ij} + \mathcal{M}_{ji} \preceq 0 \qquad 1 \leq i \neq j \leq h \tag{12}$$

which reduces the conservatism and increases the applicability of the controller.

For convex qLPV and TS systems, Equations (9) and (11) are also known as parallel distributed compensation (PDC) [68]. In this case, the feedback controller and the convex system share the same weighting functions and the LMI conditions are obtained with the direct Lyapunov method. However, the more local models the convex representation has, the greater is the conservatism of the LMI solution. This fact follows from the necessity of finding a feasible solution that employs a common matrix P for all the local models. A possible strategy to reduce the conservatism is to consider nonquadratic Lyapunov functions (NQLFs) as done, for example, in Refs. [69–72]. The solution obtained through NQLFs, which is also known as non-PDC [73,74], reduces considerably the conservatism and maintains the same weighting functions for both the convex model and the controller. However, non-PDC controllers are harder to design than PDC controllers, since the weighting functions involve time derivatives of the NLQF, leading to local results [75]. This problem does not arise in convex systems dependent on exogenous time-varying parameters, because the NLQLF would not involve time derivatives of the states, hence global solutions can be obtained for this case [72,76–78].

2.2. Convex Output-Feedback Control

A variant of the state-feedback control strategy previously described consists in using directly the output $y(t)$ for feedback, which is easier to implement in cases where the state is unavailable for measurement. Nevertheless, some conditions have to be ensured to make it possible to synthesize these controllers, by means of approaches initially developed in the robust context and later extended to the LPV framework [79–82].

The next result allows the quadratic stabilization of a convex system of the form (1)–(2) using a convex dynamic output-feedback controller defined as:

$$\dot{x}_K(t) = A_K(\theta(t))x_K(t) + B_K(\theta(t))y(t) \tag{13}$$

$$u(t) = C_K(\theta(t))x_K(t) + D_K(\theta(t))y(t) \tag{14}$$

where $x_K(t) \in \mathbb{R}^{n_x}$ is the internal state of the controller and $A_K(\theta(t))$, $B_K(\theta(t))$, $C_K(\theta(t))$, $D_K(\theta(t))$ are matrix-valued functions, such that the closed-loop system obtained by the connection of (1)–(2) and (13)–(14) is stable. In particular, the closed-loop system is described by the following autonomous convex system:

$$\begin{bmatrix} \dot{x}(t) \\ \dot{x}_K(t) \end{bmatrix} = \begin{bmatrix} A(\theta(t)) + B(\theta(t))D_K(\theta(t))C(\theta(t)) & B(\theta(t))C_K(\theta(t)) \\ B_K(\theta(t))C(\theta(t)) & A_K(\theta(t)) \end{bmatrix} \begin{bmatrix} x(t) \\ x_K(t) \end{bmatrix} \tag{15}$$

Due to the presence of $C(\theta(t))$ post-multiplying $B_K(\theta(t))$ and $D_K(\theta(t))$ in (15), the procedure to obtain LMIs for design of the controller's matrices is somehow more complex. The system (1)–(2) is quadratically stabilizable using the convex controller (13)–(14) if there exist a positive definite matrix $P \in \mathbb{S}^{2n_x}$ such that $A_{cl}(\theta)P + PA_{cl}(\theta) \prec 0$, $\forall \theta \in \Theta$, where $A_{cl}(\theta)$ is the state matrix of the autonomous system (15). Following Refs. [83,84], this condition is achieved if and only if there exist matrices $Q \succ 0$ and $S \succ 0$ and matrix-valued functions $\hat{C}_K(\theta)$ and $\hat{B}_K(\theta)$ such that the following holds $\forall \theta \in \Theta$:

$$\text{He}\{A(\theta)Q + B(\theta)\hat{C}_K(\theta)\} \prec 0 \tag{16}$$

$$\text{He}\{SA(\theta) + \hat{B}_K(\theta)C(\theta)\} \prec 0 \tag{17}$$

A possible methodology to obtain the controller's matrices after solving (16)–(17) is described hereafter [85]. If (16)–(17) and the following condition hold:

$$\begin{pmatrix} Q & I \\ I & S \end{pmatrix} \succ 0 \tag{18}$$

then, by letting M, N be non-singular matrices such that $MN^T = I - SQ$ and choosing $D_K(\theta(t)) = 0$, the controller's matrices $A_K(\theta(t))$, $B_K(\theta(t))$ and $C_K(\theta(t))$ can be computed as follows:

$$B_K(\theta(t)) = M^{-1}\left[\hat{B}_K(\theta(t)) - SB(\theta(t))D_K(\theta(t))\right] \tag{19}$$

$$C_K(\theta(t)) = \left[\hat{C}_K(\theta(t)) - D_K(\theta(t))C(\theta(t))Q\right]N^{-T} \tag{20}$$

$$A_K(\theta(t)) = M^{-1}\left[\hat{A}_K(\theta(t)) - SB(\theta(t))C_K(\theta(t))N^T - MB_K(\theta(t))C(\theta(t))Q \right. \tag{21}$$

$$\left. - S\Big(A(\theta(t)) + B(\theta(t))D_K(\theta(t))C(\theta(t))\Big)Q\right]N^{-T}$$

On the other hand, if (18) does not hold, then the matrices have to be adjusted using $Q_\lambda = \lambda Q$, $S_\lambda = \lambda S$, $\hat{B}_{K\lambda}(\theta(t)) = \lambda \hat{B}_K(\theta(t))$ and $\hat{C}_{K\lambda}(\theta(t)) = \lambda \hat{C}_K(\theta(t))$, where $\lambda > 1$, until (18) holds with these new variables and the controller's matrices can be computed.

Equation (18) guarantees the existence of the invertible matrices M and N used for controller computation. In the same way, the conditions to perform H_∞ control, control with guaranteed cost or to achieve other specifications can be obtained. Note that double polytopic sums appear due to the terms $B(\theta)\hat{C}_K(\theta)$ and $\hat{B}_K(\theta)C(\theta)$ in (16)–(17). If the controller is restricted to the case where $\hat{B}_K(\theta)$ and $\hat{C}_K(\theta)$ are constant, then the LMIs (16)–(17) can be reduced to a finite number of conditions easily, otherwise the discussion provided in the previous section about possible relaxations would apply with slight modifications. It is worth remarking that in convex systems in which $\theta(t)$ depends on

2.3. Convex Tracking Controller

Consider a convex system subject to unknown inputs ans sensor noise, described by (for the sake of simplicity, the output matrix is assumed to be constant):

$$\dot{x}(t) = A(\theta(t))x(t) + B(\theta(t))u(t) + R(\theta(t))d(t) \quad (22)$$
$$y(t) = Cx(t) + Gd(t) \quad (23)$$

where $d(t) \in \mathbb{R}^{n_d}$ is the disturbance vector comprising both unknown inputs and noise and $R(\theta(t))$ and G are matrices of appropriate dimensions. As illustrated in Figure 3, a convex tracking controller can be considered for this system, with control law:

$$u(t) = K_1(\theta(t))x(t) + K_2(\theta(t))\epsilon(t) = \mathcal{K}(\theta(t))\begin{bmatrix} x(t) & \epsilon(t) \end{bmatrix}^T \quad (24)$$

where $K_1(\theta(t))$ and $K_2(\theta(t))$ are the gains to be designed and $\epsilon(t)$ is the integration error, added to compensate steady-state errors and reach the desired output $w(t)$:

$$\dot{\epsilon}(t) = w(t) - y(t) = w(t) - Cx(t) - Gd(t) \quad (25)$$

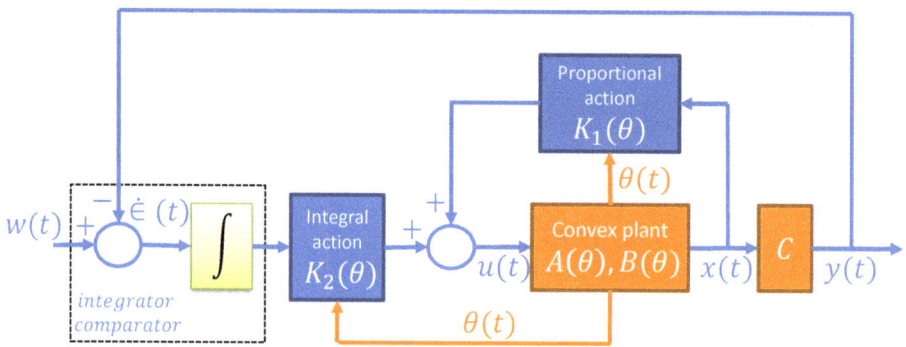

Figure 3. Convex tracking controller scheme.

The system augmented with the integrator can be rewritten in a compact form by introducing the augmented state vector $x_c(t) = [x^T(t) \ \epsilon^T(t)]^T$:

$$\dot{x}_c(t) = \tilde{A}_c(\theta(t))x_c(t) + \tilde{B}_c(\theta(t))u(t) + \tilde{R}_c(\theta(t))d(t) + \tilde{B}_w w(t) \quad (26)$$

with:

$$\tilde{A}_c(\theta(t)) = \begin{bmatrix} A(\theta(t)) & 0 \\ -C & 0 \end{bmatrix} \quad \tilde{B}_c(\theta(t)) = \begin{bmatrix} B(\theta(t)) \\ 0 \end{bmatrix} \quad \tilde{B}_w = \begin{bmatrix} 0 \\ I \end{bmatrix} \quad \tilde{R}_c(\theta(t)) = \begin{bmatrix} R(\theta(t)) \\ -G \end{bmatrix} \quad (27)$$

The closed loop system has the form:

$$\dot{x}_c(t) = \sum_{i=1}^{h} \rho_i(\theta(t)) \sum_{j=1}^{h} \rho_j(\theta(t)) \left[(\tilde{A}_{ci} - \tilde{B}_{ci}\mathcal{K}_j) x_c(t) + \tilde{B}_{Rwi} \tilde{d}_w(t) \right] \quad (28)$$

with:

$$\bar{B}_{Rwi} = \begin{bmatrix} \bar{R}_i & \bar{B}_w \end{bmatrix}, \quad \bar{d}_w(t) = \begin{bmatrix} d(t) & w(t) \end{bmatrix}^T \tag{29}$$

Sufficient conditions for the existence of the controller are given in Ref. [35] and presented hereafter. Consider the system (22)–(23), the feedback control law defined by (24), the integrator, and let the attenuation level be given by $\gamma_c > 0$. The closed loop system error (25) is globally stable with H_∞ performance if $\| x_c(t) \|_2^2 < \gamma_c^2 \| \bar{d}_w(t) \|_2^2$ and if there exists a matrix $X \succ 0$ such that $\forall i, j \in [1, 2, ..., h]$, the following holds:

$$\begin{bmatrix} \text{He}\left(X \bar{A}_{ci}^T + \Xi_j^T \bar{B}_{ci}^T \right) + \bar{B}_{Rwi} \bar{B}_{Rwi}^T & X \\ * & -\gamma_c^2 I \end{bmatrix} \prec 0 \tag{30}$$

Then, the controller gain matrices are computed by $\mathcal{K}_j = \begin{bmatrix} K_{1j} & K_{2j} \end{bmatrix} = X^{-1} \Xi_j$.

This is possible because if we consider the $\mathcal{L}_2$-gain from $\bar{d}_w(t)$ to $x_c(t)$ such that:

$$J_{x_c d} = \dot{V}(t) + x_c^T(t) x_c(t) - \gamma_c^2 \bar{d}_w^T(t) \bar{d}_w(t) < 0 \tag{31}$$

where $V(t)$ is a quadratic Lyapunov function, the LMI (30) is obtained after solving the performance criteria (31). Complete procedures are described in detail in Ref. [35].

An alternative approach is to use a reference model as originally proposed by Ref. [87] and later applied by Refs. [26,88], which has the advantage that the tracking error is described by an autonomous system, so its convergence to zero can be guaranteed even without the use of an integrator.

2.4. Model Predictive Control for Convex Systems

Model predictive control (MPC) is a control strategy that is based on the use of a mathematical model to predict the system's behavior in a future time window and then finds the optimal input sequence by minimizing a cost function [89,90]. Only the first calculated input is applied to the system and the remaining are discarded, repeating this prediction-optimization process at every sample. This control technique is popular because it can take into account systematically complex dynamics, as well as physical and process quality constraints [91].

Consider the discrete-time convex system:

$$x(k+1) = A(\theta(k)) x(k) + B(\theta(k)) u(k) \tag{32}$$

where $A(\theta(k)) = \sum_{j=1}^{l} \theta_j(k) A_j$ and $B(\theta(k)) = \sum_{j=1}^{l} \theta_j(k) B_j$. Therefore, $\theta(k)$ belongs to a convex polytope Θ defined by the values $\theta_j(k)$ such that $\sum_{j=1}^{l} \theta_j(k) = 1$, with, $0 \leq \theta_j(k) \leq 1$. On the other hand, when $\theta(k)$ varies in the polytope Θ, the system matrices vary in the polytope Ω defined as follows:

$$[A(\theta(k)), B(\theta(k))] \in \Omega = Co\{[A_1, B_1], [A_2, B_2], \ldots, [A_l, B_l]\} \tag{33}$$

where $[A_i, B_i]$ are the vertex matrices obtained when $\theta_i = 1$ and $\theta_j = 0$ for $j \neq i$. Hereafter, for illustrative purposes, it is assumed that there is no model uncertainty and that both the scheduling variable $\theta(k)$ and the state $x(k)$ are known at time k. However, the future evolution of the model is uncertain since future values of $\theta(k)$ are unknown.

Let us define the following quadratic cost function:

$$J(k) = \underbrace{x(k|k)^T Q x(k|k) + u(k|k)^T R u(k|k)}_{J_0(k)} + \underbrace{\sum_{i=1}^{\infty} x(k+i|k)^T Q x(k+i|k) + u(k+i|k)^T R u(k+i|k)}_{J_1(k)} \tag{34}$$

where Q, R are weighting matrices with appropriate dimensions and the notation $x(k+i|k)$ represents the predicted value for the state variable x at the future sample $k+i$ calculated at sample k. Hence, $J_0(k)$ correspond to the first prediction step and $J_1(k)$ correspond to the remaining of the prediction.

Let $U(k)$ be the sequence of inputs computed at sample k, that is, $U(k) = [u(k|k), U_1(k)] = [u(k|k), u(k+1|k),\ldots]$. Then, the optimal control sequence is obtained by minimizing the maximum value that the cost function (34) can take for all the possible future trajectories of the parameter $\theta(k)$, that is,

$$U^*(k) = \min_{U(k)} \max_{[A(\theta(k+i)), B(\theta(k+i))] \in \Omega, \, i \geq 0} J(k) \tag{35}$$

where $*$ denotes optimality. The first element of $U^*(k)$, that is, $u^*(k|k)$, is applied to the system, while the remaining of the sequence $U_1^*(k)$ can be proven to be equivalent to a state feedback control law whose gain does not depend on the instantaneous value of $\theta(k)$ (see Ref. [92] for further details), that is,

$$U_1^*(k) = \{u(k+i|k) = K(k)x(k+i|k), i \geq 1\} \tag{36}$$

Following Ref. [92], instead of solving (35), an upper bound for the term $J_1(k)$ can be defined, as follows:

$$\max_{[A(\theta(k+i)), B(\theta(k+i))] \in \Omega, \, i \geq 0} J_1(k) \leq V(x(k+1|k)) = x(k+1|k)^T P x(k+1|k) \quad P(k) \succ 0. \tag{37}$$

Then, an upper bound of the worst case of $J(k)$ is minimized instead of (35), as follows:

$$U^*(k) = \min_{u(k|k), P(k)} x(k|k)^T Q x(k|k)^T + u(k|k)^T R u(k|k)^T + x(k+1|k)^T P x(k+1|k) \tag{38}$$

The optimization problem (38) can be reformulated as the following minimization problem:

$$\min_{\gamma, u(k|k), \tilde{Q}(k), Y(k)} \gamma \tag{39}$$

subject to LMIs:

$$\begin{bmatrix} 1 & \hat{x}(k+1|k)^T & \hat{x}(k|k)^T Q^{\frac{1}{2}} & u(k|k)^T R^{\frac{1}{2}} \\ \hat{x}(k+1|k) & \tilde{Q}(k) & 0 & 0 \\ Q^{\frac{1}{2}} x(k|k) & 0 & \gamma I & 0 \\ R^{\frac{1}{2}} u(k|k) & 0 & 0 & \gamma I \end{bmatrix} \succeq 0 \tag{40}$$

$$\begin{bmatrix} \tilde{Q}(k) & \tilde{\Gamma}_j(k)^T & \tilde{Q}(k) Q^{\frac{1}{2}} & Y(k)^T R^{\frac{1}{2}} \\ \tilde{\Gamma}_j(k) & \tilde{Q}(k) & 0 & 0 \\ Q^{\frac{1}{2}} \tilde{Q}(k) & 0 & \gamma I & 0 \\ R^{\frac{1}{2}} Y(k) & 0 & 0 & \gamma I \end{bmatrix} \succeq 0 \quad \forall j = 1, \ldots, l \tag{41}$$

with $\hat{x}(k+1|k) = [A(\theta(k))x(k|k) + B(\theta(k))u(k|k)]$, $\tilde{\Gamma}_j(k) = A_j \tilde{Q}(k) + B_j Y(k)$ and $\tilde{Q}(k) \succeq 0$. The gain in (36) is computed as $K(k) = Y(k)\tilde{Q}^{-1}(k)$, which guarantees that the state evolves in an ellipsoidal invariant set.

Considering the system output as $y(k) = Cx(k)$, the cost function (34) may be subject to constraints [93]:

$$|u(k|k)| \leq u_{max} \tag{42}$$

$$\left\| C[A(\theta(k))x(k|k) + B(\theta(k))u(k|k)] \right\|_2 \leq y_{max} \tag{43}$$

Constraints on the inputs are satisfied if there exists a matrix $X \succeq 0$ such that:

$$\begin{bmatrix} X & Y \\ Y^T & \tilde{Q} \end{bmatrix} \succeq 0 \quad \text{with} \quad X_{ii} \leq u_{max}^2 \tag{44}$$

In a similar way, the constraints on the outputs are equivalent to the LMI:

$$\begin{bmatrix} \tilde{Q} & [A_j\tilde{Q} + B_jY]^T C^T \\ C[A_j\tilde{Q} + B_jY] & y_{max}^2 \end{bmatrix} \succeq 0 \quad j = 1, \ldots, l \tag{45}$$

The constrained MPC algorithm with control law (36) can be obtained by solving the optimization problem (39) subject to the LMIs (40), (41), (44), (45) and constraints (42) and (43). However, although this algorithm does not impose $u(k|k)$ and $y(k+1|k)$ to invariant ellipsoid constraints, still includes constraints on all future inputs and outputs. A method to improve the conservatism is to relax the future constraints (44) and (45) and bound only $u(k|k)$ and $y(k+1|k)$ [93]. To guarantee stability an additional constraint that ensures that the cost function decreases monotonously ($\phi(k) < \phi(k-1)$ with $\phi(k) = x(k|k)^T Q x(k|k)^T + u(k|k)^T R u(k|k)^T + x(k+1|k)^T P x(k+1|k))$ has to be included in the optimization:

$$\min_{\gamma, u(k|k), \hat{Q}(k), Y(k)} \gamma \tag{46}$$

subject to LMIs:

$$\begin{bmatrix} \gamma & \hat{x}(k+1|k)^T & x(k|k)^T Q^{\frac{1}{2}} & u(k|k)^T R^{\frac{1}{2}} \\ \hat{x}(k+1|k) & \hat{Q}(k) & 0 & 0 \\ Q^{\frac{1}{2}} x(k|k) & 0 & I & 0 \\ R^{\frac{1}{2}} u(k|k) & 0 & 0 & I \end{bmatrix} \succeq 0 \tag{47}$$

$$\begin{bmatrix} \phi(k-1) & \hat{\Gamma}_j(k)^T & \hat{Q}(k)Q^{\frac{1}{2}} & Y(k)^T R^{\frac{1}{2}} \\ \hat{\Gamma}_j(k) & \hat{Q}(k) & 0 & 0 \\ Q^{\frac{1}{2}}\hat{Q}(k) & 0 & I & 0 \\ R^{\frac{1}{2}} Y(k) & 0 & 0 & I \end{bmatrix} \succeq 0 \quad \forall j = 1, \ldots, l \tag{48}$$

$$\begin{bmatrix} u(k|k) - u_{max} \\ -u_{max} - u(k|k) \end{bmatrix} \leq 0 \tag{49}$$

$$\begin{bmatrix} C[A(\theta(k))x(k|k) + B(\theta(k))u(k|k)] - y_{max} \\ -y_{max} - C[A(\theta(k))x(k|k) + B(\theta(k))u(k|k)] \end{bmatrix} \leq 0 \tag{50}$$

with $\hat{\Gamma}_j(k) = A_j\hat{Q}(k) + B_jY(k)$ and $\hat{Q}(k) \succeq 0$. The gain of the control law (36) is computed as $K(k) = Y(k)\hat{Q}^{-1}(k)$. To initialize the algorithm, in $k = 0$, the Lyapunov constraint $\phi(k) < \phi(k-1)$ is not taken into account. The resulting control strategy, which is depicted in Figure 4, provides guaranteed closed-loop stability provided that a feasible solution has been found. A parameter dependant feedback law instead of (36) can also be considered as in Ref. [93].

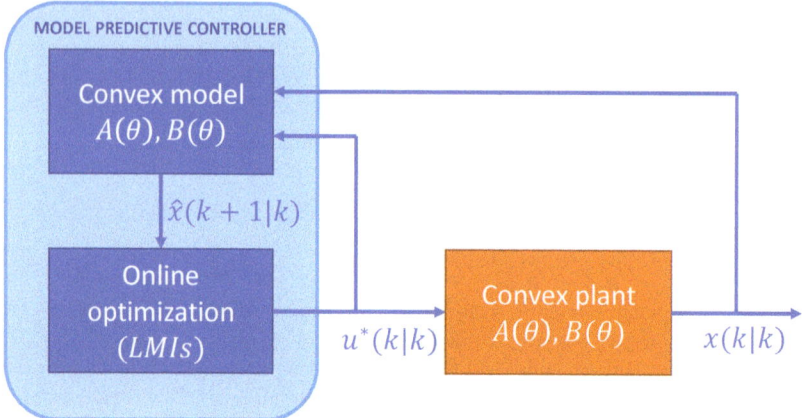

Figure 4. Convex model predictive control scheme.

Most of the MPC strategies for convex systems are based on the algorithm proposed by Ref. [92] since it stabilises robustly an LPV system for all possible parameter variations. However, such algorithm was not thought for application to LPV systems and therefore it suffers from being conservative and computationally demanding. The strategies proposed by Refs. [93,94], which consider bounds in the parameter variations, show less conservatism and a decreased computational load when compared to Ref. [92]. A modification of Ref. [93] that involves updating the polytope Ω while keeping it defined by the least possible number of vertices has been presented in Ref. [95]. This innovation is motivated by the fact that the fewer vertices are used to describe Ω, the less likely it is that infeasibility problems could occur. In Ref. [96], an extension to nonlinear systems has been presented, where a linearized model is obtained from the nonlinear model at each sample and then an LPV model that varies in a politope Ω is used for obtaining the state prediction. Other existing approaches are focused on the use of Lyapunov functions that depend on $\theta(k)$ that enlarge the feasible region [97,98]. In line with this work, an algorithm that uses closed-loop predictions with good achieved performance and low computational requirements was presented in Ref. [99]. More recently, Ref. [100] has presented a class of nonlinearly parameterized Lyapunov functions to achieve more efficient relaxed stability conditions. A robust MPC scheme for LPV systems where the varying parameters are assumed to be measured online and exploited for feedback has been derived in Ref. [101]. Explicit MPC for convex systems has been also proposed in order to avoid the need of online optimization [102,103]. In general, MPC for convex systems has been a topic that has received an intense interest by the research community in the last few years, for which the interested reader is referred to Refs. [104–107] and the references therein.

2.5. Final Comments on Control of Convex Systems

It is worth noting that what has been discussed in this section about the control of convex systems does not apply only to the problem of controller design for quadratic stabilization but also to the case of other specifications, such as $\mathcal{D}$-stabilization [108], H_∞ control [109], control with guaranteed cost [110] and many more. The described methods can be adapted to deal with convex systems with piecewise constant parameters, which provide a unifying concept lying in between the robust and the gain-scheduled perspectives, including both as extremal cases [111].

On the other hand, there are some cases in which the direct application of the convex techniques described so far would not work, for example, due to the loss of controllability for some points of the design polytope, so that alternative strategies must be employed. For instance, consider the following simplified model of a unicycle mobile robot

$$\begin{bmatrix} \dot{x}(t) \\ \dot{y}(t) \\ \dot{\alpha}(t) \end{bmatrix} = \begin{bmatrix} \cos(\alpha(t)) & 0 \\ \sin(\alpha(t)) & 0 \\ 0 & 1 \end{bmatrix} \begin{bmatrix} v \\ w \end{bmatrix} \tag{51}$$

where x, y are the spacial 2D coordinates, v and w are the mobile robot translational and angular velocities, respectively and α denotes the orientation with respect to the fixed frame. If $\alpha \in [-\pi, \pi]$ were considered, one would get a non-controllable convex models in the vertices although the underlying nonlinear system is actually controllable. For example, the solution proposed in Ref. [88] was based on dividing the parameter space in regions and use a switched approach in order to avoid the above-mentioned singularities.

Another important theoretical point to be remarked is that the LMI-based assessment of stability (or some other goal) in convex systems arising from an underlying nonlinear system could mislead to believe that global stability (or performance) would hold for the original system. This fact is in general not true, as shown remarkably by Ref. [112] with a simple second order autonomous nonlinear system, that is, the well known Van der Pol equation. Fortunately, Ref. [112] also shows that it is possible to estimate the region of attraction for the nonlinear system, based on the Lyapunov function obtained for the convex system. This fact was further studied by Refs. [113–115] and was used by Ref. [34] to create a metric to compose different convex models obtained for the same nonlinear system.

Finally, it is worth mentioning that analysis and control problems for convex systems with delays have also attracted some recent interest [116,117]. These systems belong to the intersection of convex systems and time-delay systems, so they inherit the difficulties of each one. In particular, the stability analysis of these systems must be performed using tools such as Lyapunov-Razumikhin functions and Lyapunov-Krasovskii functionals, which increase the number of decision variables [118]. The interested reader is referred to the monograph [119] and the references therein.

3. Observation of Convex Systems

In many real-world applications there are some internal state variables that cannot be measured with the available sensors. Nevertheless, many control techniques are based on the assumption that the whole state is available, which is not always true. In practice, the available information concerns the input $u(t)$ and the output $y(t)$, rather than the state $x(t)$. In such case, the observability properties state that when the system is observable, the initial state can be determined and, therefore, the state trajectory can be reconstructed from input and output measurements, by means of the so-called state observer. State observers are dynamical systems that are designed to estimate asymptotically the state vector $x(t)$. Applications of convex observers can be found in UAVs [120], electric vehicles [121], networked systems [122], DC motors [123], wind turbines [124], riderless bicycles [125], to mention a few.

3.1. Convex Observers

The most common state observer is the one named after Luenberger, which for a convex systems has the following form [126] (see Figure 5):

$$\dot{\hat{x}}(t) = A(\theta(t))\hat{x}(t) + B(\theta(t))u(t) + L(\theta(t))(y(t) - \hat{y}(t)) \tag{52}$$
$$\hat{y}(t) = C\hat{x}(t) \tag{53}$$

where $\hat{x}(t) \in \mathbb{R}^{n_x}$ denotes the state estimate, $\hat{y}(t) \in \mathbb{R}^{n_y}$ denotes the output estimate and the meaning of the remaining variables can be inferred from the previous section. The scheduled observer gains $L(\theta(t))$ are designed to guarantee closed-loop stability of the estimation error dynamics for all values of $\theta(t)$, such that the estimation error between the observer (52)–(53) and the system (1)–(2) converges towards zero. Let us define the estimation error as follows:

$$e(t) := x(t) - \hat{x}(t). \tag{54}$$

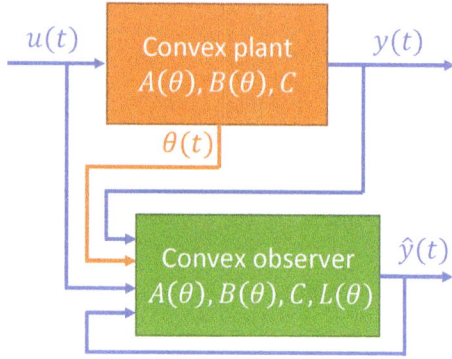

Figure 5. Convex state observer scheme.

Then, the estimation error dynamics is defined as:

$$\dot{e}(t) = (A(\theta(t)) - L(\theta(t)C))e(t) \tag{55}$$

The stability condition of the above differential equation can be obtained by means of LMI-based techniques, for example by considering the quadratic stability concept. In this case, one seeks the existence of a quadratic Lyapunov function $V(e(t)) = e(t)^T P e(t) \geq 0, P \succ 0$, whose derivative over the error dynamics is given by:

$$\dot{V}(e(t)) = \dot{e}(t)^T P e(t) + e(t)^T P \dot{e}(t) = e(t)^T \left(\text{He}\{PA(\theta(t)) - PL(\theta(t))C\}\right)e(t) < 0$$

In order to eliminate the quadratic term, the change of variable $W(\theta(t)) = PL(\theta(t))$ is considered, such that the following LMI is obtained:

$$\text{He}\{PA(\theta) - W(\theta)C\} \prec 0 \qquad \forall \theta \in \Theta \tag{56}$$

where the observer gain matrix can be computed later from its solution as $L(\theta(t)) = P^{-1}W(\theta(t))$.

Furthermore, to improve the speed convergence of the state observer, a decay rate $\alpha < 0$ can be added as requirement, by asking that:

$$\dot{V}(e(t)) + 2\alpha P < 0 \tag{57}$$

which is also known in the literature as α-stabilization. As a result, the LMI (56) is replaced by the following:

$$\text{He}\{PA(\theta) - W(\theta)C\} + 2\alpha P \prec 0 \qquad \forall \theta \in \Theta \tag{58}$$

It should be noticed that the stability of a state observer is guaranteed if the LMI (58) has a solution. Nevertheless, the approach described so far does not consider disturbances or measurement noise, which affect all physical systems.

3.2. Robust Observers

Dynamical systems can be affected by external disturbances and measurement noise and, moreover, there exists always a mismatch between the real plant and its model used for control. These effects can lead to the loss of stability or performance if not taken into account appropriately. A control system that remains stable and with none (or little) performance loss despite the disturbances/noise is said to be robust. In other words, robustness means that the system remains stable and with almost the same performance even in the presence of disturbance, model mismatches or noise. For example, a wind turbine system must keep its efficiency even in the presence of air velocity changes. In contrast, an example of unwanted disturbance amplification is the Tacoma Narrows suspension bridge, where strong winds caused resonant oscillations of increasing magnitude in the bridge structure, which ultimately led to its destruction.

In particular, the robust state observer design problem is related to finding the observer gains such that it is always possible to estimate the real states within prescribed tolerances, despite the effects of uncertainties.

A possible technique to deal with uncertainties is by means of the robust H_∞ approach, which has been developed since the beginning of the eighties and has been applied intensely, with successful results, to convex systems [11,120,127,128].

The H_∞ approach assumes that the disturbance $d(t) \in \mathbb{R}^{n_d}$ belongs to a set of norm bounded functions. The idea is to minimize the worst error that can arise from any disturbance in the following set:

$$\| d(t) \|_2 = \left(\int_0^\infty d^T(\tau)d(\tau)d\tau \right)^{1/2} \tag{59}$$

For instance, let us consider a convex model affected by the above sources of uncertainty:

$$\dot{x}(t) = A(\theta(t))x(t) + B(\theta(t))u(t) + R(\theta(t))d(t) \tag{60}$$
$$y(t) = Cx(t) + Gd(t) \tag{61}$$

where the meaning of each variable and matrix is kept as previously and the matrices $A(\theta(t))$, $B(\theta(t))$, $R(\theta(t))$ satisfy the polytopic property:

$$\begin{bmatrix} A(\theta(t)) \\ B(\theta(t)) \\ R(\theta(t)) \end{bmatrix} = \sum_{i=1}^{h} \rho_i(\theta(t)) \begin{bmatrix} A_i \\ B_i \\ R_i \end{bmatrix}, \quad \sum_{i=1}^{h} \rho_i(\theta(t)) = 1, \quad \rho_i(\theta(t)) \geq 0 \ \forall \theta \in \Theta \tag{62}$$

Let us consider the Luenberger observer given by (52)–(53), for which the dynamics of the estimation error defined as in (55), can be described after some algebraic manipulations as follows:

$$\dot{e}(t) = (A(\theta(t)) - L(\theta(t))C)\,e(t) + (R(\theta(t)) + L(\theta(t))G)d(t) \tag{63}$$

Then, the design problem can be formulated as the one of guaranteeing asymptotic stability of the estimation error (63) while at the same time minimizing, by means of the H_∞ technique, the ratio between the ℓ_2 norm of the output vector and the ℓ_2 norm of the disturbance vector against the disturbance vector $d(t)$, that is,

$$\min_{\gamma>0} \gamma : \frac{\|y\|_{\ell_2}}{\|d\|_{\ell_2}} < \gamma, \quad \|d\|_{\ell_2} \neq 0 \tag{64}$$

where $\gamma > 0$ is the prescribed attenuation level (upper bound on the above mentioned ratio). Then, by considering a bound on the $\mathcal{L}_2$ gain from $d(t)$ to $e(t)$ given by the Lyapunov function $V(e(t)) = e(t)^T P e(t)$, $P \succ 0$, the above performance criterion is satisfied if the following holds:

$$\dot{V}(e(t)) + e(t)^T e(t) - \gamma^2 d(t)^T d(t) < 0 \tag{65}$$

Similarly to the procedure described above in Section 3.1, the following LMI is obtained from manipulations on (65):

$$\begin{bmatrix} \text{He}\{PA_i - \Xi_i C\} + I & PE_i - \Xi_i G \\ * & -\tilde{\gamma} I \end{bmatrix} \prec 0 \tag{66}$$

Once solved the above LMI, the observer gain matrix can be computed as $L_i = P^{-1}\Xi_i$, which achieves an attenuation level $\gamma = \sqrt{\tilde{\gamma}}$.

3.3. Proportional-Integral Observers

Proportional-integral observers (PIOs) have become popular in recent years due to their robustness against constant or slowly varying disturbances. In a PIO an additional term, which is proportional to the integral of the output estimation error, is added in order to increase the robustness performance [129], as depicted in Figure 6. This term gives an additional degree of freedom that can be used for the estimation of unknown inputs such as disturbances [130,131], battery charge [132] and faults [133,134], among others.

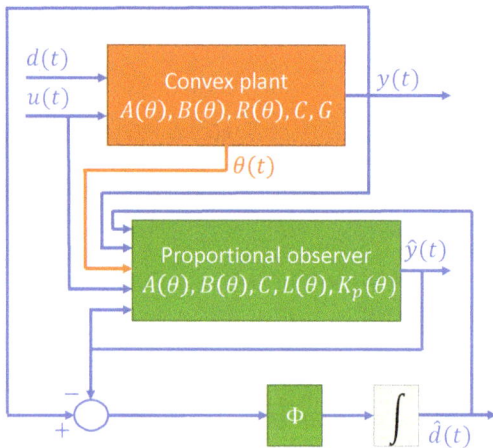

Figure 6. Convex proportional integral observer (PIO) scheme.

A PIO for a system in the form (60)–(61) is described by the following equations:

$$\dot{\hat{x}}(t) = A(\theta(t))x(t) + B(\theta(t))u(t) + L(\theta(t))(C\hat{x}(t) - y(t)) + K_p(\theta(t))\hat{d}(t) \tag{67}$$

$$\dot{\hat{d}}(t) = \Phi(y(t) - C\hat{x}(t)) \tag{68}$$

where $\hat{x}$ and $\hat{d}$ denote the estimated state and unknown input vectors, respectively and $L(\theta(t))$, $K_p(\theta(t))$, and Φ are the observer gain matrices to be computed. The addition of an integrator provides more robustness to the observer so that it can deal with measurement noise or modeling uncertainties. Let us define the estimation errors as (55) and $e_d(t) = d(t) - \hat{d}(t)$. In order to get a suitable design procedure, it can be considered that the unknown input $d(t)$ is varying slowly, which means that $\dot{d}(t) \approx 0$. Then, the dynamics of the estimation errors can be computed as:

$$\dot{e}(t) = (A(\theta(t)) - L(\theta(t))C)e(t) + (R(\theta(t)) - L(\theta(t))G)d(t) - K_p(\theta(t))\hat{d}(t) \tag{69}$$
$$\dot{e}_d(t) = -\Phi C e(t) - \Phi G d(t) \tag{70}$$

By considering:
$$R(\theta(t)) - L(\theta(t))G = K_p(\theta(t)) \tag{71}$$
$$\Phi G = 0 \tag{72}$$

and an extended error vector defined as $\bar{e}^T(t) := \left[e^T(t) e_d^T(t) \right]^T$, the overall error dynamics is rewritten as:
$$\dot{\bar{e}}(t) = (A_e(\theta(t)) - L_e(\theta(t))C_e)\bar{e}(t) \tag{73}$$

with:
$$A_e(\theta(t)) = \begin{bmatrix} A(\theta(t)) & K_p(\theta(t)) \\ 0 & 0 \end{bmatrix} \quad L_e(\theta(t)) = \begin{bmatrix} L(\theta(t)) \\ \Phi \end{bmatrix} \quad C_e = [C, \ 0] \tag{74}$$

Then, by considering a Lyapunov equation $V(\bar{e}(t)) = \bar{e}^T(t)\mathscr{P}\bar{e}(t)$, the solution is obtained in the LMI formulation, similarly to (56), as:
$$\text{He}\{\mathscr{P}A_e(\theta) - W_e(\theta)C_e\} \prec 0 \quad \forall \theta \in \Theta \tag{75}$$

where $\mathscr{P} = \text{diag}(P, Q)$, $P, Q \succ 0$ and $W_e(\theta(t)) = \mathscr{P}L_e(\theta(t))$. The observer gain matrices are obtained from the solution of (75) as $L_e(\theta(t)) = \mathscr{P}^{-1}W_e(\theta(t))$.

It is worth remarking that the slow variation assumption is very common in the literature. From a practical point of view, it can be relaxed, as done for example in Ref. [135]. It is also worth noticing that later, in Section 4.2, unknown input observers, which allow to obtain asymptotic convergence of the estimation error despite the presence of $d(t)$ without making the assumption that $d(t)$ is approximately constant, will be reviewed.

3.4. Descriptor Observers

The observers discussed above are designed for regular systems whose dynamics are represented only by ordinary differential equations. Nevertheless, some mathematical models are composed of both differential and static (algebraic) equations. Designing state observers for descriptor systems is harder than for regular systems because descriptor systems usually have three types of modes (finite dynamic, impulsive and non-dynamic [136]) and the observer must deal with all of them. Nonetheless, there are plenty of fields in which descriptor systems are applied, for example, aircraft modeling [137], complex systems [11], microgrids [138], electrical [9], mechanical and hydraulic systems [136] and biological processes [12], among others. Furthermore, there are some mechanical systems for which possible variations in time of masses and/or inertias rely on a natural descriptor model as in Refs. [139–141]. In these cases, although convex systems can be transformed into a regular system, it has been proved that it is possible to reduce the number of linear models by considering a descriptor representation, which in general makes the LMI-based constraints less conservative [142,143].

A convex descriptor system affected by disturbances is described by
$$E\dot{x}(t) = A(\theta(t))x(t) + B(\theta(t))u(t) + R(\theta(t))d(t) \tag{76}$$
$$y(t) = Cx(t) + Gd(t) \tag{77}$$

where E is a constant matrix with $\text{rank}(E) = r \leq n_x$. Note that in the particular case of $E = I$, the descriptor system becomes a regular system and the observer can be computed as described previously. On the other hand, unlike regular systems, a descriptor system has different modes that are given by differential and algebraic equations. As a result, different types of observability condition should be verified such as R-observability [144], which means that:

$$\text{rank} \begin{bmatrix} sE - A_i \\ C \end{bmatrix} = n_x \qquad \forall i \in [1, 2, ...N] \tag{78}$$

and I-observability [144], that is,

$$\text{rank} \begin{bmatrix} E & A_i \\ 0 & E \\ 0 & C \end{bmatrix} = n_x + r \qquad \forall i \in [1, 2, ..., N] \tag{79}$$

R-observability characterizes the ability to reconstruct only the reachable state from the output data. However, due to the algebraic equations, impulsive terms can appear, which are not desirable since they can saturate the state response or, in general, have negative effects on the system. On the other hand, I-observability guarantees the ability to estimate impulse terms given by the algebraic equations [12].

Then if the convex descriptor system is both R- and I-observable, the following observer can be proposed:

$$\dot{z}(t) = N(\theta(t))z(t) + J(\theta(t))u(t) + L(\theta(t))y(t) \tag{80}$$
$$\hat{x}(t) = z(t) + T_2 y(t) \tag{81}$$
$$\hat{y}(t) = C\hat{x}(t) \tag{82}$$

where $z(t)$ represents the observer state and $\hat{x}(t)$ stands for the estimated states. $N(\theta(t))$, $J(\theta(t))$, $L(\theta(t))$ and T_2 are unknown gain matrices of appropriate dimensions to be computed. Based on (76)–(77) and (80)–(82), the estimation error $e(t)$ is:

$$e(t) = x(t) - \hat{x}(t) = (I - T_2 C)x(t) - z(t) - T_2 G d(t) \tag{83}$$

Assuming that there exists a matrix $T_1 \in \mathbb{R}^{n_x \times n_x}$ such that:

$$I - T_2 C = T_1 E \tag{84}$$

the estimation error becomes:

$$e(t) = T_1 E x(t) - z(t) - T_2 G d(t). \tag{85}$$

Assuming that the disturbances is slowly varying, $\dot{d}(t) \approx 0$, the dynamics of $e(t)$ is given by:

$$\begin{aligned}
\dot{e}(t) &= T_1 E \dot{x}(t) - \dot{z}(t) \\
&= T_1 \left(A(\theta(t))x(t) + B(\theta(t))u(t) + R(\theta(t))d(t) \right) - \left(N(\theta(t))z(t) + J(\theta(t))u(t) + L(\theta(t))y(t) \right) \\
&= (T_1 A(\theta(t)) - L(\theta(t))C - N(\theta(t))T_1 E)x(t) + (T_1 B(\theta(t)) - J(\theta(t)))u(t) \\
&\quad + (T_1 R(\theta(t)) - L(\theta(t))G + N(\theta(t))T_2 G)d(t) + N(\theta(t))e(t)
\end{aligned} \tag{86}$$
$$\tag{87}$$

In order to guarantee convergence to zero of the estimation error, the following conditions are considered:

$$T_1 A(\theta(t)) - L(\theta(t))C - N(\theta(t))T_1 E = 0 \tag{88}$$
$$T_1 B(\theta(t)) - J(\theta(t)) = 0 \tag{89}$$

After some algebraic manipulations, the following equations equivalences are obtained:

$$N(\theta(t)) = T_1 A(\theta(t)) + K(\theta(t))C \tag{90}$$
$$\Gamma(\theta(t)) = L(\theta(t)) - N(\theta(t))T_2 \tag{91}$$
$$\bar{R}(\theta(t)) = T_1 R(\theta(t)) - \Gamma(\theta(t))G \tag{92}$$

A particular solution of both matrices T_1 and T_2 is computed as:

$$\begin{bmatrix} T_1 & T_2 \end{bmatrix} = \begin{bmatrix} I_{n_x} & 0 \end{bmatrix} \begin{bmatrix} E \\ C \end{bmatrix}^{\dagger} \tag{93}$$

The estimation error becomes:

$$\dot{e}(t) = N(\theta(t))e(t) + \bar{R}(\theta(t))d(t)$$

and, in order to guarantee robustness, the following H_∞ performance criterion is considered:

$$\dot{V}(e(t)) + e(t)^T e(t) - \gamma^2 d(t)^T d(t) < 0 \tag{94}$$

with attenuation level $\gamma > 0$ and quadratic Lyapunov function $V(e(t)) := e(t)^T P e(t)$, $P \succ 0$, such that the following BMI is obtained:

$$\begin{bmatrix} \text{He}\{PT_1 A(\theta(t)) + PK(\theta(t))C\} + I & PT_1 R(\theta(t)) - P\Gamma(\theta(t))G \\ * & -\gamma I \end{bmatrix} \prec 0 \tag{95}$$

Then, by considering the change of variable $\Xi(\theta(t)) = PK(\theta(t))$ and $\Omega = P\Gamma(\theta(t))$, the above LMI becomes:

$$\begin{bmatrix} \text{He}\{PT_1 A(\theta(t)) + \Xi(\theta(t))C\} + I & PT_1 R(\theta(t)) - \Omega(\theta(t))G \\ * & -\gamma I \end{bmatrix} \prec 0 \tag{96}$$

4. Safety in Convex Systems

Due to the increased demand of safety and reliability in complex systems, fault diagnosis techniques have attracted a great amount of attention in the past few decades. Concerning the recent developments of fault diagnosis for convex systems, hereafter we will review: (i) residual generation for fault detection; (ii) unknown input observers (UIO)-based fault isolation; (iii) observer-based fault estimation; and (iv) multiple model adaptive estimators (MMAEs). Fault tolerant control (FTC) systems aim at maintaining closed-loop stability and desired performances in the face of faults in some components, for example, actuators or sensors. The majority of the literature has focused the attention on LTI systems, although one can find ad-hoc approaches developed for nonlinear systems, see for example, Ref. [145]. It should not be surprising that some recent research has attempted to extend FTC strategies developed originally for LTI systems to the convex case, in order to enlarge their applicability to a wider class of nonlinear systems. This research has focused mainly on the following techniques, which will be reviewed in the following: (i) sliding mode control (SMC); (ii) control reconfiguration; and (iii) virtual actuators/sensors.

4.1. Residual Generation for Fault Detection

Fault detection aims at detecting accurately the appearance of a fault and is usually performed generating residual signals which act as fault indicators [146]. For the following convex system with unknown disturbances $d(t)$ and faults $f(t)$:

$$\dot{x}(t) = A\left(\theta(t)\right)x(t) + B\left(\theta(t)\right)u(t) + R\left(\theta(t)\right)d(t) + F\left(\theta(t)\right)f(t) \qquad (97)$$

$$y(t) = Cx(t) + Gd(t) + Hf(t) \qquad (98)$$

let us consider a detection filter with the following structure:

$$\dot{z}(t) = N\left(\theta(t)\right)z(t) + J\left(\theta(t)\right)u(t) + L\left(\theta(t)\right)y(t) \qquad (99)$$

$$\hat{x}(t) = z(t) - Ey(t) \qquad (100)$$

$$\hat{y}(t) = C\hat{x}(t) \qquad (101)$$

where $N\left(\theta(t)\right)$, $J\left(\theta(t)\right)$, $L\left(\theta(t)\right)$ are filter gains to be determined through design. By defining the estimation error signal $e(t) = x(t) - \hat{x}(t)$, the residual signal $r(t) = y(t) - \hat{y}(t)$ and the matrix $T = I + EC$, if the following constraints hold:

$$TA\left(\theta(t)\right) - N\left(\theta(t)\right)T - L\left(\theta(t)\right)C = 0 \qquad (102)$$

$$TB\left(\theta(t)\right) - J\left(\theta(t)\right) = 0 \qquad (103)$$

$$E\begin{bmatrix} G & H \end{bmatrix} = 0 \qquad (104)$$

then one obtains that the residual has a dynamics described by:

$$\dot{e}(t) = N\left(\theta(t)\right)e(t) + B_d\left(\theta(t)\right)d(t) + B_f\left(\theta(t)\right)f(t) \qquad (105)$$

$$r(t) = Ce(t) + Dd(t) + Hf(t) \qquad (106)$$

with:

$$B_d\left(\theta(t)\right) = [TR\left(\theta(t)\right) - L\left(\theta(t)\right)G - N\left(\theta(t)\right)EG)]$$

$$B_f\left(\theta(t)\right) = [TF\left(\theta(t)\right) - L\left(\theta(t)\right)H - N\left(\theta(t)\right)EH)]$$

Then, in order to achieve the fault detection goal, one must ensure the asymptotic stability of the error system while making the signal $r(t)$ as sensitive as possible to faults and as insensitive as possible to disturbances, which is usually achieved by means of a mix of H_∞ and H_- index optimization. This approach was initially proposed for filter design in the full-frequency domain, see for example, Refs. [147,148]. However, for some practical systems, fault and disturbance frequencies ranges are known beforehand, which motivated recent research on filter design in a finite-frequency domain [149–151], using the so-called generalized Kalman-Yakubovich-Popov (GKYP) lemma [152]. Another recent line of research worth of mentioning is the one that investigates the behavior of the fault detection observer when unmeasurable scheduling parameters are considered, see for example, Ref. [153].

4.2. Unknown Input Observers-Based Fault Isolation

In the last years, UIOs have shown to be a promising technique for fault detection purposes, due to their ability to provide the system state estimate even in the presence of unknown inputs, such as faults and disturbances. The approaches proposed for UIO design can be basically split into two classes: in the first one, the state estimation is decoupled from the unknown inputs, for example, by means of some structural conditions [154]. In the second case, joint estimation of the state and unknown inputs is achieved [155].

Let us consider the following convex system (for the sake of simplicity, the whole state is assumed to be measured):

$$\dot{x}(t) = A(\theta(t)) x(t) + B(\theta(t)) u(t) + F(\theta(t)) f(t) \qquad (107)$$
$$y(t) = x(t) \qquad (108)$$

and let $R(\theta(t))$ and $H(\theta(t))$ be some given matrix functions. Let us choose:

$$T(\theta(t)) = I - R(\theta(t)) \qquad (109)$$
$$S_1(\theta(t)) = R(\theta(t)) A(\theta(t)) - H(\theta(t)) \qquad (110)$$
$$S_2(\theta(t)) = H(\theta(t)) T(\theta(t)) \qquad (111)$$

then:

$$\dot{z}(t) = H(\theta(t)) z(t) + R(\theta(t)) B(\theta(t)) u(t) + [S(\theta(t)) - \dot{T}(\theta(t))] y(t) \qquad (112)$$
$$\hat{x}(t) = z(t) + T(\theta(t)) y(t) \qquad (113)$$

where $\dot{T}(\theta(t))$ is the time derivative of $T(\theta(t))$ and:

$$S(\theta(t)) = S_1(\theta(t)) + S_2(\theta(t)) \qquad (114)$$

is an unknown input observer for (107)–(108) [156], for which the dynamics of the estimation error $e(t) = x(t) - \hat{x}(t)$ is given by:

$$\dot{e}(t) = H(\theta(t)) e(t) + R(\theta(t)) F(\theta(t)) f(t) \qquad (115)$$

In fact, taking into account (107)–(108) and (112)–(113), one finds:

$$\dot{e}(t) = [A(\theta(t)) - S(\theta(t)) - T(\theta(t)) A(\theta(t))] x(t) - H(\theta(t)) z(t) \qquad (116)$$
$$+ [I - R(\theta(t)) - T(\theta(t))] B(\theta(t)) u(t) + [I - T(\theta(t))] F(\theta(t)) f(t)$$

which, using (109)–(111), can be rewritten as follows:

$$\dot{e}(t) = [H(\theta(t)) - H(\theta(t) T(\theta(t)))] x(t) - H(\theta(t)) z(t) + R(\theta(t)) F(\theta(t)) f(t) \qquad (117)$$

Then, it is easy to check that (115) follows from (117) taking into account (113).

The main feature of the estimation error dynamics in (115) is that convergence of $e(t)$ to zero when $f(t) = 0$ can be ensured by a proper choice of the matrix $H(\theta(t))$ (for example, as a diagonal matrix with strictly negative parameter-varying elements on the main diagonal). Moreover, the matrix $R(\theta(t))$ can be used to constrain the range of the matrix $R(\theta(t)) F(\theta(t))$, in such a way that different directions of $e(t)$ can be assigned to different faults, such that not only fault detection but also fault isolation can be achieved.

In the last years, one can recognize a trend in research that goes towards robustification of this technique, which was started by Ref. [157]. For instance, a few recent works have merged UIOs with interval observers [158–160], in such a way that instead of a single trajectory for the estimation error, lower and upper bounds which are compatible with the uncertainty are computed. On the other hand, other works have considered the case in which the scheduling variables are measured inexactly, see for example, Refs. [128,161,162]. Further improvements have been provided by Ref. [163], who have not restricted the parameter dependency of the UIO to mimic the one of the system, so that the decoupling conditions can be relaxed and have also considered the case in which the output equation of the convex system is not restricted to be parameter-independent.

4.3. Observer-Based Fault Estimation

As the name itself suggests, in observer-based fault estimation techniques, an observer is used to estimate the fault as though as if it were another (unmeasurable) state of the system. In order to exemplify the main idea behind these techniques, let us consider the following convex system:

$$\dot{x}(t) = A\left(\theta(t)\right)x(t) + B\left(\theta(t)\right)u(t) + F\left(\theta(t)\right)f(t) \tag{118}$$
$$y(t) = Cx(t) \tag{119}$$

and let us assume that the dynamics of $f(t)$ is described by:

$$\dot{f}(t) = A_f f(t) \tag{120}$$

with known matrix A_f (this assumption can be relaxed using, for example, an interval formulation). Then, it is possible to consider an augmented state $\bar{x}(t) = \left[x(t)^T, f(t)^T\right]^T$ such that the resulting augmented system is described by:

$$\dot{\bar{x}}(t) = \bar{A}\left(\theta(t)\right)\bar{x}(t) + \bar{B}\left(\theta(t)\right)u(t) \tag{121}$$
$$y(t) = \bar{C}x(t) \tag{122}$$

with:

$$\bar{A}\left(\theta(t)\right) = \begin{bmatrix} A\left(\theta(t)\right) & F\left(\theta(t)\right) \\ 0 & A_f \end{bmatrix} \quad \bar{B}\left(\theta(t)\right) = \begin{bmatrix} B\left(\theta(t)\right) \\ 0 \end{bmatrix} \quad \bar{C} = \begin{bmatrix} C & 0 \end{bmatrix}$$

Hence, a state observer designed to provide an estimate $\hat{\bar{x}}(t)$ of $\bar{x}(t)$ would provide an estimate $\hat{f}(t)$ of $f(t)$.

Among recent works developing further this concept, an adaptive polytopic observer which could estimated time-varying actuator faults was presented in Ref. [164] for convex descriptor systems, differing from most of other papers which assume generally that the actuator faults are constant. Sliding mode observers have been investigated by Refs. [165,166], which have considered the case of erroneous scheduling parameters. The case in which the fault's frequency content is not distributed within the whole frequency domain but in a finite interval of frequencies was addressed by Ref. [167] based on the GKYP lemma. An improvement of the design conditions has been brought by Ref. [168], which have developed a robust fault estimator via homogeneous polynomially parameter-dependent Lyapunov functions. It is worth highlighting that, although the majority of the results found in the literature consider the case of additive faults, some recent work has proposed a switched observer formulation to estimate actuator multiplicative faults in discrete-time convex systems [135]. Successful applications of observer-based fault estimation, either using high-fidelity simulations or through experimental validation, can be found in the areas of aviation [165,166], bioreactors [12], distillation columns [169], automotive suspension systems [170] and renewable microgrids [171,172].

4.4. Multiple Model Adaptive Estimators

The main idea behind MMAEs is to choose a set of models that represent the possible system behavior patterns and to obtain the state estimate as a combination of the estimates obtained from local state observers which run in parallel, each one based on the individual models that match these patterns [173,174]. The above mentioned combination is achieved as a weighted sum, where the weights represent the likelihood that the corresponding model is indeed true. Under certain conditions, the weight associated with the correct model converges to 1, while the other weights converge to 0, which allows an adaptive identification of the correct model [175]. This approach has been developed for discrete-time systems and it is exemplified hereafter.

Following Ref. [176], let us consider the discrete-time convex system:

$$x(k+1) = A\left(\zeta(k), \theta(k)\right) x(k) + B\left(\zeta(k), \theta(k)\right) u(k) \tag{123}$$
$$y(k) = C\left(\zeta(k), \theta(k)\right) x(k) \tag{124}$$

where using standard notation, $k \in \mathbb{Z}$ denotes a sample. Moreover, $\zeta(k)$ denotes an uncertain parameter, for which a finite set of candidate parameter values $\{\zeta_1, \zeta_2, \ldots, \zeta_N\}$ is considered.

For the system (123)–(124), state estimation is achieved by means of the following convex MMAE [176]:

$$\hat{x}(k) = \sum_{i=1}^{h} p_i(k) \hat{x}(k|\zeta_i) \tag{125}$$

$$\hat{y}(k) = \sum_{i=1}^{h} p_i(k) \hat{y}(k|\zeta_i) \tag{126}$$

$$\hat{\zeta}(k) = \zeta_{i^*(k)}, \quad i^*(k) = \arg\max_{i \in \{1,\ldots,h\}} p_i(k) \tag{127}$$

where $\hat{x}(k)$, $\hat{y}(k)$ and $\hat{\zeta}(k)$ denote the estimates of the state $x(k)$, the output $y(k)$ and the unknown parameter vector ζ, respectively and $p_i(k)$ are dynamic weights, which can be interpreted as a time-varying indicator of how likely it is that $\zeta = \zeta_i$. In (125)–(126), each $\hat{x}(k|\zeta_i), \hat{y}(k|\zeta_i)$ correspond to local estimates, obtained under the assumption that $\zeta = \zeta_i$.

The dynamic weights $p_i(k)$ appearing in (125)–(127) can be generated as follows:

$$p_i(k+1) = \frac{p_i(k) \beta_i(k) e^{-\omega_i(k)}}{\sum_{j=1}^{N} p_j(k) \beta_j(k) e^{-\omega_j(k)}} \tag{128}$$

where $\beta_i(k)$ is a positive weighting matrix function and $\omega_i(k)$ is the *error measuring function*, which describes how different is each local output estimate $\hat{y}(k|\zeta_i)$ from the observed output $y(k)$.

The above described convex MMAE, for which a conceptual scheme is provided in Figure 7, has some relevant properties. First of all, if $p_i(0) > 0 \,\forall i \in \{1, \ldots, h\}$, it can be proven that all the weights $p_i(k)$ generated by (128) are non-negative, uniformly bounded and contained in $[0,1]$, with $\sum_{i=1}^{h} p_i(k) = 1$, $\forall k > 0$. Moreover, it can be demonstrated that under some conditions the parameter estimate $\hat{\zeta}(k)$ will converge to a value ζ_i^* with $p_i^*(k) \to 1$ as $k \to \infty$, which corresponds to the local estimate that exhibits the smallest error measuring function.

These properties have been exploited for fault identification purposes by Ref. [177], where it was shown that a convex MMAE could be used to achieve icing diagnosis in unmanned aerial vehicles (UAVs) with the relevant feature that information about the icing location could be obtained. In this case, the idea is to assign different faulty models to different parameters ζ_i, in such a way that the dynamic weights would suggest which model is the one that fits data coming from the sensors the best. A similar idea was employed in Ref. [178], where a bank of observers, each one corresponding to a system description taking into account the presence of a particular fault, was used to address the problem of fault detection and isolation in near-space vehicles (NSVs) with actuator faults.

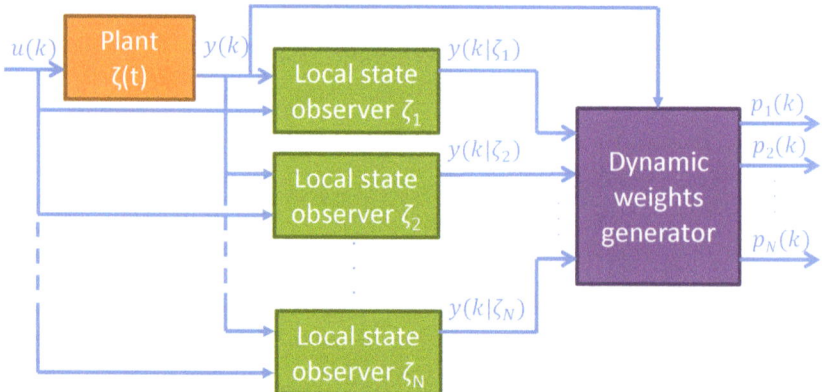

Figure 7. Multiple model adaptive estimator (MMAE) conceptual scheme.

4.5. Sliding Mode Fault Tolerant Control

The robustness of sliding mode control (SMC) against matched uncertainties makes it an interesting choice for FTC [179], although its direct application is impeded by the loss of regularity of the sliding mode when complete losses of actuators are considered. In fact, the first applications of SMC to convex systems were in a fault-free context [180]. The above mentioned problem was solved for overactuated systems by Ref. [181] by means of control allocation (CA), which provides an effective mechanism for distributing a virtual control signal among the available actuators.

More specifically, a convex plant subject to actuator faults represented by a diagonal semipositive definite matrix $W(t)$ with diagonal entries that represent the effectiveness level of actuators was considered in Ref. [181], as follows:

$$\dot{x}(t) = A\left(\theta(t)\right) x(t) + B_f E\left(\theta(t)\right) W(t) u(t) \tag{129}$$

where $E(\theta)$ is invertible for all $\theta \in \Theta$ and B_f is factored as $B_f = [B_1, B_2]^T$ with $B_2 B_2^T = I_l$, $l < n_u$ and $\|B_2\| > \|B_1\|$, so that B_2 represents the dominant contribution of the distribution of the control action within the channels of the system.

For the system (129), the control law is chosen as (see Figure 8 for an illustrative scheme):

$$u(t) = -\left(E\left(\theta(t)\right)\right)^{-1} B_2^T \left(B_2 E\left(\theta(t)\right) \hat{W}(t) \left(E\left(\theta(t)\right)\right)^{-1} B_2^T\right)^{-1} (v_l(t) + v_n(t)) \tag{130}$$

where $\hat{W}(t)$ is an estimate of $W(t)$, $v_l(t)$ is the linear component of the virtual control, chosen to be a standard state-feedback $v_l(t) = -Fx(t)$ and $v_n(t)$ is the nonlinear discontinuous part, which induces sliding and provides robustness:

$$v_n(t) = -\kappa(t) \frac{\sigma(t)}{\|\sigma(t)\|} \quad \text{for } \sigma(t) \neq 0 \tag{131}$$

where $\kappa(t)$ is an adaptive modulation function given by:

$$\kappa(t) = \|F\| \|x(t)\| \bar{\kappa}(t) + \eta \tag{132}$$

$$\dot{\bar{\kappa}}(t) = -\beta \bar{\kappa}(t) + \gamma \epsilon_0 \|F\| \|x(t)\| \|\sigma(t)\| \tag{133}$$

with $\beta, \gamma, \epsilon_0, \eta$ positive scalars and σ defines the sliding surface $\sigma(t) = 0$:

$$\sigma(t) = B_2 \left(B_f^T B_f\right)^{-1} B_f^T \left(x(t) - x(0) - \int_0^t \left(A\left(\theta(\tau)\right) - \begin{bmatrix} B_1 B_2^T \\ I_l \end{bmatrix} F\right) x(\tau) d\tau\right) \quad (134)$$

Finally, Ref. [145] shows that if F is designed such that the fault-free closed-loop system is quadratically stable, then it is possible to prove that for any faults/failures inside the set:

$$\mathcal{W}_\epsilon = \{(w_1, \ldots, w_{n_u}) \in [0,1] \times \ldots \times [0,1] : H(\theta)^T H(\theta) > \epsilon I\} \quad (135)$$

with ϵ small scalar which satisfies $0 < \epsilon \ll 1$ and $H(\theta) = B_2 E(\theta) W(t) (E(\theta))^{-1} B_2^T$, the sliding motion will be stable if:

$$\gamma_0 \gamma_1 \left(1 + \frac{c}{\sqrt{\epsilon}}\right) < 1 \quad (136)$$

where γ_0 is the $\mathcal{L}_2$ gain of $\tilde{G}(s) = F(sI - A(\theta) + B_v F)^{-1} [I_{n-l}, 0]^T$, $\gamma_1 = |B_1|$ (which, by assumption is small) and c represents the worst-case condition number (over Θ) of $E(\theta)$.

The design of the state-feedback controller in Ref. [181] was based on the assumption that all the plant states are available. This assumption was later relaxed by Ref. [182], where an unknown input observer (UIO) was used to estimate the unavailable plant states. Further research has led to develop some conditions based on the Lyapunov-Krasovskii functional approach that do not only guarantee the passivity and asymptotical stability of the closed-loop system but also cover the issue of actuator saturation and the existence of time-varying delays [183].

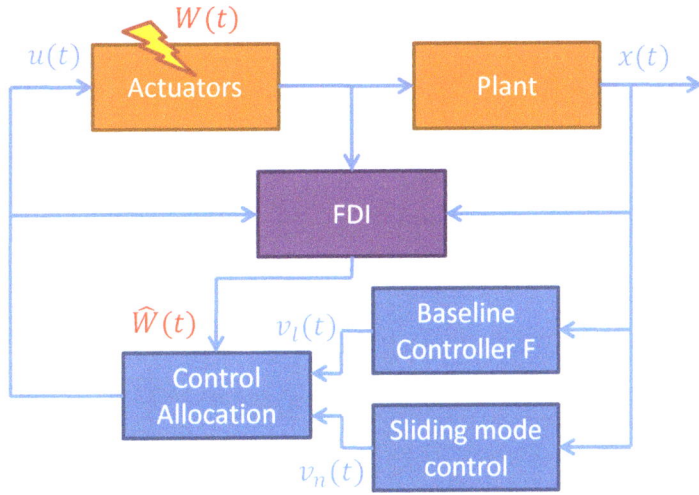

Figure 8. Sliding mode fault tolerant control (FTC) scheme.

4.6. Fault Tolerant Control Based on Controller Reconfiguration

On the other hand, as the name itself suggests, in controller reconfiguration, some modification of the control law is performed in order to compensate for the fault effect and make the faulty system behave as close as possible to the nominal system. The reconfiguration can be performed either by considering the fault estimation $\hat{f}(t)$ as an additional scheduling variable $\theta_f(t)$, as in Refs. [26,184,185] or by introducing a component in the control law, which is responsible to achieve fault tolerance, as proposed by Refs. [186–189]. In the first case, if we restrict our attention to a state-feedback control law (for the sake of simplicity), then it would have the following structure:

$$u(t) = K\left(\theta(t), \theta_f(t)\right) x(t) \quad (137)$$

with the advantage that fault tolerance would be achieved employing exactly the same LMI-based techniques employed for standard control design. On the other hand, in the second case, the control law is obtained as follows (see Figure 9):

$$u(t) = u_n(t) + u_f(t) \quad (138)$$

where $u_n(t)$ is the nominal state-feedback controller in fault-free condition and $u_f(t)$ is used to accommodate the faults. The advantage of this approach lies in that it eases the integrated design of fault estimator and fault tolerant controller, as discussed deeply in Ref. [188]. In order to illustrate this fact, let us consider the following convex system, which is a simplification of the class of systems considered in Ref. [188] by neglecting disturbances and parametric uncertainties in the state matrix:

$$\dot{x}(t) = A\left(\theta(t)\right) x(t) + B\left(\theta(t)\right) u(t) + F\left(\theta(t)\right) f_a(t) \quad (139)$$
$$y(t) = Cx(t) + F_s f_s(t) \quad (140)$$

where $f_a(t)$ and $f_s(t)$ denote actuator and sensor faults, respectively. By augmenting the state as $\tilde{x}(t) = [x(t), f_a(t), f_s(t)]^T$, the system (139)–(140) becomes:

$$\dot{\tilde{x}}(t) = \bar{A}\left(\theta(t)\right) \tilde{x}(t) + \bar{B}\left(\theta(t)\right) u(t) \quad (141)$$
$$y(t) = \bar{C}\tilde{x}(t) \quad (142)$$

for which an observer can be proposed, as follows:

$$\dot{z}(t) = M\left(\theta(t)\right) z(t) + J\left(\theta(t)\right) u(t) + L\left(\theta(t)\right) y(t) \quad (143)$$
$$\hat{x}(t) = z(t) + Hy(t) \quad (144)$$

where $z(t)$ and $\hat{x}(t)$ are the observer internal state and the estimate of $\tilde{x}(t)$, respectively. Under the assumption that:

$$L\left(\theta(t)\right) = L_1\left(\theta(t)\right) + L_2\left(\theta(t)\right) \quad (145)$$
$$L_2\left(\theta(t)\right) = \left(\Xi \bar{A}\left(\theta(t)\right) - L_1\left(\theta(t)\right) \bar{C}\right) H \quad (146)$$
$$M\left(\theta(t)\right) = \Xi \bar{A}\left(\theta(t)\right) - L_1\left(\theta(t)\right) \bar{C} \quad (147)$$
$$J\left(\theta(t)\right) = \Xi \bar{B}\left(\theta(t)\right) \quad (148)$$

with $\Xi = I - H\bar{C}$, the dynamics of the estimation error $e(t) = x(t) - \hat{x}(t)$ is described by:

$$\dot{e}(t) = M\left(\theta(t)\right) e(t) = \left[\Xi \bar{A}\left(\theta(t) - L_1\left(\theta(t)\right) \bar{C}\right)\right] e(t) \quad (149)$$

If the control law is chosen as:

$$u(t) = K\left(\theta(t)\right) \hat{x}(t) = \begin{bmatrix} K_x\left(\theta(t)\right) & K_f\left(\theta(t)\right) \end{bmatrix} \hat{x}(t) \quad (150)$$

where $K_x\left(\theta(t)\right)$ and $K_f\left(\theta(t)\right)$ are the state-feedback and actuator fault compensation gains respectively, then if $K_f\left(\theta(t)\right)$ is chosen as $K_f\left(\theta(t)\right) = -B\left(\theta(t)\right)^\dagger F\left(\theta(t)\right)$ (under the assumption that the actuator fault $f_a(t)$ is in the range space of the control input), one obtains:

$$\dot{x}(t) = [A(\theta(t)) + B(\theta(t))K_x(\theta(t))]x(t) + E(\theta(t))e(t) \tag{151}$$

with:

$$E(\theta(t)) = \begin{bmatrix} -B(\theta(t))K_x(\theta(t)) & F(\theta(t)) & 0 \end{bmatrix} \tag{152}$$

On the other hand, the sensor fault can be compensated as:

$$y_c(t) = y(t) - F_s \hat{f}_s(t) \tag{153}$$

where $y_c(t)$ is the compensated system output and $\hat{f}_s(t)$ is the sensor fault estimate. Since (149) and (151) describe an autonomous convex system, the integrated FE/FTC design can be formulated as an LMI-based stabilization problem (H_∞ optimization if there are uncertainties and/or disturbances).

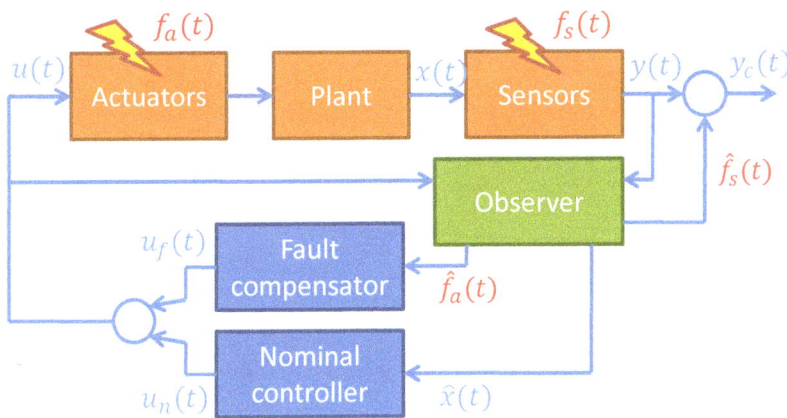

Figure 9. Controller reconfiguration FTC scheme.

4.7. Fault-Hiding via Virtual Actuators and Virtual Sensors

Contrarily to the previously described approach, the *fault-hiding* paradigm aims at reconfiguring the faulty plant instead of the controller/observer when a fault occurs [88]. The reconfiguration block hides the fault from the controller/observer point of view, such that it will see the same plant as before the fault and thus can be kept without modifying or retuning it (see Figure 10). The advantage of doing so is that fault tolerance can be added to an already existing control scheme by means of a plug-and-play philosophy. In case of actuator faults, the reconfiguration block is named *virtual actuator* [190], whereas it is named *virtual sensor* in the case of sensor faults [191]. Although virtual sensors and virtual actuators were initially considered separately, an overall scheme that employs both of them in order to tolerate simultaneous actuator and sensor faults was later developed [192,193]. Some recent work has also studied issues related to the existence of input constraints (saturations) and fault isolation delays [194].

Let us consider the following convex system:

$$\dot{x}(t) = A(\theta(t))x(t) + B(\theta(t))W(t)(u(t) + f_a(t)) \tag{154}$$
$$y(t) = V(t)Cx(t) + f_s(t) \tag{155}$$

where, consistently with the previously described approaches, $W(t)$ and $V(t)$ denote losses of effectiveness in the actuators and sensors, respectively, whereas $f_a(t)$ and $f_s(t)$ denote additive actuator/sensor faults.

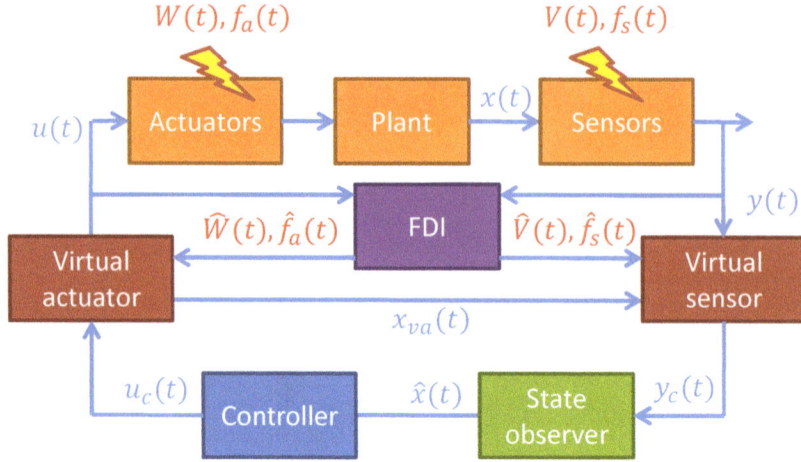

Figure 10. FTC using virtual sensors and virtual actuators.

The structure of the convex virtual actuator depends on the following rank condition ($\hat{W}$ denotes an estimation of W) [192]:

$$\text{rank}\left(B\left(\theta(t)\hat{W}(t)\right)\right) = \text{rank}\left(B\left(\theta(t)\right)\right) \quad \forall \theta \in \Theta \tag{156}$$

which describes whether fault tolerance can be achieved through a simple redistribution of the control inputs. In the first case, the reconfiguration structure is as follows:

$$u(t) = N_{va}\left(\theta(t)\right) u_c(t) - \hat{f}_a(t) \tag{157}$$

where $\hat{f}_a(t)$ is an estimation of $f_a(t)$ and $N_{va}\left(\theta(t)\right)$ is given by:

$$N_{va}\left(\theta(t)\right) = \left[B\left(\theta(t)\right)\hat{W}(t)\right]^\dagger B\left(\theta(t)\right) \tag{158}$$

In case (156) does not hold, the virtual actuator becomes a dynamical system with state equation:

$$\dot{x}_{va}(t) = \left[A\left(\theta(t)\right) + B^*\left(\theta(t)\right) M_{va}\left(\theta(t)\right)\right] x_{va}(t) + \left[B\left(\theta(t)\right) - B^*\left(\theta(t)\right)\right] u_c(t) \tag{159}$$

and output equation (reconfiguration structure):

$$u(t) = N_{va}\left(\theta(t)\right) \left[u_c(t) - M_{va}\left(\theta(t)\right) x_{va}(t) - \hat{f}_a(t)\right] \tag{160}$$

where $x_{va}(t)$ is the virtual actuator state, $M_{va}\left(\theta(t)\right)$ denotes the virtual actuator gain and the matrix $B^*\left(\theta(t)\right)$ is obtained as:

$$B^*\left(\theta(t)\right) = B\left(\theta(t)\right) \hat{W}(t) N_{va}\left(\theta(t)\right) \tag{161}$$

Similarly, the structure of the virtual sensor depends on the following rank condition ($\hat{V}$ is the estimation of V) [192]:

$$\text{rank}\left(\hat{V}(t)C\right) = \text{rank}(C) \tag{162}$$

so that if it holds, then the virtual sensor is a static block, whereas if the above condition does not hold, then it is a dynamical system with internal state $x_{vs}(t)$ and dynamics described by the equations:

$$\dot{x}_{vs}(t) = [A(\theta(t)) + M_{vs}(\theta(t))C^*] x_{vs}(t) + B(\theta(t)) u_c(t) \qquad (163)$$
$$\phantom{\dot{x}_{vs}(t) =} - M_{vs}(\theta(t)) N_{vs}(t) \left(y(t) + \hat{V}(t) C x_{va}(t) - \hat{f}_s(t) \right)$$
$$y_c(t) = N_{vs}(t) \left(y(t) + \hat{V}(t) C x_{va}(t) - \hat{f}_s(t) \right) + (C - C^*) x_{vs}(t) \qquad (164)$$

where $\hat{f}_s(t)$ denotes an estimation of $f_s(t)$, $M_{vs}(\theta(t))$ is the virtual sensor gain and:

$$N_{vs}(t) = C \left[\hat{V}(t) C \right]^{\dagger} \qquad (165)$$
$$C^* = N_{vs}(t) \hat{V}(t) C \qquad (166)$$

Then, it is possible to show that if the control loop consists of a state-feedback law with controller gain $K(\theta(t))$ and a Luenberger observer with gain $L(\theta(t))$, then thanks to the introduction of the virtual actuator/sensor in the loop, one can find an appropriate similarity transformation of the overall augmented state such that the dynamics in the new state coordinates $\check{x}$ is described by:

$$\dot{\check{x}}(t) = \begin{bmatrix} A(\theta) + M_{vs}(\theta) C^* & 0 & 0 & 0 \\ \star & A(\theta) + L(\theta) C & 0 & 0 \\ \star & \star & A(\theta) + B(\theta) K(\theta) & 0 \\ \star & \star & \star & A(\theta) + B^*(\theta) M_{va}(\theta) \end{bmatrix} \check{x}(t) \qquad (167)$$

where $\star$ denotes some generic non-zero terms, which is in a block-triangular structure. Then, since $A(\theta) + L(\theta)C$ and $A(\theta) + B(\theta)K(\theta)$ are already stable due to the stability of the faultless system, one can ensure overall stability under fault occurrence by designing the gains $M_{vs}(\theta)$ and $M_{va}(\theta)$ so that $A(\theta) + M_{vs}(\theta)C^*$ and $A(\theta) + B^*(\theta)M_{va}(\theta)$ are stable (the reader is referred to Ref. [192] for further details on the method, along with a discussion about the quadratic stability of block-triangular convex systems).

Finally, it is worth mentioning that some recent works have combined model predictive control (MPC) with the convex formulation in order to take into account possible input and state constraints associated to actuator saturation and other physical limitations [195,196]. The convex MPC framework has been used to go one step further than FTC, that is, to perform health-aware control on the basis of the information about the system reliability provided by a prognosis and health management (PHM) module [197,198]. This type of control strategy increases the overall reliability, anticipates the apparition of faults and reduces the operational costs.

5. Conclusions

In this paper, we have performed a review of the most applied techniques in control, observation and safety of convex systems. With this terminology we have wished to unify the concepts of linear parameter varying and Takagi-Sugeno systems, with the purpose of allowing the reader to taste all the flavors of techniques offered by the humongous existing literature about these classes of systems. Due to the huge amount of papers, the review is in no way meant to be exhaustive but it is meant to be a helpful document to look for any reader who wishes to locate himself/herself in this field and learn about the main used techniques. We feel that we have done our best to provide a discussion about the state-of-the-art of the topic. However, in spite of our best efforts, many publications could not be included and for this reason, we would like to apologize in advance for any omission.

In addition, it is important to mention that this paper is mainly focused on discussing the advances of polytopic convex systems. However, it is acknowledged that other approaches that lead to an LPV representation exist, such as grid-based LPV [199], linear fractional transformation (LFT)-based LPV [200,201], polynomial LPV approaches [202] and tensor model-based transformation [203,204], among others. Also, the reader should note that, throughout the review, only methods based on quadratic Lyapunov functions have been discussed. Nevertheless, less conservative solutions can be obtained based on non-quadratic Lyapunov functions, for example, the polyquadratic, as proposed in Refs. [69–71]. In general, all these topics are currently investigated and, therefore, the above references are recommended to the interested reader.

Author Contributions: Conceptualization, F.-R.L.-E., D.R., and G.V.-P.; investigation, F.-R.L.-E., D.R., and G.V.-P.; writing-original draft preparation, F.-R.L.-E., D.R., and G.V.-P.; writing-review and editing F.-R.L.-E., D.R., and G.V.-P.

Funding: This work has been supported by Tecnológico Nacional de México grants 6723.18-P and 5400.19-P, the Instituto de Ciencia Tecnológia e Innovación, Chiapas (ICTIECH) 1123/2019. This work has been partially funded by the Spanish State Research Agency (AEI) and the European Regional Development Fund (ERFD) through the projects SCAV (ref. MINECO DPI2017-88403-R) and DEOCS (ref. MINECO DPI2016-76493), and also by AGAUR ACCIO RIS3CAT UTILITIES 4.0 – P7 SECUTIL. This work has been also supported by the AEI through the Maria de Maeztu Seal of Excellence to IRI (MDM-2016-0656) and the grant Juan de la Cierva-Formacion (FJCI-2016-29019).

Conflicts of Interest: The authors declare no conflict of interest.

References

1. Shamma, J.; Athans, M. Guaranteed properties for nonlinear gain scheduled control systems. In Proceedings of the 27th IEEE Conference on Decision and Control, Austin, TX, USA, 7–9 December 1988; Volume 3, pp. 2202–2208.
2. Shamma, J.; Athans, M. Analysis of gain scheduled control for nonlinear plants. *IEEE Trans. Autom. Control* **1990**, *35*, 898–907. [CrossRef]
3. Shamma, J. An Overview of LPV Systems. In *Control of Linear Parameter Varying Systems with Applications*; Mohammadpour, J., Scherer, C.W., Eds.; Springer: Boston, MA, USA, 2012; pp. 3–26.
4. Tanaka, K.; Wang, H.O. *Fuzzy Control Systems Design and Analysis: A Linear Matrix Inequality Approach*; John Wiley & Sons: Hoboken, NJ, USA, 2004. [CrossRef]
5. Jadbabaie, A.; Jamshidi, M.; Titli, A. Guaranteed-cost design of continuous-time Takagi-Sugeno fuzzy controllers via linear matrix inequalities. In Proceedings of the IEEE International Conference on Fuzzy Systems Proceedings, IEEE World Congress on Computational Intelligence, Anchorage, AK, USA, 4–9 May 1998; Volume 1, pp. 268–273.
6. Ohtake, H.; Tanaka, K.; Wang, H.O. Fuzzy modeling via sector nonlinearity concept. *Integr. Comput.-Aided Eng.* **2003**, *10*, 333–341. [CrossRef]
7. Ichalal, D.; Marx, B.; Ragot, J.; Maquin, D. State estimation of Takagi–Sugeno systems with unmeasurable premise variables. *IET Control Theory Appl.* **2010**, *4*, 897–908. [CrossRef]
8. Lendek, Z.; Guerra, T.M.; Babuska, R.; De Schutter, B. *Stability Analysis and Nonlinear Observer Design Using Takagi-Sugeno Fuzzy Models*; Springer: Berlin/Heidelberg, Germany, 2011. [CrossRef]
9. López-Estrada, F.R.; Astorga-Zaragoza, C.M.; Theilliol, D.; Ponsart, J.C.; Valencia-Palomo, G.; Torres, L. Observer synthesis for a class of Takagi–Sugeno descriptor system with unmeasurable premise variable. Application to fault diagnosis. *Int. J. Syst. Sci.* **2017**, *48*, 3419–3430. [CrossRef]
10. Wang, Z.; Li, F.; Qin, Y.; Li, D.; Ma, G.; Ma, J. *A Novel Dual Nonlinear Observer for Vehicle System Roll Behavior With Lateral and Vertical Coupling*; SAE Technical Paper; SAE International: Warrendale, PA, USA, 2019. [CrossRef]
11. Nagy Kiss, A.M.; Marx, B.; Mourot, G.; Schutz, G.; Ragot, J. State estimation of two-time scale multiple models. Application to wastewater treatment plant. *Control Eng. Pract.* **2011**, *19*, 1354–1362. [CrossRef]
12. López-Estrada, F.R.; Ponsart, J.C.; Astorga-Zaragoza, C.M.; Camas-Anzueto, J.L.; Theilliol, D. Robust sensor fault estimation for descriptor-LPV systems with unmeasurable gain scheduling functions: Application to an anaerobic bioreactor. *Int. J. Appl. Math. Comput. Sci.* **2015**, *25*, 233–244. [CrossRef]
13. Zhao, Z.; Wang, Y.; Liu, F. A multi-way LPV modeling method for batch processes. *J. Process Control* **2018**, *65*, 56–67. [CrossRef]

14. Gómez-Peñate, S.; López-Estrada, F.R.; Valencia-Palomo, G.; Rotondo, D.; Enríquez-Zárate, J. Actuator and sensor fault estimation based on a proportional-integral quasi-LPV observer with inexact scheduling parameters. In Proceedings of the 3rd IFAC Workshop on Linear Parameter-Varying Systems, Eindhoven, The Netherlands, 4–6 November 2019.
15. Lopez-Estrada, F.R.; Astorga-Zaragoza, C.M.; Valencia-Palomo, G.; Rios-Rojas, C.; Galicia-Gonzalez, C.; Escobar-Gomez, E. Observer-based LPV stabilization system for a riderless bicycle. *IEEE Latin Am. Trans.* **2018**, *16*, 1076–1083. [CrossRef]
16. López-Estrada, F.R.; Theilliol, D.; Astorga-Zaragoza, C.M.; Ponsart, J.C.; Valencia-Palomo, G.; Camas Anzueto, J. Fault diagnosis observer for descriptor Takagi-Sugeno systems. *Neurocomputing* **2019**, *331*, 10–17. [CrossRef]
17. Pfifer, H.; Moreno, C.P.; Theis, J.; Kotikapuldi, A.; Gupta, A.; Takarics, B.; Seiler, P. Linear parameter varying techniques applied to aeroservoelastic aircraft: In memory of Gary Balas. *IFAC-PapersOnLine* **2015**, *48*, 103–108. [CrossRef]
18. Huang, B.; Lu, B.; Li, Q. A proportional–integral-based robust state-feedback control method for linear parameter-varying systems and its application to aircraft. *Proc. Inst. Mech. Eng. Part G* **2019**, *233*, 4663–4675. [CrossRef]
19. Gutjahr, B.; Gröll, L.; Werling, M. Lateral vehicle trajectory optimization using constrained linear time-varying MPC. *IEEE Trans. Intell. Transp. Syst.* **2016**, *18*, 1586–1595. [CrossRef]
20. Gómez-Peñate, S.; López-Estrada, F.R.; Valencia-Palomo, G.; Osornio-Ríos, R.; Zepeda-Hernández, J.; Rios-Rojas, C.; Camas-Anzueto, J. Sensor fault diagnosis observer for an electric vehicle modeled as a Takagi-Sugeno system. *J. Sens.* **2018**, *2018*, 3291639. [CrossRef]
21. Chadli, M.; Borne, P. *Multiple Models Approach in Automation: Takagi-Sugeno Fuzzy Systems*; Wiley Online Library: London, UK, 2013. [CrossRef]
22. Rotondo, D. *Advances in Gain-Scheduling and Fault Tolerant Control Techniques*; Springer: Cham, Switzerland, 2017. [CrossRef]
23. Bernal, M.; Estrada, V.; Márquez, R. *Diseño e Implementación de Sistemas de Control Basados en Estructuras Convexas Y Desigualdades Matriciales Lineales*; Pearson: México city, Mexico, 2019.
24. Rodrigues, M.; Hamdi, H.; BenHadj-Braiek, N.; Theilliol, D. Observer-based fault tolerant control design for a class of LPV descriptor systems. *J. Frankl. Inst.* **2014**, *351*, 3104–3125. [CrossRef]
25. Rotondo, D.; Nejjari, F.; Puig, V. Quasi-LPV modeling, identification and control of a twin rotor MIMO system. *Control Eng. Pract.* **2013**, *21*, 829–846. [CrossRef]
26. Rotondo, D.; Nejjari, F.; Puig, V. Robust quasi–LPV model reference FTC of a quadrotor UAV subject to actuator faults. *Int. J. Appl. Math. Comput. Sci.* **2015**, *25*, 7–22. [CrossRef]
27. He, D.F.; Huang, H.; Chen, Q.X. Quasi-min–max MPC for constrained nonlinear systems with guaranteed input-to-state stability. *J. Frankl. Inst.* **2014**, *351*, 3405–3423. [CrossRef]
28. He, Z.; Zhao, L. Quadrotor trajectory tracking based on quasi-LPV system and internal model control. *Math. Probl. Eng.* **2015**, *2015*, 857291. [CrossRef]
29. Rizzello, G.; Naso, D.; Turchiano, B.; Seelecke, S. Robust position control of dielectric elastomer actuators based on LMI optimization. *IEEE Trans. Control Syst. Technol.* **2016**, *24*, 1909–1921. [CrossRef]
30. Rizzello, G.; Ferrante, F.; Naso, D.; Seelecke, S. Robust interaction control of a dielectric elastomer actuator with variable stiffness. *IEEE/ASME Trans. Mech.* **2017**, *22*, 1705–1716. [CrossRef]
31. Pérez-Estrada, A.J.; Osorio-Gordillo, G.L.; Darouach, M.; Alma, M.; Olivares-Peregrino, V.H. Generalized dynamic observers for quasi-LPV systems with unmeasurable scheduling functions. *Int. J. Robust Nonlinear Control* **2018**, *28*, 5262–5278. [CrossRef]
32. Robles, R.; Sala, A.; Bernal, M. Performance-oriented quasi-LPV modeling of nonlinear systems. *Int. J. Robust Nonlinear Control* **2019**, *29*, 1230–1248. [CrossRef]
33. Baranyi, P. Extracting LPV and qLPV structures from state-space functions: A TP model transformation based framework. *IEEE Trans. Fuzzy Syst.* **2019**. [CrossRef]
34. Rotondo, D.; Puig, V.; Nejjari, F.; Witczak, M. Automated generation and comparison of Takagi–Sugeno and polytopic quasi-LPV models. *Fuzzy Sets Syst.* **2015**, *277*, 44–64. [CrossRef]
35. López-Estrada, F.R.; Ponsart, J.C.; Theilliol, D.; Zhang, Y.; Astorga-Zaragoza, C.M. LPV model-based tracking control and robust sensor fault diagnosis for a quadrotor UAV. *J. Intell. Robot. Syst.* **2016**, *84*, 163–177. [CrossRef]

36. Ramírez, M.; Villafuerte, R.; González, T.; Bernal, M. Exponential estimates of a class of time–delay nonlinear systems with convex representations. *Int. J. Appl. Math. Comput. Sci.* **2015**, *25*, 815–826. [CrossRef]
37. Arceo, J.C.; Vázquez, D.; Estrada-Manzo, V.; Márquez, R.; Bernal, M. Nonlinear convex control of the Furuta pendulum based on its descriptor model. In Proceedings of the 13th International Conference on Electrical Engineering, Computing Science and Automatic Control (CCE), Mexico City, Mexico, 26–30 September 2016; pp. 1–6.
38. Quintana, D.; Estrada-Manzo, V.; Bernal, M. A methodology for real-time implementation of nonlinear observers via convex optimization. In Proceedings of the 15th International Conference on Electrical Engineering, Computing Science and Automatic Control (CCE), Mexico City, Mexico, 5–7 September 2018; pp. 1–6.
39. Arceo, J.C.; Sánchez, M.; Estrada-Manzo, V.; Bernal, M. Convex stability analysis of nonlinear singular systems via linear matrix inequalities. *IEEE Trans. Autom. Control* **2018**, *64*, 1740–1745. [CrossRef]
40. López-Estrada, F.R.; Hernández-de León, H.R.; Estrada-Manzo, V.; Bernal, M. LMI-based fault detection and isolation of nonlinear descriptor systems. In Proceedings of the IEEE International Conference on Fuzzy Systems (FUZZ-IEEE), Naples, Italy, 9–12 July 2017; pp. 1–5.
41. Hoffmann, C.; Werner, H. A survey of linear parameter-varying control applications validated by experiments or high-fidelity simulations. *IEEE Trans. Control Syst. Technol.* **2014**, *23*, 416–433. [CrossRef]
42. Apkarian, P.; Biannic, J.M.; Gahinet, P. Self-scheduled H_∞ control of missile via linear matrix inequalities. *J. Guid. Control Dyn.* **1995**, *18*, 532–538. [CrossRef]
43. Rugh, W.J.; Shamma, J.S. Research on gain scheduling. *Automatica* **2000**, *36*, 1401–1425. [CrossRef]
44. Shamma, J.S.; Athans, M. Gain scheduling: Potential hazards and possible remedies. *IEEE Control Syst. Mag.* **1992**, *12*, 101–107.
45. Rotondo, D.; Nejjari, F.; Puig, V. Robust state-feedback control of uncertain LPV systems: An LMI-based approach. *J. Frankl. Inst.* **2014**, *351*, 2781–2803. [CrossRef]
46. Sato, M. Gain-scheduled output-feedback controllers depending solely on scheduling parameters via parameter-dependent Lyapunov functions. *Automatica* **2011**, *47*, 2786–2790. [CrossRef]
47. Sato, M.; Peaucelle, D. Gain-scheduled output-feedback controllers using inexact scheduling parameters for continuous-time LPV systems. *Automatica* **2013**, *49*, 1019–1025. [CrossRef]
48. Kajiwara, H.; Apkarian, P.; Gahinet, P. LPV Techniques for Control of an Inverted Pendulum. *IEEE Control Syst.* **1999**, *19*, 44–54.
49. Bruzelius, F.; Breitholtz, C.; Pettersson, S. LPV-based gain scheduling technique applied to a turbo fan engine model. In Proceedings of the International Conference on Control Applications, Glasgow, UK, 18–20 September 2002; pp. 713–718.
50. Scherer, C.W. *Robust Mixed Control and LPV Control with Full Block Scaling*; Technical Report; Delft University of Technology, Mechanical Engineering Systems and Control Group: Delft, The Netherlands, 2004.
51. Mohammadpour, J.; Scherer, C. *Control of Linear Parameter Varying Systems with Applications*; Springer: New York, NY, USA, 2012. [CrossRef]
52. Xu, H.E.; Jun, Z.; M, D.G.; Chao, C. Switching control for LPV polytopic systems using multiple Lyapunov functions. In Proceedings of the 30th Chinese Control Conference, Yantai, China, 22–24 July 2011; pp. 1771–1776.
53. Xu, H.E.; Jun, Z.; Dimirovski, G.M. A blending method control of switched LPV systems with slow-varying parameters and its application to an F-16 aircraft model. In Proceedings of the 30th Chinese Control Conference, Yantai, China, 22–24 July 2011; pp. 1765–1770.
54. Shin, J.; Balas, G.; Kaya, A.M. Blending methodology of linear parameter varying control synthesis of F-16 aircraft system. *J. Guid. Control Dyn.* **2002**, *25*, 1040–1048. [CrossRef]
55. Xie, W.; Kamiya, Y.; Eisaka, T. Robust control system design for polytopic stable LPV systems. *IMA J. Math. Control Inf.* **2003**, *20*, 201–216. [CrossRef]
56. Qiu, J.; Feng, G.; Gao, H. Fuzzy-Model-Based Piecewise H_∞ Static-Output-Feedback Controller Design for Networked Nonlinear Systems. *IEEE Trans. Fuzzy Syst.* **2010**, *18*, 919–934. [CrossRef]
57. Yin, X.; Zhang, L.; Zhu, Y.; Wang, C.; Li, Z. Robust control of networked systems with variable communication capabilities and application to a semi-active suspension system. *IEEE/ASME Trans. Mech.* **2016**, *21*, 2097–2107. [CrossRef]

58. Shamma, J.S. Analysis and Design of Gain Scheduled Control Systems. Ph.D. Thesis, Massachusetts Institute of Technology, Cambridge, MA, USA, 1988.
59. Shamma, J.S.; Athans, M. Guaranteed properties of gain scheduled control for linear parameter-varying plants. *Automatica* **1991**, *27*, 559–564. [CrossRef]
60. Goebel, R.; Hu, T.; Teel, A.R. Dual matrix inequalities in stability and performance analysis of linear differential/difference inclusions. In *Current Trends in Nonlinear Systems and Control*; Springer: Boston, MA, USA, 2006; pp. 103–122. [CrossRef]
61. Apkarian, P.; Gahinet, P.; Becker, G. Self-scheduled H_∞ control of linear parameter-varying systems: A design example. *Automatica* **1995**, *31*, 1251–1261. [CrossRef]
62. Pandey, A.; Sehr, M.; de Oliveira, M. Pre-filtering in gain-scheduled and robust control. In Proceedings of the American Control Conference (ACC), Boston, MA, USA, 6–8 July 2016; pp. 3698–3703.
63. Sehr, M.A.; de Oliveira, M.C. Pre-filtering and post-filtering in gain-scheduled output-feedback control. *Int. J. Robust Nonlinear Control* **2017**, *27*, 3259–3279. [CrossRef]
64. Sala, A.; Arino, C. Asymptotically necessary and sufficient conditions for stability and performance in fuzzy control: Applications of Polya's theorem. *Fuzzy Sets Syst.* **2007**, *158*, 2671–2686. [CrossRef]
65. Kruszewski, A.; Sala, A.; Guerra, T.M.; Ariño, C. A triangulation approach to asymptotically exact conditions for fuzzy summations. *IEEE Trans. Fuzzy Syst.* **2009**, *17*, 985–994. [CrossRef]
66. Sala, A.; Arino, C. Relaxed stability and performance LMI conditions for Takagi–Sugeno fuzzy systems with polynomial constraints on membership function shapes. *IEEE Trans. Fuzzy Syst.* **2008**, *16*, 1328–1336. [CrossRef]
67. Tuan, H.D.; Apkarian, P.; Narikiyo, T.; Yamamoto, Y. Parameterized linear matrix inequality techniques in fuzzy control system design. *IEEE Trans. Fuzzy Syst.* **2001**, *9*, 324–332. [CrossRef]
68. Wang, H.O.; Tanaka, K.; Griffin, M. Parallel distributed compensation of nonlinear systems by Takagi-Sugeno fuzzy model. In Proceedings of the 1995 IEEE International Conference on Fuzzy Systems, Yokohama, Japan, 20–24 March 1995; Volume 2, pp. 531–538.
69. Wang, L.K.; Zhang, H.G.; Liu, X.D. H_∞ Observer Design for Continuous-Time Takagi-Sugeno Fuzzy Model with Unknown Premise Variables via Nonquadratic Lyapunov Function. *IEEE Trans. Cybern.* **2016**, *46*, 1986–1996. [CrossRef]
70. Márquez, R.; Guerra, T.M.; Bernal, M.; Kruszewski, A. A non-quadratic Lyapunov functional for H_∞ control of nonlinear systems via Takagi-Sugeno models. *J. Frankl. Inst.* **2016**, *353*, 781–796. [CrossRef]
71. Márquez, R.; Guerra, T.M.; Bernal, M.; Kruszewski, A. Asymptotically necessary and sufficient conditions for Takagi–Sugeno models using generalized non-quadratic parameter-dependent controller design. *Fuzzy Sets Syst.* **2017**, *306*, 48–62. [CrossRef]
72. Pandey, A.; de Oliveira, M.C. Quadratic and poly-quadratic discrete-time stabilizability of linear parameter-varying systems. *IFAC-PapersOnLine* **2017**, *50*, 8624–8629. [CrossRef]
73. Lam, H.K.; Wu, L.; Zhao, Y. Linear matrix inequalities-based membership-function-dependent stability analysis for non-parallel distributed compensation fuzzy-model-based control systems. *IET Control Theory Appl.* **2014**, *8*, 614–625. [CrossRef]
74. Cherifi, A.; Guelton, K.; Arcese, L. Quadratic design of d-stabilizing non-pdc controllers for quasi-lpv/ts models. *IFAC-PapersOnLine* **2015**, *48*, 164–169. [CrossRef]
75. Guerra, T.M.; Bernal, M.; Guelton, K.; Labiod, S. Non-quadratic local stabilization for continuous-time Takagi–Sugeno models. *Fuzzy Sets Syst.* **2012**, *201*, 40–54. [CrossRef]
76. Daafouz, J.; Bernussou, J. Parameter dependent Lyapunov functions for discrete time systems with time varying parametric uncertainties. *Syst. Control Lett.* **2001**, *43*, 355–359. [CrossRef]
77. Chadli, M.; Daafouz, J.; Darouach, M. Stabilisation of singular LPV systems. *IFAC Proc. Vol.* **2008**, *41*, 9999–10002. [CrossRef]
78. Pandey, A.P.; de Oliveira, M.C. On the necessity of LMI-based design conditions for discrete time LPV filters. *IEEE Trans. Autom. Control* **2018**, *63*, 3187–3188. [CrossRef]
79. El Ghaoui, L.; Oustry, F.; AitRami, M. A cone complementarity linearization algorithm for static output-feedback and related problems. *IEEE Trans. Autom. Control* **1997**, *42*, 1171–1176. [CrossRef]
80. Prempain, E.; Postlethwaite, I. Static H_∞ loop shaping control of a fly-by-wire helicopter. *Automatica* **2005**, *41*, 1517–1528. [CrossRef]

81. Henrion, D.; Lasserre, J.B. Convergent relaxations of polynomial matrix inequalities and static output feedback. *IEEE Trans. Autom. Control* **2006**, *51*, 192–202. [CrossRef]
82. Apkarian, P.; Noll, D. Nonsmooth H_∞ synthesis. *IEEE Trans. Autom. Control* **2006**, *51*, 71–86. [CrossRef]
83. Scherer, C.; Gahinet, P.; Chilali, M. H_∞ design with pole placement constraints: An LMI approach. *IEEE Trans. Autom. Control* **1996**, *41*, 358–367.
84. Gahinet, P. Explicit controller formulas for LMI-based H_∞ synthesis. *Automatica* **1996**, *32*, 1007–1014. [CrossRef]
85. Amato, F. *Robust Control of Linear Systems Subject to Uncertain Time-Varying Parameters*; Springer: Berlin/Heidelberg, Germany, 2006; Volume 325. [CrossRef]
86. Kose, I.E.; Jabbari, F. Control of LPV systems with partly measured parameters. *IEEE Trans. Autom. Control* **1999**, *44*, 658–663. [CrossRef]
87. Abdullah, A.; Zribi, M. Model reference control of LPV systems. *J. Frankl. Inst.* **2009**, *346*, 854–871. [CrossRef]
88. Rotondo, D.; Puig, V.; Nejjari, F.; Romera, J. A fault-hiding approach for the switching quasi-LPV fault-tolerant control of a four-wheeled omnidirectional mobile robot. *IEEE Trans. Ind. Electron.* **2014**, *62*, 3932–3944. [CrossRef]
89. Valencia-Palomo, G.; Rossiter, J.A. Auto-tuned predictive control based on minimal plant information. *IFAC Proc. Vol.* **2009**, *42*, 554–559. [CrossRef]
90. Valencia-Palomo, G.; Rossiter, J.; López-Estrada, F. Improving the feed-forward compensator in predictive control for setpoint tracking. *ISA Trans.* **2014**, *53*, 755–766. [CrossRef] [PubMed]
91. Valencia-Palomo, G.; Hilton, K.; Rossiter, J.A. Predictive control implementation in a PLC using the IEC 1131.3 programming standard. In Proceedings of the 2009 European Control Conference (ECC), Budapest, Hungary, 23–26 August 2009; pp. 1317–1322.
92. Kothare, M.V.; Balakrishnan, V.; Morari, M. Robust constrained model predictive control using linear matrix inequalities. *Automatica* **1996**, *32*, 1361–1379. [CrossRef]
93. Lu, Y.; Arkun, Y. Quasi-min-max MPC algorithms for LPV systems. *Automatica* **2000**, *36*, 527–540. [CrossRef]
94. Park, P.; Jeong, S.C. Constrained RHC for LPV systems with bounded rates of parameter variations. *Automatica* **2004**, *40*, 865–872. [CrossRef]
95. Lu, Y.; Arkun, Y. Polytope updating in quasi-min-max MPC algorithms. *IFAC Proc. Vol.* **2000**, *33*, 407–412. [CrossRef]
96. Lu, Y.; Arkun, Y. A scheduling quasi–min-max model predictive control algorithm for nonlinear systems. *J. Process Control* **2002**, *12*, 589–604. [CrossRef]
97. Lee, S.; Won, S. Model predictive control for linear parameter varying systems using a new parameter dependent terminal weighting matrix. *IEICE Trans. Fund. Electron. Commun. Comput. Sci.* **2006**, *89*, 2166–2172. [CrossRef]
98. Wada, N.; Saito, K.; Saeki, M. Model predictive control for linear parameter varying systems using parameter dependent Lyapunov function. *IEEE Trans. Circuits Syst. II* **2006**, *12*, 1446–1450. [CrossRef]
99. Pluymers, B.; Rossiter, J.; Suykens, J.; De Moor, B. The efficient computation of polyhedral invariant sets for linear systems with polytopic uncertainty. In Proceedings of the American Control Conference, Portland, OR, USA, 8–10 June 2005; pp. 804–809. [CrossRef]
100. Garone, E.; Casavola, A. Receding horizon control strategies for constrained LPV systems based on a class of nonlinearly parameterized Lyapunov functions. *IEEE Trans. Autom. Control* **2012**, *57*, 2354–2360. [CrossRef]
101. Yu, S.; Böhm, C.; Chen, H.; Allgöwer, F. Model predictive control of constrained LPV systems. *Int. J. Control* **2012**, *85*, 671–683. [CrossRef]
102. Besselmann, T.; Lofberg, J.; Morari, M. Explicit MPC for LPV systems: Stability and optimality. *IEEE Trans. Autom. Control* **2012**, *57*, 2322–2332. [CrossRef]
103. Zhang, J.; Xiu, X. Kd tree based approach for point location problem in explicit model predictive control. *J. Frankl. Inst.* **2018**, *355*, 5431–5451. [CrossRef]
104. Ariño, C.; Querol, A.; Sala, A. Shape-independent model predictive control for Takagi–Sugeno fuzzy systems. *Eng. Appl. Artif. Intell.* **2017**, *65*, 493–505. [CrossRef]
105. Hanema, J.; Lazar, M.; Tóth, R. Stabilizing tube-based model predictive control: Terminal set and cost construction for LPV systems. *Automatica* **2017**, *85*, 137–144. [CrossRef]
106. Ding, B.; Wang, P.; Hu, J. Dynamic output feedback robust MPC with one free control move for LPV model with bounded disturbance. *Asian J. Control* **2018**, *20*, 755–767. [CrossRef]

107. Morato, M.M.; Nguyen, M.Q.; Sename, O.; Dugard, L. Design of a fast real-time LPV model predictive control system for semi-active suspension control of a full vehicle. *J. Frankl. Inst.* **2019**, *356*, 1196–1224. [CrossRef]
108. Ghersin, A.S.; Pena, R.S.S. Applied LPV control with full block multipliers and regional pole placement. *J. Control Sci. Eng.* **2010**, *2010*, 3. [CrossRef]
109. Ostertag, E. *Mono-and Multivariable Control and Estimation: Linear, Quadratic and LMI Methods*; Springer: Berlin/Heidelberg, Germany, 2011. [CrossRef]
110. Rotondo, D.; Puig, V.; Nejjari, F. Linear quadratic control of LPV systems using static and shifting specifications. In Proceedings of the European Control Conference (ECC), Linz, Austria, 15–17 July 2015; pp. 3085–3090.
111. Briat, C. Stability analysis and control of a class of LPV systems with piecewise constant parameters. *Syst. Control Lett.* **2015**, *82*, 10–17. [CrossRef]
112. Bruzelius, F.; Pettersson, S.; Breitholtz, C. Region of attraction estimates for LPV-gain scheduled control systems. In Proceedings of the European Control Conference (ECC), Cambridge, UK, 1–4 September 2003; pp. 892–897. [CrossRef]
113. Pitarch, J.L.; Sala, A.; Arino, C.V. Closed-form estimates of the domain of attraction for nonlinear systems via fuzzy-polynomial models. *IEEE Trans. Cybern.* **2013**, *44*, 526–538. [CrossRef]
114. Lendek, Z.; Lauber, J. Local stability of discrete-time TS fuzzy systems. *IFAC-PapersOnLine* **2016**, *49*, 7–12. [CrossRef]
115. Lendek, Z.; Nagy, Z.; Lauber, J. Local stabilization of discrete-time TS descriptor systems. *Eng. Appl. Artif. Intell.* **2018**, *67*, 409–418. [CrossRef]
116. Zhang, X.; Tsiotras, P.; Knospe, C. Stability analysis of LPV time-delayed systems. *Int. J. Control* **2002**, *75*, 538–558. [CrossRef]
117. Wu, F.; Grigoriadis, K.M. LPV systems with parameter-varying time delays: Analysis and control. *Automatica* **2001**, *37*, 221–229. [CrossRef]
118. Briat, C.; Sename, O.; Lafay, J.F. Memory-resilient gain-scheduled state-feedback control of uncertain LTI/LPV systems with time-varying delays. *Syst. Control Lett.* **2010**, *59*, 451–459. [CrossRef]
119. Briat, C. *Linear Parameter-Varying and Time-Delay Systems: Analysis, Observation, Filtering Control*; Springer: Berlin/Heidelberg, Germany, 2014; Volume 3. [CrossRef]
120. Guzmán-Rabasa, J.A.; López-Estrada, F.R.; González-Contreras, B.M.; Valencia-Palomo, G.; Chadli, M.; Pérez-Patricio, M. Actuator fault detection and isolation on a quadrotor unmanned aerial vehicle modeled as a linear parameter-varying system. *Meas. Control* **2019**. [CrossRef]
121. Zhang, H.; Zhang, G.; Wang, J. H_∞ Observer Design for LPV Systems With Uncertain Measurements on Scheduling Variables: Application to an Electric Ground Vehicle. *IEEE/ASME Trans. Mech.* **2016**, *21*, 1659–1670. [CrossRef]
122. Li, H.; Wu, C.; Yin, S.; Lam, H.K. Observer-based fuzzy control for nonlinear networked systems under unmeasurable premise variables. *IEEE Trans. Fuzzy Syst.* **2015**, *24*, 1233–1245. [CrossRef]
123. Ibrir, S.; Sabir, A. Robust observer-based stabilization and tracking of uncertain linear systems with $\mathcal{L}$2-gain performance: Application to DC motors. In Proceedings of the 2016 IEEE International Energy Conference (ENERGYCON), Leuven, Belgium, 4–8 April 2016; pp. 1–6. [CrossRef]
124. Gauterin, E.; Kammerer, P.; Kühn, M.; Schulte, H. Effective wind speed estimation: Comparison between Kalman Filter and Takagi–Sugeno observer techniques. *ISA Trans.* **2016**, *62*, 60–72. [CrossRef]
125. Brizuela-Mendoza, J.A.; Astorga-Zaragoza, C.M.; Zavala-Río, A.; Pattalochi, L.; Canales-Abarca, F. State and actuator fault estimation observer design integrated in a riderless bicycle stabilization system. *ISA Trans.* **2016**, *61*, 199–210. [CrossRef]
126. Bergsten, P.; Palm, R.; Driankov, D. Observers for Takagi-Sugeno fuzzy systems. *IEEE Trans. Syst. Man Cybern. Part B* **2002**, *32*, 114–121. [CrossRef]
127. Zhang, L.; Yin, X.; Ning, Z.; Ye, D. Robust filtering for a class of networked nonlinear systems with switching communication channels. *IEEE Trans. Cybern.* **2016**, *47*, 671–682. [CrossRef]
128. Gómez-Peñate, S.; Valencia-Palomo, G.; López-Estrada, F.R.; Astorga-Zaragoza, C.M.; Osornio-Rios, R.A.; Santos-Ruiz, I. Sensor fault diagnosis based on a sliding mode and unknown input observer for Takagi-Sugeno systems with uncertain premise variables. *Asian J. Control* **2019**, *21*, 339–353. [CrossRef]

129. Busawon, K.K.; Kabore, P. Disturbance attenuation using proportional integral observers. *Int. J. Control* **2001**, *74*, 618–627. [CrossRef]
130. Youssef, T.; Chadli, M.; Karimi, H.R.; Zelmat, M. Design of unknown inputs proportional integral observers for TS fuzzy models. *Neurocomputing* **2014**, *123*, 156–165. [CrossRef]
131. Kang, D. Design of a disturbance observer for discrete-time linear systems. In Proceedings of the 14th International Conference on Control, Automation and Systems (ICCAS), Seoul, South Korea, 22–25 October 2014; pp. 1381–1383.
132. Xu, J.; Mi, C.C.; Cao, B.; Deng, J.; Chen, Z.; Li, S. The state of charge estimation of lithium-ion batteries based on a proportional-integral observer. *IEEE Trans. Veh. Technol.* **2013**, *63*, 1614–1621.
133. Youssef, T.; Chadli, M.; Karimi, H.R.; Wang, R. Actuator and sensor faults estimation based on proportional integral observer for TS fuzzy model. *J. Frankl. Inst.* **2017**, *354*, 2524–2542. [CrossRef]
134. Rotondo, D.; Cristofaro, A.; Johansen, T.A.; Nejjari, F.; Puig, V. Detection of icing and actuators faults in the longitudinal dynamics of small UAVs using an LPV proportional integral unknown input observer. In Proceedings of the 3rd Conference on Control and Fault-Tolerant Systems (SysTol), Barcelona, Spain, 7–9 September 2016; pp. 690–697. [CrossRef]
135. Rotondo, D.; López-Estrada, F.R.; Nejjari, F.; Ponsart, J.C.; Theilliol, D.; Puig, V. Actuator multiplicative fault estimation in discrete-time LPV systems using switched observers. *J. Frankl. Inst.* **2016**, *353*, 3176–3191. [CrossRef]
136. Duan, G.R. *Analysis and Design of Descriptor Linear Systems*; Springer: New York, NY, USA, 2010. [CrossRef]
137. Masubuchi, I.; Kato, J.; Saeki, M.; Ohara, A. Gain-scheduled controller design based on descriptor representation of LPV systems: Application to flight vehicle control. In Proceedings of the 43rd IEEE Conference on Decision and Control (CDC), Nassau, Bahamas, 14–17 December 2004; Volume 1, pp. 815–820.
138. Baghaee, H.R.; Mirsalim, M.; Gharehpetian, G.B.; Talebi, H.A. A generalized descriptor-system robust H_∞ control of autonomous microgrids to improve small and large signal stability considering communication delays and load nonlinearities. *Int. J. Electr. Power Energy Syst.* **2017**, *92*, 63–82. [CrossRef]
139. González, A.; Estrada-Manzo, V.; Guerra, T.M. Gain-scheduled H_∞ admissibilisation of LPV discrete-time systems with LPV singular descriptor. *Int. J. Syst. Sci.* **2017**, *48*, 3215–3224. [CrossRef]
140. Arceo, J.C.; Villafuerte, R.; Estrada-Manzo, V.; Bernal, M. LMI-Based Controller Design for Time-Delay Nonlinear Descriptor Systems with Guaranteed Exponential Estimates. *IFAC-PapersOnLine* **2018**, *51*, 585–590. [CrossRef]
141. González, A.; Guerra, T.M. Enhanced Predictor-Based Control Synthesis for Discrete-Time TS Fuzzy Descriptor Systems With Time-Varying Input Delays. *IEEE Trans. Fuzzy Syst.* **2018**, *27*, 402–410. [CrossRef]
142. Guerra, T.M.; Estrada-Manzo, V.; Lendek, Z. Observer design for Takagi–Sugeno descriptor models: An LMI approach. *Automatica* **2015**, *52*, 154–159. [CrossRef]
143. Estrada-Manzo, V.; Lendek, Z.; Guerra, T.M. Generalized LMI observer design for discrete-time nonlinear descriptor models. *Neurocomputing* **2016**, *182*, 210–220. [CrossRef]
144. Hamdi, H.; Rodrigues, M.; Mechmeche, C.; Theilliol, D. Fault detection and isolation for linear parameter varying descriptor systems via proportional integral observer. *Int. J. Adapt. Control Signal Proc.* **2012**, *26*, 224–240. [CrossRef]
145. Alwi, H.; Edwards, C. Fault tolerant longitudinal aircraft control using non-linear integral sliding mode. *IET Control Theory Appl.* **2014**, *8*, 1803–1814. [CrossRef]
146. Gertler, J. *Fault Detection and Diagnosis*; John Wiley & Sons, Ltd.: Hoboken, NJ, USA, 2008. [CrossRef]
147. Wei, X.; Verhaegen, M. LMI solutions to the mixed H_∞/H_- fault detection observer design for linear parameter-varying systems. *Int. J. Adapt. Control Signal Proc.* **2011**, *25*, 114–136. [CrossRef]
148. Chadli, M.; Abdo, A.; Ding, S.X. H_-/H_∞ fault detection filter design for discrete-time Takagi-Sugeno fuzzy system. *Automatica* **2013**, *49*, 1996–2005. [CrossRef]
149. Chibani, A.; Chadli, M.; Shi, P.; Braiek, N.B. Fuzzy fault detection filter design for T–S fuzzy systems in the finite-frequency domain. *IEEE Trans. Fuzzy Syst.* **2016**, *25*, 1051–1061. [CrossRef]
150. Wang, Z.; Shi, P.; Lim, C.C. H_-/H_∞ fault detection observer in finite frequency domain for linear parameter-varying descriptor systems. *Automatica* **2017**, *86*, 38–45. [CrossRef]
151. Chibani, A.; Chadli, M.; Ding, S.X.; Braiek, N.B. Design of robust fuzzy fault detection filter for polynomial fuzzy systems with new finite frequency specifications. *Automatica* **2018**, *93*, 42–54. [CrossRef]

152. Iwasaki, T.; Hara, S. Generalized KYP lemma: Unified frequency domain inequalities with design applications. *IEEE Trans. Autom. Control* **2005**, *50*, 41–59. [CrossRef]
153. Estrada, F.L.; Ponsart, J.C.; Theilliol, D.; Astorga-Zaragoza, C.M. Robust H_-/H_∞ fault detection observer design for descriptor-LPV systems with unmeasurable gain scheduling functions. *Int. J. Control* **2015**, *88*, 2380–2391. [CrossRef]
154. Darouach, M.; Zasadzinski, M.; Xu, S.J. Full-order observers for linear systems with unknown inputs. *IEEE Trans. Autom. Control* **1994**, *39*, 606–609. [CrossRef]
155. Rotondo, D.; Witczak, M.; Puig, V.; Nejjari, F.; Pazera, M. Robust unknown input observer for state and fault estimation in discrete-time Takagi–Sugeno systems. *Int. J. Syst. Sci.* **2016**, *47*, 3409–3424. [CrossRef]
156. Rotondo, D.; Cristofaro, A.; Johansen, T.A.; Nejjari, F.; Puig, V. Icing detection in unmanned aerial vehicles with longitudinal motion using an LPV unknown input observer. In Proceedings of the Conference on Control Applications (CCA), Sydney, Australia, 21–23 September 2015; pp. 984–989. [CrossRef]
157. Chadli, M.; Karimi, H.R. Robust observer design for unknown inputs Takagi–Sugeno models. *IEEE Trans. Fuzzy Syst.* **2012**, *21*, 158–164. [CrossRef]
158. Meyer, L.; Ichalal, D.; Vigneron, V. Interval observer for LPV systems with unknown inputs. *IET Control Theory Appl.* **2017**, *12*, 649–660. [CrossRef]
159. Rotondo, D.; Cristofaro, A.; Johansen, T.A.; Nejjari, F.; Puig, V. State estimation and decoupling of unknown inputs in uncertain LPV systems using interval observers. *Int. J. Control* **2018**, *91*, 1944–1961. [CrossRef]
160. Rotondo, D.; Cristofaro, A.; Johansen, T.A.; Nejjari, F.; Puig, V. Robust fault and icing diagnosis in unmanned aerial vehicles using LPV interval observers. *Int. J. Robust Nonlinear Control* **2018**. [CrossRef]
161. Hassanabadi, A.H.; Shafiee, M.; Puig, V. Actuator fault diagnosis of singular delayed LPV systems with inexact measured parameters via PI unknown input observer. *IET Control Theory Appl.* **2017**, *11*, 1894–1903. [CrossRef]
162. Xu, F.; Tan, J.; Wang, Y.; Wang, X.; Liang, B.; Yuan, B. Robust fault detection and set-theoretic UIO for discrete-time LPV systems with state and output equations scheduled by inexact scheduling variables. *IEEE Trans. Autom. Control* **2019**. [CrossRef]
163. Marx, B.; Ichalal, D.; Ragot, J.; Maquin, D.; Mammar, S. Unknown input observer for LPV systems. *Automatica* **2019**, *100*, 67–74. [CrossRef]
164. Rodrigues, M.; Hamdi, H.; Theilliol, D.; Mechmeche, C.; BenHadj Braiek, N. Actuator fault estimation based adaptive polytopic observer for a class of LPV descriptor systems. *Int. J. Robust Nonlinear Control* **2015**, *25*, 673–688. [CrossRef]
165. Chandra, K.P.B.; Alwi, H.; Edwards, C. Fault detection in uncertain LPV systems with imperfect scheduling parameter using sliding mode observers. *Eur. J. Control* **2017**, *34*, 1–15. [CrossRef]
166. Chen, L.; Edwards, C.; Alwi, H. Sensor fault estimation using LPV sliding mode observers with erroneous scheduling parameters. *Automatica* **2019**, *101*, 66–77. [CrossRef]
167. Zhang, K.; Jiang, B.; Shi, P.; Xu, J. Analysis and design of robust H_-/H_∞ fault estimation observer with finite-frequency specifications for discrete-time fuzzy systems. *IEEE Trans. Cybern.* **2014**, *45*, 1225–1235. [CrossRef] [PubMed]
168. Xie, X.; Yue, D.; Zhang, H.; Xue, Y. Fault estimation observer design for discrete-time Takagi–Sugeno fuzzy systems based on homogenous polynomially parameter-dependent Lyapunov functions. *IEEE Trans. Cybern.* **2017**, *47*, 2504–2513. [CrossRef] [PubMed]
169. Lopez-Estrada, F.R.; Ponsart, J.C.; Theilliol, D.; Astorga-Zaragoza, C.; Flores-Montiel, M. Robust state and fault estimation observer for discrete-time D-LPV systems with unmeasurable gain scheduling functions. Application to a binary distillation column. *IFAC-PapersOnLine* **2015**, *48*, 1012–1017. [CrossRef]
170. Morato, M.M.; Sename, O.; Dugard, L.; Nguyen, M.Q. Fault estimation for automotive Electro-Rheological dampers: LPV-based observer approach. *Control Eng. Pract.* **2019**, *85*, 11–22. [CrossRef]
171. Morato, M.M.; Mendes, P.R.; Normey-Rico, J.E.; Bordons, C. Robustness conditions of LPV fault estimation systems for renewable microgrids. *Int. J. Electr. Power Energy Syst.* **2019**, *111*, 325–350. [CrossRef]
172. Morato, M.M.; Regner, D.J.; Mendes, P.R.; Normey-Rico, J.E.; Bordons, C. Fault analysis, detection and estimation for a microgrid via H_2/H_∞ LPV observers. *Int. J. Electr. Power Energy Syst.* **2019**, *105*, 823–845. [CrossRef]
173. Li, X.R.; Bar-Shalom, Y. Multiple-model estimation with variable structure. *IEEE Trans. Autom. Control* **1996**, *41*, 478–493.

174. Hassani, V.; Aguiar, A.P.; Athans, M.; Pascoal, A.M. Multiple model adaptive estimation and model identification usign a minimum energy criterion. In Proceedings of the American Control Conference (ACC), St. Louis, MO, USA, 10–12 June 2009; pp. 518–523. [CrossRef]
175. Xiong, K.; Wei, C.; Liu, L. Robust multiple model adaptive estimation for spacecraft autonomous navigation. *Aerosp. Sci. Technol.* **2015**, *42*, 249–258. [CrossRef]
176. Rotondo, D.; Hassani, V.; Cristofaro, A. A multiple model adaptive architecture for the state estimation in discrete-time uncertain LPV systems. In Proceedings of the American Control Conference (ACC), Seattle, WA, USA, 24–26 May 2017; pp. 2393–2398.
177. Rotondo, D.; Cristofaro, A.; Hassani, V.; Johansen, T.A. Icing diagnosis in unmanned aerial vehicles using an LPV multiple model estimator. *IFAC-PapersOnLine* **2017**, *50*, 5238–5243. [CrossRef]
178. Yang, G.H.; Wang, H. Fault detection and isolation for a class of uncertain state-feedback fuzzy control systems. *IEEE Trans. Fuzzy Syst.* **2014**, *23*, 139–151. [CrossRef]
179. Alwi, H.; Edwards, C.; Tan, C.P. *Fault Detection and Fault-Tolerant Control Using Sliding Modes*; Springer: London, UK, 2011. [CrossRef]
180. Sivrioglu, S.; Nonami, K. Sliding mode control with time-varying hyperplane for AMB systems. *IEEE/ASME Trans. Mech.* **1998**, *3*, 51–59. [CrossRef]
181. Alwi, H.; Edwards, C.; Stroosma, O.; Mulder, J.; Hamayun, M.T. Real-time implementation of an ISM fault-tolerant control scheme for LPV plants. *IEEE Trans. Ind. Electron.* **2014**, *62*, 3896–3905.
182. Hamayun, M.T.; Ijaz, S.; Bajodah, A.H. Output integral sliding mode fault tolerant control scheme for LPV plants by incorporating control allocation. *IET Control Theory Appl.* **2017**, *11*, 1959–1967. [CrossRef]
183. Selvaraj, P.; Kaviarasan, B.; Sakthivel, R.; Karimi, H.R. Fault-tolerant SMC for Takagi–Sugeno fuzzy systems with time-varying delay and actuator saturation. *IET Control Theory Appl.* **2017**, *11*, 1112–1123. [CrossRef]
184. Shin, J.Y.; Wu, N.E.; Belcastro, C. Adaptive linear parameter varying control synthesis for actuator failure. *J. Guid. Control Dyn.* **2004**, *27*, 787–794. [CrossRef]
185. Sloth, C.; Esbensen, T.; Stoustrup, J. Robust and fault-tolerant linear parameter-varying control of wind turbines. *Mechatronics* **2011**, *21*, 645–659. [CrossRef]
186. Jia, Q.; Chen, W.; Zhang, Y.; Li, H. Fault reconstruction and fault-tolerant control via learning observers in Takagi–Sugeno fuzzy descriptor systems with time delays. *IEEE Trans. Ind. Electron.* **2015**, *62*, 3885–3895. [CrossRef]
187. Li, X.; Zhu, F.; Chakrabarty, A.; Żak, S.H. Nonfragile fault-tolerant fuzzy observer-based controller design for nonlinear systems. *IEEE Trans. Fuzzy Syst.* **2016**, *24*, 1679–1689. [CrossRef]
188. Lan, J.; Patton, R.J. Integrated design of fault-tolerant control for nonlinear systems based on fault estimation and T-S fuzzy modeling. *IEEE Trans. Fuzzy Syst.* **2016**, *25*, 1141–1154. [CrossRef]
189. Li, X.; Lu, D.; Zeng, G.; Liu, J.; Zhang, W. Integrated fault estimation and non-fragile fault-tolerant control design for uncertain Takagi–Sugeno fuzzy systems with actuator fault and sensor fault. *IET Control Theory Appl.* **2017**, *11*, 1542–1553. [CrossRef]
190. Rotondo, D.; Nejjari, F.; Puig, V.; Blesa, J. Model reference FTC for LPV systems using virtual actuators and set-membership fault estimation. *Int. J. Robust Nonlinear Control* **2015**, *25*, 735–760. [CrossRef]
191. Nazari, R.; Seron, M.M.; De Doná, J.A. Fault-tolerant control of systems with convex polytopic linear parameter varying model uncertainty using virtual-sensor-based controller reconfiguration. *Annu. Rev. Control* **2013**, *37*, 146–153. [CrossRef]
192. Rotondo, D.; Nejjari, F.; Puig, V. A virtual actuator and sensor approach for fault tolerant control of LPV systems. *J. Process Control* **2014**, *24*, 203–222. [CrossRef]
193. Tabatabaeipour, S.M.; Stoustrup, J.; Bak, T. Fault-tolerant control of discrete-time LPV systems using virtual actuators and sensors. *Int. J. Robust Nonlinear Control* **2015**, *25*, 707–734. [CrossRef]
194. Rotondo, D.; Ponsart, J.C.; Nejjari, F.; Theilliol, D.; Puig, V. Virtual actuator-based FTC for LPV systems with saturating actuators and FDI delays. In Proceedings of the 3rd Conference on Control and Fault-Tolerant Systems (SysTol), Barcelona, Spain, 7–9 September 2016; pp. 831–837.
195. Witczak, P.; Luzar, M.; Witczak, M.; Korbicz, J. A robust fault-tolerant model predictive control for linear parameter-varying systems. In Proceedings of the 19th International Conference on Methods and Models in Automation and Robotics (MMAR), Miedzyzdroje, Poland, 2–5 September 2014; pp. 462–467.
196. Acevedo-Valle, J.M.; Puig, V.; Tornil-Sin, S.; Witczak, M.; Rotondo, D. Predictive Fault Tolerant Control for LPV systems using model reference. *IFAC-PapersOnLine* **2015**, *48*, 30–35. [CrossRef]

197. Pour, F.K.; Puig, V.; Cembrano, G. Health-aware LPV-MPC based on system reliability assessment for drinking water networks. In Proceedings of the IEEE Conference on Control Technology and Applications (CCTA), Copenhagen, Denmark, 21–24 August 2018; pp. 187–192.
198. Pour, F.K.; Puig, V.; Cembrano, G. Health-aware LPV-MPC based on a reliability-based remaining useful life assessment. *IFAC-PapersOnLine* **2018**, *51*, 1285–1291. [CrossRef]
199. Wu, F.; Yang, X.H.; Packard, A.; Becker, G. Induced L2-norm control for LPV systems with bounded parameter variation rates. *Int. J. Robust Nonlinear Control* **1996**, *6*, 983–998. [CrossRef]
200. Packard, A. Gain scheduling via linear fractional transformations. *Syst. Control Lett.* **1994**, *22*, 79–92. [CrossRef]
201. Veenman, J.; Scherer, C.W. Stability analysis with integral quadratic constraints: A dissipativity based proof. In Proceedings of the 52nd IEEE Conference on Decision and Control, Florence, Italy, 10–13 December 2013; pp. 3770–3775.
202. Brizuela-Mendoza, J.; Sorcia-Vázquez, F.J.; Guzmán-Valdivia, C.H.; Osorio-Sánchez, R.; Martínez-García, M. Observer design for sensor and actuator fault estimation applied to polynomial LPV systems: A riderless bicycle study case. *Int. J. Syst. Sci.* **2018**, *49*, 2996–3006. [CrossRef]
203. Baranyi, P.; Yam, Y.; Várlaki, P. *Tensor Product Model Transformation in Polytopic Model-Based Control*; CRC Press: Boca Raton, FL, USA, 2013.
204. Takarics, B.; Baranyi, P. Tensor-product-model-based control of a three degrees-of-freedom aeroelastic model. *J. Guid. Control Dyn.* **2013**, *36*, 1527–1533. [CrossRef]

© 2019 by the authors. Licensee MDPI, Basel, Switzerland. This article is an open access article distributed under the terms and conditions of the Creative Commons Attribution (CC BY) license (http://creativecommons.org/licenses/by/4.0/).

Article

Effect of Control Horizon in Model Predictive Control for Steam/Water Loop in Large-Scale Ships

Shiquan Zhao [1,2,3,*], **Anca Maxim** [1,3,4], **Sheng Liu** [2], **Robin De Keyser** [1,3] **and Clara Ionescu** [1,3,5]

1. Research Group on Dynamical Systems and Control, Department of Electrical Energy, Metals, Mechanical Constructions and Systems, Ghent University, B9052 Ghent, Belgium; anca.maxim@ac.tuiasi.ro (A.M.); Robain.DeKeyser@ugent.be (R.D.K.); ClaraMihaela.Ionescu@UGent.be (C.I.)
2. College of Automation, Harbin Engineering University, Harbin 150001, China; liu.sch@163.com
3. Core Lab EEDT-Energy Efficient Drive Trains, Flanders Make, 3920 Lommel, Belgium
4. Department of Automatic Control and Applied Informatics, Gheorghe Asachi Technical University of Iasi, Blvd. Mangeron 27, 700050 Iasi, Romania
5. Department of Automation, Technical University of Cluj-Napoca, Memorandumului Street No.28, 400114 Cluj-Napoca, Romania
* Correspondence: Shiquan.Zhao@UGent.be; Tel.: +32-048-645-2783

Received: 16 November 2018; Accepted: 11 December 2018; Published: 14 December 2018

Abstract: This paper presents an extensive analysis of the properties of different control horizon sets in an Extended Prediction Self-Adaptive Control (EPSAC) model predictive control framework. Analysis is performed on the linear multivariable model of the steam/water loop in large-scale watercraft/ships. The results indicate that larger control horizon values lead to better loop performance, at the cost of computational complexity. Hence, it is necessary to find a good trade-off between the performance of the system and allocated or available computational complexity. In this original work, this problem is explicitly treated as an optimization task, leading to the optimal control horizon sets for the steam/water loop example. Based on simulation results, it is concluded that specific tuning of control horizons outperforms the case when only a single valued control horizon is used for all the loops.

Keywords: model predictive control; control horizon; steam power plant; steam/water loop; multi-input and multi-output system; loop design

---

1. Introduction

The steam/water loop is a water supply process in a steam power plant with highly interconnected equipment. Good steam/water loop performance is a prerequisite for the steam power plant to operate properly [1]. However, due to the complicated interactions between the dynamic variables and the harsh working environment of the watercraft, there are difficulties in obtaining satisfying performance for the complex dynamics of such a steam/water loop [2]. The ever-increasing system complexity and demand for high performance of this sub-system within the broader operation system of the watercraft also pose challenges to operations. In this context, an effective control method is required to guarantee safe operation of the steam/water loop.

In order to design an effective controller for the steam/water loop, constraints such as: input saturations or rate limits have to be taken into consideration. There are several possibilities to deal with the constraints in the literature [3–6], including also model predictive control (MPC) [7,8], applied specifically in steam power plants. For example, an economic model predictive control was developed for the boiler-turbine system [9]. The economic index was utilized directly as a cost function, and the economic model predictive control realized the economic optimization as well as the dynamic tracking. In order to guarantee the stability of the closed loop system, a Sontage controller and corresponding region were designed. A stable model predictive tracking controller (SMPTC) for coordinated control of a large-scale power plant was proposed [10]. By using fuzzy clustering and a subspace identification method, a Takagi–Sugeno (TS) fuzzy model was established. Then, through the SMPTC method, the system obtained good set-point tracking performance while guaranteeing input-to-state stability and the input constraints of the system. A non-linear generalized predictive controller based on neuro-fuzzy network (NFGPC) is proposed in [11], which consists of local generalized predictive controllers (GPCs) designed using the local linear models of the neuro-fuzzy network that models the plant. Liu discussed the performance of coordinated control on the steam-boiler generation plant using two non-linear model predictive control methods [12]. One of these methods is the input output feedback linearization technique based on a suitably chosen approximated linear model. The other method is based on neuro-fuzzy networks to represent a non-linear dynamic process using a set of local models. To improve the learning ability of the MPC method, Liu proposed a non-linear model predictive controller based on iterative learning control [13]. In practice, the MPC method was also applied to the boiler control system to enable tight dynamical coordination of selected controlled variables, particularly the coordination of air and fuel flows during transients [14].

The works introduced above are mainly about the application of model predictive control on the boiler-turbine system installed on land. However, the steam power plants installed on the large-scale watercraft or ships have more differences compared to those installed on land. Some of these are: (i) receiving more disturbances from the ocean waves; (ii) of smaller capacity; (iii) used at multiple operation points with varying state processes. According to these characteristics, there is a need to develop more effective control methods for the steam/water loop.

The impact of tuning different prediction horizon sets on the steam/water loop has already been studied in our previous work, and an optimized prediction horizon set was obtained according to the specific dynamics of this complex system [15].

However, in the present paper, we summarize our findings upon the effect of tuning different control horizon sets. In [16], Rossiter analyzed the effect of varying the control horizon, and he summarized that as control horizon increases, the nominal closed-loop performance improves if the prediction horizon is large enough. However, for many models, there is not much change beyond a control horizon equal to 3 samples. For a system with an unstable equilibrium point, the sensitivity of the trajectory sometimes is very high if the input sequence and the initial state are near the unstable equilibrium point. In this case, it is necessary to reduce the sensitivity by choosing a shorter horizon length [17], while ignoring the performance increase with large control horizon length. Cortés proposed that larger values for the control horizon length will, in general, provide better performance [18]. However, the computational complexity will also increase with the horizon length.

In this paper, a comprehensive analysis was made, studying the effect of different control horizons in a linear Extended Predictive Self-Adaptive Control (EPSAC) MPC framework [19]. The results were obtained on the steam/water loop in a large-scale ship. It was found that larger control horizon values improve the loop performance, at the cost of computational complexity. Consequently, an optimization scheme was designed by minimizing an optimal performance index consisting of the tracking error and the computing time for solving the MPC problem. In the end, the best control horizon set was obtained which provides a good trade-off between the closed-loop performance and allocated or available computational complexity. According to the simulation results, there are always ripples in the system's outputs when applying different control horizon sets, with $N_c \geq 2$. Hence, a modified cost function penalizing both the control effort and the tracking error was imposed in EPSAC, which effectively removed the ripple.

The rest of the paper is structured as follows: A description of the steam/water loop is given in Section 2. In Section 3, a brief introduction of the proposed EPSAC strategy with optimized control horizon is described. The simulation results and analysis are shown in Section 4. Finally, the conclusions are given in Section 5.

2. Description of the Steam/Water Loop

In the steam/water loop, there are mainly five loops, as briefly introduced in Figure 1: (i) drum water level control loop, (ii) deaerator water level control loop, (iii) deaerator pressure control loop, (iv) condenser water level control loop, and (v) exhaust manifold pressure control loop.

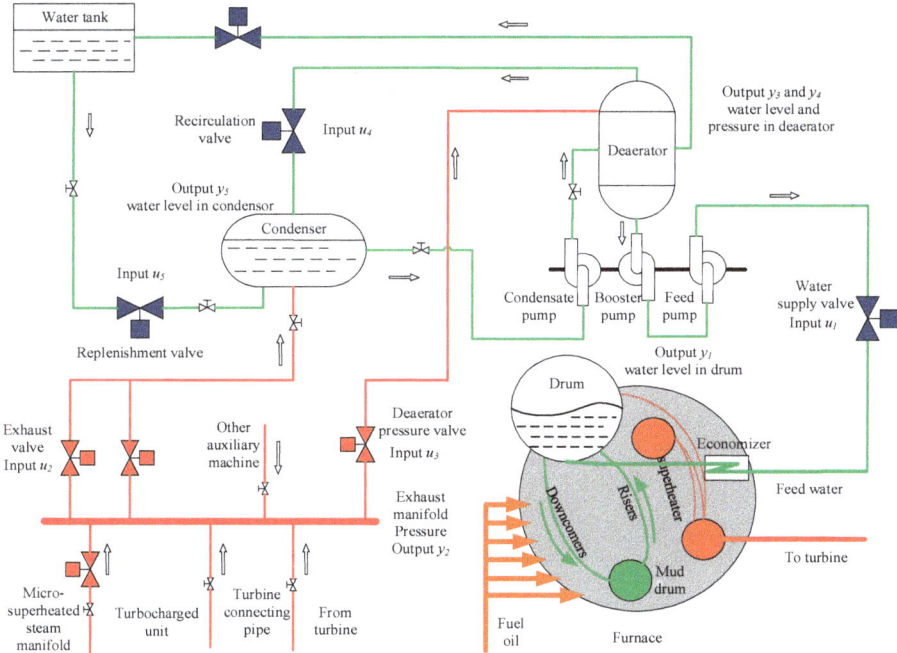

Figure 1. Scheme of complete steam/water loop investigated in this paper.

There are two main loops, one for steam indicated by red line, and another for water indicated by the green line. The system works as follows. Firstly, the water from the water tank goes to the condenser. Secondly, the water will be deoxygenated in the deaerator and be pumped to boiler. Due to a higher density of feed water, it goes into the mud drum. After being heated in the risers, the feed water turns into a mixture of steam and water. Thirdly, steam gets separated from the mixture and heated in the superheater. Finally, the steam with a certain pressure and temperature services the steam turbine. The used steam will be sent back to exhaust manifold and most of the steam gets condensed in the condenser, while the remainder services the deaerator for deoxygenation.

The references of these models for each equipment are described as follows. The model of the boiler comes from [20]; the model of exhaust manifold is approximated as a second-order model according to [21]; the models of the deaerator and condenser are obtained according to [22]. Through linearization around the operating point, the overall model shown in Equation (1) is obtained. The input vector $\mathbf{u} = [u_1, u_2, u_3, u_4, u_5]$ contains the positions of the valves that control the flow rates of feedwater to the drum (u_1), exhaust steam from the exhaust manifold (u_2), exhaust steam to the deaerator (u_3), water from the deaerator (u_4) and water to the condenser (u_5), respectively. The output vector $\mathbf{y} = [y_1, y_2, y_3, y_4, y_5]$ contains the values of the water level in drum (y_1), pressure in exhaust manifold (y_2), water level (y_3) and pressure (y_4) in the deaerator, and water level of the condenser (y_5), respectively. Table 1 includes the ranges and operating points of the output variables.

$$\begin{bmatrix} y_1 \\ y_2 \\ \vdots \\ y_5 \end{bmatrix} = \begin{bmatrix} G_{11} & G_{12} & \cdots & G_{15} \\ G_{21} & G_{22} & \cdots & G_{25} \\ \vdots & \vdots & \ddots & \vdots \\ G_{51} & G_{52} & \cdots & G_{55} \end{bmatrix} \begin{bmatrix} u_1 \\ u_2 \\ \vdots \\ u_5 \end{bmatrix} \tag{1}$$

where $G_{11} = \frac{0.0000987}{(s+0.1131)(s+0.0085+0.032j)(s+0.0085-0.032j)}$, $G_{22} = \frac{0.7254}{(s+1.2497)(s+0.0223)}$, $G_{23} = \frac{-0.5}{(s+1.9747)(s+0.0253)}$, $G_{33} = \frac{0.0132}{(s+0.0265+0.0244j)(s+0.0265-0.0244j)}$, $G_{34} = \frac{-0.009}{(s+0.0997)(s+0.0411)}$, $G_{41} = \frac{-0.0008}{(s+0.012+0.126j)(s+0.012-0.126j)}$, $G_{44} = \frac{0.0005152}{(s+0.012+0.038j)(s+0.012-0.038j)}$, $G_{54} = \frac{-0.00015}{(s+0.0175+0.0179j)(s+0.0175-0.0179j)}$, $G_{55} = \frac{0.00147}{(s+0.025+0.0654j)(s+0.025-0.0654j)}$, and other transfer functions $G_{12} = G_{13} = \ldots = G_{53} = 0$.

Table 1. Parameters used in steam/water loop.

Output Variables	Operating Points	Range	Units
Drum water level	1.79	1.39–2.19	m
Exhaust manifold pressure	100.03	87.03–133.8	MPa
Deaerator pressure	30	24.9–43.86	KPa
Deaerator water level	0.7	0.489–0.882	m
Condenser water level	0.5	0.32–0.63	m

The rates and amplitudes of the five inputs are constrained to:

$$\begin{cases} -0.007 \leq \frac{du_1}{dt} \leq 0.007 & 0 \leq u_1 \leq 1 \\ -0.01 \leq \frac{du_2}{dt} \leq 0.01 & 0 \leq u_2 \leq 1 \\ -0.01 \leq \frac{du_3}{dt} \leq 0.01 & 0 \leq u_3 \leq 1 \\ -0.007 \leq \frac{du_4}{dt} \leq 0.007 & 0 \leq u_4 \leq 1 \\ -0.007 \leq \frac{du_5}{dt} \leq 0.007 & 0 \leq u_5 \leq 1 \end{cases} \tag{2}$$

The inputs units are normalized percentage values of the valve opening (i.e., 0 represents a fully closed valve, and 1 is completely opened). Additionally, the input rates are measured in percentage per second.

3. Model Predictive Control with Optimized Control Horizon

3.1. Brief Introduction to Extended Prediction Self-Adaptive Control (EPSAC)

The following is a short summary of EPSAC and more details can be found in [23]. Consider a linear system described below:

$$y(t) = x(t) + n(t) \tag{3}$$

where $y(t)$ indicates the measured output of system; $x(t)$ is the output of model and $n(t)$ is the model/process disturbance. The output of the model $x(t)$ depends on the past outputs and inputs, and can be expressed generically as:

$$x(t) = f[x(t-1), x(t-2), \ldots, u(t-1), u(t-2), \ldots] \tag{4}$$

In EPSAC, the future input consists of two parts:

$$u(t+k|t) = u_{base}(t+k|t) + \delta u(t+k|t) \tag{5}$$

where $u_{base}(t+k|t)$ indicates basic future control scenario and $\delta u(t+k|t)$ indicates the optimizing future control actions. Then following results will be obtained by applying Equation (5) as the control effort.

$$y(t+k|t) = y_{base}(t+k|t) + y_{opt}(t+k|t) \tag{6}$$

where $y_{base}(t+k|t)$ is the effect of base future control and $y_{opt}(t+k|t)$ is the effect of optimizing future control actions $\delta u(t|t), \ldots, \delta u(t+N_c-1|t)$. The part of $y_{opt}(t+k|t)$ can be expressed as a discrete time convolution as follows:

$$y_{opt}(t+k|t) = h_k \delta u(t|t) + h_{k-1} \delta u(t+1|t) + \ldots + g_{k-N_c+1} \delta u(t+N_c-1|t) \tag{7}$$

where $h_1, \ldots h_{N_p}$ are impulse response coefficients; $g_1, \ldots g_{N_p}$ are the step response coefficients; N_c, N_p are control horizon and prediction horizon, respectively. Thus the following formulation can be obtained:

$$\mathbf{Y} = \overline{\mathbf{Y}} + \mathbf{G} \cdot \mathbf{U} \tag{8}$$

with, $\mathbf{Y} = [y(t+N_1|t) \ldots y(t+N_p|t)]^T$, $\mathbf{U} = [\delta u(t|t) \ldots \delta u(t+N_c-1|t)]^T$, $\overline{\mathbf{Y}} = [y_{base}(t+N_1|t) \ldots y_{base}(t+N_p|t)]^T$ and

$$\mathbf{G} = \begin{bmatrix} h_{N_1} & h_{N_1-1} & \cdots & g_{N_1-N_c+1} \\ h_{N_1+1} & h_{N_1} & \cdots & \cdots \\ \cdots & \cdots & \cdots & \cdots \\ h_{N_p} & h_{N_p-1} & \cdots & g_{N_p-N_c+1} \end{bmatrix} \tag{9}$$

where N_1 indicates the time-delay in the system.

The disturbance term $n(t)$ is defined as a filtered white noise signal [19]. When there is no information concerning the noise, the disturbance model used in Equation (3) can be chosen as an integrator, to ensure zero steady-state error in the reference tracking experiment:

$$n(t) = \frac{1}{1-q^{-1}} e(t) \tag{10}$$

where $e(t)$ denotes the white noise; q^{-1} is the backward shift operator.

In order to apply EPSAC for a MIMO (multiple-input and multiple-output) system, the individual error of each output is minimized separately. The cost function for the steam/water system with five sub-loops is as follows:

$$J_i = \sum_{k=N_1}^{N_p} [r_i(t+k|t) - y_i(t+k|t)]^2 (i = 1,2,\ldots,5) \tag{11}$$

By defining $\mathbf{G}_{ik}$ as the influence from kth input to ith output, Equation (11) can be rewritten as:

$$(\mathbf{R}_i - \mathbf{Y}_i)^T (\mathbf{R}_i - \mathbf{Y}_i) = (\mathbf{R}_i - \overline{\mathbf{Y}}_i - \sum_{k=1}^{5} \mathbf{G}_{ik} \mathbf{U}_k)^T (\mathbf{R}_i - \overline{\mathbf{Y}}_i - \sum_{k=1}^{5} \mathbf{G}_{ik} \mathbf{U}_k) \tag{12}$$

with $\mathbf{R}_i$ denoting the reference for loop i, and $\mathbf{Y}_i$ denotes the predicted output for loop i.

Taking constraints from inputs and outputs into account, the process to find the minimum cost function becomes an optimization problem which is called quadratic programming.

$$\min_{\mathbf{U}_i} J_i(\mathbf{U}_i (= \mathbf{U}_i^T \mathbf{H}_i \mathbf{U}_i + 2\mathbf{f}_i^T \mathbf{U}_i + c_i \text{ subject to } \mathbf{AU} \leq \mathbf{b}$$

$$\text{with} \begin{cases} \mathbf{H}_i = \mathbf{G}_{ii}^T \mathbf{G}_{ii} \mathbf{f}_i = -\mathbf{G}_{ii}^T (\mathbf{R}_i - \overline{\mathbf{Y}}_i - \sum_{k=1}^{5} \mathbf{G}_{ik} \mathbf{U}_k) \\ c_i = (\mathbf{R}_i - \overline{\mathbf{Y}}_i - \sum_{k=1}^{5} \mathbf{G}_{ik} \mathbf{U}_k)^T (\mathbf{R}_i - \overline{\mathbf{Y}}_i - \sum_{k=1}^{5} \mathbf{G}_{ik} \mathbf{U}_k) \end{cases} \quad (13)$$

where $\mathbf{A}$ is a matrix; $\mathbf{b}$ is a vector according to the constraints and $\mathbf{U}_i$ is the input for sub loop i. By solving the quadratic problem, the optimal $\mathbf{U} = [\mathbf{U}_1\ \mathbf{U}_2\ \mathbf{U}_3\ \mathbf{U}_4\ \mathbf{U}_5]$ can be obtained.

3.2. Ripple-Free Model Predictive Control (MPC)

Since MPC uses a discrete-time model, it is easy to get ripples in the system output when controlling a continuous system with periodic control effort during the sampling time. According to the simulation results of the steam/water loop in large-scale ships, there always exists ripple when applying a control horizon $N_c \geq 2$. To remove the ripple in the control effort, an alternative cost function which also penalizes the control effort imposed in the EPSAC strategy [15], obtaining:

$$J_i = J_i^e + \frac{\lambda}{1-\lambda} J_i^u \quad (14)$$

where $\lambda \in [0;\ 1)$ is a weighting parameter, and,

$$J_i^e = \sum_{k=1}^{N_p} k e_i(t+k|t)^2, \quad J_i^u = \sum_{k=1}^{N_c} (u_i(t+k|t) - u_i(t+k-1|t))^2 \quad (15)$$

are the cumulative sum that penalizes the predicted tracking error $e_i(t+k|t)$ over the prediction horizon, and the cumulative term which corrects the deviations in postulated control effort $u_i(t+k|t)$ over the control horizon, for each loop i, respectively.

In order to minimize the J_i^e in Equation (15), the tracking error must converge to zero rapidly. However, the J_i^u term has a negative impact on the tracking error. By choosing an appropriate value for λ, a good trade-off between the closed-loop performance and the control effort can be made.

3.3. Optimized Control Horizon

To our knowledge, the longer control horizon values can result in better performance, albeit the computational complexity will also increase, which makes it more difficult to realize online optimization. The relationship between performance, computational complexity and control horizon can be described by Figure 2.

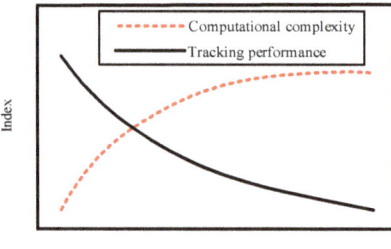

Figure 2. The relationship between tracking performance, computational complexity and control horizon.

According to Figure 2, it is possible to find a good trade-off between the tracking performance and computational complexity. In this paper, the problem is explicitly treated as an optimization problem, and the following index is applied to obtain the point of compromise for the five loops [24]:

$$I_i = T_i + \eta_i E_i \quad (i = 1, 2, \ldots, 5)$$
$$T_i = \sum_{k=1}^{N_s} t_{is}(k) \quad E_i = \sum_{k=0}^{N_s} |r_i(k) - y_i(k)|/r_i(k) \tag{16}$$

where T_i is the total simulation computation time, with t_{is} the length of time required to perform the optimization at each sampling time and N_s the number of total simulation samples; E_i is the integrated absolute normalized tracking error; η_i denotes the weighting factor.

In order to obtain the optimal control horizon N_c for loop i, experiments are required to be conducted with different control horizons. After minimizing the index I_i, the optimal control horizon can be obtained. The value of η should be chosen according to the dynamic of the system. For example, the dynamic of the system is slow in the steam/water loop, hence a large value of η should be chosen to focus more on the error than the computation time.

4. Simulation Results and Analysis

In this section, the proposed EPSAC method is applied to the steam/water loop. Firstly, the performance is shown after applying the cost function focus on penalizing the control effort by tracking several step set points in different loops. Secondly, different control horizon sets are imposed in the ripple-free EPSAC to verify their effect, and the optimal control horizon set is obtained by minimizing the index in Equation (16).

4.1. Ripple-Free Validation

According to our previous work, the parameter configuration for the EPSAC method is shown in Table 2, where the T_s is the sampling time; $N_{p1}, N_{p2}, \ldots, N_{p5}$ are prediction horizons of the five loops, respectively. (The prediction horizons were selected taking into account the specific transient dynamic for each loop). The step set points are provided in Table 3. In the experiments, the initial condition was set at the operating point of the steam/water loop.

Table 2. Parameters applied in Extended Prediction Self-Adaptive Control (EPSAC) controller.

Controllers	N_c	T_s	N_p	λ	N_1	N_s
EPSAC	10	5 s	$N_{p1} = 20; N_{p2} = 15; N_{p3} = 15;$ $N_{p4} = 20; N_{p5} = 20$	0	1	300
Ripple-free EPSAC				0.3		

Table 3. Step set points changes in the experiments.

Time (s)	2–300	300–600	600–900	900–1200	1200–1500
Drum Water Level (m)	2	2	2	2	2
Exhaust Manifold Pressure (MPa)	100.03	116	116	116	116
Deaerator Pressure (KPa)	30	30	35	35	35
Deaerator Water Level (m)	0.7	0.7	0.7	0.8	0.8
Condenser Water Level (m)	0.5	0.5	0.5	0.5	0.6

The simulation results are shown in Figure 3, including the system outputs and the corresponding control efforts. Note that the EPSAC performs better, with the cost function given in Section 3.2 that also penalizes the control effort variations, thus eliminating the severe ripples on each loop input. Also, it is noteworthy to mention that the output steady state error from loops 1, 4 and 5 is removed.

When only the tracking error is penalized in the cost function, there are ripples with $N_c \geq 2$, which means that the controller is allowed to give at least two different control values, to ensure that the predicted output reaches the imposed reference, within the prediction window. In order to minimize the cost function, the first value of control effort will be optimized as large as possible under the constraint of the system. Hence, the inputs of the system are aggressive which results in the ripples. The amplitude of the ripples is influenced by the control horizon and the sampling time. By choosing an appropriate λ value in Equation (14), the ripples can be effectively removed. It is worth mentioning that when the control horizon is $N_c = 1$, there are no ripples.

In the ship's steam/water loop, the condenser and the deaerator have smaller capacity when compared with the boiler. Therefore, as seen in Figure 3, there are large overshoot values at the condenser water level and the deaerator water level when the setpoint is changing for the drum water level. The steady errors exhibited in loops 1, 4 and 5 as shown in Figure 3, are caused by the intrinsic coupling between the respective loops. The input u_1 has a large influence on the deaerator water level y_4, which is controlled by u_4. However, input u_4 also modifies the condenser water level y_5. On the other hand, the inputs for each loop are calculated according to the cost function shown in Equation (12), where the past sample time input values for the coupling variables are used.

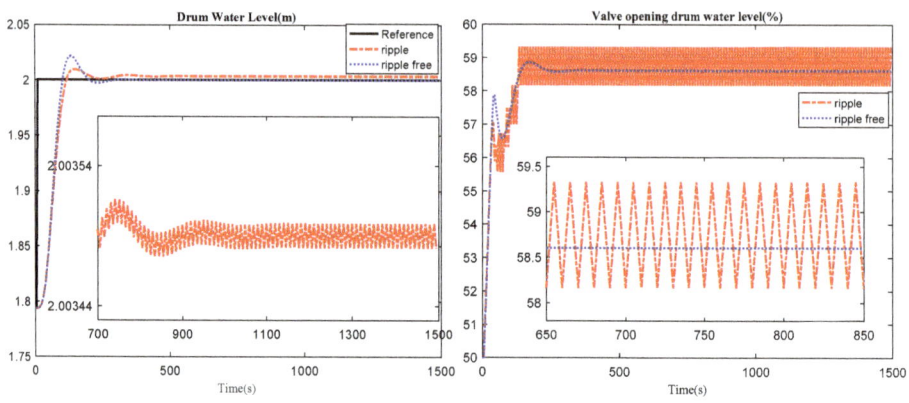

(a)

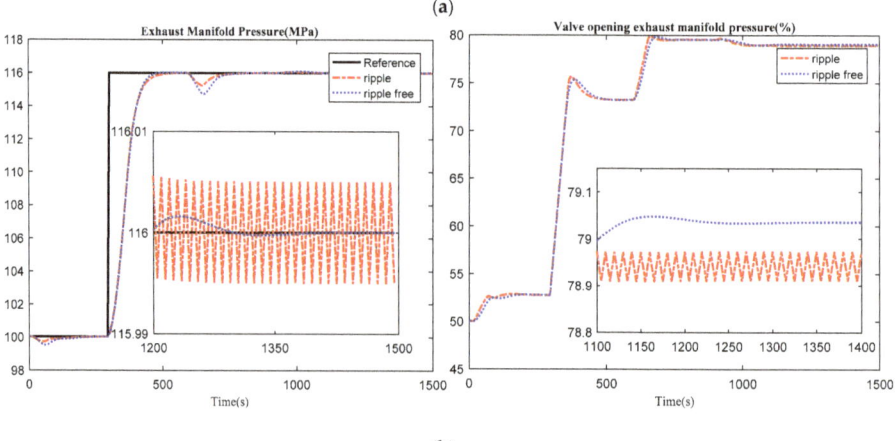

(b)

Figure 3. *Cont.*

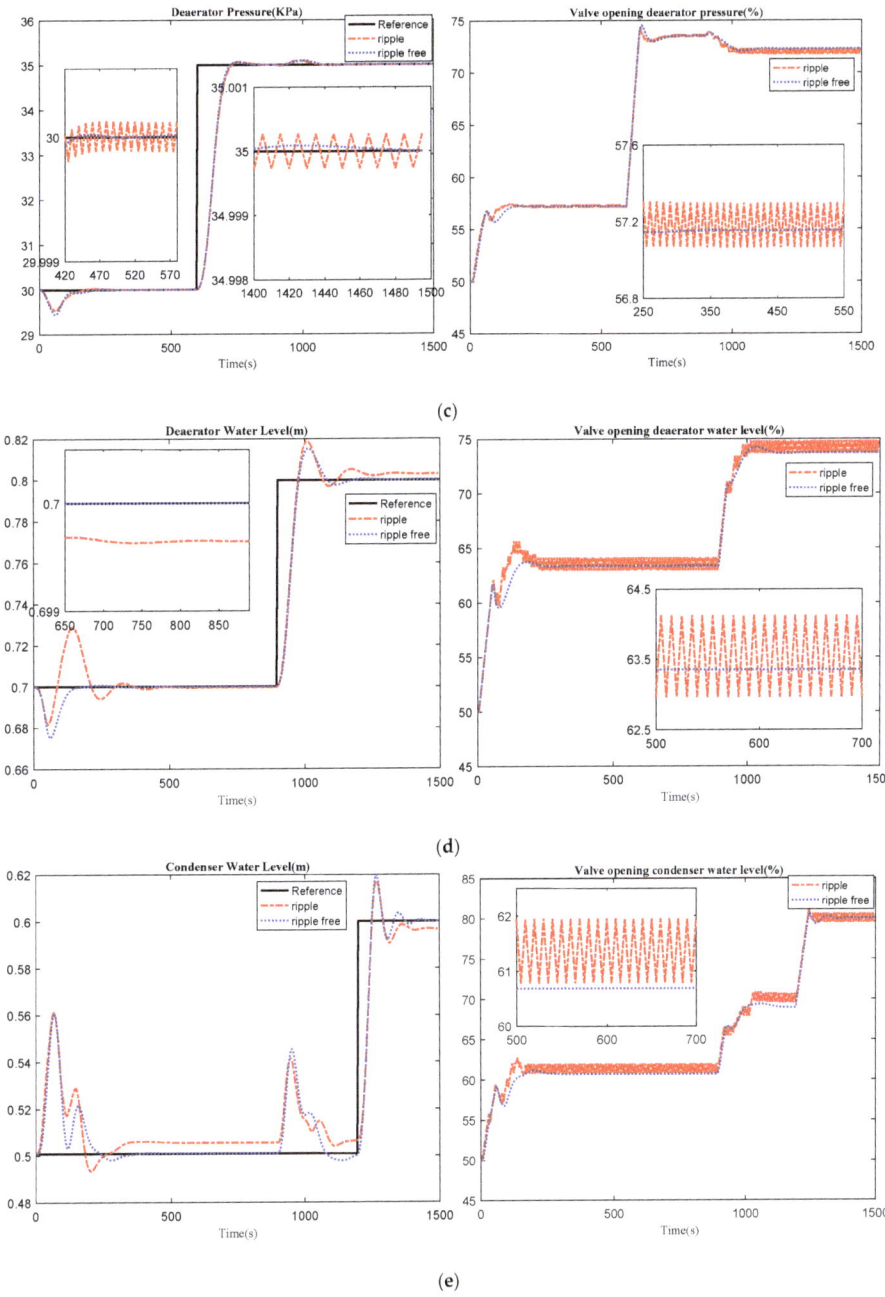

Figure 3. Responses of the steam/water loop under the EPSAC and ripple-free EPSAC for (**a**) drum water level control loop, (**b**) deaerator water level control loop, (**c**) deaerator pressure control loop, (**d**) condenser water level control loop and (**e**) exhaust manifold pressure control loop (The figures on left-hand indicate the outputs, and on the right-hand indicate the inputs).

4.2. Influence of Different Control Horizon Sets

This section summarizes the results for the five loops with different control horizon values. The simulation study cases are described as follows:

- Case 1: $N_{c1}, \ldots, N_{c5} = 1$ sample;
- Case 2: $N_{c1}, \ldots, N_{c5} = 2$ samples;
- Case 3: $N_{c1}, \ldots, N_{c5} = 5$ samples;
- Case 4: $N_{c1}, \ldots, N_{c5} = 10$ samples.

The responses of the steam/water loop with different control horizon values, are shown in Figure 4 (left-hand side), whereas the corresponding control efforts are given in Figure 4 (right-hand side). From the simulation results, one can remark that increasing the control horizon value in the proposed ripple-free EPSAC leads to better tracking performance, with a smaller overshoot and settling-time response, but with a higher control effort.

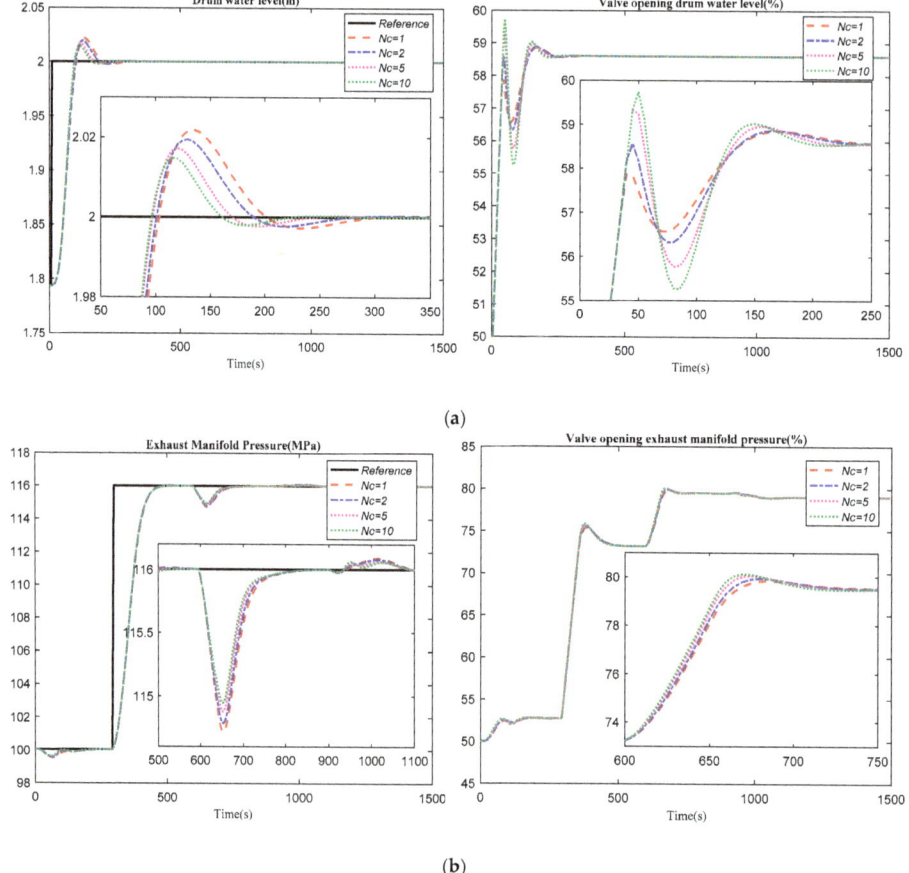

Figure 4. Cont.

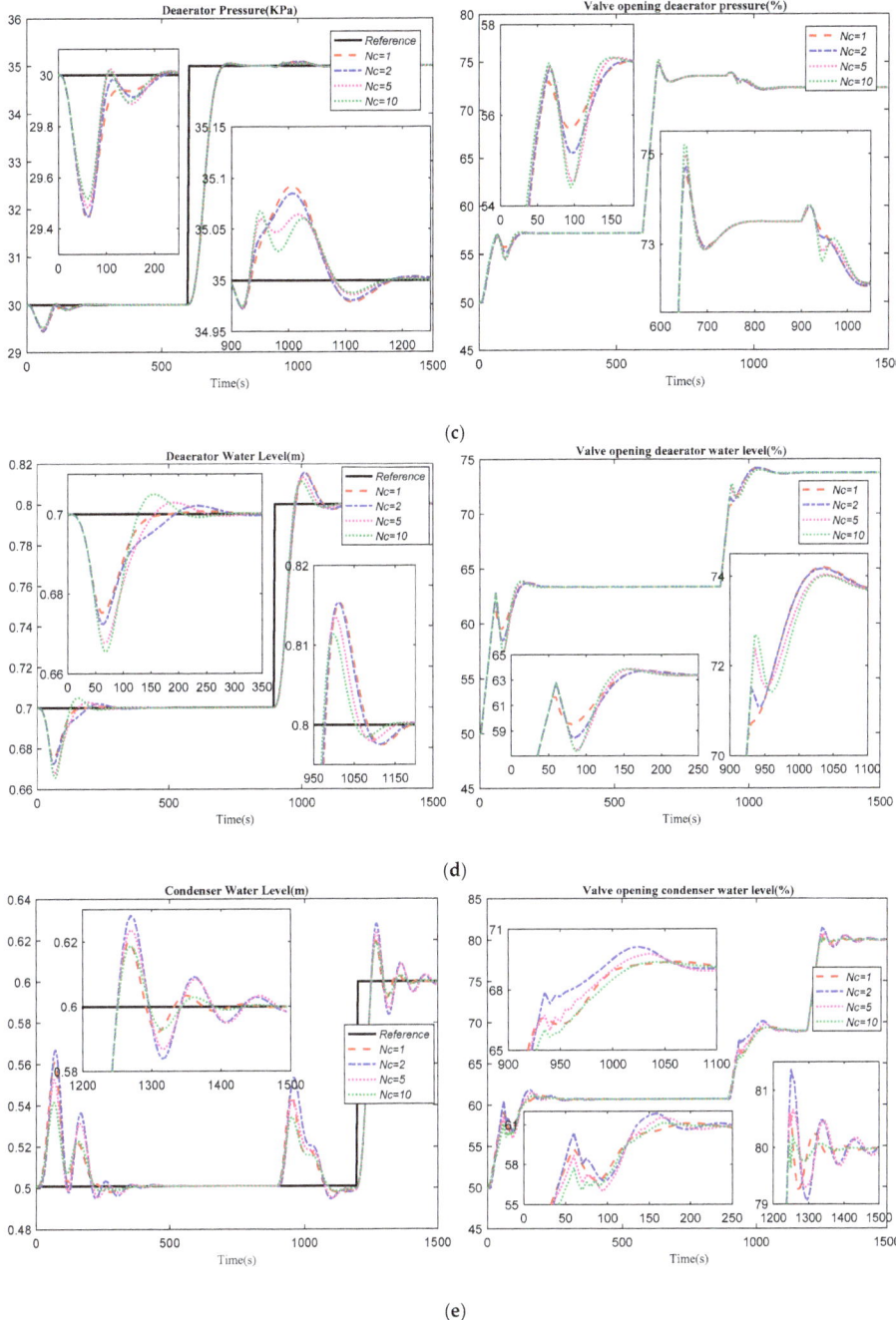

Figure 4. Responses of the steam/water loop under the ripple-free EPSAC for different control horizons for (**a**) drum water level control loop, (**b**) deaerator water level control loop, (**c**) deaerator pressure control loop, (**d**) condenser water level control loop and (**e**) exhaust manifold pressure control loop (The figures on left-hand indicate the outputs, and on the right-hand indicate the inputs).

The performance of the proposed ripple-free EPSAC algorithm was also analyzed in terms of the integrated absolute normalized tracking error (E_i) and computation time (T_i) defined in Section 3.3, in index (15). The numerical values are listed in Tables 4 and 5 respectively, and their relationship is graphically depicted in Figure 5 (for different control horizon values).

Table 4. Normalized tracking error with different control horizon sets.

	Loop 1	Loop 2	Loop 3	Loop 4	Loop 5
$N_c = 1$	1.342	2.039	2.08	1.933	4.603
$N_c = 2$	1.294	2.007	2.063	2.04	5.595
$N_c = 5$	1.242	1.976	2.038	1.999	5.012
$N_c = 10$	1.215	1.957	2.015	1.919	4.147

Table 5. Computing time in seconds with different control horizon sets.

	Loop 1	Loop 2	Loop 3	Loop 4	Loop 5
$N_c = 1$	3.384	2.687	3.182	2.584	2.79
$N_c = 2$	4.778	3.217	4.083	4.432	4.297
$N_c = 5$	6.058	4.456	5.716	5.757	5.822
$N_c = 10$	5.959	4.393	5.195	5.332	5.455

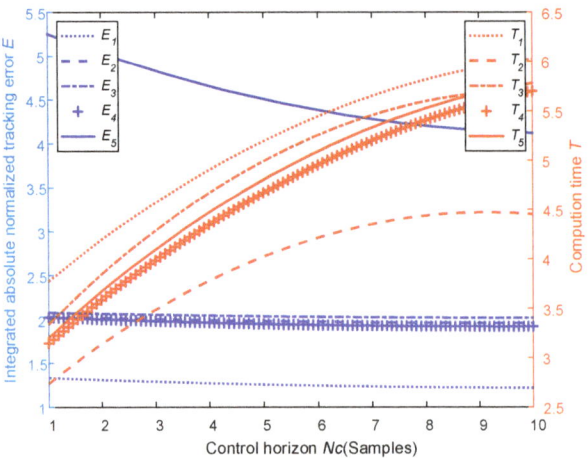

Figure 5. Computation time T_i (blue line) and integrated absolute normalized tracking error E_i (red line) in the five loops i ($i = 1,2,\ldots,5$) for different control horizon values.

Next, the information from Tables 4 and 5 is combined, and the index (16) is calculated, with $\eta_i = 0.76$, for each loop i ($i = 1,2,\ldots,5$). Note that this value compromises the computational complexity (i.e., the required computation time) in favor of a better tracking error. Given the graphical results plotted in Figure 6 and their significance, the optimal N_c set is selected as $N_{c1} = 4$, $N_{c2} = 1$, $N_{c3} = 1$, $N_{c4} = 4$, $N_{c5} = 6$ samples, which gives a good trade-off between the two components from index (15).

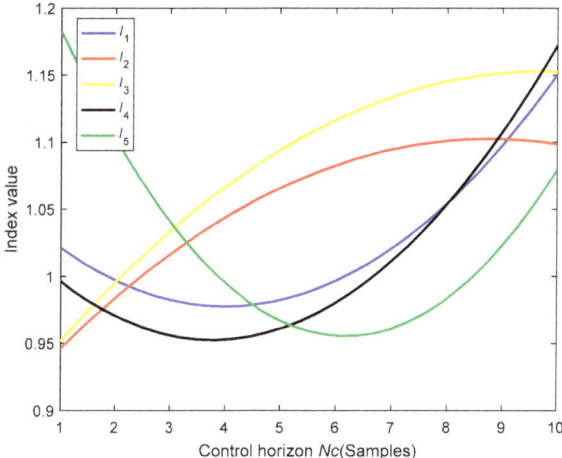

Figure 6. Optimization index for different control horizon values.

5. Conclusions

In this paper, the effect of control horizon is studied for an EPSAC model predictive control framework, and the results are validated on a complex steam/water loop process example. Since a larger control horizon improves the performance of the system at the price of a higher computational complexity and control effort, a trade-off is required. By minimizing an objective function defined as a combination between the system error and computational time, the best control horizon set of values is obtained. According to the simulation results, when applying different control horizon sets ($N_c \geq 2$) in the steam/water loop, there are always ripples in the output of the system. Hence, a cost function in terms of tracking error and deviations in the control effort was imposed in EPSAC. The simulation results show the effectiveness of the alternative cost function from EPSAC.

Author Contributions: Methodology, R.D.K., A.M. and S.Z.; software, S.Z., A.M. and S.L.; formal analysis, S.Z.; writing—original draft preparation, S.Z. and A.M.; writing—review and editing, S.Z., A.M., R.D.K., S.L., and C.I.; supervision, C.I., S.L. and R.D.K.; funding acquisition, S.Z., S.L. and C.I.

Acknowledgments: Mr Shiquan Zhao acknowledges the financial support from Chinese Scholarship Council (CSC) under grant 201706680021 and the Cofunding for Chinese PhD candidates from Ghent University under grant 01SC1918.

Conflicts of Interest: The authors declare no conflict of interest.

References

1. Sun, L.; Hua, Q.; Li, D.; Pan, L.; Xue, Y.; Lee, K.Y. Direct energy balance based active disturbance rejection control for coal-fired power plant. *ISA Trans.* **2017**, *70*, 486–493. [CrossRef] [PubMed]
2. Zhao, S.; Ionescu, C.M.; De Keyser, R.; Liu, S. A Robust PID Autotuning Method for Steam/Water Loop in Large Scale Ships. In Proceedings of the 3rd IFAC Conference in Advances in Proportional-Integral-Derivative Control, Ghent, Belgium, 9–11 May 2018; pp. 462–467.
3. Hosseinzadeh, M.; Yazdanpanah, M.J. Robust adaptive passivity-based control of open-loop unstable affine non-linear systems subject to actuator saturation. *IET Control Theory Appl.* **2017**, *11*, 2731–2742. [CrossRef]
4. Wang, W.L.A.; Mukai, H. Stability of linear time-invariant open-loop unstable systems with input saturation. *Asian J. Control* **2004**, *6*, 496–506. [CrossRef]
5. Scarciotti, G.; Astolfi, A. Approximate finite-horizon optimal control for input-affine nonlinear systems with input constraints. *J. Control Decis.* **2014**, *1*, 149–165. [CrossRef]
6. Li, Y.; Li, T.; Tong, S. Adaptive fuzzy modular backstepping output feedback control of uncertain nonlinear systems in the presence of input saturation. *Int. J. Mach. Learn. Cybern.* **2013**, *4*, 527–536. [CrossRef]

7. Maxim, A.; Copot, D.; De Keyser, R.; Ionescu, C.M. An industrially relevant formulation of a distributed model predictive control algorithm based on minimal process information. *J. Process Control* **2018**, *68*, 240–253. [CrossRef]
8. Fu, D.; Zhang, H.; Yu, Y.; Ionescu, C.M.; Aghezzaf, E.; De Keyser, R. A Distributed Model Predictive Control Strategy for Bullwhip Reducing Inventory Management Policy. *IEEE Trans. Ind. Inf.* **2018**, in press. [CrossRef]
9. Liu, X.; Cui, J. Economic model predictive control of boiler-turbine system. *J. Process Control* **2018**, *66*, 59–67. [CrossRef]
10. Wu, X.; Shen, J.; Li, Y.; Lee, K.Y. Fuzzy modeling and stable model predictive tracking control of large-scale power plants. *J. Process Control* **2014**, *24*, 1609–1626. [CrossRef]
11. Liu, X.J.; Chan, C.W. Neuro-fuzzy generalized predictive control of boiler steam temperature. *IEEE Trans. Energy Convers.* **2006**, *21*, 900–908. [CrossRef]
12. Liu, X.; Guan, P.; Chan, C.W. Nonlinear multivariable power plant coordinate control by constrained predictive scheme. *IEEE Trans. Control Syst. Technol.* **2010**, *18*, 1116–1125. [CrossRef]
13. Liu, X.; Kong, X. Nonlinear fuzzy model predictive iterative learning control for drum-type boiler–turbine system. *J. Process Control* **2013**, *23*, 1023–1040. [CrossRef]
14. Havlena, V.; Findejs, J. Application of model predictive control to advanced combustion control. *Control Eng. Pract.* **2005**, *13*, 671–680. [CrossRef]
15. Zhao, S.; Cajo, R.; De Keyser, R.; Ionescu, C.M. Nonlinear predictive control applied to steam/water loop in large scale ships. In Proceedings of the 12th IFAC Symposium on Dynamics and Control of Process Systems, including Biosystems (under review).
16. Rossiter, J.A. *Model-Based Predictive Control: A Practical Approach*; CRC Press: Boca Raton, FL, USA, 2003; pp. 85–102.
17. Ohtsuka, T. A continuation/GMRES method for fast computation of nonlinear receding horizon control. *Automatica* **2004**, *40*, 563–574. [CrossRef]
18. Cortés, P.; Kazmierkowski, M.P.; Kennel, R.M.; Quevedo, D.E.; Rodríguez, J. Predictive control in power electronics and drives. *IEEE Trans. Ind. Electron.* **2008**, *55*, 4312–4324.
19. De Keyser, R. Model based predictive control for linear systems. In *Control Systems, Robotics and Automation-Volume XI Advanced Control Systems-V*; Unbehauen, H., Ed.; UNESCO: Oxford, UK, 2003; pp. 24–58.
20. Åström, K.J.; Bell, R.D. Drum-boiler dynamics. *Automatica* **2000**, *36*, 363–378.
21. Wang, P.; Liu, M.; Ge, X. Study of the improvement of the exhaust steam maniline pressure control system of a steam-driven power plant. *J. Eng. Therm. Energy Power* **2014**, *29*, 65–70.
22. Wang, P.; Meng, H.; Dong, P.; Dai, R. Decoupling control based on pid neural network for deaerator and condenser water level control system. In Proceedings of the 34th Chinese Control Conference (CCC), Hangzhou, China, 28–30 July 2015; pp. 3441–3446.
23. De Keyser, R.; Ionescu, C.M. The disturbance model in model based predictive control. In Proceedings of the IEEE Conference on Control Applications (CCA 2003), Istanbul, Turkey, 25–25 June 2003; pp. 446–451.
24. Ribeiro, C.C.; Rosseti, I.; Souza, R.C. Effective probabilistic stopping rules for randomized metaheuristics: GRASP implementations. In *Learning and Intelligent Optimization*; Springer: Berlin, Germany, 2011; Volume 6683, pp. 146–160.

© 2018 by the authors. Licensee MDPI, Basel, Switzerland. This article is an open access article distributed under the terms and conditions of the Creative Commons Attribution (CC BY) license (http://creativecommons.org/licenses/by/4.0/).

Article

FFANN Optimization by ABC for Controlling a 2nd Order SISO System's Output with a Desired Settling Time

Aydın Mühürcü

Department of Mechatronics Engineering, Kırklareli University, Kırklareli 39000, Turkey; amuhurcu@klu.edu.tr; Tel.: +90-542-509-42-13

Received: 20 November 2018; Accepted: 19 December 2018; Published: 21 December 2018

Abstract: In this study, a control strategy is aimed to ensure the settling time of a 2nd order system's output value while its input reference value is changed. Here, Feed Forward Artificial Neural Network (FFANN) nonlinear structure has been chosen as a control algorithm. In order to implement the intended control strategy, FFANN's normalization coefficient (K), learning coefficients (η), momentum coefficients (μ) and the sampling time (Ts) were optimized by Artificial Bee Colony (ABC) but FFANN's values of weights were chosen arbitrary on start time of control system. After optimization phase, the FFANN behaves as an adaptive optimal discrete time non-linear controller that forces the system output to take the same value with the input reference for a desired settling time (ts). The success of the optimization algorithm was proved with close loop feedback control simulations on Matlab's Simulink platform based on 2nd order transfer functions. Also, the success was proved with a 2nd order physical system (buck converter) that was structured with power electronics elements on Simulink platform. Finally, the success of the control process was discussed by observing results.

Keywords: FFANN; control; optimization; ABC; modeling; buck converter; settling time

1. Introduction

Nowadays, optimization of controllers' parameters is preferred for obtaining a better cost-effective control strategies. In the last two decades, researchers have developed different types of optimization algorithms that may be used by scientists in control area. Bee [1], Firefly [2], Bat [3], Virus [4], Genetic [5], Cuckoo [6], Particle Swarm [7], Gravitation [8] and Biogeography [9] may be given for example.

The mathematical algorithms called controller are used for shaping the output variable of a physical system according to a desired behavior [10]. These mathematical algorithms are run using discrete or continuous time hardware [11]. Controller's parameters are calculated with analysis of rules sequences developed for related control algorithms [12].

Owing to the fact that the Artificial Intelligence algorithms are versatile, with the same type of structure they may be used for solving more than one type of problems such as control, prediction, estimation and modelling [13]. Using the same type of Artificial Neural Network structure, different researches have solved different type of problems. For example, Erkaymaz et al. estimate the thermal performance of a solar air collectors and predicted the modules of rupture values of oriented strand boards [14], Beg et al. proposed a discrete wavelet transform approach to classify power system transient analysis [15], Zounemat-Kermani et al. developed models to predict one day ahead stream flow of the Marion Junction station in Cahaba watershed [16] and Ardestani et al. suggested to predict contact force at the medial knee joint [17].

The most popular type of Artificial Neural Networks (ANN) are illustrated like Feed Forward, Kohonen, Radial Basis, Dynamic Neural, Multilayer Perceptron, Neural-Fuzzy, Cascading Neural and Stochastic Neural [13–17].

There are several studies in the literature that combine heuristic-based optimization algorithms with ANN based algorithms. In some studies, the optimized ANN algorithm is asked to model a system [18,19]. In another type of work, the optimized ANN is expected to work as a predictor [20,21]. There are also several studies in the estimation and control field performed by the ANN algorithm which has been optimized by heuristic algorithms [22–24]. The common point of the study types mentioned above is the optimization of the weights of the ANN algorithm. In these works, external parameters of the ANN are not put into optimization. They are fixed in arbitrary way.

In this study, the external parameters that are Kout, η, μ and Ts of the Feed Forward Artificial Neural Network (FFANN) were optimized using the Artificial Bee Colony (ABC) algorithm. The weights, which are the internal parameters of the FFANN algorithm, were randomly assigned and continuously recalculated using the Back-Propagation method in the control process. Thus, the FFANN algorithm is adapted to adaptive and optimal operation.

There are various swarm-based optimization methods in the literature. Such methods present extremely superiority in obtaining the global optimum and in handling discontinuous and non-convex objectives. However, many of these methods are not effective in managing optimization problems of integer and discrete nature. Such optimization problems can be solved by approximating the discrete and integer variables by continuous variables. Thus, the problem becomes an ordinary nonlinear programming one with continuous control parameters and the continuous values are reduced to the closest possible discrete or integer variable values. In practice, this method generally causes to the solutions that may be far from the globally optimal solution. ABC algorithm is a search method, which is inspired by the foraging behavior of honey bee swarming, and target discrete optimization problems [25].

2. The Feed Forward Artificial Neural Network (FFANN) Model

Block diagram belongs to FFANN that was used in this study is given in Figure 1. Here, the FFANN controller has 3 layers called input, hidden and output. Each layer quantity is 1.

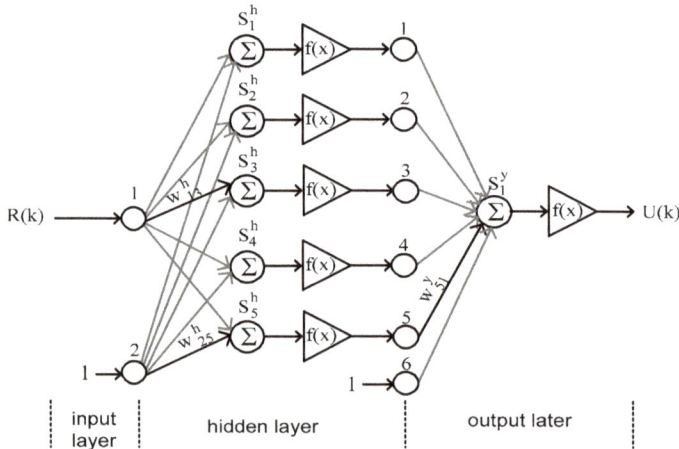

Figure 1. The Feed Forward Artificial Neural Network (FFANN) ontroller block diagram chosen for this study.

Variables in Figure 1 are described as,
$S_1^h, S_2^h, \ldots, S_5^h$: Addition centers of hidden (h) layer
$w_{11}^h, w_{12}^h, \ldots, w_{15}^h$: Weights between FFANN input and hidden layer addition centers
$w_{21}^h, w_{22}^h, \ldots, w_{25}^h$: Weights, between bias input and hidden layer addition centers, belong to input layer
$R(k)$: FFANN input
$U(k)$: FFANN output

S_1^y: Addition center of FFANN output layer

$w_{11}^y, w_{21}^y, \ldots, w_{51}^y$: Weights between output of hidden layer and output layer addition center of FFANN

w_{61}^y: Weight between bias input of output layer and addition center of output layer

$f(x)$: Activation function

In the literature, there is not any rule for numbering of hidden layers and for numbering of hidden layer's neurons [26]. On one hand, if number of hidden layers or neurons is chosen more than needed, the trading volume would be increased unnecessarily while optimizing the weights; on the other hand, if they were chosen less than needed, probability of reaching the level of acceptable minimum error would be decreased.

In this study, the control structure shown in Figure 2 was constructed. FFANN hidden layers and neurons count were determined by experimental observation. To this end, FFANN was optimized by ABC by selecting the hidden layer and the number of neurons high. Then, the optimization process was repeated by selecting the hidden layer and the number of neurons low. Fault-based cost function value was observed by running the control system after optimization process, Equation (11). The results of the experiment showed that the FFANN-based control process involved a high number of hidden layers and neurons, and the cost value of the FFANN-based control process with a low amount of hidden layer and neuron was similar, Figure 8. Considering the cost function values of the experimental processes, the numbers of hidden layers were chosen as 1 and the numbers of neurons belonging to the hidden layer were chosen as 5, Figure 1.

Another important variable is the activation function. The function type has also effects on weight optimization [27]. The FFANN controller in the closed loop negative feedback control system generates the control signal, $u(k)$, based on error signal, $e(k)$, Figure 2. The error during the control process may be greater, lower or equal to zero. Therefore, the activation function with limit values $[-1, +1]$ of tangent sigmoid is chosen in the FFANN structure, Equation (1).

$$f(x) = \frac{e^x - e^{-x}}{e^x + e^{-x}} \quad (1)$$

3. The Control System

Block diagram of close loop negative feedback control system based on FFANN controller is shown on Figure 2. Here, the controller is a discrete time algorithm but the system that is under control has continuous time structure.

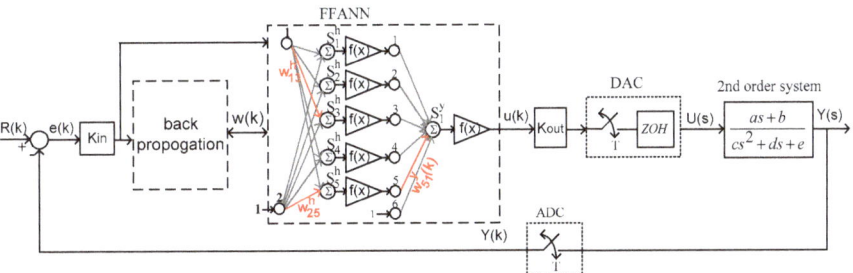

Figure 2. FFANN-based close loop control structure for controlling a 2nd order system.

For this study, a Digital to Analog Converter (DAC) was chosen, with a 1st order holder transfer function, as seen Equation (2).

$$\frac{1 - e^{sT}}{s} \quad (2)$$

Since operation range of a FFANN is between [−1, +1] [28], input signals have to be involved into FFANN algorithm only after multiply by normalization coefficient (Kin). FFANN's output signal is turned back using de-normalization coefficient (Kout) as shown in Figure 2.

FFANN algorithm is a two-step mathematical algorithm. In the first step, optimization of FFANN weights is done. In literature, this step is named as "learning process". In this study, "Steepest Descent" method has been used for weight optimization. This optimization method is simple structured and fast [28]. Renewal of weights based on this method is given in Equation (3) and Equations (3) and (4).

$$E(k) = \frac{1}{2}(R(k) - Y(k))^2 \qquad (3)$$

$$w(k+1) = w(k) - \eta \frac{\partial E(k)}{\partial w} + \mu[w(k) - w(k-1)] \qquad (4)$$

Here, η is learning coefficient and is chosen randomly in the range of (0, 1]. Another coefficient, μ, is momentum coefficient. Momentum coefficient does not only ensure to pass the local gradients but also helps to decrease the proportion of the error. The network may have an oscillation without momentum. Momentum coefficient usage prevents the network from oscillation during learning process, [28]. In literature momentum coefficient is chosen randomly between (0, 1].

At the second step of FFANN, output calculation is realized for sampling moment ($Ts \times k, k = 0, 1, 2 \ldots$) as shown in Equations (5)–(7).

$$S_j^h = \sum_{i=1}^{5}\left(R(k)w_{ji}^h\right) + w_{2j}^h, \ j = 1, 2, \ldots, 5 \qquad (5)$$

$$S_1^y = \sum_{i=1}^{5}\left(f\left(S_i^h\right)w_{i1}^y\right) + w_{61}^y \qquad (6)$$

$$U = f\left(S_1^y\right) \qquad (7)$$

4. Artificial Bee Colony (ABC) Algorithm

A bee transforms itself into a scout bee in order to find new sources when the food source is exhausted. Food source represents cost function in ABC algorithm. Therefore, the lowest cost function value represents the richest food source [29].

Food sources are created randomly in the first step of the algorithm. The bees that go directly to source of food are known as worker bees. Onlooker bees live in colony that they are directed to food sources based on the signs of worker bees. Onlooker bees go to food sources. They chose and store food like worker bees and return to the colony, Figure 3. Worker bees who consume food resources within a certain number of trials turn into scout bees to search for new food sources. When scout bees reach a random food source, process of food storage restart. These steps, continue until end criterion is satisfied [29].

"limits" and "popsize" is two fundamental variables in ABC algorithm. "popsize" is number of individuals in algorithm. The "limit" is number of trials for worker bees to leave food source. If it is not possible to develop as much as the limit value for a solution that represents a resource it is abandoned [29].

Figure 3. Bees in the hive.

The bee that abandoned food source becomes a scout bee. In the ABC algorithm, the number of onlooker bees, number of worker bees and number of food sources equal to each other. Number of food sources is half of the population [29].

First, food sources are created randomly for starting to run ABC algorithm, as seen in Equation (8).

$$x_{i,j} = x_j^{min} + rand(0,1)\left(x_j^{max} - x_j^{min}\right) \quad (8)$$

The new solution that is found by worker bees is compared with the old one. If the new solution is better, it keeps this solution and deletes the old solution from memory. If the new solution is not better than before, the previously defined abandonment counter is incremented. Searching food source by worker bees is shown in Equation (9).

$$v_{i,j} = x_{i,j} + \varphi_{i,j}\left(x_{i,j} - x_{k,j}\right) \quad (9)$$

After the worker bees have completed their food scan, onlooker bees go to random analysis so that they can select of food sources for bees, Equation (10).

$$\rho_i = \frac{fit_i}{\sum_{i=1}^{NS} fit_i} \quad (10)$$

The ρ value is obtained for each solution. Worker bees compare ρ_{ref} that are randomly selected and used as threshold value with ρ_i. If the probability of selecting ρ_i is larger than ρ_{ref} onlooker bee is moved toward this source of food and starts searching for a new solution by rerunning Equation (8).

The variables in Equations (8)–(10) are

NS: Number of food sources
D: Number of parameters that are optimized
x^{max}, x^{min}: Limit values of parameters that will be optimized
v_i: New food location in relation to x_i
x_k: Randomly food location that is different from x_i
$\varphi_{i,j}$: Random value between -1 to 1
fit_i: Normalized cost function
k: Solution in the neighborhood of i

All worker and onlooker bees check the abandonment counter for each solution after completing of food searching. If counter value reaches to limit value, the worker bee turns into a scout bee and run Equation (8). The process steps continue until the maximum number of cycles or the lowest value of the cost function value is reached.

5. Parameter Optimization by ABC

The FFANN weights in the closed-loop control system projected in Figure 2 are optimized during the control process by running the Back-Propagation algorithm. Before starting the control process, the parameters of Kout, η, μ and Ts should be optimized. They are optimized by the heuristic ABC algorithm. In the optimization process, the control system given in Figure 2 is used, too. During the process, the FFANN weights are randomly assigned for each optimization simulation, as in the beginning of the control process, Figure 2.

The implementation of the block diagram of the optimization process described in flow diagram below is given in detail in "Appendix A". As seen in the block diagram, the FFANN parameters are randomly assigned before the 1st run is performed. Then, the closed loop control system runs up to the simulation time period, Figure 5.

The input reference voltage applied to the control system has square waveform, Figure 4. The corners of the square wave are rounded off using the 1st order transfer function. The square wave is defined as the high and low time interval *ts* of the reference input voltage. In this way, the ABC algorithm optimizes FFANN parameters for the time interval of the closed-loop control system for the time period specified by *ts*.

Errors calculated by running the cost function during the simulation are collected. The total error obtained after the simulation is transferred to the ABC algorithm. The ABC algorithm calculates the new Kout, η, μ and *Ts* parameters by processing the total error within the framework of its mathematical algorithm. Next, the new parameters are replaced with the old ones for use in the next simulation. Prior to running the control system with the new parameters, the weights of FFANN are randomly assigned.

The potential solutions performed by the ABC algorithm, as in other heuristic algorithms, tries to find the closest value to the global solution by moving around instead of finding the most optimal solution available in the problem space [28]. It was also tried to be eliminated by the simulation technique in order to keep the algorithm around a certain local minimum. Therefore, FFANN weights are assigned randomly before running the required simulation for each new cost function account.

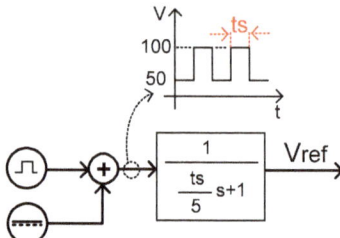

Figure 4. Vref voltage used in the optimization process as input of control system in Figure 2.

Discrete time cost function of Integral Squar Error (ISE) was benefited to determining cost belong to process of control [28]. Mathematical equation belong to discrete ISE is given in Equation (11).

$$ISE(e) = \sum_{k=0}^{t_{sim}/T_s} e_k^2 \qquad (11)$$

ABC optimization process "run ABC" belongs to flow chart in Figure 5 is summarized as:

1. Initialize the population of solutions.
2. Evaluate the population.
3. cycle = 1
4. repeat
5. Produce new solution (food-source positions) $v_{i,j}$ in the neighborhood of $x_{i,j}$ for the employed bees using Equation (9).
6. Apply the greedy selection process between x_i and v_i.
7. Calculate the probability values of p_i for the solutions x_i by means of their fitness values, Equation (10).
8. Produce the new solutions (new positions) v_i for the onlookers from the solutions x_i selected depending on p_i and evaluate them.
9. Apply the greedy selection process for the onlookers between x_i and v_i.
10. Determine the abandoned solution (source), if exists, and replace it with a new randomly produced solution x_i for the scout, Equation (8).
11. Memorize the best food source position (solution) achieved so far.
12. cycle = cycle + 1.
13. until cycle = Maximum Cycle Number.

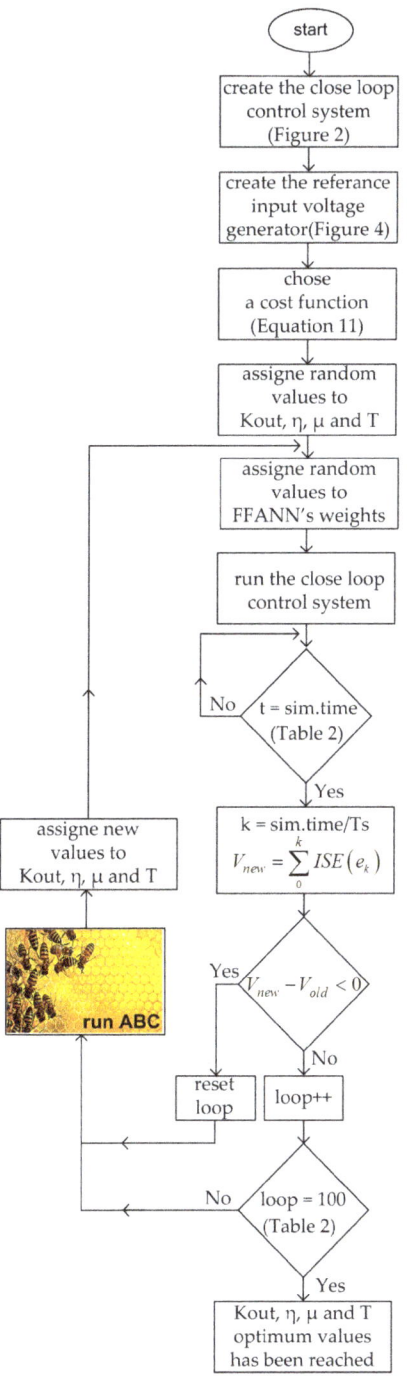

Figure 5. Artificial Bee Colony (ABC) based optimization process of system in Figure 2.

6. Control Simulations with Transfer Functions

The 2nd order transfer function belongs to a buck converter, Figure 6, can be derived using circuit theory.

$$V_{out}(s) = V_{in}(s) \frac{(Z3+Z4)//Z5}{(Z1+Z2)+(Z3+Z4)//Z5} \tag{12}$$

$$\frac{V_{out}(s)}{V_{in}(s)} = T(s) = \frac{\frac{(Z3+Z4)Z5}{Z3+Z4+Z5}}{(Z1+Z2)+\frac{(Z3+Z4)Z5}{Z3+Z4+Z5}} \tag{13}$$

where,

$Z1 = j\omega L = sL$: Impedance of coil
$Z2 = R_L$: Serial equivalent resistance of coil
$Z3 = R_c$: Serial equivalent resistance of capacitor
$Z4 = \frac{1}{j\omega C} = \frac{1}{sC}$: Impedance of capacitor
$Z5 = R_{load}$: Load resistance
denotes.

If the impedances are changed with S-domain parameters, the new transfer function takes shape as shown below:

$$T(s) = \frac{sCR_CR_{LOAD} + R_{LOAD}}{s^2(CLR_C + CLR_{LOAD}) + s(L + CR_LR_C + CR_LR_{LOAD} + CR_CR_{LOAD}) + (R_L + R_{LOAD})} \tag{14}$$

If variables of $T(s)$ are replaced with parameter values that are used for $T_2(s)$, Table 1, it will be seen that $T(s)$ output signal values are the same with the buck converter output voltage for the same input control signal, $U(s)$, Figure 6.

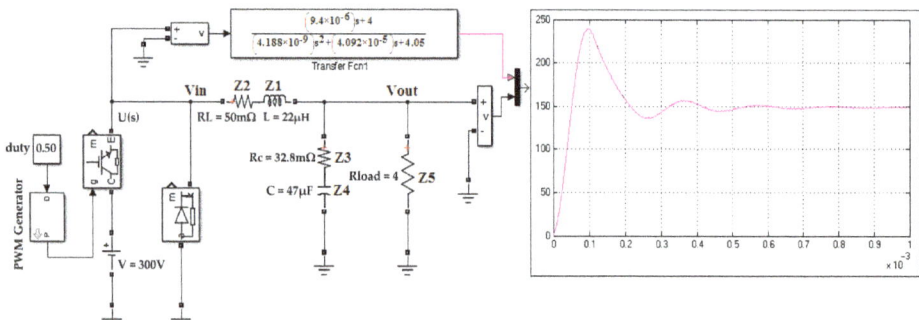

Figure 6. Simulation of models with the same input control signal, $U(s)$.

In this study, transfer functions with 3 different time constant (τ) were chosen for control application experience, Equations (15)–(18).

$$\left. \begin{array}{l} T_1(s) = \frac{(1.25\times10^{-6})s+25}{(2.505\times10^{-11})s^2+(3.502\times10^{-6})s+25.05}, \\ \text{roots} = (-0.6991 \mp 9.9755j) \times 10^5, \tau = 14.3 \times 10^{-6} \end{array} \right\} \tag{15}$$

$$\left. \begin{array}{l} T_2(s) = \frac{(9.4\times10^{-6})s+4}{(4.188\times10^{-9})s^2+(4.092\times10^{-5})s+4.05}, \\ \text{roots} = (-0.4885 \mp 3.0712j) \times 10^4, \tau = 204.69 \times 10^{-6} \end{array} \right\} \tag{16}$$

$$\left.\begin{array}{r}T_3(s) = \frac{(8.25\times 10^{-3})s+50}{0.3634s^2+2.217s+50.05},\\ roots = (-3.0500 \mp 11.3331\text{j}),\ \tau = 327.9\times 10^{-3}\end{array}\right\} \quad (17)$$

Table 1, indicates the component values of buck converter circuits belong to transfer functions of $T_1(s)$, $T_2(s)$ and $T_3(s)$.

Table 1. Buck converters' parameters.

Equation	L (μH)	R_l (mohm)	C (μF)	Rc (mohm)	Rload (ohm)
6.4	1	50	1	6.6	25
6.5	22	50	47	32.8	4
6.6	2200	50	3300	327.9	50

Chosen 3 systems have been adjusted as to have large difference between τ of each other's. The behaviors of systems' output signals based on unit step input function have been represented on Figure 7.

On Figure 7, it's seen that when a unit input signal is applied to transfer functions, overshoot of output signals is different from each other's. Here the settling times are different for about 1000 times with each other's. Control success for these 3 transfer functions, whose settling time values are very different from each other will show that the FFANN control logic based on the proposed ABC heuristic optimization method is applicable to all 2nd order systems.

Timing of ABC algorithm was defined by transfer functions' time constants, Table 2.

Table 2. Initialization ABC's timing parameters.

T. Functions	Simulation Time	Iteration
$T_1(s)$	$ts_1 \times 5$	100
$T_2(s)$	$ts_2 \times 5$	100
$T_3(s)$	$ts_3 \times 5$	100

On Figure 8, ISE(e) decreasing is shown for optimization process with ABC for $T_1(s)$ and for settling time $ts_1 = 5\tau$. On Figure 8, it is seen that after 14th iteration the cost function becomes stable and so it's found out the optimum parameters' values according to ABC algorithm.

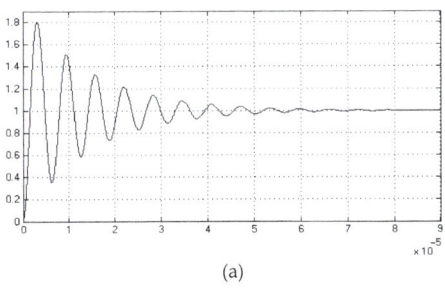

(a)

Figure 7. Cont.

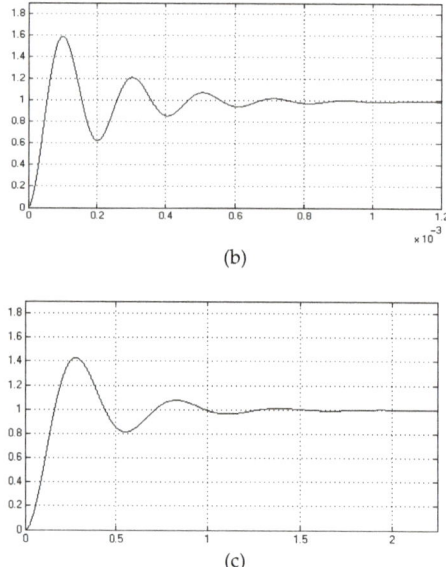

Figure 7. Systems' outputs for unit step input, (**a**) $T_1(s)$, (**b**) $T_2(s)$, (**c**) $T_3(s)$.

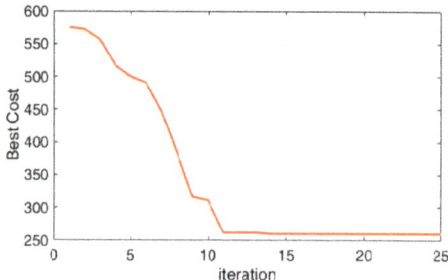

Figure 8. Decrease of Integral Squar Error (ISE (e)), optimizing FFANN's output parameters for $ts = 5\tau$ of $T_1(s)$.

In this study, ts values have been chosen according to time constant of systems, Tables 3–5. FFANN's optimized parameter values by ABC belong to three different transfer functions for 5 different settling times as shown in Tables 3–5. Parameter values of Kin for normalization had been fixed to 1/450 before optimization process was started. Value 450 is 3 times bigger than maximum input reference value ($R(k)$) that will be applied to the input of the control system, Figure 2.

Table 3. Feed Forward Artificial Neural Network (FFANN) parameter optimization for $T_1(s)$.

Setling Time	n_1, n_2	μ_1, μ_2	Kout	Tsample
5τ	0.0018	0.038	4725	286 ns
25τ	0.0022	0.052	4455	1.36 µs
250τ	0.0012	0.027	4275	12.584 µs
$10{,}000\tau$	0.0033	0.041	3825	572.16 µs

Table 4. FFANN parameter optimization for $T_2(s)$.

Setling Time	n_1, n_2	μ_1, μ_2	Kout	Tsample
5τ	0.0011	0.029	4630	4.094 µs
25τ	0.0034	0.024	4316	20.47 µs
250τ	0.0016	0.028	3804	198.7 µs
$10,000\tau$	0.0027	0.033	3710	8.26 ms

Table 5. FFANN parameter optimization for $T_3(s)$.

Setling Time	n_1, n_2	μ_1, μ_2	Kout	Tsample
5τ	0.0021	0.027	4722	6.6 ms
25τ	0.0024	0.021	4386	32.8 ms
250τ	0.0019	0.034	3854	319 ms
$10,000\tau$	0.0031	0.028	3635	13.11 s

Control simulations of Figure 2 for $T_1(s)$, $T_2(s)$ and $T_3(s)$ are shown on Figures 9–11. It's seen that success of control with optimal FFANN's parameters (red output line) is much better than those of with classical chosen FFANN's parameters (green output line). Here, $n_1 = n_2 = 0.1$, $\mu_1 = \mu_2 = 0.1$, Kout = 1/Kin = 450 and Tsample = $\tau/10$ were chosen as un-optimal FFANN's parameters.

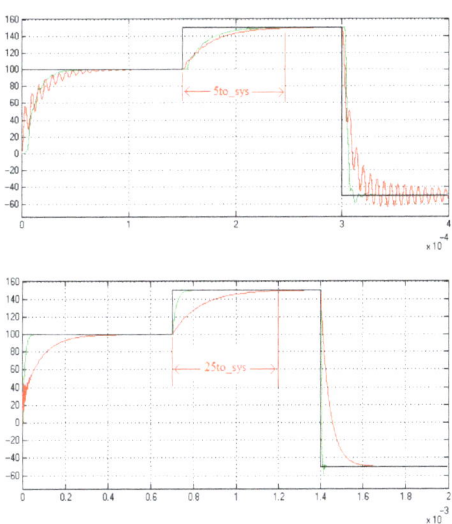

Figure 9. *Cont.*

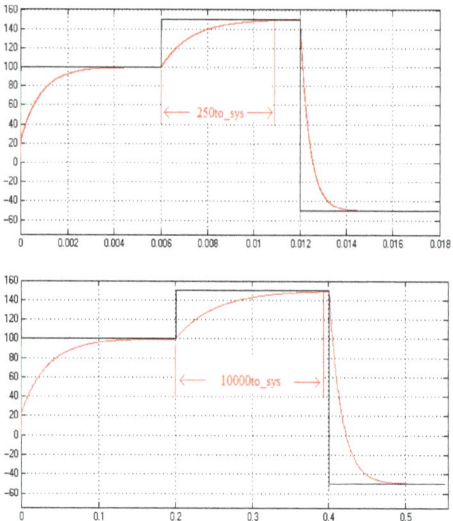

Figure 9. Control success of FFANN with optimal (red) and classical (green) chosen output parameters for $T_1(s)$.

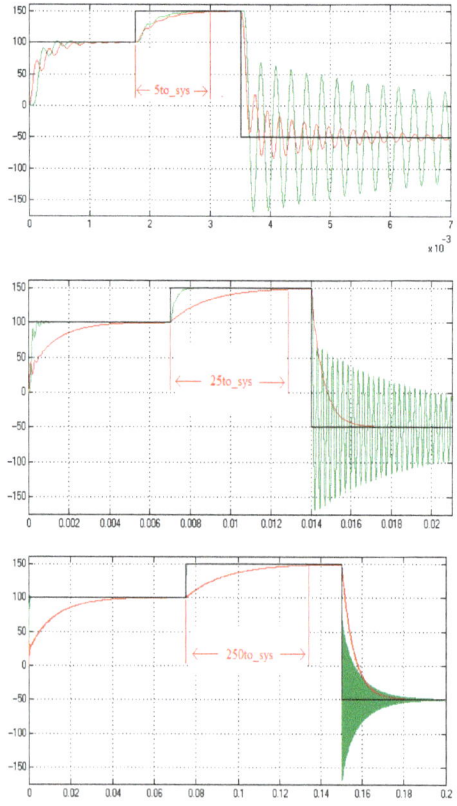

Figure 10. *Cont.*

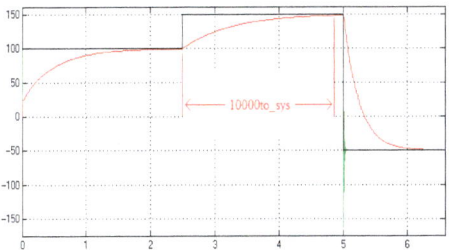

Figure 10. Control success of FFANN with optimal (red) and classical (green) chosen output parameters for $T_2(s)$.

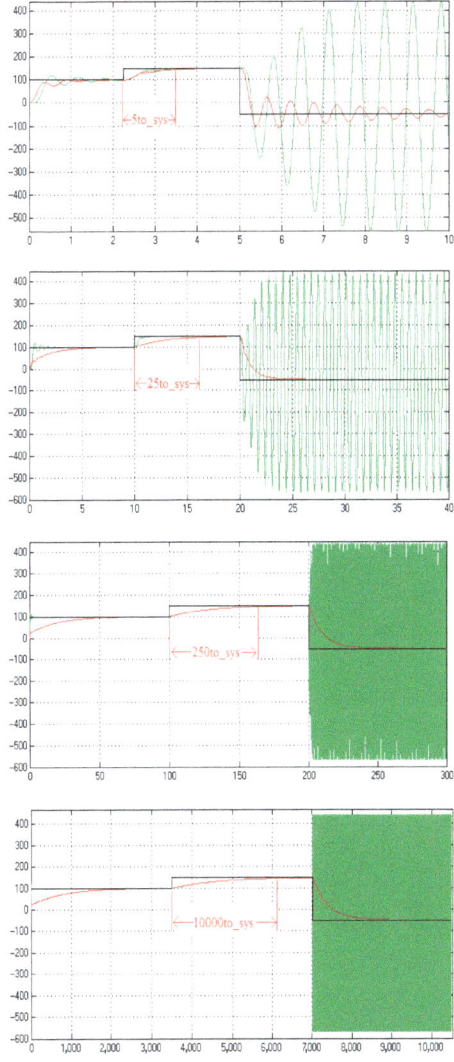

Figure 11. Control success of FFANN controller with optimal (red) and classical (green) chosen output parameters for $T_3(s)$.

The control success of the FFANN based controller assigned with the traditional method depends on the roots of the system it is going to control and therefore the time constant exists, see Figures 9–11. The response rate of the controller for some systems is sufficient and the system output is close to the input reference value as soon as possible. However, the settlement time of this convergence varies depending on whether the reference step takes up or downward value, Figure 9. The FANN based controller, whose parameters are assigned in the traditional way, can transform the controlled system output into a highly oscillating or marginal stable structure, as shown in Figures 10 and 11. However, the FFANN parameters will be optimized by ABC and the control processes to be performed will be determined in a stable manner. By running the optimization algorithm over the linking strategies of the control process, such as settlement time, more functional output signals based on simpler structured input reference value can be obtained, see Figures 9–11.

7. Control Simulations with Power Electronics Components

In Figure 12, a buck converter circuit based on feedback control system is shown. Here, the buck converter transfer function is 2nd order as shown in Section 6, Figure 6. So, Figure 12 shows FFANN based controller for a feedback control system with 2nd order model constructed by hardware on simulation platform.

Control system given in Figure 12 is similar to system given in Figure 2. The difference is that in Figure 2, the system that is under control has been given mathematically as transfer function but in Figure 12, it's been expressed using power electronic components.

Component values for 3 different bucks are given in Table 1. Optimal output parameter values of FFANN are shown on Tables 3–5. Control success of the FFANN is shown on Figures 13–15. Again, $n_1 = n_2 = 0.1$, $\mu_1 = \mu_2 = 0.1$, Kout = 1/Kin = 450 and $T_{sample} = \tau/10$ were chosen as un-optimal FFANN's parameters.

Another difference of the control system on Figure 12 from Figure 2 is that control signals are input to buck converter as Pulse Width Modulation (PWM) signals. Frequency of the PWM signals was chosen as $f_{pwm} = 2/T_{sample}$.

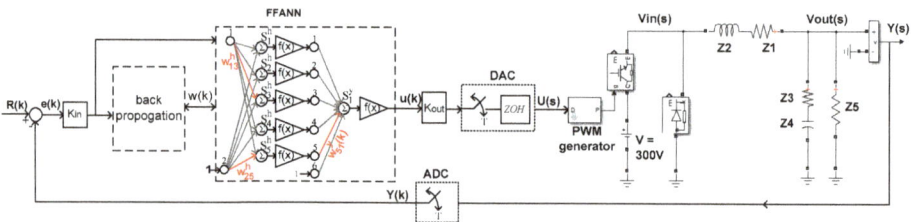

Figure 12. Hardware settings for control simulation on Matlab's Simulink.

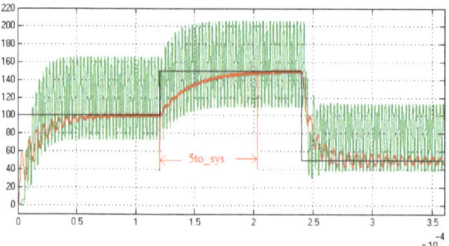

Figure 13. *Cont.*

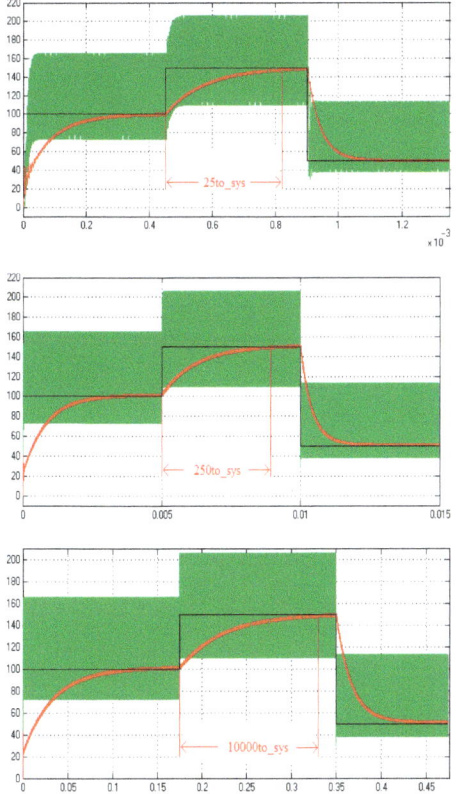

Figure 13. Control success of FFANN with optimal (red) and classical (green) chosen output parameters for $T_1(s)$.

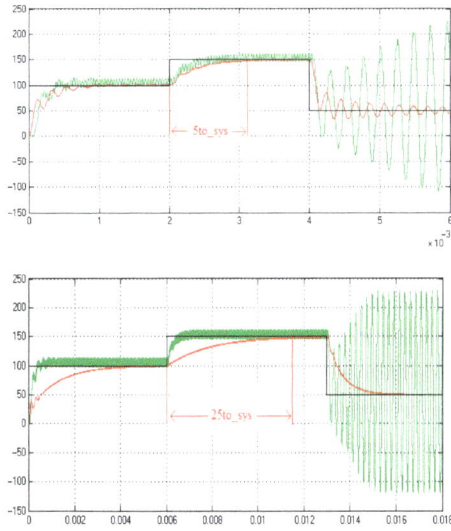

Figure 14. *Cont.*

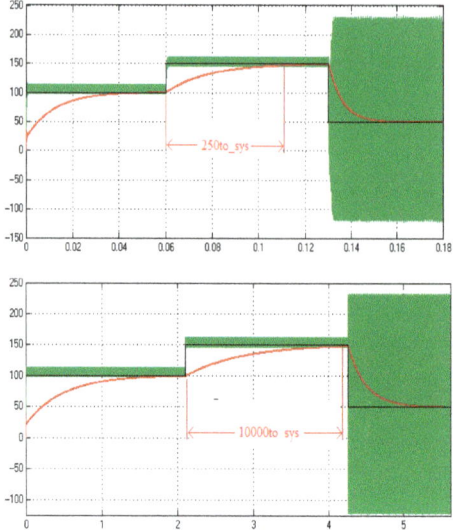

Figure 14. Control success of FFANN with optimal (red) and classical (green) chosen output parameters for $T_2(s)$.

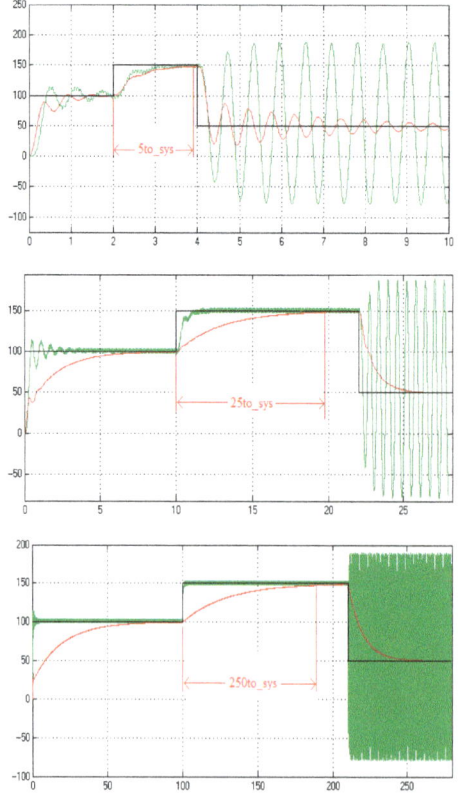

Figure 15. *Cont.*

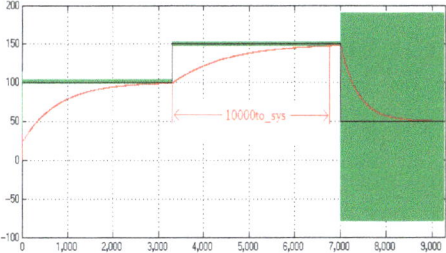

Figure 15. Control success of FFANN with optimal (red) and classical (green) chosen output parameters for $T_3(s)$.

The control success of the FFANN controller, whose parameters were determined by the conventional method, was further reduced by the conversion of the output signal to the PWM-based signal. During the control process based on Pulse Width Modulation (PWM) technique has been seen oscillations that had not been seen before in the continuous time simulations, Figures 13–15. The oscillation amplitude is related to the time constant of the controlled system. If FFANN output parameters are optimized by ABC it will be seen that the control process will be operated in a stable manner even if the output control signal is PWM structured.

8. Discussion

In this study, parameter optimization of a FFANN based controller was demonstrated. Different ANN structures should be optimized in their output parameters if they are targeted to control the 2nd order systems in an optimal and stable manner. Otherwise, high oscillation or marginal stability of the controlled system's output may be encountered, as in the case of FFANN based control. The weight of the ANN control algorithm whose parameters will be optimized can also be included in the optimization process. Thus, the ANN based controller will achieve the minimum control cost from the moment the control starts. Another optimization approach can be to change the optimization parameters. That is, the output parameters optimized in this study will be fixed based on the traditional method and only FFANN weights will be optimized. This means that during the control process, no Back-Propagation algorithm will be required and so mathematical operations would be minimized throughout the process.

9. Conclusions

In this study, artificial intelligence algorithm in FFANN structure has been transformed into a successful controller by using a heuristic algorithm. The ABC heuristic algorithm has been used in the optimization process of FFANN parameters. The reason for choosing ABC is explained in the introduction section. It has been proven by simulations that the control success of the FFANN based controller whose parameters are created by conventional methods is low and may vary even in the same control process. For a successful control process, FFANN parameters have been shown to require an optimization based on system parameters to be controlled. The optimization strategy may be developed in the form of minimum settling time or it may be improved by monitoring a desired settlement time. In this study, FFANN parameters which are optimized by observing the settlement time, in order to achieve the reference step input changes, the approximations of the input sizes of the outputs of the 2nd order systems have been achieved. The FFANN based controller has succeeded to converge of the system output value for the reference step input changes inside the prescribed time. The control simulations with transfer function and hardware-based control have proved the accuracy of the FFANN parameters to be optimized with a heuristic optimization algorithm such as ABC.

Funding: This research received no external funding.

Conflicts of Interest: The author declare no conflict of interest.

Appendix A

ABC optimization process for the system that is shown on Figure 2 is given below as a Matlab's function (*.m) file. The optimization process optimizes output parameters of FFANN.

%ABC optimization process for the system shown in Figure 2.

%Problem Definition

CostFunction = @(x) **Run_Fig2(x);** % first simulate Figure 3 to find out cost function
nVar = 4;% number of decision variables K_1 for $n_{1\,2}$, K_2 for $\mu_{1\,2}$, K_3 for Tsmpl, K_4 for Kout
VarSize = [1 nVar];% decision variables matrix size
VarMin = 0.001; % decision variables lower bound
VarMax = 10000; %decision variables upper bound, chosen acording to τ
% ABC Settings
MaxIt = 40; % maximum number of iterations
nPop = 40; % population size (colony size)
nOnlooker = nPop; %number of onlooker bees
L = round(0.0025 × nVar × nPop); % abandonment limit parameter (trial limit)
H = 0.025; % acceleration coefficient upper bound
% Initialization
empty_bee.Position = [];
empty_bee.Cost = []; % empty bee structure
Pop = repmat(empty_bee,nPop,1); % initialize population array
BestSol.Cost = inf; % initialize best solution ever found
for i = 1:nPop % create initial population, **start1**
pop(i).Position = unifrnd(VarMin,VarMax,VarSize);
pop(i).Cost = **Run_Fig2 (pop(i).Position);**
  if pop(i).Cost <= BestSol.Cost
    BestSol = pop(i);
  end
end % create initial population, **end1**
C = zeros(nPop,1); % abandonment counter
BestCost = zeros(MaxIt,1); % hold best cost values
% ABC Main Loop
for it = 1:MaxIt % abc main loop, **start2**
  for i= 1:nPop % recruited bees, **start3**
    % Choose k randomly, not equal to i
    K = [1:i-1 i+1:nPop];
    K = K(randi([1 numel(K)]));
    % Define Acceleration Coeff.
    phi = h × unifrnd(−1,+1,VarSize);
% New Bee Position
    newbee.Position = pop(i).Position+
    phi. × (pop(i).Position-pop(k).Position);
    % Evaluation
    newbee.Cost = **Run_Fig2(newbee.Position);**

```
    % Comparision
    if newbee.Cost <= pop(i).Cost
        pop(i) = newbee;
    else
        C(i) = C(i)+1;
    end
 end % recruited bees, end3
% Calculate Fitness Values and Selection Probabilities
 F = zeros(nPop,1);
 MeanCost = mean([pop.Cost]);
 for i = 1:nPop    % convert cost to fitness
    F(i) = exp( −pop(i).Cost/MeanCost );
 end
 P = F/sum(F); % probability calculation
 for m = 1:nOnlooker % onlooker bees, start4
    % Select Source Site
    i = RouletteWheelSelection(P);
    % Choose k randomly, not equal to i
    K = [1:i-1 i+1:nPop];
    k = K(randi([1 numel(K)]));
    % Define Acceleration Coeff.
    phi = h × unifrnd(−1,+1,VarSize);
    % New Bee Position
    newbee.Position=
    pop(i).Position+phi. × (pop(i).Position-pop(k).Position);
    % Evaluation
    newbee.Cost = Run_Fig2 (newbee.Position);
% Comparision
if newbee.Cost <= pop(i).Cost
        pop(i) = newbee;
    else
        C(i) = C(i)+1;
    end
 end % onlooker bees, end4
 for I = 1:nPop % scout bees
    if C(i) >= L          pop(i).Position = unifrnd(VarMin,VarMax,VarSize);
        pop(i).Cost = Run_Fig2 (pop(i).Position);
        C(i) = 0;
    end
 end
 for I = 1:nPop % update best solution ever found
    if pop(i).Cost <= BestSol.Cost
        BestSol = pop(i);
    end
 end
```

```
% Store Best Cost Ever Found
BestCost(it) = BestSol.Cost;
end % abc main loop, stop2
% Results
figure(1);
xlabel('Iteration');
ylabel('Best Cost');
plot(BestCost);
hold on;
grid on;
semilogy(BestCost, 'LineWidth',2);
K1=BestSol.Position(1) % n_1 2
K2=BestSol.Position(2) % μ_1 2
K3=BestSol.Position(3) % Tsample
K4=BestSol.Position(4) % Kout
```
Inside the algorithm it is used "Roulette Wheel Selection" function that is described below.
```
function i = RouletteWheelSelection (P)
   r = rand;
   C = cumsum(P);
   i=find(r <= C,1,'first');
end
```

References

1. Gholipour, R.; Khosravi, A.; Mojallali, H. Multi-objective optimal backstepping controller design for chaos control in a rod-type plasma torch system using Bees algorithm. *Appl. Math. Modell.* **2015**, *39*, 4432–4444. [CrossRef]
2. Sekhar, G.C.; Sahu, R.K.; Baliarsingh, A.K.; Panda, S. Load frequency control of power system under deregulated environment using optimal firefly algorithm. *Int. J. Electr. Power Energy Syst.* **2016**, *74*, 195–211. [CrossRef]
3. Veysi, M.; Soltanpour, M.R.; Khooban, M.H. A novel self-adaptive modified bat fuzzy sliding mode control of robot manipulator in presence of uncertainties in task space. *ROBOTICA* **2015**, *33*, 2045–2064. [CrossRef]
4. Liang, Y.C.; Cuevas Juarez, J.R. A novel metaheuristic for continuous optimization problems: Virus optimization algorithm. *Eng. Optim.* **2016**, *48*, 73–93. [CrossRef]
5. Kose, E.; Abaci, K.; Kizmaz, H.; Aksoy, S.; Yalçin, M.A. Sliding mode control based on genetic algorithm for WSCC systems include of SVC. *Elektron. Elektrotech.* **2013**, *19*, 25–28. [CrossRef]
6. Sekhar, P.; Mohanty, S. An enhanced cuckoo search algorithm based contingency constrained economic load dispatch for security enhancement. *Int. J. Electr. Power Energy Syst.* **2016**, *75*, 303–310. [CrossRef]
7. Moharam, A.; El-Hosseini, M.A.; Ali, H.A. Design of optimal PID controller using hybrid differential evolution and particle swarm optimization with an aging leader and challengers. *Appl. Soft Comput.* **2016**, *38*, 727–737. [CrossRef]
8. Das, S.; Chatterjee, D.; Goswami, S.K. A Gravitational Search Algorithm Based Static VAR Compensator Switching Function Optimization Technique for Minimum Harmonic Injection. *Electr. Power Compon. Syst.* **2015**, *43*, 2297–2310. [CrossRef]
9. Kumar, A.R.; Premalatha, L. Optimal power flow for a deregulated power system using adaptive real coded biogeography-based optimization. *Int. J. Electr. Power Energy Syst.* **2015**, *73*, 393–399. [CrossRef]
10. Ang, K.H.; Chong, G. PID control system analysis, design, and technology. *IEEE Trans. Control Syst. Technol.* **2005**, *13*, 559–576.
11. Hallworth, M.; Shirsavar, S.A. Microcontroller based peak current mode control using digital slope compensation. *IEEE Trans. Power Electron.* **2012**, *27*, 3340–3351. [CrossRef]
12. Aksoy, S.; Mühürcü, A. PI Elman neural network based nonlinear state estimation for induction motors. *IREE* **2011**, *6*, 706–718.

13. Okan, E.; Mahmut, Ö.; Nejat, Y. Impact of small-world topology on the performance of a feed-forward articial neural network based on 2 different real-life problems. *Turk. J. Elect. Eng. Comp Sci.* **2014**, *22*, 708–718.
14. Beg, M.A.; Khedkar, M.K.; Paraskar, S.R.; Dhole, G.M. Feed-forward Artificial Neural Network–Discrete Wavelet Transform Approach to Classify Power System Transients. *Electr. Power Compon. Syst.* **2013**, *41*, 586–604. [CrossRef]
15. Kermani, M.Z.; Kisi, O.; Rajaee, T. Performance of radial basis and LM-feed forward artificial neural networks for predicting daily watershed runoff. *Appl. Soft Comput.* **2013**, *13*, 4633–4644. [CrossRef]
16. Nabiyev, V. *Yapay Zeka*, 1st ed.; Seçkin: Istanbul, Turkey, 2012.
17. Nourmohammadzadeh, A.; Hartmann, S. Fault Classification of a Centrifugal Pump in Normal and Noisy Environment with Artificial Neural Network and Support Vector Machine Enhanced by a Genetic Algorithm. In *International Conference on Theory and Practice of Natural Computing*; Springer: Cham, Switzerland, 2015; pp. 58–70.
18. Yu, Y.; Li, Y.; Li, J. Nonparametric modeling of magneto rheological elastomeric base isolator based on artificial neural network optimized by ant colony algorithm. *J. Intell. Mater. Syst. Struct. Rep.* **2015**, *26*. [CrossRef]
19. Zhu, C.; Zhao, X.; Zhou, J. ANN based on PSO for Surface Water Quality Evaluation Model and Its Application. *Chin. Control Decision Conf. Rep.* **2009**, *6*, 3264. [CrossRef]
20. Chang, J.; Xu, X. Applying Neural Network with Particle Swarm Optimization for Energy Requirement Prediction. In Proceedings of the 7th World Congress on Intelligent Control and Automation, Chongqing, China, 25–27 June 2008.
21. Ma, L.; Lee, K.Y.; Ge, G. An Improved Predictive Optimal Controller with Elastic Search Space for Steam Temperature Control of Large-Scale Supercritical Power Unit. In Proceedings of the 51st IEEE Conference on Decision and Control, Maui, HI, USA, 10–13 December 2012.
22. Deepa, P.; Sivakumar, R. Synthesis of Heuristic Control Strategies for Liquid Level Control in Spherical Tank. In Proceedings of the 3rd International Conference on Advances in Electrical, Electronics, Information, Communication and Bio-Informatics, Chennai, India, 27–28 February 2017.
23. Ma, L.; Cao, P.; Gao, Z.; Lee, K.Y. ANN and PSO Based Intelligent Model Predictive Optimal Control for Large-Scale Supercritical Power Unit. In Proceedings of the 2016 IEEE International Conference on Information and Automation, Ningbo, China, 1–3 August 2016.
24. Lin, X.; Li, A.; Zhang, W. Application of PSO-based ANN in Knowledge Acquisition for the Selection of Optimal Milling Parameters. In Proceedings of the 6th World Congress on Intelligent Control and Automation, Dalian, China, 21–23 June 2006.
25. Ayan, K.; Kılıç, U. Artificial bee colony algorithm solution for optimal reactive power flow. *Appl. Soft Comput. Rep.* **2012**, *12*, 1477. [CrossRef]
26. Wani, S.M.A. Comparative study of back propagation learning algorithms for neural networks. *Int. J. Res. Comput. Commun. Eng.* **2013**, *3*, 1151–1156.
27. Smola, A.; Vishwanathan, S.V.N. *Introduction to Machine Learning*, 1st ed.; Cambridge University Press: Cambridge, UK, 2008.
28. Kose, E.; Muhurcu, A. The control of a non-linear chaotic system using genetic and particle swarm based on optimization algorithms. *Int. J. Intell. Syst. Appl. Eng.* **2016**, *4*, 145–149. [CrossRef]
29. Karaboga, D. *An Idea Based on Honey Bee Swarm for Numerical Optimization*; Computers Engineering Department, Engineering Faculty, Erciyes University: Kayseri, Turkey, 2005.

© 2018 by the author. Licensee MDPI, Basel, Switzerland. This article is an open access article distributed under the terms and conditions of the Creative Commons Attribution (CC BY) license (http://creativecommons.org/licenses/by/4.0/).

Article

On the Boundary Conditions in a Non-Linear Dissipative Observer for Tubular Reactors

Irandi Gutierrez-Carmona [1], Jaime A. Moreno [2] and H.F. Abundis-Fong [3],*

1. Departamento de Control Automático, Cinvestav, Mexico City 07360, Mexico; igutierrez@ctrl.cinvestav.mx
2. Instituto de Ingeniería, Universidad Nacional Autónoma de México, Mexico City 04510, Mexico; jmorenop@iingen.unam.mx
3. Tecnológico Nacional de México/I.T. Laguna, Torreón, Coah. 27000, Mexico
* Correspondence: habundis@correo.itlalaguna.edu.mx; Tel.: +52-871-705-1324

Received: 28 November 2018; Accepted: 21 December 2018; Published: 28 December 2018

Abstract: The modal injection mechanism ensures the exponential convergence of an observer in a continuous tubular reactor in dependence with the system parameters, the sensor location, and the observer gains. In this paper, it is shown that by simple considerations in the boundary conditions, the observer convergence is improved regardless of the presence of perturbations, the sensor locations acquire a meaningful physical meaning, and by simple numerical manipulations, the perturbations in the inflow can be numerically estimated.

Keywords: distributed observers; sensor position; perturbation estimation; PDE

1. Introduction

Tubular reactors are of great importance in chemical and biochemical processes, specially those with non-monotonic kinetics [1], e.g., catalytic reactors with Langmuir–Hinshelwood kinetics [2,3] or bioreactors with Haldane kinetics [4]. The tubular reactors are continuous systems where the mass concentration in some inner point depends on the spatial and temporal coordinates (see Figure 1).

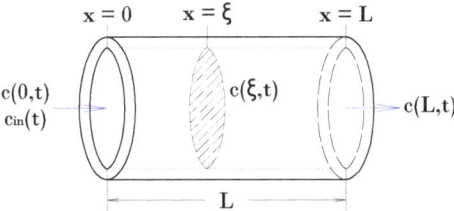

Figure 1. Simplified model of a tubular reactor.

In this kind of reactors, it is almost impossible to measure the concentration along the reactor; it is usually found that only a finite set of points can be measured, and the system states must be reconstructed from this information. The necessity to measure or estimate the system states has motivated the design of observers for this distributed parameter system, including absolute stability results [5], adaptive switching observers [6], Lyapunov-based approaches [7], backstepping designs [8], sliding modes observers [9], kalman schemes [10,11], interval observers [12], and finally (the main interest of this work) dissipative approaches [13].

Dissipative observers deal with a Luenberger-type observer; this is, the observer may be understood as a copy of the original system, plus correction terms to adjust the system response. The observer dynamic in the infinite–dimensional space is studied using the Garlekin's method, where the orthonormal basis is defined by the eigenfunctions, which in turn may be divided into slow eigenfunctions and fast eigenfunctions that describe, correspondingly, the slow and fast dynamics of the system [14–16].

The main idea of the dissipative observers is, through a modal injection mechanism, to move the slow eigenfunctions sufficiently far into the left-half complex plane to ensure that the potentially destabilising effects of the non-linear reaction terms are compensated [13,16]. The effect of the fast eigenfunctions, corresponding to fast dynamics, is assumed stable and disappears rapidly.

In the non-linear dissipative observer [17], three measurements of the concentration are made in the reactor: In some inner point and in both boundaries. The observer behaviour depends on the position of the inner measurement point but not explicitly on the boundary measurements [18]; thus, the boundary measurements can be used for other purposes rather than stability—for example, to provide further information for the sensor allocation or improve the observer performance in the presence of inflow uncertainties.

In this paper, we propose a simple but meaningful way to select the boundary gains in order to improve the observer convergence, provide a physical meaning for the sensor position, and allow the estimation of the input uncertainties in the inflow. The results are shown in a numerical example.

This paper is organised as follows: In Section 2, the previous results and inconvenience of neglecting the effects of the boundary gains are described; in Section 3, the advantages of a correct selection of the boundary gains are proposed; in Section 4, the numerical results are shown; and in Section 5, the conclusions are presented.

2. Problem Formulation

Consider the tubular reactor depicted in Figure 1, where $c(x,t)$ is the mass concentration at the spatial coordinate $x \in [0,1]$ at time t. For this tubular reactor, the dynamical equations are given as:

$$\begin{aligned} \frac{\partial c(x,t)}{\partial t} &= \frac{1}{P_{ec}} \frac{\partial^2 c(x,t)}{\partial x^2} - \frac{\partial c(x,t)}{\partial x} - D_a r\left(c(x,t)\right), \\ \frac{1}{P_{ec}} \frac{\partial c(x,t)}{\partial x}\bigg|_{x=0} &= c(0,t) - c_{in}(t), \\ \frac{1}{P_{ec}} \frac{\partial c(x,t)}{\partial x}\bigg|_{x=1} &= 0, \end{aligned} \quad (1)$$

where P_{ec} is the system Peclet number, $r(x,t)$ is the non-linear reaction rate, D_a is a constant reaction rate, and $c_{in}(t)$ is the inflow mass concentration.

The mass concentration $c(x,t)$ can be measured by sensors located at the positions $x = \{0, \xi, 1\}$, for some $\xi \in (0,1)$; this is, the mass concentration is measured in the inflow, some inner point of the reactor and outflow. To estimate the complete mass concentration in the distributed system, the Luenberger-type observer may be used [13]:

$$\begin{aligned} \frac{\partial \hat{c}(x,t)}{\partial t} &= \frac{1}{P_{ec}} \frac{\partial^2 \hat{c}(x,t)}{\partial x^2} - \frac{\partial \hat{c}(x,t)}{\partial x} - D_a r\left(\hat{c}(x,t)\right) \\ &\quad \cdots - l_\xi(x)\left(\hat{c}(\xi,t) - c(\xi,t)\right), \\ \frac{1}{P_{ec}} \frac{\partial \hat{c}(x,t)}{\partial x}\bigg|_{x=0} &= \hat{c}(0,t) - c_{in}(t) - l_0\left(\hat{c}(0,t) - c(0,t)\right), \\ \frac{1}{P_{ec}} \frac{\partial \hat{c}(x,t)}{\partial x}\bigg|_{x=1} &= -l_1\left(\hat{c}(1,t) - c(1,t)\right). \end{aligned} \quad (2)$$

Note that the observer is a copy of the original system (Equation (1)), plus the distributed correction term $l_\xi(x)$ and the boundary correction terms $\{l_0, l_1\}$. The observation error

$$e(x,t) := \hat{c}(x,t) - c(x,t), \quad (3)$$

is the difference between the real and the estimated mass concentration, with a dynamical evolution given as:

$$\frac{\partial e(x,t)}{\partial t} = \frac{1}{P_{ec}}\frac{\partial^2 e(x,t)}{\partial x^2} - \frac{\partial e(x,t)}{\partial x} - D_a \rho(x,t) - l_\xi(x)e(\xi,t)$$
$$\left.\frac{1}{P_{ec}}\frac{\partial e(x,t)}{\partial x}\right|_{x=0} = (1-l_0)\,e(0,t),$$
$$\left.\frac{1}{P_{ec}}\frac{\partial e(x,t)}{\partial x}\right|_{x=1} = -l_1 e(1,t), \quad (4)$$

where the non-linear term $\rho(x,t) = r(\hat{c}(x,t)) - r(c(x,t))$ is the difference between the reaction rate in the system and the observer. In Reference [18], the following theorem is described.

Theorem 1. *If in the observer (Equation (2)), the boundary correction terms are set to zero, and using the correction term:*

$$l_\xi(x) = \sum_{k=1}^{N} l_{\xi,k} \Phi_k(x), \quad (5)$$

where $\phi_k(x)$ are the solutions of the Sturm–Liouville problem:

$$\left(\frac{1}{P_{ec}}\frac{\partial^2}{\partial x^2} - \frac{\partial}{\partial x}\right)\Phi_k(x) = \lambda_k \Phi_k(x), \quad (6)$$

then the weighted error norm:

$$\|e(x,t)\|_{\mathcal{L}_2^\omega} = \int_0^1 \exp^{-P_{ec}x} e^2(x,t)dx =: E(t), \quad (7)$$

converges exponentially to zero; this is:

$$E(t) \leq E(0)\exp^{-\tilde{\Lambda}t}, \quad (8)$$

for some positive constant $\tilde{\Lambda}$, if the following conditions are met:

(i) *The non-linear term $\rho(x,t)$ satisfies the sector condition:*

$$S_h := \int_0^1 \omega(x) \begin{bmatrix} e(x,t) \\ \rho(x,t) \end{bmatrix}^T \begin{bmatrix} -s_u s_l & \frac{1}{2}(s_u+s_l) \\ \frac{1}{2}(s_u+s_l) & -1 \end{bmatrix} \begin{bmatrix} e(x,t) \\ \rho(x,t) \end{bmatrix} dx \geq 0, \quad (9)$$

where $s_l = \min \frac{\partial r}{\partial c}$, and $s_u = \max \frac{\partial r}{\partial c}$ are, respectively, the minimal and maximal slope of the reaction rate with respect to the concentration;
(ii) *the sensor location $x = \xi$ does not correspond to any root of the first N eigenfunctions $\Phi_k(x)$, this is, $\Phi_k(\xi) \neq 0$;*
(iii) *noticing that the eigenvalues λ_k, given as:*

$$\lambda_k = \frac{P_{ec}^2 + 4\omega_k^2}{4P_{ec}}, \quad (10)$$

are real, negative, and form a discrete monotonically decreasing series [19], $\lambda_1 > \lambda_2 > \ldots > \lambda_N > \lambda_{N+1} > \ldots$, for some $(k-1)\pi \leq \omega_k \leq \pi$, see Equation (20). Then, the modal correction dimension N is chosen such that:

$$-2\lambda_{N+1} > \frac{(2D_a - [s_u + s_l])^2}{4} - s_u s_l + 2\Lambda, \quad (11)$$

and finally;
(iv) *the maximal eigenvalue of the matrix $(A_N - LC^s)$, where:*

$$A_N = \begin{bmatrix} \lambda_1 & 0 & \cdots & 0 \\ 0 & \lambda_2 & & 0 \\ \vdots & & \ddots & \vdots \\ 0 & 0 & \cdots & \lambda_N \end{bmatrix}, \quad L = \begin{bmatrix} l_{\xi,1} \\ l_{\xi,2} \\ \vdots \\ l_{\xi,N} \end{bmatrix}, \quad C^s = \begin{bmatrix} \phi_1(\xi) \\ \phi_2(\xi) \\ \cdots \\ \phi_N(\xi) \end{bmatrix}, \quad (12)$$

is smaller than λ_{N+1}.

Remark 1. *The eigenvalues λ_k are functions of the Peclet number P_{ec}, determining the convergence rate of the weighted error norm (Equation (7)), and the dimension N of the modal correction mechanism (Equation (11)). A small Peclet number produces a high diffusion term, whereas a big Peclet number produces a small diffusion term.*

The basic idea of the observer is to accelerate the convergence rate of the slowest $N-$ eigenfunctions. Noticing that the observer stability proof does not depend on the boundary conditions (see Appendix A), the pair $\{l_0, l_1\}$ is selected to improve the observer performance, without seemingly any restriction on the pair $\{l_0, l_1\}$. In similar works [16], solely the boundary conditions in the observer convergence are studied, leading to restrictive conditions.

In this work, as an extension of the previous theorem, we show that the gains $\{l_0, l_1\}$ can be selected to:

(a) Improve the observer convergence;
(b) provide more information about the sensor position;
(c) facilitate tuning the observer parameters;
(d) and allow the estimation of the inflow perturbation.

3. Main Results

The gains $\{l_0, l_1\}$ are not required directly in the proof of Theorem 1, but they certainly modify the eigenvalues $\Phi_k(x)$ used to design the correction term (Equation (5)) and play an important role in the observation error (see Equations (3) and (16)). In the following corollary, we show how the correct gains $\{l_0, l_1\}$ simplify the observer design and constrain the error behaviour in a suitable way.

Corollary 1. *Assume conditions of Theorem 1 are fulfilled, but consider the boundary conditions:*

$$|l_0| \gg 1 \quad \text{and} \quad l_1 = l_0 - 2, \quad (13)$$

then:

(i) *The correction term $l_\xi(x)$ in Equation (5) simplifies to:*

$$l_\xi(x) = \sqrt{2} \exp^{P_{ec} x/2} \sum_{k=1}^{N} l_{\xi,k} \sin(k\pi x), \quad (14)$$

(ii) *the feasible sensor positions are given by all points in the set:*

$$\xi = \left\{ y \in (0,1) : y \neq \frac{n}{k} \quad \text{for} \quad k = 1, \cdots, N \quad \text{and} \quad n \in \mathbb{N} \right\}, \quad (15)$$

(iii) *and the observation error in the boundaries is close to zero or converges rapidly to zero.*

The proof of this corollary is given along this section. First recall that the solution for the error Equation (4) may be decomposed as:

$$e(x,t) = \sum_{k=1}^{\infty} e_k(t) \Phi_k(x), \tag{16}$$

where the set $\{\phi_k(x)\}|_{k\in\mathbb{N}}$ defines a basis for the spatial distribution and $\{e_k(t)\}|_{k\in\mathbb{N}}$ defines the time evolution of the system. From Equation (6), it follows that the eigenfunctions $\{\phi_k(x)\}|_{k\in\mathbb{N}}$ are of the form:

$$\Phi_k(x) = \exp^{P_{ec}x/2}\left(A_k \sin(\omega_k x) + B_k \cos(\omega_k x)\right). \tag{17}$$

The eigenfrequencies $\{\omega_k\}|_{k\in\mathbb{N}}$ and the amplitudes $\{A_k, B_k\}|_{k\in\mathbb{N}}$ are obtained, substituting the eigenfunctions from Equation (17) in the Sturm–Liouville boundary conditions:

$$\begin{aligned} \left.\frac{1}{P_{ec}}\frac{\partial \phi_k(x)}{\partial x}\right|_{x=0} &= (1-l_0)\phi_k(0), \\ \left.\frac{1}{P_{ec}}\frac{\partial \phi_k(x)}{\partial x}\right|_{x=1} &= -l_1\phi_k(1), \end{aligned} \tag{18}$$

as:

$$(1-2l_0)P_{ec}B_k - 2\omega_k A_k = 0, \tag{19}$$

and:

$$(\tan(\omega_k))^{-1} = \frac{1}{(l_1 - l_0 + 1)}\left(\frac{1}{P_{ec}}\omega_k + \frac{(1+2l_1)(2l_0-1)}{4}\frac{P_{ec}}{\omega_k}\right). \tag{20}$$

From Equations (19) and (20), it follows that numerical approximations are required to build the inyection term (Equation (5)). From Equation (13), for example, $l_1 - l_0 + 1 = -1$ and $|l_0| \gg 1$, Equation (20) becomes:

$$(\tan(\omega_k))^{-1} \approx -\left(\frac{1}{P_{ec}}\omega_k + |l_0|^2\frac{P_{ec}}{\omega_k}\right). \tag{21}$$

The right-hand side is a concave function with upper bound:

$$-\left(\frac{1}{P_{ec}}\omega_k + |l_0|^2\frac{P_{ec}}{\omega_k}\right) \leq -2|l_0|, \tag{22}$$

and Equation (20) simply becomes:

$$(\tan(\omega_k))^{-1} \leq -2|l_0|, \tag{23}$$

and using $|l_0| \gg 1$, we find:

$$\omega_k \approx k\pi \quad \text{for all} \quad k \in \mathbb{N}. \tag{24}$$

In Figure 2, the left and right-hand sides of Equation (20) are plotted for $l_0 = 12$ and $l_1 = 10$, where the intersection points are the solutions of the equation, which verifies Equation (24).

Once the eigenfrequencies ω_k are fixed, Equation (19) becomes:

$$B_k = \frac{2k\pi A_k}{(1-2l_0)P_{ec}}, \tag{25}$$

and for $|l_0|$ large enough, the first N terms may be neglected, this is:

$$\{B_k\}|_{k=1,\cdots,N} \approx 0. \tag{26}$$

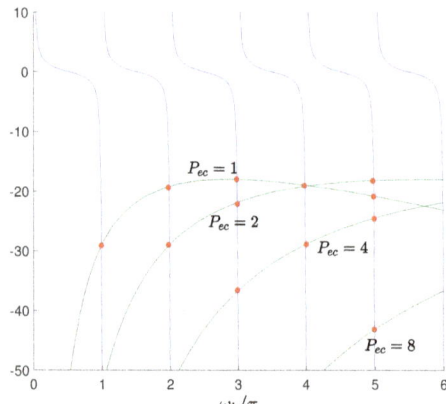

Figure 2. Numerical approximation for Equation (20).

Therefore, the first N eigenfunctions are:

$$\phi_k(x) \approx A_k \exp^{P_{ec}x/2} \sin(k\pi x) \qquad k = \{1, \cdots, N\}. \qquad (27)$$

To satisfy the orthogonal condition depicted in the Appendix A (see Equation (A3)), $A_k = \sqrt{2}$ is selected for $k = \{1, \cdots, N\}$, and the first N eigenfunctions become:

$$\phi_k(x) \approx \sqrt{2} \exp^{P_{ec}x/2} \sin(k\pi x) \quad \text{for} \quad k = \{1, \cdots, N\}, \qquad (28)$$

From where Equation (14) follows. In Figure 3, the first four eigenvalues for $l_0 = 12$ and $l_1 = 10$ are shown. Increasing $|l_0|$ will make Equation (28) a better approximation to the real eigenvalues.

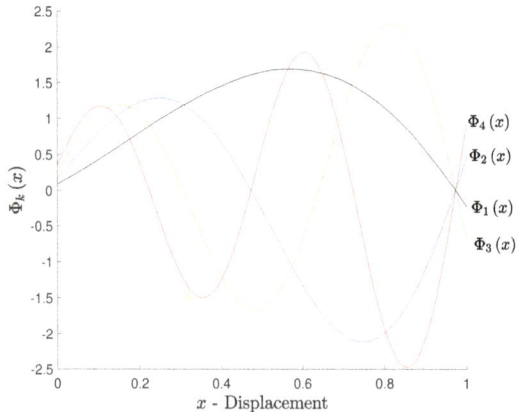

Figure 3. First four eigenvalues $\phi_k(x)$.

From Equation (28), it is immediately obvious that the sensor should avoid any position $\xi \in (0,1)$ such that:

$$\sin(\xi \pi x) = 0 \quad \text{for} \quad k = \{1, \cdots, N\}, \qquad (29)$$

and Equation (15) follows.

Now, noticing that the slowest N eigenfunctions are approximated by sine functions, the error decomposition $e(x,t) = \sum_{k=1}^{\infty} e_k(t) \phi_k(x)$ may be written as:

$$e(x,t) = \sum_{k=1}^{N} e_k(t) A_k \sin(k\pi x) + \sum_{k=N+1}^{\infty} e_k(t) \phi_k(x). \quad (30)$$

Therefore, at the boundaries:

$$e(x,t)|_{x=\{0,1\}} = \sum_{k=N+1}^{\infty} e_k(t) \phi_k(x), \quad (31)$$

the error depends only of fastest modes $\{\phi_k(x)\}|_{k=\{N+1,\ldots\}}$ that converges rapidly to zero.

Remark 2. *If the values of B_k for $k = \{1, 2, ..., N\}$ are not negligible, this may occur for small Peclet numbers or l_0 close to $\frac{1}{2}$ (see Equation (19)); then, a peaking phenomenon may arise. This is exemplified in the numerical simulation section.*

Remark 3. *Equation (13), say $l_1 = l_0 - 2$ is an algebraic condition, not the only one, proposed to preserve the right-hand side of Equation (20) as a concave function, keeping valid the approximation $\omega_k \approx k\pi$.*

The precise sensor position is something that should be discussed more carefully; however, noticing that the sensor position ξ should be selected to increase the effect of the correction mechanism, given by the product $l_\xi(x) e(\xi, t)$ in Equation (4), the sensor position can be proposed to satisfy the relation:

$$\xi = \left\{ y : \sum_{j=1}^{N} |\phi_j(y)| = \max_{x \in (0,1)} \sum_{j=1}^{N} |\phi_j(x)| \right\}. \quad (32)$$

4. Numerical Simulation

In this section, a tubular reactor with a non-monotonic Langmuir–Hinshelwood type kinematics is considered [2,3], where the reaction rate is given by:

$$r(c(x,t)) = \frac{c(x,t)}{(1 + \sigma c(x,t))^2}, \quad (33)$$

where the constant coefficient σ denotes some inhibition parameter. Simulation studies were carried out, considering a diffusion dominated behaviour corresponding to $(P_{ec}, \sigma, D_a) = (6, 3, 4)$, and an inflow as the sum of a nominal and a perturbation term:

$$c_{in}(t) = \underbrace{0.3}_{\text{nominal}} + \underbrace{0.1 \left(\sum_{m=0}^{5} \cos(6m\pi t) \right)}_{\text{perturbation}}. \quad (34)$$

Figure 4 shows the error surface when no correction mechanism is applied, this is, $l_0 = l_1 = l_\xi = 0$. The behaviour at $x = 0$ is due to the inflow perturbations.

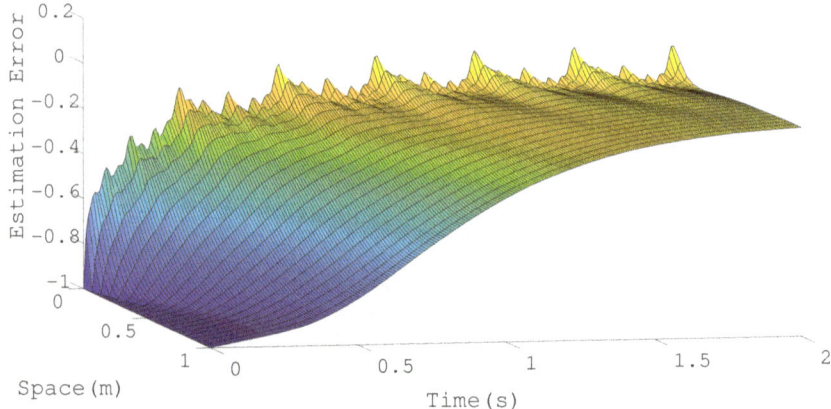

Figure 4. Error surface without boundary feedback.

By setting $\{l_0, l_1\} = \{102, 100\}$, the error in the boundaries can be brought to values close to zero rapidly (see Figure 5); even the effect of the inflow perturbations is reduced. Note that without the modal injection mechanism $l_{\tilde{\zeta}}(x)$, at $t = 0.6$ (s), the spatial behaviour of the error resembles the behaviour of the first and slowest eigenfunction $\phi_1(x) = \sqrt{2} \exp^{Pec x/2} \sin(\pi x)$.

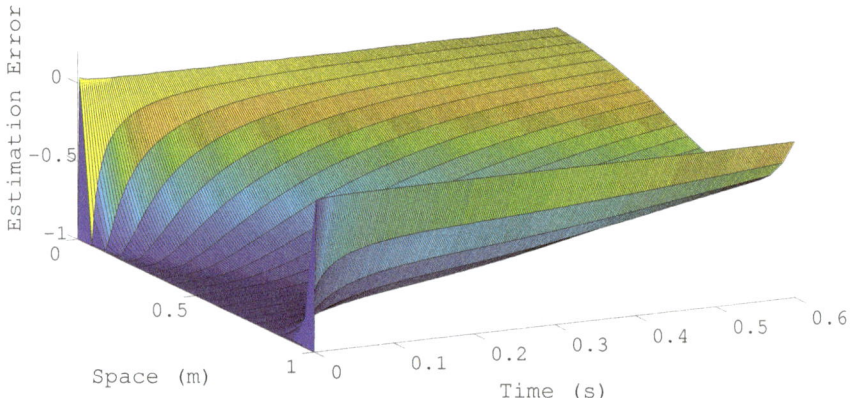

Figure 5. Error surface with only boundary feedback.

Now, it is straightforward to verify condition *(iii)* of Theorem 1; this is, the eigenvalues form the decreasing series:

$$\lambda_1 = -3.15, \quad \lambda_2 = -8.08, \quad \lambda_3 = -16.30, \quad \cdots \quad (35)$$

and Equation (11) is fulfilled for $N = 1$:

$$-2\lambda_{N+1} > \frac{(2D_a - [s_u + s_l])^2}{4} - s_u s_l + 2\Lambda = 12.41 + 2\Lambda, \quad (36)$$

and any $\Lambda \in (0, 1.87)$. It is proposed the modal correction mechanism:

$$l_{\tilde{\zeta}}(x) = l_{\tilde{\zeta},1} \phi_1(x) = l_{\tilde{\zeta},1} \sqrt{2} \exp^{Pec x/2} \sin(\pi x), \quad (37)$$

that will affect the slowest eigenfunction, allowing a better convergence of the error surface to zero. Using Equation (32) to fix the sensor position to $\xi = 0.74$, and by setting $l_{\xi,1} = 2.87$, condition *(iv)* of the Theorem 1 is fulfilled:

$$\lambda_1 - l_{\xi,1}\phi(\xi) = -30 < \lambda_2. \tag{38}$$

In Figure 6, the error surface with the the boundary feedback and the correction mechanism is shown. The effect of the modal injection mechanism is immediate.

Figure 7 shows the error norm $E(t)$ for all different feedback conditions:

(a) No feedback;
(b) only boundary feedback;
(c) only modal correction mechanism;
(d) both boundary and modal correction mechanism.

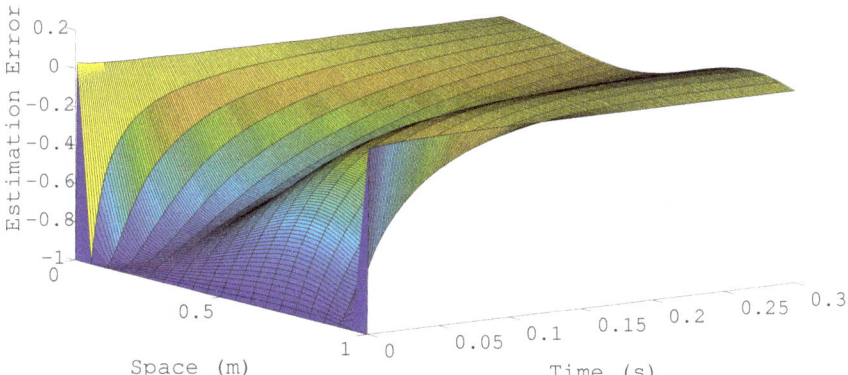

Figure 6. Error surface with boundary and the first mode feedback.

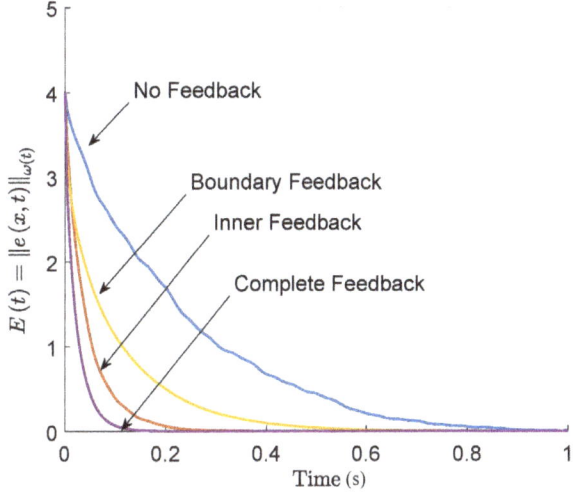

Figure 7. Error comparison.

The combination of a modal injection mechanism with boundary feedback increases the convergence rate without compromising stability. To improve the observer convergence, more modes can be added to the modal correction mechanism; for example, consider the modal correction mechanism:

$$l_{\xi}(x) = \sum_{k=1}^{3} l_{\xi,k} \Phi_k(x) = \sqrt{2} \exp^{P_{ec}x/2} \sum_{k=1}^{3} l_{\xi,k} \sin(k\pi x). \tag{39}$$

It is immediate to verify that setting:

$$l_{\xi,1} = 12.6, \quad l_{\xi,2} = 6.2, \quad l_{\xi,3} = 0.3, \tag{40}$$

then:

$$\max \left\{ \text{eig} \left(\begin{bmatrix} -3.15 & 0 & 0 \\ 0 & -8.08 & 0 \\ 0 & 0 & -16.30 \end{bmatrix} - \begin{bmatrix} l_{\xi,1} \\ l_{\xi,2} \\ l_{\xi,3} \end{bmatrix} \begin{bmatrix} \phi_1(\xi) & \phi_2(\xi) & \phi_3(\xi) \end{bmatrix} \right) \right\} = -20 < \lambda_3, \tag{41}$$

and the convergence rate of the three slowest modes is increased. In Figure 8, the corresponding error surface is shown. Comparing Figures 6 and 8, the effect of adding more modes in the modal correction mechanism is immediate.

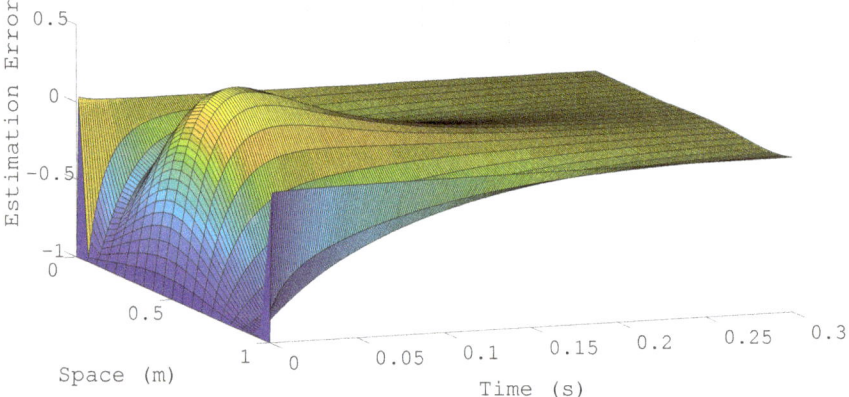

Figure 8. Error surface with boundary and three modal correction mechanism.

Now, using the fact that the error in the boundaries is close to zero, this is:

$$e(x,t)|_{x=0} \approx 0 \quad \Rightarrow \quad \hat{c}(0,t) \approx c(0,t), \tag{42}$$

simple numeric manipulations will allow the estimation of the inflow perturbation. From the boundary conditions:

$$\frac{1}{P_{ec}} \frac{\partial c(0,t)}{\partial x} = c(0,t) - \{c_{in}(t) + \theta_{per}(t)\}, \tag{43}$$

and the corresponding discrete approximation:

$$\frac{1}{P_{ec} \Delta x} (c(\Delta x, t) - c(0,t)) = c(0,t) - \{c_{in}(t) + \theta_{per}(t)\}, \tag{44}$$

a non-rigorous estimation of the perturbation is obtained by replacing the actual concentration with the estimated concentration, this is, $c(\Delta x, t) \to \hat{c}(\Delta x, t)$, and solving for θ_{per} as:

$$\theta_{\text{per}}(t) \approx \frac{1}{P_{ec}\Delta x}\left((1 + P_{ec}\Delta x)c(0,t) - \hat{c}(\Delta x, t)\right) - c_{\text{in}}(t). \tag{45}$$

The estimation of the inflow perturbation is shown in Figure 9.

Finally, and for completeness, an example is presented of the peaking phenomenon that commonly occurs when high gains are implemented.

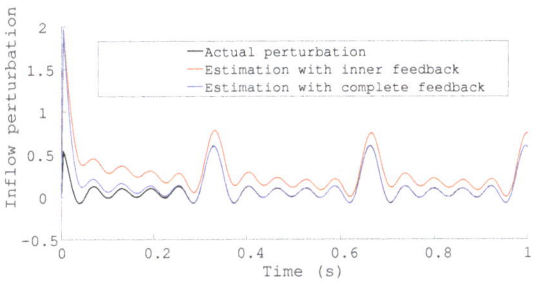

Figure 9. Perturbation estimation in the concentration input.

Peaking Phenomenon

Consider a tubular reactor with a small Peclet number, for example, $(P_{ec}, \sigma, D_a) = (2, 3, 4)$, and the boundary gains $(l_0, l_1) = (12, 10)$. The corresponding error surface is depicted in Figure 10, where a peak in the spatial boundary appears. Contrary to what is thought, this peak is reduced when the boundary gains are increased, making this peaking phenomenon something that should be more carefully analysed, especially when the observer data will be used for feedback control [20].

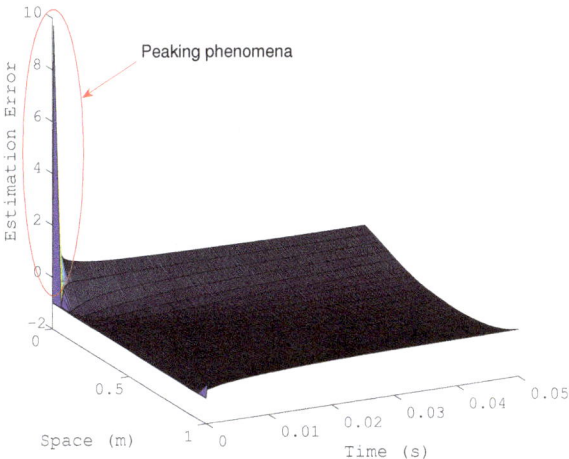

Figure 10. Peaking phenomena is the error surface.

5. Conclusions

In this work, we extend some results of the non-linear dissipative observer to show that correctly chosen boundary gains allow a simple tuning of the observer and its parameters, improving the observer convergence, and allowing the estimation of the inflow perturbations. Numerical validation of the presented algorithm shows the validity of the proposed approach.

Author Contributions: Conceptualisation, I.G.-C. and J.A.M.; Formal analysis, I.G.-C.; Investigation, J.A.M.; Visualisation, writing—review and editing, H.F.A.-F.

Funding: This research received no external funding.

Acknowledgments: Special thanks the National Autonomous University of México, UNAM, for the support.

Conflicts of Interest: The authors declare no conflict of interest.

Appendix A

Proof of Theorem 2.1. In this section, a simplified proof is provided. First recall that the error norm (see Equation (7)) may be equivalently written as:

$$\|e(x,t)\|_{\mathcal{L}_2^\omega} = E(t) = \int_0^1 \omega(x) e^2(x,t) \, dx, \tag{A1}$$

where $\omega(x) = \exp^{P_{ec} x/2}$ [21,22]. Combining Equations (16) and (A1), we have:

$$E(t) = \sum_{k=1}^{\infty} \sum_{j=1}^{\infty} e_k(t) e_j(t) \int_0^1 \omega(x) \phi_k(x) \phi_j(x) \, dx, \tag{A2}$$

selecting the eigenfuntions $\{\phi_k(x)\}|_{k \in \mathbb{N}}$ in such a way that:

$$\int_0^1 \omega(x) \phi_k(x) \phi_j(x) \, dx = \delta_{k,j}, \tag{A3}$$

where:

$$\delta_{k,j} = \begin{cases} 0 & \text{if } k \neq j \\ 1 & \text{if } k = j \end{cases}, \tag{A4}$$

then the error norm can be written as:

$$E(t) = \sum_{k=1}^{\infty} \sum_{j=1}^{\infty} e_k(t) e_j(t) \delta_{k,j} = \sum_{k=1}^{\infty} e_k^2(t), \tag{A5}$$

and deriving the error norm $E(t)$, we obtain:

$$\frac{d}{dt} E(t) = \frac{d}{dt} \left(\int_0^1 \omega(x) e^2(x,t) \, dx \right) = 2 \int_0^1 \omega(x) e(x,t) \frac{\partial e(x,t)}{\partial t} \, dx, \tag{A6}$$

and substituting (4) we have:

$$\frac{d}{dt} E(t) = D_T + D_K, \tag{A7}$$

with:

$$\begin{aligned} D_T &= 2 \int_0^1 \omega(x) e(x,t) \left(\left(\frac{1}{P_{ec}} \frac{\partial^2}{\partial x^2} - \frac{\partial}{\partial x} \right) e(x,t) - l_\xi(x) e(\xi,t) \right) dx \\ D_K &= -2 \int_0^1 \omega(x) e(x,t) D_a \rho(x,t) \, dx \end{aligned} \tag{A8}$$

Now, using Equations (16) and (6), we rewrite D_T as:

$$D_T = 2 \sum_{k=1}^{\infty} \sum_{j=1}^{\infty} \lambda_k e_j(t) e_k(t) \int_0^1 \omega(x) \Phi_k(x) \Phi_j(x) dx \\ -2 \sum_{k=1}^{\infty} \sum_{j=1}^{N} l_{\xi,j} e(\xi,t) e_k(t) \int_0^1 \Phi_j(x) \Phi_k(x) dx, \quad (A9)$$

and using the orthogonality condition (Equation (A3)), we have:

$$D_T = 2 \sum_{k=1}^{\infty} \lambda_k e_k^2(t) - 2 e(\xi,t) \sum_{k=1}^{N} l_{\xi,k} e_k(t): \quad (A10)$$

since $e(\xi,t) = \sum_{k=1}^{\infty} e_k(t) \phi_k(\xi)$, then:

$$D_T = 2 \sum_{k=1}^{\infty} \lambda_k e_k^2(t) - 2 \sum_{k=1}^{\infty} \sum_{j=1}^{N} e_k(t) \Phi_k(\xi) l_{\xi,j} e_j(t), \quad (A11)$$

which can be written in the quadratic form:

$$D_T = 2 e^T(t) \left[\begin{array}{c|c} A_N - LC^s & LC^f \\ \hline 0 & A_{N+1} \end{array} \right] e(t) \quad (A12)$$

where:

$$e(t) = \begin{bmatrix} e_1(t) & \cdots & e_N(t) & e_{N+1}(t) & \cdots \end{bmatrix}^T, \quad (A13)$$

and:

$$A_N = \begin{bmatrix} \lambda_1 & & 0 \\ & \ddots & \\ 0 & & \lambda_N \end{bmatrix}, \quad A_{N+1} = \begin{bmatrix} \lambda_{N+1} & 0 & \\ 0 & \lambda_{N+2} & \\ & & \ddots \end{bmatrix} \quad (A14)$$

$$L = \begin{bmatrix} l_{\xi,1} \\ \cdots \\ l_{\xi,N} \end{bmatrix}, \quad C^s = \begin{bmatrix} \Phi_1(\xi) \\ \cdots \\ \Phi_N(\xi) \end{bmatrix}, \quad C^f = \begin{bmatrix} \Phi_{N+1}(\xi) \\ \Phi_{N+2}(\xi) \\ \cdots \end{bmatrix}. \quad (A15)$$

If $\phi_k(\xi) \neq 0$ for $k = \{1, \cdots, N\}$, then the pair $\{A_N, C^s\}$ is observable and there exists a vector L such that:

$$\lambda_{N+1} \leq \max \sigma(A_N - LC^s), \quad (A16)$$

so D_T is bounded by:

$$D_T \leq \lambda_{N+1} e^T(t) e(t) = \lambda_{N+1} E(t). \quad (A17)$$

The perturbation term D_K is bounded using the sector condition $S_h \geq 0$ (see Equation (9)) so:

$$D_K \leq -2 \int_0^1 \omega(x) e(x,t) D_a \rho(x,t) dx + S_h, \quad (A18)$$

or, regrouping terms:

$$D_K \leq -\int_0^1 \omega(x) \begin{bmatrix} e(x,t) \\ \rho(c,e) \end{bmatrix}^T \begin{bmatrix} s_u s_l & D_a - \frac{1}{2}(s_u + s_l) \\ D_a - \frac{1}{2}(s_u + s_l) & 1 \end{bmatrix} \begin{bmatrix} e(x,t) \\ \rho(c,e) \end{bmatrix} dx. \quad (A19)$$

Equation (A7) is then bounded as:

$$\frac{d}{dt}E(t) \leq -\int_0^1 \omega(x) \begin{bmatrix} e(x,t) \\ \rho(c,e) \end{bmatrix}^T P \begin{bmatrix} e(x,t) \\ \rho(c,e) \end{bmatrix} dx, \tag{A20}$$

where $P = P^T \in \mathbb{R}^{2\times 2}$ is given by:

$$P = \begin{bmatrix} s_u s_l - 2\lambda_{N+1} & D_a - \frac{1}{2}(s_u + s_l) \\ D_a - \frac{1}{2}(s_u + s_l) & 1 \end{bmatrix}. \tag{A21}$$

Now, P is positive definite if for some positive scalar $\Lambda > 0$:

$$s_u s_l - 2\lambda_{N+1} - \left(D_a - \frac{1}{2}(s_u + s_l)\right)^2 = \Lambda, \tag{A22}$$

or, equivalently:

$$-2\lambda_{N+1} = \frac{(2D_a - [s_u + s_l])^2}{4} - s_u s_l + \Lambda, \tag{A23}$$

Therefore:

$$\frac{d}{dt}E(t) \leq -\tilde{\Lambda}\int_0^1 \omega(x) \begin{bmatrix} e(x,t) \\ \rho(x,e) \end{bmatrix}^T \begin{bmatrix} e(x,t) \\ \rho(x,e) \end{bmatrix} dx \quad \tilde{\Lambda} = \min \sigma(P) \tag{A24}$$

$$\frac{d}{dt}E(t) \leq -\tilde{\Lambda}\int_0^1 \omega(x) e^2(x,t) dx = -\tilde{\Lambda}E(t), \tag{A25}$$

and:

$$E(t) \leq E(0)\exp^{-\tilde{\Lambda}t}. \tag{A26}$$

□

References

1. Lapidus, L.; Amundson, N. *Chemical Reactor Theory*; Printice-Hall: Upper Saddle River, NJ, USA, 1977.
2. Carberry, J.J. *Chemical and Catalytic Reaction Engineering*; Courier Corporation: Chelmsford, MA, USA, 2001.
3. Elnashaie, S.S.; Abashar, M.E. The implication of non-monotonic kinetics on the design of catalytic reactors. *Chem. Eng. Sci.* **1990**, *45*, 2964–2967. [CrossRef]
4. Bailey, J.E.; Ollis, D.F. Biochemical engineering fundamentals. *Chem. Eng. Educ.* **1976**, *10*, 162–165.
5. Curtain, R.F.; Demetriou, M.A.; Ito, K. Adaptive compensators for perturbed positive real infinite-dimensional systems. *Int. J. Appl. Math. Comput. Sci.* **2003**, *13*, 441–452.
6. Mercorelli, P. An adaptive and optimized switching observer for sensorless control of an electromagnetic valve actuator in camless internal combustion engines. *Asian J. Control* **2014**, *16*, 959–973. [CrossRef]
7. Mercorelli, P. A motion-sensorless control for intake valves in combustion engines. *IEEE Trans. Ind. Electron.* **2017**, *64*, 3402–3412. [CrossRef]
8. Krstic, M.; Smyshlyaev, A. *Boundary Control of PDEs: A Course on Backstepping Designs*; Siam: Philadelphia, PA, USA, 2008; Volume 16.
9. Orlov, Y.V. *Discontinuous Systems: Lyapunov Analysis and Robust Synthesis under Uncertainty Conditions*; Springer Science & Business Media: Berlin, Germany, 2008.
10. Schaum, A.; Alvarez, J.; Meurer, T.; Moreno, J. State-estimation for a class of tubular reactors using a pointwise innovation scheme. *J. Process Control* **2017**, *60*, 104–114. [CrossRef]
11. Chen, L.; Mercorelli, P.; Liu, S. A Kalman estimator for detecting repetitive disturbances. In Proceedings of the 2005 IEEE American Control Conference, Portland, OR, USA, 8–10 June 2005; pp. 1631–1636.
12. Kharkovskaia, T.; Efimov, D.; Fridman, E.; Polyakov, A.; Richard, J.P. On design of interval observers for parabolic PDEs. *IFAC-PapersOnLine* **2017**, *50*, 4045–4050. [CrossRef]

13. Schaum, A.; Moreno, J.A.; Alvarez, J. Spectral dissipativity observer for a class of tubular reactors. In Proceedings of the 47th IEEE Conference on Decision and Control, CDC 2008, Cancun, Mexico, 9–11 December 2008; pp. 5656–5661.
14. Pourkargar, D.B.; Armaou, A. Design of APOD-based switching dynamic observers and output feedback control for a class of nonlinear distributed parameter systems. *Chem. Eng. Sci.* **2015**, *136*, 62–75. [CrossRef]
15. Pourkargar, D.B.; Armaou, A. Dynamic shaping of transport–reaction processes with a combined sliding mode controller and Luenberger-type dynamic observer design. *Chem. Eng. Sci.* **2015**, *138*, 673–684. [CrossRef]
16. Schaum, A.; Moreno, J.; Meurer, T. Dissipativity-based observer design for a class of coupled 1-D semi-linear parabolic PDE systems. *IFAC-PapersOnLine* **2016**, *49*, 98–103. [CrossRef]
17. Schaum, A.; Moreno, J.A.; Díaz-Salgado, J.; Alvarez, J. Dissipativity-based observer and feedback control design for a class of chemical reactors. *J. Process Control* **2008**, *18*, 896–905. [CrossRef]
18. Schaum, A. On the Design of Nonlinear Dissipative Observers for Agitated and Tubular Reactors. Ph.D. Thesis, Instituto de Ingeniería, Universidad Nacional Autónoma de México, Mexico City, Mexico, 2009.
19. Delattre, C.; Dochain, D.; Winkin, J. Sturm-Liouville systems are Riesz-spectral systems. *Int. J. Appl. Math. Comput. Sci.* **2003**, *13*, 481–484.
20. Sussmann, H.; Kokotovic, P. The peaking phenomenon and the global stabilization of nonlinear systems. *IEEE Trans. Autom. Control* **1991**, *36*, 424–440. [CrossRef]
21. Dettman, J.W. *Mathematical Methods in Physics and Engineering*; Courier Corporation: Chelmsford, MA, USA, 2013.
22. Jeffreys, H.; Jeffreys, B. *Methods of Mathematical Physics*; Cambridge University Press: Cambridge, UK, 1999.

© 2018 by the authors. Licensee MDPI, Basel, Switzerland. This article is an open access article distributed under the terms and conditions of the Creative Commons Attribution (CC BY) license (http://creativecommons.org/licenses/by/4.0/).

Article

Model-Based Stochastic Fault Detection and Diagnosis of Lithium-Ion Batteries

Jeongeun Son and Yuncheng Du *

Department of Chemical & Biomolecular Engineering, Clarkson University, Potsdam, NY 13676, USA; son@clarkson.edu
* Correspondence: ydu@clarkson.edu; Tel.: +1-315-268-2284

Received: 6 December 2018; Accepted: 9 January 2019; Published: 13 January 2019

Abstract: The Lithium-ion battery (Li-ion) has become the dominant energy storage solution in many applications, such as hybrid electric and electric vehicles, due to its higher energy density and longer life cycle. For these applications, the battery should perform reliably and pose no safety threats. However, the performance of Li-ion batteries can be affected by abnormal thermal behaviors, defined as faults. It is essential to develop a reliable thermal management system to accurately predict and monitor thermal behavior of a Li-ion battery. Using the first-principle models of batteries, this work presents a stochastic fault detection and diagnosis (FDD) algorithm to identify two particular faults in Li-ion battery cells, using easily measured quantities such as temperatures. In addition, models used for FDD are typically derived from the underlying physical phenomena. To make a model tractable and useful, it is common to make simplifications during the development of the model, which may consequently introduce a mismatch between models and battery cells. Further, FDD algorithms can be affected by uncertainty, which may originate from either intrinsic time varying phenomena or model calibration with noisy data. A two-step FDD algorithm is developed in this work to correct a model of Li-ion battery cells and to identify faulty operations in a normal operating condition. An iterative optimization problem is proposed to correct the model by incorporating the errors between the measured quantities and model predictions, which is followed by an optimization-based FDD to provide a probabilistic description of the occurrence of possible faults, while taking the uncertainty into account. The two-step stochastic FDD algorithm is shown to be efficient in terms of the fault detection rate for both individual and simultaneous faults in Li-ion batteries, as compared to Monte Carlo (MC) simulations.

Keywords: fault detection and classification; uncertainty analysis; lithium-ion battery; optimization; thermal management; polynomial chaos expansion

1. Introduction

Lithium-ion (Li-ion) batteries are widely used in many applications, such as cell phones, electric and hybrid electric vehicles, since they exhibit a higher energy density and have a relatively longer life compared to other batteries [1]. In these systems, Li-ion batteries must possess a high reliability and pose no safety threats [2]. However, the thermal behavior can greatly affect the safety, durability, and performance of Li-ion batteries [3]. For example, fire and explosions caused by thermal runaway were reported [4]. Thus, reliable battery management systems are essential to mitigate negative effects (e.g., thermal runaway) and avoid catastrophic failures [5]. As a key component of the battery management system, fault detection and diagnosis play an important role in the management of Li-ion batteries [6].

Fault detection and diagnosis (FDD) methods generally can be classified into two major groups, i.e., first-principle model-based methods and data driven (or empirical) methods [7]. For the former, models describing the physical mechanisms of the fault dynamics are oftentimes used, while historical data are typically collected for data driven methods to derive empirical models. Each of these approaches has its own advantage and drawback depending on the specific problems. It is recognized that first-principle model-based methods exhibit a better extrapolation ability, whereas data-driven methods are easier to design [8]. This work focuses on the use of the first-principle models for FDD, since these models provide a fundamental understanding of the thermal physics of batteries [9].

Several first-principle thermal models have been previously developed for Li-ion batteries. For example, a three-dimensional thermal finite element model was developed to investigate the cell behavior under abnormal events such as overheating and external short circuits [10]. This model requires high computational capabilities, and its application is limited to stationary storage [11]. Compared to the three-dimensional models, the one-dimensional model of Li-ion batteries, developed using the average lumped temperature of the cell, is viable for real-time applications and can enable online battery management [12]. However, such a model may fail to provide insights into the thermal (fault) dynamics due to its simplicity [13]. As a trade-off, a two-dimensional thermal model was developed, which can predict the core and the surface temperature of Li-ion battery cells [3,13]. Since the two-dimensional model can provide a better understanding of the thermal dynamics of battery cells, while maintaining the computational complexity, it is used in this work for the design of a stochastic FDD scheme.

Measurements of temperatures such as surface and core temperatures are often used for FDD in Li-ion batteries, but there is no direct measurement of the core temperature. To take the core temperature into account, estimation techniques are often required. In the literature, several estimation techniques have been developed. For example, an adaptive observer based on the lumped thermal model [14] and state observer using partial differential algebraic equations [15] were proposed to estimate the temperature. Compared to these estimation techniques, the real-time monitoring and diagnosis of faults in batteries are less explored. Although there have been several proposed works related to diagnostic algorithms for internal faults in Li-ion batteries [3,16,17], it is important to note that previously reported FDD work mostly investigated sensor or actuator fault detection problems [18–20].

In this work, we propose to estimate the core temperature and use the estimation results to identify and classify two sets of faults. That is, faults that can introduce dynamic changes in core temperatures and faults that can affect the surface temperatures. The FDD scheme in this work can potentially provide more information about the thermal dynamics of batteries and enable an internal thermal fault detection to improve the performance of the Li-ion battery.

For FDD, the available algorithms compare the observed behavior to the corresponding model results, estimated from first-principle models [21]. When a fault is detectable, the FDD scheme will generate fault signatures, which in turn can be referred to an FDD scheme to identify the root cause of faults using a threshold [22]. However, the main restrictive factor for the first-principle model-based FDD is the model uncertainty [23]. The accuracy of the fault detection algorithm can be affected by any uncertainty in the model parameters. Such an uncertainty may result from intrinsic time varying phenomena or originate from the model calibration with noisy measurements [24]. The uncertainty can be quantitatively approximated by a calibration with experimental data, which include principles such as least squares errors or the Delphi method [25,26].

The procedures that firstly quantify the uncertainty and then propagate the uncertainty onto the FDD scheme are typically omitted in previously reported works. This subsequently may lead to a loss of information about the effect of uncertainty on FDD performance. Recently, several techniques, such as the adaptive observer [27,28] and the sliding mode observer [29], were developed for FDD in the presence of uncertainty. However, most of these methods cannot provide information, such as the probability that a fault has occurred. In addition, since the faults in the batteries may happen in a stochastic fashion, the use of fixed thresholds to identify the root cause of faults may not be effective.

There are differences between the actual thermal dynamics of Li-ion batteries and fundamental models derived from physical phenomena. For example, to make models tractable and useful, it is common to make simplifications during the model development, which will introduce a mismatch between the model and the Li-ion battery system of interest. Thus, the first principle model-based FDD scheme should be designed to compensate the mismatch. Specifically, a set of fixed model parameters may not be accurate enough for estimating the core temperature in the presence of a model mismatch. Consequently, any inaccuracy in the temperature estimation may potentially lead to a low fault detection rate. To ensure the accuracy of FDD, it is essential to simultaneously calibrate the model parameters and adjust the FDD scheme. However, this is generally challenging due to the presence of uncertainty such as the measurement noise and an unknown model mismatch.

In this work, we propose to address these aforementioned limitations by developing an FDD scheme for Li-ion batteries described by a two-dimensional first-principle thermal dynamic model, for which both model parameters and faults are of a stochastic nature. Specifically, the faults considered in this work, such as the thermal runaway, are stochastic perturbations superimposed on step changes in the specific thermal dynamic parameter and electric current. The objective is to identify the changes in the mean values of the thermal dynamic parameter and current in the presence of random perturbations, the measurement noise, and a model mismatch. As compared to other existing thermal diagnostic techniques, the main feature of the FDD scheme is the efficient quantification of the effect of stochastic changes in model parameters on fault detection, and the rapid propagation of the stochasticity onto the estimation of temperatures that are required for FDD.

Note that one possible way to propagate uncertainty in model parameters onto temperature estimates is the use of Monte Carlo (MC) simulations [30]. However, methods such as MC may be computationally demanding, since they often require a larger number of simulations in order to obtain accurate results. It is worth mentioning that although the calibration of an FDD scheme can be performed offline, the online re-calibration of the model in the presence of a model mismatch with MC as shown later in current work is computationally prohibitive. Recently, the uncertainty propagation with generalized Polynomial Chaos (gPC) expansion has been studied in different modelling [31], optimization [32], and fault detection problems [24]. As compared to MC, the advantage of gPC is that it can propagate a complex probability distribution of uncertainty in model parameters onto model predictions rapidly and can analytically approximate the statistical moments of model predictions in a computationally efficient manner [31]. The improvement in computational time may facilitate its application in the real-time model adjustment for improved FDD.

The FDD algorithm in this work is specifically targeted to identify and diagnose stochastic thermal faults consisting of uncertainty around a set of mean values of thermal properties in the presence of a model mismatch. In summary, the contributions in this work include: (i) The use of an intrusive gPC model for stochastic FDD of Li-ion batteries by approximating the uncertainty in thermal dynamics with gPCs and by propagating the uncertainty directly onto temperatures that can be used for FDD; (ii) the identification and classification of a fault based on the probability information of temperatures other than a single point estimate or threshold; (iii) the formulation of an optimization to account for a model mismatch and adjust the thermal dynamic models by incorporating the discrepancy between model predictions and measurements.

This paper is organized as follows. Section 2 presents the theoretical background and the principal methodologies in this work, including a two-dimensional thermal dynamic model, the introduction of generalized polynomial chaos (gPC) expansion, and the formulation of the stochastic fault detection and diagnosis (FDD) problem. The methodology for FDD and the formulation of an optimization for model correction to account for the model mismatch is presented in Section 3. The analysis and discussion of the results are given in Section 4, followed by conclusions in Section 5.

2. Theoretical Backgrounds

2.1. Thermal Model of Lithium-ion Battery

The two-dimensional deterministic thermal dynamic model is used to describe a cylindrical Li-ion battery cell in this work [3,13]. A schematic diagram of the Li-ion battery cell is shown in

Figure 1. This model can provide information about the heat source of the battery and estimate the core temperature based on measurements of the surface temperature. The surface temperature T_s and the core temperature T_c can be defined as:

$$C_c \dot{T}_c = I^2 R_e + \frac{T_s - T_c}{R_c} \tag{1}$$

$$C_s \dot{T}_s = \frac{T_f - T_s}{R_u} - \frac{T_s - T_c}{R_c} \tag{2}$$

$$R_e = \beta_0 + \beta_1 SOC + \beta_2 T_c \tag{3}$$

where I is the current, T_f represents the surrounding air temperature, R_e is the internal (or electrical) resistance, R_c is the thermal resistance between the surface and core of the battery, R_u denotes the convection resistance between the surface and the surroundings of the battery, C_c and C_s represent the heat capacity of the internal battery material and the surface battery material, respectively. The internal resistance R_e is given in Equation (3) which consists of state of charge (SOC), core temperature T_c, and parameters $\beta_0, \beta_1, \beta_2$ that can be pre-estimated by an offline estimation scheme [3].

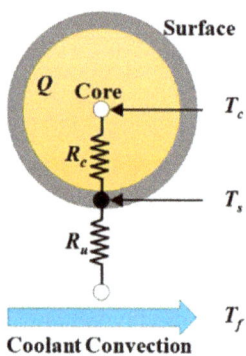

Figure 1. Schematic of thermal model of Li-ion battery cell.

For the Li-ion battery cell model given in Equations (1) and (2), model parameters including R_e are generally assigned with constant values. A set of parameters used in the two-dimensional thermal dynamic model is given in Table 1 [33].

Table 1. Parameter declaration for the thermal model of Li-ion battery cell.

Model Parameters	C_c	C_s	R_e	R_c	R_u
Units	JK^{-1}	JK^{-1}	mΩ	KW^{-1}	KW^{-1}
Value	268	18.8	10	2	1.5

It is important to note that the model of the battery and the model parameters may involve uncertainty. For example, the thermal dynamics of a Li-ion battery cell can change with respect to time, which may be caused by factors such as the surrounding temperature and the state of charge. In addition, the estimates of model parameters can be affected by noisy data used for model calibration. These possible sources of uncertainty can be briefly categorized into three groups as follows.

1. Observational uncertainty: This includes measurement errors in experimental data, such as the measurements of voltage, current, and surface temperatures.
2. Parametric uncertainty: This refers to uncertainty in parameters, which may originate from the observational uncertainty or result from lack of information. It may be advantageous to represent

a model parameter, e.g., R_e in Equation (1), as a random variable with a distribution other than a fixed value.
3. Structural uncertainty: This describes the differences between a model and the actual Li-ion battery system. For example, models in Equations (1) and (2) may not be an exact representation of the thermal dynamics of a Li-ion battery cell.

In this current work, we focus on the development of FDD algorithms in the presence of these uncertainties. Specifically, the conduction resistance R_c in Equations (1) and (2) is considered as an uncertain parameter and changes in R_c are defined as stochastic faults. The conduction resistance R_c is often used to incorporate conduction and thermal resistance across materials with compact and inhomogeneous properties. It is difficult to accurately estimate the exact parameter value of R_c, since the rolled electrodes consist of the cathode, anode, separator, and current collectors, which may complicate the parameter estimation and reduce the estimation accuracy [14]. Any variations in R_c, may significantly affect the performance of the battery. In addition, it is assumed that the current I in Equation (1) is the second uncertainty in this work, since the internal state of the battery can be affected by the current [34]. For example, as previously reported [14] current variations may lead to the fluctuation in temperatures of the battery. Furthermore, the electric current of the battery can be time-varying in practice and can be corrupted by measurement errors. Thus, the exact value of current can be an unknown prior.

Since the convection resistance R_u is related to the surrounding coolant flowrate [35], which is oftentimes tightly controlled to maintain a consistent battery temperature, R_u is assumed to be a constant rather than a parametric uncertainty. For the internal resistance R_e in Equation (1), it can be affected by various conditions such as the state of charge of battery, temperature, and drive cycle [14,36,37] leading to the changes in model predictions such as temperature. However, this thermal parameter in Li-ion battery has been investigated by many researchers and is well formulated with the state of charge and temperatures as shown in Equation (3) [3,14,38]. For example, it can be estimated offline with experimental data or determined online with SOC estimation based on an equivalent circuit model (ECM) [38]. In this work, it is assumed that R_e is a constant rather than a time-varying parameter and it is not considered as a parametric uncertainty for simplicity. However, the proposed uncertainty propagation and diagnostic scheme can be extended to R_u and R_e according to their intrinsic properties when there is evidence to support a significant variation in R_u and R_e.

In this work, sudden changes of temperatures in the Li-ion battery caused by the current I and resistance R_c will be diagnosed and classified by the proposed method. Additionally, to introduce structural uncertainty, it is assumed that the exact statistical moments of uncertainties, such as the actual mean value of R_c is unknown to the modelers, which will be corrected by incorporating the differences between model predictions and the measurement of temperatures. Further, it should be noted that only the surface temperature of the battery can be directly measured, thus the estimations of the core temperatures will be used in the model correction.

2.2. Generalized Polynomial Choas Expansion

The generalized polynomial chaos (gPC) expansion approximates a random variable with an arbitrary probability density function (PDF) of another random variable (e.g., ξ) with a known prior distribution. For brevity, suppose that the battery thermal models in Equations (1) and (2) can be described by a set of ordinary differential equations (ODEs) as:

$$\dot{x} = f(t, x, u, p) \qquad (4)$$

where the vector $x = \{x_j\}$ ($j = 1, 2, \ldots, n$) represents the core and the surface temperatures, i.e., T_c and T_s, with initial values x_0 at $t = 0$, u is deterministic parameters, i.e., fixed constant values, while p is a vector of uncertainties, i.e., I and R_c in this work, which will be approximated with PDFs. To evaluate the effect of uncertainty on temperatures, a key step is to approximate each parameter in $p = \{p_i\}$ ($i = 1, 2, \ldots, n_p$) as a function of a set of the independent random variable $\xi = \{\xi_i\}$ as:

$$p_i = p_i(\xi_i) \tag{5}$$

where ξ_i denotes the ith independent random variable following a standard PDF [31]. Based on the definition of gPC expansion, each parametric uncertainty $\{p_i\}$ and the model predictions x can be defined using the orthogonal polynomial basis functions $\{\phi_k(\xi)\}$ as:

$$p_i(\xi_i) = \sum_{k=0}^{\infty} \hat{p}_{i,k} \phi_k(\xi_i) \tag{6}$$

$$x_j(t, \xi) = \sum_{k'=0}^{\infty} \hat{x}_{j,k'}(t) \varphi_{k'}(\xi) \tag{7}$$

where $\{\hat{p}_{i,k}\}$ denote the gPC coefficients of the ith parametric uncertainty, $\{\hat{x}_{j,k'}\}$ are the gPC coefficients of the jth model predictions at time instant t, and $\{\varphi_{k'}(\xi)\}$ are the orthogonal polynomial basis functions of random variables ξ [31]. When the PDFs of p are a given prior, a set of coefficients $\{\hat{p}_{i,k}\}$ in Equation (6) can be determined such that $p_i(\xi_i)$ follows a prior known distribution. Otherwise, optimization techniques can be used to estimate $\{\hat{p}_{i,k}\}$. As compared to p, the gPC coefficients of x are unknown and have to be calculated. To calculate $\{\hat{x}_{j,k'}\}$, Equations (6) and (7) are firstly substituted into Equation (4), which is followed by applying a Galerkin projection and by projecting Equation (4) onto each of the polynomial chaos basis function $\{\varphi_{k'}(\xi)\}$ as:

$$\langle \dot{x}_j(t, \xi), \varphi_{k'}(\xi) \rangle = \langle f(t, x_j(t, \xi), u, p(\xi)), \varphi_{k'}(\xi) \rangle \tag{8}$$

For practical application, truncation, i.e., a finite number of terms, is often used other than infinite terms in Equations (6) and (7). For example, the total number of approximation terms (i.e., Q) that can be used for $\{x_j\}$ in Equation (7) can be calculated as:

$$Q = ((n_p + q)! / (n_p! q!)) - 1 \tag{9}$$

where q is the number of terms that is necessary to approximate an arbitrary uncertainty with a prior known PDF in Equation (6), and n_p is the total number of parametric uncertainties in p. As seen in Equation (9), the number of terms required for the gPC approximation of $x = \{x_j\}$ depends on the order of polynomial q and/or the number of unknown parametric uncertainty n_p.

The inner product between any two vectors in Equation (8) can be calculated as [31]:

$$\langle \psi(\xi), \psi'(\xi) \rangle = \int \psi(\xi) \psi'(\xi) W(\xi) d\xi \tag{10}$$

where the integral is calculated over the entire domain defined by random variables ξ in the Wiener-Askey framework, $W(\xi)$ is the PDF of ξ that is defined as a weighting function in gPC theory. For example, Hermite polynomial basis functions can be used for normal distributions [31]. Using gPC coefficients of model predictions x in Equation (7), the statistical moments of x at a given time t can be quickly estimated as follows:

$$E(x_j(t)) = E\left[\sum_{k'=0}^{Q} \hat{x}_{j,k'}(t) \varphi_{k'}\right] = \hat{x}_{j,0}(t) E(\varphi_0) + \sum_{k'=1}^{Q} E[\varphi_i] = \hat{x}_{j,0}(t) \tag{11}$$

$$\text{Var}(x_j(t)) = E\left[(x_j(t) - E[x_j(t)])^2\right] = E\left[\left(\sum_{k'=0}^{Q} \hat{x}_{j,k'}(t)\varphi_{k'} - \hat{x}_{j,k'=0}(t)\right)^2\right]$$
$$= E\left[\left(\sum_{k'=1}^{Q} \hat{x}_{j,k'}(t)\varphi_{k'}\right)^2\right] = \sum_{k'=1}^{Q} \hat{x}_{j,k'}(t)^2 E(\varphi_{k'}^2)$$
(12)

In addition, the PDF of model predictions x can be estimated by sampling from the PDF of ξ and by substituting samples into the gPC expressions of x in Equation (7). The calculation of statistical moments with the analytical formulae in Equations (11) and (12) and the rapidly approximation of the PDF of x are the main rationale of using the gPC in this current work, since it can reduce the computational burden involved in the model correction in the presence of structural and parametric uncertainty. Note that the FDD procedure in this work consists of the inverse of the procedures summarized above, i.e., the identification of the PDFs (e.g., mean values) of parametric uncertainty using the measurements and model predictions of x. The details concerning the FDD will be discussed in Section 3.

2.3. Formulation of FDD Problem

The faults considered in this work consist of stochastic perturbations superimposed on a particular set of mean values of these two aforementioned uncertainties, i.e., current I and conduction resistance R_c. For example, Figure 2 shows a possible fault profile (Figure 2a) and the resulting noise-free temperature responses (Figure 2b). For clarity, two mean values of each faults in Figure 2 are presented. As can be seen, any changes in the mean values of faults can induce variations in temperatures. The objective is to use the measurements of the temperature to identify the step changes between different mean values of the current (I) and the thermal resistance R_c.

A mathematical description of stochastic faults is defined as:

$$p_i = \overline{p_i} + \Delta p_i (i = 1, \ldots, n_p)$$
(13)

where $p_i \in p$ ($i = 1, 2, \ldots, n_p$), $\{\overline{p_i}\}$ denotes a set of mean values, and $\{\Delta p_i\}$ represents the variation around each mean value of the ith uncertainty. For example, the solid bold lines (blue and red) in Figure 2a are the mean values of current (I) and thermal resistance R_c, while the purple and green lines are the perturbations around each of the mean values. It is assumed in this work that the statistical moment of $\{\Delta p_i\}$ is *time-invariant* for simplicity and can be estimated with offline model calibration algorithms. In addition, the total number of possible mean values of p_i can be experimentally inferred from the constancy of measured quantities such as the surface temperature as shown in Figure 2b, but the exact mean values can be unknown to the modelers.

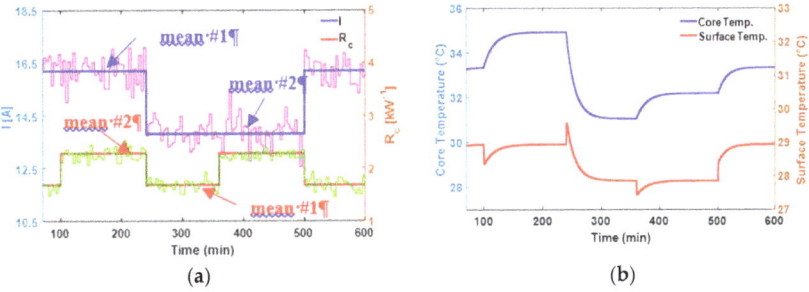

Figure 2. Profiles of faults (**a**) and the corresponding noise-free temperature (**b**). Note that the purple and green lines in (**a**) represent the perturbations around the mean values of possible faults, and noise free measurements of temperatures are used in (**b**) for clarity.

As seen in Figure 2b, the core temperature is higher, when the mean values of I and R_c are larger. Since any significant changes in the core temperatures are harmful and may cause catastrophic failures in Li-ion batteries [4], the smaller mean values of I and R_c are used to represent the normal operating mode of Li-ion battery in this work, while the larger mean value in either I and R_c is used to represent the faulty operating modes. Thus, the objective is to identify the mean value (or mean value changes) of I and R_c in the presence of uncertainty.

To summarize, two types of faults are considered. (i) Fault 1: Current fault (I), representing the switch between two mean values of I, which can affect the core temperature dynamics and further induce thermal runaway faults. (ii) Fault 2: Thermal resistance fault (R_c), representing a significant deviation in the mean value of thermal resistance R_c, which may result from battery aging and can affect both the core and temperatures. Based on the definition of the faults, the setting of normal and faulty operating modes in this work is given in Table 2, respectively.

Table 2. Faults definition and description.

Modes	Description	Type
Normal	$I = \bar{I}^1, R_c = \bar{R}_c^1$	No fault
Faulty 1	$I = \bar{I}^2, R_c = \bar{R}_c^1$	Individual fault
Faulty 2	$I = \bar{I}^1, R_c = \bar{R}_c^2$	Individual fault
Faulty 3	$I = \bar{I}^2, R_c = \bar{R}_c^2$	Simultaneous faults

3. Methodology of Fault Detection and Diagnosis

The objective of the FDD algorithm is to identify a change in the mean values of I and R_c and classify an operating condition as a normal or faulty mode described in Table 2, using measurements of temperatures. A Joint Confidence Region (JCR) based FDD algorithm is first presented in Section 3.1, which is followed by an optimization-based model correction method in Section 3.2 for improved FDD in the presence of a model mismatch.

3.1. Fault Detection Algorithm Using JCR Profiles

In Section 2, the propagation of uncertainty onto model predictions was discussed, from which the PDF profile of each model prediction can be approximated using the gPC models. The main idea of the FDD algorithm in this work is to solve the inverse problem, i.e., to identify the mean values of uncertainty with gPC models. The FDD method consists of three steps. (*a*) The stochasticity in faults (i.e., I and R_c) is propagated onto model predictions, thus producing a family of gPC models of the core and surface temperatures around each mean value of faults considered in this work. (*b*) Since two uncertainties (faults) are studied, a set of joint confidence region (JCR) profiles of the core and surface temperatures is used to infer the possible mean values or any changes in mean values of faults. The generation of the JCR, which predicts the probability that a pair of measurements belongs to a particular JCR, will be discussed later. (*c*) Because of the measurement noise and the overlaps among JCRs, the JCR-based FDD may provide a lower fault detection rate. Thus, a gPC model-based minimum distance optimization is developed to improve the FDD performance.

Step a

The formulation of the gPC models for the core and surface temperatures follows the procedures as outlined in Section 2. It is assumed that the stochastic perturbations in faults I and R_c are independent stochastic events, thus a two-dimensional random space is used, i.e., $\xi = \{\xi_1, \xi_2\}$. Consequently, the predictions of temperatures obtained from Equation (7) are functions of $\xi = \{\xi_1, \xi_2\}$, i.e., any changes in faults can affect both the core and surface temperatures.

Step b

Since two faults are studied, JCR profiles of the core and surface temperatures are used to infer mean value changes in faults I and R_c. Figure 3 shows a schematic of generated JCRs from gPC models. The generation of JCRs proceeds as follows.

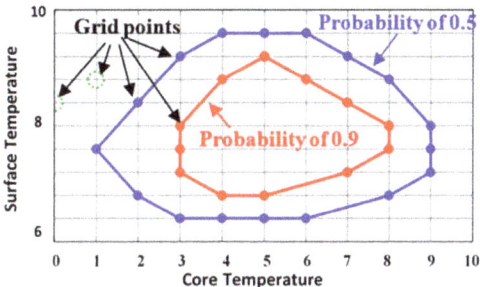

Figure 3. Schematic of joint confidence regions (JCRs) with different probabilities. Note that the units of temperatures in this work is Celsius degree (°C).

(i) In the case of stochastic perturbations in both I and R_c, the maximum variations of core and surface temperatures are first estimated. (ii) A two-dimensional discrete domain made of combinations of core and surface temperature values can be generated based on the temperature estimations in Step i. (iii) Random samples of ξ_1 and ξ_2 are substituted into the gPC models of the core and surface temperatures as defined in Equation (7), which can provide the temperatures values. (iv) Each pair of the core and surface temperatures is assigned to a particular grid generated in Step ii, and the total number of temperature pairs can be calculated when all the samples from Step iii have been assigned. (v) The probability at each discrete grid is calculated as the ratio between the number of temperature pairs at a particular grid point and the total number of temperature pairs (i.e., the number of combinations of ξ_1 and ξ_2 that are used in Step iii). (vi) A JCR can be generated by connecting discrete grid points with the same probability (see Figure 3).

Step c

Following the procedures above, a family of JCR profiles can be generated for each pair of mean values of I and R_c, as shown in Table 2, which can be used for FDD. However, as seen in Figure 4a, the JCRs used to infer faults can be misleading, when a pair of measurements (red star) is found to be in the overlap of JCRs. In addition, the measurements may lay outside of JCR profiles due to the measurement noise, as shown in Figure 4b. Thus, a gPC model-based minimum distance criterion is used to improve the FDD performance, which is explained below.

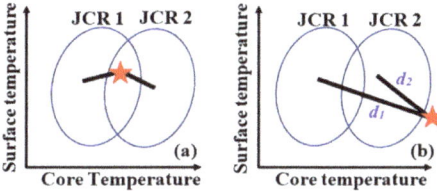

Figure 4. Visual interpretation of fault detection and diagnosis (FDD) algorithm using JCRs. Note that (a) represents that a pair of measurements can be found in the overlap of the JCRs, and (b) represent that a pair of measurements can be found outside the JCRs due to measurement noise. In addition, d_1 and d_2 in (b) represent the distance between the measurements and the centers of JCRs, which can be used for FDD with a minimum distance criterion as defined in Equations (14) and (15).

107

As seen in Equation (7), the gPC models of the core and surface temperatures are functions of random variables $\xi = \{\xi_1, \xi_2\}$, which can provide the statistical information of temperatures resulting from stochasticity in faults I and R_c. The combination of gPC models of the core and surface temperatures can provide the mathematical description of JCRs. When a pair of temperatures is available, e.g., red star in Figure 4, it is possible to calculate the distance between a pair of temperatures and the center of a JCR. For a prescribed confidence region (or specific probability), the shortest distance between the measurements and a specific JCR can then be used to infer the mean values of faults. For example, as seen in Figure 4b, the distance d_2 is smaller than d_1, thus indicating that the mean values of faults, used to generate JCR-2, are the most probable operating mode. To analytically decide the Euclidean distance between a pair of measurements and a JCR, an optimization problem is developed as:

$$\min_{\lambda} J_i = (T_{c,i} - T_{c,p})^2 + (T_{s,i} - T_{s,p})^2 \tag{14}$$

$$Operating mode: M_{FCR} = \arg \min\{J_i\} \tag{15}$$

where i is the total number of combination of mean values of faults I and R_c as shown in Table 2, $T_{c,i}$, and $T_{s,i}$ are the gPC models for a particular set of mean value I and R_c, which are functions of ξ given in Equation (7), $T_{c,p}$, and $T_{s,p}$. are the core and surface temperatures that are used for FDD. Note that M_{FCR} in Equation (15) is the identified operating mode defined in Table 2 based on the minimum distance criterion. It should be noted that there is no direct measurement of the core temperatures of the battery, thus models, i.e., Equations (1) and (2), are used to estimate the core temperature with the measurement of the surface temperature. The decision variable λ is a vector of random samples of $\xi = \{\xi_1, \xi_2\}$ from the sample domain defined by the three-sigma rules [39]. This optimization problem in Equation (14) will be performed for each pair of core and surface temperature measurements and combination of mean values of faults I and R_c that are defined in Table 2. Then, the minimum distance as defined in Equation (15) can be used to identify an operating mode as defined in Table 2.

3.2. Optimization-Based Model Correction

The FDD algorithm in Section 3.1 assumes that the exact statistical moments of I and R_c are given priors, which can be propagated onto the temperatures to formulate the JCR profiles of temperatures. However, it cannot account for the discrepancy between the model and the actual thermal dynamics of the Li-ion battery. For example, a model calibration with noisy data can introduce model uncertainty. Further, model assumptions and simplifications are often made to make a model tractable, which may result in structural uncertainty. To account for uncertainty (and/or mismatch) between the model and the actual battery cells, we propose to correct the model by incorporating the error between model predictions and available measurements. The correction criterion is formulated as follows:

$$\dot{\tilde{x}} = f(t, \tilde{x}, u, p) + \mu(\hat{x} - \tilde{x}) \tag{16}$$

where $\mu = \{\mu_j\}$ (j = 1, 2, ..., n) is a vector of correction gains, $\tilde{x}$ is model predictions, and $\hat{x}$ is the measurements of temperatures. To implement Equation (16), it is assumed that the measurements of the surface temperature are available, and the core temperature can be estimated with the model that is being corrected. It is also assumed that the exact statistical information, such as mean value of the uncertainty, is not available for the user, in order to represent a model involving model mismatch. Such a difference will be compensated using correction gains μ in Equation (16).

To calculate the correction gains, a set of measurements inside a sliding time window will be used in this work. A schematic of the sliding time moving window is shown in Figure 5, where L represents the size of the moving window and M is the moving rate, i.e., L determines a total number of required temperatures and M decides the overlap between the windows. A smaller window size can be less accurate and may be time consuming, but it can be sensitive as it would better capture the thermal dynamics of battery. A larger window size can reduce the computational burden, but it may lead to a coarse estimation. The moving rate decides the number of measurements changed at a time. For example, when 1 is used for M, which means that the one measurement is changed at a time, i.e., the first measurement in L will be removed and one new measurement will be appended to L. When M is larger, it may produce poor model correction result, while it will increase the computational load when M is smaller. The choice of L and M is problem specific and requires a trade-off, which can be determined with insights of the dynamic natures of batteries.

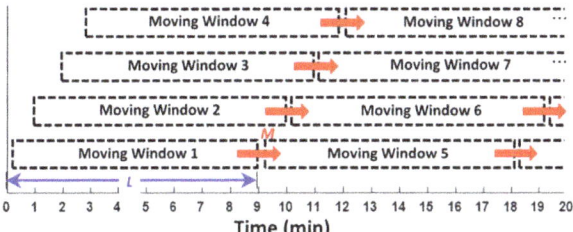

Figure 5. Schematic of sliding time moving windows for model correction.

For a sliding time moving window with temperature measurements, the correction gains μ can be optimized with an optimization as:

$$\min_{\lambda=\mu} J = \sum_{i=1}^{L} (T_{c,i} - T_{c,p})^2 + \sum_{i=1}^{L} (T_{s,i} - T_{s,p})^2 \qquad (17)$$

where $T_{c,i}$, and $T_{s,i}$ are gPC model predictions of core and surface temperatures obtained from Equation (16), $T_{c,p}$ and $T_{s,p}$ denotes the temperatures inside moving windows that are used for the model correction. Note that core temperatures are estimated from the deterministic models that are being corrected based on the measurements of the surface temperatures. The decision variable λ in Equation (17) is the correction gain that can be recursively updated with moving time windows. It will be shown in the results section that the model correction can be executed at each time interval in a real-time fashion, and the fault detection results can be greatly improved with the recursively-updated gPC model.

3.3. Summary of FDD Algorithm

An overview of the proposed model correction and FDD is shown in Figure 6. In summary, the algorithm proceeds as follows.

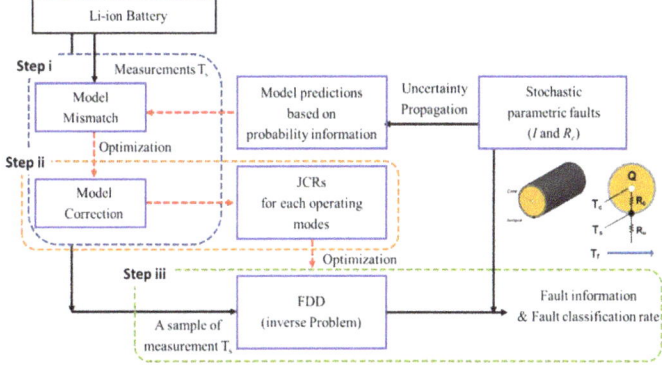

Figure 6. Overview of the proposed FDD algorithm.

Step i—Collect measurements of surface temperatures as a training set when the battery is operated at normal and faulty operating modes, described in Table 2. Using the optimization defined as Equation (17), the models of Li-ion battery cells can be corrected around each pair of the mean values of I and R_c. Note that the measurements of the temperatures for faults can be obtained from either a historical database or designed experiments.

Step ii—Using the corrected models, the JCR profiles of the core and surface temperatures for each operating mode can be generated following the procedures described in Section 3.1.

Step iii—When a sample of surface temperature is available, the core temperature will be firstly estimated, and the minimum distance can be calculated with Equations (14) and (15), which can be used to infer a particular set of mean values of I and R_c.

To evaluate the performance of the proposed FDD approach, the fault classification rate (r_{FCR}) defined as below is used:

$$r_{FCR} = \frac{n_{id}}{n_{total}} \quad (18)$$

where n_{total} represents the total number of testing samples used for algorithm verification, and n_{id} is the number of samples that have been correctly identified and classified.

4. Results and Discussion

4.1. Uncertainty Propagation and Model Predictions

The FDD algorithm is applied to the Li-ion battery cells as explained in Section 2.1. For clarity, two mean values of fault I and R_c are considered, respectively. For the current fault, I, these mean values are $\bar{I}^1$ = 16.2 A and $\bar{I}^2$ = 13.8 A, respectively. It is assumed that the stochastic perturbations in I around each of these mean values follow a normal distribution with a mean of zero and a standard deviation of 0.45 A. For the conduction resistance R_c, two mean values are $\bar{R}_c^1$ =1.68 KW^{-1} and $\bar{R}_c^2$ = 2.28 KW^{-1}, respectively. In addition, the random variations around each mean value are normally distributed, which has a mean value of zero and a standard deviation of 0.066 KW^{-1}, i.e., a 5% variation with respect to the average of two mean values. Since the perturbations around the mean values follow a normal distribution, Hermite polynomial basis functions are used for gPC models in this work. It is important to note that for arbitrary distributions, the polynomial basis functions from the Askey-Wiener scheme other than Hermite polynomial basis functions can be used to improve the convergence of the gPC approximation in Equation (6) [31].

Following the uncertainty propagation procedures described in Section 2.2, Figure 7 shows the mean of temperatures and the corresponding variance around the mean values at each time interval, when the battery is operated at the normal mode. Since two sources of uncertainty are studied (i.e., n_p = 2 in Equation (9)), and two terms can be used to approximate a normally distributed I or R_c (i.e., p = 1), six terms are required to approximate each temperature (i.e., Q = 5 in Equation (9)). The gPC coefficients of the temperatures can be solved by substituting the gPC models of uncertainties and temperatures into the Li-ion battery model (Equations (1) and (2)), which can then be solved by a Galerkin projection as explained in Section 2.2. This will produce a set of coupled equations to describe the stochastic thermal dynamics of Li-ion battery cells. The resulting gPC models of the core and the surface temperatures are given by Equations (A1)–(A12) in Appendix A for brevity.

As seen in Figure 7, T_{c0} and T_{s0} represent the mean values of the rcore and surface temperatures, and the *bar-plots* represent the variances around the mean values which can be calculated from the higher order gPC coefficients, using Equation (12) in Section 2.2. Additionally, it was found that the core temperature can be significantly affected by variations in I and R_c, as compared to the surface temperature, i.e., a larger variance as seen in Figure 7.

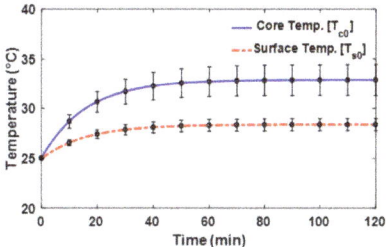

Figure 7. Uncertainty propagation in the lumped thermal models of the Li-ion battery cell at the normal operating mode, (mean value of temperatures and the variance at a few particular time intervals).

4.2. FDD Using JCR Profiles and Computational Efficiency

Based on the gPC model developed with each pair of the mean values of I and R_c, a family of JCRs can be generated following the procedures as explained in Section 3. Figure 8 shows the JCRs for a set of specific confidence regions, where 1000 pairs of temperature samples are used. Based on the JCRs profile, the mean values of I and R_c can be inferred by solving the optimization problem defined in Equations (14) and (15) for a pair of temperatures. Taking a pair of temperatures as given in Figure 8 (the star) as an example, it can be concluded that the battery system is operated around the second set of mean values of I and R_c, since the distance between the given samples of temperatures and JCR-2 is minimal. It should be noted that the JCR profiles can not only distinguish a specific faulty operating mode from the normal operation, but also provide the probability information of occurred faults.

In addition, comparison studies were conducted to compare the gPC-based FDD with Monte Carlo (MC) simulations-based method. For MC, a similar optimization problem as done for the gPC is defined as:

$$\min_{\lambda'} J = \sum_{j=1}^{N} (T_c^j - T_{c,p})^2 + \sum_{j=1}^{N} (T_s^j - T_{c,p})^2 \qquad (19)$$

where λ' is the decision variables, i.e., the mean and the standard deviation of I and R_c that have to be determined with respect to a given pair of measurements of temperature, i.e., $T_{c,p}$ and $T_{s,p}$. Also, N is the total number of samples used in the MC simulations in each iteration of the optimization, T_c^j and T_s^j are a particular set of core and surface temperatures simulated with respect to the decision variables. When the optimization of Equation (19) is finished, the optimization results λ' are compared with mean values defined in Table 2 based on a minimum distance criterion, which can identify a corresponding operating mode.

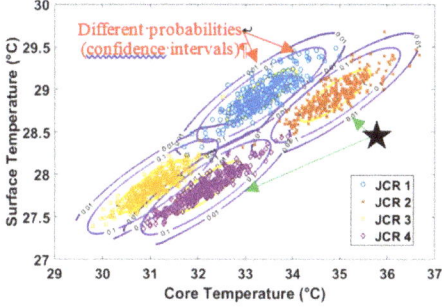

Figure 8. JCRs generated with a set of specific mean values of I and R_c, which are summarized in Table A1 in Appendix B. (i) JCR 1: 16.2 and 1.68 for I and R_c; (ii) JCR 2: 16.2 and 2.28 for I and R_c; (iii) JCR 3: 13.8 and 1.68 for I and R_c; (iv) JCR 4: 13.8 and 2.28 for I and R_c.

For the gPC-based FDD, it was found that the optimization problem described in Equations (14) and (15) can be finished within an average of 5 seconds. However, for the MC-based method, the calculation of the mean values of I and R_c on average requires approximately 321 seconds, when 100 pairs of samples of I and R_c were used to simulate T_c^j and T_s^j in each optimization iteration. This clearly shows the computational efficacy of gPC, compared with that of MC. In addition, it was found that MC with 100 samples cannot provide as accurate results as gPC. For example, it was found that the fault classification rate r_{FCR} of gPC and MC is ~0.94 and ~0.75, respectively. To improve the FDD performance, a larger number of samples are required in each iteration of the optimization with MC. However, this may significantly increase the computational burden. Especially, for the real-time model correction that will be discussed in next section, it can be computationally prohibitive with MC. A summary of the comparison between gPC and MC is given in Appendix C.

4.3. FDD Results Using JCRs in Combination with Model Correction

In previous case studies, it is assumed that the models of a battery are accurate, and JCR profiles are used for FDD. In this section, the JCR profiles-based FDD algorithm is integrated with a model correction procedure to deal with the FDD problem in the presence of a model mismatch. For clarity, it is assumed that the exact mean values of I and R_c for each operating modes (JCRs) are unknown to the modeler, thus a set of correction gains will be used to compensate the effect of a model mismatch on FDD. Since the exact mean values of faults are unknown, the mean values in the gPC models of the core and surface temperature are corrected using model predictions and measurements collected at each time interval inside the time moving windows, which can be described as:

$$\frac{dT_{c0}}{dt} = \frac{1}{C_c}\left(I_0^2 R_e + I_1^2 R_e + \frac{1}{R_{c0}}((T_{s0}-T_{c0})A + (T_{s2}-T_{c2})B + (T_{s4}-T_{c4})C)\right) + \mu_1(T_{c0}-\hat{T_c}) \quad (20)$$

$$\frac{dT_{s0}}{dt} = \frac{1}{C_s}\left(\frac{1}{R_u}(T_f-T_{s0}) - \frac{1}{R_{c0}}((T_{s0}-T_{c0})A + (T_{s2}-T_{c2})B + (T_{s4}-T_{c4})C)\right) + \mu_2(T_{s0}-\hat{T_s}) \quad (21)$$

where T_{c0} and T_{s0} are the first coefficients (i.e., mean values) in gPC models of the core and surface temperatures, I_0 and R_{c0} are the gPC coefficients in Equation (6) used to approximate the mean values of I and R_c, $\hat{T_c}$ and $\hat{T_s}$ are the measurements of temperatures. Note that μ_1 and μ_2 are correction gains which will be recursively optimized with the optimization defined in Equation (17), T_{s2}, T_{c2}, T_{s4}, and T_{c4} are higher order gPC coefficients of the core and surface temperatures, which can be determined with gPC models as given in Appendix A. In addition, A, B, and C are constants calculated using gPC models with the Galerkin projection. For illustration, Figure 9 shows the model correction results of μ_1 and μ_2, when the system is operated at different operating modes as defined in Tables 2 and A1 in Appendix B. To introduce the model mismatch, a ±10% change was randomly added to these mean values given in Table A1.

For different JCR profiles, the first column in Figure 9 represents the correction gains of the core temperature calculated at each time instant, whereas the second column is the correction gain of the surface temperature. As can be seen in Figure 9, the profiles of correction gains μ_1 and μ_2 fluctuated within a certain range when the optimization of Equation (17) was executed, and eventually reached a plateau. For example, the correction gain of the core temperature, i.e., μ_1, varied significantly when the optimization was initially executed, e.g., 0 to ~80 min. In contrast, the changes in correction gains appear to be smaller after approximately 80 min of simulations. It is important to note that the perturbations in correction gains may either result from measurement noises or stochasticity in the current I and conduction resistance R_c. In addition, it was found that the correction gain μ_2 of the surface temperature stabilizes faster than the correction gain of core temperature μ_1. This is due to the fact that random variations in I and R_c can significantly affect core temperatures as previously discussed in Section 4.1 (see Figure 7). Note that the size of moving time window (L) is set to 80 for simulations as shown in Figure 9, i.e., 80 measurements were used to optimize the correction gains at each time instant. The moving rate M is set to 1 in this case study. In addition, random noise was added to the surface temperatures, which was further used to estimate core temperatures for optimization as defined in Equation (17).

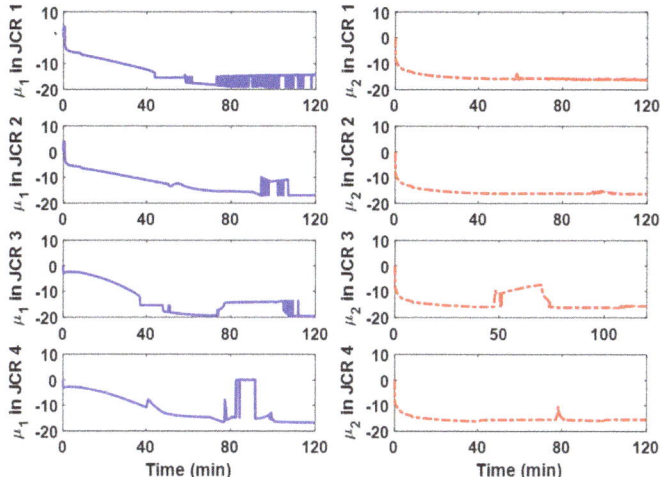

Figure 9. Correction gains μ_1 and μ_2 for different operating modes, where solid lines (blue) are the correction gain used for core temperatures and the dash-dotted line (red) are the results of surface temperature.

Using these correction gains and the gPC coefficients, the distributions of the core and surface temperatures as each time interval can be rapidly estimated. For example, Figure 10 shows the simulation results of temperatures for the normal operation. Based on the corrected gPC models and the distributions of temperatures, a set of JCR profiles can be formulated and used for FDD following the steps as explained in Section 3.1.

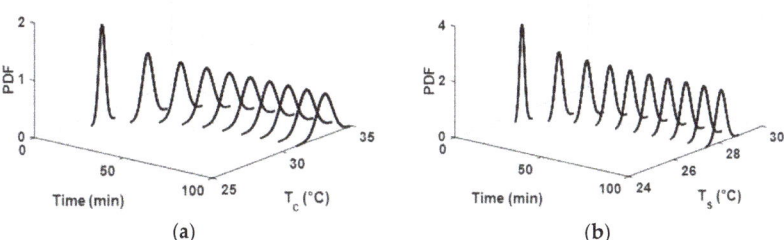

Figure 10. Distribution of temperatures for a few time intervals estimated with the generalized polynomial chaos (gPC) coefficients and the correction gains, which can be used to define a two-dimensional domain to generate JCR profiles for FDD: (**a**) Core temperature approximated with gPC and correction gains and (**b**) surface temperature approximated with gPC and correction gains.

To evaluate the efficiency of the correction and its effect on FDD, two case scenarios were investigated. For the first one, JCR profiles generated with the inaccurate mean values of I and R_c were used, whereas the correction algorithm was combined with the JCR-based FDD in the second case scenario. Table 3 shows the results of FDD for both case studies.

As seen in Table 3, the fault classification rate r_{FCR} can be improved approximately by 25% on average with the correction algorithm defined in Equation (17). In addition, study was conducted to investigate the effect of measurement noise on the accuracy of FDD, and Table 4 shows the results of r_{FCR} with respect to different levels of measurement noise. It can be seen that the measurement noise can significantly affect the accuracy of FDD. For instance, the fault classification rate r_{FCR} is about 73% with a 5% measurement noise in the surface temperatures, which has been decreased about 22%, as compared with the case where the measurement noise is 1%.

Table 3. Faults classification rate with different joint confidence region (JCR) profiles.

r_{FCR} (%)	JCR 1	JCR 2	JCR 3	JCR 4
without correction	59.1	62.3	59.9	69.7
with correction	89.6	89.7	82.7	88.4

Table 4. Faults classification rate of the model corrected by optimization-based model correction.

	1%	2%	3%	4%	5%
r_{FCR} (%)	95	89.6	84.5	78.2	72.7

Using the gPC models, it was found that the optimization of Equation (17) for one function evaluation can be completed in ~1 second on average and the optimum can be achieved in about 30 iterations, which results in an overall simulation time of about ~30 seconds. On the other hand, it was found that if Monte Carlo simulations were used for updating the correction gains with 100 samples, ~5 min were required for one evaluation of the optimization in Equation (17). Thus, 30 iterations would take ~2.5 h. This is significantly higher than the gPC-based FDD method, which may be computationally prohibitive for a real-time application of model correction with MC.

5. Conclusions

Lithium-ion (Li-ion) batteries are widely used due to their higher energy density and longer life as compared to other batteries. However, the thermal behavior can greatly affect the safety, durability, and performance of Li-ion batteries. Fault detection and diagnosis (FDD), as a key component of the battery management system, play an important role in the management of Li-ion batteries. This paper presents a stochastic FDD algorithm to identify thermal dynamic faults such as the thermal runaway fault in a Li-ion battery using generalized polynomial chaos (gPC) expansion models. The proposed algorithm consists of three consecutive procedures: (i) Uncertainty propagation with gPC models to evaluate the effect of uncertainty on measured quantities, which can be used for FDD; (ii) accurate fault diagnosis with JCR profiles, which can provide the probabilistic information of being in a faulty operating mode; (iii) recursive optimization to adjust the FDD algorithm to account for a mismatch between models and thermal dynamics of Li-ion battery cells. It was found that the gPC-based FDD method can outperform sampling-based techniques such as Monte Carlo (MC) simulations in terms of computational efficiency and FDD accuracy. This ensures its on-line applications in Li-ion battery systems such as electric and hybrid electric vehicles. However, the application of the proposed FDD algorithm in complex systems is not pursued for brevity and left for future study. In addition, it is assumed that the uncertainty in this work follows the standard distribution in the Askey–Wiener scheme for algorithm clarity. For other distributions, the arbitrary gPC algorithm as explained in our previous work can be used to improve the computational efficiency [40].

Author Contributions: Y.D. designed the stochastic models and the computational framework, J.S. carried out the computer experiments. Both Y.D. and J.S. contributed to the final version of the manuscript. Y.D. also supervised the project.

Conflicts of Interest: The authors declare no conflict of interest.

Appendix A. Results of the gPC Expansion for the Lumped Thermal Model of Li-Ion Battery

$$\frac{dT_{c0}}{dt} = \frac{1}{C_c}\left(I_0^2 R_e + I_1^2 R_e + \frac{1}{R_{c0}}((T_{s0} - T_{c0})A + (T_{s2} - T_{c2})B + (T_{s4} - T_{c4})C)\right) \quad (A1)$$

$$\frac{dT_{s0}}{dt} = \frac{1}{C_s}\left(\frac{1}{R_u}(T_f - T_{s0}) - \frac{1}{R_{c0}}((T_{s0} - T_{c0})A + (T_{s2} - T_{c2})B + (T_{s4} - T_{c4})C)\right) \quad (A2)$$

$$\frac{dT_{c1}}{dt} = \frac{1}{C_c}\left(2I_0 I_1 R_e + \frac{1}{R_{c0}}((T_{s1} - T_{c1})A + (T_{s5} - T_{c5})B)\right) \quad (A3)$$

$$\frac{dT_{s1}}{dt} = \frac{1}{C_s}\left(\frac{1}{R_u}(-T_{s1}) - \frac{1}{R_{c0}}((T_{s1} - T_{c1})A + (T_{s5} - T_{c5})B)\right) \quad (A4)$$

$$\frac{dT_{c2}}{dt} = \frac{1}{C_c}\left(\frac{1}{R_{c0}}((T_{s0} - T_{c0})B + (T_{s2} - T_{c2})C + (T_{s4} - T_{c4})D)\right) \quad (A5)$$

$$\frac{dT_{s2}}{dt} = \frac{1}{C_s}\left(\frac{1}{R_u}(-T_{s2}) - \frac{1}{R_{c0}}((T_{s0} - T_{c0})B + (T_{s2} - T_{c2})C + (T_{s4} - T_{c4})D)\right) \quad (A6)$$

$$\frac{dT_{c3}}{dt} = \frac{1}{C_c}\left(I_1^2 R_e E + \frac{1}{R_{c0}}((T_{s3} - T_{c3})FA)\right) \quad (A7)$$

$$\frac{dT_{s3}}{dt} = \frac{1}{C_s}\left(\frac{1}{R_u}(-T_{s3})F - \frac{1}{R_{c0}}((T_{s3} - T_{c3})FA)\right) \quad (A8)$$

$$\frac{dT_{c4}}{dt} = \frac{1}{C_c}\left(\frac{1}{R_{c0}}((T_{s0} - T_{c0})C + (T_{s2} - T_{c2})D + (T_{s4} - T_{c4})G)\right) \quad (A9)$$

$$\frac{dT_{s4}}{dt} = \frac{1}{C_s}\left(\frac{1}{R_u}(-T_{s4})H - \frac{1}{R_{c0}}((T_{s0} - T_{c0})C + (T_{s2} - T_{c2})D + (T_{s4} - T_{c4})G)\right) \quad (A10)$$

$$\frac{dT_{c5}}{dt} = \frac{1}{C_c}\left(\frac{1}{R_{c0}}((T_{s1} - T_{c1})B + (T_{s5} - T_{c5})C)\right) \quad (A11)$$

$$\frac{dT_{s5}}{dt} = \frac{1}{C_s}\left(\frac{1}{R_u}(-T_{s5}) - \frac{1}{R_{c0}}((T_{s1} - T_{c1})B + (T_{s5} - T_{c5})C)\right) \quad (A12)$$

where A, B, C, D, E, F, G, and H are all constants calculated with the Galerkin Projection.

Appendix B. Definition and Description of Faults and Their Mean Values

Table A1. Faults Definition and Description.

JCRs (Mode)	Mean Values	Type
JCR 1 (Faulty 1)	I = 16.2, R_c = 1.68	Individual fault
JCR 2 (Faulty 3)	I = 16.2, R_c = 2.28	Simultaneous faults
JCR 3 (Normal)	I = 13.8, R_c = 1.68	No fault
JCR 4 (Faulty 2)	I = 13.8, R_c = 2.28	Individual fault

Appendix C. Summary of Comparison between gPC and MC

Table A2. Comparison results between gPC and MC.

Method	Classification Rate	Computational Time
gPC	0.94	5 s
MC (100 samples)	0.75	324 s *

* Per optimization iteration of Equation (17).

References

1. Tarascon, J.-M.; Armand, M. Issues and challenges facing rechargeable lithium batteries. *Nature* **2001**, *414*, 359–367. [CrossRef] [PubMed]
2. Yan, W.; Zhang, B.; Zhao, G.; Weddington, J.; Niu, G. Uncertainty Management in Lebesgue-Sampling-Based Diagnosis and Prognosis for Lithium-Ion Battery. *IEEE Trans. Ind. Electron.* **2017**, *64*, 8158–8166. [CrossRef]
3. Dey, S.; Biron, Z.A.; Tatipamula, S.; Das, N.; Mohon, S.; Ayalew, B.; Pisu, P. Model-based real-time thermal fault diagnosis of Lithium-ion batteries. *Control Eng. Pract.* **2016**, *56*, 37–48. [CrossRef]
4. Wang, Q.; Ping, P.; Zhao, X.; Chu, G.; Sun, J.; Chen, C. Thermal runaway caused fire and explosion of lithium ion battery. *J. Power Sources* **2012**, *208*, 210–224. [CrossRef]
5. Charkhgard, M.; Farrokhi, M. State-of-Charge Estimation for Lithium-Ion Batteries Using Neural Networks and EKF. *IEEE Trans. Ind. Electron.* **2010**, *57*, 4178–4187. [CrossRef]
6. Zhang, J.; Lee, J. A review on prognostics and health monitoring of Li-ion battery. *J. Power Sources* **2011**, *196*, 6007–6014. [CrossRef]
7. Ding, S.X. *Model-Based Fault Diagnosis Techniques: Design Schemes, Algorithms and Tools*, 2nd ed.; Springer: Berlin, Germany, 2013. [CrossRef]
8. Du, Y.; Budman, H.; Duever, T.A. Comparison of stochastic fault detection and classification algorithms for nonlinear chemical processes. *Comput. Chem. Eng.* **2017**, *106*, 57–70. [CrossRef]
9. Venkatasubramanian, V.; Rengaswamy, R.; Yin, K.; Kavuri, S.N. A review of process fault detection and diagnosis: Part I: Quantitative model-based methods. *Comput. Chem. Eng.* **2003**, *27*, 293–311. [CrossRef]
10. Guo, G.; Long, B.; Cheng, B.; Zhou, S.; Xu, P.; Cao, B. Three-dimensional thermal finite element modeling of lithium-ion battery in thermal abuse application. *J. Power Sources* **2010**, *195*, 2393–2398. [CrossRef]
11. Chen, S.C.; Wan, C.C.; Wang, Y.Y. Thermal analysis of lithium-ion batteries. *J. Power Sources* **2005**, *140*, 111–124. [CrossRef]
12. Smith, K.; Wang, C.-Y. Power and thermal characterization of a lithium-ion battery pack for hybrid-electric vehicles. *J. Power Sources* **2006**, *160*, 662–673. [CrossRef]
13. Doughty, D.H.; Butler, P.C.; Jungst, R.G.; Roth, E.P. Lithium battery thermal models. *J. Power Sources* **2002**, *110*, 357–363. [CrossRef]
14. Lin, X.; Perez, H.E.; Siegel, J.B.; Stefanopoulou, A.G.; Li, Y.; Anderson, R.D.; Ding, Y.; Castanier, M.P. Online Parameterization of Lumped Thermal Dynamics in Cylindrical Lithium Ion Batteries for Core Temperature Estimation and Health Monitoring. *IEEE Trans. Control Syst. Technol.* **2013**, *21*, 1745–1755. [CrossRef]
15. Klein, R.; Chaturvedi, N.A.; Christensen, J.; Ahmed, J.; Findeisen, R.; Kojic, A. Electrochemical Model Based Observer Design for a Lithium-Ion Battery. *IEEE Trans. Control Syst. Technol.* **2013**, *21*, 289–301. [CrossRef]
16. Dey, S.; Ayalew, B. A Diagnostic Scheme for Detection, Isolation and Estimation of Electrochemical Faults in Lithium-Ion Cells. In Proceedings of the ASME 2015 Dynamic Systems and Control Conference, Columbus, OH, USA, 28–30 October 2015. [CrossRef]
17. Muddappa, V.S.; Anwar, S. Electrochemical Model Based Fault Diagnosis of Li-Ion Battery Using Fuzzy Logic. In Proceedings of the ASME 2014 International Mechanical Engineering Congress and Exposition, Montreal, QC, Canada, 14–20 November 2014. [CrossRef]
18. Dey, S.; Mohon, S.; Pisu, P.; Ayalew, B. Sensor Fault Detection, Isolation, and Estimation in Lithium-Ion Batteries. *IEEE Trans. Control Syst. Technol.* **2016**, *24*, 2141–2149. [CrossRef]
19. Lombardi, W.; Zarudniev, M.; Lesecq, S.; Bacquet, S. Sensors fault diagnosis for a BMS. In Proceedings of the 2014 European Control Conference (ECC), Strasbourg, France, 24–27 June 2014. [CrossRef]
20. He, H.; Liu, Z.; Hua, Y. Adaptive Extended Kalman Filter Based Fault Detection and Isolation for a Lithium-Ion Battery Pack. *Energy Procedia* **2015**, *75*, 1950–1955. [CrossRef]
21. Gao, Z.; Ding, S.X.; Cecati, C. Real-time fault diagnosis and fault-tolerant control. *IEEE Trans. Ind. Electron.* **2015**, *62*, 3752–3756. [CrossRef]
22. Izadian, A.; Khayyer, P.; Famouri, P. Fault Diagnosis of Time-Varying Parameter Systems with Application in MEMS LCRs. *IEEE Trans. Ind. Electron.* **2009**, *56*, 973–978. [CrossRef]
23. Du, Y.; Duever, T.A.; Budman, H. Generalized Polynomial Chaos-Based Fault Detection and Classification for Nonlinear Dynamic Processes. *Ind. Eng. Chem. Res.* **2016**, *55*, 2069–2082. [CrossRef]
24. Du, Y.; Duever, T.A.; Budman, H. Fault detection and diagnosis with parametric uncertainty using generalized polynomial chaos. *Comput. Chem. Eng.* **2015**, *76*, 63–75. [CrossRef]

25. Guo, H.; Wang, X.; Wang, L.; Chen, D. Delphi Method for Estimating Membership Function of Uncertain Set. *J. Uncertain. Anal. Appl.* **2016**, *4*, 3. [CrossRef]
26. Liu, B. *Uncertainty Theory: A Branch of Mathematics for Modeling Human Uncertainty*; Springer: Berlin, Germany, 2010. [CrossRef]
27. Patton, R.J.; Putra, D.; Klinkhieo, S. Friction compensation as a fault tolerant control problem. *Int. J. Syst. Sci.* **2010**, *41*, 987–1001. [CrossRef]
28. Sidhu, A.; Izadian, A.; Anwar, S. Adaptive Nonlinear Model-Based Fault Diagnosis of Li-Ion Batteries. *IEEE Trans. Ind. Electron.* **2015**, *62*, 1002–1011. [CrossRef]
29. Yan, X.-G.; Edwards, C. Nonlinear robust fault reconstruction and estimation using a sliding mode observer. *Automatica* **2007**, *43*, 1605–1614. [CrossRef]
30. Spanos, P.D.; Zeldin, B.A. Monte Carlo Treatment of Random Fields: A Broad Perspective. *Appl. Mech. Rev.* **1998**, *51*, 219–237. [CrossRef]
31. Xiu, D. *Numerical Methods for Stochastic Computation: A Spectral Method Approach*; Princeton University Press: Princeton, NJ, USA, 2010.
32. Mandur, J.; Budman, H. Robust optimization of chemical processes using Bayesian description of parametric uncertainty. *J. Process Control* **2014**, *24*, 422–430. [CrossRef]
33. Lin, X.; Fu, H.; Perez, H.E.; Siege, J.B.; Stefanopoulou, A.G.; Ding, Y.; Castanier, M.P. Parameterization and Observability Analysis of Scalable Battery Clusters for Onboard Thermal Management. *Oil Gas Sci. Technol.* **2013**, *68*, 165–178. [CrossRef]
34. Savoye, F.; Venet, P.; Millet, M.; Groot, J. Impact of Periodic Current Pulses on Li-Ion Battery Performance. *IEEE Trans. Ind. Electron.* **2012**, *59*, 3481–3488. [CrossRef]
35. Lin, X.; Stefanopoulou, A.G.; Perez, H.E.; Siegel, J.B.; Li, Y.; Anderson, R.D. Quadruple Adaptive Observer of the Core Temperature in Cylindrical Li-ion Batteries and their Health Monitoring. In Proceedings of the 2012 American Control Conference (ACC), Montreal, QC, Canada, 27–29 June 2012. [CrossRef]
36. Mahamud, R.; Park, C. Reciprocating air flow for Li-ion battery thermal management to improve temperature uniformity. *J. Power Sources* **2011**, *196*, 5685–5696. [CrossRef]
37. Smith, K.; Kim, G.; Darcy, E.; Pesaran, A. Thermal/electrical modeling for abuse-tolerant design of lithium ion modules. *Int. J. Energy Res.* **2010**, *34*, 204–215. [CrossRef]
38. Mathew, M.; Janhunen, S.; Rashid, M.; Long, F.; Fowler, M. Comparative Analysis of Lithium-Ion Battery Resistance Estimation Techniques for Battery Management Systems. *Energies* **2018**, *11*, 1490. [CrossRef]
39. Pukelsheim, F. The Three Sigma Rule. *Am. Stat.* **1994**, *48*, 88–91. [CrossRef]
40. Du, Y.; Budman, H.; Duever, T. Parameter estimation for an inverse nonlinear stochstic problem: Reactivity ratio studies in copolymerization. *Macromol. Simul.* **2017**, *26*, 1600095. [CrossRef]

© 2019 by the authors. Licensee MDPI, Basel, Switzerland. This article is an open access article distributed under the terms and conditions of the Creative Commons Attribution (CC BY) license (http://creativecommons.org/licenses/by/4.0/).

Article

Profile Monitoring for Autocorrelated Reflow Processes with Small Samples

Shu-Kai S. Fan [1,*], Chih-Hung Jen [2] and Jai-Xhing Lee [1]

1. Department of Industrial Engineering and Management, National Taipei University of Technology, Taipei City 10608, Taiwan; simpleyearning@hotmail.com
2. Department of Information Management, Lunghwa University of Science and Technology, Guishan, Taoyuan County 33306, Taiwan; f7815@mail.lhu.edu.tw
* Correspondence: morrisfan@ntut.edu.tw; Tel.: +886-2-2771-2171

Received: 12 January 2019; Accepted: 12 February 2019; Published: 15 February 2019

Abstract: The methodology of profile monitoring combines both the model fitting and statistical process control (SPC) techniques. Over the past ten years, a variety of profile monitoring methods have been proposed and extensively investigated in terms of different process profiles. However, monitoring tasks still exhibit a primary problem in that the errors surrounding the functional relationship are frequently assumed to be independent within every single profile. However, the assumption of independence is an unrealistic assumption in many practical instances. In particular, within-profile autocorrelation often occurs in the profile data. To mitigate the within-profile autocorrelation, a monitoring method incorporating an autoregressive (AR)(1) model to cope with autocorrelation is proposed. In this paper, the reflow process with small samples in surface mount technology (SMT) is investigated. In Phase I, three different process models are compared in combination with the first-order autoregressive model, while an appropriate profile model is sought. The Hotelling T^2 and exponentially weighted moving average (EWMA) control charts are used together to monitor the parameter estimates (i.e., profile shape) and residuals (i.e., profile variability), respectively.

Keywords: profile monitoring; polynomial regression model; sum of sine function; Hotelling's T^2 control chart; EWMA control chart

1. Introduction

Statistical process control (SPC) has globally been applied for dealing with process monitoring in a variety of manufacturing processes [1]. The control charting technique is typically designed to monitor a univariate statistic, e.g., the sample average, standard deviation, range of a sequence of sample data, among others. However, several productive processes (e.g., reflow oven, heat treatment, etc.) have proven difficult to manage with a traditional SPC operation. The difficulty in these cases arises because a quality characteristic cannot be suitably characterized. If the quality characteristic of a product or process can be represented by a functional form between the quality characteristic and the input variable, then effective monitoring can be established. This scenario is the so-called "profile monitoring".

A major problem for many profile monitoring models lies in the dependence of within-profile residuals, i.e., within-profile autocorrelation. This problem may cause the parameter estimates of the fitted model to be unstable or it might make monitoring performance unsatisfactory. By this account, within-profile autocorrelation is often present and it should not be intentionally ignored. Jensen et al. [2] applied a mixed model to monitor nonlinear profiles in order to account for the correlation structure. Chicken et al. [3] has proposed a semiparametric wavelet method for monitoring the changes in sequences of nonlinear profiles. In their paper, no assumptions are made on the nature of form or the changes between the profiles other than finite square-integrability. Based on a likelihood ratio test involving a change-point model, the method uses the spatial adaptability properties of wavelets to detect the profile changes. Qiu et al. [4] proposed a new control chart to deal with the within-profile autocorrelation. Hung et al. [5] used support vector regression to describe the within-profile relationship. In [6], a B-spline approach was presented for process profile modeling. To mitigate the dependency of the process data, the bootstrap method was utilized. Ghahyazi et al. [7] used a multistage process in phase II to monitor a simple linear profile. In that paper, a first-order autoregressive correlation model was first modeled. Subsequently, a U statistic is utilized to eliminate the cascade effect and the control scheme is modified accordingly. Zhang et al. [8] proposed that a Gaussian process model be applied to the characterization of the within-profile correlation. Herein, two multivariate control charts (Hotelling T2 and multivariate EWMA) were proposed to monitor the linear trend term and the within-profile correlation separately in phase II. Khedmati and Niaki [9] proposed using the U statistic for the general linear profiles to eliminate the effect of between-profile autocorrelation of error terms in phase-II monitoring. Based on the simulation results, this proposed method could provide a significantly better result in detecting shifts in the regression parameters. Jensen et al. [10] used a nonlinear model for fitting the profiles, thus reducing the profiles to a smaller set of parameter estimates. In that paper, a T^2 control chart using the difference-covariance matrix is employed to perform profile monitoring. The proposed statistic that was based on the differences was modified to account for the correlation between the profiles in phase I and phase II analysis.

The main objective of this research is to construct a monitoring system that can compensate for the one-step-ahead residuals, particularly for the reflow process with small samples in surface mount technology (SMT). In the reflow process, it is of critical importance to monitor the oven temperature condition and to identify potential process irregularity before the product quality becomes worse. In this paper, 15 profiles of the reflow process from [11] will be investigated.

The three different parametric models will be considered as the modeling candidates. Afterwards, the different fitted models are evaluated by means of R^2_{adj}, Akaike information criterion (AIC), and Schwarz information criterion (SIC). Next, in terms of the best-fitted model, phase I and II process monitoring is performed.

The remainder of this paper is organized, as follows. Section 2 presents the three different fitting models, together with the autocorrelation effect. The basic engineering details of the reflow process will be elaborated in Section 3. Simulation results of profile monitoring are presented to demonstrate the performance of different fitting models. Lastly, the conclusions of the paper and the summary of our findings are remarked in Section 4.

2. The Proposed Method for Monitoring Process Profiles

In the proposed framework, three different models are investigated to seek an appropriate profile model. Additionally, the Hotelling T^2 and the EWMA control charts are employed in order to monitor the profile shape and the profile residual, respectively. The flowchart is shown in Figure 1. First, the different process models are compared on the basis of R^2_{adj}, AIC_C, and SIC_C. According to the parameter estimates of the profile model, the nonlinear profile can be monitored and analyzed. In phase I, the Hotelling T^2 control chart is used to evaluate the process stability and remove any outlying profiles. The Hotelling T^2 control chart is also considered for phase II analysis via the out-of-control average run length (ARL_{OUT}). An EWMA control chart is utilized to check the residuals of the fitted model if the autocorrelation effect is changed or not.

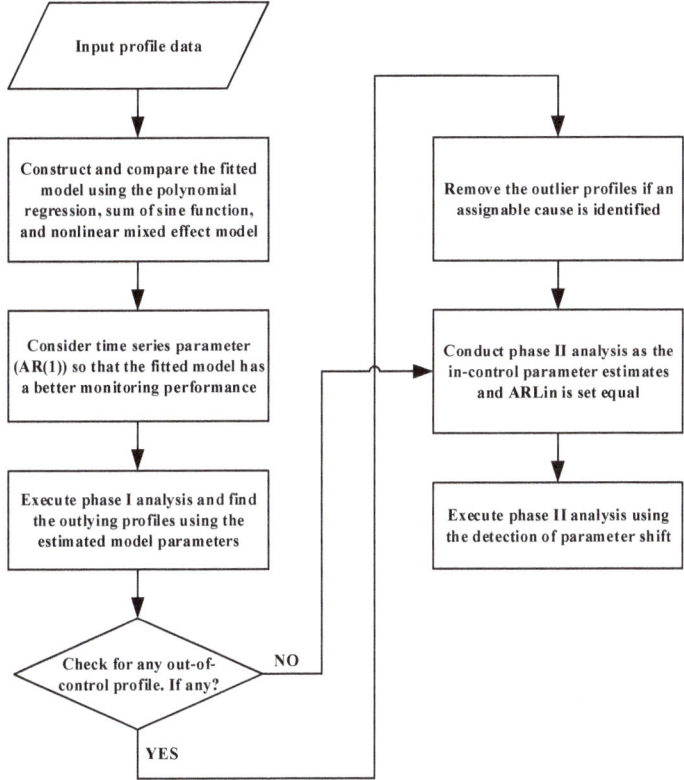

Figure 1. Flow chart for the proposed framework.

2.1. Constructing the Profile Model

To evaluate and determine an appropriate process model, R^2_{adj} is firstly considered to be an immediate measure so that the fitted performance can be more quickly compared. However, the performance evaluation of model fitting merely considers R^2_{adj} that can cause the problem of overfitting. Hurvich and Tsai [12] had pointed out that the AIC would generate the overfitting problem when the fitting samples belong to a smaller number. Although Hurvich and Tsai [12] claimed that the AIC_C could enhance the accuracy of model selection, the overfitting problem can still occur to circumvent better estimation solutions. When referring to [13], we can find that the SIC_C seems to be able to deal with the overfitting problem for the small sample case. To deal with the accuracy of model selection and the overfitting problem, in this paper, the AICc and SICc are simultaneously adopted, with the expectation of obtaining adequate results. The small sample SIC_c and AIC_c criteria derived by [13] are described, as follows:

$$\text{SIC}_c = \log(\hat{\sigma}_k^2) + \frac{\log(n)k}{n-k-2} \tag{1}$$

$$\text{AIC}_c = \log(\hat{\sigma}_k^2) + \frac{n+k}{n-k-2} \tag{2}$$

In (1) and (2), the variance estimate is denoted by $\hat{\sigma}_k^2 = \sum_{j=1}^{n}(y_j - \hat{y}_j)^2/(n-1)$. Here, k is the number of the parameters in the process model. The number of measurement points that are in the profile is denoted by n. It is of particular importance to note that the variance estimate $\hat{\sigma}_k^2$ is calculated after the autocorrelation effect has been discounted by using the first-order autoregressive model, which will be addressed shortly.

2.1.1. Polynomial Model

The polynomial model with one input variable can be defined by

$$y_{jp} = \boldsymbol{\beta}'_p \mathbf{x}_j + \eta_{jp}, p = 1, \ldots, q; j = 1, \ldots, n \tag{3}$$

where the first-order autoregressive model is defined by $\eta_{jp} = \phi \eta_{j-1,p} + \varepsilon_{jp}$, which is logical in deeming the profile as a time series data set. Since the autocorrelation effect often occurs in profile monitoring, the first-order autoregressive model (i.e., AR(1)) is considered as the disturbance term, η, and is incorporated into the profile model. Therefore, the profile model will include the first-order autoregressive parameter to compensate. The parameter ϕ is the first-order autoregressive (AR(1)) coefficient. The noise ε_{jp} is the error term of white noise and its variance estimate is indicated by $\hat{\sigma}_k^2$, as in Equations (1) and (2). Also, $\boldsymbol{\beta}'_p = [\beta_{0p}, \beta_{1p}, \ldots, \beta_{rp}]$ is the vector of unknown parameters in the polynomial function, the vector of regressors is denoted by $\mathbf{x}'_j = [1, x_j, x_j^2, \ldots, x_j^r]$ and r denotes the order of the model in (3). Note that all of the parameters in Equation (3) are estimated by using the ordinary least squares estimation method (see [11]).

2.1.2. Model of the Modified Sum of Sine Functions in two Different Forms

The modified sum of sine functions is represented in the original form as

$$y_{jp} = \sum_{r=1}^{k} a_{rp} \sin(b_{rp} x_j + c_{rp}) + \eta_{jp}, r = 1, \ldots, k; p = 1, \ldots, q; j = 1, \ldots, n \tag{4}$$

where a_r is the amplitude, b_r is the frequency, and c_r is the horizontal phase constant at each sine wave term. For example, when the profile model is considered as the modified sum of two-sine functions, then the model can be defined by

$$y_{jp} = a_{1p} \sin(b_{1p} x_j + c_{1p}) + a_{2p} \sin(b_{2p} x_j + c_{2p}) + \eta_{jp} \tag{5}$$

where x_j denotes the input variable for the jth measurement, $\boldsymbol{\beta}_p$ denotes the parameter vector in profile p ($\boldsymbol{\beta}'_p = [a_{1p}, a_{2p}, b_{1p}, b_{2p}, c_{1p}, c_{2p}]$), and the η_{jp} term is defined, as in Equation (3). As mentioned above, the parameters a_1 and a_2 are the amplitude of the function, b_1 and b_2 determine the period, and c_1 and c_2 influences the horizontal shift. The parameter estimation is performed by using the nonlinear least squares estimation method (see [11]).

To strengthen the fitting of nonlinear models by means of the modified sum of sine functions, we also use the nonlinear mixed effects model (NLME) to test the fitted performance, which then is extended into a nonlinear model with random effects. The generic form of NLME is given by the following equation:

$$y_{jp} = f(\boldsymbol{\beta}_{jp}, \mathbf{x}_j) + \varepsilon_{jp}, \boldsymbol{\beta}_{jp} = \mathbf{A}_{jp} \boldsymbol{\theta} + \mathbf{B}_{jp} \boldsymbol{\gamma}_{jp}, p = 1, \ldots, q, j = 1, \ldots, n \tag{6}$$

In (6), f is the function governing within-profile behavior, $\boldsymbol{\beta}_{jp}$ is a vector of group-specific model parameters, $\mathbf{A}_{jp}$ is a design matrix for combining fixed effects, $\boldsymbol{\theta}$ is a vector of fixed effects, $\mathbf{B}_{jp}$ is a design matrix for combining random effects, $\boldsymbol{\gamma}_{jp}$ is a vector of multivariate normally distributed random effects with $\gamma_{jp} \sim N(0, \mathbf{D})$, where $\mathbf{D}$ is a covariance matrix for the random effects, and ε_{jp} is a vector of errors, which is assumed to be independent, identically, normally distributed, and independent of γ_{jp}, $\varepsilon_{jp} \sim N(0, \sigma^2)$.

According to the mixed model that $\beta = \mathbf{A} \cdot Fixed\ effect + \mathbf{B} \cdot Random\ effect$, the estimated profile parameters (β) of the two-sine function in terms of the NLME model can be described in Equation (7):

$$\begin{bmatrix} a_{1p} \\ b_{1p} \\ c_{1p} \\ a_{2p} \\ b_{2p} \\ c_{2p} \end{bmatrix} = \mathbf{A} \cdot \begin{bmatrix} \bar{a}_1 \\ \bar{b}_1 \\ \bar{c}_1 \\ \bar{a}_2 \\ \bar{b}_2 \\ \bar{c}_2 \end{bmatrix} + \mathbf{B} \cdot \begin{bmatrix} a_{1p} - \bar{a}_1 \\ b_{1p} - \bar{b}_1 \\ c_{1p} - \bar{c}_1 \\ a_{2p} - \bar{a}_2 \\ b_{2p} - \bar{b}_2 \\ c_{2p} - \bar{c}_2 \end{bmatrix} \quad (7)$$

In Equation (7), the **A** and **B** are assumed to be the 1 matrix, the bar symbol refers to an average. The NLME form can then be represented, as follows:

$$\begin{aligned} y_{jp} &= [\bar{a}_1 + (a_{1p} - \bar{a}_1)] \sin[(\bar{b}_1 + (b_{1p} - \bar{b}_1)x_j) + (\bar{c}_1 + (c_{1p} - \bar{c}_1))] + \\ &\quad [\bar{a}_2 + (a_{2p} - \bar{a}_2)] \sin[(\bar{b}_2 + (b_{2p} - \bar{b}_2)x_j) + (\bar{c}_2 + (c_{2p} - \bar{c}_2))] + \varepsilon_{jp} \\ &= (a_{1,fixed} + a_{1p,random}) \sin((b_{1,fixed} + b_{1p,random})x_{ij} + (c_{1,fixed} + c_{1p,random})) + \\ &\quad (a_{2,fixed} + a_{2p,random}) \sin((b_{2,fixed} + b_{2p,random})x_{ij} + c_{2,fixed}) + \varepsilon_{jp} \end{aligned} \quad (8)$$

The parameter estimates are obtained by using the maximum likelihood estimation method (see [2]). Herein, it should be noted that the NLME model does not include the AR(1) term. In previous literature (see [2]), the NLME model has been used to solve the problem of autocorrelation. Therefore, the two fitting models together with AR(1) and the NLME model are compared in phase I analysis.

2.2. Phase I and II Monitoring and Analysis

In phases I and II, the parametric T^2 control chart is used to check whether the process is in the statistical control status and to identify potential outliers. Here, $\hat{\beta}_p$ is the estimate of the parameter vector. Over the entire profile data, the sample mean vector $\overline{\hat{\beta}}$ and the sample variance-covariance matrix $\mathbf{S} = s^2\{\hat{\beta}\}$ can be computed by using the parameter estimates that were obtained from different fitting models. For example, the estimate of the parameter vector for the fourth order polynomial with AR(1) is defined as $\begin{bmatrix} \hat{\beta}_0 & \hat{\beta}_1 & \hat{\beta}_2 & \hat{\beta}_3 & \hat{\beta}_4 & \hat{\phi} \end{bmatrix}$; the estimate of the parameter vector for the sum of two-sine functions with AR(1) is defined as $\begin{bmatrix} \hat{a}_1 & \hat{a}_2 & \hat{b}_1 & \hat{b}_2 & \hat{c}_1 & \hat{c}_2 & \hat{\phi} \end{bmatrix}$.

According to the aforementioned parameter estimates, the T2 control chart (Brill, 2001) is described by

$$T^2_{c,p} = (\hat{\beta}_p - \overline{\hat{\beta}})' \mathbf{S}_c^{-1} (\hat{\beta}_p - \overline{\hat{\beta}}), p = 1, 2, \ldots, q, \quad (9)$$

where $\mathbf{S}_c$ is the sample variance-covariance estimator, as defined by

$$\mathbf{S}_c = \frac{1}{q-1} \sum_{p=1}^{q} (\hat{\beta}_p - \overline{\hat{\beta}})(\hat{\beta}_p - \overline{\hat{\beta}})' \quad (10)$$

In (10), q is the number of profiles in the process data. The approximate upper control limit (UCL), as derived by [14], is as follows:

$$\text{UCL}_c = \frac{(q-1)^2}{q} B_{\alpha,k/2,(q-k-1)/2} \quad (11)$$

In Equation (11), $B_{\alpha,k/2,(q-k-1)/2}$ is the upper α percentage point of a beta distribution with parameters $k/2$ and $(q-k-1)/2$, where k is the number of parameter estimates. According to [15], the T^2 control chart in (9) is shown to be ineffective in detecting sustained shifts in the mean vector.

In this regard, the alternative T² control chart that was proposed by [16] is also considered. The control chart is defined by

$$T_{D,p}^2 = (\hat{\boldsymbol{\beta}}_p - \bar{\bar{\boldsymbol{\beta}}})' \mathbf{S}_D^{-1} (\hat{\boldsymbol{\beta}}_p - \bar{\bar{\boldsymbol{\beta}}}), p = 1, 2, \ldots, q \tag{12}$$

In (12), the variance-covariance estimator ($\mathbf{S}_D$) is calculated by using successive differences in the following:

$$\mathbf{S}_D = \frac{\hat{\mathbf{V}}'\hat{\mathbf{V}}}{2(q-1)} \tag{13}$$

where $\hat{\mathbf{v}}_p = \hat{\boldsymbol{\beta}}_{p+1} - \hat{\boldsymbol{\beta}}_p$ for $p = 1, \ldots, q-1$ and the transpose of these $q-1$ difference vectors are stacked into the $(q-1) \times k$ matrix $\hat{\mathbf{V}}$, as follows:

$$\hat{\mathbf{V}} = [\hat{\mathbf{v}}_1' \hat{\mathbf{v}}_2' \cdots \hat{\mathbf{v}}_{q-1}']' \tag{14}$$

In [17], the approximate UCL_D for a large sample size ($q > k^2 + 3k$) can be estimated according to

$$\text{UCL}_D = \chi^2(1 - \alpha, k), \tag{15}$$

where k denotes the degrees of freedom and α denotes the significance level. Sullivan and Woodall [15] argue that the simulation results can be used to discover that the T_C^2 control chart (see Equation (9)) performs worse in detecting the step change and the ramp shift in the mean vector during phase I than the T_D^2 chart, as shown in (12). Based on this fact, in this paper, the T_D^2 control chart is employed to evaluate the different fitting models while identifying the outlying profiles.

While phase I is executed, the process should be able to achieve a stable situation. Subsequently, the data of the in-control profiles is employed to estimate the unknown parameters. In phase II, the exponentially weighted moving average (EWMA) chart is additionally used for detecting the autoregressive (AR) effect in residuals in order to determine whether the AR parameter in the process model should be re-estimated. In sum, the T_D^2 control chart is used to monitor the parameters of the model (i.e., profile shape). In the meantime, the EWMA control chart is used to monitor the residuals (i.e., profile variability). The EWMA statistic is computed by

$$\text{EWMA}_\varepsilon(j) = \theta e_j + (1-\theta)\text{EWMA}_\varepsilon(j-1), j = 1, 2, \ldots, n, \tag{16}$$

where e_j is the jth residual; $\theta(0 < \theta \leq 1)$ is a smoothing constant and the starting value is assumed $\text{EWMA}_\varepsilon(0) = 0$. An out-of-control signal is issued as soon as $\text{EWMA}_\varepsilon(j) < LCL$ or $\text{EWMA}_\varepsilon(j) > UCL$, where

$$LCL = 0 - L_\varepsilon \hat{\sigma}_\varepsilon \sqrt{\frac{\theta}{2-\theta}}, \quad UCL = 0 + L_\varepsilon \hat{\sigma}_\varepsilon \sqrt{\frac{\theta}{2-\theta}} \tag{17}$$

In (17), $\hat{\sigma}_\varepsilon$ denotes the standard error of the residual as $\hat{\sigma}_k$ in Equations (1) and (2); $L_\varepsilon(>0)$ is a half-length that is designed to generate a specific in-control ARL. Under this monitoring framework, the ARL_OUT performance of the T^2 control chart, together with the EWMA chart, is evaluated for phase II analysis in the next section. In terms of the aforementioned methods, the proposed monitoring framework can be formalized as pseudo-code 1 in the appendix.

3. Experimental Results for Profile Monitoring

In this section, the proposed monitoring framework is illustrated and evaluated while using the simulated reflow process in SMT. The application domains of the monitoring system and some implementation issues are discussed. In terms of the simulation results, the analysis of profile monitoring can be done in three parts: (i) making a comparison of the fitting performance between the polynomial regression with AR(1), the modified sum of sine functions with AR(1), and the NLME model; (ii) screening the outlying profiles by means of the T_D^2 control chart for phase I analysis; and,

(iii) testing the proposed T_D^2 control chart with the EWMA control chart, while also monitoring the process parameters and the residual in phase II analysis.

3.1. Fundamentals of Reflow Process

The operation of the reflow process is the heating sequence for assembling printed circuit boards (PCB) using solder paste at successively higher temperatures. As an assembly moves through a soldering system, it will perform a controlled temperature curve in order to achieve the required quality. Such a temperature curve is also called a "temperature profile". The temperature profile is often measured along a variety of technical dimensions, such as slope, soak, time above liquidus, and peak. In general, reflow soldering processes contain four stages. Each operation presents a unique temperature profile: preheat, thermal soak (dwell), reflow (liquidus), and cooling. Figure 2 shows a typical example of a schematic temperature profile.

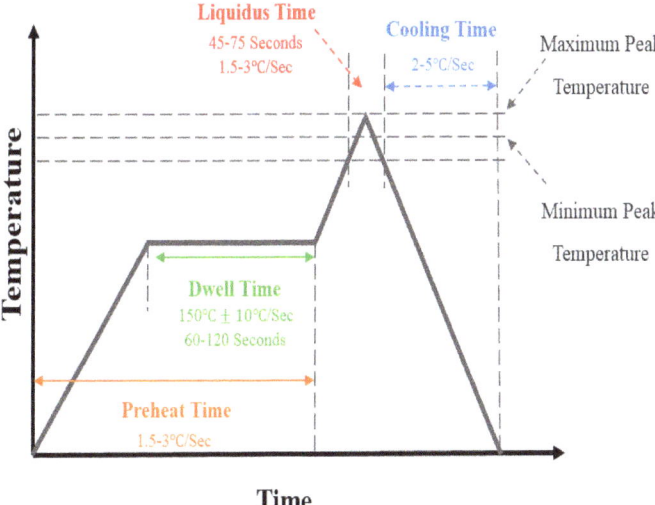

Figure 2. A typical temperature profile.

In the preheat zone step, the changes of the temperature curve can be described as an ascending tendency from normal temperature to approximately 150 °C. In this step, the ascending temperature facilitates the removal of solvent and water vapor in the solder paste. Rapid heating helps the flux softening temperature to be reached quickly, so the flux can spread quickly and cover the maximum area of solder joints. It also integrates some activator into the actual alloy liquid. Furthermore, because some parts of the motherboard cannot deal with the sharp temperature changes, the rate of temperature change in the preheating zone is set to between 1.5 °C/s and 3 °C/s.

When the operation approaches the thermostatic zone, the temperature is usually maintained in a region of 150 ± 10 °C. This operational zone is a flat temperature profile to enhance the effect of the soldering, and it especially prevents tombstoning. The reflow zone is also called Time Above Liquidus (TAL). The TAL is the period of time above the maximum temperature at which crystals can coexist with the melt in thermodynamic equilibrium. The peak of reflow temperature usually depends on the melting temperature of the solder, while also taking into account the temperature that assembled components can endure. For instance, a typical lead-free manufacturing process must not exceed the limit of 260 °C, which is the highest temperature that tantalum capacitors can endure.

Following the reflow zone, the product is cooled and the solder joints are solidified so that it can rejoin the assembly process. Note that Figure 2 only provides an overall, schematic diagram of the temperature profile of the reflow process. In fact, the temperature control in the reflow process belongs to a nonlinear profile pattern. Hence, any linear approximation approach will not be suited to this research. The high nonlinearity and curvature somewhat warrants the need for a new profile monitoring approach. In this paper, the practical data of the same product type is gathered so as to form individual profiles for process monitoring. The production line in the SMT practice is essentially constructed with high flexibility to deal with the different types of products. Hence, to perform profile monitoring of a wide variety of low-volume products in small-to-moderate batches is our research target. Note that the data set that was used in this research is available upon request.

3.2. Comparing and Evaluating the Different Profile Models

In this section, the polynomial regression model, the modified sum of sine functions, and the NLME model are first used to fit the reflow process data. The polynomial models of orders 3–5 and the modified sum of 1–3 sine functions are selected for model fitting. In every profile, n measurements in the ith random profile are collected over time, as indicated by (x_{jp}, y_{jp}), for $p = 1, 2, \ldots, q$ and $j = 1, 2, \ldots, n$. The polynomial models are as shown in Equation (3). The modified sum of sine functions and the nonlinear mixed effects models are as shown in Equations (4)–(7).

In here, seven models, including the polynomial model of different orders with AR(1), the modified sum of sine functions with AR(1), and the NLME model, are tested. Fifteen profiles of 48 data points that were collected each in the reflow process are individually modelled by using the seven different models, and the parameter estimates are utilized for phase I monitoring. To compare the fitting results, four performance measures (R^2_{adj}, RSS, SIC_c, and AIC_c) are selected as the performance measures. Moreover, the number of times that each model is chosen best over 15 profiles is also reported. The computational results are displayed in Table 1 and Figures 3–5. From the fitting results in Table 1, it can be seen that the modified sum of two-sine functions exhibits a better fitting performance (with less SIC and AIC) than the other fitted models. Typically, using a large sample of profiles for parameter estimation in the phase I analysis is necessary, especially for nonlinear profiles. In this study, only fifteen profiles can be obtained due to a technical limitation. The excellent fitting performance that is shown in Table 1 and Figures 3–5 must be attributed to the appropriateness of model selection and the flexibility of the chosen models under investigation. If the fitting performance is not satisfactory, then more profile data need to be collected for the estimation purpose before proceeding to the phase II analysis.

The particularly high performance based on the four measures arises from the suitability of the modified sum of two-sines with AR(1) for the reflow process data. Thus, the model previously mentioned is considered to be the best model to undertake research. The polynomial model of order 4 with AR(1) outperforms the other polynomial models, thus being considered as the benchmark model. These two process models with AR(1) show great flexibility in dealing with complex model-building situations, and therefore they are also expected to be extensively applied in a wide variety of nonlinear processes. They will be selected for evaluation in phase I and II monitoring.

Table 1. The fitting performances for polynomial, sum of sine, and nonlinear mixed effects model (NLME) models. (**a**) The fitting performances of the polynomial with first-order autoregressive (AR(1)) model; (**b**) The fitting performances of the sum of sine with AR(1) model; (**c**) The number of times each model was chosen best over 15 profiles; (**d**) The fitting performances of NLME model based on two-sine function.

The Different Order	The Fitting Performance for Polynomial with AR(1) Model			
	$\overline{R}^2_{adj}$	$\overline{AIC}_C$	$\overline{SIC}_C$	$\overline{RSS}$
3rd order	0.9873	4.6003	3.7309	1522.7104
4th order	0.9904	4.3485	3.5279	1126.4767
5th order	0.9908	4.3158	3.5465	1048.1086

The Different Order	The Fitting Performance for Sum of Sine with AR(1) Model			
	$\overline{R}^2_{adj}$	$\overline{AIC}_C$	$\overline{SIC}_C$	$\overline{RSS}$
One-sine model	0.9863	4.6576	3.7417	1677.0915
Two-sine model	0.9955	3.1722	2.4029	517.3413
Three-sine model	0.9909	3.4961	2.8972	994.5594

Models	R^2_{adj}	AIC_C	SIC_C	RSS
3rd order polynomial with AR(1)	0	0	0	0
4th order polynomial with AR(1)	0	0	0	0
5th order polynomial with AR(1)	0	0	0	0
One-sine with AR(1)	0	0	0	0
Two-sine with AR(1)	9	9	10	7
Three-sine with AR(1)	6	6	5	8

	$\overline{R}^2_{adj}$	$\overline{AIC}_C$	$\overline{SIC}_C$	$\overline{RSS}$
NLME model	0.9929	3.9730	3.5039	718.9822

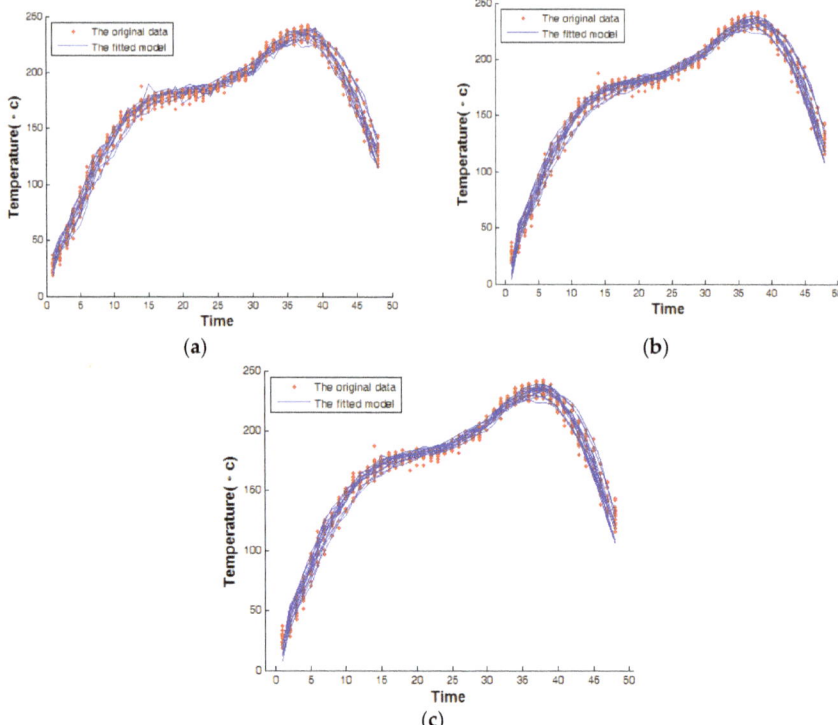

Figure 3. The fitting profile using the polynomial with AR(1) model. (**a**) The third-order model; (**b**) The fourth-order model; and, (**c**) The fifth-order model.

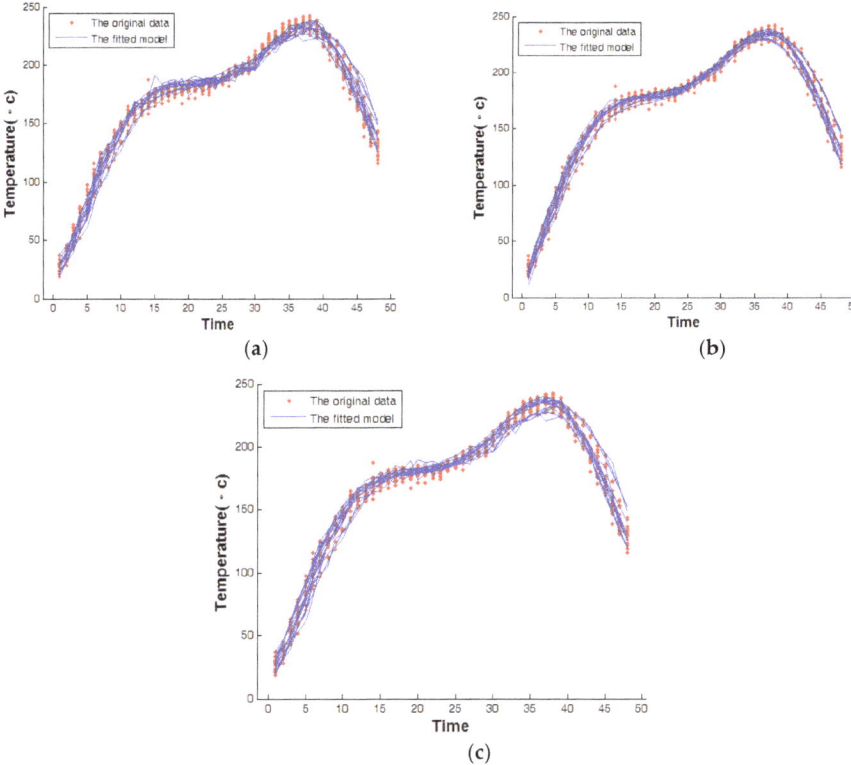

Figure 4. The fitting profile using the sum of sine model with AR(1) model. (**a**) The one-sine model; (**b**) The two-sine model; a d, (**c**) The three-sine model.

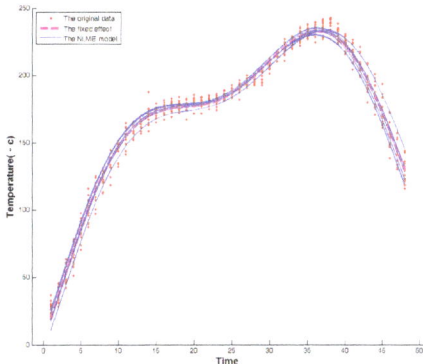

Figure 5. The fitted profiles of the nonlinear mixed effects model (NLME).

3.3. Simulations for Phase I Analyses

In phase I, following [18] and [16], the T_D^2 chart is used for identifying the outlying profiles. First, the statistic T_D^2 of the parameter estimates (including the AR(1) parameter) is calculated based on Equation (12). Based on the polynomial regression model of order 4 and the modified sum of two-sines, the control limit of T_D^2 control chart is plotted in Figure 6. Since no outlying profiles are found in the 15 phase I runs, as in the common SPC practice, no profiles need to be removed prior to the phase II.

On the other hand, three types of the hypothetical process abnormalities are also tested in order to assess and compare the performance of phase I monitoring. The first scenario assumes that the preheat zone has a lower temperature slope. The maintenance of temperature in the dwell zone is assumed to be unstable for the second scenario, and in the third scenario the temperature is set to over-heating in the reflow zone. Figure 7 shows the three outlying profiles along with the average baseline using 15 in-control profiles. In common practice, these three types of abnormality will not happen at the same time. Accordingly, in our paper, we test each abnormal profile separately.

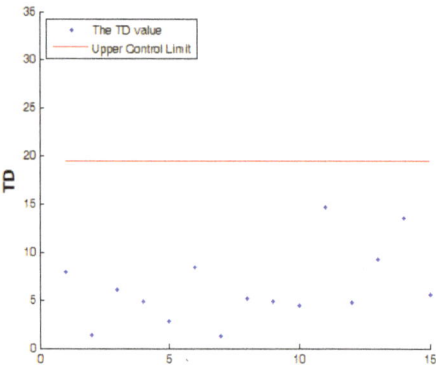

Figure 6. T_D^2 control chart of the sum of two-sine function by using the fifteen in-control profiles.

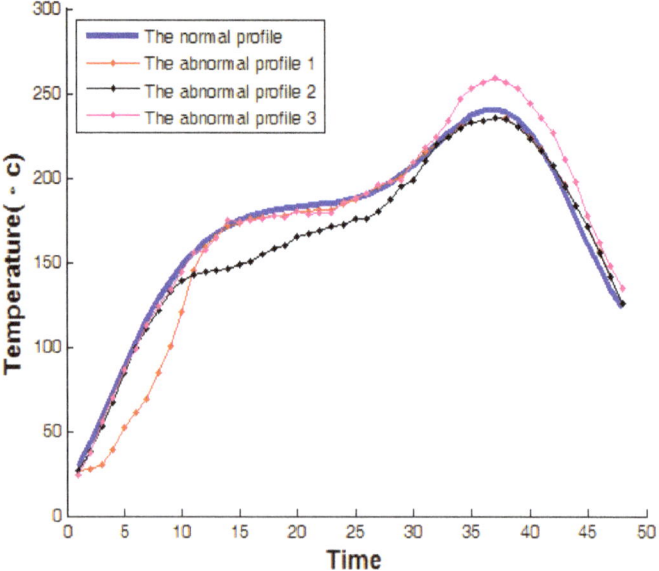

Figure 7. The simulations of abnormal profiles.

In this simulation, the significance level of $\alpha = 0.05$ is used as the significance level for any individual profile, i.e., $T_p^2, p = 1, \ldots, 15$, and to construct the control limits [14,17]. The three simulated abnormal profiles are individually added to the 15 in-control profiles. The T_D^2 statistic is computed for the polynomial regression and the modified sum of two-sine functions with the AR(1) model. Figures 8 and 9 show the monitoring results of the T_D^2 control charts for abnormal profiles 1, 2, and 3. The simulated results reveal that the T_D^2 control charts are able to identify the outlying situations if the abnormal profiles are present. It is worth to note that the T_D^2 control chart is like a moving range with individual observations, and it is not affected by shifts in the mean vector, and as a result of which it has greater power.

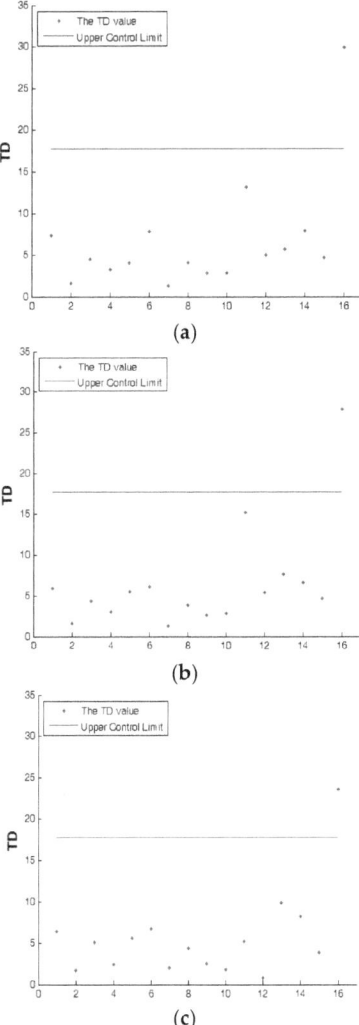

Figure 8. Detect the abnormal profiles using the polynomial model with the 16 profiles (fifteen in-control and one out-of-control for each scenario). (**a**) Detecting the abnormal profile 1; (**b**) Detecting the abnormal profile 2; and, (**c**) Detecting the abnormal profile 3.

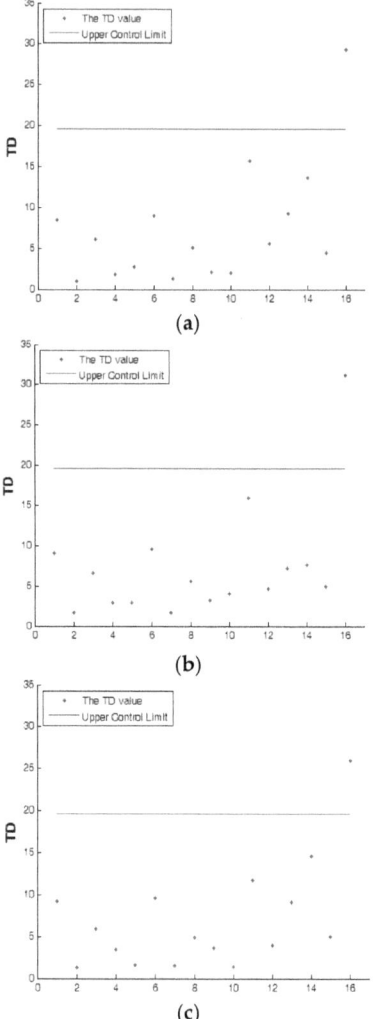

Figure 9. Detect the abnormal profiles using the sum of sine function with the 15 in-control profiles. (**a**) Detecting the abnormal profile 1; (**b**) Detecting the abnormal profile 2; and, (**c**) Detecting the abnormal profile 3.

The polynomial model of order 4 with AR(1) and the modified sum of two-sine functions with AR(1) can account for the autocorrelation effects appropriately, so the simulated abnormal profiles can be successfully detected. From a fitting performance viewpoint based on Table 1, the modified sum of two-sine functions with AR(1) takes the lead between the compared models. From an operational point of view, the polynomial model of order 4 can be adopted, since it contains fewer unknown parameters to be estimated and it is easier to implement in practice than the modified sum of two-sine functions. By contrast, the modified sum of two-sine functions explains the data variance better and is more powerful than the polynomial model of order 4. Therefore, during the pre-production stage where the process insights should be fully gained for the purpose of process adjustment/optimization, the modified sum of two-sine functions can be considered instead.

3.4. Simulation Results for Phase II Monitoring

The objective of phase II is to detect shifts quickly in the process. The reliable control limits can be established to achieve effective on-line monitoring in phase II when the in-control process data is stable. The proposed monitoring framework uses the composite method to monitor two possible process irregularities. First, the T_D^2 control chart is applied, so that the profile shape of the fitted model can be monitored. Second, the autocorrelation of residuals is monitored by using the EWMA control chart. If the process has been confirmed to be stable, then the 15 in-control profiles will be employed in order to estimate the parameters of the fitted model, the mean vector, and the variance-covariance. The parameter estimates are used to replace the unknown parameters in Equations (3)–(7), plus an error term to simulate the profile data. The practical process is assumed to follow the modified sum of two-sine functions with the error term to simulate the process changes. The variance of the fitted model is estimated by using the mean square error (MSE), over the 15 in-control profiles, from the model-building stage. The error term of the process model is assumed to be normally distributed with zero mean and constant variance. Next, the T_D^2 and EWMA control charts are employed to implement the phase II analysis, evaluating the detection performance as the process parameters shift. The control limits are constructed using the parameter estimates of the three different fitting models (the polynomial regression model of order 4 with AR(1), the modified sum of two-sine functions with AR(1), and the traditional polynomial regression model of order 4) to evaluate and compare the monitoring performances. The traditional polynomial regression model of order 4 is only used as a benchmark. To monitor the process shape on a fair basis, in each fitting model the different control limits should be particularly designed to have an approximately equal in-control ARL (i.e., ARL_{IN} = 100) when using 10,000 simulation cycles. Moreover, the smoothing constant (θ) in the EWMA chart is set to 0.02, as in [11] and [19]. Here, in each experiment, 20,000 profiles are simulated for ARL_{OUT} evaluation in terms of the shifts of different scales in the six parameters of the modified sum of two-sine functions. Note that, based on an earlier experiment, using a typical ARL_{IN} in our study, like 370 or even larger, will cause indistinguishable performance in ARL_{OUT} in the presence of a small scale of parameter change. The formal procedure in the Phase II analysis is represented as pseudo-code 2 in the Appendix A.

In the ordinary SMT operation, if the process recipe is suitably tuned before volume production, then the profile specification should be pre-determined and only subjected to a minuscule adjustment as a result of product changes. Therefore, seven different types of shifts are considered in our simulation study. The shifts of these process parameters are applied to the amplitude, the frequency, and the horizontal phase constant in the model, as shown in Equation (5). The simulation results of ARL_{OUT} can be used to evaluate the on-line monitoring capability. These ARL_{OUT} values are calculated by setting equally spaced parameter shifts every 100 simulated profiles (maximum $ARL_{IN} = 100$) and then averaging across 20,000 simulation cycles. The parameter shift in the scale of σ_β ranges from 0.5 to 3 for the six process parameters. Any parameter shift will cause the change of curve shape that is closest to the type of sustained shift in SPC practice. Figure 10a shows the comparative profiles under various shifts of different process parameters. In order to detect the autocorrelation effect in the residuals of each profile, the EWMA control chart is used to implement the related monitoring tasks using the different scales of the autocorrelation coefficient, from 0.1 to 0.9 (see Figure 10b).

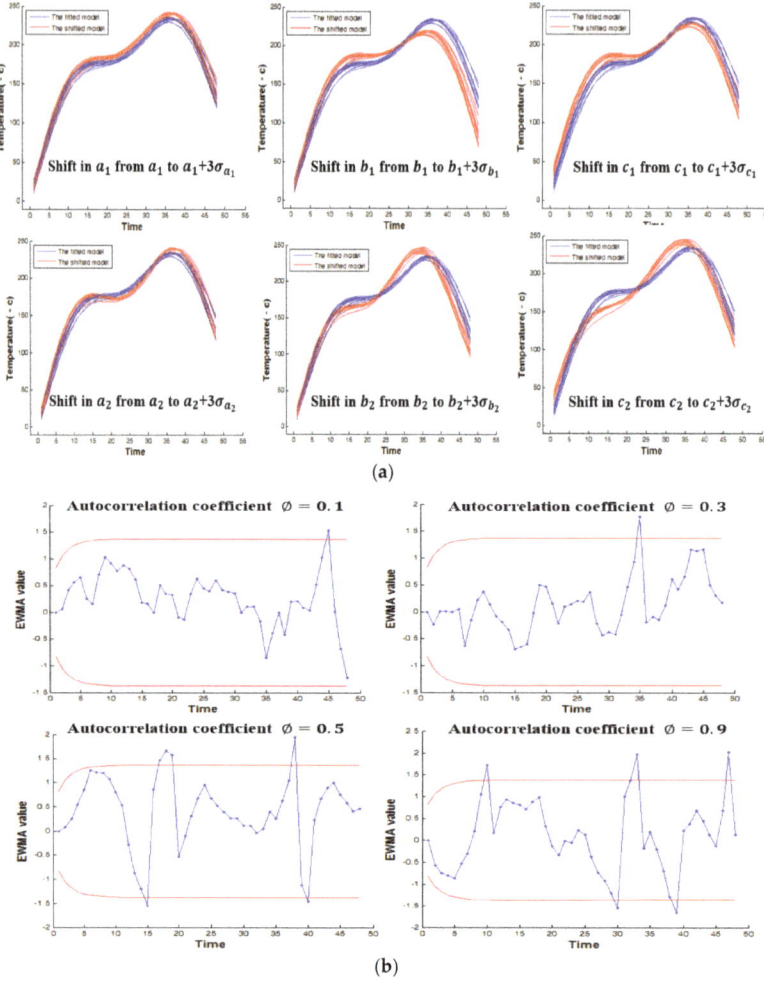

Figure 10. Phase II analysis. (**a**) Detecting the curve changes using different shifts of process parameters; and, (**b**) Detecting the autocorrelation effect of residuals using exponentially weighted moving average (EWMA) control chart.

Table 2 gives the ARL$_{OUT}$ estimates for shifts of the six parameters. The experimental results indicate that, for monitoring the shape of the model, both of the composite models perform reasonably well regardless of whether the data is modeled by the modified sum of two-sine functions or the polynomial regression model. We also use the EWMA control chart to monitor residuals if any autocorrelation effect in addition to the AR(1) already included in the model is exhibited. The results show that the modified sum of two-sine functions, combined with AR(1), performs much better than the pure polynomial regression model as $\lambda \geq 1.5$. Even so, it is very difficult to compare the modified sum of two-sine functions and the polynomial regression model in the composite approach, although the former performs slightly better than the latter. In a word, it is reasonable to allege that the modified sum of two-sine functions can be a viable modeling option for nonlinear profiling monitoring circumstances where only small samples are available for the reflow process.

Table 2. The average run length (ARL) comparison for different parameter shifts using three different models. (a) a_1 from a_1 to $a_1 + \lambda\sigma_{a_1}$; (b) b_1 from b_1 to $b_1 + \lambda\sigma_{b_1}$; (c) c_1 from c_1 to $c_1 + \lambda\sigma_{c_1}$; (d) a_2 from a_2 to $a_2 + \lambda\sigma_{a_2}$; (e) b_2 from b_2 to $b_2 + \lambda\sigma_{b_2}$; (f) c_2 from c_2 to $c_2 + \lambda\sigma_{c_2}$; (g) The ARL for monitoring the autocorrelation of residuals.

Control Chart Based on the Polynomial Regression Model of Order 4 with AR(1)						
λ	0.5	1	1.5	2	2.5	3
$ARL_{OUT}(T_D^2)$	72.7286	40.0651	18.9343	9.2455	4.8773	2.8661
Control Chart Based on the Sum of Two-sine Functions with AR(1) Model						
λ	0.5	1	1.5	2	2.5	3
$ARL_{OUT}(T_D^2)$	71.0655	40.0492	18.6020	9.0533	4.7834	2.8495
Control Chart Based on the Polynomial Regression of Order 4						
λ	0.5	1	1.5	2	2.5	3
$ARL_{OUT}(T_D^2)$	75.8772	45.8895	21.2343	10.5677	7.0577	3.0632
Control Chart Based on the Polynomial Regression Model of Order 4 with AR(1)						
λ	0.5	1	1.5	2	2.5	3
$ARL_{OUT}(T_D^2)$	73.5965	41.9621	19.6273	8.9772	4.8677	2.9916
Control Chart Based on the Sum of Two-sine Functions with AR(1)						
λ	0.5	1	1.5	2	2.5	3
$ARL_{OUT}(T_D^2)$	73.5521	40.9701	18.2644	8.7256	4.7681	2.8211
Control Chart Based on Polynomial Regression of Order 4						
λ	0.5	1	1.5	2	2.5	3
$ARL_{OUT}(T_D^2)$	74.9043	52.3352	21.2921	10.8889	7.8225	4.5232
Control Chart Based on the Polynomial Regression Model of Order 4 with AR(1)						
λ	0.5	1	1.5	2	2.5	3
$ARL_{OUT}(T_D^2)$	74.8687	40.2647	18.5716	9.0064	4.9945	2.8466
Control Chart Based on the Sum of Two-sine Functions with AR(1)						
λ	0.5	1	1.5	2	2.5	3
$ARL_{OUT}(T_D^2)$	74.4160	40.1462	18.3835	8.8046	4.6832	2.7475
Control Chart Based on the Polynomial Regression of Order 4						
λ	0.5	1	1.5	2	2.5	3
$ARL_{OUT}(T_D^2)$	78.3815	55.6741	22.7029	15.3900	6.3135	3.6904
Control Chart Based on the Polynomial Regression Model of Order 4 with AR(1)						
λ	0.5	1	1.5	2	2.5	3
$ARL_{OUT}(T_D^2)$	71.2188	40.5620	19.4892	9.1471	5.0632	2.9617
Control Chart Based on the Sum of Two-sine Functions with AR(1)						
λ	0.5	1	1.5	2	2.5	3
$ARL_{OUT}(T_D^2)$	70.7991	39.8413	19.2414	8.7882	4.7695	2.8921
Control Chart Based on the Polynomial Regression of Order 4						
λ	0.5	1	1.5	2	2.5	3
$ARL_{OUT}(T_D^2)$	79.4577	59.0987	21.8773	9.9247	6.9499	3.1433

Table 2. Cont.

Control Chart Based on the Polynomial Regression Model of Order 4 with AR(1)						
λ	0.5	1	1.5	2	2.5	3
$ARL_{OUT}(T_D^2)$	77.9581	40.7159	17.8598	9.5453	5.2117	2.9759
Control Chart Based on the Sum of Two-sine Functions with AR(1)						
λ	0.5	1	1.5	2	2.5	3
$ARL_{OUT}(T_D^2)$	77.6478	40.7232	17.7346	9.0055	4.9806	2.8821
Control Chart Based on the Polynomial Regression of Order 4						
λ	0.5	1	1.5	2	2.5	3
$ARL_{OUT}(T_D^2)$	80.9116	51.3598	25.5898	11.8557	6.0081	4.3123
Control Chart Based on the Polynomial Regression Model of Order 4 with AR(1)						
λ	0.5	1	1.5	2	2.5	3
$ARL_{OUT}(T_D^2)$	72.7417	41.3423	18.8105	9.0889	4.9321	2.8128
Control Chart Based on the Sum of Two-sine Functions with AR(1)						
λ	0.5	1	1.5	2	2.5	3
$ARL_{OUT}(T_D^2)$	72.0051	40.9314	18.6684	9.0452	4.8423	2.7555
Control Chart Based on the Polynomial Regression of Order 4						
λ	0.5	1	1.5	2	2.5	3
$ARL_{OUT}(T_D^2)$	74.3391	45.2862	22.9480	11.3352	5.8202	3.4110
$\varepsilon_{ij}=\phi\varepsilon_{ij-1}+a_{ij}, a_{ij}\sim N(0,\sigma^2)$						
Autocorrelation coefficient ϕ	0.1	0.3	0.5	0.7	0.9	
$ARL_{OUT}(EWMA)$	76.4314	50.9765	26.2965	13.8156	8.7692	

4. Conclusions

This paper presents a new monitoring framework for dealing with the autocorrelation effect that exists in the errors around the functional relationship when only small samples are available. The research framework includes model building and phase I and II analyses. The central idea of the proposal is how to construct an appropriate profile model that is capable of dealing with the time series effect. Using different profile models (the polynomial regression model, the modified sum of two-sine functions, and the nonlinear mixed effects model), the phase I and II analyses of reflow process data can be conducted. In phase I, the Hotelling T_D^2 control chart is utilized to screen the outlying profiles. When the outlying profiles are investigated and removed, then the same control charts with the EWMA control chart for monitoring autocorrelation are used for phase II monitoring, where the detectability of parameter shifts in terms of ARL$_{OUT}$ is evaluated. According to the comparison results, some concluding remarks and suggestions can be provided:

1. If the profile pattern exhibits a significant autocorrelation effect, then the proposed framework can use a different profile model with AR(1) and the proposed model selection procedure to strengthen the fitting performance. Furthermore, we feel safe to conclude that the sum of two-sine functions with AR(1) can be a viable modelling option for nonlinear profiling monitoring instances where only small samples are available for the reflow process.
2. In phase I of the reflow process that is investigated in this paper, two types of composite models all have good monitoring ability for identifying outlying profiles. However, the nonlinear mixed effects model cannot resolve the problem of autocorrelation in the residuals. This situation will cause difficulties in monitoring when autocorrelation is present.

3. According to the phase I results of the reflow process that was investigated in this study, the Hotelling T^2 control chart can produce satisfactory performance for monitoring of the process profile.
4. On the whole, the proposed monitoring framework displays better detecting performances than the traditional polynomial regression model in phase II analysis for the reflow process that is discussed in this paper. In addition, the proposed EWMA control chart is also effective in detecting changes of the autocorrelation effect in residuals. This study pinpoints a major finding, a fact that the modified sum of two-sine functions is able to statistically fit the nonlinear profile of the reflow process data extremely well. In the proposed framework, the Hotelling T^2 control chart and the EWMA control chart work in harmony to simultaneously monitor the parameter estimates (i.e., profile shape) and residuals (i.e., profile variability), respectively. The simulation results in phases I and II illustrate the proposed monitoring framework. Therefore, the practitioner can follow the guidelines of model building and process monitoring that are demonstrated in this paper, as the nonlinear profile monitoring task of the reflow process is necessary.
5. To achieve desirable monitoring performances for other potential applications, the parameter setting of the control chart bears further scrutiny. A real-data examination of phase II analysis should be further conducted to complement the research outcomes that are delivered in this paper.

Author Contributions: This is a joint work of the three authors; nevertheless, each author was especially in charge of his expert and capability: S.-K.S.F. and J.-X.L. for conceptualization, methodology, investigation and formal analysis, S.-K.S.F. and C.-H.J. for validation, original draft preparation and writing.

Funding: Shu-Kai S. Fan was partially funded by the Ministry of Science and Technology Project MOST 105-2221-E-027-071-MY3.

Conflicts of Interest: The authors declare no conflict of interest.

Appendix A

1. The pseudo-code for the proposed monitoring framework:

Input the reflow process data
Do

Use three nonlinear models to fit the data;
Calculate (R^2_{adj}, SSE, SIC_c, AIC_c);

While (the goodness of fit test is satisfied)
If (autocorrelation in the residuals)
{
Incorporate the time series model;
}
Construct the fitted model for each profile data
Calculate the T^2 statistics using the vector of parameter estimates
Calculate the control limits for the T^2 statistics
If ($T^2_C > UCL_C$) or ($T^2_D > UCL_D$)
{
Do
Remove the out-of-control profiles;
Recalculate the T^2 and its upper control limit to check for any out-of-control profile;
While (all out-of-control profiles removed)
}

Calculate ARL_{IN} and ARL_{OUT} for phase II analysis

2. The pseudo-code for Phase II analysis

```
For (the number of executions = 1:10,000)
Count = 0;
    For (the number of simulated profiles = 1:20,000)
        Count = count + 1;
        Index = 0;
        If (T² > UCL_T²)
            RL (the number of simulations) = the number of simulated profiles;
            Break;
        Else
            For (the sampling number of each profile = 1:48)
                Calculate EWMA Z(the sampling number of each profile)
                If (Z > UCL_EWMA or Z < LCL_EWMA)
                    Index = 1;
                    Break;
                End
            End
            If (index = 1)
                RL (the number of executions) = count;
                Break;
            End
        End
    End
End
Calculate ARL;
```

References

1. Montgomery, D.C. *Introduction to Statistical Quality Control*, 5th ed.; Wiley: New York, NY, USA, 2005.
2. Jensen, W.A.; Birch, J.B.; Woodall, W.H. Monitoring correlation within linear profiles using mixed models. *J. Qual. Technol.* **2008**, *40*, 167–185. [CrossRef]
3. Chicken, E.; Pignatiello, J.J., Jr.; Simpson, J.R. Statistical process monitoring of nonlinear profiles using wavelets. *J. Qual. Technol.* **2009**, *41*, 198–212. [CrossRef]
4. Qiu, P.; Zou, C.; Wang, Z. Nonparametric profile monitoring by mixed effects modeling. *Technometrics* **2010**, *52*, 265–277. [CrossRef]
5. Hung, Y.C.; Tsai, W.C.; Yang, S.F.; Chuang, S.C.; Tseng, Y.K. Nonparametric profile monitoring in multi-dimensional data spaces. *J. Process Control* **2012**, *22*, 397–403. [CrossRef]
6. Chuang, S.C.; Hung, Y.C.; Tsai, W.C.; Yang, S.F. A framework for nonparametric profile monitoring. *Comput. Ind. Eng.* **2013**, *64*, 482–491. [CrossRef]
7. Ghahyazi, M.E.; Niaki, S.T.A.; Soleimani, P. On the monitoring of linear profiles in multistage processes. *Qual. Reliab. Eng. Int.* **2014**, *30*, 1035–1047. [CrossRef]
8. Zhang, Y.; He, Z.; Zhang, C.; Woodall, W.H. Control charts for monitoring linear profiles with within-profile correlation using Gaussian process models. *Qual. Reliab. Eng. Int.* **2014**, *30*, 487–501. [CrossRef]
9. Khedmati, M.; Niaki, S.T.A. Phase II monitoring of general linear profiles in the presence of between-profile autocorrelation. *Qual. Reliab. Eng. Int.* **2016**, *32*, 443–452. [CrossRef]
10. Jensen, W.A.; Grimshaw, S.D.; Espen, B. Nonlinear profile monitoring for oven-temperature data. *J. Qual. Technol.* **2016**, *48*, 84–97.
11. Fan, S.-K.S.; Chang, Y.J.; Aidara, N. Nonlinear profile monitoring of reflow process data based on the sum of sine functions. *Qual. Reliab. Eng. Int.* **2013**, *29*, 743–758. [CrossRef]
12. Hurvich, C.M.; Tsai, C.L. Regression and time series model selection in small samples. *Biometrika* **1989**, *76*, 297–307. [CrossRef]

13. McQuarrie, A.D. A small-sample correction for the Schwarz SIC model selection criterion. *Stat. Prob. Lett.* **1999**, *44*, 79–86. [CrossRef]
14. Tracy, N.D.; Young, J.C.; Mason, R.I. Multivariate control charts for individual observations. *J. Qual. Technol.* **1992**, *24*, 88–95. [CrossRef]
15. Sullivan, J.H.; Woodall, W.H. A comparison of multivariate control charts for individual observations. *J. Qual. Technol.* **1996**, *28*, 398–408. [CrossRef]
16. Holmes, D.S.; Mergen, A.E. Improving the performance of the T2 control chart. *Qual. Eng.* **1993**, *5*, 619–625. [CrossRef]
17. Williams, J.D.; Woodall, W.H.; Birch, J.B. Statistical monitoring of nonlinear product and process quality profiles. *Qual. Reliab. Eng. Int.* **2007**, *23*, 925–941. [CrossRef]
18. Brill, R.V. A case study for control charting a product quality measure that is a continuous function over time. Presented at the 45th Annual Fall Technical Conference, Toronto, ON, Canada, 18–19 October 2001.
19. Kim, K.; Mahmoud, M.A.; Woodall, W.H. On the monitoring of linear profile. *J. Qual. Technol.* **2003**, *35*, 317–328. [CrossRef]

© 2019 by the authors. Licensee MDPI, Basel, Switzerland. This article is an open access article distributed under the terms and conditions of the Creative Commons Attribution (CC BY) license (http://creativecommons.org/licenses/by/4.0/).

Article

Availability Assessment of IMA System Based on Model-Based Safety Analysis Using AltaRica 3.0

Haiyong Dong [1,*,†], Qingfan Gu [2,†], Guoqing Wang [1,2,3,†], Zhengjun Zhai [1,†], Yanhong Lu [1,†] and Miao Wang [3,†]

1. School of Computer Science, Northwestern Polytechnical University, Xi'an 710072, China; wang_guoqing@careri.com (G.W.); zhaizjun@nwpu.edu.cn (Z.Z.); yanhonglu@nwpu.edu.cn (Y.L.)
2. China National Aeronautical Radio Electronics Research Institute, Shanghai 200233, China; gu_qingfan@careri.com
3. School of Aeronautics and Astronautics, Shanghai Jiao Tong University, Shanghai 200240, China; wang_miao@careri.com
* Correspondence: donghaiyong@mail.nwpu.edu.cn; Tel.: +86-182-2050-7569
† These authors contributed equally to this work.

Received: 18 January 2019; Accepted: 20 February 2019; Published: 25 February 2019

Abstract: The integrated modular avionics (IMA) system is widely used in all classes of aircraft as a result of its high functional integration and resource utilization in developing advanced avionics systems. However, a series of challenges related to safety assessment exist in the background of the logical architecture for multi-message interactions of the IMA system. Traditional safety assessment methods are mainly based on engineering experience, and are difficult to reuse, incomplete, and even error-prone. Here we propose a method to assess the availability of the IMA system based on the thinking of model-based safety analysis. To aid the proposed method, we implement a tool to generate a AltaRica 3.0 file used to assess the IMA system model. The simulation results show that the proposed method makes the availability assessment fast, efficient, and effective. Moreover, we apply this method to the modification analysis of the IMA system under the condition of satisfying the safety requirement. Our study can enhance the safety assessment of safety-critical systems effectively, assist the design of IMA systems, and reduce the amount of errors during the programming process of the safety model.

Keywords: availability assessment; integrated modular avionics; model-based safety analysis; AltaRica 3.0

1. Introduction

Integrated modular avionics (IMA) is the state-of-the-art methodology in the real-time computer network airborne system domain, which consists of a number of computing modules capable of supporting numerous hosted applications with different criticality levels [1,2]. Up to now, IMA has been widely used in large, civil aircraft, such as the Airbus A380 and Boeing B787, due to the remarkable improvement in system efficiency, with weight and power consumption reductions by means of comprehensive resources integration or high resources sharing [3]. Different from the federated digital architecture, the IMA system can be divided into three levels: the functional layer, logical layer, and physical layer. The visual objects in the logical layer work together to provide services for hosted applications in the functional layer by utilizing the resources in the physical layer. In addition, some IMA systems, like that in the A380, use two redundant avionics full duplex switched ethernet (AFDX) networks to guarantee the required availability [4]. However, at the same time, the reuse of the traditional safety assessment will become more difficult. Virtual link (VL), the central feature of an AFDX network, is a unidirectional logic path from the source end-system to all the destination end-systems [5]. In this way, VLs are mapped onto visual objects in the logical layer, AFDX switches and end-systems are mapped onto the resource in the physical layer, and functions are mapped onto the application in the functional layer. In practice, the system engineer utilizes the IMA configure tool to obtain a specific VL configuration, whose network performance meets the needs of hosted applications. However, the above VL configuration needs to be further analyzed to verify that the availability of specific applications appropriate to a required criticality level is satisfied. Availability is the qualitative or quantitative attribute that signals that a system is in a functioning state at a given point in time, and it is sometimes expressed in terms of the probability that a system does not provide its output(s) (i.e., unavailability) [6]. It is an important factor in the area of reliability and safety, especially for the safety-critical system. Traditionally, the safety assessment and hazard analysis are modeled on fault tree analysis (FTA) by analysts based on engineering experience, which is easy to understand, but hard to reflect in real designs. Even more important for complex avionics systems, the FTA model is too huge to modify with any minor change by manual operation [7]. In addition, traditional safety analyses (FTA, etc.) are usually based on informal system models, which are always regarded as incomplete, inconsistent, and error-prone [8]. Moreover, a consistent formal model is needed in both system design procedure and safety analysis procedure. To solve these problems, model-based safety analysis (MBSA) is proposed.

Up to now, MBSA has been widely used in the fields of aviation [9], railways [10], automotives [11], and other safety-critical systems [12]. During the process of MBSA, system engineers and safety analysts share a common system model. It extends the system model with a fault model as well as relevant portions of the physical system, and is recommended to model complex systems in ARP 4761A draft [13]. In addition, Laboratoire Bordelais de Recherche en Informatique (LaBRI) developed a free formal language, AltaRica, to model both functional and dysfunctional behaviors of systems. Models in AltaRica 3.0 are described by guarded transition systems (GTS), which consists of state variables, flow variables, events, transitions, and assertions [14]. AltaRica 3.0 can support the modeling of event driven systems based on MBSA, and the model described can be hierarchical and compositional [15]. Thus, AltaRica 3.0 has been widely used to model these safety-critical systems [16,17].

Some researchers have investigated the safety assessment of avionics systems based on MBSA. Morel used MBSA to validate several IMA architectures with three levels, and suggested that MBSA is a good method for safety assessment in early validation to support flexible and rapid prototyping of integrated systems, and expressed that his study needed to do some quantity analysis to verify whether the availability further met the requirements [9]. Li used MBSA to study the safety assessment of complex aircraft products, proposed a safety modeling approach based on AltaRica, and proved its validity through simple hydraulic system verification [18]. The safety analysis of IMA based on MBSA have also been studied, while the model described by AltaRica was totally coded by hand, this makes it difficult to reuse and easy to make mistakes with [19].

In this paper, to study the impact of using the effective procedure and tool to analyze the safety of IMA systems, a method based on MBSA using AltaRica 3.0 to assess the availability of the IMA system is proposed and a tool to aid the assessment method is implemented. An IMA system case is modeled to verify the validity of the proposed method. In addition, we do some research on design optimization of the IMA system. Finally, the advantages and disadvantages of the different assessment methods are analyzed. This provides new insights into the safety assessment and hazard analysis in an IMA system operating within an acceptable safety level.

2. Assessment Method

Model-based safety analysis (MBSA) is able to build a complete, accurate, and consistent safety model for complex, safety-critical systems [20]. Generally, there are seven steps in a MBSA process: "Gather the most complete system data available at the time", "Define the goal and the granularity of the analysis", "Define the failure conditions to be studied", "Build the failure propagation model (FPM) according to the collected data", "Build the failure condition logic", "Verification of the FPM and failure condition logic", and "Failure condition evaluation & analysis". However, not every MBSA process suits the above steps. In addition, there is no limit to the model languages in a MBSA process. Considering this, AltaRica 3.0, an available, high-level language for event driven modeling of complex systems, is especially well suited for safety analyses and performance analyses. AltaRica 3.0 defines the block by representing the component with failure mode, which is composed of the declaration of variables and events, and the definition of transitions and assertions [18].

As shown in Figure 1, the system designer is responsible for the system model while the safety engineer is responsible for the safety model. An IMA system model consists of three layers: the physical layer, logical layer, and functional layer. These three layers have one-to-one correspondence with the failure modes, failure propagation models, and failure conditions, which constitute an IMA safety model. According to the requirement, the system designer utilizes the IMA configure tool to generate the xml file for data exchange. The file in xml format contains the failure rate and configured VLs of every component to describe the logical relationship between the components in the IMA system. In this paper, the data about safety of failure modes and failure propagation models can be collected from the system model and translated into the description of the alt file, which is used to describe the safety model. The safety engineer obtains the information about the function from the system designer and defines the invalid function as the studied failure condition. The configuration xml file is used as interface control document (ICD) for components in IMA system. Note that the ICD file is built by the system designer, and it is always regarded as the input for the safety engineer.

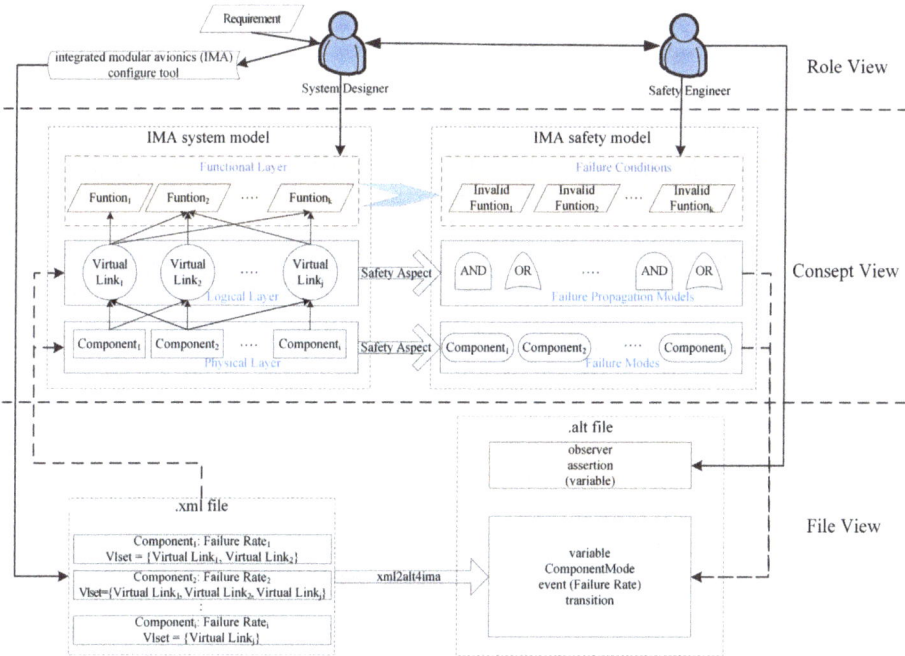

Figure 1. Generated mapping from system model to safety model for a integrated modular avionics (IMA) system

In this paper, a method is proposed to assess the availability of the typical IMA system based on MBSA using AltaRica 3.0. On the basis of the system model described in the xml file, we build the safety model in the alt file (AltaRica 3.0 format) by utilizing a tool named xml2alt4ima which was developed by our team. In the xml file, there is one root element named "VirtualLinks" after XML declaration. Each component is an element of the VirtualLinks. Every component has three attributes to represent "Name", "GuID", and failure rate of "Loss". Every component also contains several elements to represent the configured VLs. Every VL consists of 10 attributes: "Name", "GuID", "BAG", "MaxFrameSize", "MinFrameSize", "Priority", "Captain", "ActualPath", "Source", and "Destination". Note that not every attribute is utilized in our method, for example, "BAG" is used to set the time gap between two packages, and it's used to assess the performance in real-time. In our method, "Name" and failure rate of "Loss" of each component, and "Name" and "ActualPath" of each VL are the safety properties and they are utilized in describing the safety model.

The xml file contains all components with thier configurations, and these configurations can be mapped onto VLs in the physical layer and components in the logical layer. In addition, since the fault may occur in every component in the physical layer, each component has a responding failure mode in safety model. Each VL in the logical layer has its own working status, which can be changed when fault occurs in the related components. The failure propagation models and failure modes are described by "variable", "ComponentMode", "event", and "transition" in the alt file. The "variable" is used to describe the state of the system or subsystem, the "ComponentMode" is used to describe the status of component, the "event" is used to describe the event that may occur in the system, and the "transition" describes how the system evolves. The variable with brackets in the alt file generated by the functions means it is not required, but it can assist in the process of calculating the observer, especially when the observer is a complex function or failure condition. Moreover, the functions in the functional layer are defined by the system designer, and it is the basis of the failure condition in the safety model during the process of MBSA. The failure condition is a condition having an effect on the aircraft, which is usually caused by one or more failures or errors associated with the flight phase, relevant adverse operational or environmental conditions, or external events [21]. To sum up, the system model described in the xml file can be mapped onto the physical layer and logical layer, and it contains the basis data of the safety model which is described in the alt file. The safety engineer needs to understand the system model, extract valid information of the safety assessment, build the failure condition with the help of the system designer, and thus, realize the safety model.

Therefore, the process to assess the availability of the IMA system can be concluded in a method as follows:

- **Step 1. Define the failure condition of the IMA system and their safety requirements.**
  The failure condition means an unexpected state. It is always a logical combination of some unexpected states. For the IMA system, it means an invalid function.
- **Step 2. Utilize the special generation tool (xml2alt4ima) to generate an alt file based on the configuration xml file.**
  The xml2alt4ima tool is designed to aid the construction of the alt file according to the xml configure file.
- **Step 3. Manually add the observer, assertion, and variables if needed.**
  Complete the alt file manually. The observer is used to represent the failure condition and complex function. The assertion contains some sentences to represent the logical relationship. Variables provide assistance in understanding the logical relationship between the failure condition and failure mode. In addition, we need to add a variable named "failed", which is used to represent the top event of the fault tree.
- **Step 4. Utilize the AltaRica 3.0 assessment tool to compile the alt file, and obtain the cut set, probability, contribution, and so on.**
  The AltaRica 3.0 compiler can explain the meaning of the alt file. We recommend the free OpenAltaRica tool [22], which integrates many analysis functions.

The xml2alt4ima tool is developed in Matlab 2016a, and the core algorithm is illustrated in Algorithm 1.

Algorithm 1 The algorithm to generate the alt file from the configuration xml file of the IMA system.

Input: xml file, including m ($m \geq 1$) components with failure rate and configured VLs;
Output: alt file, including file structure, event, transition and required variable;
1: Begin initialization
2:     Define the domain of ComponentMode for all components
3: End initialization
4: $m \Leftarrow$ the quantity of the components in the xml file
5: For component i ($1 \leq i \leq m$)
6:     $Rate_i \Leftarrow$ the failure rate of component i
7:     Define the state of component i based on ComponentMode
8:     Define the event for component i with $Rate_i$
9:     Define the transition of component i based on event
10:     n $\Leftarrow$ the quantity of the VLs configured in component i
11:     For VL j ($1 \leq j \leq n$)
12:       Define variables for VL j configured in component i
13:       $p \Leftarrow$ he quantity of components in the actual path through VL j
14:       For component k ($1 \leq k \leq p$)
15:         Add action for VL j in the transition configured in component i
16:       End component k
17:     End VL j
18: End component i
19: Delete redundant variables for VL
20: Begin modification
21:     Add assertion for the failure condition
22:     Add block for the whole model
23: End modification

3. Case Study

In this section, a typical example of the IMA system model is introduced in Section 3.1, some general assumptions and failure condition are presented in Section 3.2, the results based on our proposed method is calculated, and it is also verified by other methods in Section 3.3. On the basis of the results, we try to optimize the system model and propose advice for the system designer in Section 3.4. We also try to explore the efficiency of different safety assessment methods in Section 3.5.

3.1. IMA System Model

As a result of the high requirement of performance and availability, the utilization of the existing resource becomes the most difficult point in the structural design of IMA and the core architecture of civil avionics systems [23,24]. For example, to avoid a single-point failure, all AFDX networks and end-systems are designed to be double or triple module redundant. Figure 2 shows a typical IMA system model with two redundant AFDX networks, three general processing modules (GPM), three remote data concentrators (RDC), and two hosted functions (HF). The RDC is designed for data acquisition from the sensor (SEN) and other signal sources, the GPM is designed for data calculation and procession, the HF is designed for data display and upper application, and the switch (SW) is designed for transferring data through the IMA system. It is assumed that HF1 is used by the captain and HF2 is used by the copilot. Every HF needs data processed by the GPM from both SEN1 and SEN2.

SEN1 BU denotes the backup of SEN1, and SEN2 BU denotes the backup of SEN2. Every RDC obtains the data from the connected sensors through the ARINC 429 bus, and transfers these data to three GPMs through the redundant network. After processing these data, every GPM transfers the processed data to two HFs through the redundant network. The GPM and HF are able to utilize the effective data and drop the redundant data. In addition, since the sensors are connected with the RDC through the ARINC 429 bus, instead of the AFDX switch, there is no VL configuration in the sensors.

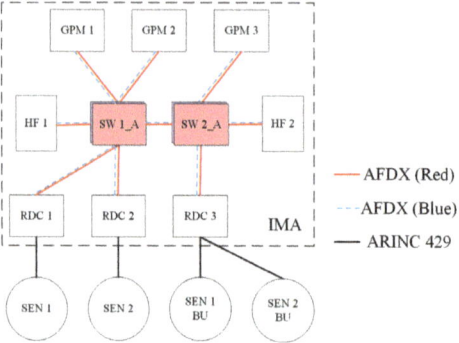

Figure 2. The model of a integrated modular avionics (IMA) system

3.2. Assumption and Failure Definition

There are 12 components in the IMA system model, as shown in Figure 2. To simplify the model, there are six general assumptions [24–26].

Assumptions

- Faults are modeled as statistically independent distributed events;
- The failure rate of each component is a constant;
- A fault occurs instantaneously and at most one fault event in a minimum time slice;
- The system and its components have two states: normal and failure;
- The system and its components are unrepairable while in use;
- The cable between two components keeps working.

Note that the failure distribution of the components is assumed to be a λ-exponential distribution where λ is equal to the failure rate per flight hour. The mean time between failure (MTBF) and failure rate per flight hour of these components are shown in Table 1. The failure rate of the sensor comes from the book written by Jukes [27], the MTBF of the switch and GPM come from Reference [4], and the MTBF of the RDC comes from a booklet published by a RDC manufacturer [28]. The HF exists in a specific line replaceable unit (LRU), so these data vary with different LRU. The MTBF of the HF refers to the devices in the display system designed by the China National Aeronautical Radio Electronics Research Institute (CARERI). Components fail instantaneously without any common cause effect. Since the sensors do not belong to the IMA system, their failures are not calculated in Section 3.3.

Table 1. Failure rate of components in the IMA system.

Component	Mean Time Between Failure (MTBF)	Failure Rate per Flight Hour
Sensor (SEN)	20,000 h	5.00×10^{-5}
Switch (SW)	100,000 h	1.00×10^{-5}
Remote data concentrator (RDC)	14,000 h	7.14×10^{-5}
Hosted function (HF)	16,000 h	6.25×10^{-5}
General processing module (GPM)	50,000 h	2.00×10^{-5}

In this paper, the meaning of the failure condition is similar to the functional failure mode. According to their severity, failure conditions can be classified into catastrophic, hazardous, major, minor, and of no safety effect. On the basis of the design experience of civil aircraft projects, a safety-critical function is defined in the model: at least one HF can get both sets of sensor data processed by the GPM. In other words, a hazardous failure condition defined and denoted as LOSS_SEN_HF. LOSS_SEN_HF means the crew, both HFs, cannot get either set of sensor data processed by the GPM. Development assurance level (DAL) is defined in aerospace recommended practice (ARP) 4754A [6] and the above function should satisfy with DAL B, which means this failure condition may cause the hazardous effect and its failure rate must be lower than 10^{-7} per flight hour [29].

3.3. Results

We added observers and assertions for a special failure condition based on the alt file generated by xml2alt4ima tool. Then, we ran its program in OpenAltaRica tool, the aim of which was to develop a complete set of tools for the high-level modeling language AltaRica 3.0 [30]. Then, we generated a fault tree in open probabilistic safety assessment (OPSA) format from AltaRica 3.0 model as below.

As shown in Figure 3, the top event of the fault tree is LOSS_SEN_HF, and the basic events are the failures of these components. The size of the generated OPSA file was 8318 KB. There were thousands of automatic defined gates, including all the combinations of different basic events.

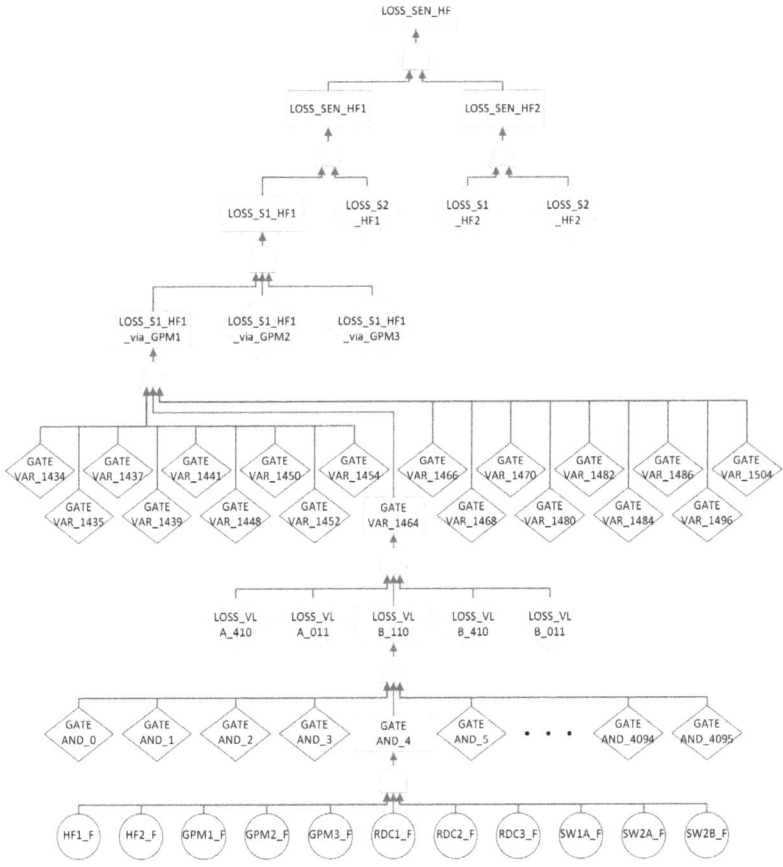

Figure 3. The fault tree of the defined failure condition

In addition, we modified the file (XFTA.xml) to meet with the failure conditions defined above [31,32]. Finally, we got the minimal cut set of the failure condition, as shown in Table 2.

Table 2. Minimal cut set of LOSS_SEN_HF in the IMA system.

Rank	Minimal Cut Set	Probability
1	rdc1_f, rdc3_f	5.10×10^{-9}
2	rdc2_f, rdc3_f	5.10×10^{-9}
3	hf1_f, hf2_f	3.91×10^{-9}
4	gpm1_f, gpm2_f, gpm3_f	8.00×10^{-15}
5	rdc1_f, sw2A_f, sw2B_f	7.14×10^{-15}
6	rdc2_f, sw2A_f, sw2B_f	7.14×10^{-15}
7	rdc3_f, sw1A_f, sw1B_f	7.14×10^{-15}
8	hf1_f, sw2A_f, sw2B_f	6.25×10^{-15}
9	hf2_f, sw1A_f, sw1B_f	6.25×10^{-15}
10	gpm3_f, sw1A_f, sw1B_f	2.00×10^{-15}
11	hf2_f, rdc1_f, sw1A_f, sw2B_f	4.46×10^{-19}
⋮	⋮	⋮

Since switches are configured as redundant devices, the IMA configure tool denotes sw1A and sw1B to represent switches with the same location. There are hundreds of minimal cut sets generated by OpenAltaRica, while three second-order cut sets and seven third-order cut sets make up the majority of the top event.

The probability of LOSS_SEN_HF per flight hour is 1.41022×10^{-8}, which complies with the safety requirements.

Besides, we used other assessment methods to verify the proposed method. We utilized Simfia (software developed by APSYS) to build the model of the above system, generate the fault tree, and calculate the availability. The model and the fault tree of Simfia are shown in Figure 4.

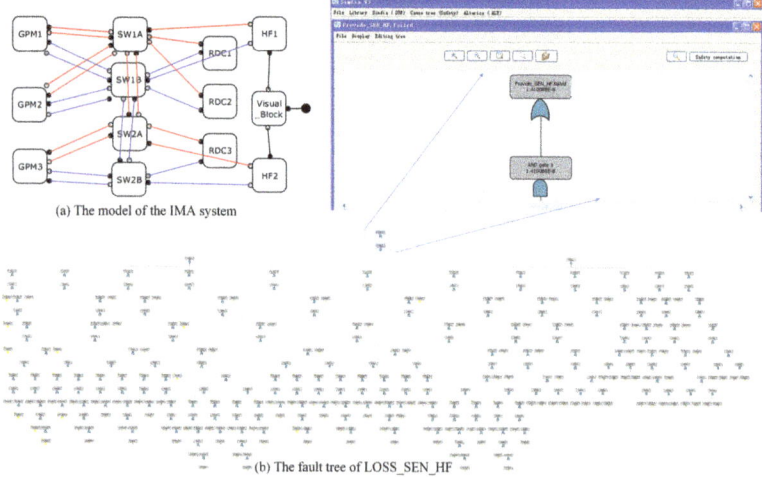

Figure 4. The model and the fault tree of Simfia. (a) Description of the IMA model. (b) The fault tree generated by the Simfia.

The availabilities of the above two methods were a little different (1.41088×10^{-8} calculated by Simfia), and the deviation comes from the computational accuracy of Simfia. We also used the Monte Carlo simulation to verify the proposed method. The failure condition "LOSS_SEN_HF" occurred 141 times over 10,000,000,000 simulation runs. The result of the Monte Carlo simulation is approximately the same with the above two methods. In summary, the Monte Carlo simulation proves the correctness of our method.

3.4. Optimization of the IMA System

The GPM in the IMA system is always configured with many applications, thus it is capable of supporting numerous upper functions of differing critical levels. In this paper, we only define one function: the HF obtains sensor data from the RDC processed by GPMs. All devices are redundant in improving the availability of the above function.

As shown in Table 2, the cut set rank first and second are the combination failures of the RDCs, which means that the failures of both data sources can lead to a top event. The cut set rank third, meaning the failures of both data destinations, can lead to a top event as well. The third-order cut set is a combination of different types of components, excluding the cut set rank fourth, which is the failure set of all GPMs. As is well known, three redundant devices failing per flight hour is an event with a small probability, unless these devices are designed with the same unknown bug.

For the function defined in this paper, it seems that there is no need to use three GPMs to satisfy the safety requirement, which should be researched further. Then, we tried to modify the logical architecture with only two GPMs, and generate the alt file referring to the prior subsection. The probability of LOSS_SEN_HF per flight hour with two GPMs is shown in Table 3.

Table 3. Probability of LOSS_SEN_HF in the IMA system with two GPMs.

Configuration	Probability of LOSS_SEN_HF
without GPM1	1.45022×10^{-8}
without GPM2	1.45022×10^{-8}
without GPM3	1.46022×10^{-8}

Note that the original IMA system includes three GPMs: GPM1, GPM2, and GPM3. Configuration of "without GPM1" denotes the current defined function in the IMA system model only employing GPM2 and GPM3.

Table 3 shows that the probability of LOSS_SEN_HF in the IMA system without GPM1 or GPM2 is lower than that without GPM3. This is due to the fact that in the latter configuration there exists another second-order cut set "sw1A_f, sw1B_f", more than in the first two configurations. As Figure 2 shows, switches are designed to connect with part of the GPMs to balance the communication load. In the above model, it can be concluded that the load balancing designation can reduce the risk of common cause failure, especially for same zone risk.

3.5. Efficiency of Safety Assessment Process

The advantage and disadvantage of the three safety assessment methods are analyzed in this section. The first one uses the proposed method based on the MBSA, the second one uses the safety assessment tool (Simfia) to build the safety model manually, and the third one uses the traditional tool (PTC Windchill Quality Solutions, also known as Relex) to build the fault tree based on engineering experience directly. The main steps of obtaining the availability of the failure condition with the different methods are summarized in Table 4. "Auto" means the corresponding step can be operated by software, while "manually" means that the step should be done by a safety engineer, and the context after [source] of each [auto] signifies the data source of corresponding step. It is clear to see that the MBSA requires less manual operations.

With the help of the senior safety engineers in CARERI who have five years experience in using both Simfia and Relex, we tried to analyze the availability of different IMA cases for analyzing the efficiency of the above three methods. All our experiments were conducted on a single core of a 2.7 GHz Intel Core2Duo processor with 2 GB RAM running on Windows XP. The statistical data are shown in Table 5.

Table 4. Main steps of obtaining the availability of failure condition.

Main Steps	MBSA	Simfia	Relex
1. Definition of components	[auto] Define the ComponentMode by xml2alt4ima [source] system model (xml file)	[manually] Create the blocks as the components of system	[manually] Define the basic event of the fault tree
2. Configuration of components	[auto] Get the failure rate for every component; Define the event with corresponding failure rate by xml2alt4ima [source] system model (xml file)	[manually] Set the failure rate for every block; Create the connector to link the blocks; Set the state types of IN/OUT port of each connector	[manually] Set the failure rate for every basic event; Define the gate; Set the gate type
3. Modeling of logical causes	[auto] Add the transition as the relationship between Visual Links and Components by xml2alt4ima [source] system model (xml file)	[manually] Set the logical causes of every state type of each OUT connector	[manually] Connect the gate with other gates and basic events
4. Definition of failure condition	[manually] Add an observer for the failure condition	[manually] Define a new block for the failure condition	[manually] Define the gate as the top event of the fault tree
5. Configuration of failure condition	[manually] Build the assertion as the relationship between Failure condition and Visual Links	[manually] Create the connector to link the new blocks; Set the state types of IN/OUT port of each connector; Set the logical causes of every state type of each OUT connector	[manually] Connect the top event gate with other gates and basic events
6. Obtain the fault tree	[auto] Generated by OpenAltaRica [source] safety model (alt file)	[auto] Generate the fault tree [source] safety model	[manually] Build the fault tree from top to down
7. Obtain the cut set of failure condition	[auto] Obtain the cut set of failure condition [source] fault tree (opsa file)	[auto] Obtain the cut set of failure condition [source] safety model	[auto] Obtain the cut set of failure condition [source] fault tree
8. Calculate the availability	[auto] Calculate the availability [source] fault tree (opsa file)	[auto] Calculate the availability [source] safety model	[auto] Calculate the availability [source] fault tree

Table 5. Comparison of different methods for the IMA cases.

	Case 1 [a]			Case 2 [b]			Case 3 [c]		
	MBSA	Simfia	Relex	MBSA	Simfia	Relex	MBSA	Simfia	Relex
Quantity of components in model	10 Components			12 Components			15 Components		
Time for modeling	≈1 h	≈4 h	≈6 h	≈1 h	≈6 h	≈10 h	≈1 h	≈8 h	≈12 h
Time for generating fault tree	<1 min	<10 s	none	<10 min	<10 s	none	<1 h	<1 min	none
Quantity of gates and basic events [d]	1254	167	84	5828	232	115	45,844	282	136
Time for calculating the cut set	<1 min	<10 s	<10 s	1 min	<10 s	<10 s	1 min	<10 s	<10 s
Time for remodeling [e]	≈40 min	≈1 h	≈3 h	≈1 h	≈2 h	≈5 h	≈1 h	≈3 h	≈6 h

[a] The IMA system with two HFs, two GPMs, four switches, and two RDCs. [b] The IMA system in the prior subsection with two HFs, three GPMs, four switches, and three RDCs. [c] The IMA system with four HFs, three GPMs, four Switches, and four RDCs. [d] The fault tree contains many useless gates generated by the OpenAltaRica and Simfia. The repeated basic event of the fault tree was only counted once. [e] The modification mainly refers to the change of logical architecture of the IMA model.

Table 4 shows that our proposed method based on the MBSA costs less time to model than the traditional method or Simfia. When the system changes, our proposed method can be changed more efficiently than the other two methods. However, the quantity of gates and basic events of the fault tree generated by OpenAltaRica is larger than that of the other two methods, and the time for calculating the cut set of our proposed method is longer. In addition, different analysts have different styles to define the gate of the fault tree, which makes it difficult for other analysts to understand the fault tree generated by Relex. When utilizing Simfia to analyze the availability, analysts need to remodel the system according to their comprehension, which is inefficient and prone to error. On the contrary, our proposed method enables the safety analyst to devote their time to the safety analysis and designation advice rather than on duplicate work.

4. Conclusions

IMA is recommended because of its high utility as regards resources and hierarchical architecture, as well as its ease of use for the engineer. However, the traditional availability assessment of the IMA system with the feature of the redundant AFDX network is time-consuming and error-prone. For a safety-critical system, the common way to analyze the availability is through modeling the fault tree based on engineering experience. In this paper, we propose an availability assessment method for the IMA system based on MBSA using AltaRica 3.0 and implement a tool to generate an alt file based on the configuration xml file of the IMA system. In this way, the availability assessment becomes faster and can be modified effectively according to the change of system. Taking a typical IMA system model as an example, the results indicate that the application in the IMA system satisfies the safety requirements. In addition, we also find that the load balancing designation of the IMA system is advantageous in reducing the risk of common cause failure. Our method can also be used in the availability assessment of a safety-critical system with hierarchical architecture with a functional–logical–physical layer.

In the future, we will research the safety analysis of the IMA system in the following two aspects: On the one hand, considering that different GPMs can process respective functions, it is necessary to study the feature of the fault propagation process in the IMA system; on the other hand, it is of great importance to study a process which can handle the availability assessment considering errors or the confusing of basic components and functions, rather than the loss of these modules.

Author Contributions: Conceptualization, H.D.; Data curation, Z.Z.; Formal analysis, G.W.; Funding acquisition, G.W.; Investigation, Z.Z.; Methodology, H.D., G.W., Z.Z., Y.L. and M.W.; Project administration, Q.G.; Resources, Q.G.; Software, H.D.; Validation, Q.G.; Writing—original draft, H.D.; Writing—review & editing, Y.L. and M.W.

Funding: This research was funded by National Key Basic Research Program of China grant number 2014CB744900.

Acknowledgments: The authors would like to thank Hao Rong for his technical assistance in the construction of the case model, Zilong He, Ning Guo and Ruoxing Mei for their advice and assistance in proofreading the article.

Conflicts of Interest: The authors declare no conflict of interest.

Abbreviations

The following abbreviations are used in this manuscript:

IMA	Integrated Modular Avionics
AFDX	Avionics Full Duplex Switched Ethernet
VL	Visual Link
FTA	Fault Tree Analysis
MBSA	Module Based Safety Analysis
ARP	Aerospace Recommended Practice
GTS	Guarded Transition Systems
FPM	Failure Propagation Model
CARERI	China National Aeronautical Radio Electronics Research Institute

ICD	Interface Control Document
GPM	General Processing Module
RDC	Remote Data Concentrator
HF	Hosted Function
SW	Switch
SEN	Sensor
BU	Backup
ARINC	Aeronautical Radio Inc.
MTBF	Mean Time Between Failure
LRU	Line Replaceable Unit
DAL	Development Assurance Level
KB	Kilo Byte
OPSA	Open Probabilistic Safety Assessment

References

1. Windsor, J.; Deredempt, M.H.; de Ferluc, R. Integrated modular avionics for spacecraft spacecraft-user requirements, architecture and role definition. In Proceedings of the IEEE/AIAA 30th Digital Avionics Systems Conference (DASC 2011), Seattle, WA, USA, 16–20 October 2011; pp. 8A6:1–8A6:16.
2. DO297. *Integrated Modular Avionics*; RTCA, Inc.: Washington, DC, USA, 2005.
3. Watkins, C.B.; Walter, R. Transitioning from federated avionics architectures to integrated modular avionics. In Proceedings of the IEEE/AIAA 26th Digital Avionics Systems Conference (DASC 2007), Dallas, TX, USA, 21–25 October 2007; pp. 2.A.1:1–2.A.1:10.
4. Itier, J.B. A380 Integrated Modular Avionics. Available online: http://www.artist-embedded.org/docs/Events/2007/IMA/Slides/ARTIST2_IMA_Itier.pdf (accessed on 21 December 2018).
5. Alena, R.L.; Ossenfort, J.P.; Laws, K.I.; Goforth, A.; Figueroa, F. Communications for Integrated Modular Avionics. In Proceedings of the 2007 IEEE Aerospace Conference, Big Sky, MT, USA, 3–10 March 2007; pp. 1–18.
6. ARP4754A. *Guidelines for Development of Civil Aircraft and Systems*; SAE International: Warrendale, PA, USA, 2010.
7. Güdemann, M. Qualitative and Quantitative Formal Model-Based Safety Analysis: Push the Safety Button. Ph.D. Thesis, Otto von Guericke University Magdeburg, Magdeburg, Germany, 2011.
8. Hönig, P.; Lunde, R.; Holzapfel, F. Model Based Safety Analysis with smartIflow. *Information* 2017, *8*, 7. [CrossRef]
9. Morel, M. Model-Based Safety Approach for Early Validation of Integrated and Modular Avionics Architectures. In Proceedings of the 4th International Model-Based Safety and Assessment (IMBSA 2014), Munich, Germany, 27–29 October 2014; pp. 57–69.
10. Issad, M.; Kloul, L.; Rauzy, A. A Model-Based Methodology to Formalize Specifications of Railway Systems. In Proceedings of the 4th International Model-Based Safety and Assessment (IMBSA 2014), Munich, Germany, 27–29 October 2014; pp. 28–42.
11. Papadopoulos, Y.; Grante, C. Evolving car designs using model-based automated safety analysis and optimisation techniques. *J. Syst. Softw.* 2005, *76*, 77–89. [CrossRef]
12. Lisagor, O.; Kelly, T.; Niu, R. Model-based safety assessment: Review of the discipline and its challenges. In Proceedings of the IEEE 9th International Conference on Reliability, Maintainability and Safety (ICRMS 2011), Guiyang, China, 12–15 June 2011, pp. 625–632.
13. ARP 4761A (Draft) Associated Appendix. *Model Based Safety Analysis*; SAE International: Warrendale, PA, USA, 201X.
14. Prosvirnova, T.; Batteux, M.; Brameret, P.A.; Cherfi, A.; Friedlhuber, T.; Roussel, J.M.; Rauzy, A. The altarica 3.0 project for model-based safety assessment. In Proceedings of the 4th IFAC Workshop on Dependable Control of Discrete Systems (DCDS 2013), York, UK, 4–6 September 2013, pp. 1–7.
15. Prosvirnova, T. Altarica 3.0: A Model-Based Approach for Safety Analyses. Ph.D. Thesis, Ecole Polytechnique, Palaiseau, France, 2014.

16. Mortada, H.; Prosvirnova, T.; Rauzy, A. Safety assessment of an electrical system with AltaRica 3.0. In Proceedings of the 4th International Model-Based Safety and Assessment (IMBSA 2014), Munich, Germany, 27–29 October 2014; pp. 181–194.
17. Brameret, P.A.; Rauzy, A.; Roussel, J.M. Automated generation of partial Markov chain from high level descriptions. *Reliab. Eng. Syst. Saf.* **2015**, *139*, 179–187. [CrossRef]
18. Li, S.; Duo, S. A practicable mbsa modeling process using Altarica. In Proceedings of the 4th International Model-Based Safety and Assessment (IMBSA 2014), Munich, Germany, 27–29 October 2014; pp. 1–13.
19. Gu, Q.; Wang, G.; Zhai, M. Model-based safety analysis for integrated avionics system. In Proceedings of the 14th AIAA Aviation Technology, Integration, and Operations Conference (ATIO 2014), Atlanta, GA, USA, 16–20 June 2014; pp. 2226:1–2226:8.
20. Joshi, A.; Miller, S.P.; Whalen, M.; Heimdahl, M.P.E. A proposal for model-based safety analysis. In Proceedings of the 24th IEEE Digital Avionics Systems Conference (DASC 2005), Washington, DC, USA, 30 October–3 November 2005; pp. 13:1–13:12.
21. *Design Assurance Guidance for Airborne Electronic Hardware*; DO254; RTCA, Inc.: Washington, DC, USA, 2005.
22. OpenAltaRica web page. Available online: http://openaltarica.fr/ (accessed on 21 December 2018).
23. Ananda, C.M.; Venkatanarayana, K.G.; Preme, M.; Raghu, M. Avionics systems, integration, and technologies of the light transport aircraft. *Def. Sci. J.* **2011**, *61*, 289–298. [CrossRef]
24. Moir, I.; Seabridge, A.; Jukes, M. *Civil Avionics Systems*; John Wiley & Sons Ltd.: West Sussex, UK, 2013; ISBN 978-1-118-34180-3.
25. Tu, J.; Cheng, R.; Tao, Q. Reliability analysis method of safety critical avionics system based on dynamic fault tree under fuzzy uncertainty. *Mt. Reliab.* **2015**, *17*, 156–163. [CrossRef]
26. Dugan, J.B.; Bavuso, S.J.; Boyd, M.A. Dynamic fault-tree models for fault-tolerant computer systems. *IEEE Trans. Reliab.* **1992**, *41*, 363–377. [CrossRef]
27. Jukes, M. *Aircraft Sisplay Systems*; American Institute of Aeronautics and Astronautics: Reston, VA, USA, 2004; ISBN 978-1-56347-657-0.
28. RDC booklet of Flight Data Systems. Available online: https://www.flightdata.aero/static/uploads/files/fds-rdc-final-wfvctuhozukm.pdf (accessed on 21 December 2018).
29. *Guidelines and Methods for Conducting the Safety Assessment Process on Civil Airborne Systems and Equipment*; ARP4761; SAE International: Warrendale, PA, USA, 1996.
30. Aupetit, B.; Batteux, M.; Rauzy, A.; Roussel, J.M. Improving performances of the altarica 3.0 stochastic simulator. In Proceedings of the 25th European Safety and Reliability Conference (ESREL 2015), Zurich, Switzerland, 7–10 September 2015; pp. 1815–1823.
31. Open-PSA Format Web Page. Available online: http://www.open-psa.org/ (accessed on 21 December 2018).
32. Yakymets, N.; Jaber, H.; Lanusse, A. Model-based system engineering for fault tree generation and analysis. In Proceedings of the 1st International Conference on Model-Driven Engineering and Software Development (MODELSWARD 2013), Barcelona, Spain, 19–21 February 2013; pp. 210–214.

© 2019 by the authors. Licensee MDPI, Basel, Switzerland. This article is an open access article distributed under the terms and conditions of the Creative Commons Attribution (CC BY) license (http://creativecommons.org/licenses/by/4.0/).

Article

A Novel Method for Gas Turbine Condition Monitoring Based on KPCA and Analysis of Statistics T² and SPE

Li Zeng [1], Wei Long [2] and Yanyan Li [2,*]

[1] School of Aeronautics & Astronautics, Sichuan University, Chengdu 610065, China; zengli_sichuan@163.com
[2] School of Manufacturing Science and Engineering, Sichuan University, Chengdu 610065, China; longwei@scu.edu.cn
* Correspondence: Liyanyan@scu.edu.cn

Received: 24 January 2019; Accepted: 24 February 2019; Published: 27 February 2019

Abstract: Gas turbines are widely used all over the world, in order to ensure the normal operation of gas turbines, it is necessary to monitor the condition of gas turbine and analyze the tested parameters to find the state information contained in parameters. There is a problem in gas turbine condition monitoring that how to locate the fault accurately if failure occurs. To solve the problem, this paper proposes a method to locate the fault of gas turbine components by evaluating the sensitivity of tested parameters to fault. Firstly, the tested parameters are decomposed by the kernel principal component analysis. Then construct the statistics of T² and SPE in the principal elements space and residual space, respectively. Furthermore, the thresholds of the statistics must be calculated. The influence of tested parameters on faults is analyzed, and the degree of influence is quantified. The fault location can be realized according to the analysis results. The research results show that the proposed method can realize fault diagnosis and location accurately.

Keywords: KPCA; T² statistical model; SPE statistical model; kernel function

1. Introduction

Gas turbines provide power for generators, ships, aircraft, etc. Gas turbines need to withstand the influence of high temperature and high pressure when working. Obviously, the harsh working condition of turbines will definitely lead to the performance degradation of components. Fault occurs when performance degradation is severe. It is essential to locate the fault in time after a fault happens [1–4]. Currently, there are four categories of fault diagnosis—the turbine model-based method, the knowledge-based method, the data-driven-based method, and the techniques fusion-based method [5,6]. The model-based method of the diagnosed object must establish an accurate turbine model and on-line input parameters are employed [5]. Silvio Simani and Farsoni Saverio [7] established an identified fuzzy model which based on the Takagi–Sugeno prototype to detect and isolate the fault. Hector Sanchez and Teresa Escobet [8] established a model and proposed a method to check whether the measurements fall inside the output interval. A diagnosis was proposed based on this model. Method based on knowledge is to essentially formulate the diagnostic problem solving as a pattern recognition problem [9,10]. Zhang, Bingham, and Gallimore [11] proposed two techniques to detect the fault. They promoted the concept of y indices based on a transposed formulation of data matrix, and residual errors (REs) and faulty sensor identification indices (FSIIs) are introduced in another method. A large number of data must be available if the method based on data-driven is adopted. The potential relationships between these data need to be extracted. Zhu, Ge, and Song [12] proposed a robust variable model driven by the hidden Markov model and a probabilistic model with Student's t mixture output was designed to tolerate outliers. Furthermore, Zhang Peng [13] studied the Kalman filter and applied it in the location of a fault. He focused on how to establish the linear and nonlinear models of turbine. Based on the models, two faulty location algorithms which apply to the steady working state and dynamic working state respectively were constructed. Vasile Palade, Ron J. Patton, and Faisel J. Uppal [14] applied a neuro-fuzzy technique in an actuator fault location of a gas turbine. Based on learning and adaptation of the TSK fuzzy model, a neuro-fuzzy model was used to generate he residual, and a neuro-fuzzy classifier for the Mamdani model is used to evaluate the residual. Che Changchang, Wang Huawei, and Ni Xiaomei [15] proposed a fault fusion diagnosis model which is based on deep learning. The model analyzes a large number of performance data and obtains fault classification confidence by extracting hidden features from the performance data, then conducts the decision fusion of multiple fault classification results. Tayarani-Bathaie and Khorasani [16] constructed two types of dynamic neural networks to learn the turbine dynamic state. For the measurable variables of the turbine, different neural networks are trained to capture the dynamic relationships. Then, construct a multilayer perception network function to isolate the fault. All model-based methods need to build models that accurately reflect the turbine state. However, due to the large number of turbine parts and the bad working environment, there are too many factors that affect the working state of gas turbine. Thus, it is very difficult to build high-precision models. In addition, the data-driven approaches require sufficient samples to be obtained to locate the fault. Furthermore, the algorithm designer must know the fault generation mechanism and the relationship between these samples. All the above conditions are difficult to meet at the same time.

To avoid the problems mentioned above, and to locate the faults successfully, this paper proposes a fault location approach based on the sensitivity analysis of tested aerodynamic parameters. This approach belongs to the category of data-driven method and the faulty samples are not needed. Firstly, when the turbine is testing, collects the measured data in real-time. Then decomposes the measured data based on the kernel principal component analysis, constructs the Hotelling-T^2 (T^2) statistic, which is the application of the T-statistic in multivariate analysis in the principal space and squared prediction error (SPE) statistics in residual space after data decomposition. Further, the thresholds of statistics must be calculated, determining whether the fault occurs by comparing the relationship between the T^2 statistic and its threshold. If a fault occurs during detection, we calculate the partial derivatives of the T^2 and SPE statistics to the measured parameters. The greater the values of the partial derivatives, the greater the impacts of the measured parameters to the statistics. According to the working principle of gas turbines, it can be known that the parameters at the outlet of a component will fluctuate firstly and then the fluctuation spreads to other components if a component is faulty. The amplitude of the fluctuation at the outlet of the failed component is the greatest. Obviously, partial derivatives can be used to indicate the degree of influence of the measured parameters when a component fails.

2. Materials and Methods

Principal component analysis is a method of data processing which is suitable for linear system and transforms the correlated data into uncorrelated ones by a series of orthogonal changes. Gas turbine is a typical nonlinear system and great error may be caused if PCA is directly used to diagnose the fault of turbine. This paper adopts the kernel principal component analysis to detect the gas turbine fault. By using the kernel function, KPCA has strong nonlinear system processing ability [17,18]. The processes of KPCA are shown below [19–23].

For a given data sample collection, $x_1, x_2, x_3, \ldots, x_q \in R^n$, a nonlinear transformation $\varnothing$ maps the samples into a higher feature space F:

$$x \in R^n \xrightarrow{\varnothing} \varnothing(x) \in F \tag{1}$$

where $\varnothing(x)$ is the expression of samples in feature space. The covariance of $\varnothing(x)$ can be expressed as:

$$C_F = \frac{1}{n}\sum_{i=1}^{n}\varnothing(x_i)\varnothing(x_i)^T \qquad (2)$$

$$\lambda v = C_F v \qquad (3)$$

where λ is the eigenvalue of C_F and v is the eigenvector of C_F. Calculating the inner product of $\varnothing(x_i)$ with λv and $C_F v$ respectively:

$$\lambda(\varnothing(x_j)\cdot v) = \varnothing(x_j)\cdot C_F v \qquad (4)$$

The eigenvectors can be represented by a series of constants α_i, as follows:

$$v = \sum_{i=1}^{n}\alpha_i\varnothing(x_i) \qquad (5)$$

Combine Equations (2)–(5):

$$\begin{aligned}\lambda \sum_{i=1}^{n}\alpha_i[\varnothing(x_j)\cdot\varnothing(x_i)] &= \varnothing(x_j)\frac{1}{n}\sum_{i=1}^{n}\varnothing(x_i)\varnothing(x_i)^T\sum_{k=1}^{n}\alpha_k\varnothing(x_k)\\ &= \frac{1}{n}\sum_{k=1}^{n}[[\sum_{i=1}^{n}\varnothing(x_i)\varnothing(x_i)^T][\alpha_k\varnothing(x_j)\cdot\varnothing(x_k)]]\end{aligned} \qquad (6)$$

Simplify Equation (6) into Equation (7):

$$n\lambda\alpha = K\alpha \qquad (7)$$

In Equation (7), $K = [\varnothing(x_j)\cdot\varnothing(x_i)]$. K is a kernel function which calculates the inner product of vectors in high-dimensional feature space. To strengthen the ability of KPCA to deal with nonlinear problem, the Gauss radial basis function is adopted and its expression is:

$$K(x_i, x) = \exp[-\frac{\|x_i - x\|^2}{\sigma^2}] \qquad (8)$$

Normalize the eigenvectors v by Equation (8):

$$\langle v_k, v_k \rangle = 1 \qquad (9)$$

It can be seen that the vector α is normalized by Equation (9). Representing the mapped data in the feature space as t_k, there is:

$$t_k = \langle v_k, \varnothing(x)\rangle = \sum_{i=1}^{n}\alpha_i^k[\varnothing(x_i)\cdot\varnothing(x)] = \sum_{i=1}^{n}\alpha_i^k K(x_i, x) \qquad (10)$$

where α_i^k is the i-th coefficient of the k-th eigenvalue of matrix K to eigenvector. The cumulative contribution rate of variance is used to determine the number of principal components which mapped to the feature space. The calculation equation is as follow:

$$\frac{\sum_{k=1}^{l}\lambda_k}{\sum_{k=1}^{n}\lambda_k} > \varepsilon \qquad (11)$$

where l is the number of principal component and ε is a constant. The value of ε reflect the influence of noise. Usually, the value of ε is between 0 and 1.

Equations (2)–(11) are the steps to conduct the kernel principal component analysis. To achieve the fault detection of gas turbine components, the statistics of T^2 and SPE must be constructed, as shown below:

$$T^2 = [t_1, t_2, \ldots, t_l]\Lambda^2[t_1, t_2, \ldots, t_l]^T \qquad (12)$$

where Λ is a diagonal matrix consisting of principal component eigenvalues. t_k is the mapped data of samples in the feature space.

$$T_{th}^2 = \frac{1(n^2-1)}{n(n-1)} F_\alpha(1, n-1) \tag{13}$$

T_{th}^2 is the threshold of the T^2 statistic and $F_\alpha(1, n-1)$ is upper limit value of F-ditribution with confidence level α:

$$SPE = \|\varnothing(x) - \varnothing_k(x)\|^2 = \sum_{i=1}^{n} t_i^2 - \sum_{i=1}^{1} t_i^2 \tag{14}$$

$$SPE_{th} = \theta_1 \left[\frac{C_a \left(2\theta_2 h_0^2\right)^{\frac{1}{2}}}{\theta_1} + 1 + \frac{\theta_2 h_0 (h_0 - 1)}{\theta_1^2} \right]^{1/h_0} \tag{15}$$

$$\theta_1 = \sum_{i=k+1}^{n} \lambda_i, \theta_2 = \sum_{i=k+1}^{n} \lambda_i^2, \theta_3 = \sum_{i=k+1}^{n} \lambda_i^3, h_0 = 1 - 2\theta_1\theta_2/\left(3\theta_2^2\right) \tag{16}$$

Based on the KPCA introduced from Equations (2)–(15), a fault diagnosis algorithm is designed to determine whether the turbine component is fault. The tested parameters include all the values of total temperature and total pressure at the outlet of gas turbine components. Decomposing these parameters by kernel principal component analysis, construct the T^2 statistic, SPE statistic, and their corresponding thresholds. Determine whether a fault occurs by comparing the relationship of the T^2 statistic and its threshold.

This section focuses on how to locate the fault when the failure occurs. By calculating the partial derivatives of statistics T^2 and SPE to the tested parameters, the sensitivity of tested parameters to the fault can be expressed, and the location of the fault can be determined according to the sensitivity. For the T^2 statistic, the greater the value of sensitivity is, the more likely it is the location of the fault. Kernel function analysis is the most important step in sensitivity calculation, so we make the following changes to the kernel function:

$$K(x_i, x_j) = \exp[-\|x_i - x_j\|^2/\sigma^2] = K(v \cdot x_i, v \cdot x_j) = \exp[-\|v \cdot x_i - v \cdot x_j\|^2/\sigma^2] \tag{17}$$

$v = [v_1, v_2, \ldots, v_n], v_i = 1$, n is the number of categories of measured parameters, x_i is the i-th measurement vector consisting of different measured parameters. Calculating the partial derivative of kernel function to v_k, there is:

$$\begin{aligned}\frac{\partial K(x_i, x_j)}{\partial v_k} = \frac{\partial K(v \cdot x_i, v \cdot x_j)}{\partial v_k} &= -\frac{1}{\sigma^2}\left(x_{i,k} - x_{j,k}\right)^2 K(x_i, x_j) \\ &= -\frac{1}{\sigma^2}\left(v_k \cdot x_{i,k} - v_k \cdot x_{j,k}\right)^2 K(v \cdot x_i, v \cdot x_j)\end{aligned} \tag{18}$$

The value of partial derivative indicates the effect of parameters to kernel function. $x_{j,k}$, $x_{i,k}$ are the k-th elements of the i-th and j-th measured parameters. The partial derivative of the product between kernels can be expressed as:

$$\frac{\partial K(x_i, x_{new}) K(x_j, x_{new})}{\partial v_k} = -\frac{1}{\sigma^2}\left[\left(x_{i,k} - x_{new,k}\right)^2 + \left(x_{j,k} - x_{new,k}\right)^2\right] \times K(x_i, x_{new}) K(x_j, x_{new}) \tag{19}$$

x_{new} is a vector consisting of measured parameters. Define the partial derivatives of statistics as $C_{T^2,i,new}$ and $C_{SPE,i,new}$, there are:

$$C_{T^2,i,new} = \left|\frac{\partial T_{new}^2}{\partial v_k}\right| C_{SPE,i,new} = \left|\frac{\partial SPE_i}{\partial v_k}\right| \tag{20}$$

The values of $C_{T^2,i,new}$ and $C_{SPE,i,new}$ indicate the sensitivity level of the i-th element of the statistics. Steps of calculate the $C_{T^2,i,new}$ are as follows:

$$T^2 = [t_1, t_2, \ldots, t_k] \Lambda^{-1} [t_1, t_2, \ldots, t_k]^T = \overline{K}_{new}^T \alpha \Lambda^{-1} \alpha^T \overline{K}_{new} = tr(\alpha^T \alpha^T \overline{K}_{new} \overline{K}_{new}^T \alpha \Lambda^{-1}) \quad (21)$$

$$\overline{K}_{new} = \begin{matrix} K(x_1, x_{new}) - \frac{1}{n}\sum_{j=1}^{n} K(x_1, x_j) - \frac{1}{n}\sum_{j=1}^{n} K(x_{new}, x_j) + \frac{1}{n^2}\sum_{j=1}^{n}\sum_{j=1}^{n} K(x_j, x_j) \\ K(x_2, x_{new}) - \frac{1}{n}\sum_{j=1}^{n} K(x_2, x_j) - \frac{1}{n}\sum_{j=1}^{n} K(x_{new}, x_j) + \frac{1}{n^2}\sum_{j=1}^{n}\sum_{j=1}^{n} K(x_j, x_j) \\ \ldots \\ K(x_n, x_{new}) - \frac{1}{n}\sum_{j=1}^{n} K(x_2, x_j) - \frac{1}{n}\sum_{j=1}^{n} K(x_{new}, x_j) + \frac{1}{n^2}\sum_{j=1}^{n}\sum_{j=1}^{n} K(x_j, x_j) \end{matrix} \quad (22)$$

There is:

$$C_{T^2,i,new} = \left| \frac{\partial T_{new}^2}{\partial v_k} \right| = \left| \frac{\partial(tr(\alpha^T \overline{K}_{new} \overline{K}_{new}^T \alpha \Lambda^{-1}))}{\partial v_k} \right| = \left| tr[\alpha^T (\frac{\partial \overline{K}_{new} \overline{K}_{new}^T}{\partial v_k}) \alpha \Lambda^{-1}] \right| \quad (23)$$

The calculation steps of SPE are as follows:

$$SPE = K(x_{new}, x_{new}) - \frac{2}{n}\sum_{i=1}^{n} K(x_i, x_{new}) + \frac{1}{n^2}\sum_{i=1}^{n}\sum_{i=1}^{n} K(x_i, x_i) - t_{new}^T t_{new} \quad (24)$$

$$\begin{aligned} C_{SPE,i,new} &= \left| \frac{\partial SPE}{\partial v_i} \right| = \left| -\frac{1}{\sigma^2}[-\frac{2}{n}\frac{\partial \sum_{i=1}^{n} k(x_i, x_{new})}{\partial v_i} - \frac{\partial t_{new}^T t_{new}}{\partial v_i}] \right| \\ &= \left| \frac{1}{\sigma^2}[\frac{2}{n}\frac{\partial \sum_{i=1}^{n} k(x_i, x_{new})}{\partial v_i} + \frac{\partial t_{new}^T t_{new}}{\partial v_i}] \right| \\ &= \left| \frac{1}{\sigma^2}[\frac{2}{n}\frac{\partial \sum_{i=1}^{n} k(x_i, x_{new})}{\partial v_i} + tr(\alpha^T \frac{\partial t_{new}^T t_{new}}{\partial v_i} \alpha)] \right| \end{aligned} \quad (25)$$

Figure 1 shows an algorithm for fault diagnosis and location based on above research.

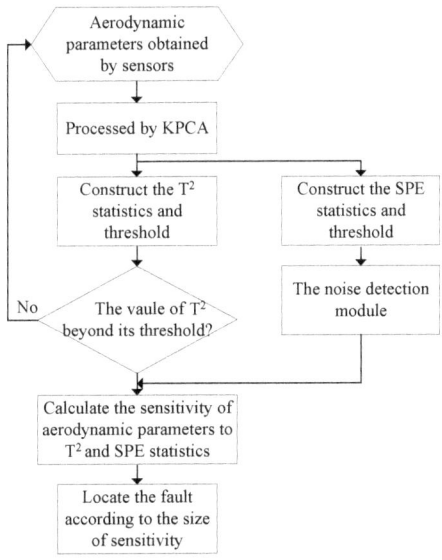

Figure 1. Process of fault diagnosis and location.

3. Results

In order to verify the effectiveness of the method proposed in this paper, certain of twin-spool aviation gas turbine is adopted as the research object. It is widely known that the working condition of the engine is very bad [13,24] (suffering from high pressure, high temperature, high stress, etc.) and the performances of gas turbine components (such as the compressor, rotator, turbine, etc.) are decreasing as working hours increase [25–30]. The initial working parameters of this turbine are shown in Table 1. When these working parameters are determined, the state parameters of the engine are shown in Figures 2 and 3 when the flight altitude and speed are different.

Table 1. Initial working parameters of turbine.

Efficiency of LPC	0.868	Pressure ratio of LPC	3.8
Efficiency of HPC	0.878	Pressure ratio of HPC	4.474
Fuel low calorific value	42,900	Total temperature of combustor outlet	1600 K
Efficiency of HPR	0.98	Efficiency of LPR	0.98
Combustion chamber efficiency	0.98	Engine room air entrainment coefficient	0.01
Cooling parameter of HPT	0.03	Efficiency of HPT	0.89
Cooling parameter of LPT	0.01	Efficiency of LPT	0.91
Design speed of LPR	10,000 (r/m)	Design Speed of HPR	16,000 (r/m)

LPC—Low Pressure Compressor; HPC—High Pressure Compressor; HPR—High Pressure Rotor; LPR—Low Pressure Rotor; HPT—High Pressure Turbine; LPT—Low Pressure Turbine.

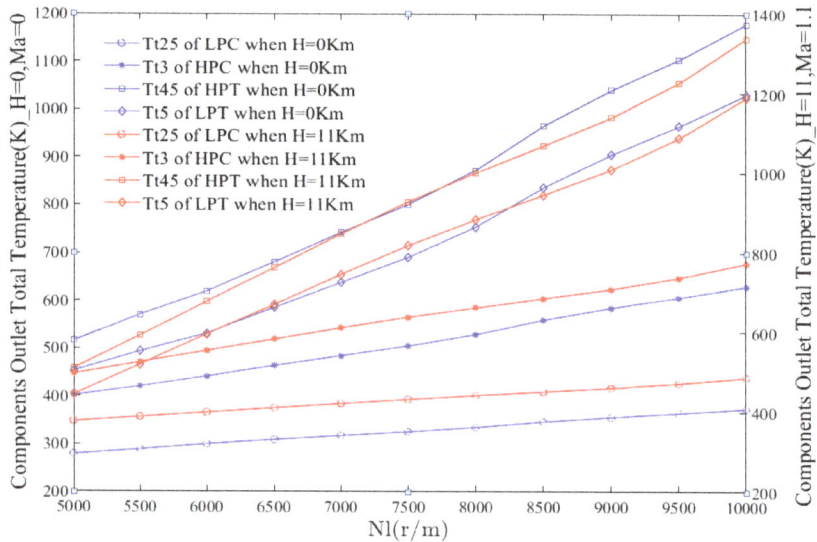

Figure 2. Total temperatures of components.

In the experiment, the measured parameters include the total temperature at the outlets of low pressure compressor LPC (Tt25), total temperature of how pressure compressor HPC (Tt3), total temperature of the high pressure turbine HPT (Tt45), and total temperature of the low pressure turbine LPT (Tt5). In addition, the total pressure at the outlets of the low pressure compressor LPC (Pt25), total pressure of the high pressure compressor HPC (Pt3), total pressure of the high pressure turbine HPT (Pt45), and the total pressure of the low pressure turbine LPT (Pt5) are included. Two faults occurred at the 2600th sampling moment: one is the misalignment of the LPC rotor, and another one is the crack generation of the LPT blade. The proposed method is adopted to detect and locate the faults. Figures 4–7 are the diagrams of fault diagnosis. At the 2600th sampling time, the faults of the LPC and LPT are generated, respectively. In Figure 2, the value of the T^2 statistic is smooth and lower than its threshold before the 2600th sampling time. Due to the occurrence of fault at the 2600th sample, the curve takes a large jump and exceeds its threshold. In Figure 5, the SPE statistic approaches the threshold at some time before the occurrence of fault. Since the SPE statistic mainly contains noise information, KPCA processing cannot eliminate the noise completely. When the noise amplitude increases, the value of the SPE statistic may exceed its threshold, which has been introduced in Equations (13) and (15). This does not affect the fault diagnosis of the components. Figures 6 and 7 show the fault detection of the LPT and the detection results are similar with those of the LPC. The fault location algorithm mentioned above is used to locate the fault of the LPC, HPC, LPT, and HPT. The location of the results are shown in Table 2.

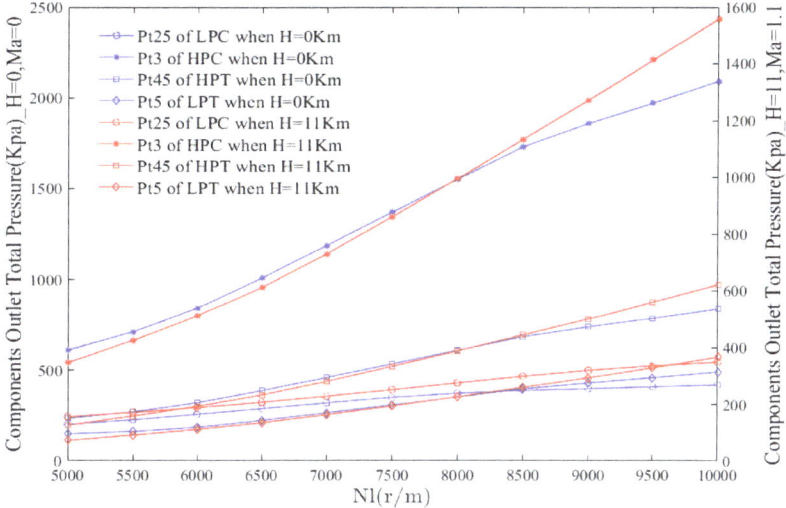

Figure 3. Total pressures of components.

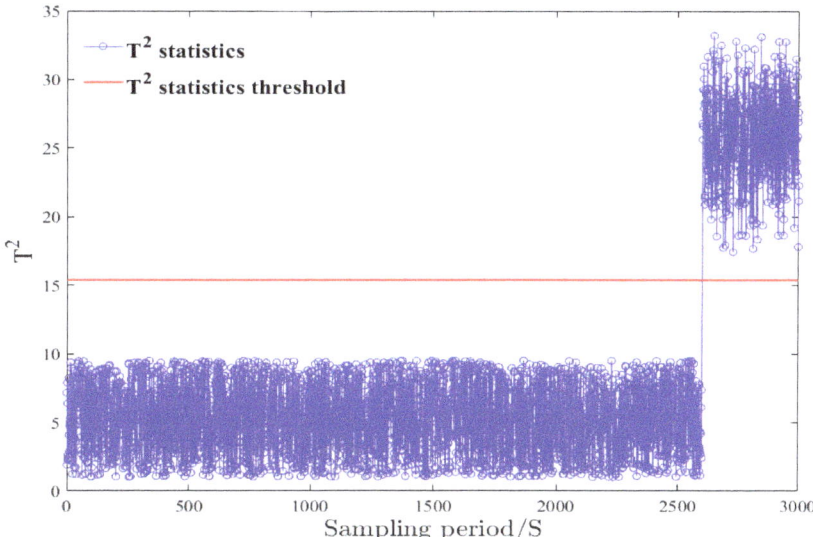

Figure 4. Value of the T^2 statistic under the condition of an LPC (low pressure compressor) fault.

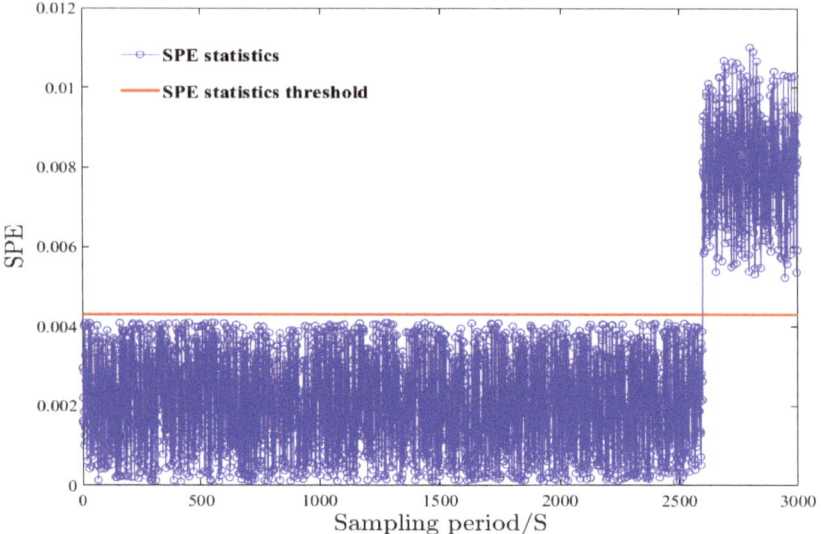

Figure 5. Value of SPE statistic under the condition of LPC fault.

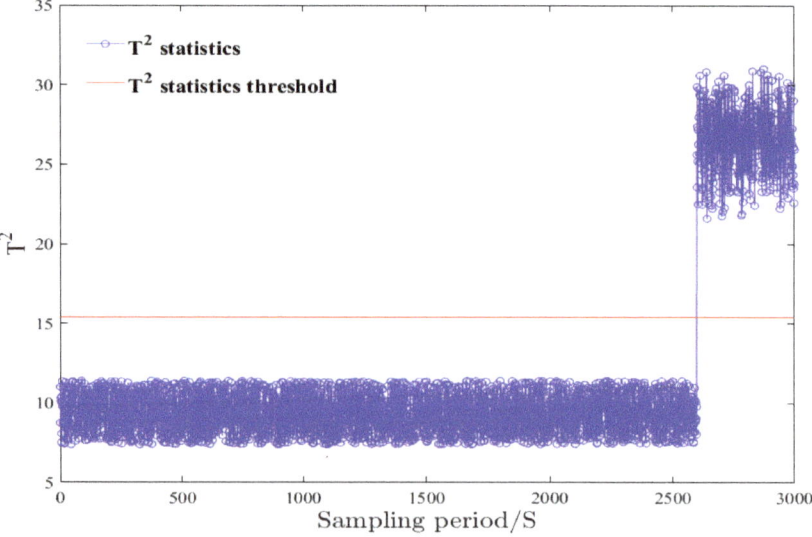

Figure 6. Value of the T^2 statistic under the condition of an LPT fault.

Table 2 shows that when any part of gas turbine components is fault, the sensitivity of measured parameters of faulty part to the statistics of T^2 and SPE is greater than that of normal ones. Take LPC as an example to illustrate the result. If the efficiency coefficient of LPC decreased by 1% due to the misalignment of LPC rotor, the measured parameters at the outlet of the LPC fluctuates firstly. Sensitivities of total temperature and total pressure at the outlet of LPC measured by the sensors to T^2 statistic are 0.3239 and 0.6271, which are obviously higher than those of other measured parameters. The sensitivities to the SPE statistic are both 0.125, which are also higher than those of other parameters. The fault location method can locate the fault to the each component. Figure 8 shows the sensitivity distribution spectrums of the measured parameters when the gas turbine is working.

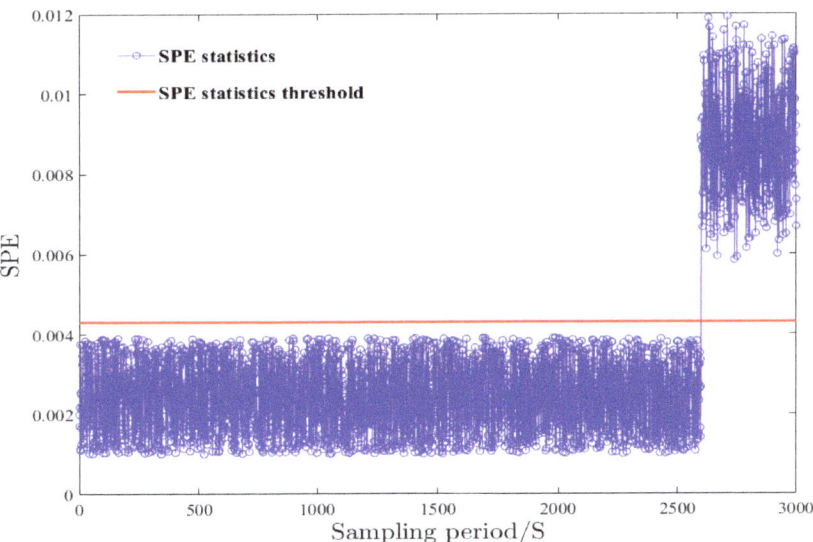

Figure 7. Value of the SPE (Squared Prediction Error) statistic under the condition of an LPT fault.

Table 2. The sensitivity of measured parameters.

	Sensitivity	F_{LPC}	F_{HPC}	F_{LPT}	F_{HPT}
	$C_{T^2\_T25}$	0.3239	0.0397	0.037	0.0279
	$C_{T^2\_T3}$	0.0045	0.6043	0.0215	0.0133
	$C_{T^2\_T45}$	0.0144	0.0789	0.0584	0.1764
	$C_{T^2\_T5}$	0.005	0.0248	0.2073	0.0177
	$C_{T^2\_P25}$	0.6271	0.063	0.0554	0.0448
	$C_{T^2\_P3}$	0.0042	0.1014	0.0246	0.0195
Sensitivity of	$C_{T^2\_P45}$	0.0156	0.0739	0.0713	0.6866
measured	$C_{T^2\_P5}$	0.0049	0.014	0.5244	0.0138
parameters to	C_{SPE_T25}	0.1250	0.1249	0.1249	0.1249
statistics	C_{SPE_T3}	0.1249	0.125	0.1249	0.1249
	C_{SPE_T45}	0.1249	0.1249	0.1249	0.125
	C_{SPE_T5}	0.1249	0.1249	0.125	0.1249
	C_{SPE_P25}	0.1250	0.1249	0.1249	0.1249
	C_{SPE_P3}	0.1249	0.125	0.1249	0.1249
	C_{SPE_P45}	0.1249	0.1249	0.1249	0.125
	C_{SPE_P5}	0.1249	0.1249	0.1245	0.1249

It can be seen that before the fault occurs in this figure, the sensitivity distribution curves are gentle and the differences between the sensitivity curves are not obvious. When a low pressure compressor failure occurs, the curves representing the sensitivity of the LPC increased sharply in a short time, and the values are significantly higher than others. In addition, according to the rule of failure caused by the degradation of gas turbine components, if the performance of a component degrades to a certain extent and is about to fail, the degradation speed will be accelerated until the failure occurs. In this process, the measured parameters at the outlet of deteriorating components will deviate from the real value as the deterioration of performance. In the sensitivity distribution spectrum, the sensitivity of the measured values of the deteriorating components will increase continuously, and the fault prediction can be realized by comparing the changes in sensitivity.

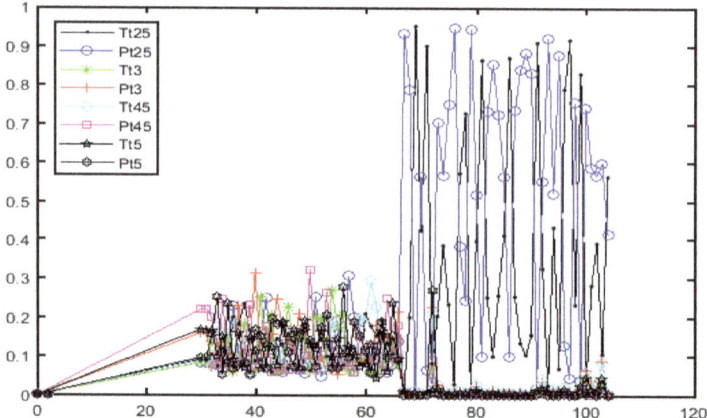

Figure 8. Sensitivity distribution spectra of the tested parameters.

In addition, due to the influence of harsh working circumstance, the sensor outputs may seriously deviate from their actual values and this may lead to the misdiagnosis. It is essential to keep output within a reasonable range. According to the working principle of gas turbines (taking an aero gas engine as an example), the power and flow balance conditions must be observed when the turbine is under normal working conditions and all parameters remain unchanged or fluctuate in a small range. If the state of turbine changes due to the variation of control parameters, all the aerodynamic parameters will bound to change greatly, reflecting an anomaly of sensor measurements. Another case is that if only a few measurements are abnormal, according to the working principle of gas turbines and the balance conditions, it can be known that the anomalies are caused by noise or the fault of the sensors and the measurements must be restored. The process to detect the abnormal value and restore the measurements is shown in Figure 9.

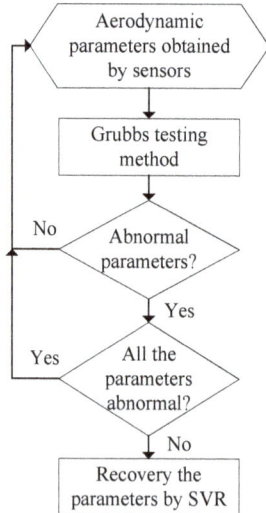

Figure 9. Processing steps for outliers.

Firstly, Grubb's method for testing is adopted to check if the parameters of sensors are abnormal or not. It is important to note that the value of detection level α must be determined according to the variation of the aerodynamic parameters. If the checked parameters are abnormal, all the parameters should be tested by Grubb's method. Then count the number of parameters which are abnormal. If the number is less than the quantity of aerodynamic parameters for fault diagnosis, it indicates that the sensors are faulty or the influence of noise increases. In such a situation, the abnormal parameters must be restored by a support vector machine (SVR) [31] to ensure the fault location proceeds smoothly.

4. Conclusions

In this paper, a novel method to locate the fault of a gas turbine is proposed. Kernel principal component analysis is adopted to detect the occurrence of fault. Based on the analysis of the fault indicator and aerodynamic parameters, the partial derivative of the T^2 and SPE statistics to aerodynamic parameters are calculated. The results are used to represent the influence degree of fault to these parameters and the fault location can be realized by different influence degrees. There are four conclusions can be drawn, as follows:

(1) The KPCA is an effective way to detect the fault of a gas turbine. T^2 and SPE statistics, and their corresponding thresholds must be constructed. By comparing the size of the T^2 statistic and its threshold, the on-line fault diagnosis of a gas turbine can be realized. Furthermore, both statistics help to locate the fault.

(2) The fault location of gas turbine is realized by calculate the partial derivatives of T^2 to aerodynamic parameters. The size of partial derivatives represent the sensitivity degrees of aerodynamic parameters to fault. Based on the balance working condition of gas turbine, fault location can be achieved according to the size of partial derivatives.

(3) Sensitivity distribution spectra can be used to represent the performance degradation of the gas turbine and identify a potential fault. When a fault which resulted by performance degradation is about to occur, the partial derivatives of the aerodynamic parameters associated with this fault will change dramatically and this change is easily reflected by the sensitivity distribution spectra.

(4) The method proposed in this paper is used to locate the fault of a gas turbine and how to recognize the fault is not discussed. Whether this method can be used in fault identification needs to be verified in the follow-up work.

Author Contributions: Data curation: L.Z.; funding acquisition: W.L.; methodology: L.Z.; project administration: W.L. and Y.L.; supervision: Y.L.; writing—original draft, L.Z.; writing—review and editing: L.Z.

Funding: This research was funded by the National Natural Science Foundation of China, grant number 51875371 and National Green Manufacturing System Plan Number: [2017] 327.

Conflicts of Interest: The authors declare no conflict of interest.

References

1. Siddiqui, M.E.; Almitani, K.H. Energy Analysis of the S-CO$_2$ Brayton Cycle with Improved Heat Regeneration. *Processes* **2019**, *7*, 3. [CrossRef]
2. Lee, H.G.; Huh, J.H. A Cost-Effective Redundant Digital Excitation Control System and Test Bed Experiment for Safe Power Supply for Process Industry 4.0. *Sensors* **2018**, *7*, 85. [CrossRef]
3. Lu, F.; Ju, H.F.; Huang, J.Q. An improved extended Kalman filter with inequality constraints for aero engine health monitoring. *Aerosp. Sci. Technol.* **2016**, *58*, 36–47. [CrossRef]
4. Zeng, L.; Long, W.; Li, Y.Y. Performance Degradation Diagnosis of Gas Turbine Based on Improved FUKF. *J. Southwest Univ.* **2018**, *53*, 873–878. [CrossRef]
5. Gao, Z.W.; Ding, S.X.; Cecati, C. Real-time fault diagnosis and Fault-toulent Control. *IEEE Trans. Ind. Electron.* **2015**, *62*, 3752–3756. [CrossRef]
6. Gao, Z.W.; Sheng, S.W. Real-time monitoring, prognosis, and resilient control for wind turbine systems. *IEEE Trans. Ind. Electron.* **2018**, *116*, 1–4. [CrossRef]
7. Simani, S.; Farsoni, S.; Castaldi, P. Fault Diagnosis of a Wind Turbine Benchmark via Identified Fuzzy Models. *IEEE Trans. Ind. Electron.* **2015**, *62*, 3775–3782. [CrossRef]

8. Sanchez, S.; Escobet, T.; Puig, V.; Odgaard, P.F. Fault Diagnosis of an Advanced Wind Turbine Benchmark Using Interval-Based ARRs and Observers. *IEEE Trans. Ind. Electron.* **2015**, *62*, 3783–3792. [CrossRef]
9. Gao, Z.; Cecati, C.; Ding, S.X. A survey of fault diagnosis and fault-tolerant techniques—Part I: Fault diagnosis with model-based and signal-based approaches. *IEEE Trans. Ind. Electron.* **2015**, *62*, 3757–3767. [CrossRef]
10. Gao, Z.; Cecati, C.; Ding, S.X. A Survey of Fault Diagnosis and Fault-tolerant Techniques—Part II: Fault diagnosis with knowledge-based and hybrid/active approaches. *IEEE Trans. Ind. Electron.* **2015**, *62*, 3768–3774. [CrossRef]
11. Zhang, Y.; Bingham, C.M.; Gallimore, M. Fault Detection and Diagnosis Based on Extensions of PCA. *Adv. Mil. Technol.* **2013**, *8*, 1–15. [CrossRef]
12. Zhu, J.L.; Ge, Z.Q.; Song, Z.H. HMM-Driven Robust Probabilistic Principal Component Analyzer for Dynamic Process Fault Classification. *IEEE Trans. Ind. Electron.* **2015**, *62*, 3814–3821. [CrossRef]
13. Zhang, P. Aero Engine Fault Diagnostics Based on Kalman Filter. Ph.D. Thesis, Nanjing University of Aeronautics ans Astronautics, Nanjing, China, 2008.
14. Vasile, P.; Ron, J.P.; Faisel, J.U.; Joseba, Q. Fault Diagnosis of An Industrial Gas Turbine Using Neuro-fuzzy Methods. *World Congr.* **2002**, *35*, 1630–1636. [CrossRef]
15. Che, C.C.; Wang, H.W.; Ni, X.M.; Hong, J.Y. Fault Fusion Diagnosis of Aero-engine Based on Deep Learning. *J. Beijing Univ. Aeronaut. Astronaut.* **2018**, *44*, 621–628. [CrossRef]
16. Tayarani-Bathaie, S.S.; Khorasani, K. Fault Detection and Isolation of Aero Engines Using a Bank of Neural Networks. *Process Control* **2015**, *36*, 22–41. [CrossRef]
17. Lei, W.; Zhang, J.X.; Xie, S.S.; Ren, L.T.; Hu, J.H. Fault Estimation for Actuators in Aero engine Control System Based on Fault Decomposition. *Fire Control Command Control* **2014**, *39*, 127–130.
18. Fuat, D.; Jose, A.R.; Ahmet, P. A strategy for Detection and Isolation of Sensor Failures and Process Upsets. *Chemom. Intell. Lab. Syst.* **2001**, *55*, 109–123.
19. Wold, S. Exponentially weighted moving principal component analysis and projection to latent structures. *Chemom. Intell. Lab. Syst.* **1994**, *21*, 149–161. [CrossRef]
20. Navi, M.; Meskin, N.; Davoodi, M. Sensor Fault Detection and Isolation of An Industrial Gas Turbine Using Partial Adaptive KPCA. *J. Process Control* **2015**, *2*, 37–48. [CrossRef]
21. Chen, S.G.; Wu, X.J.; Yin, H.F. KPCA Method Based on Within-class Auxiliary Training Samples and Its Application to Pattern Classification. *Pattern Anal. Appl.* **2016**, *3*, 749–767. [CrossRef]
22. Alain, R. Variable Selection Using SVM-based Criteria. *J. Mach. Learn. Res.* **2003**, *3*, 1357–1370. [CrossRef]
23. Jiang, W.L.; Wu, S.X.; Liu, S.Y. Exponentially Weighted Dynamic Kernel Principal Component Analysis Algorithm and Its Application in Fault Diagnosis. *J. Mech. Eng.* **2011**, *47*, 63–67. [CrossRef]
24. Zhang, K.Y.; Li, Z.G.; Li, J. Effect of gas turbine endwall misalignment with slashface leakage on the blade endwall aerothermal performance. *Int. J. Heat Mass Transf.* **2018**, *131*, 1247–1259. [CrossRef]
25. Dewallef, P.; Romessis, C.; Leonard, O.; Mathioudakis, K. Combining classification techniques with Kalman filters for aircraft engine diagnostics. *J. Eng. Gas Turbines Power* **2006**, *128*, 595–603. [CrossRef]
26. Huang, M.Y.; Dey, S. Stability of Kalman filtering with Markovian packet losses. *Automatica* **2007**, *43*, 598–607. [CrossRef]
27. Chen, Y. Gas Path Fault Diagnosis for Turbojet Engine Based on Nonlinear Model. Ph.D. Thesis, Nanjing University of Aeronautics and Astronautics, Nanjing, China, 2014.
28. Zhang, P.; Huang, J.Q. SRUKF Research on Aero engine for Gas Path Component Fault Diagnosis. *J. Aeros. Power* **2008**, *23*, 169–172.
29. Kurzke, J. Gas turbine cycle design methodology: A comparison of parameter variation with numerical optimization. *J. Eng. Gas Turbines Power* **1999**, *121*, 6–11. [CrossRef]
30. Lu, F.; Huang, J.Q. Contracted Kalman Filter Estinmation for Turbofan Engine Gas-path Health. *Control Theory Appl.* **2012**, *29*, 1543–1550. [CrossRef]
31. Zeng, L.; Long, W. Prediction Method of Metal Content of Aero Engine Lubricating Oil Based on SVR. *J. Sichuan Univ.* **2016**, *48*, 161–164. [CrossRef]

© 2019 by the authors. Licensee MDPI, Basel, Switzerland. This article is an open access article distributed under the terms and conditions of the Creative Commons Attribution (CC BY) license (http://creativecommons.org/licenses/by/4.0/).

Article

Rare Event Chance-Constrained Optimal Control Using Polynomial Chaos and Subset Simulation

Patrick Piprek [1,*,†,‡], **Sébastien Gros** [2,‡] **and Florian Holzapfel** [1]

1. Institute of Flight System Dynamics, Technical University of Munich, 85748 Garching bei München, Germany; florian.holzapfel@tum.de
2. Department of Eng. Cybernetics, Norwegian University of Science and Technology, 7034 Trondheim, Norway; sebastien.gros@ntnu.no
* Correspondence: patrick.piprek@tum.de; Tel.: +49-89-289-16530
† Current address: Boltzmannstrasse 15, 85748 Garching bei München, GER.
‡ These authors contributed equally to this work.

Received: 19 February 2019; Accepted: 26 March 2019; Published: 30 March 2019

Abstract: This study develops a chance–constrained open–loop optimal control (CC–OC) framework capable of handling rare event probabilities. Therefore, the framework uses the generalized polynomial chaos (gPC) method to calculate the probability of fulfilling rare event constraints under uncertainties. Here, the resulting chance constraint (CC) evaluation is based on the efficient sampling provided by the gPC expansion. The subset simulation (SubSim) method is used to estimate the actual probability of the rare event. Additionally, the discontinuous CC is approximated by a differentiable function that is iteratively sharpened using a homotopy strategy. Furthermore, the SubSim problem is also iteratively adapted using another homotopy strategy to improve the convergence of the Newton-type optimization algorithm. The applicability of the framework is shown in case studies regarding battery charging and discharging. The results show that the proposed method is indeed capable of incorporating very general CCs within an open–loop optimal control problem (OCP) at a low computational cost to calculate optimal results with rare failure probability CCs.

Keywords: robust open-loop optimal control; generalized polynomial chaos; chance constraints; subset simulation; open-loop optimal control; battery charge–discharge

1. Introduction

In the context of open–loop optimal control (OC), the calculation of robust trajectories, i.e., trajectories that remain safe despite model uncertainties, is crucial for safety-critical applications. This type of problem is often treated by a chance constraint (CC) formulation, which must generally be approached via sampling techniques. Here, the method of generalized polynomial chaos (gPC), introduced by Xiu and Karniadakis in 2002 [1], allows for calculating arbitrarily good approximations of the system response due to uncertainties, which can consequently be used to generate samples of the system trajectories.

In this work, we introduce gPC in open–loop optimal control problems (OCPs) with CCs, combined with a subset simulation (SubSim) sampling technique, to calculate trajectories subject to probabilistic constraints imposing very rare failure events. To make the CC formulation applicable for Newton-type optimization algorithms, a differentiable approximation of the indicator function is used. Additionally, a homotopy strategy is implemented to gradually approximate the exact CC failure domain.

The formulation of OCPs with uncertainties by chance–constrained open–loop optimal control (CC–OC) techniques is a commonly used approach, although CC–OC can be computationally expensive and difficult to solve. The following research has been conducted within the field of CC–OC: In [2], a general overview of different methods for handling CCs is given. The author introduces both analytical (e.g., ellipsoid relaxation) and sampling-based methods (e.g., mixed integer programming). Both methods are subsequently combined in a hybrid approach. Additionally, feedback control is used to satisfy system constraints. In [3], a strategy to approximate a CC based on split Bernstein polynomials is introduced. Here, pseudo-spectral methods are applied and a single optimization run is used on the transformed CC–OC. Therefore, joint CCs are decomposed and a Markov–chain Monte–Carlo (MCMC) algorithm is used to evaluate the samples.

In general, CCs are also common in model predictive control (MPC) applications [4]. A very popular choice for this robust MPC are so-called min-max algorithms, which try to achieve a worst-case design in the presence of uncertainties to increase the robustness [5–7]. As online applicability is very important in MPC, CC algorithms in MPC often try to transform CCs in algebraic constraints [8]. Further methods comprise maximizing the feasible set with CCs [9] or randomization [10]. It should be noted that none of these methods is specifically tailored to treat rare events, which is desired in this study. These rare events are of special importance in reliability engineering and safety critical applications and are thus very prominent in the engineering domain. In addition, the developed method in this study should not be too conservative as it would e.g., be the case with transformations to algebraic constraints.

In order to deal with rare events, we use the method of SubSim, proposed in [11,12]. This methodology begins the probability estimation with a general Monte–Carlo analysis (MCA) solution and then gradually explores different samples within the failure domain. This MCMC algorithm converges to a series of conditional probabilities that yield the failure probability of the rare event. In the context of CC–OC, we use the samples generated from the MCMC as evaluation samples for the solution of the OCP. Here, we use a homotopy strategy to adapt the samples after each OCP solution until the SubSim, as well as the OCP, fulfill the desired rare-event failure probability.

To give an overview of the development of a CC–OC framework, the paper is organized as follows. In Section 2, some theoretical background and fundamentals of OC and gPC are introduced. Section 3 introduces the proposed incorporation of CCs in the OCP and the combination with the gPC expansion and SubSim. The model for the CC–OC case studies is presented in Section 4, while the results are given in Section 5. Conclusive remarks and an outlook are looked at in Section 6.

2. Theoretical Background

This section gives an overview of the methods used within this paper as well as some characteristics of their implementation. Here, Section 2.1 introduces the general OCP formulation, while Section 2.2 gives an overview of the gPC method and how to calculate statistics for the OCP from it.

2.1. Open-Loop Optimal Control

The OCP in this paper is given as follows (extended form of [13]):

$$\min_{\mathbf{x}, \mathbf{u}, \mathbf{p}, t_f} J = e\left(\mathbf{x}\left(t_f\right), \mathbf{u}\left(t_f\right), \mathbf{p}, t_f\right) + \int_{t_0=0}^{t_f} L(\mathbf{x}, \mathbf{u}, \mathbf{p}) \, dt,$$

$$\text{s.t.} \quad \mathbf{c}_{lb}(\mathbf{x}, \mathbf{u}, \mathbf{p}) \leq \mathbf{c}(\mathbf{x}, \mathbf{u}, \mathbf{p}) \leq \mathbf{c}_{ub}(\mathbf{x}, \mathbf{u}, \mathbf{p}),$$
$$\mathbf{f}(\mathbf{x}, \mathbf{u}, \mathbf{p}; \theta) = \dot{\mathbf{x}},$$
$$\boldsymbol{\psi}(\mathbf{x}, \mathbf{u}, \mathbf{p}) = \mathbf{0},$$
$$\mathbb{P}_{OCP}(\mathbf{y} \notin \mathcal{F}) \geq \eta$$

(1)

In Equation (1), the lower and upper bounds of box constraints are denoted by lb and ub respectively and the output variables $\mathbf{y} \in \mathbb{R}^{n_y}$ are defined by the following nonlinear function:

$$\mathbf{y} = \mathbf{g}(\mathbf{x}, \mathbf{u}, \mathbf{p}) \quad (2)$$

It should be noted that we deliberately distinguish between probabilistic and deterministic ("hard") constraints in Equation (1). This is done due to e.g., the fact that the state integration is generally carried out in the deterministic rather than the probabilistic domain (in our case, we use specific evaluation nodes provided by the gPC theory), while we also have constraints that are specifically designed in the probabilistic domain (i.e., our CCs).

The optimization/decision variables of the OCP include the states of the system $\mathbf{x} \in \mathbb{R}^{n_x}$, the controls $\mathbf{u} \in \mathbb{R}^{n_u}$, the time-invariant parameters $\mathbf{p} \in \mathbb{R}^{n_p}$ (these might be design parameters of the model, e.g., a surface area or a general shape parameter), and the final time $t_f \in \mathbb{R}$. We combine these variables within the vector $\mathbf{z} = \left[t_f, \mathbf{p}^T, \mathbf{x}^T, \mathbf{u}^T\right]^T$. The external parameters $\theta \in \mathbb{R}^{n_\theta}$ are considered uncertain, but of known probability density function (pdf). The set $\mathcal{F}$, labeled *failure set* hereafter, is the set of states, controls, and parameters, i.e., the outputs, which lead to a failure of the system. Note that Equation (1) is the probability to not hit the failure set with the desired probability. This choice creates a better conditioned nonlinear programming problem (NLP) within the Newton-type optimization. In this paper, we assume that the probability $\eta = \mathbb{P}_{des}(\mathbf{y} \notin \mathcal{F})$ of not encountering a failure is selected arbitrarily close to 1.

It should be noted that we can assume without loss of generality that the initial time t_0 is zero. The objective is to minimize the cost functional J consisting of the final time cost index e and the running cost index L. The OCP is subject to the following constraints:

- the state dynamics $\dot{\mathbf{x}}$ that ensure a feasible trajectory,
- the inequality path and point constraints $\mathbf{c}$ that ensure limits of the trajectory to be feasibly enforced by box constraints (i.e., by lower and upper bound)
- the equality path and point constraints ψ that ensure a specific condition during the flight, e.g., the initial and final state condition.

Generally, when the state dimension is not trivially small, OCPs as in Equation (1) are best solved using direct methods. Direct methods first discretize the problem into a NLP, which is then solved by classic NLP solvers. In the following, we use the trapezoidal collocation method for the discretization [13], which is readily implemented in the OC software FALCON.m [14]. This software tool is also used to implement the proposed CC–OC approach. Furthermore, the primal-dual interior-point solver Ipopt [15] is used to solve the discretized NLP.

2.2. Generalized Polynomial Chaos

This section gives an overview on the gPC method. Here, Section 2.2.1 introduces the basics of the gPC method. The calculation of the statistical moments is then presented in Section 2.2.2.

2.2.1. Definition of Expansion and Incorporation in the Optimal Control Problem

The gPC method was originally developed by Xiu and Karniadakis in 2002 [1] and is an extension of the Wiener polynomial chaos, which was only valid for Gaussian uncertainties [16]. It can be construed as a Fourier-like expansion with respect to the uncertain parameters, which approximates the response of the output variables **y** and reads as follows (it is reminded that the output variables are defined as a nonlinear function of states, controls, and parameters in Equation (2)) [1,17]:

$$\mathbf{y}(\mathbf{z};\boldsymbol{\theta}) \approx \sum_{m=0}^{M-1} \hat{\mathbf{y}}^{(m)}(\mathbf{z}) \Phi^{(m)}(\boldsymbol{\theta}), \quad (M-1) = \binom{N+D}{N}, \tag{3}$$

where the multivariate expansion polynomials $\Phi^{(m)} \in \mathbb{R}$ are orthogonal with m as their highest polynomial exponent [1]. The order of the gPC expansion is given by M, the number of uncertain parameters by N, and the highest order of the orthogonal polynomials by D. The Wiener–Askey scheme provides general rules to select the orthogonal polynomials Φ based on the pdf $\rho(\boldsymbol{\theta})$ of the uncertain parameters $\boldsymbol{\theta}$. For some specific pdfs of the gPC expansion, these polynomial relations are summarized in Table 1. Take into account that extensions to general pdfs are also available [18].

The expansion coefficients $\hat{\mathbf{y}}^{(m)} \in \mathbb{R}^{n_y}$ in Equation (3) are given by a Galerkin projection [19]:

$$\hat{\mathbf{y}}^{(m)}(\mathbf{z}) = \int_{\Omega} \mathbf{y}(\mathbf{z};\boldsymbol{\theta}) \Phi^{(m)}(\boldsymbol{\theta}) \rho(\boldsymbol{\theta}) \, d\boldsymbol{\theta}, \tag{4}$$

where Ω is the support of the pdf $\rho(\boldsymbol{\theta}) \in \mathbb{R}$ (Table 1).

Table 1. Continuous density function-orthogonal polynomial connection for standard generalized polynomial chaos (after [1]) for a scalar parameter θ.

Distribution	Probability Density Function	Support	Symbol	Orthogonal Polynomial
Gaussian/Normal	$\frac{1}{\sqrt{2\pi}} \exp\left(\frac{\theta^2}{2}\right)$	$]-\infty, \infty[$	$\mathcal{N}(\mu, \sigma)$	Hermite
Gamma	$\frac{\theta^\alpha \exp(-\theta)}{\Gamma(\alpha+1)}$	$[0, \infty[$	$\gamma(\mu, \sigma, \alpha)$	Laguerre
Beta	$\frac{\Gamma(\alpha+\beta+2)}{2^{\alpha+\beta+1}\Gamma(\alpha+1)\Gamma(\beta+1)}(1-\theta)^\alpha(1+\theta)^\beta$	$]-1, 1[$	$\mathcal{B}(a, b, \alpha, \beta)$	Jacobi
Uniform	$\frac{1}{2}$	$]-1, 1[$	$\mathcal{U}(a, b)$	Legendre

To connect the expansion coefficients with the physical trajectories of the system, the stochastic collocation (SC) method is used [19]. This is also done to constrain the viable domain of the expansion coefficients based on the physical system response in the OCP. Generally, the SC method tries to approximate the integral in Equation (4) by Gaussian quadrature using a finite sum, discrete expansion at a set of nodes $\boldsymbol{\theta}^{(j)} \in \mathbb{R}^{n_\theta}$ with corresponding integration weights $\alpha^{(j)} \in \mathbb{R}$. These are specifically chosen in order to have a high approximation accuracy [19]. This yields the following approximation formula for Equation (4) [19]:

$$\hat{\mathbf{y}}^{(m)}(\mathbf{z}) = \int_{\Omega} \mathbf{y}(\mathbf{z};\boldsymbol{\theta}) \Phi^{(m)}(\boldsymbol{\theta}) \rho(\boldsymbol{\theta}) \, d\boldsymbol{\theta} \approx \sum_{j=1}^{Q} \underbrace{\mathbf{y}\left(\mathbf{z};\boldsymbol{\theta}^{(j)}\right)}_{\mathbf{y}^{(j)}} \Phi^{(m)}\left(\boldsymbol{\theta}^{(j)}\right) \alpha^{(j)}. \tag{5}$$

Here, Q is the number of specifically selected nodes according to the Gaussian quadrature rules (zeros of orthogonal polynomial), defining the accuracy of the integral approximation [19]. It should be noted that the Gaussian quadrature approach is subject to the curse of dimensionality for a large number of uncertain parameters because it is generally evaluated on a tensor grid [19]. Thus, sparse grids [19] must be employed in higher dimensions, e.g., starting from $n_\theta > 5$. For the sake of simplicity, we use the tensor grid in this study, but the methods can directly be extended to sparse grids as well.

The continuous OCP (Equation (1)), is discretized into a NLP using the gPC expansion for states and controls, and then solved using a Newton-type optimization algorithm. Here, we use a trapezoidal collocation scheme. It should be noted that we apply the state integration and the state constraint to each of the SC nodes in this context. This makes it possible to calculate any desired output variable gPC expansion for the CC using the SC expansion in Equation (5). Here, the physical state trajectories and the output equation (Equation (2)) must be applied to calculate the required output expansion coefficients for Equation (10). In addition, we ensure feasible, physical trajectories by constraining the physical states at each of the SC nodes as this task might not be trivial by merely constraining the expansion states that are part of the decision variables (Equation (7)). Take into account that it is crucial in this context to ensure the constraint qualifications/regularity conditions for the NLP, e.g., linear independence, such that the optimization is well-behaved ([20], p. 45).

The basic form of the NLP is as follows (after [13]):

$$\min_{\hat{z}} \quad J = e(\hat{z}_N) + \frac{\hat{t}_f^{(0)}}{2} h_\tau \sum_{i=1}^{N-1} [L(\hat{z}_i) + L(\hat{z}_{i+1})]$$

s.t. $\hat{z}_{lb} \leq \hat{z} \leq \hat{z}_{ub}$,

$$\mathbf{x}_{lb} \leq \begin{bmatrix} \mathbf{x}_1^{(j)} = \sum_{m=0}^{M-1} \hat{\mathbf{x}}_1^{(m)} \Phi^{(m)}\left(\boldsymbol{\theta}^{(j)}\right) \\ \vdots \\ \mathbf{x}_N^{(j)} = \sum_{m=0}^{M-1} \hat{\mathbf{x}}_N^{(m)} \Phi^{(m)}\left(\boldsymbol{\theta}^{(j)}\right) \end{bmatrix} \leq \mathbf{x}_{ub}, \quad \forall j,$$

$$\boldsymbol{\psi}(\hat{z};\boldsymbol{\theta}) = \begin{bmatrix} \mathbf{x}_2^{(j)} - \mathbf{x}_1^{(j)} - \frac{t_f^{(j)}}{2} h_\tau \left(\dot{\mathbf{x}}_2^{(j)} + \dot{\mathbf{x}}_1^{(j)}\right) \\ \vdots \\ \mathbf{x}_N^{(j)} - \mathbf{x}_{N-1}^{(j)} - \frac{t_f^{(j)}}{2} h_\tau \left(\dot{\mathbf{x}}_N^{(j)} + \dot{\mathbf{x}}_{N-1}^{(j)}\right) \end{bmatrix} = 0, \quad \forall j,$$

$$\begin{bmatrix} \mathbb{P}_{OCP}(\mathbf{y}_1 \notin \mathcal{F}) \\ \vdots \\ \mathbb{P}_{OCP}(\mathbf{y}_N \notin \mathcal{F}) \end{bmatrix} \geq \eta.$$

(6)

Take into account that the differential equation, used for the model dynamics in Equation (1), is directly included in the equality constraints ψ using the trapezoidal integration scheme. Additionally, the deterministic equality and inequality constraints of the NLP must be fulfilled in our framework at each SC nodes to ensure feasibility.

The discretization step is depicted by h_τ and generally comprises N discretized time steps. The decision variable vector $\hat{z}$, with corresponding lower and upper bounds depicted by $\hat{z}_{lb}$ and $\hat{z}_{ub}$, respectively, is defined using the gPC expansion coefficients for states, controls, and parameters as follows:

$$\hat{z} = [\hat{t}_f^{(0)}, \ldots, \hat{t}_f^{(M-1)}, \hat{\mathbf{p}}^{(0)}, \hat{\mathbf{x}}_1^{(0)}, \ldots, \hat{\mathbf{x}}_1^{(M-1)}, \hat{\mathbf{u}}_1^{(0)}, \ldots, \hat{\mathbf{x}}_N^{(0)}, \ldots, \hat{\mathbf{x}}_N^{(M-1)}, \hat{\mathbf{u}}_N^{(0)}]^T. \quad (7)$$

Take into account that the control history is not expanded in Equation (7). This is due to the fact that expanding the control history would yield a set of optimal control histories. As we want to calculate a robust trajectory, i.e., a trajectory that is robust considering that the control history is not adapted, we only use the mean value in the decision vector. Still, an extension of Equation (7) to a distributed control history is possible. It should be noted that the same argumentation applies for the time-invariant parameters, as it is also generally desired to calculate a single robust value for these.

Further note that the outputs **y**, required for the CC in Equation (6), can be calculated directly using the decision variables in Equation (7), the gPC expansion in Equation (3) (to calculate the physical trajectories), the output equation in Equation (2), and the SC method in Equation (5).

It should be noted that the cost function in Equation (6) is depending on the decision variables directly, which are the expansion coefficients (Equation (7)). This is done to be able to optimize statistical moments (e.g., mean value and variance; Section 2.2.2) in the OCP. Further take into account that the inequality path constraints are box constraints enforced at each discretization point for the physical trajectories with the same lower and upper bound and independent of the uncertainty. This is done as the state limits normally do not vary over time and should also not change depending on the uncertainty. In addition, we enforce physical trajectories calculated by the NLP optimizer using this procedure.

Further take into account that Equation (6) is a deterministic version of the uncertain OCP in Equation (1) except for the CCs. Further note that the inequality as well as equality constraints are evaluated at the physical SC nodes (Equation (5)) using the gPC expansion in Equation (3).

2.2.2. Statistical Moments

Statistical moments, such as mean or variance, can be calculated directly from the gPC expansion in Equation (3), if the expansion coefficients are known. For instance, the mean is given by [19]:

$$\mathbb{E}\left[\mathbf{y}(\mathbf{z};\boldsymbol{\theta})\right] \approx \int_\Omega \left(\sum_{m=0}^{M-1} \hat{\mathbf{y}}^{(m)}(\mathbf{z}) \Phi^{(m)}(\boldsymbol{\theta})\right) \rho(\boldsymbol{\theta}) \, d\boldsymbol{\theta} = \hat{\mathbf{y}}^{(0)}(\mathbf{z}). \tag{8}$$

Equation (8) shows that the mean is only depending on the first deterministic expansion coefficient. The variance of the outputs **y** calculated as [19]:

$$\sigma^2\left[\mathbf{y}(\mathbf{z};\boldsymbol{\theta})\right] = \mathbb{E}\left[\mathbf{y}(\mathbf{z};\boldsymbol{\theta}) - \mathbb{E}[\mathbf{y}(\mathbf{z};\boldsymbol{\theta})]^2\right] \approx \sum_{m=1}^{M-1} \left[\hat{\mathbf{y}}^{(m)}(\mathbf{z})\right]^2 \tag{9}$$

is only dependent on the deterministic expansion coefficients $\hat{\mathbf{y}}^{(1,\dots,M-1)}$.

3. Chance Constraints in the Polynomial Chaos Optimal Control Framework

Within this section, we look at the CC framework based on the gPC approximation within the OCP that should approximate the probability of not being the failure event, i.e., $\mathbb{P}_{OCP}(\mathbf{y}_i^{(j)} \notin \mathcal{F}), \forall i$. In Section 3.1, the general formulation of CCs in the deterministic OCP is introduced. Afterwards, Section 3.2 introduces a differentiable approximation of the sharp CCs and a homotopy strategy to iteratively sharpen the differentiable CC representation. The SubSim method and its incorporation within CC–OC to calculate rare-event failure probabilities are described in Section 3.3.

3.1. Derivation of Chance Constraint Formulation

Sampling techniques such as the Metropolis–Hastings algorithm (MHA) [21] or importance sampling [22] are frequently used to approximate the probability of an event (in this case: fulfilling a CC) when its pdf is difficult to sample from or integrate. A drawback of these methods is a non-deterministic evaluation procedure of the probability. Generally, this study still tries to apply sampling-based algorithms to estimate the probability $\mathbb{P}_{OCP}(\mathbf{y}_i \notin \mathcal{F})$ in the OCP (Equation (6)). Additionally, rare events should be covered, which makes SubSim a viable choice [12]. The basic SubSim method uses a modified Metropolis–Hastings algorithm (MMHA), i.e., random sampling, to explore the failure region and calculate the failure probability. Here, a further issue arises when using direct methods to solve OCPs with

Newton-type NLP solvers: the samples cannot be redrawn in each iteration of the NLP solution process as this would yield a stochastic Newton-type optimization procedure. Generally, this would be necessary in the context of sampling techniques, such as SubSim, which ultimately results in problems defining accurate step-sizes and exit criteria in the NLP. Thus, this study uses a homotopy strategy to cope with these issues that move the creation of new samples from the NLP iteration to a homotopy step.

In order to apply the mentioned sampling techniques, we need a good approximation for the probabilistic quantity, i.e., the quantity with respect to whom the CC is defined, depending on the stochastic disturbance. When applying gPC, the gPC expansion in Equation (3) provides this approximation. Thus, in cases where the expansion coefficients are available within the NLP, as e.g., in Equation (6) (remember that Equations (2), (3) and (5) can be applied to calculate the expansions coefficients for any output quantity based on the known physical trajectories at the SC nodes for the states used in Equation (6)), we can sample the gPC expansion for thousands of samples via a matrix-vector operation in an MCA-type way, but with improved efficiency due to the simple evaluation as follows: consider n_s random samples obtained from the pdf of θ, labeled $\theta^{(1)}, \ldots, \theta^{(n_s)}$. It should be noted that these samples can now be drawn randomly in contrast to the SC method as we are not trying to approximate the integral in Equation (4), but the probability of the CC. These samples for the uncertain parameters yield corresponding samples for the output $\mathbf{y}$, given by:

$$\underbrace{\left[\mathbf{y}\left(\mathbf{z};\theta^{(1)}\right) \ \cdots \ \mathbf{y}\left(\mathbf{z};\theta^{(n_s)}\right)\right]}_{\mathbb{R}^{n_y \times n_s}} = \underbrace{\left[\hat{\mathbf{y}}^{(0)}(\mathbf{z}) \ \cdots \ \hat{\mathbf{y}}^{(M-1)}(\mathbf{z})\right]}_{\mathbb{R}^{n_y \times M}} \underbrace{\begin{bmatrix} \Phi^{(0)}\left(\theta^{(1)}\right) & \cdots & \Phi^{(0)}\left(\theta^{(n_s)}\right) \\ \vdots & \ddots & \vdots \\ \Phi^{(M-1)}\left(\theta^{(1)}\right) & \cdots & \Phi^{(M-1)}\left(\theta^{(n_s)}\right) \end{bmatrix}}_{\mathbb{R}^{M \times n_s}}, \quad (10)$$

such that the output samples are provided from a simple matrix-vector multiplication operating on the expansion coefficients $\hat{\mathbf{y}}$, which are part of the OCP formulation due to Equations (2), (3) and (5). With the samples available from Equation (10), the general equation for fulfilling, i.e., not being in the failure set, a CC is given as follows:

$$\mathbb{P}\left[\mathbf{y}(\mathbf{z};\theta) \notin \mathcal{F}\right] = \int_{\Omega} \mathcal{I}(\mathbf{y}(\mathbf{z};\theta)) \rho(\theta) \, d\theta \approx \frac{1}{n_s} \sum_{i=1}^{n_s} \mathcal{I}\left(\mathbf{y}\left(\mathbf{z};\theta^{(i)}\right)\right). \quad (11)$$

Here, $\mathcal{I}(\mathbf{y}(\mathbf{z};\theta))$ is the indicator function, defined as:

$$\mathcal{I}\left(\mathbf{y}\left(\mathbf{z};\theta^{(i)}\right)\right) = \begin{cases} 1, & \text{for } \mathbf{y}\left(\mathbf{z};\theta^{(i)}\right) \notin \mathcal{F}, \\ 0, & \text{else.} \end{cases} \quad (12)$$

It should be noted that the indicator function $\mathcal{I}$ is trivial to evaluate but non-differentiable, and can therefore create difficulties when used in the context of a Newton-type NLP solver. Thus, we introduce a smooth approximation s of the indicator functions having the following properties:

$$\begin{aligned} & s(\mathbf{y}(\mathbf{z};\theta)) \in [0;1], \\ & s(\mathbf{y}(\mathbf{z};\theta)) \approx 1 \quad \mathbf{y}(\mathbf{z};\theta) \notin \mathcal{F}, \\ & s(\mathbf{y}(\mathbf{z};\theta)) \approx 0 \quad \mathbf{y}(\mathbf{z};\theta) \in \mathcal{F}. \end{aligned} \quad (13)$$

3.2. Approximation of Chance Constraints by Sigmoids

A group of functions that can be used for the approximation of an indicator function that must fulfill the conditions given in Equation (13) are the logistics functions. An example for this class of functions is the sigmoid function, which is defined as follows for a scalar output y:

$$s(y; a, b) = \frac{1}{\exp\left[-a \cdot (y - b)\right] + 1} \in \mathbb{R}. \tag{14}$$

The parameters $a \in \mathbb{R}$ and $b \in \mathbb{R}$ are the scaling and offset parameter of the sigmoid, respectively. These are used to shape the sigmoid in order to suitably approximate the desired CC domain. Their design using a homotopy strategy while solving the CC–OC problem is illustrated in Algorithm 1.

Algorithm 1 Implemented homotopy strategy for sigmoid scaling and offset parameter in CC–OC framework.

Require: Define the homotopy factor a_{hom} and the desired final sigmoid parameter $a_{desired}$.
1: Initialize sigmoid parameter a and confidence level CL.
2: Define the bound value of the sigmoid y_{bound} (i.e., the bound value of the CC)
3: **while** $a < a_{desired}$ **do**
4:     Calculate the sigmoid values:
5:     $c = -\frac{\ln\left(\frac{1}{CL} - 1\right)}{a}$
6:     $b = y_{bound} - c$
7:     Solve the CC–OC problem including SubSim in Algorithm 4.
8:     Increase a by homotopy factor: $a = a_{hom} \cdot a$.
9: **end while**
10: **return** Robust optimal trajectory.

Furthermore, the sigmoid in Equation (14) has a very simple derivative that can be used to efficiently calculate the gradient that is necessary for OC. It is given as follows:

$$\frac{ds(y; a, b)}{dy} = a \cdot s(y; a, b) \cdot [s(y; a, b) - 1]. \tag{15}$$

The sigmoid in Equation (14) can be combined by multiplication in order to approximate the indicator function for $\mathcal{F}$ being an interval in $\mathbb{R}$. This is depicted in Figure 1, which shows the multiplication of two sigmoids (solid blue) with one gradual descend (dashed green; number 1) and one steep ascend (dashed red; number 2) to approximate a box constraint on a scalar output. Here, one sigmoid $s(y; a_{lb}, b_{lb})$ describes the lower bound, while the other sigmoid $s(y; a_{ub}, b_{ub})$ describes the upper bound.

For the sake of simplicity, we further assume box constraints on all our CCs, i.e., we assume that $\mathcal{F}$ is a hyper-rectangle with lower and upper bounds lb_i, ub_i on dimension $i = 1, \ldots, n_y$. This is a viable assumption for most OCP applications, as box constraints are very prominent in OC. The proposed approach can be trivially extended to any set $\mathcal{F}$ that can be described by a set of smooth inequality constraints.

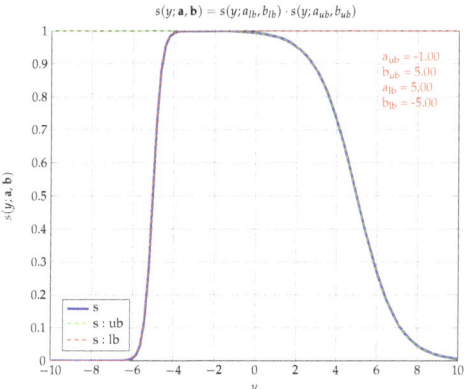

Figure 1. Product of two sigmoids with different scaling and offset parameters to approximate an uncertainty domain by box constraints.

We can then form the hyper-rectangle by applying the basic sigmoid in Equation (14) as follows:

$$s(\mathbf{y};\mathbf{a},\mathbf{b}) = \prod_{i=1}^{n_y} s\left(y_i; a_{lb_i}, b_{lb_i}\right) \cdot s\left(y_i; a_{ub_i}, b_{ub_i}\right) \in \mathbb{R}. \quad (16)$$

Here, $\mathbf{a} = \left[a_{lb_1}, a_{ub_1}, \ldots, a_{lb_{n_y}}, a_{ub_{n_y}}\right]^T$ and $\mathbf{b} = \left[b_{lb_1}, b_{ub_1}, \ldots, b_{lb_{n_y}}, b_{ub_{n_y}}\right]^T$ are simplifying notations. Take into account that the derivative of Equation (16) can be formed using the chain rule and Equation (15).

In order to calculate the probability, we must only sum up the function values of the multidimensional sigmoid in Equation (16) and divide it by the number of samples as follows:

$$\mathbb{P}[\mathbf{y} \notin \mathcal{F}] \approx \frac{1}{n_s} \sum_{i=1}^{n_s} s\left(\mathbf{y}^{(i)}; \mathbf{a}, \mathbf{b}\right). \quad (17)$$

This approximation can now be used within the OCP (Equation (6)). In order to include rare failure events, the next subsection introduces the SubSim method that elaborates on the CC modeling of this subsection.

3.3. Subset Simulation in Chance-Constrained Optimal Control

The probability approximation in Equations (11) and (17) converges for reasonably low choices of n_s only if rather loose bounds on the probability (e.g., domain of $\eta = 99\%$) are considered. For tighter bounds typically used for rare events, as often required in e.g., reliability engineering (where $\eta = 99.9999\%$ is common), better suited algorithms to calculate and sample the probability are required. Indeed, a reliable estimation of the probability of rare events normally requires a very large number of samples. A classical approach to circumvent this difficulty is the use of SubSim, which is tailored to evaluate the probability of rare events [11,12,23].

SubSim methods are based on an MCMC algorithm typically relying on a MMHA, which ensures that the failure region is properly covered by the samples. To that end, it stratifies the choice of samples iteratively in order to draw significantly more samples from the failure region than a classical MCA sampling would. SubSim methods are based on an expansion of the failure probability as a series of conditional probabilities:

$$\mathbb{P}\left(\mathcal{F}\right) = \mathbb{P}\left(\mathcal{F}_m\right) = \mathbb{P}\left(\bigcap_{i=1}^{m} \mathcal{F}_i\right) = \mathbb{P}\left(\mathcal{F}_1\right) \prod_{i=2}^{m} \mathbb{P}\left(\mathcal{F}_i | \mathcal{F}_{i-1}\right). \quad (18)$$

In Equation (18), $\mathcal{F}$ is the set of failure events and $\mathcal{F}_1 \supset \mathcal{F}_2 \supset \ldots \supset \mathcal{F}_m = \mathcal{F}$ is a sequence of set of events with decreasing probability of occurrence. The conditional probability $\mathbb{P}\left(\mathcal{F}_i | \mathcal{F}_{i-1}\right)$ describes the probability that an event in $\mathcal{F}_i \subset \mathcal{F}_{i-1}$ occurs assuming that an event in $\mathcal{F}_{i-1}$ has already occurred. Thus, instead of evaluating the rare event $\mathbb{P}\left(\mathcal{F}\right) = \mathbb{P}\left(\mathcal{F}_m\right)$, one can evaluate a chain of relatively likely conditional probabilities $\mathbb{P}\left(\mathcal{F}_i | \mathcal{F}_{i-1}\right)$, each of which is relatively easy to evaluate via sampling.

The evaluation of the conditional probabilities is the main task in SubSim, achieved using e.g., the MMHA approach. The MMHA, working on each component of a random vector (i.e., vector of one–dimensional random variables) (RVec), is introduced in Algorithm 2 [11,12,24].

Algorithm 2 Modified Metropolis–Hastings algorithm for subset simulation for each component of the random vector θ_i to create new sample for random parameter (after [11]).

Require: Current sample for the random parameter: θ_i.

1: Define a symmetric proposal pdf, $\rho_i^*\left(\tilde{\theta} | \theta_i\right) = \rho_i^*\left(\theta_i | \tilde{\theta}\right)$, centered around the current random sample θ_i.
2: Generate candidate sample $\tilde{\theta}$ from $\rho_i^*\left(\tilde{\theta} | \theta_i\right)$ and calculate the system response (e.g., by Equations (1), (3), or (6)) of this candidate.
3: Calculate the ratio between the proposed and the candidate sample evaluated at the target pdf, e.g., $r = \frac{\rho(\tilde{\theta})}{\rho(\theta_i)}$ with $\rho \propto N\left(0,1\right)$ (this is the stationary pdf according to the central limit theorem ([17], p. 23)).
4: Set the acceptance ratio of the candidate sample $\tilde{\theta}$ as follows: $a = \min\{1, r\}$ and accept it with this probability.
5: Draw a random sample from the uniform distribution as follows: $s \propto \mathcal{U}\left(0,1\right)$.
6: **if** $s \geq a$ **then**
7:     Set: $\theta_p = \tilde{\theta}$
8: **else**
9:     Set: $\theta_p = \theta_i$
10: **end if**
11: Update the new sample by the rule:
12: **if** $\theta_p \in \mathcal{F}_i$ **then**
13:     Set: $\theta_{new,i} = \theta_p$
14: **else**
15:     Set: $\theta_{new,i} = \theta_i$
16: **end if**
17: **return** New sample for the random parameter: $\theta_{new,i}$.

Normally, the MCMC algorithm based on the MMHA in Algorithm 2 is fast converging as especially the sampling of new candidates is done locally around the current sample point. Thus, the acceptance rate is normally very high and progress is made quite fast. An issue of the MMHA is the choice of an appropriate proposal distribution ρ_i^*. Here, generally a Uniform or a Gaussian pdf is chosen, in order to have a simple evaluation and the symmetric property. The general behavior of the MMHA in SubSim can be visualized as in Figure 2: it can be seen that after the random sampling by MCA in level 0, the samples get shifted to the failure domain, which are the arcs in the corners of the domain. This is done until a sufficient amount of samples is located in the failure domain.

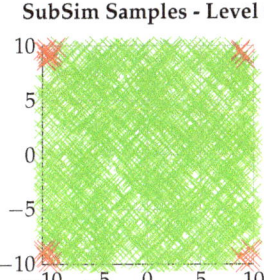
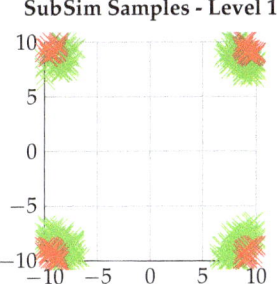
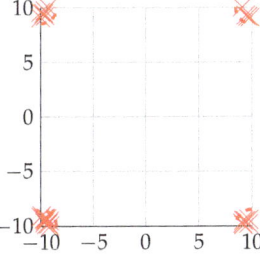

Figure 2. General behavior of subset simulation with modified Metropolis–Hastings algorithm and $p_0 = 0.1$ showing movement of the samples (red: in failure domain, green: not in failure domain) over the different subset levels with arc failure domains at edges.

We detail the SubSim method as in Algorithm 3 [11,12]. The SubSim starts with a general MCA and afterwards subsequently evaluates the failure region yielding the chain of conditional probabilities. It should be noted that the choice of the intermediate failure events $\mathcal{F}_{1,...,m}$ is critical for the convergence speed of the SubSim. In [11], the "critical demand-to-capacity ratio" is introduced that is based on normalized intermediate threshold values. Based on this ratio, it is common to choose the thresholds adaptively such that the conditional probabilities are equal to a fixed, pre-defined value ([12], p. 158). This is done by appropriately ordering the previous samples and their result. An often and a normally very efficient conditional probability value is $p_0 = 0.1$ [11].

Finally, we can estimate the failure probability of the SubSim regarding the desired threshold b and the $(m-1)-th$ Markov chain element, which is the last one of the SubSim as follows ([12], p. 179) (see Algorithm 3 line 12):

$$1 - \mathbb{P}_{ss}\left(\mathbf{y}\left(\mathbf{z};\boldsymbol{\theta}\right) \notin \mathcal{F}\right) = \mathbb{P}\left(\mathcal{F}\right) = \frac{1}{n_s} p_0^{m-1} \sum_{j=1}^{n_c} \sum_{k=1}^{n_{sc}} \tilde{\mathcal{I}}\left(\mathbf{y}_{jk}^{(m-1)} > b\right), \quad b > b_{m-1}. \tag{19}$$

Algorithm 3 General algorithm used for a subset simulation in connection with generalized polynomial chaos (after [12], p. 158ff).

Require: Define the number of samples per level n_s, the conditional probability p_0, and the critical threshold b.

1: Calculate the number of Markov chains $n_c = p_0 \cdot n_s$ and the number of samples $n_{sc} = p_0^{-1}$ for each of the chains.
2: Initialize the SubSim by creating the random sample set $\{\theta_k^{(0)} : k = 1, \ldots, n_s\}$.
3: Calculate the output set $\{y_k^{(0)}\left(z; \theta_k^{(0)}\right) : k = 1, \ldots, n_s\}$ by Equation (3) related to $\{\theta_k^{(0)} : k = 1, \ldots, n_s\}$.
4: Sort $\{y_k^{(0)}\left(z; \theta_k^{(0)}\right) : k = 1, \ldots, n_s\}$ in ascending order to create $\{b_k^{(0)} : k = 1, \ldots, n_s\}$. Here, $b_k^{(0)}$ is an estimate of the exceedance probability $\mathbb{P}[y(z;\theta) > b] = \frac{n_s - k}{n_s}$.
5: Set $b_1 = b_{n_s - n_c}^{(0)}$ and $\{\theta_{j0}^{(1)} : j = 1, \ldots, n_c\}$ corresponding to $\{b_{n_s - n_c + j}^{(0)} : j = 1, \ldots, n_c\}$ as the threshold and the seeds for the next level.
6: **for** i=1...m-1 **do**
7: Use e.g., the MMHA (Algorithm 2) to generate the samples $\{\theta_{jk}^{(i)} : k = 1, \ldots, n_{sc}\}$ of the conditional pdf $\rho_i^*(.|\mathcal{F}_i)$ for each seed $\{\theta_{j0}^{(i-1)} : j = 1, \ldots, n_c\}$. This creates n_c Markov chains with n_{sc} samples.
8: Calculate the output set $\{y_{jk}^{(i)}\left(z; \theta_{jk}^{(i)}\right) : j = 1, \ldots, n_c, k = 1, \ldots, n_{sc}\}$ by Equation (3) related to $\{\theta_{jk}^{(i)} : j = 1, \ldots, n_c, k = 1, \ldots, n_{sc}\}$.
9: Sort $\{y_{jk}^{(i)}\left(z; \theta_{jk}^{(i)}\right) : j = 1, \ldots, n_c, k = 1, \ldots, n_{sc}\}$ in ascending order to create $\{b_k^{(i)} : k = 1, \ldots, n_s\}$. Here, $b_k^{(i)}$ is an estimate of the exceedance probability $\mathbb{P}[y(z;\theta) > b] = p_0^i \frac{n_s - k}{n_s}$.
10: Set $b_{i+1} = b_{n_s - n_c}^{(i)}$ and $\{\theta_{j0}^{(i+1)} : j = 1, \ldots, n_c\}$ corresponding to $\{b_{n_s - n_c + j}^{(i)} : j = 1, \ldots, n_c\}$ as the threshold and the seeds for the next level.
11: **end for**
12: Calculate the failure probability $\mathbb{P}_{ss}(y(z;\theta) \notin \mathcal{F})$ based on Equation (19)
13: **return** Failure probability $\mathbb{P}_{ss}(y(z;\theta) \notin \mathcal{F})$.

It should be noted that, for the OCP in Equation (1) or Equation (6), the calculated probability in Equation (19) must be subtracted from 1 as the CC in Equation (6) is defined for not being in the failure set. Here, $\tilde{\mathcal{I}}(y(z;\theta))$ is the complementary indicator function from Equation (12) defined for the failure region:

$$\tilde{\mathcal{I}}\left(y\left(z;\theta^{(i)}\right)\right) = \begin{cases} 1, & \text{for } y\left(z;\theta^{(i)}\right) \in \mathcal{F}, \\ 0, & \text{else.} \end{cases} \quad (20)$$

Take into account that the accuracy of Equation (19) can be quantified using the coefficient of variation (c.o.v.):

$$\nu = \frac{\sigma[\mathbb{P}(\mathcal{F})]}{\mathbb{E}[\mathbb{P}(\mathcal{F})]}. \quad (21)$$

Here, $\mathbb{E}[\mathbb{P}(\mathcal{F})]$ is given by Equation (19), while the standard deviation of the failure probability can be calculated by a Beta pdf fit as proposed in ([25], p. 293). Overall, we can compare the resulting c.o.v. with literature values [24] to access the viability. Generally, a small c.o.v. indicates that the standard deviation of our failure probability estimation is smaller than our expected/mean value. Thus, the goal is to have a small c.o.v. as then the dispersion of the data is small and we can be certain about the CC being fulfilled.

In this study, we propose to introduce the SubSim algorithm in the CC–OC algorithm. Our procedure is to calculate the subset samples based on the analytic response surface of the gPC expansion (Equation (3)), which is based on the initial solution of the OCP (Equation (1)) by MCA. The samples are then used to run a new optimization fulfilling the desired rare event probability. A new response surface is calculated from which new samples are generated using a SubSim. This procedure is repeated until both the SubSim probability $\mathbb{P}_{ss}(\mathbf{y}(\mathbf{z};\theta) \notin \mathcal{F})$ (Equation (19)) as well as the probability level assigned to the constraint $\mathbb{P}_{OCP}(\mathbf{y}(\mathbf{z};\theta) \notin \mathcal{F})$ (Equation (6)) in the OCP fulfill the desired rare event probability $\mathbb{P}_{des}(\mathbf{y}(\mathbf{z};\theta) \notin \mathcal{F})$. The procedure is described in Algorithm 4. It should be noted that this procedure can generally be applied as long as the underlying OCP in Equation (6) can be solved.

Algorithm 4 Basic strategy of the subset simulation algorithm within CC–OC framework.

Require: OCP as in Equation (6) with initial guess for decision variables **z**.

1: Calculate an optimal solution for a likely failure (e.g., $\mathbb{P}(\mathbf{y}(\mathbf{z};\theta) \notin \mathcal{F}) = 99\%$) using MCA.
2: Obtain the subset probability $\mathbb{P}_{ss}(\mathbf{y}(\mathbf{z};\theta) \notin \mathcal{F})$ and samples, based on the analytic gPC response surface (Equation (3)) and by applying Algorithm 3.
3: **while** $\mathbb{P}_{ss}(\mathbf{y}(\mathbf{z};\theta) \notin \mathcal{F}) > \mathbb{P}_{des}(\mathbf{y}(\mathbf{z};\theta) \notin \mathcal{F})$ and $\mathbb{P}_{OCP}(\mathbf{y}(\mathbf{z};\theta) \notin \mathcal{F}) > \mathbb{P}_{des}(\mathbf{y}(\mathbf{z};\theta) \notin \mathcal{F})$ **do**
4:     Assign the SubSim samples to the evaluation routine of the CC within the OCP.
5:     Solve the CC–OC problem (Equation (6)).
6:     **if** Optimization not successful **then**
7:         Reduce the probability of the constraint $\mathbb{P}_{OCP}(\mathbf{y}(\mathbf{z};\theta) \notin \mathcal{F})$ to relax the OCP (e.g., by factor of 10).
8:     **else**
9:         **if** $\mathbb{P}_{OCP}(\mathbf{y}(\mathbf{z};\theta) \notin \mathcal{F}) \neq \mathbb{P}_{des}(\mathbf{y}(\mathbf{z};\theta) \notin \mathcal{F})$ **then**
10:             Increase the constraint probability within the OCP (e.g., factor of 10).
11:         **end if**
12:     **end if**
13:     Obtain the new subset probability $\mathbb{P}_{ss}(\mathbf{y}(\mathbf{z};\theta) \notin \mathcal{F})$ from Equation (19) and samples based on the new analytic gPC response surface and using Algorithm 3.
14: **end while**
15: **return** Optimal decision variables **z**.

Regarding the homotopy strategy, it should be noted that, by using the SubSim samples calculated from the last optimal solution within the new optimization, we might introduce a bias as the samples drawn from the Markov chain are based on the optimal results created by the last NLP solution. Generally, they would have to adapted in each iteration of the NLP as the system response changes. As we do not update the samples within the NLP, but within the homotopy step after the optimal solution has been calculated, we technically solve the CC and its rare event probability using biased samples compared to the ones that would be calculated within the SubSim. We cope with this issue in this paper by checking the fulfillment of the CC both in the OCP as well as after the OCP is solved by the SubSim, i.e., with the new response surface. Thus, the CC–OC is only solved if both results show that the CC is fulfilled to the desired level. Within this paper, the OCP converges fine, but further studies should explore the effects and influences of this bias and how to reduce it (e.g., by importance sampling).

4. Optimization Model

The implemented optimization model is based on the work in [26]: at first, the dynamic model is introduced in Section 4.1. The OCP setup, including constants and parameters, is afterwards defined in Section 4.2.

4.1. Battery Dynamic Equations

The following section summarizes the dynamic equations for a battery modeled by an extended equivalent circuit model (XECM) as depicted in Figure 3. Here, the equations of motion are introduced in Section 4.1.1. Afterwards, Section 4.1.2 introduces a battery heating model.

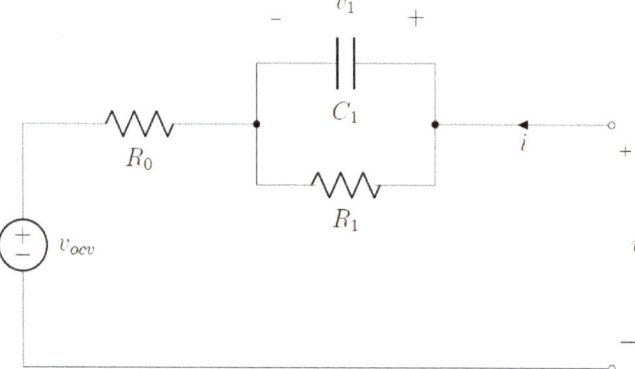

Figure 3. Schematics of Extended Equivalent Circuit Model (XECM) [26].

4.1.1. Local Voltage and Ion Concentration Dynamic Equations

The local voltage v_1 and ion concentration Δz_1 equations of motion are based on the parallel resistor–capacitor arrangement in Figure 3. The equations are given by first order lags with the current i as the control variable:

$$\dot{v}_1 = -\frac{1}{R_1 \cdot C_1} \cdot v_1 + \frac{1}{2 \cdot C_1} i, \tag{22a}$$

$$\Delta \dot{z}_1 = -\frac{1}{R_1 \cdot C_1} \cdot \Delta z_1 + \frac{1}{2 \cdot C_1} i. \tag{22b}$$

Here, R_1 and C_1 are the resistance and capacity, respectively. It should be noted that R_1 and C_1 are functions of the battery temperature (Section 4.1.2).

In addition, we have the total ion concentration z dynamic equation, also called state of charge (SoC), that is only dependent on the current:

$$\dot{z} = \frac{1}{Q} i. \tag{23}$$

Here, Q is the battery capacity.

4.1.2. Battery Heating

As an extension to the standard XECM in Section 4.1.1, we also model the heating of the battery when a current is applied. This heating can again be formulated by an equation of motion for the battery temperature T_{batt} that is mainly influenced by the square of the applied current:

$$\dot{T}_{batt} = k_1 \cdot i^2 + k_2 \cdot (T_{amb} - T_{batt}). \tag{24}$$

The coefficients $k_1 = \frac{k_{R_0} \cdot R_0}{m_{batt} c}$ as well as k_2 are again parameters of the battery, while T_{amb} is the ambient temperature. It should be noted that k_{R_0} is the considered uncertainty and is a scaling factor for the lumped resistance term R_0. It is uniformly distributed as follows:

$$k_{R_0} \in \mathcal{U}(0.8; 1.2), \quad \mu = 1, \quad \sigma \approx 0.1155. \tag{25}$$

Thus, the lumped resistance term can vary up to ±20%, which refers to the uncertainty that is introduced to the system when identifying the parameter. Take into account that we choose this parameter as the uncertain value as it is also the main contributor to the battery temperature increase, which we want a robust trajectory against. Additionally, it should be noted that the uncertainty definition in Equation (25) implies using a Uniform pdf as the proposal distribution in Algorithm 2.

For the CC optimization, we want to achieve that the following probability for the battery temperature is always fulfilled:

$$\mathbb{P}\left[0°C \leq T_{batt} \leq 40°C\right] \geq 0.999999. \tag{26}$$

This CC is implemented using the SubSim approach presented in Section 3.3 and the sigmoid approximation of the CC with the homotopy strategy presented in Section 3.2. We use this kind of probability as we want to assure that the battery is not damaged by a temperature that is too high, but also charges as optimally as possible without being too conservative. In addition, it might not be possible in general applications, due to other system constraints, to calculate a fully robust trajectory, which makes the use of CCs viable.

4.2. Problem Setup

The problem consists of two phases that model one charge and one discharge of the battery. The following initial boundary conditions (IBC; Table 2) for the states $\mathbf{x} = \begin{bmatrix} v_1 & \Delta z_1 & z & T_{batt} \end{bmatrix}$ that define these conditions in the beginning of the first phase are as follows (these are assigned as inequality constraints in Equation (6)):

Table 2. Initial boundary condition (IBC) of the optimization problem.

Phase	IBC_{lb}	IBC_{ub}
1	[0, 0, 0.15, 20]	[0, 0, 0.15, 20]

Furthermore, the final boundary conditions (FBC; Table 3) for the same states in all phases are defined as follows (again assigned as inequality constraints in Equation (6)):

Table 3. Final boundary conditions (FBC) of the optimization problem.

Phase	FBC_{lb}	FBC_{ub}
1	$[0, 0, 0.95, -10]$	$[0.15, 0.5, 1.00, 40]$
2	$[0, 0, 0.00, -10]$	$[0.15, 0.5, 0.15, 40]$

It should be noted here that trying to e.g., enforce an equality constraint for the final boundary condition with only the mean robust control, as in Equation (7), might yield an infeasible OCP. In this case, a CC should be considered to model the final boundary condition or an inequality constraint (as used in this study) can be applied.

The states with their respective lower and upper bounds x_{lb}, x_{ub}, and scaling x_S are as given in Table 4.

Table 4. States upper and lower bounds as well as scalings.

State	Description	x_{lb}	x_{ub}	x_S
v_1	Local voltage	0	0.15	10^0
Δz_1	Local concentration	0	1	10^0
z	Total concentration	0.05	0.95	10^0
T_{batt}	Battery temperature	-10	40	10^{-1}

The controls with their respective lower and upper bounds u_{lb}, u_{ub}, and scaling u_S are defined as in Table 5.

Table 5. Control upper and lower bounds as well as scalings.

Control	Description	u_{lb}	u_{ub}	u_S
i_{charge}	Charge current	0	$3 \cdot Q$	10^{-1}
$i_{discharge}$	Discharge current	$-3 \cdot Q$	0	10^{-1}

Finally, the parameters and the constants of the optimization model are defined in Table 6.

Table 6. Parameters and constants of the optimization model.

Value	Description	Reference
T_{amb}	Ambient Temperature	$20\,°C$
$m_{batt}c$	battery mass times specific heat capacity	$260\,\frac{J}{°C}$
k_2	convection coefficient of battery with ambient	$0.00001\,\frac{1}{s}$
R_1	Local Resistance	$fcn\,(T_{batt})$
R_0	Lumped Resistance	$fcn\,(T_{batt})$
C_1	Local Capacity	$fcn\,(T_{batt})$
Q	Battery Capacity	$26\,Ah$

Finally, we consider the following cost function to minimize the cycle time:

$$J = \hat{t}_f^{(0)}. \tag{27}$$

This is a parameter cost that actually requires a trade-off between fulfilling the CC and finishing the cycle as fast as possible, due to the fact that a fast charging/discharging with large current yields a fast temperature increase.

5. Optimization Cases

This section covers the test cases for the CC–OC framework. At first, a single phase with only charging is looked at in Section 5.1 to get an overview of the problem characteristics. Then, Section 5.2 looks at a charge–discharge cycle. Generally, each phase has $N = 125$ time discretization steps yielding NLP problem sizes of around 2000 optimization variables and constraints.

For the final results, which are depicted in the following, the scaling factor of the sigmoid CC approximation is $a = \pm 50$. This could be achieved in a single homotopy step for the results in Section 5.1 and with two homotopy steps for the results in Section 5.2. The homotopy begins with $a = \pm 1$ and has an intermediate step, in the second example, at $a = \pm 25$. In general, we use $n_s = 10,000$ random samples to approximate the probability of the CC using the methods introduced in Section 3. The gPC expansion order is chosen to be three as this has shown to be viable for these kind of problems. The CC is defined as given in Equation (26). Take into account that the initial MCA solution fulfills the CC in Equation (26) with a probability of 97.5%. After this initial solution, we directly assign the desired probability, given in Equation (26), to the CC–OC and thus require one homotopy step.

In the following, we show the results obtained for the different SubSim level ("SSLevel") runs with $p_0 = 0.1$ and $n_s = 2500$ during the homotopy procedure. Here, the zeroth level is the basic MCA solution. The gPC order is chosen to be three, which was determined to be sufficient by comparing the accuracy of the expansion with MCA optimization runs.

5.1. Battery Charging Optimization

This section introduces the optimization of a single battery charge by looking at the general time-optimal OCP (Equation (27)).

In Figure 4, which depicts the probability of not fulfilling the CC, it can be seen that our desired probability level is fulfilled after the first SubSim level. This probability is calculated in a post-processing step using 1 million samples and the analytic indicator function in Equation (12): here, we get a level close to 0% failures and thus the CC is, based on our sampling, fulfilled with a very high certainty. Indeed, the SubSim evaluates the failure probability to be $\mathbb{P}(\mathcal{F}) = 2.0088 \cdot 10^{-5}$% with a c.o.v. of $\nu = 1.0052$. Thus, we fulfill our desired failure probability of 10^{-4}% even though we have a slightly high c.o.v..

This can also be seen looking at Figure 5 that shows the fitted marginal distribution of the failure probability at the final point in time (i.e., the end time). This is the point where the violation is most likely to occur. Here, the already mentioned method of fitting a Beta pdf for the c.o.v. estimation by applying the theory in study [25] is used. The pdf is depicted in solid blue and plotted in the range of $[0, \mathbb{P}_{des}(\mathcal{F})]$, which covers the range of the pdf and its probability until the maximal allowed failure probability. The mean value is depicted in dashed black. It is evident that the pdf fulfills our rare-event CC with high probability and thus we can be confident in the certainty of our result. To be specific, according to the SubSim and the Beta pdf of [25], the CC is fulfilled with almost 97% certainty for our application.

In Figure 6, the current for the robust optimal result is shown. We can see that the current is mainly at its maximum bound for the MCA optimization (SSLevel: 0), while it decreases linearly from a general lower level to fulfill the tighter bounds of the desired rare-event probability in the SSLevel 1 run. The optimal time is slightly increased for the SSLevel 1.

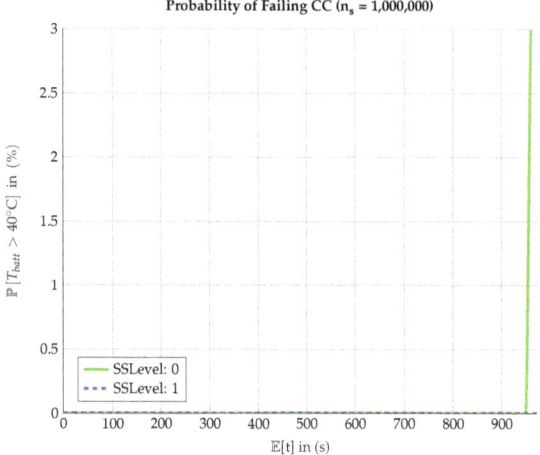

Figure 4. Probability of fulfilling the chance constraint over time for charging.

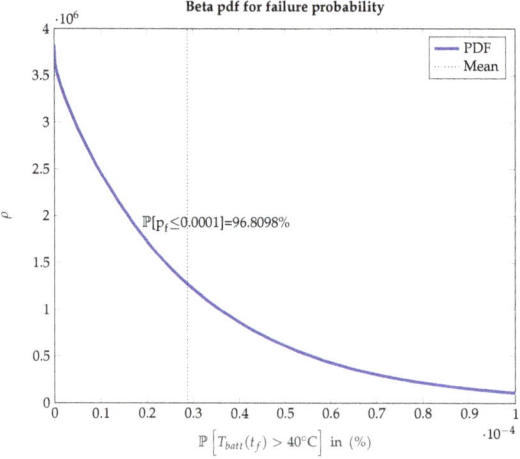

Figure 5. Estimation of failure probability density function from subset simulation for a final point on the optimization time horizon using a Beta distribution with mean value and probability density function area estimation until the allowed failure probability.

Finally, Figure 7 shows the development of the SoC for the mean value and the standard deviation. We observe a basically linear increase in the mean value reaching the desired charging level and a small standard deviation. Overall, the SoC is only subject to minor influences by the defined uncertainty that are mainly based on the temperature variations. Thus, although there is an uncertainty, we can still reach a similar charging level with the proposed robust open–loop optimal control (ROC) method. Furthermore, there are only minor differences between the different SubSim levels.

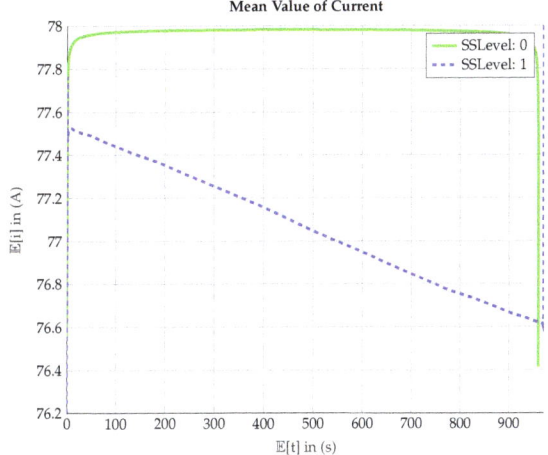

Figure 6. Robust control history for the current as the command variable over time for charging.

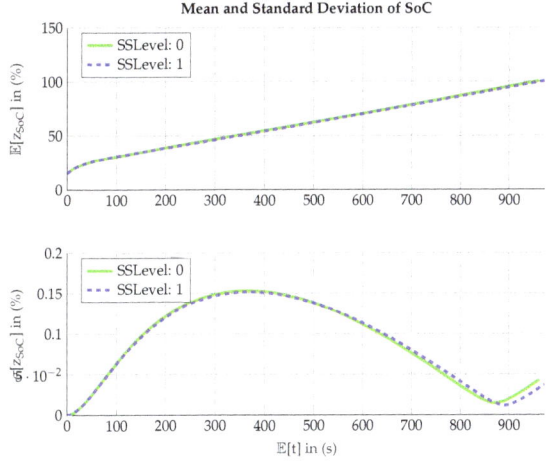

Figure 7. Mean and standard deviation value for state-of-charge over time for charging.

5.2. Battery Charge-Discharge Cycle Optimization

Within this section, we are looking at the full charge–discharge cycle: For this, Figure 8 shows the optimal current histories. In contrast to the single cycle, we can see that the current now is not reaching the maximal value anymore (78A; Section 4.2) and is also gradually decreasing over time. The differences between the SubSim levels are overall quite minimal but the same trend as for the results in Figure 6 can be observed, i.e., that the SubSim levels after the MCA have an overall lower level and are slightly longer.

Then, Figure 9 shows the fitted marginal distribution of the failure probability at the final point in time (i.e., the end time) once more. Once again, the pdf is plotted in the range of $[0, \mathbb{P}_{des}(\mathcal{F})]$, i.e., we cover the part of the pdf and its probability until the allowed maximal failure probability. Although we can see that the failure probability is now twice as large in the mean as for only the charge cycle (Figure 5), we can still be certain regarding our confidence in fulfilling the CC (around 91.5%).

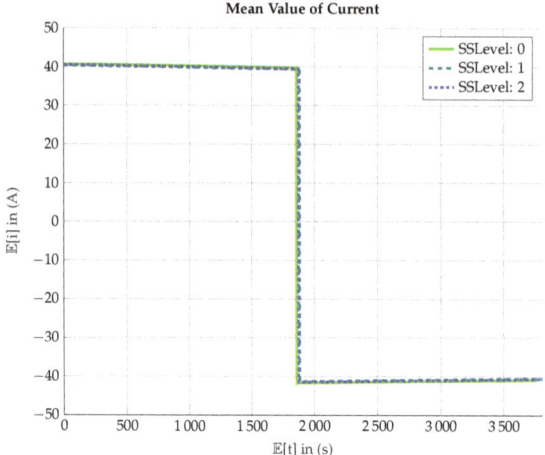

Figure 8. Robust control history for the current as the command variable over time for charge–discharge cycle.

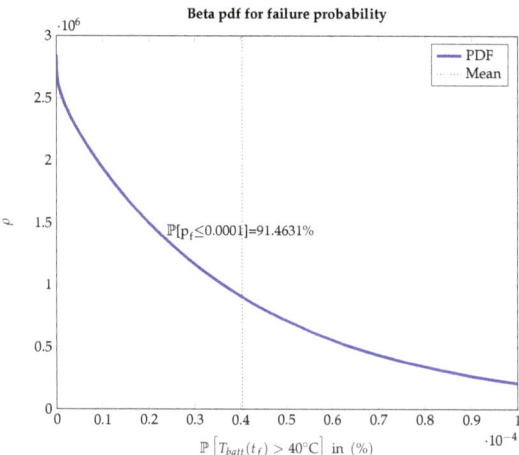

Figure 9. Estimation of failure probability density function for charge–discharge cycle from subset simulation for the final point on an optimization time horizon using a Beta distribution with mean value and probability density function area estimation until the allowed failure probability.

Now, Figure 10 shows the mean value of the SoC and its standard deviation. We can observe that the SoC is virtually independent of the SubSim level, but a major increase in the standard deviation can be seen, which is a consequence of the robust charging and the uncertainty. Generally, this increase is based on an increased standard deviation of the optimal time.

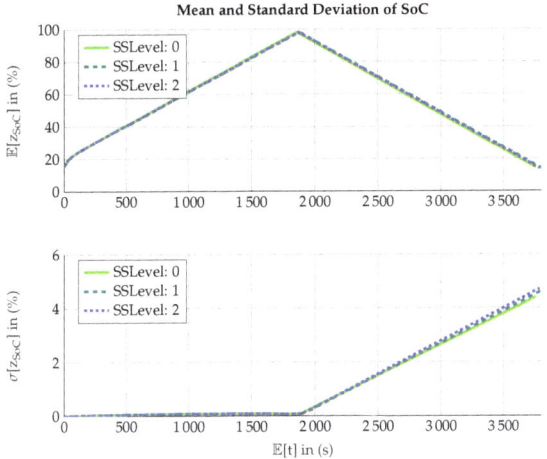

Figure 10. Mean and standard deviation value for state-of-charge over time for charge–discharge cycle.

Finally, Figure 11 depicts the battery temperature mean value and mean value with an added standard deviation. It should be noted that, in this figure, the standard deviation is added with a factor of $k_\sigma = 1.7321$: this value is chosen as it is exactly the value that yields the boundary value of the original uniform pdf if adding the standard deviation in Equation (25). As the optimization model is linear in the uncertainty, we can expect that the propagated uncertainty at the output is thus also almost linear. This is strengthened by Figure 11 as the solution that is offset by the factor $k_\sigma = 1.7321$ is just violating the constraint at the end, which is based on the CC fulfillment that must not be 100%. Again, the level 1 SubSim yields a reduced exceedance with a longer charge time.

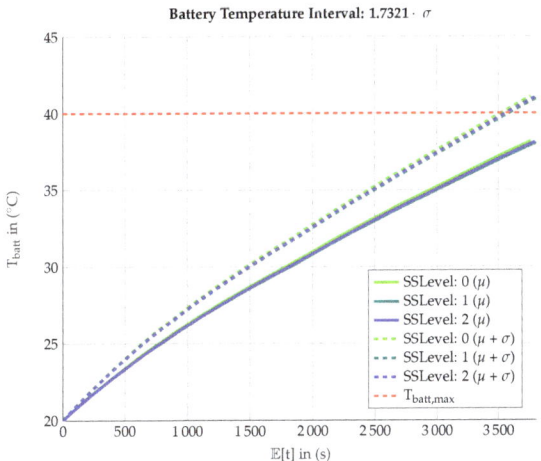

Figure 11. Mean battery temperature including the standard deviation interval provided by the underlying uniform parameter uncertainty over time for the charge–discharge cycle.

6. Conclusions

This paper presented an efficient method for ROC that relies on a CC–OC framework. In this framework, CCs are approximated by efficiently sampling from the gPC expansion. Therefore, a direct transcription method using the gPC expansion is applied for solving the OCP: by this, the gPC expansion can be evaluated in each optimization step as a matrix-vector operation and efficient sampling of the CC value, and thus its probability, is possible. In order to make the sharp CC bounds usable for Newton-type OC applications, the paper additionally introduced an approximation of CCs by means of sigmoids. This sigmoid approximation is gradually adapted using a homotopy strategy to reach a good approximation of the original sharp bound. Finally, the method of SubSim is applied to get the capability of calculating rare event failure probabilities. Overall, the applicability of the framework is shown using battery charge and discharge examples.

In the future, developments can be made combining the proposed method with distributed open–loop optimal control (DOC). The DOC framework can then be used to handle smaller sub-problems of the original OCP, which can be solved more efficiently. Here, also a distribution of the SubSim can be considered for improved efficiency.

Furthermore, more research should be directed into the suitable choice of the gPC expansion order to find a good response surface approximation for the evaluation of the CC. Here, especially the truncation errors of the gPC expansion as well as the SC evaluation must be considered. This is necessary to accurately calculate the probability of fulfilling the CC. In addition, e.g., a moment-matching optimization [27], which tries to find the best nodes-weights combination for the given problem, could be applied. Especially with a highly nonlinear model, larger orders of the gPC expansion must also be used, which decreases the efficiency. In the context of this large expansion order as well as a high-dimensional uncertainty space, further research must also be directed to efficient (adaptive) sparse grid implementations for the SC evaluation [19,28].

Additionally, in future applications, general pdfs can be introduced in the proposed method using the theory of arbitrary polynomial chaos [18,29] or Gaussian mixture models [30]. This could also be of special interest for the accurate choice of the SubSim evaluation points.

Finally, research can be directed into controlling the c.o.v. of the SubSim using the Beta pdf introduced in [25]. This can be beneficial to have a small dispersion of the data and thus high confidence in the calculated failure probability estimation and, consequently, the calculated robust optimal trajectories.

Author Contributions: The concept and methodology of this article was developed by P.P. and S.G. The formal analysis was carried out by P.P., who also wrote the original draft of this paper. S.G., P.P., and F.H. provided review and editing to the original draft. F.H. was responsible for funding acquisitions as well as supervision.

Funding: This work was supported by the Deutsche Forschungsgemeinschaft (DFG) through the Technical University of Munich (TUM) International Graduate School of Science and Engineering (IGSSE).

Acknowledgments: The authors want to acknowledge the German Research Foundation (DFG) and the Technical University of Munich (TUM) for supporting the publication in the framework of the Open Access Publishing Program.

Conflicts of Interest: The authors declare no conflict of interest.

References

1. Xiu, D.; Karniadakis, G.E. The Wiener-Askey Polynomial Chaos for Stochastic Differential Equations. *SIAM J. Sci. Comput.* **2002**, *24*, 619–644. [CrossRef]
2. Vitus, M.P. Stochastic Control via Chance Constrained Optimization and its Application to Unmanned Aerial Vehicles. Ph.D. Thesis, Stanford University, Stanford, CA, USA, 2012.

3. Zhao, Z.; Liu, F.; Kumar, M.; Rao, A.V. A novel approach to chance constrained optimal control problems. In Proceedings of the American Control Conference (ACC), Chicago, IL, USA, 1–3 July 2015; pp. 5611–5616. [CrossRef]
4. Mesbah, A. Stochastic Model Predictive Control: An Overview and Perspectives for Future Research. *IEEE Control Syst.* **2016**, *36*, 30–44. [CrossRef]
5. Diehl, M.; Bjornberg, J. Robust Dynamic Programming for Min–Max Model Predictive Control of Constrained Uncertain Systems. *IEEE Trans. Autom. Control* **2004**, *49*, 2253–2257. [CrossRef]
6. Lu, Y.; Arkun, Y. Quasi-Min-Max MPC algorithms for LPV systems. *Automatica* **2000**, *36*, 527–540. [CrossRef]
7. Villanueva, M.E.; Quirynen, R.; Diehl, M.; Chachuat, B.; Houska, B. Robust MPC via min–max differential inequalities. *Automatica* **2017**, *77*, 311–321. [CrossRef]
8. Gavilan, F.; Vazquez, R.; Camacho, E.F. Chance-constrained model predictive control for spacecraft rendezvous with disturbance estimation. *Control Eng. Pract.* **2012**, *20*, 111–122. [CrossRef]
9. Schaich, R.M.; Cannon, M. Maximising the guaranteed feasible set for stochastic MPC with chance constraints. *IFAC-PapersOnLine* **2017**, *50*, 8220–8225. [CrossRef]
10. Zhang, X.; Georghiou, A.; Lygeros, J. Convex approximation of chance-constrained MPC through piecewise affine policies using randomized and robust optimization. In Proceedings of the 2015 54th IEEE Conference on Decision and Control (CDC), Osaka, Japan, 15–18 December 2015; pp. 3038–3043. [CrossRef]
11. Au, S.K.; Beck, J.L. Estimation of small failure probabilities in high dimensions by subset simulation. *Probab. Eng. Mech.* **2001**, *16*, 263–277. [CrossRef]
12. Au, S.K. *Engineering Risk Assessment with Subset Simulation*; Wiley: Hoboken, NJ, USA, 2014. [CrossRef]
13. Betts, J.T. *Practical Methods for Optimal Control and Estimation Using Nonlinear Programming*, 2nd ed.; Advances in Design and Control; Society for Industrial and Applied Mathematics (SIAM 3600 Market Street Floor 6 Philadelphia PA 19104): Philadelphia, PA, USA, 2010.
14. Rieck, M.; Bittner, M.; Grüter, B.; Diepolder, J.; Piprek, P. FALCON.m User Guide, Institute of Flight System Dynamics, Technical University of Munich, 2019. Available online: www.falcon-m.com (accessed on 30 March 2019).
15. Wächter, A.; Biegler, L.T. On the implementation of an interior-point filter line-search algorithm for large-scale nonlinear programming. *Math. Program.* **2006**, *106*, 25–57. [CrossRef]
16. Wiener, N. The Homogeneous Chaos. *Am. J. Math.* **1938**, *60*, 897. [CrossRef]
17. Xiu, D. *Numerical Methods for sTochastic Computations: A Spectral Method Approach*; Princeton University Press: Princeton, NJ, USA, 2010.
18. Witteveen, J.A.; Bijl, H. Modeling Arbitrary Uncertainties Using Gram-Schmidt Polynomial Chaos. In Proceedings of the 44th AIAA Aerospace Sciences Meeting and Exhibit, Reno, Nevada, 9–12 January 2006.
19. Xiu, D. Fast Numerical Methods for Stochastic Computations: A Review. *Commun. Comput. Phys.* **2009**, *5*, 242–272.
20. Bittner, M. Utilization of Problem and Dynamic Characteristics for Solving Large Scale Optimal Control Problems. Dissertation, Technische Universität München, Garching, Germany, 2016.
21. Sherlock, C.; Fearnhead, P.; Roberts, G.O. The Random Walk Metropolis: Linking Theory and Practice Through a Case Study. *Stat. Sci.* **2010**, *25*, 172–190. [CrossRef]
22. Bucklew, J.A. *Introduction to Rare Event Simulation*; Springer Series in Statistics; Springer: New York, NY, USA, 2004. [CrossRef]
23. Au, S.K.; Patelli, E. Rare event simulation in finite-infinite dimensional space. *Reliab. Eng. Syst. Saf.* **2016**, *148*, 67–77. [CrossRef]
24. Papaioannou, I.; Betz, W.; Zwirglmaier, K.; Straub, D. MCMC algorithms for Subset Simulation. *Probab. Eng. Mech.* **2015**, *41*, 89–103. [CrossRef]
25. Zuev, K.M.; Beck, J.L.; Au, S.K.; Katafygiotis, L.S. Bayesian post-processor and other enhancements of Subset Simulation for estimating failure probabilities in high dimensions. *Comput. Struct.* **2012**, *92–93*, 283–296. [CrossRef]

26. Skötte, J. Optimal Charging Strategy for Electric Vehicles. Master's Thesis, Chalmers University of Technology, Gothenburg, Sweden, May 2018.
27. Paulson, J.A.; Mesbah, A. Shaping the Closed-Loop Behavior of Nonlinear Systems Under Probabilistic Uncertainty Using Arbitrary Polynomial Chaos. In Proceedings of the 2018 IEEE Conference on Decision and Control (CDC), Miami Beach, FL, USA, 17–19 December 2018; pp. 6307–6313. [CrossRef]
28. Blatman, G.; Sudret, B. Adaptive sparse polynomial chaos expansion based on least angle regression. *J. Comput. Phys.* **2011**, *230*, 2345–2367. [CrossRef]
29. Paulson, J.A.; Buehler, E.A.; Mesbah, A. Arbitrary Polynomial Chaos for Uncertainty Propagation of Correlated Random Variables in Dynamic Systems. *IFAC-PapersOnLine* **2017**, *50*, 3548–3553. [CrossRef]
30. Piprek, P.; Holzapfel, F. Robust Trajectory Optimization combining Gaussian Mixture Models with Stochastic Collocation. In Proceedings of the IEEE Conference on Control Technology and Applications (CCTA), Mauna Lani, HI, USA, 27–30 August 2017; pp. 1751–1756.

© 2019 by the authors. Licensee MDPI, Basel, Switzerland. This article is an open access article distributed under the terms and conditions of the Creative Commons Attribution (CC BY) license (http://creativecommons.org/licenses/by/4.0/).

Article

Global Evolution Commended by Localized Search for Unconstrained Single Objective Optimization

Rashida Adeeb Khanum [1,†], **Muhammad Asif Jan** [2,*,†], **Nasser Tairan** [3,†], **Wali Khan Mashwani** [2,†], **Muhammad Sulaiman** [4,†], **Hidayat Ullah Khan** [5,†] and **Habib Shah** [3,†]

1. Jinnah College for Women, University of Peshawar, Peshawar 25000, Pakistan; rakhan@uop.edu.pk
2. Institute of Numerical Sciences, Kohat University of Science & Technology, Kohat 26000, Pakistan; mashwanigr8@gmail.com
3. College of Computer Science, King Khalid University, Abha 61321, Saudi Arabia; nmtairan@kku.edu.sa (N.T.); habibshah.uthm@gmail.com (H.S.)
4. Department of Mathematics, Abdul Wali Khan University Mardan, Mardan 23200, Pakistan; sulaiman513@gmail.com
5. Department of Economics, Abbottabad University of Science & Technology, Abbottabad 22010, Pakistan; masmaleo@yahoo.com
* Correspondence: majan@kust.edu.pk or majan.math@gmail.com; Tel.: +92-313-998-6123
† These authors contributed equally to this work.

Received: 27 April 2019; Accepted: 27 May 2019; Published: 11 June 2019

Abstract: Differential Evolution (DE) is one of the prevailing search techniques in the present era to solve global optimization problems. However, it shows weakness in performing a localized search, since it is based on mutation strategies that take large steps while searching a local area. Thus, DE is not a good option for solving local optimization problems. On the other hand, there are traditional local search (LS) methods, such as Steepest Decent and Davidon–Fletcher–Powell (DFP) that are good at local searching, but poor in searching global regions. Hence, motivated by the short comings of existing search techniques, we propose a hybrid algorithm of a DE version, reflected adaptive differential evolution with two external archives (RJADE/TA) with DFP to benefit from both search techniques and to alleviate their search disadvantages. In the novel hybrid design, the initial population is explored by global optimizer, RJADE/TA, and then a few comparatively best solutions are shifted to the archive and refined there by DFP. Thus, both kinds of searches, global and local, are incorporated alternatively. Furthermore, a population minimization approach is also proposed. At each call of DFP, the population is decreased. The algorithm starts with a maximum population and ends up with a minimum. The proposed technique was tested on a test suite of 28 complex functions selected from literature to evaluate its merit. The results achieved demonstrate that DE complemented with LS can further enhance the performance of RJADE/TA.

Keywords: optimization; evolutionary computation; population minimization; hybridization; local search; global search; adaptive differential evolution; external archives; metaheuristics

---

1. Introduction

Nonlinear unconstrained optimization is an active research area, since many real-life challenges/problems can be modeled as a continuous nonlinear optimization problem [1]. To deal with this kind of optimization problems, various nature-inspired population based search mechanisms have been developed in the past [2]. A few of those are Differential Evolution (DE) [3,4], Evolution Strategies (ES) [2,5], Partical Swarm Optimization (PSO) [6–9], Ant Colony Optimization (ACO) [10–13], Bacterial Foraging Optimization (BFO) [14,15], Genetic Algorithm (GA) [16–18], Genetic Programming (GP) [2,19–21], Cuckoo Search (CS) [22,23], Estimation of Distribution Algorithm (EDA) [24–28] and Grey Wolf Optimization (GWO) [29,30].

DE does not need specific information about the complicated problem at hand [31]. That is why DE is implemented to solve a wide variety of optimization problems in the past two decades [30,32–34]. DE has merits over PSO, GA, ES and ACO, as it depends upon few control parameters. Its implementation is very easy and user friendly, too [2]. Due to these advantages, we selected DE to perform global search in the suggested hybrid design. In addition, because of its easy nature, DE is implemented widely [35–42] on practical optimization problems [35–42]. However, its convergence to known optima is not guaranteed [2,31,43]. Stagnation of DE is another weakness identified in various studies [31].

Traditional search approaches, such as Nelder–Mead algorithm, Steepest Descent and DFP [44] may be hybridized with DE to improve its search capability. Implementing LS into a global search for enhancing the solution quality is called Memetic Algorithms (MAs) [31,45]. Some of the recent MAs can be found in [1,31]. Very recently, Broyden–Fletcher–Goldfarb–Shanan LS was merged with an adaptive DE version, JADE [46], which produced the MA, Hybridization of Adaptive Differential Evolution with an Expensive Local Search Method [47]. In the majority of the established designs, LS is implemented to the overall best solutions, while in our design it is applied to the migrated elements of the archive. In addition, the population is adaptively decreased.

In this work, we propose a hybrid algorithm that combines DFP [44,48,49] with a recently developed algorithm, RJADE/TA [50], to enhance RJADE/TA's performance in local regions. The main idea is to operate DFP on the elements that are shifted to archive and record the information from both solutions, the previously brought forward and the new potential solutions to discourage the chance of losing the globally best solution. For this purpose, firstly, DFP is implemented to the archived information. Secondly, a decreasing population mechanism is suggested. The new algorithm is denoted by RJADE/TA-ADP-LS.

The structure of this work is as follows. Section 2 presents primary DE, DFP, and RJADE/TA methods. Section 3 describes the literature review. In Section 4, the suggested hybrid algorithm is outlined. Section 5 is devoted to the validation of results achieved by RJADE/TA-ADP-LS. At the end, the conclusions are summarized in Section 6.

2. Primary DE, DFP, and RJADE/TA

We reviewed in detail traditional DE and JADE in our previous works [47,50]. Here, we briefly review primary DE, DFP and RJADE/TA for ready reference.

2.1. Primary DE

DE [3,4] starts with a random population in the given search region. After initialization, a mutation strategy, where three different individuals from population are randomly selected and the scaled difference of the two individuals to the third one, target vector is added to produce a mutant vector. Following mutation, the mutant and the target vectors are combined through a crossover operator to produce a trial vector. At last, the target and trial vectors are compared based on a fitness function to select the better one for the next generation (see Lines 7–20 of Algorithm 1).

Algorithm 1 Outlines of RJADE/TA Procedure.

1: To form the primary population P_p produce $N^{[pop]}$ vectors uniformly and randomly, $\mathbf{w}_{[j,s_1]}^{\{y\}}, \mathbf{w}_{[j,s_2]}^{\{y\}}, \ldots, \mathbf{w}_{[j,s_{N^{[pop]}}]}^{[y]}$;
2: $M^{[first]} = M^{[sec]} = \varnothing$;
3: Initialize $\lambda CR = \lambda F = 0.5; p = 5\%; c = 0.1$;
4: Set $S_{CR} = S_F = \varnothing$;
5: Evaluate P_p;
6: **while** $FEs < MaxFEs$ **do**
7:     $F_j = rand(\lambda F, 0.1)$;
8:     Randomly sample $\mathbf{w}_{(best)}^{[p,y]}$ in $100p\%$ pop;
9:     Choose $\mathbf{w}_{[i,s_1]}^{\{y\}} \neq \mathbf{w}_{[i,s]}^{\{y\}}$ in P_p;
10:     Choose $\tilde{\mathbf{w}}_{[i,s_2]}^{\{y\}} \neq \mathbf{w}_{[i,s_2]}^{\{y\}}$ in $P_p \cup M^{[first]}$ do random selection;
11:     Produce the mutant vector $\mathbf{w}_{[i,mut]}^{\{y\}}$ as $\mathbf{w}_{[i,mut]}^{\{y\}} = \mathbf{w}_{[i,s]}^{\{y\}} + F_j(\mathbf{w}_{(best)}^{\{p,y\}} - \mathbf{w}_{[i,s]}^{\{y\}}) + F_j(\mathbf{w}_{[i,s_1]}^{\{y\}} - \tilde{\mathbf{w}}_{[i,s_2]}^{\{y\}})$;
12:     Produce the trial vector $\mathbf{q}_{[i,j]}^{\{y\}}$ as follows.
13:     **for** $i = 1$ to n **do**
14:         **if** $i < i_{rand}$ or $rand(0,1) < CR_j$ **then**
15:             $q_{[i,j]}^{\{y\}} = w_{[i,mut_j]}^{\{y\}}$;
16:         **else**
17:             $q_{[i,j]}^{\{y\}} = w_{[i,s_j]}^{\{y\}}$;
18:         **end if**
19:     **end for**
20:     Best selection $\{\mathbf{w}_{[i,s]}^{\{y\}}, \mathbf{q}_{[i,s]}^{\{y\}}\}$;
21:     **if** $\mathbf{q}_{[i,s]}^{\{y\}}$ is the best **then**
22:         $\mathbf{w}_{[i,s]}^{\{y\}} \to M^{[first]}, CR_j \to S_{CR}, F_j \to S_F$;
23:     **end if**
24:     If size of $M^{[first]} > N^{[pop]}$, delete extra solutions from $M^{[first]}$ randomly;
25:     Update $M^{[sec]}$ as follows.
26:     **if** $y = \kappa$ **then**
27:         $\mathbf{w}_{[j,best]}^{\{y\}} \to M^{[sec]}$;
28:         $P_p - \mathbf{w}_{[j,best]}^{\{y\}}$;
29:         Centroid calculation $\to \mathbf{w}_{[j,c]}^{\{y\}} = \frac{1}{N^{[pop]}-1}\sum_{i=2}^{N^{[pop]}} \mathbf{w}_{[j,c]}^{\{y\}}$;
30:         Reflection mechanism $\to \mathbf{w}_{[j,r]}^{\{y\}} = \mathbf{w}_{[j,c]}^{\{y\}} + (\mathbf{w}_{[j,c]}^{\{y\}} - \mathbf{w}_{[j,best]}^{\{y\}})$;
31:     **end if**
32:     $\lambda CR = (1-c) \cdot \lambda CR + c \cdot mean_A(S_{CR})$;
33:     $\lambda F = (1-c) \cdot \lambda F + c \cdot mean_L(S_F)$;
34: **end while**
35: **Result:** The best solution $\mathbf{w}_{(best)}^{\{y\}}$ corresponding to minimum function $f(\mathbf{w})$ value from $P_p \cup M^{[sec]}$ in the optimization.

2.2. Reflected Adaptive Differential Evolution with Two External Archives (RJADE/TA)

RJADE/TA [50] is an adaptive DE variant. Its main idea is to archive comparatively best solutions of the population at regular interval of optimization process and reflect the overall poor solutions. RJADE/TA inserts the following techniques in JADE. The techniques are presented in Table 1.

Table 1. Algorithmic parameters.

$M^{[first]}$	First archive	$M^{[sec]}$	Second archive
P_p	Primary population	$N^{[pop]}$	Population size
FEs	Function evaluations	$MaxFEs$	Maximum function evaluations
λ	FEs of RJADE/TA	κ	Gap between two successive updates of $M^{[sec]}$
λCR	Crossover probability	λF	Mutation scaling factor
S_{CR}	Set of successful crossover probabilities	S_F	Set of successful mutation factors
w	No. of iterations of DFP	r	Number of migrated solutions to $M^{[sec]}$
$\mathbf{w}_{[j,new]}^{\{y\}}$	j^{th} New candidate/solution at iteration y	$\mathbf{w}_{[j,best]}^{\{y\}}$	j^{th} Ever best candidate/solution at iteration y

To prevent premature convergence and stagnation, the best solution, $\mathbf{w}_{[j,best]}^{\{y\}}$ is replaced by its reflection in RJADE/TA and is then shifted to the second archive $M^{[sec]}$.

The reflected solution replaces $\mathbf{w}_{[j,best]}^{\{y\}}$ in the population and the ever best candidate $\mathbf{w}_{[j,best]}^{\{y\}}$ by itself is migrated to the second archive $M^{[sec]}$. RJADE/TA maintains two archives, termed as $M^{[first]}$ and $M^{[sec]}$ for convenience. After half of available resources are utilized ($MaxFEs$), the first archive update of the second archive, $M^{[sec]}$, is made. Afterwards, $M^{[sec]}$ is updated adaptively with a continuing intermission of generations (see Algorithm 1).

The overall best candidates are transferred to $M^{[sec]}$, whereas $M^{[first]}$ records the recently explored poor solutions. The size of $M^{[first]}$ is fixed, equal to population size $N^{[pop]}$, while the size of $M^{[sec]}$ may exceed $N^{[pop]}$. As $M^{[sec]}$ keeps information of all best solutions found, no solution is deleted from it. $M^{[sec]}$ records only one solution of the current iteration, it may be a child or a parent, whereas $M^{[first]}$ makes a history of more than one inferior "parent solutions" only. $M^{[first]}$ is updated at every iteration and $M^{[sec]}$, initialized as $\emptyset$, is updated with a gap of κ iterations adaptively. The recorded history of $M^{[first]}$ is utilized in reproduction later on. In contrast, in $M^{[sec]}$, the recorded best individual is reflected with a new solution, which is then sent to the population. Once a candidate solution is posted to $M^{[sec]}$, it remains passive during the whole optimization. When the search procedures are terminated, then the recoded information contributes towards the selection of the best candidate solution.

2.3. Davidon–Fletcher–Powell (DFP) Method

The DFP method is a variable metric method, which was first proposed by Davidon [51] and then modified by Powell and Fletcher [52]. It belongs to the class of gradient dependent LS methods. If a right line search is used in DFP method, it will assure convergence (minimization) [49]. It calculates the difference between the old and new points, as given in Equation (1). Then, it finds the difference of the gradients at these points as calculated in Equation (2).

$$\mathbf{t}_{\{w\}} = \mathbf{w}_{[j+1]}^{\{y\}} - \mathbf{w}^{[j]} \qquad (1)$$

$$\mathbf{t}_{\{g\}} = \nabla f(\mathbf{w}^{[j+1]}) - \nabla f(\mathbf{w}^{[j]}). \qquad (2)$$

It then updates the Hessian matrix $\mathbf{H}$ as presented in Equation (3). Afterwards, it locates the optimal search direction $\mathbf{s}^{[j]}$ with the help of the Hessian matrix information as calculated in Equation (4). Finally, the output solution $\mathbf{w}^{[j+1]}$ is computed by Equation (5), where $\alpha^{[j]}$ is calculated by a line search method; golden section search method is used in this work.

$$\mathbf{H}^{[j+1]} = \mathbf{H}^{[j]} + \frac{(\mathbf{t}_{\{w\}}^{'} \mathbf{t}_{\{w\}})}{\mathbf{t}_{\{w\}}^{'} \mathbf{t}_{\{g\}}} - \frac{(\mathbf{H}^{[j]} \mathbf{t}_{\{g\}}^{'} \mathbf{t}_{\{g\}} \mathbf{H}^{[j]})}{\mathbf{t}_{\{g\}} \mathbf{H}^{[j]} \mathbf{t}_{\{g\}}^{'}} \qquad (3)$$

$$\mathbf{s}^{[j]} = -\mathbf{H}^{[j]} \nabla f(\mathbf{w}^{[j]}) \tag{4}$$

$$\mathbf{w}^{[j+1]} = \mathbf{w}^{[j]} + \alpha^{[j]} \mathbf{s}^{[j]} \tag{5}$$

3. Related Work

To fix the above-mentioned weaknesses of DE, many researchers merged various LS techniques in DE. Nelder–Mead LS is hybridized with DE [53] to improve the local exploitation of DE. Recently, two new LS strategies are proposed and hybridized iteratively with DE in [1,31]. These hybrid designs show performance improvement over the algorithms in comparison. Two LS strategies, Trigonometric and Interpolated, are inserted in DE to enhance its poor exploration. Two other LS techniques are merged in DE along with a restart strategy to improve its global exploration [54]. This algorithm is statistically sound, as the obtained results are better than other algorithms. Furthermore, alopex-based LS is merged in DE [55] to improve its diversity of population. In another experiment, DE's slow convergence is enhanced by combining orthogonal design LS [56] with it. To avert local optima in DE, random LS is hybridized [57] with it. On the other hand, some researchers borrowed DE's mutation and crossover in traditional LS methods (see, e.g., [58,59]).

To the best of our knowledge, none of the reviewed algorithms in this section integrate DFP into DE's framework. Further, the proposed work here maintains two archives: the first one stores inferior solutions and the second one keeps information of best solutions migrated to it by the global search. Furthermore, the second archive improves the solutions quality further by implementing DFP there. Hence, our proposed work has the advantage that the second archive keeps complete information of the solution before and after LS. This way, any good solution found is not lost. It also adopts a population decreasing mechanism.

4. Developed Algorithm

As discussed in the literature review, LS techniques, due to their demerits, should not be used alone to solve optimization problems [2]. The global optimality of global evolution techniques is very high, but they can get stuck in local regions and cannot fine tune the solution at hand. Thus, motivated by above issues of global/local techniques, we hybridize a global optimizer RJADE/TA with DFP to enhance the convergence in both regions. The new design is named as RJADE/TA-ADP-LS. We specifically handle unconstrained, nonlinear, continuous, and single objective optimization problems in the current work.

RJADE/TA-ADP-LS

The initial population is evolved globally by RJADE/TA [50] until $\lambda\%$ of the function evaluations; that is, after RJADE/TA's iterative mutation, crossover, selection and $M^{[first]}$ process, as shown in Algorithm 1, the population is sorted and the current best solution $\mathbf{w}^{[k]}_{(i,best)}$ is translated to $M^{[sec]}$. This best solution may be a parent or a child solution. The DFP is applied to the shifted elements for w iterations. After implementation of DFP, a new improved solution $\mathbf{w}^{[k]}_{(i,new)}$ is produced from an old migrant. Then, the previously explored best solution and this new solution are posted to archive $M^{[sec]}$. Unlike our perviously proposed archive $M^{[sec]}$ in RJADE/TA, where the archive keeps the record of best solutions only and no LS is implemented, $M^{[sec]}$, as mentioned above, in this method maintains information of both solutions, i.e., the migrated best solution and its improved version, if any, after implementation of DFP.

The archive $M^{[sec]}$ is updated after regular intervals of κ generations (20 here). The migrated solutions and those explored by DFP remain there during the entire evolution process. When the evolution process completes, the overall best candidate is selected from $P_p \cup M^{[sec]}$. The novelty of RJADE/TA-ADP-LS is that it employs DFP to the archived solutions only, unlike all hybrid designs reviewed in Section 3.

In the proposed hybrid mechanism, we implement DFP to the migrated best solution to obtain its improved form, but without reflection, as displayed in the flowchart given in Figure 1, unlike in our recently proposed work [60]. Moreover, in this model, we propose adaptively decreasing population (ADP) mechanism different from the fixed population approach of Khanum et al. [60]. We refer to this new hybrid as RJADE/TA-ADP-LS throughout this work. The idea of RJADE/TA-ADP-LS is novel in proposing the ADP approach, because, in the literature, majority of the evolutionary algorithms (as reviewed in Section 3) maintain a fixed population throughout the searching process.

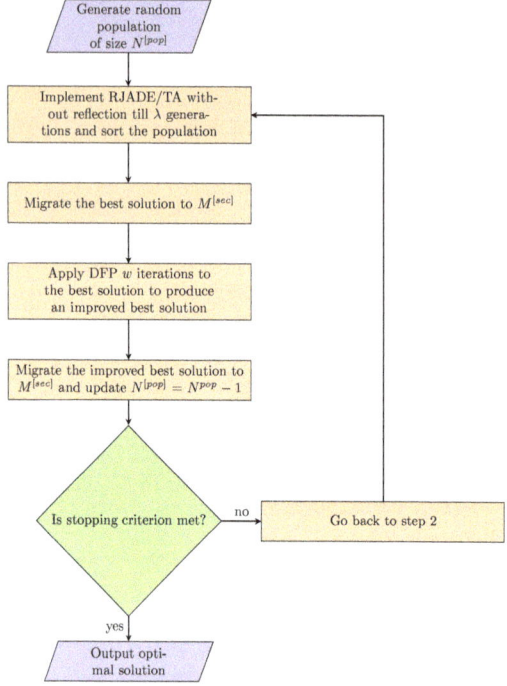

Figure 1. Flowchart of RJADE/TA-ADP-LS.

In this design, when the first update of $M^{[sec]}$ is made after half of the available resources are spent, DFP is applied to the archive members. The implementations of DFP and ADP are shown in Algorithm 2. Both the previously located best solution, $\mathbf{w}_{[j,best]}^{\{y\}}$, and the one exploited by DFP, $\mathbf{w}_{[j,new]}^{\{y\}}$, are propagated to $M^{[sec]}$. No reflection is made here to compensate the decreasing population. The ADP approach (Algorithm 2, Lines 6–8) is implemented as:

$$P_p = P_p - \mathbf{w}_{[j,best]}^{\{y\}}. \tag{6}$$

Hence,

$$P_p = \{\mathbf{w}_{[j,s_1]}^{\{y\}}, \mathbf{w}_{[j,s_2]}^{\{y\}}, \ldots, \mathbf{w}_{[j,s_{N^{[pop]}-1}]}^{\{y\}}\} \tag{7}$$

$$f(P_p) = \{f(\mathbf{w}_{[j,s_1]}^{\{y\}}), f(\mathbf{w}_{[j,s_2]}^{\{y\}}), \ldots, f(\mathbf{w}_{[j,s_{N^{[pop]}-1}]}^{\{y\}})\} \tag{8}$$

Every time $M^{[sec]}$ is updated, the migrated element is removed from the current population P_p (see Equation (6)), and the population is decreased by one. Thus, after each break of κ generations, $r(=$ the number of times the κ breaks occur) solutions are removed from $N^{[pop]}$, and the population size is updated to $N^{[pop]} - r$, as demonstrated on Line 11 of Algorithm 2. Furthermore, the function values are updated accordingly (see Equations (7) and (8)). In ADP approach, the algorithm begins with a maximum population and terminates with a minimum population.

Algorithm 2 RJADE/TA-ADP-LS.

1: Update $M^{[sec]}$ as follows.
2: **if** $k = \kappa$ **then**
3: $\quad \mathbf{w}_{[j,best]}^{\{y\}} \to M^{[sec]}$;
4: $\quad$ Apply DFP to $\mathbf{w}_{[j,best]}^{\{y\}}$ to prduce $\mathbf{w}_{[j,new]}^{\{y\}}$;
5: $\quad \mathbf{w}_{[j,new]}^{\{y\}} \to M^{[sec]}$;
6: $\quad P_p = \{\mathbf{w}_{([j,s_1])}^{\{y\}}, \mathbf{w}_{[j,s_2]}^{\{y\}}, \ldots, \mathbf{w}_{[j,s_{N^{[pop]}-1}]}^{\{y\}}\}$;
7: $\quad f(P_p) = \{f(\mathbf{w}_{[j,s_1]}^{\{y\}}), f(\mathbf{w}_{[j,s_2]}^{\{y\}}), \ldots, f(\mathbf{w}_{[j,s_{N^{[pop]}-1}]}^{\{y\}})\}$;
8: $\quad N^{[pop]} = N^{[pop]} - 1$;
9: **end if**
10: Terminate the iteration;
11: Repeat the process r number of times and update $N^{[pop]} = N^{[pop]} - r$.

5. Validation of Results

In this section, first we briefly illustrate the five algorithms used for comparison and then the experimental results are presented.

5.1. Global Search Algorithms in Comparison

Among the five algorithms for comparison, the first two, RJADE/TA and RJADE/TA-LS, are our recently proposed hybrid algorithms, while the remaining three, jDE, jDEsoo and jDErpo, are non-hybrid, but adaptive and popular DE variants.

5.1.1. RJADE/TA

RJADE/TA [50], similar to RJADE/TA-ADP-LS, utilizes two archives for information. One of the archives stores inferior solutions, while the other keeps a record of superior solutions. However, in RJADE/TA-ADP-LS, the second archive stores elite solutions, which are then improved by DFP. Further details of RJADE/TA can be seen in Section 2.2.

5.1.2. RJADE/TA-LS

RJADE/TA-LS [60] is a very recently proposed hybrid version of global and local search. However, it is different from RJADE/TA-ADP-LS in the sense that it utilizes reflection mechanism and a fixed population, while RJADE/TA-ADP-LS uses DFP as LS without reflection and a population decreasing approach.

5.1.3. jDE

jDE [61] is an adaptive version of DE, which is based on self-adaption of control parameters F and CR. In jDE, the parameters F and CR keep changing during the evolution process, while the population size $N^{[pop]}$ is kept unchanged. Every solution in jDE has its own F and CR values. Better individuals are produced due to better values of F and CR. Such parameter values translate to upcoming generations of jDE. Because of its unique mechanism and simplicity, jDE has gained popularity among researchers in the field of optimization. Since its establishment, people use it to compare with their own algorithms.

5.1.4. jDEsoo and jDErpo

jDEsoo [62] is a new version of DE that deals with single-objective optimization. jDEsoo subdivides the population and implements more than one DE strategies. To enhance diversity of population, it removes those individuals from population that remain unchanged in the last few generations. It was primarily developed for CEC 2013 competition.

jDErpo [61] is an improvement of jDE. It is based on the following mechanisms. Firstly, it incorporates two mutation strategies, different from jDE, DE and RJADE/TA. Secondly, it uses adaptively increasing strategy for adjusting the lower bounds of control parameters. Thirdly, it utilizes two pairs of control parameters for two different mutation strategies in contrast to one pair of parameters used in jDE, classic DE and RJADE/TA. jDErpo was also specially designed for solving CEC 2013 competition problems.

5.2. Parameter Settings/Termination Criteria

Experiments were performed on 28 benchmark test problems of CEC 2013 [63]. They are referred as BMF1–BMF28. The parameters' settings were kept the same as demanded in [63]. The dimension n of each problem was set to 10, population size $N^{[pop]}$ to 100, and the $MaxFEs$ to $10,000 \times n$. The number of elite solutions r was kept as 1. The iterations number w of DFP was set to 2. The reduction of population per archive update r was also chosen as 1. The gap κ between successive updates of $M^{[sec]}$ was kept as 20. The optimization was terminated if either $MaxFEs$ were reached or the difference between the means of function error values was less than 10^{-8}, as suggested in [50,63].

Table 2. Comparison of RJADE/TA-ADP-LS with Well Established Algorithms.

Bench Marks	jDE	jDEsoo	jDErpo	RJADE/TA	RJADE/TA-ADP-LS
BMF1	0.0000e + 0=	0.0000e + 0=	0.0000e + 0=	0.0000e + 0=	0.0000e + 0
BMF2	7.6534e − 05−	1.7180e + 03−	0.0000e + 0=	0.0000e + 0=	0.0000e + 00
BMF3	1.3797e + 0+	1.6071e + 0+	3.7193e − 05+	1.2108e + 02+	2.0350e + 02
BMF4	3.6639e − 08+	1.2429e − 01+	0.0000e + 0+	1.1591e + 02+	2.9749e + 02
BMF5	0.0000e + 0=	0.0000e + 0=	0.0000e + 0=	0.0000e + 0=	0.0000e + 00
BMF6	8.6581e + 0−	8.4982e + 04−	5.3872e + 0+	7.8884e + 0−	5.4656e + 00
BMF7	2.7229e − 03+	9.4791e − 01−	1.6463e − 03−	1.5927e − 01+	2.3707e − 01
BMF8	2.0351e + 01=	2.0348e + 01+	2.0343e + 01+	2.0366e + 01−	2.0352e + 01
BMF9	2.6082e + 0+	2.7464e + 0+	6.4768e − 01+	4.4593e + 0+	4.6182e + 00
BMF10	4.5263e − 02−	7.0960e − 02−	6.4469e − 02−	3.5342e − 02−	3.2488e − 02
BMF11	0.0000e + 0=	0.0000e + 0=	0.0000e + 0=	0.0000e + 0=	0.0000e + 0
BMF12	1.2304e + 01−	6.1144e + 0+	1.3410e + 01−	7.7246e + 0−	7.0574e + 00
BMF13	1.3409e + 01−	7.8102e + 0+	1.4381e + 01−	6.7571e + 0+	9.7072e + 00
BMF14	0.0000e + 0+	5.0208e − 02−	1.9367e + 01−	1.1994e − 02−	5.3105e − 03
BMF15	1.1650e + 03−	8.4017e + 02−	1.1778e + 03−	6.6660e + 02+	7.3411e + 02
BMF16	1.0715e + 0−	1.0991e + 0−	1.0598e + 0+	1.1336e + 0−	1.0545e + 00
BMF17	1.0122e + 01=	9.9240e + 0+	1.0997e + 01−	1.0122e + 01=	1.0122e + 01
BMF18	3.2862e + 01−	2.7716e + 01−	3.2577e + 01−	2.2715e + 01+	2.4399e + 01
BMF19	4.3817e − 01−	3.1993e − 01−	7.4560e − 01−	4.4224e − 01−	4.2674e − 01
BMF20	3.0270e + 0−	2.7178e + 0−	2.5460e + 0+	2.5317e + 0+	2.6153e + 00
BMF21	3.7272e + 02+	3.5113e + 02+	3.7272e + 02+	3.9627e + 02+	4.0019e + 02
BMF22	7.9231e + 01+	9.1879e + 01−	9.7978e + 01−	2.7022e + 01−	1.3178e + 01
BMF23	1.1134e + 03−	8.1116e + 02−	1.1507e + 03−	7.0015e + 02−	4.8553e + 02
BMF24	2.0580e + 02−	2.0851e + 02−	1.8865e + 02−	2.0217e + 02−	1.0823e + 02
BMF25	2.0471e + 02−	2.0955e + 02−	1.9885e + 02−	2.0314e + 02−	1.7732e + 02
BMF26	1.8491e + 02−	1.9301e + 02−	1.1732e + 02+	1.2670e + 02−	1.2096e + 02
BMF27	4.7470e + 02−	4.9412e + 02−	3.0000e + 02+	3.0351e + 02−	3.0514e + 02
BMF28	2.9216e + 02−	2.8824e + 02−	2.9608e + 02−	2.8824e + 02−	2.8500e + 02
−	17	17	14	13	
+	6	8	10	10	
=	5	3	4	5	

5.3. Comparison of RJADE/TA-ADP-LS against Established Global Optimizers

The mean of function error values, the difference between known and approximated values, for jDE, jDEsoo, jDErpo, RJADE/TA and RJADE/TA-ADP-LS, are presented in Table 2. In Table 2,

+ indicates that the algorithm won against our algorithm, RJADE/TA-ADP-LS; − indicates that the particular algorithm lost against our algorithm; and = indicates that both algorithms obtained the same statistics. The comparison of RJADE/TA-ADP-LS with other competitors showed its outstanding performance against all of them. RJADE/TA-ADP-LS achieved higher mean values than jDE and jDEsoo on 17 out of 28 problems; the many − signs in columns 2 and 3 of Table 2 support this fact. In contrast, jDE and jDEsoo performed better on six and eight problems, respectively.

RJADE/TA-ADP-LS showed performance improvement against jDErpo and RJADE/TA algorithms as well. In general, RJADE/TA-ADP-LS performed better than all algorithms in comparison, especially in the category of multimodal and composite functions. The proposed mechanism is not only based on LS for local tuning with no reflection, but it also implements an ADP approach, which could be the reasons for its good performance.

5.4. Performance Evaluation of RJADE/TA-ADP-LS Versus RJADE/TA-LS

We empirically studied the performance of RJADE/TA-ADP-LS against RJADE/TA-LS. Table 3 presents the mean results achieved by both methods in 51 runs. The best results are shown in bold face. It is very clear from the results in Table 3 that the proposed RJADE/TA-ADP-LS performed higher than RJADE/TA-LS on 13 out of 28 problems. Furthermore, on five problems, they obtained the same results. RJADE/TA-LS showed performance improvement on 10 test problems.

Table 3. Comparing RJADE/TA-ADP-LS with RJADE/TA-LS.

		BMF1	BMF2	BMF3	BMF4	BMF5	BMF6	BMF7
RJADE/TA-LS		**0.0000e+00**	**0.0000e+00**	2.5750e + 02	**3.9511e+01**	**0.0000e+00**	6.9264e + 00	**2.3707e−01**
RJADE/TA-ADP-LS	Mean	**0.0000e+00**	**0.0000e+00**	**2.0350e + 02**	2.9749e+02	**0.0000e+00**	**5.4656e+00**	**2.3707e − 01**
		BMF8	BMF9	BMF10	BMF11	BMF12	BMF13	BMF14
RJADE/TA-LS		**2.0342e+01**	4.4888e + 00	**3.2488e-02**	**0.0000e+00**	6.8613e+00	7.9039e+00	7.3105e − 003
RJADE/TA-ADP-LS	Mean	2.0352e + 01	**4.6182e+00**	**3.2488e-02**	**0.0000e+00**	**7.0574e + 00**	**9.7072e + 00**	**5.3105e-03**
		BMF15	BMF16	BMF17	BMF18	BMF19	BMF20	BMF21
RJADE/TA-LS		**6.6733e+02**	1.0855e + 00	**1.0122e+01**	**1.0122e+01**	4.4752e − 01	2.5707e+00	**3.9627e+02**
RJADE/TA-ADP-LS	Mean	7.3411e + 02	**1.0545e+00**	**1.0122e+01**	2.4399e + 01	**4.2674e-01**	**2.6153e + 00**	4.0019e + 02
		BMF22	BMF23	BMF24	BMF25	BMF26	BMF27	BMF28
RJADE/TA-LS		2.0589e + 01	6.7549e + 02	1.9809e + 02	2.0190e + 02	1.3596e + 02	**3.0033e+02**	3.0000e + 02
RJADE/TA-ADP-LS	Mean	**1.3178e+01**	**4.8553e+02**	**1.0823e+02**	**1.7732e+02**	**1.2096e+02**	3.0514e + 02	**2.8500e+02**

It is interesting to note that RJADE/TA-ADP-LS showed outstanding performance in the category of composite functions, where it solved BMF22–BMF28 better than RJADE/TA-LS. Again, the two different mechanisms, the ADP approach and the LS search with out reflection, of RJADE/TA-ADP-LS could be the reasons for its better performance. Among 28 problems, RJADE/TA-LS was better on 10 functions. Further, Table 4 presents the percentage performance of RJADE/TA-ADP-LS and RJADE/TA-LS. Since on five test problems, both algorithms showed equal results, thus we compared the percentage for the remaining 23 problems. As shown in Table 4, RJADE/TA-ADP-LS was able to solve 57% of problems against 43% of problems solved by RJADE/TA-LS out of 23 test instances.

Table 4. Comparing RJADE/TA-ADP-LS with RJADE/TA-LS.

Algorithms	RJADE/TA-ADP-LS	RJADE/TA-LS
Number of Problems solved in total of 23	13 of 23	10 of 23
% age	57%	43%

Furthermore, box plots were plotted from all means obtained in 25 runs of RJADE/TA, RJADE/TA-LS and RJADE/TA-ADP-LS. Figures 2 and 3 plot one function from each three functions.

Box plots are very good tools to show the spread of the data. Figure 2b–d shows that the boxes obtained by RJADE/TA-ADP-LS were lower than the other two boxes, indicating its better performance. Figure 2a presents the plot of BMF3, in which the two boxes in comparison were lower than RJADE/TA-ADP-LS, thus they were better.

Figure 3b,d,f shows that the boxes obtained by RJADE/TA-ADP-LS on BMF19, BMF25 and BMF27 were lower than the boxes of RJADE/TA and RJADE/TA-LS, indicating higher performance of RJADE/TA-ADP-LS. Figure 3a,c,e shows that the two other algorithms were better on the respective test instances.

5.5. Analysis/Discussion of Various Parameters Used

The number of solutions r to be migrated to archive and undergo DFP was kept as 1, since DFP is an expensive method due to gradient calculation. Further, its application to more than one solution might slow down the algorithm. The users may take two, but at most three is suggested. The number of iteration w of DFP to archive elements was kept as 2. DFP is a very good method; it could fine tune the solutions in only two iterations. Moreover, the decreasing number r of population per archive update was also chosen as 1. Since the archive was updated after regular gap of global evolution, each time population was decreased by one. However, if we reduced it by more than one solutions, then a stage would come where the diversity of the population would be decreased and the algorithm would either stop at local optima or converge prematurely. We suggest that the decreasing number be at most 3. In general, these parameters are user defined but should be chosen wisely to compliment the global and local search together, instead of premature convergence or stagnation.

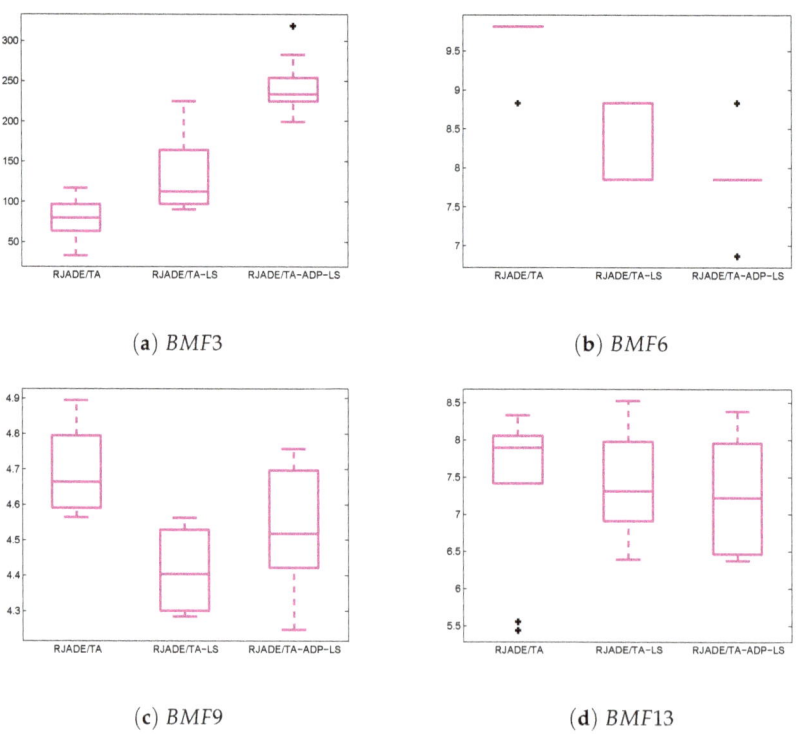

Figure 2. Box plots of various algorithms in comparison.

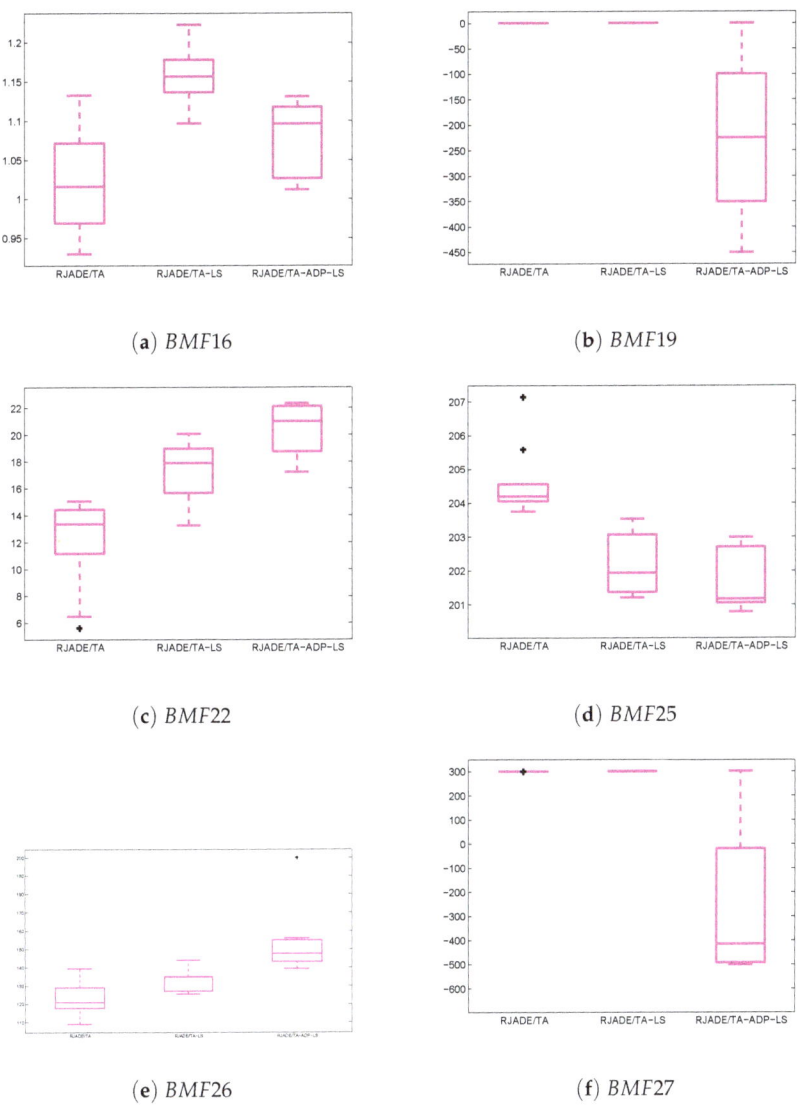

Figure 3. Box plots of various algorithms in comparison.

6. Conclusions

This paper proposed a new hybrid algorithm, RJADE/TA-ADP-LS, where a LS mechanism, DFP is combined with a DE based global search scheme, RJADE/TA to benefit from their searching capabilities in local and global regions. Further, a population decreasing mechanism is also adopted. The key idea is to shift the overall best solution to archive at specified regular intervals of RJADE/TA, where it undergoes DFP for further improvement. The archive stores both the best solution and its improved form. Furthermore, the population is decreased by one solution at each archive update. We evaluated and compared our hybrid method with five established algorithms on test suit of CEC 2013. The results demonstrated that our new algorithm is better than other competing algorithms on majority of the tested problems, particularly our algorithm showed superior performance on hard multimodal and composite problems of CEC 2013. In future, the present work will be extended to constrained optimization. As a second task, some other gradient free LS methods, global optimizers and archiving strategies will be tried to design more efficient algorithms for global optimization.

Author Contributions: Conceptualization, R.A.K. and M.A.J.; methodology, R.A.K., M.A.J., and W.K.M.; software, R.A.K., N.T. and H.S.; validation, H.U.K., M.S. and H.S.; formal analysis, R.A.K., M.A.J., and W.K.M.; investigation, R.A.K., M.A.J. and M.S.; resources, N.T. and H.S.; writing—original draft preparation, R.A.K., M.A.J.; writing—review and editing, H.U.K. and M.S.; project administration, N.T.; and funding acquisition, N.T. and H.S.

Funding: The authors would like to thank King Khalid University of Saudi Arabia for supporting this research under the grant number R.G.P.2/7/38.

Conflicts of Interest: The authors declare no conflict of interest.

References

1. Price, K.V. Eliminating drift bias from the differential evolution algorithm. In *Advances in Differential Evolution*; Springer: Berlin, Germany, 2008; pp. 33–88.
2. Xiong, N.; Molina, D.; Ortiz, M.L.; Herrera, F. A walk into metaheuristics for engineering optimization: principles, methods and recent trends. *Int. J. Comput. Intell. Syst.* **2015**, *8*, 606–636. [CrossRef]
3. Storn, R.; Price, K.V. Differential Evolution—A simple and efficient heuristic for global optimization over continuous spaces. *J. Glob. Optim.* **1997**, *11*, 341–359. [CrossRef]
4. Storn, R. Differential evolution research—Trends and open questions. In *Advances in Differential Evolution*; Springer: Berlin, Germany, 2008; pp. 1–31.
5. Engelbrecht, A.; Pampara, G. Binary Differential Evolution Strategies. In Proceedings of the IEEE Congress on Evolutionary Computation (CEC 2007), Singapore, 25–28 September 2007; pp. 1942–1947.
6. Kennedy, J.; Eberhart, R.C. Particle Swarm Optimization. In Proceedings of the IEEE International Conference on Neural Networks, Perth, WA, Australia, 27 November–1 December 1995; pp. 1942–1948.
7. Kennedy, J.; Eberhart, R. A Discrete Binary Version of the Partical Swarm Algorithm. In Proceedings of the World Multiconference on Systemics, Cybernetics and Informatics, Orlando, FL, USA, 12–15 October 1997; pp. 4104–4109.
8. Eberhart, R.C.; Kennedy, J. A New Optimizer using Particle Swarm Theory. In Proceedings of the 6th International Symposium on Micromachine and Human Science, Nagoya, Japan, 4–6 October 1995; pp. 39–43.
9. Eberhart, R.C.; shi, Y. Guest Editorial: Special Issue on Particle Swarm Optimization. *IEEE Trans. Evol. Comput.* **2004**, *8*, 201–203. [CrossRef]
10. Dorigo, M. Ant colony optimization. *Scholarpedia* **2007**, *2*, 1461. [CrossRef]
11. Dorigo, M.; Birattari, M. Ant colony optimization. In *Encyclopedia of Machine Learning*; Springer: Berlin, Germany, 2011; pp. 36–39.
12. Al-Salami, N.M. System evolving using ant colony optimization algorithm. *J. Comput. Sci.* **2009**, *5*, 380. [CrossRef]
13. Cui, L.; Zhang, K.; Li, G.; Wang, X.; Yang, S.; Ming, Z.; Huang, J.Z.; Lu, N. A smart artificial bee colony algorithm with distance-fitness-based neighbor search and its application. *Future Gener. Comput. Syst.* **2018**, *89*, 478–493. [CrossRef]
14. Passino, K.M. Bacterial foraging optimization. *Int. J. Swarm Intell. Res. (IJSIR)* **2010**, *1*, 1–16. [CrossRef]
15. Gazi, V.; Passino, K.M. Bacteria foraging optimization. In *Swarm Stability and Optimization*; Springer: Berlin, Germany, 2011; pp. 233–249.
16. Moscato, P. On evolution, search, optimization, genetic algorithms and martial arts: Towards memetic algorithms. *Caltech Concurr. Comput. Prog. C3P Rep.* **1989**, *826*, 1989.
17. Fan, S.K.S.; Zahara, E. A hybrid Simplex Search and Partical Swarm optimization for unconstrained optimization. *Eur. J. Oper. Res.* **2007**, *181*, 527–548. [CrossRef]
18. Yuen, S.Y.; Chow, C.K. A Genetic Algorithm that Adaptively Mutates and Never Revisits. *IEEE Trans. Evol. Comput.* **2009**, *13*, 454–472. [CrossRef]
19. Koza, J.R. *Genetic Programming II, Automatic Discovery Of Reusable Subprograms*; MIT Press: Cambridge, MA, USA, 1992.

20. Koza, J.R. Genetic programming as a means for programming computers by natural selection. *Stat. Comput.* **1994**, *4*, 87–112. [CrossRef]
21. Koza, J.R. *Genetic Programming: On the Programming of Computers by Means of Natural Selection*; MIT Press: Cambridge, MA, USA, 1992.
22. Yang, X.S.; Deb, S. Cuckoo search via Lévy flights. In Proceedings of the IEEE World Congress on Nature & Biologically Inspired Computing, Coimbatore, India, 9–11 December 2009; pp. 210–214.
23. Yang, X.S.; Deb, S. Engineering optimisation by cuckoo search. *arXiv* **2010**, arXiv:1005.2908.
24. Larrañaga, P.; Lozano, J.A. *Estimation of Distribution Algorithms: A New Tool for Evolutionary Computation*; Springer: Berlin, Germany, 2001; Volume 2.
25. Zhang, Q.; Sun, J.; Tsang, E.; Ford, J. Hybrid estimation of distribution algorithm for global optimization. *Eng. Comput.* **2004**, *21*, 91–107. [CrossRef]
26. Zhang, Q.; Muhlenbein, H. On the convergence of a class of estimation of distribution algorithms. *IEEE Trans. Evol. Comput.* **2004**, *8*, 127–136. [CrossRef]
27. Lozano, J.A.; Larrañaga, P.; Inza, I.; Bengoetxea, E. *Towards a New Evolutionary Computation: Advances on Estimation of Distribution Algorithms*; Springer: Berlin, Germany, 2006; Volume 192.
28. Hauschild, M.; Pelikan, M. An introduction and survey of estimation of distribution algorithms. *Swarm Evol. Comput.* **2011**, *1*, 111–128. [CrossRef]
29. Mirjalili, S.; Mirjalili, S.M.; Lewis, A. Grey wolf optimizer. *Adv. Eng. Softw.* **2014**, *69*, 46–61. [CrossRef]
30. Gupta, S.; Deep, K. Hybrid Grey Wolf Optimizer with Mutation Operator. In *Soft Computing for Problem Solving*; Springer: Berlin, Germany, 2019; pp. 961–968.
31. Leon, M.; Xiong, N. Eager random search for differential evolution in continuous optimization. In *Portuguese Conference on Artificial Intelligence*; Springer: Berlin, Germany, 2015; pp. 286–291.
32. Maučec, M.S.; Brest, J.; Bošković, B.; Kačič, Z. Improved Differential Evolution for Large-Scale Black-Box Optimization. *IEEE Access* **2018**, *6*, 29516–29531. [CrossRef]
33. Biswas, P.P.; Suganthan, P.; Wu, G.; Amaratunga, G.A. Parameter estimation of solar cells using datasheet information with the application of an adaptive differential evolution algorithm. *Renew. Energy* **2019**, *132*, 425–438. [CrossRef]
34. Sacco, W.F.; Rios-Coelho, A.C. On Initial Populations of Differential Evolution for Practical Optimization Problems. In *Computational Intelligence, Optimization and Inverse Problems with Applications in Engineering*; Springer: Berlin, Germany, 2019; pp. 53–62.
35. Wu, G.; Shen, X.; Li, H.; Chen, H.; Lin, A.; Suganthan, P. Ensemble of differential evolution variants. *Inf. Sci.* **2018**, *423*, 172–186. [CrossRef]
36. Awad, N.H.; Ali, M.Z.; Mallipeddi, R.; Suganthan, P.N. An improved differential evolution algorithm using efficient adapted surrogate model for numerical optimization. *Inf. Sci.* **2018**, *451*, 326–347. [CrossRef]
37. Al-Dabbagh, R.; Neri, F.; Idris, N.; Baba, M. Algorithm Design Issues in Adaptive Differential Evolution: Review and taxonomy. *Swarm Evol. Comput.* **2018**, *43*, 284–311. [CrossRef]
38. Betzig, L.L. *Despotism, Social Evolution, and Differential Reproduction*; Routledge: Abingdon, UK, 2018.
39. Opara, K.R.; Arabas, J. Differential Evolution: A survey of theoretical analyses. *Swarm Evol. Comput.* **2018**, *44*, 546–558. [CrossRef]
40. Das, S.; Mullick, S.S.; Suganthan, P. Recent advances in differential evolution An updated survey. *Swarm Evol. Comput.* **2016**, *27*, 1–30. [CrossRef]
41. Cui, L.; Huang, Q.; Li, G.; Yang, S.; Ming, Z.; Wen, Z.; Lu, N.; Lu, J. Differential Evolution Algorithm With Tracking Mechanism and Backtracking Mechanism. *IEEE Access* **2018**, *6*, 44252–44267. [CrossRef]
42. Cui, L.; Li, G.; Zhu, Z.; Ming, Z.; Wen, Z.; Lu, N. Differential evolution algorithm with dichotomy-based parameter space compression. *Soft Comput.* **2018**, *23*, 1–18. [CrossRef]
43. Meng, Z.; Pan, J.S.; Zheng, W. Differential evolution utilizing a handful top superior individuals with bionic bi-population structure for the enhancement of optimization performance. *Enterpr. Inf. Syst.* **2018**, 1–22. [CrossRef]
44. Fletcher, R. *Practical Methods of Optimization*, 2nd ed.; Wiley: Hoboken, NJ, USA, 1987; pp. 80–87.
45. Lozano, M.; Herrera, F.; Krasnogor, N.; Molina, D. Real-Coded Memetic Algorithms with Crossover Hill-Climbing. *Evol. Comput.* **2004**, *12*, 273–302. [CrossRef] [PubMed]
46. Zhang, J.; Sanderson, A.C. JADE: adaptive differential evolution with optional external archive. *IEEE Trans. Evol. Comput.* **2009**, *13*, 945–958. [CrossRef]

47. Khanum, R.A.; Jan, M.A.; Tairan, N.M.; Mashwani, W.K. Hybridization of Adaptive Differential Evolution with an Expensive Local Search Method. *J. Optim.* **2016**, *1016*, 1–14. [CrossRef]
48. Davidon, W.C. Variable metric method for minimization. *SIAM J. Optim.* **1991**, *1*, 1–17. [CrossRef]
49. Antoniou, A.; Lu, W.S. *Practical Optimization: Algorithms and Engineering Applications*; Springer: Berlin, Germany, 2007.
50. Khanum, R.A.; Tairan, N.; Jan, M.A.; Mashwani, W.K.; Salhi, A. Reflected Adaptive Differential Evolution with Two External Archives for Large-Scale Global Optimization. *Int. J. Adv. Comput. Sci. Appl.* **2016**, *7*, 675–683.
51. Spedicato, E.; Luksan, L. Variable metric methods for unconstrained optimization and nonlinear least squares. *J. Comput. Appl. Math.* **2000**, *124*, 61–95.
52. Mamat, M.; Dauda, M.; bin Mohamed, M.; Waziri, M.; Mohamad, F.; Abdullah, H. Derivative free Davidon-Fletcher-Powell (DFP) for solving symmetric systems of nonlinear equations. *IOP Conf. Ser. Mater. Sci. Eng.* **2018**, *332*, 012030. [CrossRef]
53. Ali, M.; Pant, M.; Abraham, A. Simplex Differential Evolution. *Acta Polytech. Hung.* **2009**, *6*, 95–115.
54. Khanum, R.A.; Jan, M.A.; Mashwani, W.K.; Tairan, N.M.; Khan, H.U.; Shah, H. On the hybridization of global and local search methods. *J. Intell. Fuzzy Syst.* **2018**, *35*, 3451–3464. [CrossRef]
55. Leon, M.; Xiong, N. A New Differential Evolution Algorithm with Alopex-Based Local Search. In *International Conference on Artificial Intelligence and Soft Computing*; Springer: Berlin, Germany, 2016; pp. 420–431.
56. Dai, Z.; Zhou, A.; Zhang, G.; Jiang, S. A differential evolution with an orthogonal local search. In Proceedings of the IEEE Congress on Evolutionary Computation, Cancun, Mexico, 20–23 June 2013; pp. 2329–2336.
57. Ortiz, M.L.; Xiong, N. Using random local search helps in avoiding local optimum in differential evolution. In Proceedings of the IASTED, Innsbruck, Austria, 17–19 February 2014; pp. 413–420.
58. Khanum, R.A.; Zari, I.; Jan, M.A.; Mashwani, W.K. Reproductive nelder-mead algorithms for unconstrained optimization problems. *Sci. Int.* **2015**, *28*, 19–25.
59. Zari, I.; Khanum, R.A.; Jan, M.A.; Mashwani, W.K. Hybrid (N)elder-mead algorithms for nonlinear numerical optimization. *Sci. Int.* **2015**, *28*, 153–159.
60. Khanum, R.A.; Jan, M.A.; Mashwani, W.K.; Khan, H.U.; Hassan, S. RJADETA integrated with local search for continuous nonlinear optimization. *Punjab Univ. J. Math.* **2019**, *51*, 37–49.
61. Brest, J.; Zamuda, A.; Fister, I.; Boskovic, B. Some Improvements of the Self-Adaptive jDE Algorithm. In Proceedings of the IEEE Symposium on Differential Evolution (SDE), Orlando, FL, USA, 9–12 December 2014; pp. 1–8.
62. Brest, J.; Boskovic, B.; Zamuda, A.; Fister, I.; Mezura-Montes, E. Real Parameter Single Objective Optimization using self-adaptive differential evolution algorithm with more strategies. In Proceedings of the IEEE Congress on Evolutionary Computation (CEC), Cancun, Mexico, 20–23 June 2013; pp. 377–383.
63. Liang, J.; Qu, B.; Suganthan, P.; Hernández-Díaz, A.G. Problem definitions and evaluation criteria for the CEC 2013 special session on real-parameter optimization, 2013. Available online: http://al-roomi.org/multimedia/CEC_Database/CEC2013/RealParameterOptimizationCEC2013_RealParameterOptimization_TechnicalReport.pdf (accessed on 22 April 2019).

© 2019 by the authors. Licensee MDPI, Basel, Switzerland. This article is an open access article distributed under the terms and conditions of the Creative Commons Attribution (CC BY) license (http://creativecommons.org/licenses/by/4.0/).

Article

PEM Fuel Cell Voltage Neural Control Based on Hydrogen Pressure Regulation

Andrés Morán-Durán, Albino Martínez-Sibaja *, José Pastor Rodríguez-Jarquin, Rubén Posada-Gómez and Oscar Sandoval González

Division of Postgraduate and Research studies, Tecnológico Nacional de México/Instituto Tecnológico de Orizaba, Orizaba, Veracruz 92670, México
* Correspondence: albino3_mx@yahoo.com; Tel.: +52-272-183-3256

Received: 30 May 2019; Accepted: 24 June 2019; Published: 10 July 2019

Abstract: Fuel cells are promising devices to transform chemical energy into electricity; their behavior is described by principles of electrochemistry and thermodynamics, which are often difficult to model mathematically. One alternative to overcome this issue is the use of modeling methods based on artificial intelligence techniques. In this paper is proposed a hybrid scheme to model and control fuel cell systems using neural networks. Several feature selection algorithms were tested for dimensionality reduction, aiming to eliminate non-significant variables with respect to the control objective. Principal component analysis (PCA) obtained better results than other algorithms. Based on these variables, an inverse neural network model was developed to emulate and control the fuel cell output voltage under transient conditions. The results showed that fuel cell performance does not only depend on the supply of the reactants. A single neuro-proportional–integral–derivative (neuro-PID) controller is not able to stabilize the output voltage without the support of an inverse model control that includes the impact of the other variables on the fuel cell performance. This practical data-driven approach is reliably able to reduce the cost of the control system by the elimination of non-significant measures.

Keywords: feature selection; PEM fuel cell; control; neural network; principal component analysis; modeling; system identification

---

1. Introduction

The constant increase in energy consumption, environmental issues, and the rapid exhaustion of fossil fuel reservoirs have motivated researchers around the world to design renewable solutions to this global challenge [1]. Hydrogen is a potential energy renewable source, and it could be the clean fuel of the future [2]; its main characteristics are as follows:

- Hydrogen has the highest energy content per unit weight (142 kJ g^{-1}) [3];
- It is a carbon-free fuel due to its combustion product being water [4];
- Hydrogen can be used as a direct fuel or as an energy carrier for a fuel cell [4].

"One of the most promising hydrogen energy conversion technologies is the fuel cell" [5]. However, fuel cells need an operational control strategy supported by a fault detection and isolation method which can reconfigure the energy system to overcome potential faults and increase both the reliability and useful life of the fuel cell [6].

1.1. Fuel Cell Operation Principles

Fuel cells are devices that transform chemical energy into electricity. A fuel stack is made up of a group of single fuel cells placed in series. Each cell is formed by a proton exchange membrane (PEM) placed between two electrodes (anode and cathode) which are coated with a catalyst layer, usually platinum (see Figure 1).

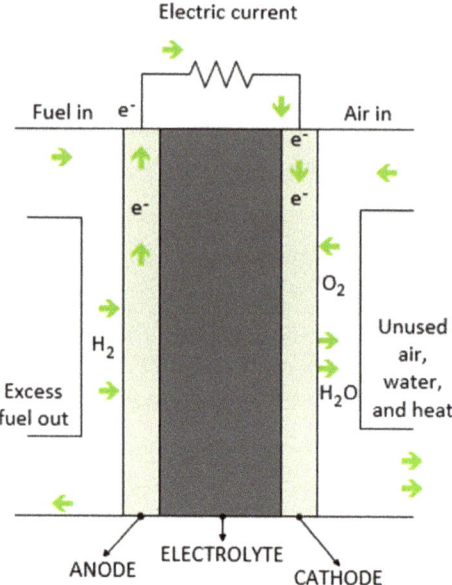

Figure 1. Proton exchange fuel cell diagram [7]. Reproduced with permission from Daud, W.R.W., *Renew. Energy*; published by Elsevier, 2007.

The fuel (hydrogen) is supplied at the anode, and the oxidant (oxygen, generally taken from the air) is supplied at the cathode. At the anode, hydrogen in the presence of a platinum catalyst is ionized into positively charged hydrogen ions and negatively charged electrons. At the cathode, electrons which come from the anode and protons that have crossed the membrane combine with oxygen from the air to form water that flows out of the fuel cell [8,9].

The overall reaction is described as follows.

$$Hydrogen + Oxygen \Rightarrow Electricity + Water + Heat$$

1.2. PEM Fuel Cell System Control

In [10] are mentioned the main components that form a PEM fuel cell system. Below, these four principal sub-systems are described:

- Reactant Flow Subsystem

This subsystem consists of a hydrogen and air supply loop; its objective is to maintain an adequate stoichiometry of the reactants according to the operating conditions of the cell. The air supply loop in a high-pressure fuel cell system uses a compressor to feed the air, while in a low-pressure system, a low-speed blower is used to feed the air.

- Temperature Subsystem

A low-power PEM fuel cell only needs a blower to regulate its operation temperature, which is around 80 °C. A high-power fuel cell cannot dissipate heat by air convection and radiation through the surface of the stack; it needs to be cooled down by the flow of deionized water.

- Water Management Subsystem

The objective of this system is to maintain good hydration of the membrane while balancing the use/consumption of water in the cell. Dry membranes and flooded fuel cells cause high polarization losses.

- Power Management Subsystem

This subsystem controls the power drawn from the fuel cell stack. The load current is considered as a disturbance that has a direct impact on other subsystems.

If the reactant flow system is controlled correctly, the main variables of the stack, such as the temperature and water concentration, will be indirectly controlled. This subsystem has a major impact on the other subsystems; because of this, its control is critical for the performance of the stack.

Polarization phenomena at the PEM fuel cell reduce the voltage that can be delivered by the system whenever more current is drawn by the load, which may affect the performance of equipment that requires a fixed voltage to work correctly. Therefore, the output voltage of the stack must be controlled by adjusting the flow rates of hydrogen and air. Another option to control the output voltage is by using outside means such as a battery or a supercapacitor or both [7].

The stack must operate with maximum efficiency most of the time to achieve profitable operation. Optimizing the hydrogen supply is a priority control objective to achieve cost-effective operation since, at this moment, the hydrogen production cost is still too high [11].

It is possible that a fixed-parameter electrochemical model does not offer a reliable prediction in transient conditions for a conventional controller. For this reason, systems identification techniques seem to be more appropriate to control complex nonlinear systems.

Following the above mentioned, this paper is focused on the control of the reactant system. It is organized as follows: Section 2 cites works related to the data-driven control of PEM fuel cells. Section 3 describes the dataset characteristics and briefly describes the types of feature selection algorithms and some regression algorithms used for systems modeling. Section 4 presents the results obtained by the application of the algorithms of feature selection and regression, and it also shows the control scheme proposed.

2. Related Works

This section presents papers related to the modeling and control of PEM fuel cells using artificial intelligence techniques. In [12], a methodology was presented for systems identification using NARX (nonlinear autoregressive network with exogenous inputs) and NOE (nonlinear output error) neural networks. The control-oriented black box model obtained was implemented in embedded hardware with limited capacity for memory and processing. In [13], the performance of classical neural network (NN) models and stacked models was compared. The stacking approach using partial least squares as a combining algorithm obtained the best prediction. In [14], the authors compared an NN model against a dynamic model using three statistical indices to validate their performance: the absolute mean error (AME), the root-mean-square error (RMSE), and the standard deviation error (SDE). The maximum value of the three indices indicated that the NN model is more precise and accurate but has bigger variation in predicting the outputs when compared with a dynamic model. Different methods have been tested to construct nonlinear empirical models. In [15], the performance of an artificial neural network (ANN) and a support vector machine (SVM) in predicting fuel cell output voltage was compared. The NN model presented excellent performance in predicting the polarization curves of the stack with $R^2 = 0.999$; the SVM model exhibited a slightly inferior performance with $R^2 = 0.980$. However, Kheirandish et al. [16] proposed a different approach for predicting the performance of an electric bicycle using SVM and ANN. Their results showed that SVM has better accuracy in predicting the power curve, approximately 99%, whereas ANN reached an accuracy of 97%. This difference is mainly due to the selection of the hyperparameters. Parametric neural network (PNN) and group method of data handling (GMDH) techniques were used to predict and control the output voltage of a PEM fuel cell of 25 W. The system inputs were gas pressure, fuel cell temperature, and input current. Both methods presented high accuracy in predicting the voltage. However, the GMDH model had less deviation [17]. Some parameters are difficult to measure, or it is very expensive to measure them, especially in fuel cell stacks. Chávez-Ramírez et al. [18] developed a simulator, based on ANN, to predict the stack voltage and cathode output temperature. They concluded that simulators based on ANN are reliably able to predict voltage and temperature behavior, saving time and resources. Recurrent neural networks were used to develop degradation prognostic models. In [19], a grid long short-term memory (G-LSTM) recurrent neural network (RNN) was used to predict the lifetime of fuel cells.

A detailed description of the neural control techniques applied to PEM fuel cells is provided in [20]. In Figure 2 are shown these different approaches. A feed-forward control system, including a neural network together with a proportional–integral–derivative controller, was presented in [21]. The control objective was maintaining a proper stack voltage using an inverse model of the plant to calculate the control signal (air pressure). In [22], a neural network adaptive control with feedback linearization was developed. The control variables were the pressure values of hydrogen and oxygen. The model presented excellent disturbance rejection, even under load variations.

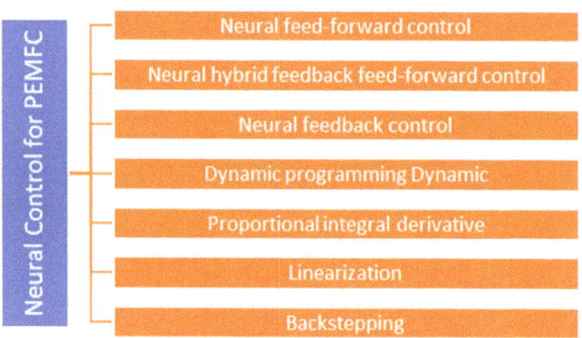

Figure 2. Neural control techniques applied to PEM fuel cells.

However, other artificial intelligence techniques have been applied to fuel cell systems to control airflow rate, temperature, and mass flow, among others. In [23], an interval type-2 fuzzy proportional–integral–derivative (IT2FPID) controller was designed to regulate the air flow. The results were compared with those of conventional PID and type-1 fuzzy PID controllers. IT2FPID presented a better performance in terms of transient response. In [24], a fuzzy cognitive map (FCM) was used to model an electric bicycle powered by a fuel cell. The Hebbian algorithm was proposed for the FCM to self-learn from its own data.

3. Materials and Methods

The development stages of the proposed control scheme are described below.

1. Apply a feature selection algorithm to determine the variables needed to model and control the fuel cell voltage;
2. Define the system inputs from the subset formed by the feature selection algorithm and try different regression algorithms to predict the output variable;
3. Develop the inverse model of the fuel cell, turning the system inputs into outputs. The output of the regression model will become a system input.
4. Integrate the inverse model with a PID neuro control to track the errors and tune the control signal to achieve the reference value of the system output. The reason why these two types of

control are integrated is to modify the control signal by not only considering the error between the output variable and the reference value but also considering the state of the other variables in the transient state.

3.1. Experimental Setup

The proposed approach was applied to the test data from IEEE 2014 [25]. Experiments were carried out on a testbench that allows running the PEM stacks under constant or variable operating conditions while controlling and recording operation data like power loads, temperatures, and stoichiometry rates of hydrogen and air. The variables monitored are presented in Table 1.

Table 1. Variables monitored.

Variable	Description	Unit
Time	Time aging	H
Vout	Stack output voltage	V
I	Current	A
J	Current density	A/cm^2
Tin, Tout H2	Inlet and outlet H$_2$ temperature	°C
Tin, Tout Air	Inlet and outlet air temperature	°C
Pin, Pout H2	Inlet and outlet H$_2$ pressure	mBar
Fin, Fout H2	Inlet and outlet H$_2$ flow	L/min
Fin, Fout Air	Inlet and outlet air flow	L/min
Fwat	Flow rate of cooling water	L/min
HrAIR	Inlet Hygrometry (Air)	%

The stack was formed by five cells. Each cell had an active area of 100 cm^2. The nominal current density of the cells was 0.70 A/cm^2, and their maximum current density was 1 A/cm^2. The test was carried out under dynamic changes in the load current (around 1020 h). The load current connected was of 70 A with oscillations of 10% at a frequency of 5 kHz. The ranges of the operating parameters are shown in Table 2.

Table 2. Range of parameters controlled.

Parameter	Range
Air flow	0 to 100 L/min
H$_2$ flow	0 to 30 L/min
Gas pressure	0 to 2 bars
Temperature	20 to 80 °C
Cell current	0 to 300 A

3.2. Feature Selection Algorithms and Data-Driven Models for Fuel Cells

Dimensionality reduction techniques can be classified into two groups: *feature selection* and *feature extraction*. Each one has its characteristics, and its accuracy depends on the characteristics of the database to be analyzed. Feature extraction techniques achieve dimensionality reduction by combining the variables. In this way, they can generate a set of new components, reducing the data dimensionality while maintaining enough information to describe the system. In some applications, such as image analysis, where model accuracy is more important than model interpretability, these techniques are very useful. Instead, feature selection reduces data dimensionality by removing irrelevant and redundant variables. Feature selection techniques aim to obtain a subset of variables that describes with accuracy the system characteristics with minimum performance degradation. Feature selection can be grouped into three main categories: Filters, Wrappers, and Embedded. A brief description of their main characteristics is given below:

- Filter methods measure the relevance of the variables by their correlations with the output variable;

- Wrapper methods create a subset of the original dataset using a training algorithm;
- Filter methods are much faster than wrapper and embedded methods;
- Wrapper methods can fall into overfitting;
- Embedded and wrapper methods capture feature dependencies while filters methods do not.

The operating principles of PEM fuel cells include electrochemistry and thermodynamics principles that are frequently very hard to model mathematically. One alternative to overcome this issue is the use of modeling methods based on artificial intelligence techniques. In this work, neural networks were used to model and control PEM fuel cells because deep learning techniques, in general, present better performance in modeling highly nonlinear systems than do machine learning algorithms. Section 4.2 compares the performance of different algorithms against dynamic neural networks. These data-driven models can be used as an emulator to detect possible failures in fuel cell systems or to develop an inverse neural control system, as is shown in [26] (see Figure 3).

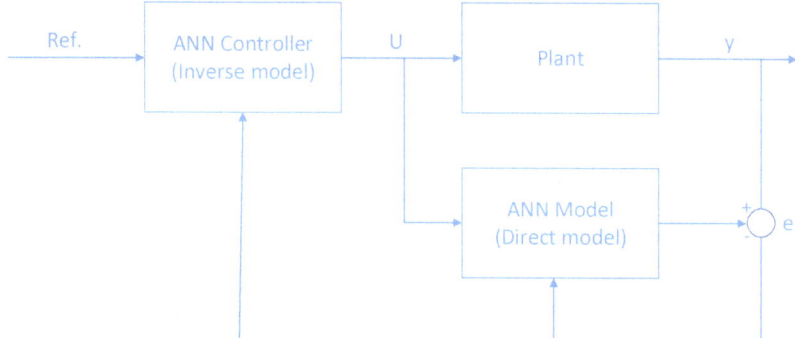

Figure 3. Direct inverse neural control.

4. Results and Discussion

For research purposes, all the information collected during the test is useful to understand and improve the material quality and the design performance; these improvements can lead to increasing the lifetime and thus reducing the cost of operation, which at the moment is still too high. However, for control purposes, in real applications, it would be very expensive to install all of these sensors and actuators. The control objective is to identify the critical operating variables and reduce the cost of the control system using Feature Selection.

4.1. Fuel Cell Feature Selection

An attempt was made to train a regression algorithm without applying feature selection. The poor results obtained were due to the noise generated by the low correlation of some variables. This section presents the results of the application of various feature selection algorithms to the original dataset. The best results were obtained using a feature extraction algorithm: PCA analysis. For this reason, Section 4.1.4 was extended to describe how the variables were selected.

4.1.1. Filter Methods

The Pearson correlation method selected the next variables: current, current density, and the output flow rates of hydrogen and air. These variables were selected due to their correlation grades being superior to 0.5. However, although these variables could model the fuel cell voltage, none of them can be considered as a system input useful to controlling the fuel cell.

4.1.2. Wrapper Methods

Two wrapper methods were applied to perform feature selection: Recursive Feature Elimination (RFE) and Backward Elimination (BE).

RFE selected 16 variables with a model precision of 0.85. The removed variables were the current density and hydrogen output temperature. BE selected 17 variables according to a p value of 0.05 (statistically significant). The removed variable was the inlet hydrogen flow.

The dimensionality reduction achieved by both algorithms, RFE and BE, was nonsignificant.

4.1.3. Embedded Methods

The selection was made using lasso regularization. If the variable is irrelevant, lasso penalizes its coefficient by changing it to zero. The best score using built-in LassoCV was 0.8617. Lasso picked 11 variables and eliminated the other 7 variables. The reduction achieved by this algorithm was highly significant. However, the fit was barely acceptable.

4.1.4. Principal Component Analysis (PCA)

PCA analysis is a statistical method used to reduce the dimensionality of a dataset while retaining as much as possible of the variation present in the data. For more details about this technique and its applications to fuel cells, refer to [27].

The first step to performing a PCA analysis is to make a descriptive statistic that summarizes the central tendency and dispersion of the values; the next step is to make a correlation matrix, which allows us to observe which variables have a solid relationship, as shown in Figure 4.

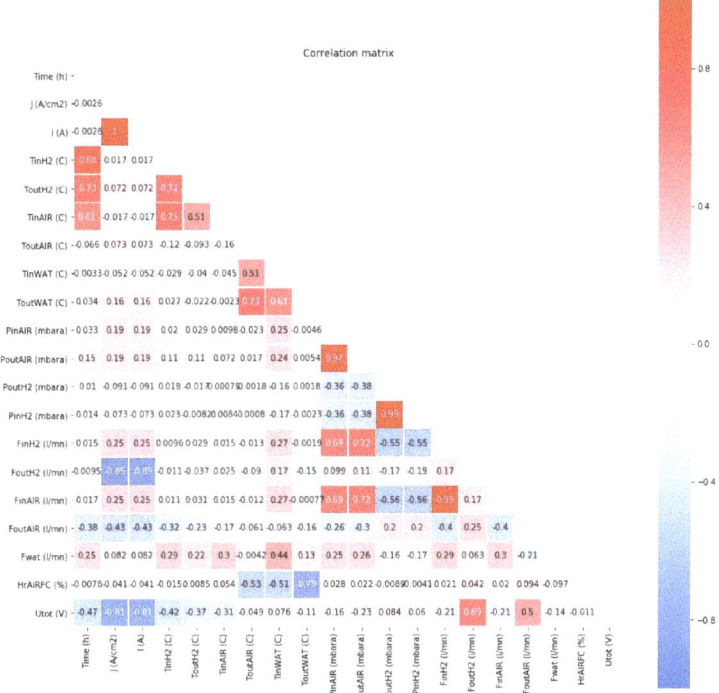

Figure 4. Fuel cell correlation matrix.

As can be seen, due to the low correlation between variables, it is very difficult to determine the parameters for operating a fuel cell system [17]. This low correlation is mainly due to PEM fuel cells having complex electrochemical reactions with multiple nonlinear input/output variables [28]. Systems with these characteristics are complicated to model accurately and, therefore, to optimize.

The main relationships are between the following:

- The air inlet and outlet pressure;
- The hydrogen inlet and outlet pressure; and
- The air inlet flow rate and hydrogen inlet flow rate.

However, as can be seen in Figure 4, there is a negative correlation between time and current (−0.81), as well as between time and current density (−0.81); these relationships are not so significant because they only reflect the natural wear of the membrane. The time variable was not considered during the feature selection process.

After that, it is necessary to determine the number of components which explain the main variance of the data. This number is obtained by trial and error. In this case, five components describe the variance of the data correctly (see Figure 5).

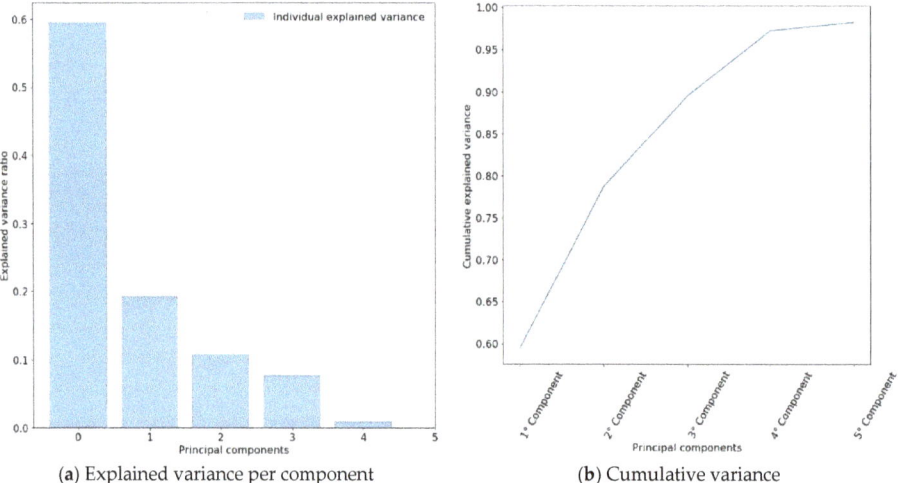

(a) Explained variance per component (b) Cumulative variance

Figure 5. Explained variance per component.

However, four components explain more than 97% of the variance. The fifth component is not relevant, so it can be omitted. In Figure 6 can be seen which variables have a major impact on each one of the four components (see also Table 3).

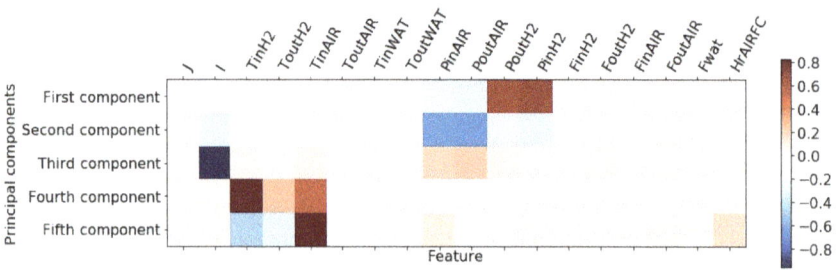

Figure 6. Fuel cell correlation matrix.

Table 3. Selected variables from PCA analysis.

Variables	Type
Current	State variable
Hydrogen inlet temp.	State variable
Air inlet temp.	State variable
Air inlet pressure	Input system
Air outlet pressure	State variable
Hydrogen inlet pressure	Input system
Hydrogen outlet pressure	State variable

However, during the regression process, it is necessary to add the time variable due to the natural wear of the membrane depending on the work hours, which reduce in an almost linear way the output voltage. Once the main variables of the fuel cell have been identified, it is possible to create a control-oriented model to track the output voltage.

4.2. Data-Driven Control-Oriented Models for PEM Fuel Cells

This section is divided into two parts. Section 4.2.1 describes the results obtained by some of the most robust regression algorithms used in machine learning. Section 4.2.2 is extended to show in more detail the neural network training process. Neural networks achieved better results than the algorithms tested in Section 4.2.1, mainly due to their ability to track nonlinear variables and system delays.

4.2.1. Fuel Cell Modeling Using Machine Learning Regression Algorithms

Different regression algorithms were tested to create a robust control-oriented model, and their performance was compared with the Explained Variance score ratio. The *k*-fold method was used for the cross-validation of the model using five folds, and a fixed seed was established to ensure reproducibility. The methods compared were ridge (RID), Bayesian ridge (BYR), decision tree regressor (DTR), gradient boosting regressor (GBR), and random forest regressor (RFR). The results show the averages and standard deviations of the Explained Variance.

- RID: 0.840495 (0.075010)
- BYR: 0.840494 (0.075011)
- DTR: 0.815885 (0.131130)
- GBR: 0.860727 (0.124138)
- RFR: 0.830844 (0.120877)

Figure 7 compares via a box plot the performance of the algorithms tested. In the graph, it can be seen that gradient boosting regressor is the algorithm that presents less variation and better accuracy. However, the gradient boosting regressor only reaches a score of 0.86, which is slightly low for control purposes.

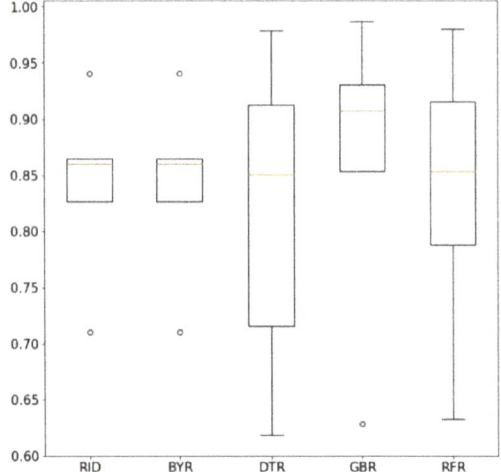

Figure 7. Algorithm comparison.

4.2.2. Fuel Cell Modeling Based on Neural Networks

Neural networks can be classified according to their behavior in time as either static or dynamic. A static neural network can model with high accuracy the performance of a PEM fuel cell. However, as can be seen in Figure 4, the time variable impacts negatively on the output voltage and current, even in steady-state conditions (see also Figure 8). A dynamic neural network takes into consideration the time variable, and its structure can be used as a generic model for system control [29].

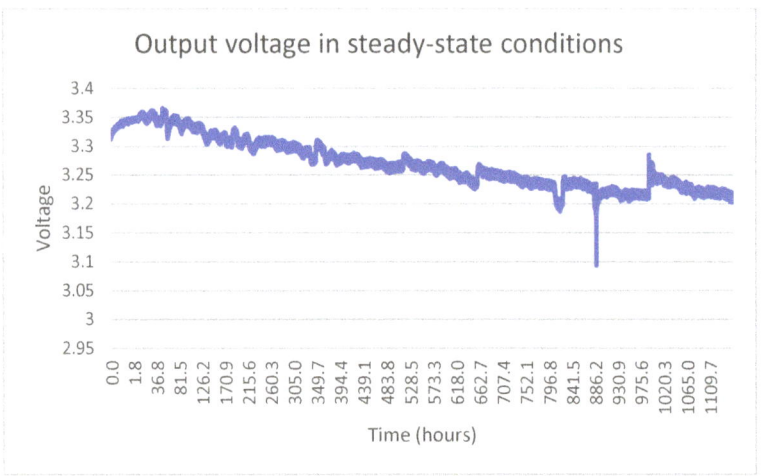

Figure 8. Fuel cell output voltage as a function of time.

A nonlinear autoregressive with external exogenous input (NARX) network was used to model the fuel cell. The validation process was done by a cross-validation technique (k-fold) with ten splits.

The dataset was divided into training and validation sets. The input layer consisted of eight inputs (the variables selected in the PCA analysis, see Table 3), the hidden layer had ten neurons with a log-sigmoid activation function with two delays (sampling time 30 seg.), and the output layer used the purelin activation function to calculate the voltage. The dynamic neural network (DNN) configuration is shown in Figure 9. The training algorithm selected was Levenberg–Marquardt because, in general, it has the fastest convergence and reduces the computational cost.

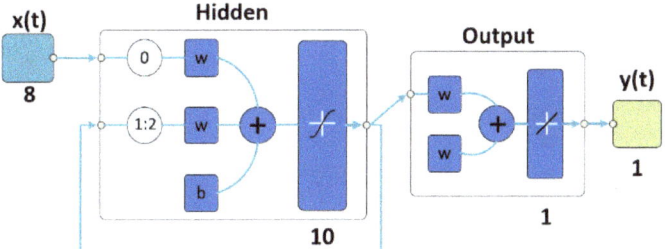

Figure 9. Dynamic neural network structure.

The high regression accuracy ($R^2 = 0.96$) and the fast convergence are mainly due to the fact that in the PCA analysis, the irrelevant and redundant variables which have no impact on the output voltage were eliminated. The eliminated variables do not have value for control purposes. In Table 4 are presented the scores of each fold. In Figure 10, a comparison of the actual values against the predicted values is presented.

Table 4. Regression score function of each fold.

Fold	Score
1	0.955929840882997
2	0.953074662444409
3	0.959505398134269
4	0.957813889951952
5	0.958048357252375
6	0.958116362163286
7	0.959355574634461
8	0.960026345242201
9	0.966767102061810
10	0.971974506261939
Ave.	0.960061203902970

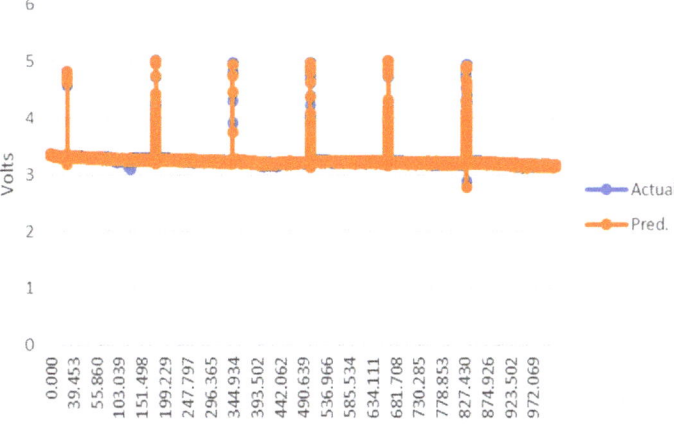

Figure 10. Actual values and predicted values.

4.3. Hybrid Control Scheme

According to the eight main variables identified in the PCA analysis, only two can be considered as system inputs: the inlet pressures of hydrogen and oxygen. However, when analyzing Figure 6, it can be seen that hydrogen pressure is the variable that most affects the fuel cell performance. Variations in air inlet pressure can be considered non-representative for control purposes if they are kept within a specific range of operation.

Keeping constant or following a reference is not the objective of this control approach for the fuel cell output voltage; this is because the output voltage does not depend only on the supply of the reactants. The load has a delayed negative correlation on the voltage level and in transient-state conditions is the variable that impacts it the most. The load (current) can be considered as an external disturbance. For the abovementioned, a MISO (multiple inputs, single output) control is needed to supply the optimal hydrogen pressure to the cell according to the operating conditions, such as temperature, current, or air pressure.

The neuro-PID controller is an already proven control approach in cases of system fault recovery, such as flooding, drying out, and auxiliary failures, such as of a compressor [20]. A PID-series neuro control scheme (with an inverse model of the fuel cell) was proposed to supply the optimal hydrogen pressure by taking into account the values of the main variables under transient conditions (see Figure 11). The self-autotuning of the PID control was done according to the method proposed by Omatu et al. [30].

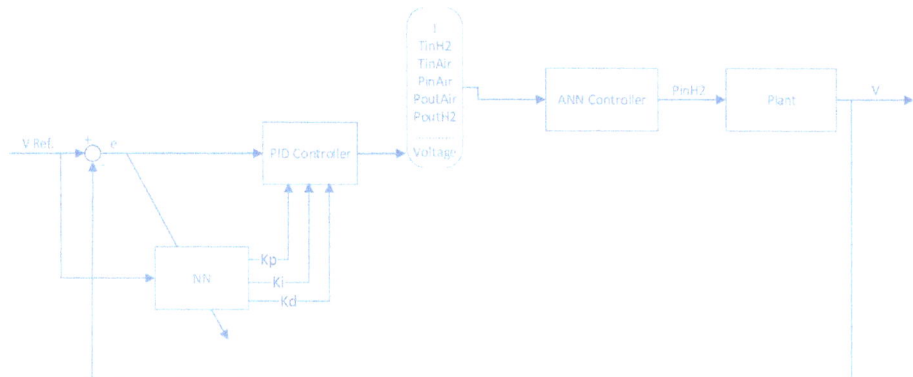

Figure 11. Proportional–integral–derivative (PID)-series neuro controller.

The ANN controller is the inverse model of the plant; this means that the output voltage of the plant was turned into input, and the hydrogen pressure became the system output. The nominal voltage is 3.3 volts in steady-state conditions; however, this nominal value depends on the changes in the load and its effect on the temperature. The ANN controller not only considers the error between the nominal voltage value and the actual value but also considers the values of the variables selected in the PCA analysis to estimate the control signal—in this case, the hydrogen pressure. The training was done following the same approach described in Section 4.2.2.

In Figure 12 are compared the voltage, current, and hydrogen pressure for both controllers, the conventional and neuro PID-series. Both controllers achieved similar performance in steady-state and transient conditions. The main difference is the reduction in the hydrogen pressure in the steady state. This reduction in pressure causes a decrease in the flow of hydrogen, which in turn decreases hydrogen consumption. It is necessary to recall that the difference is only about 45 mbar. In practical applications, a high-precision actuator (expensive) would be needed to control these small differences.

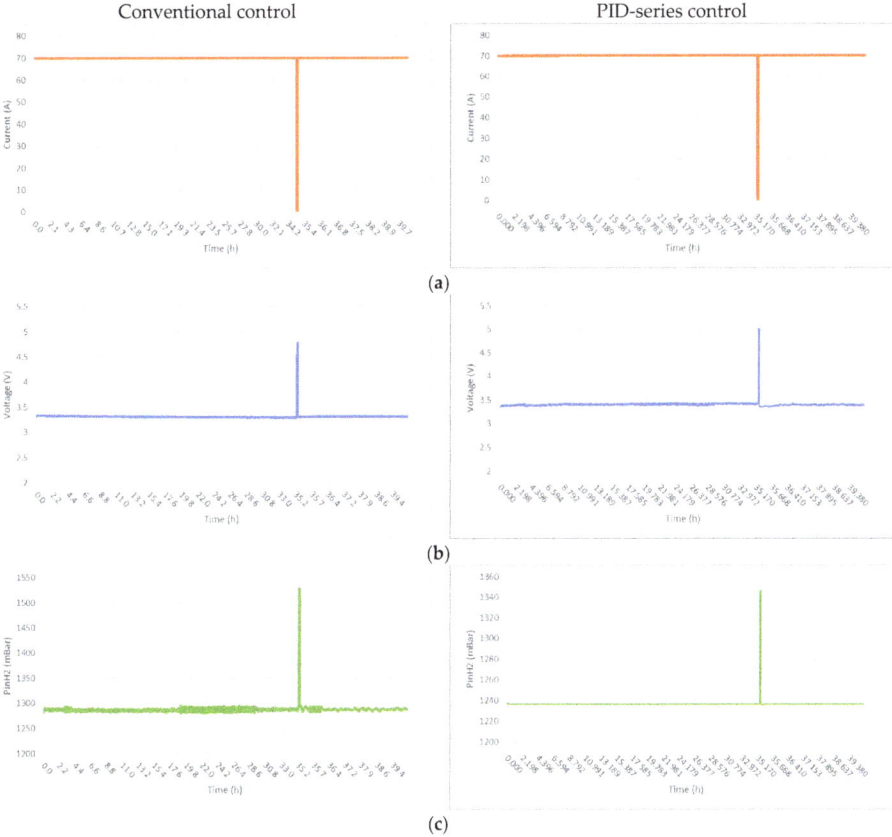

Figure 12. Performance analysis of PID-series and conventional control. (**a**) The same load was connected to both controllers. At time 35.42 h the load was changed to evaluate the effect on the output voltage and the supply hydrogen pressure; (**b**) The output voltage reached the same level in steady state in both cases practically. When the load connected was reduced (until open circuit) the output voltage increased; (**c**) Hydrogen pressure in the conventional control (left) oscillated (in the steady state) mainly between 1280 and 1295 mBar, whereas in the neuro-control (right), the hydrogen pressure remained practically constant at 1238 mBar in the steady state.

The training algorithm derives the error partially, so each neuron updates its weight according to its proportion in that error. If the neuro-PID controller only considers the gap between the desired voltage and the actual value without taking into account the changes in the variables selected in the PCA analysis, the control signal, the hydrogen inlet pressure, will not stabilize the fuel cell performance. PEM fuel cells must operate in steady-state conditions in order to avoid premature failure, such as starvation due to improper gas supply or an excessive transient load demand [31]. An energy management system is required to deliver a fixed voltage to equipment so it can work correctly. This paper proposes a practical approach to stabilize the fuel cell performance in transient conditions at minimum control cost, focusing attention on the variables that impact the most on the performance of the cell and eliminating unnecessary measurements. However, this control can be improved if the air inlet pressure is also regulated. An incorrect startup/shutdown process can cause accelerated or permanent damage to the catalyst layer. These considerations have to be included in the control process to improve the proposed control.

5. Conclusions

In this paper, we developed a data-driven control approach for PEM fuel cells to minimize the cost of control. Several feature selection algorithms were used for dimensionality reduction. Principal component analysis (PCA) obtained the best results by removing irrelevant and redundant variables. The selected variables (Table 3) can describe with high accuracy the PEM fuel cell performance. Some of the most powerful regression algorithms were compared to predict the output voltage of the cell. However, neural networks obtained the highest accuracy (R^2 = 0.96) due to their capacity to map complex nonlinear relationships. With the selected variables, an inverse model of the fuel cell was developed using neural networks in order to develop a neuro-PID controller. A PID-series control was integrated with the inverse model to regulate the system input (hydrogen inlet pressure) by considering the values of the other variables. The fuel cell voltage level does not depend only on the supply of the reactants, and in transient conditions, the load is the variable that impacts the most on fuel cell performance. This method is a practical way to save mathematical modeling time and reduce the number of sensors in the control system.

In the future, this study will be improved via experimental tests in a real PEM fuel cell system which includes the measurements detected in the PCA analysis. Later, an intelligent fault diagnosis and isolation scheme will be developed to prevent permanent damage in the catalyst layer.

Author Contributions: Conceptualization, A.M.-S.; Data curation, O.S.G.; Formal analysis, A.M.-D. and A.M.-S.; Investigation, A.M.-D.; Methodology, J.P.R.-J.; Project administration, R.P.-G.; Supervision, J.P.R.-J.; Validation, O.S.G.; Visualization, R.P.-G.; Writing – original draft, A.M.-D.; Writing – review & editing, A.M.-S.

Funding: This research received no external funding.

Conflicts of Interest: The authors declare no conflicts of interest.

References

1. Oncel, S.; Vardar-Sukan, F. Application of proton exchange membrane fuel cells for the monitoring and direct usage of biohydrogen produced by Chlamydomonas reinhardtii. *J. Power Sources* **2011**, *196*, 46–53. [CrossRef]
2. Manish, S.; Banerjee, R. Comparison of biohydrogen production processes. *Int. J. Hydrogen Energy* **2008**, *33*, 279–286. [CrossRef]
3. Rahman, S.N.A.; Masdar, M.S.; Rosli, M.I.; Majlan, E.H.; Husaini, T. Overview of Biohydrogen Production Technologies and Application in Fuel Cell. *Am. J. Chem.* **2015**, *5*, 13–23.
4. Kotay, S.M.; Das, D. Biohydrogen as a renewable energy resource—Prospects and potentials. *Int. J. Hydrogen Energy* **2008**, *33*, 258–263.
5. Lucia, U. Overview on fuel cells. *Renew. Sustain. Energy Rev.* **2014**, *30*, 164–169. [CrossRef]
6. Gao, F.; Member, S.; Blunier, B.; Sim, M.G. PEM Fuel Cell Stack Modeling for Real-Time Emulation in Hardware-in-the-Loop Applications. *IEEE Trans. Energy Convers.* **2011**, *26*, 184–194.
7. Daud, W.R.W.; Rosli, R.E.; Majlan, E.H.; Hamid, S.A.A.; Mohamed, R.; Husaini, T. PEM fuel cell system control: A review. *Renew. Energy* **2017**, *113*, 620–638. [CrossRef]
8. Gao, F.; Blunier, B.; Miraoui, A. *Proton Exchange Membrane Fuel Cells Modeling*; Wiley: Hoboken, NJ, USA, 2012.
9. Puranik, S.V.; Keyhani, A.; Khorrami, F. State-Space Modeling of proton exchange membrane fuel cell. *IEEE Trans. Energy Convers.* **2010**, *25*, 804–813. [CrossRef]
10. Pukrushpan, J.T.; Stefanopoulu, A.G.; Peng, H. *Control of Fuel Cell Power Systems*, 2nd ed.; Springer-Verlag: London, UK, 2005.
11. Ainscough, C.; Peterson, D.; Miller, E. Hydrogen Production Cost From PEM Electrolysis. *DOE Hydrog. Fuel Cells Progr. Rec.* **2014**. Available online: https://www.hydrogen.energy.gov/pdfs/14004_h2_production_cost_pem_electrolysis.pdf (accessed on 30 May 2019).
12. da Costa Lopes, F.; Watanabe, E.H.; Rolim, L.G.B. A Control-Oriented Model of a PEM Fuel Cell Stack Based on NARX and NOE Neural Networks. *IEEE Trans. Ind. Electron.* **2015**, *62*, 5155–5163. [CrossRef]
13. Napoli, G.; Ferraro, M.; Sergi, F.; Brunaccini, G.; Antonucci, V. Data driven models for a PEM fuel cell stack performance prediction. *Int. J. Hydrogen Energy* **2013**, *38*, 11628–11638. [CrossRef]

14. Sisworahardjo, N.S.; Yalcinoz, T.; El-Sharkh, M.Y.; Alam, M.S. Neural network model of 100 W portable PEM fuel cell and experimental verification. *Int. J. Hydrogen Energy* **2010**, *35*, 9104–9109. [CrossRef]
15. Han, I.S.; Chung, C.B. Performance prediction and analysis of a PEM fuel cell operating on pure oxygen using data-driven models: A comparison of artificial neural network and support vector machine. *Int. J. Hydrogen Energy* **2016**, *41*, 10202–10211. [CrossRef]
16. Kheirandish, A.; Shafiabady, N.; Dahari, M.; Kazemi, M.S.; Isa, D. Modeling of commercial proton exchange membrane fuel cell using support vector machine. *Int. J. Hydrogen Energy* **2016**, *41*, 11351–11358. [CrossRef]
17. Pourkiaei, S.M.; Ahmadi, M.H.; Hasheminejad, S.M. Modeling and experimental verification of a 25W fabricated PEM fuel cell by parametric and GMDH-type neural network. *Mech. Ind.* **2015**, *17*, 105. [CrossRef]
18. Chávez-Ramírez, A.U.; Muñoz-Guerrero, R.; Duron-Torres, S.M.; Ferraro, M.; Brunaccini, G.; Sergi, F.; Antonucci, V.; Arriaga, L.G. High power fuel cell simulator based on artificial neural network. *Int. J. Hydrogen Energy* **2010**, *35*, 12125–12133. [CrossRef]
19. Ma, R.; Yang, T.; Breaz, E.; Li, Z.; Briois, P.; Gao, F. Data-driven proton exchange membrane fuel cell degradation predication through deep learning method. *Appl. Energy* **2018**, *231*, 102–115. [CrossRef]
20. Lin-kwong-chon, C.; Grondin-pérez, B.; Kadjo, J.A.; Damour, C.; Benne, M. A review of adaptive neural control applied to proton exchange membrane fuel cell systems. *Annu. Rev. Control* **2019**, *47*, 133–154. [CrossRef]
21. Rakhtala, S.M.; Ghaderi, R.; Noei, A.R. Proton exchange membrane fuel cell voltage-tracking using artificial neural networks. *J. Zhejiang Univ. Sci. C* **2011**, *12*, 338–344. [CrossRef]
22. Abbaspour, A.; Khalilnejad, A.; Chen, Z. Robust adaptive neural network control for PEM fuel cell. *Int. J. Hydrogen Energy* **2016**, *41*, 20385–20395. [CrossRef]
23. Aliasghary, M. Control of PEM Fuel Cell Systems Using Interval Type-2 Fuzzy PID Approach. *Fuel Cells* **2018**, *18*, 449–456. [CrossRef]
24. Kheirandish, A.; Motlagh, F.; Shafiabady, N.; Dahari, M.; Wahab, A.K.A. Dynamic fuzzy cognitive network approach for modelling and control of PEM fuel cell for power electric bicycle system. *Appl. Energy* **2017**, *202*, 20–31. [CrossRef]
25. IEEE PHM 2014 Data Challenge. 2014. Available online: http://eng.fclab.fr/ieee-phm-2014-data-challenge/ (accessed on 22 May 2019).
26. Almeida, P.E.M.; Simões, M.G. Neural optimal control of PEM fuel cells with parametric CMAC networks. *IEEE Trans. Ind. Appl.* **2005**, *41*, 237–245. [CrossRef]
27. Placca, L.; Kouta, R.; Candusso, D.; Blachot, J.F.; Charon, W. Analysis of PEM fuel cell experimental data using principal component analysis and multi linear regression. *Int. J. Hydrogen Energy* **2010**, *35*, 4582–4591. [CrossRef]
28. Al-Othman, A.K.; Ahmed, N.A.; Al-Fares, F.S.; AlSharidah, M.E. Parameter Identification of PEM Fuel Cell Using Quantum-Based Optimization Method. *Arab. J. Sci. Eng.* **2015**, *40*, 2619–2628. [CrossRef]
29. Hatti, M.; Tioursi, M. Dynamic neural network controller model of PEM fuel cell system. *Int. J. Hydrogen Energy* **2009**, *34*, 5015–5021. [CrossRef]
30. Omatu, S.; Yoshioka, M.; Kosaka, T.; Yanagimoto, H. Neuro-PID Control of Speed and Torque of Electric Vehicle. *Int. J. Adv. Syst. Meas.* **2010**, *3*, 82–91.
31. Rosli, R.E.; Sulong, A.B.; Daud, W.R.W.; Zulkifley, M.A.; Rosli, M.I.; Majlan, E.H.; Haque, M.A. Reactant Control System for Proton Exchange Membrane Fuel Cell. *Procedia Eng.* **2016**, *148*, 615–620. [CrossRef]

© 2019 by the authors. Licensee MDPI, Basel, Switzerland. This article is an open access article distributed under the terms and conditions of the Creative Commons Attribution (CC BY) license (http://creativecommons.org/licenses/by/4.0/).

Article

The Bilinear Model Predictive Method-Based Motion Control System of an Underactuated Ship with an Uncertain Model in the Disturbance

Huu-Quyen Nguyen, Anh-Dung Tran and Trong-Thang Nguyen *

Faculty of Electrical-Electronic Engineering, Vietnam Maritime University, Haiphong 181810, Vietnam
* Correspondence: thangnt.ddt@vimaru.edu.vn; Tel.: +84-38-846-8555

Received: 18 June 2019; Accepted: 10 July 2019; Published: 12 July 2019

Abstract: Ship transportation plays an increasingly important role in and accounts for a large proportion of cargo transport. Therefore, it is necessary to improve the quality of the trajectory control system of the ship for improving the transport efficiency and ensuring maritime safety. This paper deals with the advanced control system for the three-degrees-of-freedom model of the underactuated ship in the condition of uncertain disturbance. Based on the three-degrees-of-freedom model of the underactuated ship, the authors built a bilinear model of the ship by linearizing each nonlinear model section. Then, the authors used the state estimator to compensate for uncertain components and random disturbances in the model. Finally, the authors built the output-feedback predictive controller based on the channel-separation principle combined with direct observation of the continuous model for controlling the motion of the underactuated ship in the case of uncertain disturbance and the bound control signals. The result is that the movement quality of the underactuated ship is very good in the context of uncertain disturbance and bound control signals.

Keywords: underactuated ship; bilinear model predictive controller; directly observer; uncertain estimator

1. Introduction

Maritime transport plays a particularly important role in international trade because about 80% of imports and exports are transported through the sea. Maritime transport is a large market because of its essential advantages, such as its wide transport range, large carrying capacity, low shipping cost, etc. Therefore, it is necessary to conduct research for improving the trajectory control system of the ship in order to improve the transport efficiency and ensure maritime safety. However, controlling the ship movement with a high-quality is a challenge for scientists because the ship is a complex object, with large nonlinearity and unknown structures, and works in dynamic environments with complex noise.

The dynamic model of a ship is an uncertain nonlinear model, and the model parameters depend on the control states. The equation used to describe ship motion is a high-order differential equation. Considering the kinetic properties, ships have the following characteristics: The oscillation and time constant are large, and the stabilization margin is small [1]. Therefore, controlling the ship motion is always a challenge for scientists, especially controlling the underactuated ship that has fewer control signals than the state variables to be controlled [2]. Studies on ship control with the model of a lack of actuators have been presented in the literature [3–7].

A ship is a large nonlinear object, so the use of simplified or linear control models does not give us the expected results. In recent years, the development of electronic and informatics technology has allowed us to apply modern control theory for ship motion, such as adaptive control, backstepping, sliding mode control, model predictive control, etc. The research [8] uses the backstepping technique to control an underactuated ship following the set trajectory. However, this research assumes that the ship only moves along a straight trajectory and the speed is constant and positive, and the uncertainties in the kinematic model and the disturbance are not considered. The research [9] has provided a trajectory ship controller based on the Lyapunov function and backstepping technique on the condition that the control signals are not bound. Additionally, the research [10] has proposed a control method based on linear algebra, where the controller is designed based on linearizing the nonlinear model of the control object. However, this research does not address the problem of disturbance and uncertainties in the dynamic model of the ship.

Therefore, difficult problems when designing the ship motion controller are as follows: the state variables cannot be measured, and the coefficients in the model matrices are changed, depending on the control states, so it is difficult to accurately determine the coefficients in the model matrices. For simplicity, some previous research has ignored many factors, leading to models of the ship that are very different to the actual situation.

In order to solve the problems caused by uncertain parameters and disturbance, the research [11,12] has used the neural network to update and estimate the uncertain components in the model. In the study [13], the author used the coordinate transformation method to overcome the third-order uncertain component in the derivative of the Lyapunov function caused by the centrifugal and Coriolis forces. To update the uncertain parameters and the time-varying parameters of the inertial mass matrix of the model, the study [14] proposed a motion controller of the self-driver ship using the unscented Kalman filter to compensate for the hidden noise in the model. The work [15] estimated the uncertain components based on the finite-time disturbance observer. The works [16,17] proposed the nonlinear disturbance observer based on the kinetic model to estimate the disturbance that is compensated for the controller.

Another problem faced when designing the trajectory controllers of ship motion is that the studies often ignore the limit of control signals. In fact, the angle of the steer is always limited from 35 degrees left to 35 degrees right. The research [18] also addressed this problem when designing the nonlinear model predictive controller for a ship following a set trajectory.

The review of previous studies shows that it is very complicated to design the trajectory control system of ships in the condition of uncertain disturbance. There are some works that have studied this topic in an attempt to solve these problems, but each work has only solved a specific problem in the case of assumptions to simplify the object. There is no research that has simultaneously solved all the problems, such as designing the advanced trajectory controller, eliminating the disturbance in the system, control in the case of the object lacking the actuator, and the limit of control signals.

To solve all of the above limitations, this study will perform the following tasks. It will compensate for the disturbance components in the object model and the disturbance components from the environment by the state observers based on the continuous object model and the difference between the object model and the reference model. Additionally, it will design the trajectory controller of the underactuated ship based on the model predictive controller (MPC) combined with segment-linearization techniques of the nonlinear object in the time-axial.

2. The Model of the Ship

2.1. The Motion of the Ship

Considering the ship motion on the sea surface, the motion of the ship is described as Figure 1. It is characterized by the following motion components: the straight slide motion (u), horizontal slide motion (v), and rotary motion (r). This ignores the following motion components: the roll rotary motion ($p = 0$), pitch rotary motion ($q = 0$), and yaw rotary motion ($\omega = 0$).

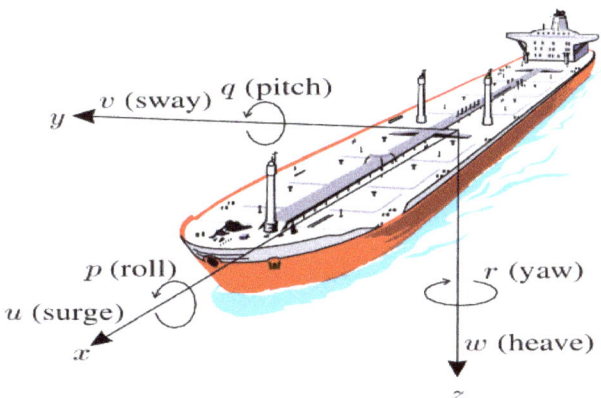

Figure 1. The motion components of the ship.

To describe the ship motion in the three-degrees-of-freedom space, we can employ the speed vector $\underline{v} = [u\, v\, r]^T$ and the position vector $\underline{\eta} = [x\, y\, \psi]^T$.

Here, u, v, r denote the straight slide speed, the horizontal slide motion speed, and the rotation speed, respectively.

2.2. The Equations of Ship Motion

The equations of ship motion in the three-degree freedom space are as follows [19]:

$$\begin{cases} \dot{\underline{\eta}} = J(\underline{\eta})\underline{v} \\ M\dot{\underline{v}} + \bar{C}(\underline{v})\underline{v} + D(\underline{v})\underline{v} + g(\underline{\eta}) = \underline{\tau} + \underline{\tau}_w \end{cases} \quad (1)$$

where:

M denotes the inertial matrix;
$C(\underline{v})$ denotes the centrifugal and Coriolis forces;
$D(\underline{v})$ denotes the hydrodynamic damping matrix;
$J(\underline{\eta})$ denotes the orthogonal matrix;
$g(\underline{\eta})$ represents the gravity forces;
$\underline{\tau}$ is the vector of control torques, including the propeller force and the rudder force;
$\underline{\tau}_w$ represents disturbances from the environment.

In the mathematical model (1), if the control force includes all of the components $\underline{\tau} = [\tau_u\, \tau_v\, \tau_r]^T$, then the ship is called a fully-actuated ship. This model of a ship has many actuators, such as the main propeller for creating the straight-slide force, the horizontal propeller on both sides for creating the horizontal slide force, and the rudder for controlling the ship direction. This model often appears in types of ship such as the serving-ship, the special-task ship, and the ship for researching marine dynamic stability control.

If the control force $\underline{\tau} = [\tau_u\, 0\, \tau_r]^T$ means that there is no horizontal slide force, then the ship is called an underactuated ship. This ship only has two actuators, such as the main propeller for creating the straight-slide force and the rudder for controlling the ship direction. This model often appears in types of ship such as cargo ships and container ships with a long transport journey.

The motion equations of an underactuated ship in the three-degree-freedom space are as follows [2]:

$$\begin{cases} \dot{\underline{\eta}} = J(\underline{\eta})\underline{v} \\ M\dot{\underline{v}} + \bar{C}(\underline{v})\underline{v} + D(\underline{v})\underline{v} + g(\underline{\eta}) = \underline{\tau} + \underline{\tau}_w \\ \underline{\tau} = [\tau_u \; 0 \; \tau_r]^T \end{cases} \quad (2)$$

It is very difficult to fully define the coefficients of M, $C(\underline{v})$, $D(\underline{v})$. These coefficients can be determined by driving the ship at different speeds in different directions and measuring the response signals. However, we still have to assume that the high-order nonlinear components are zero. Moreover, the coefficients in the above matrices also depend on other factors, such as the cargo weight on the ship and the waters in which the ships move.

In order for the model equations to fully express the dynamics of the underactuated ship in the space of three-degrees-of-freedom, this research proposes merging all of the components that are difficult to identify, uncertain components, and the environmental disturbance into the undefined vector $\underline{\Delta}(\underline{\eta}, \underline{v})$.

$$\begin{cases} \dot{\underline{\eta}} = J(\underline{\eta})\underline{v} \\ M\dot{\underline{v}} + \bar{C}(\underline{v})\underline{v} + D(\underline{v})\underline{v} + g(\underline{\eta}) = F\underline{\tau} + \underline{\Delta}(\underline{\eta}, \underline{v}) \end{cases} \quad (3)$$

where:

1. F denotes the force distribution matrix, $F = \begin{bmatrix} 1 & 0 \\ 0 & 0 \\ 0 & 1 \end{bmatrix}$;

2. $\underline{\Delta}(\underline{\eta}, \underline{v})$ is the force and torque vector that is synthesized from the uncertainty components of the ship model and disturbance from the external environment.

The matrices in Equation (3) are as follows:

$$M = \begin{bmatrix} m_{11} & 0 & 0 \\ 0 & m_{22} & 0 \\ 0 & 0 & m_{33} \end{bmatrix}, \; C(\underline{v}) = \begin{bmatrix} 0 & 0 & -m_{22}v \\ 0 & 0 & m_{11}u \\ m_{22}v & -m_{11}u & 0 \end{bmatrix}$$

$$D(\underline{v}) = \begin{bmatrix} d_{11} & 0 & 0 \\ 0 & d_{22} & 0 \\ 0 & 0 & d_{33} \end{bmatrix}, \; J(\underline{\eta}) = \begin{bmatrix} \cos\psi & -\sin\psi & 0 \\ \sin\psi & \cos\psi & 0 \\ 0 & 0 & 1 \end{bmatrix}$$

To solve the uncertainty component $\underline{\Delta}(\underline{\eta}, \underline{v})$ in Equation (3), this research will propose the estimator and compensate for the uncertainty component in the controller.

3. Building the Control System

3.1. The Diagram of the Control System

The targets of this study are to build the output-feedback MPC according to the separation principle and combine a state-feedback MPC and a state observer to control an underactuated ship in case the model contains an uncertain component $\underline{\Delta}(\underline{\eta}, \underline{v})$. The mission includes building the MPC controller based on a bilinear model of the ship combined with the direct state observer that is built from a continuous model of the ship, and building the estimator of uncertain components to estimate and compensate for the uncertain components in the model. The proposed control system structure is described in Figure 2.

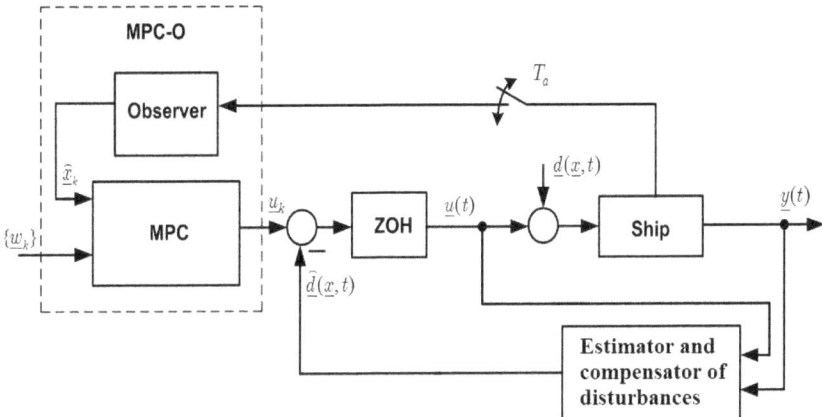

Figure 2. The model predictive controller (MPC) control system structure of the underactuated ship with the uncertain component.

In the figure, $\underline{d}(\underline{x},t)$ is an uncertain component that arises in the model, $\widehat{\underline{d}}(\underline{x},t)$ is the uncertain component which is estimated from the estimator, $\underline{\omega}_k$ is the set trajectory, $\underline{u}(t)$ is the control signal, and $\underline{y}(t)$ is the output signal. MPC-O denotes the output-feedback MPC controller. ZOH denotes the zero-order hold component.

3.2. The Bilinear Model of the Underactuated Ship in the Three-Freedom Space

Considering Equation (2), $\underline{x}_1 = \underline{\eta}$, $\underline{x}_2 = \underline{v}$, $\underline{u} = \underline{\tau}$, and $\underline{x} = \text{col}(\underline{x}_1, \underline{x}_2)$. We thus have the following:

$$\begin{pmatrix} \dot{\underline{\eta}} \\ \dot{\underline{v}} \end{pmatrix} = \begin{pmatrix} 0_{3\times 3} & J(\underline{\eta}) \\ -G(\underline{\eta}) & -M^{-1}[C(\underline{v}) + D(\underline{v})] \end{pmatrix} \begin{pmatrix} \underline{\eta} \\ \underline{v} \end{pmatrix} + \begin{pmatrix} 0_{3\times 2} \\ M^{-1}F \end{pmatrix} \underline{\tau} \tag{4}$$

Then,

$$A(\underline{x}) = \begin{pmatrix} 0_{3\times 3} & J(\underline{x}_1) \\ -G(\underline{x}_1) & -M^{-1}[C(\underline{x}_2) + D(\underline{x}_2)] \end{pmatrix} \tag{5}$$

$$B = \begin{pmatrix} 0_{3\times 2} \\ M^{-1}F \end{pmatrix}, \ C = (I_3, 0_{3\times 3})$$

Equation (4) is transformed into the following equation:

$$\begin{cases} \dot{\underline{x}} = A(\underline{x})\underline{x} + B\underline{u} \\ \underline{y} = \underline{\eta} = (I_3, 0_{3\times 3})\underline{x} = C\underline{x} \end{cases} \tag{6}$$

Equation (6) is the bilinear model of the underactuated ship in the three-freedom space.

Since the MPC controller is discrete, Equation (6) must be discrete with the sample time T_a. Performing the approximate equation $\dot{\underline{x}}(t) \approx [x((k+1)T_a) - x(kT_a)]/T_a$, Equation (6) is transformed into the following equation:

$$\begin{cases} \underline{x}_{k+1} = [I_6 + T_a A(\underline{x}_k)]\underline{x}_k + T_a B \underline{u}_k \\ \underline{y}_k = C\underline{x}_k \end{cases} \tag{7}$$

Linearizing each segment $\mathcal{H}_k$ of the model (6) along the time-axis with the assumption of a small amount of time, model (7) is approximated by the linear time-invariant model as follows [20]:

$$\mathcal{H}_k : \begin{cases} \underline{z}_{k+1} = \overline{A}_k \underline{z}_k + \overline{B} \Delta \underline{u}_k \\ \underline{y}_k = \overline{C} \underline{z}_k \end{cases} \tag{8}$$

where

$$\underline{z}_k = \text{col}(\underline{x}_k, \underline{u}_{k-1}) \in \mathbb{R}^8, \ \Delta \underline{u}_k = \underline{u}_k - \underline{u}_{k-1} \in \mathbb{R}^2$$
$$\overline{C} = (C, 0_{3 \times 2}) \in \mathbb{R}^{3 \times 8}$$

$$\overline{A}_k = \begin{pmatrix} I_6 + T_a A(\underline{x}_k) & T_a B \\ 0_{2 \times 6} & I_2 \end{pmatrix} \in \mathbb{R}^{8 \times 8}, \ \overline{B} = \begin{pmatrix} T_a B \\ I_2 \end{pmatrix} \in \mathbb{R}^{8 \times 2}$$

3.3. Building the Direct State Observer Based on the Continuous Model

Since the state variables in model (6) cannot be measured, the authors propose the direct observer from the continuous model (2) or (6) based on the measured output signal $\underline{y}(t) = \underline{\eta}(t) = [x \ y \ \psi]^T$. Here, x, y, ψ denote the x-axis coordinate, y-axis coordinate, and the direction of the ship, respectively. These values can be measured by the Global Position System (GPS) and the compass on board.

The observer's task is to identify the state vector of the continuous model (2) or (6) $\underline{x} = \text{col}(\underline{\eta}, \underline{v})$. In the status vector $\underline{x}(t) = \text{col}(\underline{\eta}(t), \underline{v}(t))$, the component $\underline{\eta}(t) = \underline{y}(t)$ has been measured, so we only need to define the second component $\underline{v}(t)$.

From Equation (2), we have $\underline{\dot{\eta}} = J(\underline{\eta}) \underline{v}$, with the orthogonal matrix $J(\underline{\eta})$ shown as follows:

$$J(\underline{\eta}) = \begin{bmatrix} \cos \psi & -\sin \psi & 0 \\ \sin \psi & \cos \psi & 0 \\ 0 & 0 & 1 \end{bmatrix} \tag{9}$$

Therefore, we can determine the state component as follows: $\underline{v}(t) = J(\underline{\eta})^{-1} \underline{\dot{\eta}} = J(\underline{y})^{-1} \underline{\dot{y}}$.

To determine the derivative value $\underline{\dot{y}}(t)$ of the output signal $\underline{y}(t)$, we can use the first-order inertia derivative stage $D_T(s) = \frac{s}{1+Ts}$, where $T > 0$ is tiny. The input is $\underline{y}(t)$ and the output is $\widehat{\underline{\dot{y}}}(t)$, so we have the following: $\widehat{\underline{\dot{y}}} + T \widehat{\underline{\ddot{y}}} = \underline{\dot{y}}$. Since $T \approx 0$, the output $\widehat{\underline{\dot{y}}}(t)$ is defined as follows: $\widehat{\underline{\dot{y}}} \approx \underline{\dot{y}} = J(\underline{\eta}) \underline{v}$.

Finally, we can obtain the observed signal as follows:

$$\underline{v}(t) = J(\underline{\eta})^{-1} \underline{\dot{\eta}} = J(\underline{y})^{-1} \underline{\dot{y}} \approx J(\underline{y})^{-1} \widehat{\underline{\dot{y}}} \tag{10}$$

3.4. Building the Estimator to Compensate for the Uncertainty Component in the Model

The incorrect model of a ship containing an uncertainty component (3) can be rewritten as follows:

$$\begin{cases} \underline{\dot{x}} = A(\underline{x})\underline{x} + B[\underline{u} + \underline{d}(\underline{x}, t)] \\ \underline{y} = C\underline{x} \end{cases} \tag{11}$$

The matrixes of the model $A(\underline{x}), B, C$ are inferred from the original model (3). Compared with the exact model in Equation (6), the model Equation (11) has an uncertain component $\underline{d}(\underline{x}, t)$, which is smooth. The article will identify this uncertain component.

The identification value is named $\widehat{\underline{d}}(\underline{x}, t) \approx \underline{d}(\underline{x}, t)$, with the tiny error $\|\widehat{\underline{d}}(\underline{x}, t) - \underline{d}(\underline{x}, t)\| \leq \delta_e$. After compensating for $\widehat{\underline{d}}(\underline{x}, t)$ in the input for the uncertainty model (11), this model is equivalent to the correct model (6).

The research will propose an approximate estimation method for $\widehat{\underline{d}}(x,t) \approx \underline{d}(x,t)$ based on the discontinuous model. This method is reasonable because the ships are not a fast-changing system, and in a control cycle, the uncertain components seem constant. This means that $\underline{d}(x,t)$ is a constant uncertain function in each cycle.

Equation (3) can be re-written according to the discontinuous form at the time (k) by linearizing each segment according to the time-axis. In the present control cycle, the uncertain component of the input $\underline{d}$ has been compensated for by $\widehat{\underline{d}}_{k-1}$, which is defined in the previous control cycle. We have the model at period k as follows:

$$\underline{v}_k = \widehat{A}(\underline{v}_{k-1})\underline{v}_{k-1} + \widehat{B}\left[\underline{u}_{k-1} + \underline{d} - \widehat{\underline{d}}_{k-1}\right] \quad (12)$$

where

$$\widehat{A}(\underline{v}_k) = I_3 - T_a M^{-1}\left[C(\underline{v}_k) + D(\underline{v}_k)\right] \in \mathbb{R}^{3\times 3},$$
$$\widehat{B} = T_a M^{-1} F \in \mathbb{R}^{3\times 2}, \underline{u}_k = \underline{\tau}_k = \underline{\tau}(kT_a) \quad (13)$$

The exact reference model at the time of k is

$$\underline{v}_k = \widehat{A}(\underline{v}_{k-1})\underline{v}_{k-1} + \widehat{B}\left(\underline{u}_{k-1} - \widehat{\underline{d}}_{k-1}\right) \quad (14)$$

where $\underline{u}_{k-1}$ is the input. The error of two models is $\underline{e}_k = \underline{v}_k - \underline{v}_k$. This error completely depends on the uncertain component, so we can determine $\widehat{\underline{d}}_k \approx \underline{d}$ from $\underline{e}_k$. Then, $\widehat{\underline{d}}_k$ is used to compensate for the uncertain component in the next control loop (k + 1).

From (13) and (14), we have

$$\underline{e}_k = \widehat{A}(\underline{v}_{k-1})\underline{v}_{k-1} + \widehat{B}\left[\underline{u}_{k-1} + \widehat{\underline{d}}_k\right] - \widehat{A}(\underline{v}_{k-1})\underline{v}_{k-1} - \widehat{B}\underline{u}_{k-1}$$
$$= \widehat{A}(\underline{v}_{k-1})\underline{v}_{k-1} - \widehat{A}(\underline{v}_{k-1})\underline{v}_{k-1} + \widehat{B}\underline{d}$$

Therefore, if the matrix $\widehat{B}$ has a rank of 2, then

$$\underline{d} \approx \widehat{\underline{d}}_k = \left(\widehat{B}^T\widehat{B}\right)^{-1}\widehat{B}^T\left[\underline{e}_k - \widehat{A}(\underline{v}_{k-1})\underline{v}_{k-1} + \widehat{A}(\underline{v}_{k-1})\underline{v}_{k-1}\right] \quad (15)$$

Equation (15) is used for approximating $\widehat{\underline{d}}_k \approx \underline{d}$ from $\underline{e}_k = \underline{v}_k - \underline{v}_k$, $\underline{v}_{k-1}$, and $\underline{v}_{k-1}$. $\underline{v}_k$, $\underline{v}_{k-1}$ can be measured or observed from the system. $\underline{v}_{k-1}$, $\underline{v}_k$ are defined by Equation (14). After defining $\widehat{\underline{d}}_k$, we can compensate for the input of the system at time $k+1$ as Figure 3.

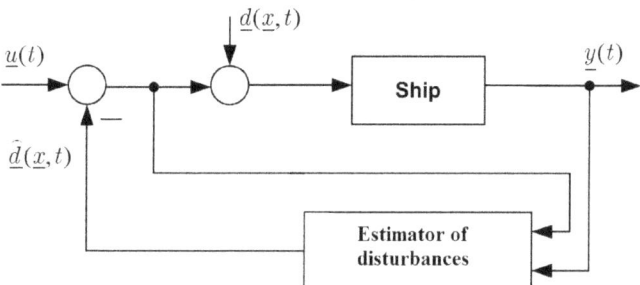

Figure 3. Compensating for the uncertain component for the incorrect model.

3.5. Building the MPC Controller with Output-Feedback Based on the Separation Principle

The separation principle-based output-feedback MPC controller is the combination of the state feedback MPC controller and the state observer directly from the continuous model. Unlike the conventional controller of linear MPC or nonlinear MPC, the proposed MPC controller is a linear controller used for a nonlinear object based on the linearization of each nonlinear model section. Along the time axis, the object includes countless linear models (8).

Based on the linear approximation model of each segment $\mathcal{H}_k(k = 0, 1, \ldots)$ and the predictive control principle, we can obtain the future outputs $\underline{y}_k$ for the entire forecasting window (N) as follows [21]:

$$\underline{y}_k = \begin{pmatrix} \underline{y}_{k+1} \\ \underline{y}_{k+2} \\ \vdots \\ \underline{y}_{k+N} \end{pmatrix} = \begin{pmatrix} \overline{CB} & 0_{3\times 2} & \cdots & 0_{3\times 2} \\ \overline{CA_k B} & \overline{CB} & \cdots & 0_{3\times 2} \\ \vdots & \vdots & \ddots & \vdots \\ \overline{CA_k^{N-1} B} & \overline{CA_k^{N-2} B} & \cdots & \overline{CB} \end{pmatrix} \begin{pmatrix} \Delta \underline{u}_k \\ \Delta \underline{u}_{k+1} \\ \vdots \\ \Delta \underline{u}_{k+N-1} \end{pmatrix} + \begin{pmatrix} \overline{CA_k} \\ \overline{CA_k^2} \\ \vdots \\ \overline{CA_k^N} \end{pmatrix} \underline{z}_k = H_k \underline{p} + \underline{b}_k \quad (16)$$

$$H_k = \begin{pmatrix} \overline{CB} & 0_{3\times 2} & \cdots & 0_{3\times 2} \\ \overline{CA_k B} & \overline{CB} & \cdots & 0_{3\times 2} \\ \vdots & \vdots & \ddots & \vdots \\ \overline{CA_k^{N-1} B} & \overline{CA_k^{N-2} B} & \cdots & \overline{CB} \end{pmatrix}, \underline{p} = \begin{pmatrix} \Delta \underline{u}_k \\ \Delta \underline{u}_{k+1} \\ \vdots \\ \Delta \underline{u}_{k+N-1} \end{pmatrix}, \underline{b}_k = \begin{pmatrix} \overline{CA_k} \\ \overline{CA_k^2} \\ \vdots \\ \overline{CA_k^N} \end{pmatrix} \underline{z}_k \quad (17)$$

where p is the vector of future input signals that need to be defined.

After having obtained the future output signals $\underline{y}_{k+i}$ $(i = 1, 2, \ldots, N)$ in the current forecasting window that depend on the future input signals $\Delta \underline{u}_{k+j} (j = 0, 1, \ldots, N-1)$ through Equation (8), we must define the future input signals so that the output signals follow the set signals $\{\underline{\eta}_k\}$.

Setting the target function is the sum of the squared deviation of the errors in the current forecasting window. In order to improve the response speed, this study proposed a technique to adjust the set signal [22]. The set-signal after adjustment is as follows: $\underline{r}_k = \begin{pmatrix} \underline{\eta}_{k+1} - K' \underline{e}_k \\ \vdots \\ \underline{\eta}_{k+N} - K' \underline{e}_k \end{pmatrix}$.

Here, $\underline{\eta}_k = \eta(kT_a)$, $\underline{e}_k = \underline{y}'_k - \underline{\eta}_k$ is the error at the previous time and K' is the calibration parameter of the set signal $(0 < K' < 1)$.

The objective function is set as follows:

$$J'_k(\underline{p}) = \underline{p}^T \left(H_k^T Q_k H_k + R_k \right) \underline{p} + 2 \left(\underline{b}_k - \underline{r}_k \right)^T Q_k H_k \underline{p} \underset{\underline{p}}{\to} \min$$

The solution is as follows:

$$\underline{p}* = \underset{\underline{p} \in P}{\operatorname{argmin}} J'_k(\underline{p}) \quad (18)$$

P is bound because of the mandatory requirements for the steer-angle α that is as follows:

$$-35^0 \leq \alpha \leq +35^0$$

$$P = \Delta U^N$$

$$\Delta U = \left\{ \Delta \underline{u} \in R^2 \mid \underline{b}_1(\underline{x}_k) - \underline{b}_1(\underline{x}_{k-1}) \leq \Delta \underline{u} \leq \underline{b}_2(\underline{x}_k) - \underline{b}_2(\underline{x}_{k-1}) \right\}$$

The common methods employed to find the optimal solution with bound conditions are the sequential quadratic programming (SQP) method, interior-point method, or evolutionary methods for optimal control, such as genetic algorithms (GA) and the particle swarm optimization (PSO) method. This research will use the GA method [23] to find the optimal solution of Equation (18).

The genetic algorithm (GA) is based on two basic rules of natural adaptations that hybridize 'good' elements $\underline{p}_{-i}$, $\underline{p}_{-j}$ together to get better elements and transform the 'bad' elements $\underline{p}_{-k}$. The 'bad' and 'good' valuations of these elements are evaluated through their objective function $J_i = J(\underline{p}_{-i})$.

Firstly, N elements $\underline{p}_{-i}$, $i = 1, 2, \ldots, N$ are randomly chosen in the constraint set p. Then, the values $J_i = J(\underline{p}_{-i})$ will be calculated respectively. Elements with $\delta_i \leq \delta_c$ are 'good', and elements with $\delta_k \geq \delta_m$ are 'bad'.

Here,

$$\delta_i = |J_i| / \sum_{k=1}^{N} |J_k|; 0 < \delta_c < \delta_m < 1 \text{ are optional.}$$

Next, a good pair of elements is hybridized to form new pairs, and bad elements are mutated. Therefore, a new generation formed from the old generation contains better elements. This process, with two calculations of the hybrid and mutant, is repeated many times until the end condition is satisfied. When the end condition of the algorithm is satisfied, the element $\underline{p}_{-i}$ of the current generation with the smallest J_i is chosen as the solution.

Currently, the GA has been installed into the command ga (·) in MATLAB, and we can use this command to build the proposed MPC controller. The parameters, such as the number of generations, the population size, and the type of selection, will automatically be defined. The syntax detail of the command is as follows:

$$[x, \text{fval}] = \text{ga}(\text{FUN}, \text{NVARS}, A, B, AE, BE, \text{OPTIONS})$$

where FUN is the objective function; NVARS is the number of variables of the objective function; and A, B, AE, and BE are the boundary conditions.

In this research, the authors execute the command as follows:
The objective function:

$$\text{FUN} = f = J_k'(\underline{p}) = \underline{p}^T\left(H_k^T Q_k H_k + R_k\right)\underline{p} + 2\left(\underline{b}_k - \underline{r}_k\right)^T Q_k H_k \underline{p}$$

The number of the variables: $NVARS = 2$.
The boundary conditions: $-2 \leq u_1 \leq 2$, $-3 \leq u_2 \leq 3$.
Finally, the details of the command are as follows:

$$[x, \text{fval}] = \text{ga}(f, 2, [\,], [\,], [-2; 2], [-3; 3], [\,], [\,])$$

After defining the solution $\underline{p}*$ of (18), we get the control signal $\underline{u}_k$ for controlling the ship motion (3) in the present cycle, as follows:

$$\underline{u}_k = \underline{u}_{k-1} - \left(I_2, 0_{2 \times 2(N-1)}\right)\underline{p}* \tag{19}$$

where $\underline{u}_k$ is the control signal during one sampling period T_a.
The control algorithm is as follows:

- Step 1: Initialing and setting the forecasting window width $N \geq 2$, the sample period T_a, and the calibration parameter of the set signal $0 < K' < 1$. Calculating B, C according to (5), $\overline{B}, \overline{C}$ according to (8), and $\widehat{B}$ according to (13). Setting $k = 1$, $\underline{\widehat{d}}_0 = 0$;
- Step 2: Setting the two positive symmetric matrices $Q_k \in \mathbb{R}^{3N \times 3N}$, $R_k \in \mathbb{R}^{2N \times 2N}$;
- Step 3: Measuring $\underline{\eta}_k$ and estimating $\underline{\widehat{v}}_k$. Calculating $\overline{A}_k$ according to (8) and $\widehat{A}(\underline{\widehat{v}}_{k-1})$ according to (13). Determining $H_k, \underline{b}_k, \underline{r}_k$ according to (17);
- Step 4: Determining $\underline{v}_k$ according to (12) and $\underline{\widehat{d}}_k$ according to (15) for the next cycle;

- Step 5: Determining the optimal solution $p*$ according to (18);
- Step 6: Calculating $\underline{u}_k$ from $p*$ according to (12);
- Step 7: Using $\underline{u}_k - \underline{\hat{d}}_{k-1}$ for controlling the continuous model (3), which is also the object (4) and (6);
- Step 8: Setting $k := k + 1$. If it needs to update G, go back to the step (2); otherwise, set $Q_{k+1} = Q_k$, $R_{k+1} = R_k$ and go back to step (3).

The observer is considered a continuous-time model and the control input is calculated by using the discrete-time model. Therefore, in the proposed controller algorithm, after observing the state of the continuous model, we have to make the observed signal of the continuous model discrete with the sample time Ta and then feed it into the controller. This is shown in step 3, where, after measuring $\underline{\eta}_k$ and estimating $\underline{\hat{v}}_k$, we calculate $\overline{A}_k$ according to (8).

4. The Results and Discussion

To verify the quality of the proposed control system, the authors have run the system on Matlab software (R2014b, MathWorks Inc., Natick, MA, USA). The control object is an underactuated ship with three degrees-of-freedom. The specifications of the ship given by Do K. D and J. Pan in the document [2], with the length of 32 m, the weight of 118×10^3 kg, the minimum radius of curvature of 150 m, and other parameters, are shown in Table 1.

Table 1. The parameters of the ship.

	τ_{umax}	τ_{rmax}	m_{11}	m_{22}	m_{33}	d_{11}	d_{22}	d_{33}
Unit	N	N.M	Kg	Kg	Kgm^2	Kgs^{-1}	Kgs^{-1}	Kgm^{2}s^{-1}
Value	5,2.10^9	8,5.10^8	120.10^3	177,9.10^3	636.10^5	215.10^2	177.10^3	802.10^4

The set trajectory is a straight and circular one, and the details are as follows: The ship moves straight for a period of 300 s with a distance of 1200 m, and the ship then moves in a circle with a radius of 200 m for a period of 325 s. The uncertain signals are assumed according to the document [5] as follows:

$$d_1 = 0.05\sin(0.1t) - 0.01, \quad d_2 = 0.2\sin(0.2t) + 0.4\cos(0.3t)$$

The simulation results are shown in Figures 4–7. Figure 4a shows the ability of the ship to follow the set trajectory, with the error shown in Figure 4b. Figure 5a shows the actual direction error. The errors obtained from the observer are shown in Figure 5b. The qualities of estimating the uncertain components d1 and d2 are shown in Figure 6a,b, respectively.

During the control process, the control signals of the straight slip force and the torque force are bound in the defined range, and they are shown as Figure 7a,b. The shapes of the control signals in Figure 7 are feasible because the control signal depends on the disturbance if the disturbance becomes large and the control becomes large. The simulation time of 800 s is long, so the number of times the control signal is changed in the simulation time is small. This number conforms to the Maritime regulation on the number of signal changes of the steering angle.

The quality of the control system has been evaluated based on the following factors: The error of trajectory, the actual direction error, the errors obtained from the observer, and the error of the estimator. The results show that these errors are very small, so the quality of the control system is very good.

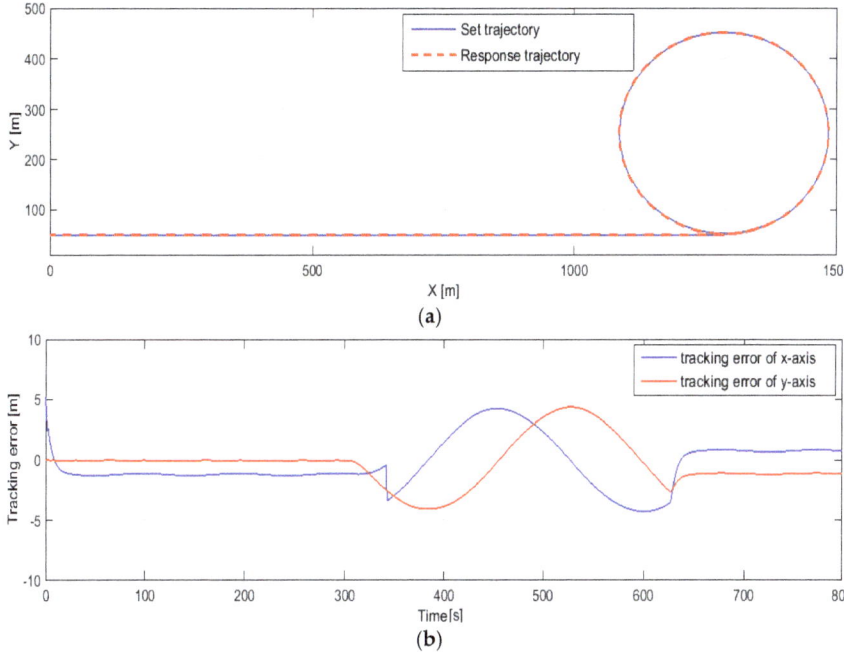

Figure 4. The ability of the control system to follow the set trajectory: (**a**) The response trajectory; (**b**) the error of the trajectory.

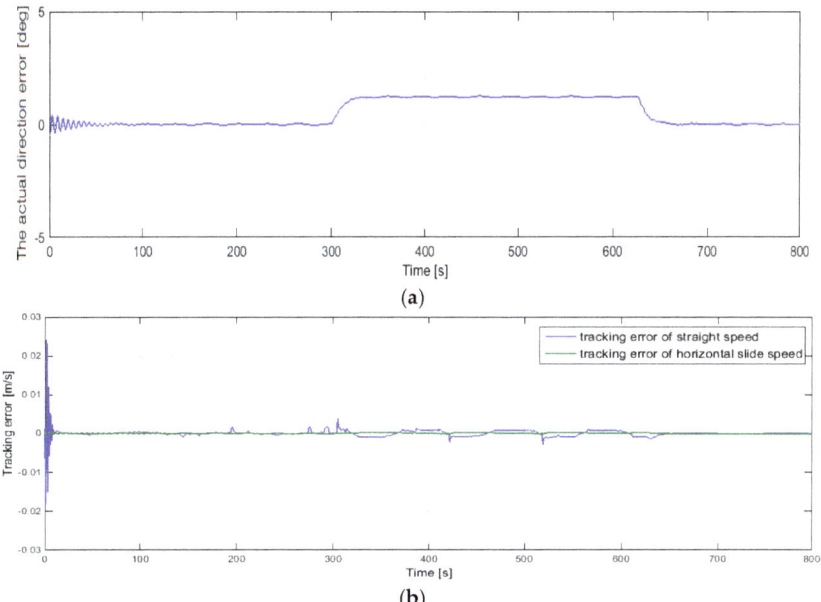

Figure 5. The simulation results of the observer: (**a**) The actual direction error; (**b**) the errors obtained from the observer.

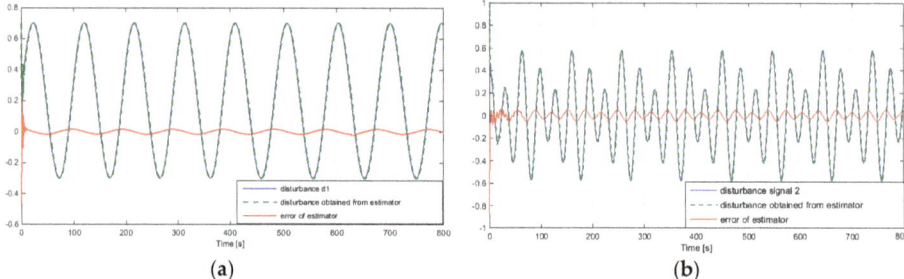

Figure 6. The results of estimating the uncertain components: (**a**) $d_1 = 0.05\sin(0.1t) - 0.01$; (**b**) $d_2 = 0.2\sin(0.2t) + 0.4\cos(0.3t)$.

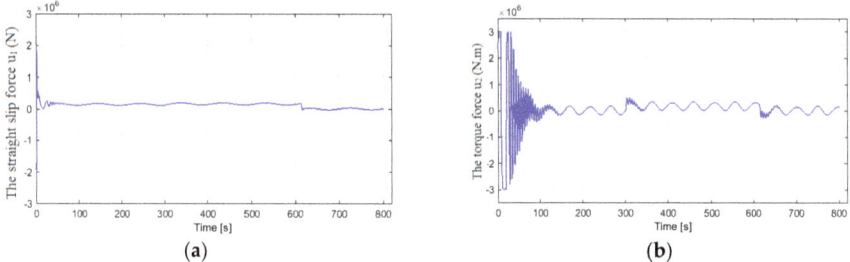

Figure 7. The control signals: (**a**) The straight slip force; (**b**) the torque force.

Although the control object has undefined components and the control signals are bound within limits, the quality of the entire control system is still very good: The response trajectory is very close to the set trajectory, and the error of the trajectory and the error of direction are very small. In addition, the quality of the observer depends on the receding horizon of the controller and the time needed to calculate the differential stage by the direct observer.

The estimator performed the mission well, compensating for the uncertain signals d_1, d_2 in the system. The difference between the estimated signal and the uncertainty signal is very small. The estimation error also depends on the receding horizon of the controller and the estimated value at the initial time.

In addition, the quality of the entire control system depends on the forecasting window N and the two positive symmetric matrices $Q_k \in R^{3N \times 3N}$, $R_k \in R^{2N \times 2N}$. In fact, we can choose matrixes Q_k, R_k that match the control target.

5. Conclusions

This article was successful in building the MPC control system with the output-feedback based on the separation principle, and building the direct observer based on the continuous model in order to control the trajectory of the underactuated ship. The controller was built based on the optimal control combined with the linearizing technique of each nonlinear model section. Disturbances from the environment and uncertainty components in the model were estimated and compensated for by the estimator. The simulation results show that although the control object has undefined components and the control signals are bound, the quality of the entire control system is still very good. The response values follow the set values, with tiny error. The success of this research is the basis for the authors to apply it to actual ships in further studies.

Author Contributions: H.-Q.N. proposed the initial idea. A.-D.T., H.-Q.N., and T.-T.N. developed the research, analyzed the results, and wrote the article together. T.-T.N. edited and finalized the article.

Funding: This research received no external funding.

Conflicts of Interest: The authors declare no conflict of interest.

References

1. Fossen, T.I. *Marine Control System-Guidance, Navigation and Control of Ships, Rigs and Underwater Vehicles*; Marine Cybernetics: Trondheim, Norway, 2002.
2. Do, K.D.; Pan, J. *Control of Ships and Underwater Vehicles: Design for Underactuated and Nonlinear Marine Systems*; Springer: London, UK, 2009.
3. Ashrafiuon, H.; Kenneth, R.M.; Lucas, C.M.; Reza, A.S. Sliding-mode tracking control of surface vessels. *IEEE Trans. Ind. Electron.* **2008**, *55*, 4004–4012. [CrossRef]
4. Do, K.D.; Pan, J.; Jiang, Z.P. Robust Adaptive Control of Underactuated Ships on a Linear Course with Comfort. *Ocean Eng.* **2003**, *30*, 2201–2225. [CrossRef]
5. Chwa, D. Global Tracking Control of Underactuated Ships with Input and Velocity Constraints Using Dynamic Surface Control Method. *IEEE Trans. Control Syst. Technol.* **2010**, *19*, 1357–1370.
6. Do, K.D.; Jiang, Z.P.; Pan, J. Robust adaptive path following of underactuated ships. *Automatica* **2004**, *40*, 929–944. [CrossRef]
7. Liu, Y.; Bu, R.; Gao, X. Ship Trajectory Tracking Control System Design Based on Sliding Mode Control Algorithm. *Polish Marit. Res.* **2018**, *25*, 26–34. [CrossRef]
8. Do, K.D.; Jiang, Z.P.; Pan, J. Underactuated Ship Global Tracking Under Relaxed Conditions. *IEEE Trans. Autom. Control* **2002**, *47*, 1529–1536. [CrossRef]
9. Godhavn, J.M. Nonlinear tracking of underactuated surface vessels. In Proceedings of the 35th IEEE Conference on Decision and Control, Kobe, Japan, 13 December 1996; pp. 975–980.
10. Serrano, M.E.; Gustavo, J.E.; Vicente, M.; Oscar, A.O.; Mario, J. Tracking trajectory of underactuated surface vessels: a numerical method approach. *Control Eng. Appl. Inform.* **2013**, *15*, 15–25.
11. Dai, S.L.; He, S.; Wang, M.; Yuan, C. Adaptive neural control of underactuated surface vessels with prescribed performance guarantees. *IEEE Trans. Neural Netw. Learn. Syst.* **2018**. [CrossRef] [PubMed]
12. Liu, C.; Zou, Z.; Yin, J. Trajectory tracking of underactuated surface vessels based on neural network and hierarchical sliding mode. *J. Mar. Sci. Technol.* **2015**, *20*, 322–330. [CrossRef]
13. Ngongi, W.E.; Du, J. Controller Design for Tracking Control of an Under-Actuated Surface Ship. *Int. J. Comput. Theory Eng.* **2015**, *7*, 469. [CrossRef]
14. Peng, Y.; Han, J.; Song, Q. Tracking control of underactuated surface ships: Using unscented Kalman filter to estimate the uncertain parameters. In Proceedings of the 2007 International Conference on Mechatronics and Automation, Harbin, China, 5–8 August 2007; pp. 1884–1889.
15. Wang, N.; Sun, Z.; Yin, G.; Sun, S.F.; Sharma, S. Finite-time observer based guidance and control of underactuated surface vehicles with unknown sideslip angles and disturbances. *IEEE Access* **2018**, *6*, 14059–14070. [CrossRef]
16. Lin, X.; Jiang, H.; Nie, J.; Jiao, Y. Adaptive-sliding-mode trajectory tracking control for underactuated surface vessels based on NDO. In Proceedings of the 2018 International Conference on Mechatronics and Automation, Changchun, China, 5–8 August 2018; pp. 1043–1049.
17. Liu, Z. Ship adaptive course keeping control with nonlinear disturbance observer. *IEEE Access* **2017**, *5*, 17567–17575. [CrossRef]
18. Abdelaal, M.; Franzle, M.; Hahn, A. Nonlinear model predictive control for tracking of underactuated vessels under input constraints. In Proceedings of the European Modelling Symposium (EMS), Madrid, Spain, 6–8 October 2015; pp. 13–318.
19. Fossen, T.I. *Guidance and Control of Ocean Vehicles*; John Wiley Sons Inc.: Chichester, UK, 1994.
20. Phuoc, N.D.; Le, T.T. Constrained output tracking control for time-varying bilinear systems via RHC with infinite prediction horizon. *J. Comput. Sci. Cybern.* **2015**, *31*, 97–106. [CrossRef]
21. Camacho, B. *Model Predictive Control*; Springer: London, UK, 2004.
22. Moore, K.L. *Iterative Learning Control for Deterministic Systems*; Springer: London, UK, 1993.
23. Liu, J. *Intelligent Control Design and MATLAB Simulation*; Springer: Singapore, 2018.

 © 2019 by the authors. Licensee MDPI, Basel, Switzerland. This article is an open access article distributed under the terms and conditions of the Creative Commons Attribution (CC BY) license (http://creativecommons.org/licenses/by/4.0/).

Article
Fine-Tuning Meta-Heuristic Algorithm for Global Optimization

Ziyad T. Allawi [1], Ibraheem Kasim Ibraheem [2] and Amjad J. Humaidi [3,*]

[1] College of Engineering, Department of Computer Engineering, University of Baghdad, Al-Jadriyah, Baghdad 10001, Iraq; ziyad.allawi@coeng.uobaghdad.edu.iq
[2] College of Engineering, Department of Electrical Engineering, University of Baghdad, Al-Jadriyah, Baghdad 10001, Iraq; ibraheemki@coeng.uobaghdad.edu.iq
[3] Department of Control and Systems Engineering, University of Technology, Baghdad 10001, Iraq
* Correspondence: 601116@uotechnology.edu.iq; Tel.: +964-7901227676

Received: 20 August 2019; Accepted: 10 September 2019; Published: 26 September 2019

Abstract: This paper proposes a novel meta-heuristic optimization algorithm called the fine-tuning meta-heuristic algorithm (FTMA) for solving global optimization problems. In this algorithm, the solutions are fine-tuned using the fundamental steps in meta-heuristic optimization, namely, exploration, exploitation, and randomization, in such a way that if one step improves the solution, then it is unnecessary to execute the remaining steps. The performance of the proposed FTMA has been compared with that of five other optimization algorithms over ten benchmark test functions. Nine of them are well-known and already exist in the literature, while the tenth one is proposed by the authors and introduced in this article. One test trial was shown to check the performance of each algorithm, and the other test for 30 trials to measure the statistical results of the performance of the proposed algorithm against the others. Results confirm that the proposed FTMA global optimization algorithm has a competing performance in comparison with its counterparts in terms of speed and evading the local minima.

Keywords: global optimization; meta-heuristics; swarm intelligence; benchmark functions; exploration; exploitation; global minimum; local minimum

---

1. Introduction

Meta-heuristic optimization describes a broad spectrum of optimization algorithms that need only the relevant objective function along with key specifications, such as variable boundaries and parameter values. These algorithms can locate the near-optimum, or perhaps the optimum values of that objective function. In general, meta-heuristic algorithms simulate the physical, biological, or even chemical processes that happen in nature. Of the meta-heuristic optimization algorithms, the following are the most widely used:

1. Genetic algorithms (GAs) [1], which simulate Darwin's theory of evolution;
2. Simulated annealing (SA) [2], which emerged from the thermodynamic argument;
3. Ant colony optimization (ACO) algorithms [3], which mimic the behavior of an ant colony foraging for food;
4. Particle swarm optimization (PSO) algorithms [4], which simulates the behavior of a flock of birds;
5. Artificial bee colony (ABC) algorithms [5], which mimic the behavior of the honeybee colony; and
6. Differential evolution algorithms (DEAs) [6], for solving global optimization problems.

Xing and Gao collected more than 130 state-of-the-art optimization algorithms in their book [7], and these swarm-based optimizations are applied in different applications and study cases [8–14].

Some algorithms start from a single point, such as SA, but the majority begin from a population of initial solutions (agents) like GAs, PSO, and DEAs, most of which is referred to as "swarm intelligence" in their mimicry of animal behaviors [15]. In these algorithms, every agent shares its information with other agents through a system of simple operations. This information sharing results in improvements to the algorithm performance and helps find the optimum or near-optimum solution(s) more quickly [3].

In any meta-heuristic optimization algorithm, there are three significant types of information exchange between a particular agent with other agents in the population. The first is called exploitation, which is a local search for the latest, and the best solution found so far. The second is called exploration, which is a global search using another agent existing in the problem space [16]. The third is called randomization, which is rarely used in some algorithms or may not be used at all. This last procedure is similar to exploration, but instead of an existing agent, a randomly-generated agent is used. For instance, ABC algorithms use randomization for the scout agent; therefore, it often succeeds in evading many local minima. Many algorithms begin with exploration and gradually shift to exploitation after several generations to avoid falling into local optimum values. Meta-heuristic algorithms then maintain trade between exploration and exploitation [17]. However, the different types demonstrate variations in how they perform this trade; by using this trade, these algorithms may get close to near-optimum or even optimum solutions.

All agents compete with themselves to stay alive inside the population. Every agent that improves its performance replaces any agent that did not promote itself. Therefore, in the fourth stage (i.e., selection) a variable selection method, such as greedy selection or roulette wheel, is used to choose the best agent to replace the worst one [1]. Meta-heuristic algorithms may find near-optimum solutions for some objective functions, but it may fall into local minima for other ones. This fact will be apparent in the results of this article. To date, an optimization algorithm that offers a superior convergence time and avoids local minima for objective functions has yet to be developed. Therefore, the area is open to improving the existing meta-heuristic algorithms or inventing new ones to fulfill these requirements [18].

In this article, a novel algorithm called the fine-tuning meta-heuristic algorithm (FTMA) is presented. It utilizes information sharing among the population agents in such a way that it finds the global optimum solution faster without falling into local ones; this is accomplished by performing the necessary optimization procedures sequentially. In the next section, the proposed algorithm is described in detail. Then, five well-known optimization algorithms are presented to compete with FTMA over a ten-function benchmark. The results and discussion are shown in the final section, along with the conclusions.

2. Literature Review

In the scope of the recent trends in nature-based meta-heuristic optimization algorithms, since the genetic algorithms [1] and simulated annealing [2] has been presented, the race begins in inventing many algorithms thanks to the rapid advances in computer speed and efficiency, especially in the new millennia. From these algorithms, we mention the firefly algorithm (FA) [19], cuckoo search (CS) [20], bat algorithm (BA) [21], flower pollination algorithm (FPA) [22], and many others mentioned in [23].

Many optimization algorithms were invented over the past five years. Some of them are new, and the others are modifications and enhancements to the already-existing ones. One of the recent and widely-used algorithms is grey wolf optimization (GWO) [24]; it is inspired by the grey wolves and their hunting behaviors in nature. Four types of leadership hierarchy of the grey wolves as well as three steps of prey hunting strategies are implemented. Mirjalili continued to invent other algorithms. The same authors presented moth–flame optimization (MFO) [25]. This algorithm mimics the navigation method of moths in nature which is called "traverse orientation". The main path which the moths travel along is towards the Moon. However, they may fall into a useless spiral path around artificial lights if they encounter these in their way. Ant lion optimizer (ALO) has been proposed in [26], which simulated the hunting mechanism of antlions in nature. Five main steps of hunting are implemented in this algorithm. Moreover, the same authors of [24] proposed a novel population-based optimization algorithm, namely, the sine–cosine algorithm (SCA) [27], it fluctuates the solution agents towards, or outwards, the best solution using a model based on sine and cosine functions. It uses random and adaptive parameters to emphasize the search steps like exploration and exploitation. Another proposed algorithm in the literature is the whale optimization algorithm (WOA) [28]. This algorithm mimics the social behavior of humpback whales, using a hunting strategy called bubble-net, as well as three operators to simulate this hunting strategy. All these algorithms mentioned above are developed, enhanced, and modified through the years, hopefully to make them suitable for every real problem which needs solving. However, no-free-lunch theorems state that there is no single universal optimization method that can deal with every realistic problem [18].

3. Fine Tuning Meta-Heuristic Algorithm (FTMA)

The FTMA is a meta-heuristic optimization algorithm used to search for optimum solutions for simple and/or complex objective functions. The fundamental feature of FTMA is the fine-tuning meta-heuristic method used when searching for the optimum.

FTMA performs the fundamental procedures of solution update, which are exploration, exploitation, randomization, and selection in sequential order. In FTMA, the first procedure of exploration is undertaken concerning an arbitrarily-selected solution in the solution space. If the solution is not improved according to the probability, the second procedure of exploitation is performed concerning the best global solution found so far. Again, if the solution is not enhanced according to probability, then the third procedure of randomization is performed concerning a random solution generated in the solution space. The fourth procedure of selection is performed by comparing the new solution and the old one and choosing the best according to the objective function. The FTMA procedure steps are:

1) Initialization: FTMA begins with initialization. Its equation is shown below:

$$x_i^0(k) = lb(k) + rand \times (ub(k) - lb(k)); \; k = 1, 2, \ldots d; i = 1, 2, \ldots, N. \qquad (1)$$

At this point in the process, all the solutions x_i^t are initialized randomly at the iteration counter $t = 0$ according to the lower bound lb and the upper bound ub for each solution space index k inside the solution space dimension d. A random number $rand$, its value is between 0 and 1, is used to place the solution value randomly somewhere between the lower and upper bounds. The space dimension, along with the number of solutions N must be specified prior to the process. Then, the fitness f_{xi}^0 is evaluated for each solution x_i^0 using the objective function. The values of the best objective fitness f_b^0 and its associated best solution x_b^0 are initially obtained from the fitness and solutions vectors, respectively. Additionally, the probabilities of exploitation and randomization, p, and r, respectively, are initialized.

After incrementing the iteration counter inside of the generation iteration loop, the four steps in each iteration are performed in the FTMA core, as follows:

2) Exploration: The general formula of this step is as follows:

$$y(k) = x_i^t(k) + rand \times \left(x_j^t(k) - x_i^t(k)\right). \qquad (2)$$

In this step, every solution x_i^t is moved with respect to another existing solution vector x_j^t, where $j \neq i$. The value of the objective function for the temporary solution y is then evaluated as a temporary fitness g.

3) Exploitation: Its equation is presented as follows:

$$if \; g > f_{xi}^t \; \&\& \; p > rand, \; y(k) = x_i^t(k) + rand \times \left(x_b^t(k) - x_i^t(k)\right). \qquad (3)$$

If the fitness g is not improved compared with f_{xi}^t and the probability of exploitation p is greater than a random number $rand$; then the exploitation step will be initiated. In this step, the temporary solution vector y is calculated by moving the solution x_i^t with respect to the best global solution, x_b^t. The value of the objective function for the temporary solution y is re-evaluated and stored in the temporary fitness g.

4) Randomization: The formula of this step is as follows:

$$if\ g > f_{xi}^t\ \&\&\ r > rand,\ y(k) = x_i^t(k) + rand \times \left(lb(k) + rand \times (ub(k) - lb(k)) - x_i^t(k)\right). \tag{4}$$

If the fitness g is not improved again in comparison with f_{xi}^t and the probability of randomization r is higher than a random number *rand*, then the randomization step will be initiated. In this step, the solution x_i^t moves with respect to a randomly-generated solution. The value of the objective function for the temporary solution y is again re-evaluated and then stored in the temporary fitness g.

5) Selection: The final step of the FTMA iteration process is the selection step, which is summarized as:

$$if\ g < f_{xi}^t,\ x_i^{t+1} = y;\ f_{xi}^{t+1} = g, \tag{5}$$

$$if\ g < f_b^t,\ x_b^{t+1} = y;\ f_b^{t+1} = g. \tag{6}$$

6) Stopping Condition: The search ends if the global fitness value f_b^{t+1} reaches zero or below a specified tolerance value ε, or if the iteration counter t reaches its previously-specified maximum value R. The pseudocode of FTMA is summarized as in Algorithm 1 below.

4. Methodology

To check the validity of the proposed FTMA, it should be tested with different well-known optimization algorithms that were used widely in the literature. Five algorithms are chosen, although there are many.

4.1. Well-Known Optimization Algorithms

(1) Genetic algorithm (decimal form) (DGA): This is similar to a conventional GAs with the exception that the chromosomes are not converted to binary digits. It has the same steps as GAs, selection, crossover, and mutation. Here, the crossover or mutation procedures are performed upon the decimal digits as they are performed upon the bits in a binary GA. The entire procedure of the DGA is taken from [29].
(2) Genetic algorithm (real form) (RGA): In this algorithm, the vectors are used in optimization as real values, without converting them to integers or binary numbers. As a binary GA, it performs the same procedures. The complete steps of DGA are taken from [30].
(3) Particle swarm optimization (PSO) with optimizer: The success of this famous algorithm is down to its simplicity. It uses the velocity vector to update every solution, using the best solution of the vector along with the best global solution found so far. The core formula of PSO is taken from [4].
(4) Differential evolution algorithm (DEAs): This algorithm chooses two (possibly three) solutions other than the current solution and searches stochastically, using selected constants to update the current solution. The whole algorithm is shown in [6].
(5) Artificial bee colony (ABC): This algorithm gained use for its distributed behavior simulating the collaborative system of a honeybee colony. The system is divided into three parts, the employed bees which perform exploration, the onlooker which shows exploitation, and the scout which performs randomization. The algorithm is illustrated in [5].

Algorithm 1: Fine-Tuning Meta-Heuristic Algorithm

Input: No. of solution population N, Maximum number of iterations R;
Tick;
for $i = 1$ to N
    Initialize x_i^0 using Equation (1);
    Evaluate f_{xi}^0 for every x_i^0;
end for
Search for x_b^0 and f_b^0;
Initialize $t = 0$, set p and r;
while $t < R$ && $f_b^t > \varepsilon$
    $t = t + 1$;
    for $i = 1$ to N
        Choose x_j^t such that $j \neq i$;
        Compute y using Exploration (Equation (2));
        Evaluate g for y;
        if $g > f_{xi}^t$ && $p > rand$
            Compute y using Exploitation (Equation (3));
            Evaluate g for y;
            if $g > f_{xi}^t$ && $r > rand$
                Compute y using Randomization (Equation (4));
                Evaluate g for y;
            end if
        end if
        if $g < f_{xi}^t$
            Update x_i^{t+1} and f_{xi}^{t+1} using Equation (5);
            if $g < f_b^t$
                Update x_b^{t+1} and f_b^{t+1} using Equation (6);
            end if
        end if
    end for
end while
Output: x_b^{t+1}, f_b^{t+1}, t, and the computation time.

4.2. Benchmark Test Functions

The optimization algorithms mentioned above, along with the proposed algorithm, will be tested on ten unimodal and multimodal benchmark functions. These functions have been used widely as alternatives to real-world optimization problems. Table 1 illustrates nine of these functions.

where x_i represents one of the solution parameters that $i = 1, 2, 3 \ldots d$ where d is the solution space dimension. The bold **0** represents a solution vector of zeros, whereas the bold **1** represents a solution vector of ones. The tenth benchmark function is proposed by the authors and introduced for the first time in this article, which is a multimodal function with multiple local and one global minimum, as shown in Table 2.

This function has $3^d - 1$ local minima which are located on points whose coordinates equal either 0 or ±1 except for the global minimum which is located precisely at the origin. The positive real parameter ε should be slightly higher than zero to trick the optimization algorithm to fall into the local minima.

Table 1. List of nine benchmark test functions used in global optimization.

Fn.Sym.	Function	Formula	$|x_i|$	Optimum
F1	SPHERE	$\sum_{i=1}^{d} x_i^2$	<5	$f(0) = 0$
F2	ELLIPSOID	$\sum_{i=1}^{d} i x_i^2$	<5	$f(0) = 0$
F3	EXPONENTIAL	$1 - \exp\left(-0.5 \times \sum_{i=1}^{d} x_i^2\right)$	<5	$f(0) = 0$
F4	ROSENBROCK	$\sum_{i=1}^{d-1} 100\left(x_{i+1} - x_i^2\right)^2 + (x_i - 1)^2$	<2	$f(1) = 0$
F5	RASTRIGIN	$10d + \sum_{i=1}^{d} \left(x_i^2 - 10 \cos 2\pi x_i\right)$	<5	$f(0) = 0$
F6	SCHWEFEL	$418.983d - \sum_{i=1}^{d} (x_i + 420.968) \sin \sqrt{|x_i + 420.968|}$	<100	$f(0) = 0$
F7	GREIWANK	$\sum_{i=1}^{d} \frac{x_i^2}{4000} - \prod_{i=1}^{d} \cos \frac{x_i}{\sqrt{i}} + 1$	<600	$f(0) = 0$
F8	ACKLEY	$-20 \exp\left(-0.2 \sqrt{\frac{\sum_{i=1}^{d} x_i^2}{d}}\right) - \exp\left(\frac{\sum_{i=1}^{d} \cos 2\pi x}{d}\right) + e + 20$	<32	$f(0) = 0$
F9	SCHAFFER	$\sum_{i=1}^{d-1} 0.5 + \frac{\sin^2\left(x_i^2 - x_{i+1}^2\right) - 0.5}{\left(1 + 0.001\left(x_i^2 + x_{i+1}^2\right)\right)^2}$	<100	$f(0) = 0$

Table 2. The introduced benchmark test function.

Fn.Sym.	Function	Formula	$|x_i|$	Optimum
F10	ALLAWI	$\sum_{i=1}^{d} \left(x_i^6 - 2(\varepsilon + 1)x_i^4 + (4\varepsilon + 1)x_i^2\right), 0 < \varepsilon \ll 1$	<2	$f^*(0) = 0$

Figure 1 illustrates that function for $d = 2$ and for $\varepsilon = 2.22 \times 10^{-16}$, which is the default constant called eps used in MATLAB® package (MathWorks, Natick, MA, USA). There are eight local minima distributed in a square space around the global minimum. The value of the function at these minima may be represented as $f(x) = 2\varepsilon \sum_{i=1}^{d} |x_i|$.

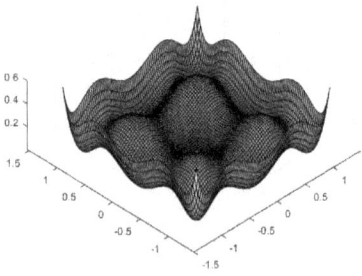

Figure 1. Graph of ALLAWI test function for $d = 2$ and $\varepsilon = 2.22 \times 10^{-16}$.

5. Results and Discussion

The two most essential requirements for an optimization algorithm are fast convergence and reaching the global minimum without falling into the local minima. Therefore, the judge for which of the optimization algorithms is the best will be taken according to these two criteria. The optimization algorithms were used to find the optimum values for the ten benchmark functions through 30 trials, to check for the mean error and the standard deviation for statistical comparison purposes. The parameters of the optimization algorithm FTMA, p and r were set to be 0.7 to make the flow control probably bypass the exploitation and randomization steps, even if the fitness is not improved in the exploration step. For all the algorithms, the number of dimensions was $d = 2$, the number of solution population agents was $N = 1000$, and the maximum number of iterations was $R = 1000$. The results of a sample trial are illustrated in Table 3.

Table 3. Results of the global fitness and computation time (s) for a sample trial. DGA: Genetic algorithm (decimal form); RGA: Genetic algorithm (real form); PSO: Particle swarm optimization; DEA: Differential evolution algorithms; ABC: Artificial bee colony; FTMA: fine-tuning meta-heuristic algorithm.

Fn.	DGA Fitness	DGA Time	RGA Fitness	RGA Time	PSO Fitness	PSO Time	DEA Fitness	DEA Time	ABC Fitness	ABC Time	FTMA Fitness	FTMA Time
F1	1.06×10^{-16}	1.05	1.64×10^{-16}	0.34	1.66×10^{-16}	0.46	3.01×10^{-17}	0.27	7.50×10^{-17}	0.30	**5.8×10^{-17}**	**0.12**
F2	1.07×10^{-16}	1.39	5.10×10^{-16}	8.88	1.16×10^{-17}	0.69	9.28×10^{-17}	0.40	2.15×10^{-17}	0.45	**1.51×10^{-17}**	**0.13**
F3	2.22×10^{-16}	102	2.22×10^{-16}	0.28	2.22×10^{-16}	0.38	1.11×10^{-16}	0.25	1.11×10^{-16}	0.35	**1.11×10^{-16}**	**0.10**
F4	1.95×10^{-6}	6.96	1.20×10^{-5}	4.84	1.97×10^{-16}	0.46	1.63×10^{-16}	0.46	4.37×10^{-7}	2.52	**7.05×10^{-17}**	**0.41**
F5	0	1.18	0	0.50	0	0.67	5.22×10^{-6}	4.85	0	0.65	0	0.18
F6	0	1.00	0	0.46	0	0.58	0	0.28	0	0.633	0	0.15
F7	2.22×10^{-16}	1.23	0	0.53	0	0.71	5.33×10^{-9}	5.41	1.11×10^{-16}	2.15	0	0.21
F8	0	2.67	1.34×10^{-12}	7.02	0	0.83	0	0.65	0	0.75	0	0.30
F9	0	0.58	2.22×10^{-16}	0.20	0	0.31	0	0.43	2.22×10^{-16}	0.66	0	0.09
F10	4.44×10^{-16}	8.01	1.18×10^{-16}	0.29	1.76×10^{-16}	0.51	1.70×10^{-13}	4.19	1.54×10^{-16}	0.67	**2.12×10^{-16}**	**0.11**

The data represent the output fitness value and the time taken by the optimization algorithm to drive its optimum global fitness below the minimum tolerance error $\varepsilon = 2.22 \times 10^{-16}$. The data in bold represents the algorithm that simultaneously scored the fastest time and found the global minimum for a specific benchmark function. The underlined data represents the algorithms that failed to pass the tolerance and completed all 1000 generation cycles. The following ten figures in Figure 2 represent the ten benchmark functions, illustrating the process of the optimization. All charts contain six lines which differ in pattern, one for each optimization algorithm.

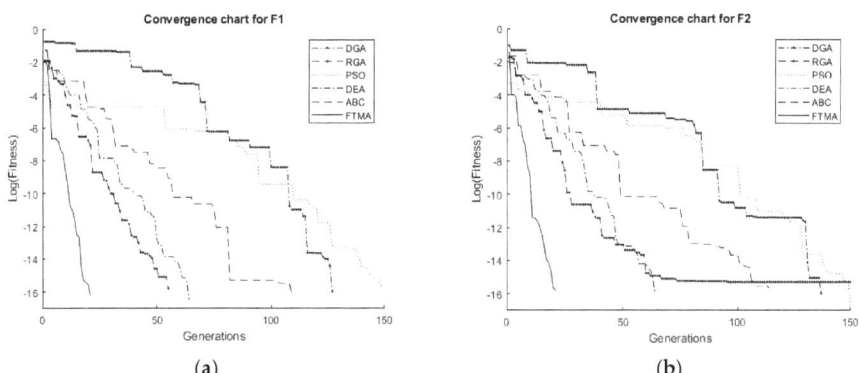

Figure 2. Cont.

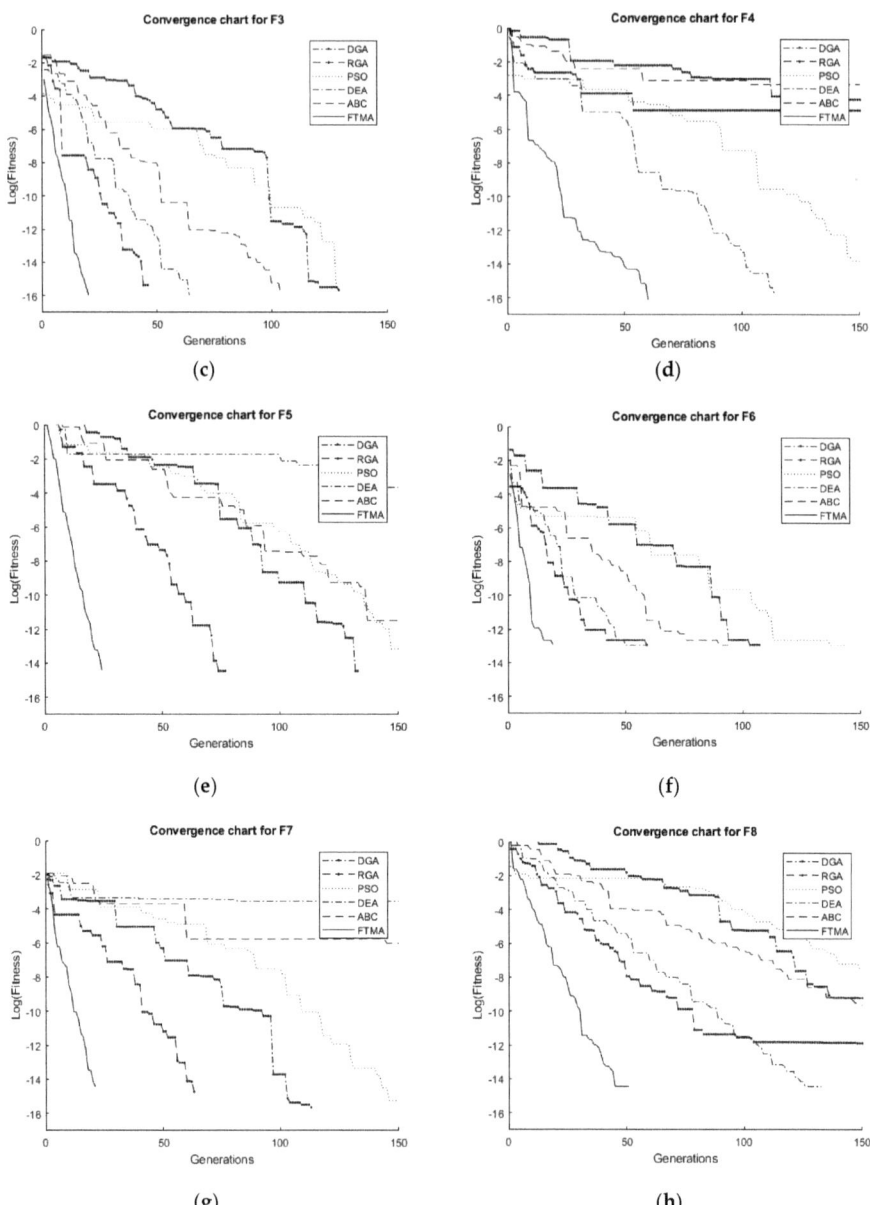

Figure 2. *Cont.*

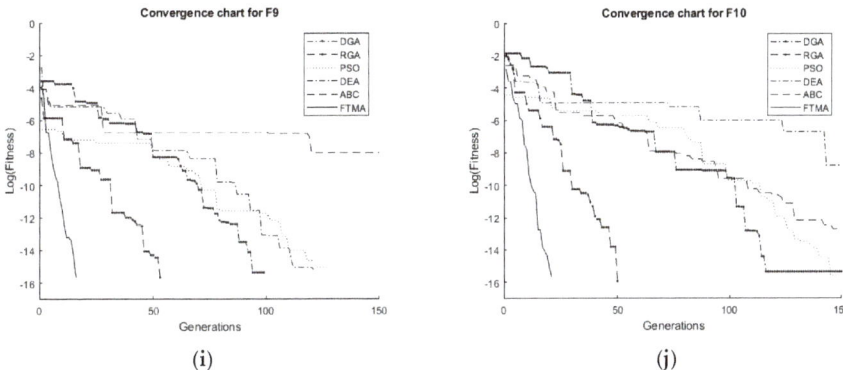

Figure 2. Convergence charts for the ten benchmark functions. (**a**) F1, (**b**) F2, (**c**) F3, (**d**) F4, (**e**) F5, (**f**) F6, (**g**) F7, (**h**) F8, (**i**) F9, and (**j**) F10.

Concerning the computation time, it is evident from Table 3 that the FTMA outperforms all other algorithms. Furthermore, we can see that the DGA failed to reach the optimum in F4 and barely in F7. For RGA, F2, F4, and F8 also failed. PSO evaded all local minima in all the benchmark functions. Furthermore, DEA failed in F10 along with F5 and F7. The ABC algorithm succeeded in avoiding local minima except for F4. In F5, F6, F8, and F9, most of the algorithms succeeded in capturing the zero global optimum value. However, FTMA never fell into the local minima, scoring the best convergence time out of all the optimization algorithms. Additionally, it reaches zero optimum value in the functions from F5 through F9. One can see that some of the optimization algorithms are suitable for some problems and not ideal for others. For example, DGA, RGA, and ABC failed in F4, but DEA succeeded; the situation is in contrary to F5. This confirms the no-free-lunch theorems of the absence of a universal algorithm for every problem. PSO, as well as FTMA, have both succeeded in evading the local maxima and converging to the global one. However, the time taken by PSO to reach the optimum is three to four times the time taken by FTMA. If we look at the ten subgraphs, which represent the search progression of the algorithms for one trial (its results are illustrated in Table 3), we find that the FTMA line (solid black) is the closest line to the vertical axis, which is the logarithmic scale of the global fitness against the number of generations. Although the maximum number of iterations is 1000, the maximum number of iterations displayed in the plots is set to be 150, because most of the algorithms catch the global optimum at or before this generation. In all figures, FTMA is the best-performing function. PSO and ABC are next best in most of the graphs. DGA, RGA, and DEA failed on many occasions. If we take the time which FTMA reached the critical error tolerance, the best of the other functions barely reached the fitness value at the same time. It can be seen from the plots that some of the algorithms have trapped in local minima, especially in F4. This implies that FTMA has the fastest convergence speed among the identified optimization algorithms. The values of the mean and standard deviation for the 30 trials are evaluated for each optimization algorithm and benchmark function. Table 4 illustrates the distribution of the output error, and Table 5 shows the distribution of computation time. The bold and underlined values represent the fastest and failed sets of trials, respectively. The trial sets are presented in ten sub-figures in Figure 3, one for each benchmark function.

Table 4. Statistical results of the output error for 30 trials.

Fn.	DGA m.	DGA Std.	RGA m.	RGA Std.	PSO m.	PSO Std.	DEA m.	DEA Std.	ABC m.	ABC Std.	FTMA m.	FTMA Std.
F1	5.94×10^{-17}	5.84×10^{-17}	5.10×10^{-15}	2.43×10^{-14}	1.03×10^{-16}	6.23×10^{-17}	9.13×10^{-17}	5.75×10^{-17}	1.00×10^{-16}	6.80×10^{-17}	7.84×10^{-17}	5.35×10^{-17}
F2	6.29×10^{-17}	5.46×10^{-17}	2.20×10^{-16}	7.04×10^{-16}	7.74×10^{-17}	5.98×10^{-17}	1.05×10^{-16}	6.76×10^{-17}	8.78×10^{-17}	6.24×10^{-17}	9.95×10^{-17}	6.52×10^{-17}
F3	1.29×10^{16}	9.96×10^{-17}	2.77×10^{-15}	1.31×10^{-14}	1.36×10^{-16}	7.94×10^{-17}	1.25×10^{-16}	7.97×10^{-17}	1.14×10^{-16}	7.29×10^{-17}	1.03×10^{-16}	8.56×10^{-17}
F4	2.69×10^{-06}	8.17×10^{-06}	3.51×10^{-05}	0.0001	1.07×10^{-16}	6.38×10^{-17}	1.09×10^{-16}	6.92×10^{-17}	1.05×10^{-06}	1.85×10^{-06}	1.25×10^{-16}	6.43×10^{-17}
F5	0	0	8.52×10^{-15}	4.52×10^{-14}	0	0	3.99×10^{-07}	4.84×10^{-07}	0	0	0	0
F6	0	0	0	0	0	0	0	0	0	0	0	0
F7	1.18×10^{-16}	1.03×10^{-16}	1.33×10^{-16}	8.78×10^{-17}	1.03×10^{-16}	8.56×10^{-17}	1.50×10^{-07}	1.96×10^{-07}	1.11×10^{-16}	9.50×10^{-17}	1.03×10^{-16}	8.56×10^{-17}
F8	2.36×10^{-16}	8.86×10^{-16}	1.59×10^{-08}	4.54×10^{-08}	0	0	0	0	0	0	0	0
F9	1.03×10^{-16}	1.10×10^{-16}	1.55×10^{-16}	1.01×10^{-16}	1.25×10^{-16}	1.1×10^{-16}	1.55×10^{-16}	1.01×10^{-16}	1.48×10^{-16}	1.04×10^{-16}	1.40×10^{-16}	1.07×10^{-16}
F10	6.56×10^{-16}	9.14×10^{-17}	1.19×10^{-16}	1.11×10^{-16}	3.73×10^{-16}	3.01×10^{-16}	1.41×10^{-09}	6.34×10^{-09}	1.04×10^{-16}	6.07×10^{-17}	9.80×10^{-17}	5.83×10^{-17}

Table 5. Statistical results of the computation time (seconds) for 30 trials.

Fn.	DGA m.	DGA Std.	RGA m.	RGA Std.	PSO m.	PSO Std.	DEA m.	DEA Std.	ABC m.	ABC Std.	FTMA m.	FTMA Std.
F1	1.085	0.221	0.566	1.091	0.546	0.072	0.315	0.047	0.364	0.040	0.125	0.021
F2	1.423	0.221	2.389	3.780	0.754	0.103	0.413	0.052	0.490	0.063	0.159	0.026
F3	1.021	0.219	1.215	2.224	0.525	0.075	0.291	0.034	0.353	0.067	0.166	0.015
F4	6.299	2.660	5.747	0.658	0.518	0.087	0.500	0.076	2.920	0.410	0.449	0.126
F5	1.275	0.247	1.788	2.561	0.687	0.091	5.200	0.500	0.733	0.084	0.203	0.033
F6	1.122	0.202	0.365	0.037	0.617	0.030	0.312	0.032	0.372	0.032	0.129	0.018
F7	1.361	0.290	0.686	0.110	0.714	0.030	5.783	0.111	2.100	0.241	0.213	0.028
F8	5.055	2.518	7.832	0.124	0.984	0.147	0.691	0.035	0.811	0.059	0.307	0.026
F9	0.773	0.139	0.404	0.761	0.343	0.054	0.478	0.051	0.765	0.092	0.119	0.022
F10	2.288	3.154	0.785	1.188	2.460	1.585	4.583	0.300	0.643	0.110	0.124	0.055

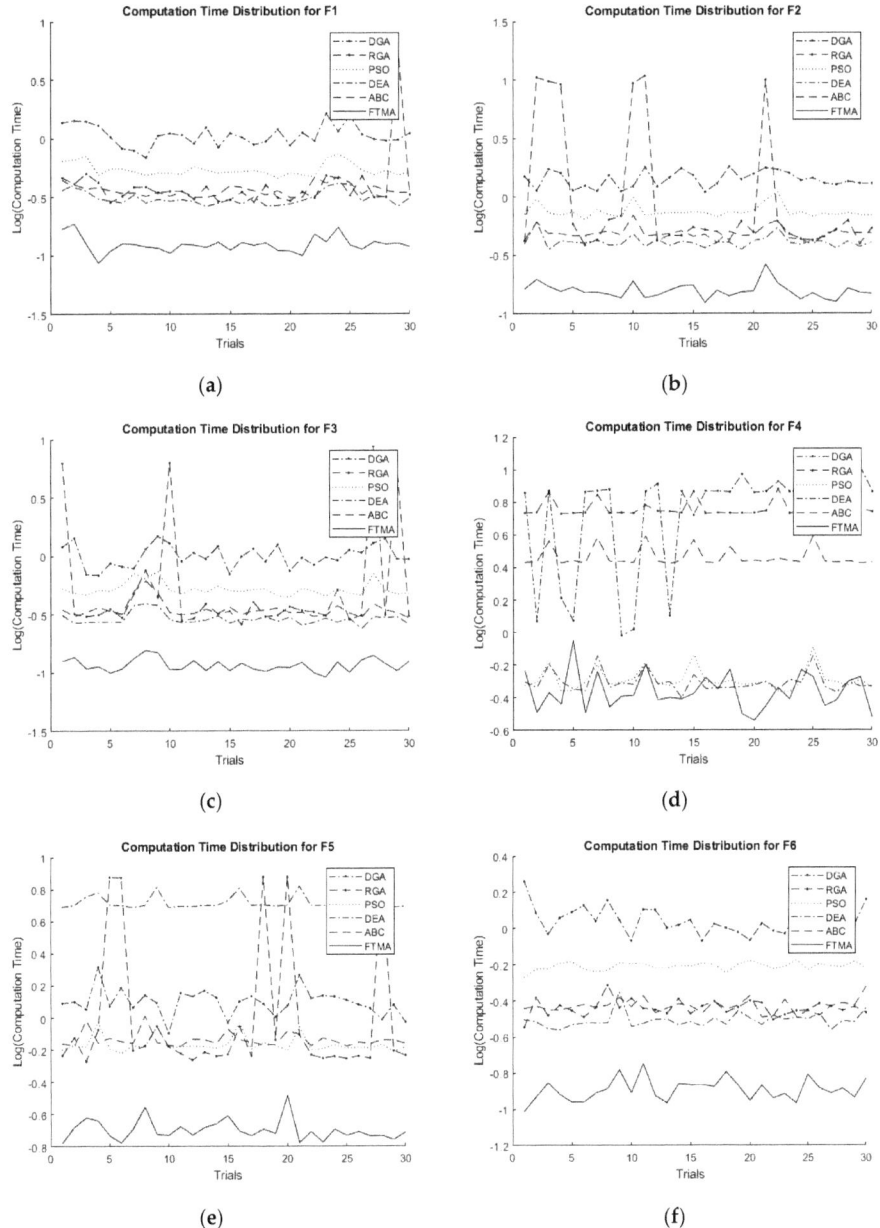

Figure 3. *Cont.*

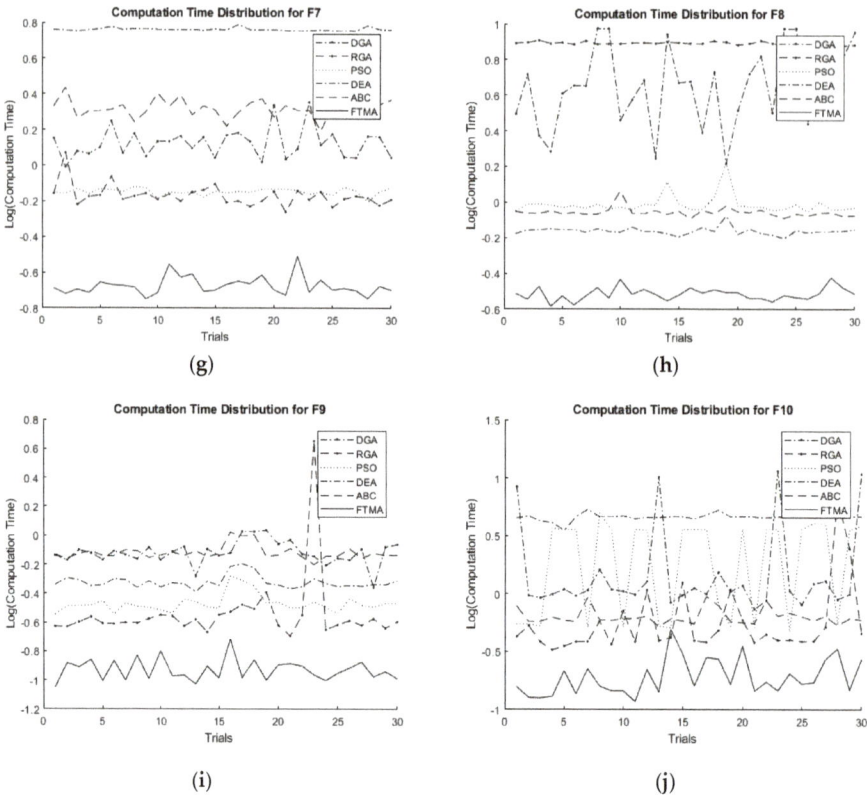

Figure 3. Computation time distribution for the ten benchmark functions. (**a**) F1, (**b**) F2, (**c**) F3, (**d**) F4, (**e**) F5, (**f**) F6, (**g**) F7, (**h**) F8, (**i**) F9, and (**j**) F10.

In Table 4, the overall trials show that some algorithms that succeeded in one of the tests might not achieve well in another one. It can be concluded that RGA failed in F2; DGA, RGA, and ABC failed in F4. Moreover, DEA failed in F5; DEA, and ABC failed in F7; DGA and RGA failed in F8; while PSO failed in the proposed benchmark function F10, which succeeded in all the other functions. Although DGA average error is slightly less than the mean error of FTMA in F1, F2, and F9, the average computation time is about eight times the computation time of the proposed algorithm. This implies that the proposed algorithm succeeded in reaching the global minimum before DGA. It can be seen that the computation time for the proposed algorithm is the best for all the benchmark functions. In Figure 3, the plots contain six lines with different patterns, one for each optimization algorithm. The figures show the logarithm of the computation time against computation trials. One can determine from these plots that some optimization functions are suitable for some algorithms and not for another. For instance, DEA is suitable for F4 but not for F5. The proposed algorithm always has the best computation time among all the remaining algorithms. Its solid line lies in the bottom near the horizontal axis. In F4, it is accompanied by PSO and DEA; in the other plots, it was alone in the bottom. For the proposed benchmark system, DEA was the worst. PSO fell in local optima many times, and DGA a few times. ABC and RGA performed well, but FTMA was the best.

6. Conclusions

This paper proposed a new global optimization named the fine-tuning meta-heuristic algorithm (FTMA). From the simulation results, it can be concluded that the FTMA reaches the optimum value faster than any other optimization algorithm used in the comparison. Its performance is competing with state-of-the-art methods, namely, RGA, DEA, ABC, PSO, and DGA. It accomplishes this in real-time and, unlike other optimization algorithms, evading any local optima. Moreover, it maintains the accuracy and robustness at the least runtime. Therefore, the FTMA offers a promising approach which, thanks to its rapid convergence time, could be applied in more complicated real-time systems where the time is a crucial factor. This result does not mean that this algorithm can solve any real problem we may encounter in practice, as it stated in the no-free-lunch theorems, there may be processes that this algorithm struggles to solve. So, there are possible opportunities to enhance the FTMA and/or its counterparts. Future studies include using the FTMA in combinatorial optimization or integrating the FTMA in control applications as an online or offline tuning algorithm for finding the optimal parameters of the feedback controllers. Moreover, because the lack of resources (supercomputers, etc.), the computation time of more than two parameters in the algorithm takes hours or sometimes days. So, it is intended to make the problem space higher if these resources become available. Finally, checking multi-dimensional spaces and using multi-objective problem scenarios are possible aspects for future research.

Author Contributions: Conceptualization, Z.T.A.; validation, I.K.I.; methodology, Z.T.A., I.K.I.; writing—review and editing, I.K.I., and A.J.H.; investigation, Z.T.A., I.K.I., and A.J.H.; formal analysis, A.J.H.

Funding: This research did not receive any specific grant from funding agencies in the public, commercial, or not-for-profit sectors.

Conflicts of Interest: The authors declare no conflicts of interest.

References

1. Goldberg, D. *Genetic Algorithms in Search, Optimization and Machine Learning*, 1st ed.; Addison-Wesley Longman Publishing Co., Inc.: Boston, MA, USA, 1989.
2. Kirkpatrick, S.; Gelatt, C.; Vecchi, M. Optimization by simulated annealing. *Science* **1983**, *220*, 671–680. [CrossRef] [PubMed]
3. Dorigo, M.; Maniezzo, V.; Colorni, A. Ant system: Optimization by a colony of cooperating agents. *IEEE Trans. Syst. Man Cybern. Part B (Cybern.)* **1996**, *26*, 29–41. [CrossRef] [PubMed]
4. Eberhart, R.; Kennedy, J. A new optimizer using particle swarm theory. In Proceedings of the Sixth International Symposium on Micro Machine and Human Science, MHS'95, Nagoya, Japan, 4–6 October 1995; IEEE: Piscataway, NJ, USA, 1995; pp. 39–43.
5. Karaboga, D.; Basturk, B. A powerful and efficient algorithm for numerical function optimization: Artificial bee colony (ABC) algorithm. *J. Glob. Optim.* **2007**, *39*, 459–471. [CrossRef]
6. Storn, R.; Price, K. Differential evolution—A simple and efficient heuristic for global optimization over continuous spaces. *J. Glob. Optim.* **1997**, *11*, 341–359. [CrossRef]
7. Xing, B.; Gao, W. *Innovative Computational Intelligence: A Rough Guide to 134 Clever Algorithms*, 1st ed.; Springer: Berlin/Heidelberg, Germany, 2016.
8. Ibraheem, I.K.; Ajeil, F.H. Path Planning of an autonomous Mobile Robot using Swarm Based Optimization Techniques Technique. *Al-Khwarizmi Eng. J.* **2016**, *12*, 12–25. [CrossRef]
9. Ibraheem, I.K.; Al-hussainy, A.A. Design of a Double-objective QoS Routing in Dynamic Wireless Networks using Evolutionary Adaptive Genetic Algorithm. *Int. J. Adv. Res. Comput. Commun. Eng.* **2015**, *4*, 156–165.
10. Ibraheem, I.K.; Ibraheem, G.A. Motion Control of an Autonomous Mobile Robot using Modified Particle Swarm Optimization Based Fractional Order PID Controller. *Eng. Technol. J.* **2016**, *34*, 2406–2419.
11. Humaidi, A.J.; Ibraheem, I.K.; Ajel, A.R. A Novel Adaptive LMS Algorithm with Genetic Search Capabilities for System Identification of Adaptive FIR and IIR Filters. *Information* **2019**, *10*, 176. [CrossRef]
12. Humaidi, A.; Hameed, M. Development of a New Adaptive Backstepping Control Design for a Non-Strict and Under-Actuated System Based on a PSO Tuner. *Information* **2019**, *10*, 38. [CrossRef]
13. Allawi, Z.T.; Abdalla, T.Y. A PSO-Optimized Reciprocal Velocity Obstacles Algorithm for Navigation of Multiple Mobile Robots. *Int. J. Robot. Autom.* **2015**, *4*, 31–40. [CrossRef]

14. Allawi, Z.T.; Abdalla, T.Y. An ABC-Optimized Type-2 Fuzzy Logic Controller for Navigation of Multiple Mobile Robots Ziyad. In Proceedings of the Second Engineering Conference of Control, Computers and Mechatronics Engineering, Baghdad, Iraq, February 2014; pp. 239–247.
15. Tarasewich, P.; McMullen, P. Swarm intelligence: Power in numbers. *Commun. ACM* **2002**, *45*, 62–67. [CrossRef]
16. Crepinsek, M.; Liu, S.; Mernik, M. Exploration and exploitation in evolutionary algorithms: A survey. *ACM Comput. Surv.* **2013**, *45*, 35. [CrossRef]
17. Chen, J.; Xin, B.; Peng, Z.; Dou, L.; Zhang, J. Optimal contraction theorem for exploration and exploitation tradeoff in search and optimization. *IEEE Trans. Syst. Man Cybern. Part A Syst. Hum.* **2009**, *39*, 680–691. [CrossRef]
18. Wolpert, D.; Macready, W. No free lunch theorems for optimization. *IEEE Trans. Evol. Comput.* **1997**, *1*, 67–82. [CrossRef]
19. Yang, X.S. Firefly algorithms for multimodal optimisation. In Proceedings of the Fifth Symposium on Stochastic Algorithms, Foundations and Applications, Sapporo, Japan, 26–28 October 2009; Watanabe, O., Zeugmann, T., Eds.; Springer: Berlin/Heidelberg, Germany, 2009; Volume 5792, pp. 169–178, Lecture notes in computer, science.
20. Yang, X.S.; Deb, S. Cuckoo Search via Lévy flights. In Proceedings of the 2009 World Congress on Nature & Biologically Inspired Computing (NaBIC), Coimbatore, India, 9–11 December 2009; IEEE: Piscataway, NJ, USA, 2009; pp. 210–214.
21. Yang, X.S. A new Meta-heuristic bat-inspired algorithm. In *Nature Inspired Cooperative Strategies for Optimization (NISCO 2010)*; Studies in computational intelligence; Springer: Berlin, Germany, 2010; pp. 65–74.
22. Yang, X.S. Flower pollination algorithm for global optimization. In *Unconventional Computation and Natural Computation*; Lecture notes in computer science; Springer: Berlin/Heidelberg, Germany, 2012; Volume 7445, pp. 240–249.
23. Yang, X.S. *Nature-Inspired Optimization Algorithms*, 1st ed.; Elsevier: London, UK, 2014.
24. Mirjalili, S.A.; Mirjalili, S.M.; Lewis, A. Grey Wolf Optimizer. *Adv. Eng. Softw.* **2014**, *69*, 46–61. [CrossRef]
25. Mirjalili, S.A. Moth-flame optimization algorithm: A novel nature-inspired heuristic paradigm. *Knowl. Based Syst.* **2015**, *89*, 228–249. [CrossRef]
26. Mirjalili, S.A. The Ant Lion Optimizer. *Adv. Eng. Softw.* **2015**, *83*, 80–98. [CrossRef]
27. Mirjalili, S.A. SCA: A Sine Cosine Algorithm for solving optimization problems. *Knowl. Based Syst.* **2016**, *96*, 120–133. [CrossRef]
28. Mirjalili, S.A.; Lewis, A. The Whale Optimization Algorithm. *Adv. Eng. Softw.* **2016**, *95*, 51–67. [CrossRef]
29. Lee, Y.; Marvin, A.; Porter, S. Genetic algorithm using real parameters for array antenna design optimization. In *MTT/ED/AP/LEO Societies Joint Chapter UK and Rep. of Ireland Section. 1999 High Frequency Postgraduate Student Colloquium*; Cat. No.99TH840; IEEE: Piscataway, NJ, USA, 1999; pp. 8–13.
30. Bessaou, M.; Siarry, P. A genetic algorithm with real-value coding to optimize multimodal continuous functions. *Struct. Multidiscip. Optim.* **2001**, *23*, 63–74. [CrossRef]

© 2019 by the authors. Licensee MDPI, Basel, Switzerland. This article is an open access article distributed under the terms and conditions of the Creative Commons Attribution (CC BY) license (http://creativecommons.org/licenses/by/4.0/).

Article

A Holonic-Based Self-Learning Mechanism for Energy-Predictive Planning in Machining Processes

Seung-Jun Shin [1,*], Young-Min Kim [2] and Prita Meilanitasari [2]

1 Division of Interdisciplinary Industrial Studies, Hanyang University, Seoul 04763, Korea
2 Graduate School of Technology and Innovation Management, Hanyang University, Seoul 04763, Korea; yngmnkim@hanyang.ac.kr (Y.-M.K.); dintamio@hanyang.ac.kr (P.M.)
* Correspondence: sjshin@hanyang.ac.kr; Tel.: +82-2-2220-2358

Received: 30 August 2019; Accepted: 11 October 2019; Published: 14 October 2019

Abstract: The present work proposes a holonic-based mechanism for self-learning factories based on a hybrid learning approach. The self-learning factory is a manufacturing system that gains predictive capability by machine self-learning, and thus automatically anticipates the performance results during the process planning phase through learning from past experience. The system mechanism, including a modeling method, architecture, and operational procedure, is structured to agentize machines and manufacturing objects under the paradigm of Holonic Manufacturing Systems. This mechanism allows machines and manufacturing objects to acquire their data and model interconnection and to perform model-driven autonomous and collaborative behaviors. The hybrid learning approach is designed to obtain predictive modeling ability in both data-existent and even data-absent environments via accommodating machine learning (which extracts knowledge from data) and transfer learning (which extracts knowledge from existing knowledge). The present work also implements a prototype system to demonstrate automatic predictive modeling and autonomous process planning for energy reduction in milling processes. The prototype generates energy-predictive models via hybrid learning and seeks the minimum energy-using machine tool through the contract net protocol combined with energy prediction. As a result, the prototype could achieve a reduction of 9.70% with respect to energy consumption as compared with the maximum energy-using machine tool.

Keywords: cyber-physical production systems; self-learning factory; holonic manufacturing systems; machine learning; transfer learning; predictive analytics

1. Introduction

Manufacturing intelligence reinforces real-time understanding, reasoning, planning, and management of manufacturing processes with the pervasive use of sensor-based data analytics and modeling [1]. Such intelligence is nothing new in manufacturing; however, it is not mature despite much effort related to its implementation and utilization over the past decades [2]. Implementing manufacturing intelligence is becoming more important than ever due to the evolution of manufacturing technology (MT) itself and the convergence of MT with Internet of things and cyber-physical systems (CPS).

CPS have been recognized as a cutting-edge technology in implementing machine intelligence in various domains, as CPS are "physical and engineered systems whose operations are monitored, coordinated, controlled and integrated by a computing and communicating core" [3]. The concept of CPS is naturally being deployed to industrial automation in the manufacturing realm, and the manufacturing version of CPS is known as cyber-physical production systems (CPPS). CPPS seek to realize intelligence, connectedness, and responsiveness through autonomous and cooperative objects and sub-systems based on context awareness within and across all levels of production [4].

CPPS can be categorized based on their maturity levels, which consist of: visibility, transparency, predictive capability, and self-optimization [5]. CPPS ultimately pursue the acquisition of self-optimizing ability so that manufacturing machines are directly involved in problem-solving or optimization through autonomous and collaborative decision-making and communication with minimization of human intervention. This self-optimizing ability obviously requires predictive capability at the precedent level. Machines can achieve self-optimization only after they can predict their performance by themselves through learning algorithms and use this ability to enhance the accuracy and robustness of their decision-making through evolutionary learning. Such predictive capability can be realized if the machine can self-learn. Self-learning endows manufacturing systems (especially manufacturing objects like machines, material handling equipment, workpieces, work-in-process and products) with the ability to learn from history for future decisions [6].

From the perspective of CPPS implementation, CPPS require control architecture suites fit for autonomous and collaborative operation and control on manufacturing objects. Holonic Manufacturing Systems (HMS) represent one of the most promising architecture suites, with the same goals as those of CPPS [4]. This coincidence can be demonstrated in the Product–Resource–Order–Staff Architecture (PROSA) reference architecture. This referential architecture is structured to achieve both hierarchical and heterarchical control by employing holons (autonomous and cooperative objects in manufacturing systems) and their holarchy (a system of holons) for efficiency in resource utilization, stability against disturbances, and flexibility during changes [7]. To pursue manufacturing intelligence, we suggest that holonic-based systems should be reshaped to obtain learning ability within the complex and dynamic nature of manufacturing environments. Even good stationary structures and mechanisms can hardly accommodate huge numbers of manufacturing conditions which are rapidly changing. Without learning, it is extremely difficult to identify concrete behaviors and activities that will improve the performance of manufacturing systems [8].

Traditionally, self-learning largely depends on creating predictive models derived by machine-learning techniques (e.g., regression, decision tree, Artificial Neural Network (ANN), support vector machine, and genetic algorithms). Machine-learning techniques are used to acquire the knowledge needed to make future decisions from historical training examples [9]. It is known that they enhance the validity of machine-specific models by using real and historical data even in dynamic and complex manufacturing environments. Machine-learning determine cause-and-effect relationships from the training datasets that have been collected from previous manufacturing operations. As cause-and-effect relationships are derived into mathematical representation under certain conditions and constraints, machine-learned models can faithfully work as predictive models by anticipating an effect from an input of cause values.

However, traditional machine learning has a drawback. It does not work unless training datasets exist. Manufacturing environments cannot always create or keep training datasets due to difficult data collection, data loss, data becoming outdated, or even data missing from manufacturing operations that have not been run. Collecting new datasets by performing additional manufacturing operations is desirable; however, it is time-consuming and is sometimes impossible. Nevertheless, the self-learning ability should be obtained and maximized even in such data-absent environments, and transfer learning can be a complementary means of achieving the machine's self-learning. Transfer learning is a technique to extract knowledge from source tasks and apply the knowledge to a target task to reduce the effort required to collect training datasets [10]. As transfer learning involves knowledge extraction from existing models, it allows machines to create knowledge-transferred models in the data-absent environment. Eventually, the adaptive convergence of machine learning and transfer learning enables machines to implement their self-learning ability in both data-existent and even data-absent environments.

In the metal cutting industry, energy consumption becomes a major metric for improving energy-efficiency and environmental performance. According to a survey [11], the manufacturing sub-sectors of fabricated metal products and machinery where the metal cutting industry involves as a part consume 41,869 and 23,424 million kWh of net electricity in the United States of America, respectively. As these values respectively occupy 5.5% and 3.1% of the entire manufacturing sector, reducing energy consumption in the metal-cutting industry is important for improving environmental performance. In this context, machining in a machine tool affects energy consumption and varies the energy difference in terms of machining power and time by about 66% [12]. Thus, process planning for energy efficiency works as a useful means for reducing the energy consumed during the execution of machine tools because machining sequences and process parameters decided during process planning significantly influence the performance of machine tool operations [13].

Accordingly, much of the literature has elucidated the relational models between process planning decisions and energy consumption based on theoretical and experimental modeling approaches [14]. A theoretical approach uses the theory of metal-cutting with some coefficient assumptions; however, it has limits in predicting energy values correctly due to the gap between assumptive and real coefficient values. An experimental approach can be subdivided into statistical and learning approaches. A statistical approach generates statistical models based on Design of Experiments, which aims at generating response surfaces with a small set of experimental data. This approach derives polynomial equations for energy prediction; however, it only works within the restricted experimental condition. A learning approach uses real data from machining operations for creating machine-learned energy models and shows high accuracy of energy prediction; however, it is limited to creating such models in a data-absent environment, as mentioned above.

In view of the above, a holonic-based approach is necessary to gain the predictive capability through self-learning for reducing energy in machining processes. As holons result from the application of object-oriented concept, they can work as decentralized individuals who independently operate for how-to-create and how-to-use models. These object-oriented holons can adaptively and evolutionarily create learned models based on their associated data and thus can cope with the variability of data, which frequently take place in manufacturing systems due to the changes in manufacturing setup, condition and environment. Furthermore, holons' mutual cooperation via their message exchanges pursues performance optimization centralized on a holarchy. A plausible scenario for energy reduction in machining is that the machines abstracted by holons automatically create energy-predictive models through learning techniques, and predict their energy values using the models. In succession, the machines autonomously and cooperatively make an optimal decision for reducing energy consumption during the process planning phase.

For such purposes, we designed a holonic-based mechanism for self-learning factories based on a hybrid-learning approach. We also implemented a prototype system to perform predictive process planning for energy reduction in milling processes. The hybrid-learning approach is proposed to obtain the ability of self-learned predictive modeling in both data-existent and even data-absent environments via accommodating traditional machine learning and transfer learning. The holonic-based mechanism, consisting of a modeling method, system architecture, and operational procedure, is designed to provide an autonomous and collaborative decision-making environment through the virtual agentization of machines and their associated objects under the paradigm of HMS. This mechanism provides interconnections between data/models and virtual agents. Thus, we can create and apply energy-predictive models automatically on machines with minimal human intervention. The implementation demonstrates how individual machine tools utilize real data or existing models for creating their learned models, predict energy based on their own models, and automatically negotiate between themselves to find the best machine tool that can minimize energy consumption in milling machining.

Section 2 reviews the relevant literature, and Section 3 introduces the concepts of a self-learning factory and hybrid learning. Section 4 presents the holonic-based mechanism. Section 5 demonstrates a prototype system with discussions, and Section 6 summarizes our conclusions.

2. Related Works

This section reviews the literature relevant to HMS and learning ability. Our mechanism builds upon the concept of HMS from a systematic perspective and the application of learning-based analytics from a methodological perspective, respectively.

2.1. Holonic Manufacturing Systems

HMS originated from PROSA, which adopted a holonic organization to achieve stability against disturbances, flexibility in changes, and efficiency in resource utilization [7]. The PROSA identified major keywords as defined below [8].

- Holon: An autonomous and cooperative building block for transforming, transporting, storing, and validating information and physical manufacturing objects. Basic holons consist of product, resource, and order holons, whereas staff holons assist the basic holons.
- Holarchy: A system of holons that cooperates to achieve a goal. It defines the basic rules for cooperation of holons, thereby limiting their autonomy.
- Autonomy: The capability of a holon to create and control the execution of its own plans and strategies.
- Cooperation: A process whereby a set of holons develops and executes mutually acceptable plans.

Extensive knowledge of HMS can be found in outstanding reviews including [8,15–17]. The reviews interestingly imply that Multi-Agent Systems (MAS) are a commonly-used and efficient technology to implement HMS due to the suitability of implementing the modularity, decentralization, and complexity of holons and their holarchy [17]. Here, an agent is a computational system situated in a dynamic environment with the capability of exhibiting autonomous and intelligent behavior, while MAS operate the community of interacting agents as a whole [16]. Thus, it makes sense that the conceptual frame of HMS needs to be transformed to programmable outcomes using MAS technology. Note that the present work also adopts this view due to the reasons given above.

Previous literature has attempted to develop and apply holonic-based systems to enhance target Key Performance Indicators (KPI) in broad applications. The following describes the purposes of HMS implementations in individual applications [15].

- Automation: The low- and (or) high-level control architecture to synchronize physical and software control units for flexibility at the machine or shop floor levels.
- Task allocation: Task assignment involving the distribution of tasks to available resources with the use of Contract Net Protocol (CNP), a negotiation procedure between a manager and a set of candidate contractors about the assignment of a task [18]. Task allocation can be a part of planning and scheduling in some sense.
- Fault-tolerance: Detection of failure, diagnosis of failure, and determination of reasonable recovery actions.
- Real-time control: The system control that reacts within precise time constraints, being classified into hard (missing deadline results in catastrophic consequences) or soft (meeting deadline is desirable but missing a deadline will not cause serious damage).
- Planning and scheduling (the application of the present work): Optimal planning and scheduling of available resources in the production or process level.

Table 1 lists the studies that have endeavored to develop holonic-based systems for enhancing the target KPI (e.g., productivity, flexibility, reconfigurability and fault-free operation). Staff holons are designed to efficiently carry out evaluation, mediation, management, and coordination to assist the basic holons, which concentrate on achieving goals. However, the previous studies did not much focus on data-driven modeling methodologies where the process of acquiring data and creating models becomes critical in structuring holons' functionalities and behaviors. As presented in Table 1, the previous studies are limited in identifying pivotal mediators that can interconnect historical data and create data-driven models with basic holons for predictive process planning. These studies are also limited in specifying operational and negotiating procedures between holons based on their recognition of model-based prediction.

MAS-driven studies have recently contributed to improve energy efficiency in manufacturing in accordance with the increase of energy reduction requirements. Alotaibi et al. developed a MAS prototype to optimize bi-objective functions (energy and tardiness) in a flexible job shop [19]. Marchiori et al. presented a dynamical approach for energy trades in steel production with the use of autonomous software agents [20]. Giret et al. proposed a software engineering approach for designing sustainable intelligent control systems based on multi-agent and holonic principles [21]. However, their studies depend on a deterministic or discrete event method and do not deal with energy-predictive models based on a data-driven method, which can deliver better predictability and adaptability of models at the machine level.

Table 1. A summary of previous literature on Holonic Manufacturing Systems (HMS).

Application	Citation	Objective	Holon Characteristics (Basic/Staff Holon)
Automation	[22]	A logic control system that supports autonomous and cooperative actions on Automated Guided Vehicles (AGV)	- Resource (B)[1]: the set of devices consisting of robots, workpiece suppliers, and conveyors - Order (B): contains the data relevant to top-level holon orders - Product (B): represents the abstraction of constituent parts - Scheduling (S)[1]: provides scheduling data to other holons - Traffic controller (S): surveys AGV traffic and advises AGV holons to sort out conflicts and avoid congestion
	[23]	An adaptive control system that operates using a role-based mechanism to maintain performance even when faced with perturbations	- Holon agent (B): a resource or an intelligent product, which can inform, communicate, decide and act independently - Optimization mechanism (S): a societal or environmental optimization mechanism to ensure global performance
	[24]	A high-level architecture with integration of both behavioral and structural self-organization for adaptive production control	- Product (B): represents products and the knowledge for production - Task (B): manages real-time execution of orders - Operational (B): represents resources (e.g., robots and operators) - Supervisor (S): introduces optimization into the system
Task allocation	[25]	A machine allocation algorithm that employs the genetic algorithm control for schedule generation and the shortest processing time-based Contract Net Protocol (CNP) for schedule negotiation	- Job (B): represents single jobs - Machine (B): represents machines - Job management (S): manages job holons and informs decomposition into decomposition holons - Decomposition (S): allows for cluster generation and scheduling assignment - Cluster (S): groups a number of machine holons - Genetic Algorithm (S): determines job process sequences in a meta-heuristic method
	[26]	A software design methodology that formulates workflow adaptation problems based on the CNP	- Task (B): represents tasks to be processed - Actor (B): represents workers or customers in the system
Fault-tolerance	[27]	A distributed architecture that performs disturbance handling and predicts mean time between failures	- Resource (B): monitors the device's status for predictive maintenance - Quality control (S): performs detection through quality verification - Task (S): monitors production orders - Supervisor (S): performs meta-monitoring by aggregating the information under holon clusters
	[28]	A service-oriented architecture that operates mechanisms for fault treatment with reconfiguration of dispersed manufacturing	- Product (B): manages requests of products, searches for corresponding holons, and creates a work order - Task (B): manages recipes - Operation (B): represents equipment and humans - Supervisor (S): coordinates the services of all holons

Table 1. *Cont.*

Application	Citation	Objective	Holon Characteristics (Basic/Staff Holon)
Real-time control	[29]	An adaptive control and distributed approach using IEC 61499 function blocks in machining groups	- Product (B): requests manufacturing to planning holons - Planning (B): assigns product operations into machines - Main Control (B): sequences operations and calculates the processing time of production - Operation (B): consists of: control layers to encapsulate agent and function blocks, interface layers to contain virtual simulation and a logical adapter, and machine layers to encapsulate a real machine and its virtual model
	[30]	A holonic hybrid control system that interacts discrete planning and continuous processes	- Product (B): stores and communicates recipes of products - Order (B): determines future planning and holarchy of the system - Resource (B): performs device online control
Planning/scheduling	[31]	An integrated planning and scheduling algorithm that dynamically generates optimal process plans and schedules	- Job (B): represents jobs to be manufactured - Machine tool (B): represents machine tools - Machining process (B): represents machining processes by machine tools - Production engineering (S): initializes the basic holons and modifies job specifications for disturbance handling - Job order (S): represents manufacturing tasks - Coordination (S): determines suitable assignment of job holons using mathematical models
	[32]	A Petri-net based methodology that computes sequential flexibility for dead-lock free planning and scheduling	- Product (B): possesses product's process information - Resource (B): provides manufacturing capability - Order (B): coordinates the execution of production plans - Directory facilitator (S): provides static information on the system as an information server
	[33]	A hybrid metaheuristic-based holonic multi-agent model for flexible job shop scheduling and robot routing	- Robot-system (B): represents robots used for operation - Order-system (B): executes orders within each job - Machine-system (B): represents machine system tools - Scheduler (B): prepares the best promising reasons of the search space - Cluster (B): guides the search to the global optimum solution of the problem
	[34]	A rescheduling mechanism for unpredicted orders and unavailable resource appearance	- Product/Order (B): issues goals according to orders to be solved as managers - Resource (B): represents machines to carry out operations - Contractor (S): performs negotiations among resource holons within the CNP

[1] (B): Basic holon, (S): Staff holon.

2.2. Learning-Based Analytics

Learning ability is recognized as an indispensable feature of manufacturing intelligence [35]. As machine learning inherits the ability to learn, the applications of machine learning dramatically increased in manufacturing domains over the last two decades, and proved suitable in prediction, optimization, control, maintenance, and troubleshooting. This suitability stems from the advantages of machine-learning techniques that handle high dimensional problems, increase the usability of machine-learning practice, discover unknown knowledge, and adapt automatically to dynamic and complex environments [36].

The learning ability has also been incorporated into agent-based manufacturing systems [37]. Kadar and Monostori presented resource/system-level learning to improve the performance of distributed systems by expanding the adaptive characteristics of agents [38]. Shen et al. proposed a learning mechanism for identifying organizational knowledge and selective interaction propagation from emergent system behavior, and it was used for adjusting distributed schedules and planning dynamically [6].

The learning ability has been widely applied into energy-efficient machining as well. Previous works have demonstrated that machine-learning techniques are powerful for predicting and optimizing energy consumption through utilizing the prior knowledge of a concerned system [39]. For example, Garg et al. applied a multi-gene genetic programming approach to generate the model structure and coefficients automatically for energy prediction and optimization in milling machining [40]. Bhinge et al. presented a data-driven approach for energy prediction in milling machining through the application of Gaussian process regression [41], and Liu et al. used a tree-based gradient boosting method, which is a machine learning method to combine weak models into a single strong model in an iterative fashion, to predict specific cutting energy in milling [39].

Despite such efforts, a common problem remains in that values of certain attributes are not available or are missing in the dataset [36]. The recent emergence of transfer learning appears to overcome this problem in manufacturing. Transfer-learning applications are increasing and include fault detection and condition causality in product quality management, fault diagnosis and condition-based maintenance in machine maintenance, and tool tip dynamics prediction in machine chatter [42,43].

Consequently, the motivation of the present work is to develop a HMS mechanism for gaining learning ability, where basic holons can interconnect data, create data-driven models, and determine their behaviors autonomously and collaboratively for energy-efficient machining through predictive process planning. The convergence of machine learning and transfer learning provides a basis for proactive decision-making about the future behaviors of agent-based manufacturing systems, thereby resulting in learning ability in complex and dynamic environments.

3. Self-Learning Factory and Hybrid Learning

This section introduces the concept of a self-learning factory and a hybrid-learning approach, respectively. Section 3.1 explains the conceptual structure and process of a self-learning factory. Section 3.2 describes the theoretical methodology of the hybrid-learning approach.

3.1. Self-Learning Factory

Manufacturing systems operate in dynamic and real-time environments and are frequently confronted with unexpected events such as machine failure. In this circumstance, MAS have been applied to facilitate adaptive, flexible and efficient use of manufacturing resources. However, determining concrete behaviors and activities in MAS a priori is challenging because the following things should be known: the environmental requirements that will emerge in the future, which agents are available, and how those agents need to interact in response to these requirements. Such challenges should be overcome by endowing the agents with the ability to improve the future performance of manufacturing systems through experience [6]. In the present work, a self-learning factory is the manufacturing system that allows manufacturing objects themselves to learn from past experience, perform predictive simulations and analytics based on the learned-experience, and thus proactively determine their behaviors and activities for improving, sustaining or recovering their target KPI.

Figure 1 presents the concept of a self-learning factory. It consists of a physical and cyber pairwise factory, which mirrors the physical factory and uses virtual agents for representing their physical objects. The cyber factory collects manufacturing data acquired from physical objects. It processes data to generate training datasets and manufacturing context information. Here, the manufacturing context means a machining condition that specifies which machine, material, machining feature, operation and strategy are applied when a certain dataset is generated. The manufacturing context information can be used a model identifier for categorizing the entire training dataset into individual datasets because different process conditions create disparate models. For example, models for a machine need to be different from those of another machine because both machines have different capabilities and performances. It then creates models from training datasets using learning techniques and stores them in a knowledge database (model repository). Here, it can adaptively choose machine learning or transfer learning, depending on whether training datasets exist or not (more details in Section 3.2). The cyber factory makes predictive planning and control decisions based on learned knowledge and models, and eventually feeds such decisions forward to the physical objects located in the physical factory. This cycle repeats, and the cyber factory evolutionarily improves the robustness of knowledge and models, thereby allowing for more accurate planning and control in physical factories.

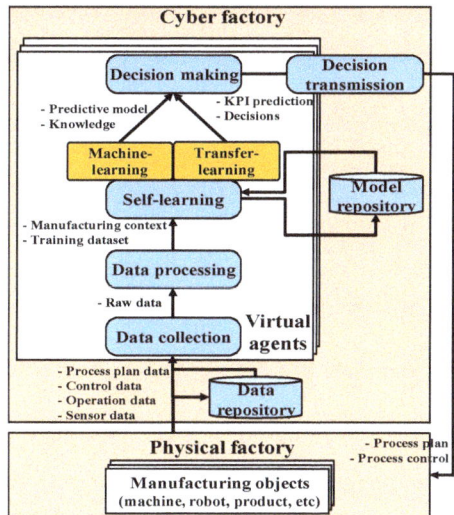

Figure 1. The concept of a self-learning factory. KPI: Key Performance Indicators.

3.2. Hybrid Learning

Manufacturing data are very important because data-driven knowledge creation is the foundation for the self-learning factory. Figure 2 presents data flow on a computer-aided chain in machining processes. Machining processes require part geometries, production plans (macro-level plans for managing a shop floor), process plans (micro-level plans for machining a part), and Numerical Control (NC) programs. Supplementary information like part libraries and machine and cutting tool specifications aid in efficient planning and control by providing technical requirements about products and resources. Here, the data associated with specifications, planning, and control work as causative data because they characterize commands and instructions by which machines must operate. While machine tools run the machining, they generate machine-monitoring data to represent their actions and movements along with timestamps. After or during machining, inspection equipment records values that are used to check whether machining was satisfactorily completed as designed or not. Machine monitoring and inspection data can be resultant data because they result from the machine's actual operations commanded and instructed by the causative data. Specifically, process plan and NC program data significantly influence machine-monitoring data because a machine tool takes the actions designated by the NC programs, which are outcomes of process plans [44].

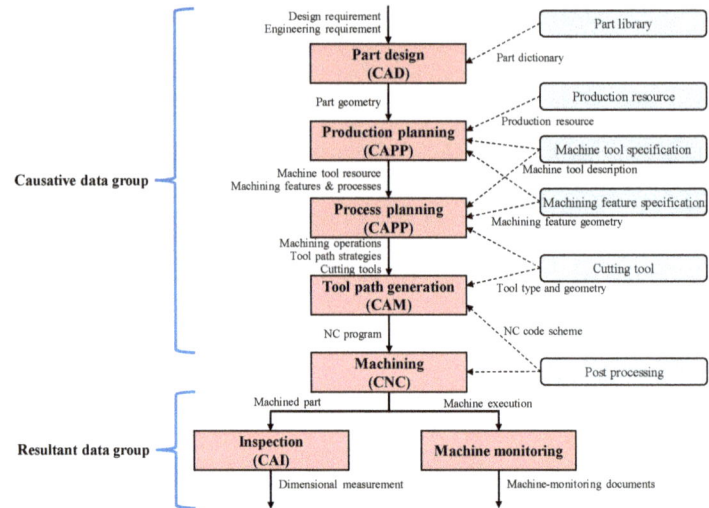

Figure 2. A computer-aided process chain and its data flow. CAD: Computer-Aided Design; CAPP: Computer-Aided Process Planning; CAM: Computer-Aided Manufacturing; CNC: Computerized Numerical Control; CAI: Computer-Aided Inspection.

From a learning perspective, causative data can correspond to input variables (x variables) as training datasets consist of x–y pairwise data instances; meanwhile, the resultant data are included as output variables (y variables) [37]. Additionally, certain causative data are involved in identifying the manufacturing contexts because they specify machining conditions. Hence, process plan and NC program data configure manufacturing context information or x variable data instances, whereas machine-monitoring data are related to y variable data instances. Training datasets can be constructed by integrating machine-monitoring data instances with their corresponding process plan and NC program data instances. Such training datasets are the primary requirements for machine learning and are used to compute their causal relationship by learning techniques.

When implementing the self-learning factory, it is necessary to achieve the self-learning ability through creating predictive models by means of appropriate learning approaches. Predictive models allow machines to forecast KPI under uncertainties, thereby helping the KPI optimization through their self-aware abilities [45]. The traditional learning that uses machine-learning techniques shows excellence at creating predictive models, as reviewed in Section 2.2. However, it does not work unless training datasets exist. To overcome this limitation of the traditional learning, we apply hybrid learning. Hybrid learning can be defined as a learning method where traditional machine learning creates predictive models in a data-existent environment; on the other hand, transfer learning does in a data-absent environment.

Figure 3 presents the concept of hybrid learning for creating energy prediction models in machining processes. Note that our problem is supervised learning because the x and y variables are supervised by humans and desired outputs are supplied during training. When training datasets exist, traditional learning computes a mathematical function, $y = f(X) + \varepsilon$ (ε: error term), based on learning x–y pairwise training datasets. The upper part of Figure 3 shows how traditional learning is used to create an energy model using an ANN, which is useful for energy prediction in machining [46]. This model can calculate an anticipated energy value based on the input of process parameters (feedrate, spindle speed, and cutting depth) in a certain manufacturing context because ANN makes the x–y relationship numerically known. Such models can provide reliable prediction capability because they build on training datasets that come from real data.

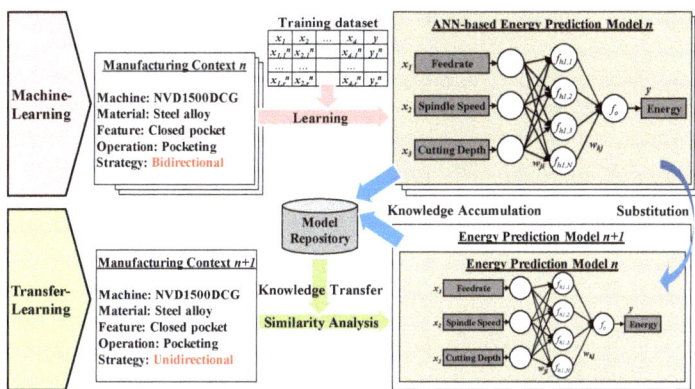

Figure 3. The concept of hybrid learning. ANN: Artificial Neural Network.

Transfer-learning can work when training datasets are unavailable. This transfer learning can create substituent models by transferring learned knowledge (existing models) as it builds upon the similarity between models. A target manufacturing context (target task) captures a substituent model that has the best similarity among existing models (source tasks), as presented in the lower part of Figure 3. Transfer learning unavoidably requires prior knowledge, where the similarity between models has been investigated in a certain manufacturing system (domain). The prior knowledge can be obtained from a preliminary analysis of the target KPI (here, energy). Table 2 shows an example of the similarity of strategies in 2.5 dimensional pocketing machining. This similarity comes from the previous work [47], which observes that unidirectional x-axis up/down milling and unidirectional y-axis strategies have the similar energy pattern in pocket machining due to the dependency of cycle time; on the other hand, bidirectional x-axis, contour, and spiral strategies do. Model similarities can be graded in terms of high, middle or low levels, depending on their energy pattern likeness. When creating a substituent model, one of several models that have a high-level of similarity can be selected and then be substituted for the model that needs to be created (the selection method explained in Section 4.1.2). For example, a contour or spiral strategy model can be substituted for the model of bidirectional strategy due to their high-level of similarity.

The adaptive convergence of machine learning and transfer learning enables self-learning ability regardless of the degrees of freedom in the data. While models are continuously created by hybrid learning, enormous knowledge can be accumulated to ensure predictive capability in a huge number of manufacturing contexts.

Table 2. A similarity matrix of energy patterns in 2.5 dimensional pocketing strategies.

Strategy.	Unidirectional x-axis down Milling	Unidirectional x-axis up Milling	Bidirectional x-axis	Unidirectional y-axis	Contour	Spiral
Unidirectional x-axis down milling		High	Low	High	Low	Low
Unidirectional x-axis up milling	High		Low	High	Low	Low
Bidirectional x-axis	Low	Low		Low	High	High
Unidirectional y-axis	High	High	Low		Low	Low
Contour	Low	Low	High	Low		High
Spiral	Low	Low	High	Low	High	

4. Mechanism

This section presents the mechanism for implementing the self-learning factory based on the hybrid-learning approach. The mechanism includes a modeling method, system architecture and operational procedure, and it focuses on predictive process planning for energy reduction in the cyber part of the self-learning factory.

4.1. Modeling Method

Predictive process planning requires models so that machines make proactive and autonomous decisions through model-based anticipation. The hybrid-learning approach needs to be fully specified because it should be implemented to compile the knowledge needed for automatic creation and use of models. Figure 4 shows high-level methods of the hybrid-learning approach. Sections 4.1.1 and 4.1.2 explain the methods of machine learning and transfer learning, respectively.

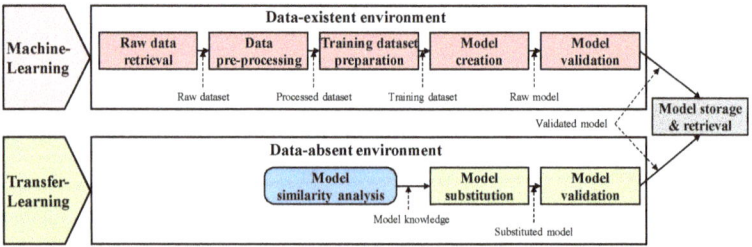

Figure 4. The modeling method of a hybrid-learning approach

4.1.1. Machine-Learning Method

The machine-learning method handles manufacturing data and data-driven models with the use of machine-learning techniques. It consists of: (1) raw data retrieval, (2) data pre-processing, (3) training dataset preparation, (4) model creation, (5) model validation, and (6) model storage and retrieval.

(1) Raw data retrieval involves the search and retrieval of raw data stored in a data repository for collecting data instances in training datasets. As explained in Section 3.2, the process plan, NC program and machine-monitoring data need to be searched and retrieved through a certain search method. A metadata-based search is useful as the metadata indicate the data about the data and serve as a map for locating data instances [48]. Once data instances are tagged with metadata attributes as header information, they can be effectively detected through mapping metadata-tagged data instances with the data queries utilizing the metadata attributes. We design the attributes of the metadata by

considering generality and accessibility. Generality assures that basic information about data instances will be represented across various data formats and dispersed data sources. Accessibility increases the availability of data searches even when some attributes are null. The attributes of the metadata of data can be identified as follows.

Metadata of data = {UUID, Group ID, Creator, Source, Duration, Means, Purpose, Creation}.

- UUID: Universally Unique Identifier.
- Group ID: an identifier for grouping instances.
- Creator: an identifier indicating who creates instances.
- Source: an identifier indicating where instances are stored.
- Duration: a period of time for gathering instances.
- Means: an identifier indicating how instances are obtained.
- Purpose: data attributes to be requested.
- Creation: a timestamp of data creation.

Figure 5 shows an example of the data retrieval using the metadata of data. When a set of raw data associated with process planning (formalized as ISO14649 [49]), NC programming (conforming to Fanuc codes), and machine monitoring (represented by MTConnect [50]) needs to be retrieved, 'O9131' (the NC program name) can work as 'group ID'. In Figure 5a, if the metadata contain 'group ID' and 'purpose', the relevant data instances can be retrieved because 'O9131' (red italic letters) is encoded at 'FILE_DESCRIPTION' in the header section and 'purpose' corresponds to the entity of 'PROJECT' in the data section. In Figure 5b, 'purpose' can request a list of 'CODE BLOCKS' in the NC program named 'O9131'. Figure 5c illustrates the data retrieval from an MTConnect document when data instances regarding 'position' and 'wattage' attributes during a period of time are necessary.

(2) Data pre-processing re-produces high-quality data from raw data and handles them as designated for preparing training datasets. The data pre-processing basically includes data cleaning, integration, transformation, and reduction [51].

Data cleaning resolves missing, noisy, outlying, duplicate, or incorrect data. Raw data unavoidably include sparse, imprecise, faulty, missing, or null data due to the dynamics of manufacturing systems and the limited capability of measurement devices [52]. These uncleaned data cause an increase in data uncertainty and result in negative impact on data-driven learning. Data cleaning produces so-called good data by keeping the data uncertainty under control, thereby increasing the reliability of data-sensitive learning.

Data integration combines heterogeneous data sources or separate formats into a single dataset for the desired learning analysis. For example, data instances retrieved from three different data formats in Figure 5 should be integrated into a tabular training dataset to connect the x and y variables. Data integration can be achieved by the backward tracing that scans from an MTConnect document and an NC program to an ISO14649 program. Here, a key attribute should be identified as the linking point for backward tracing, for which 'position' can be chosen as this key. Since 'position' indicates the coordinates of a cutting tool, a value of 'power' matched with a certain position can be obtained. A NC code block associated with the given position can be traced because the block obviously commands cutting tool movement involving the position. In turn, a machining operation associated with the NC block can be found because the former creates a group of NC blocks where the latter gets involved.

Data transformation converts data instances into the desired format, scale or unit that is more useful for the learning analysis. For example, real data values about feedrate, spindle speed, cutting depth, and power (blue and underlined letters in Figure 5a) need to be adjusted to a 0–1 scale through minimum–maximum normalization due to their different scales. In addition, it is necessary to convert a power unit to an energy unit. This is because a power meter typically measures power values, as shown in Figure 5c, while the y variable in our model uses energy units, which are scalar quantities. We adopt the delta-energy unit, which can be calculated by multiplying power with a sampling rate of measured power [41]. For example, the sampling rate is given by 0.365 s (the average sampling time of the power meter used in our case study).

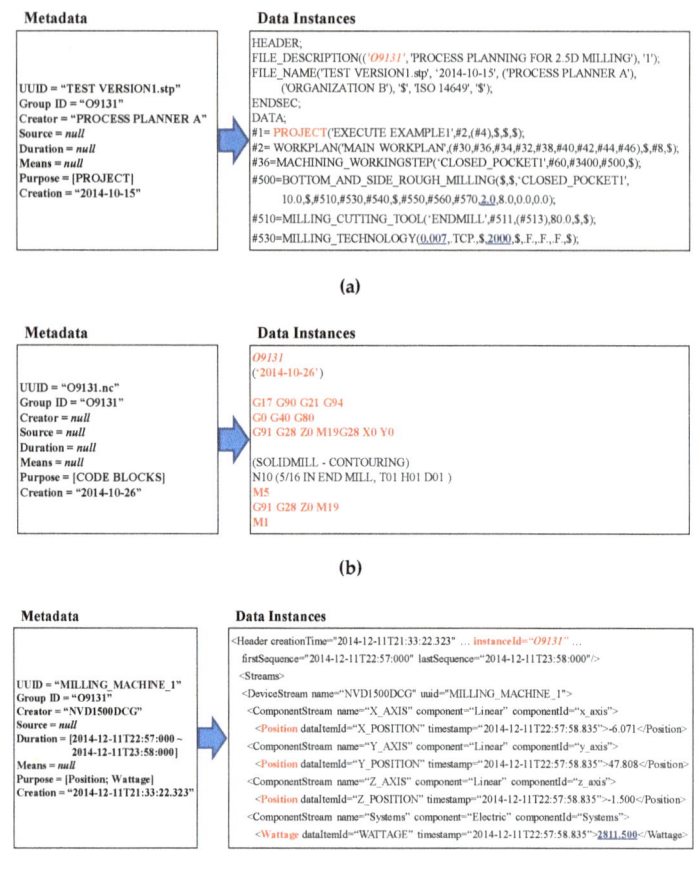

Figure 5. Data retrieval using a metadata-based search (**a**) ISO14649 program; (**b**) Numerical Control (NC) program; (**c**) MTConnect document. UUID: Universally Unique Identifier.

Data reduction may involve the removal of redundant data instances or a reduction in data dimensions to alleviate computational burdens or obtain straightforward learning results. The present dataset forms a data tuple of {feedrate, spindle speed, cutting depth}-{delta energy} at every sampling rate. We need to reduce the data dimension to {feedrate, spindle speed, cutting depth}-{energy} in terms of a manufacturing context because our models seek to output an energy value using the input of certain process parameters in a manufacturing context, as described in Section 3.2. For this purpose, all delta energy values within a manufacturing context are aggregated into a single energy value in the manufacturing context.

(3) Training dataset preparation decomposes the entire pre-processed dataset into individual training datasets separated by manufacturing contexts. Different manufacturing contexts require different models. This comes from the disparate power patterns caused by different cutting force distributions. For example, the power pattern for unidirectional strategy is different from that for bidirectional strategy due to their different tool movements and their different cutting forces. Table 3 presents an example of training datasets in two different manufacturing contexts (bidirectional and

contour machining strategies) within the same pocketing operation. These two datasets are used to create two different models. For example, when the feedrate is set to 0.333, spindle speed to 0.5, and cutting depth to 1, its corresponding energy value equals to 0.188, which is an aggregated value of individual delta energy values consumed by operating the bidirectional strategy for the pocketing. Note that the numerical values are normalized to a 0–1 scale based on original values.

Table 3. An example of training datasets.

Manufacturing Context					x Variables			y Variable
Machine	Material	Feature	Operation	Strategy	Feedrate	Spindle speed	Cutting depth	Energy
NVD1500DCG	Steel alloy	Closed pocket	Pocketing	Bi-directional	0.333	0.5	1	0.188
					0.667	0	0	0.546
					0.667	1	0	0.227
					0.667	1	1	0.000
					0.333	0.5	0	0.796
				Contour	1	0.5	0	0.256
					0.333	0.5	1	0.269
					0.667	0	0	0.386

(4) Model creation involves the generation of predictive models through learning training datasets by machine-learning techniques. As noted in Section 3.2, our model is supervised learning and thus machine-learning techniques can be used to derive mathematical functions that determine the relationship between the x and y variables. Equation (1) expresses an ANN-based function for energy prediction [53]. Figure 6 shows the structure of an energy prediction model (the graphical structure is presented in Figure 3) and its example where the attributes of an ANN function are instantiated. The manufacturing context enrolls model identification, and the numerical function performs the energy calculation based on the input of the process parameters.

$$y = f_O(\sum_{j=0}^{p} w_{Oi} f_h(\sum_{i=0}^{q} w_{ji} x_i)) + \varepsilon \quad (1)$$

where y: energy, x: process parameter (feedrate, spindle speed, and cutting depth), p and q: the numbers of neurons at each layer, w_{oi} and w_{ji}: weight values, f_o and f_h: activation functions, and ε: learning error

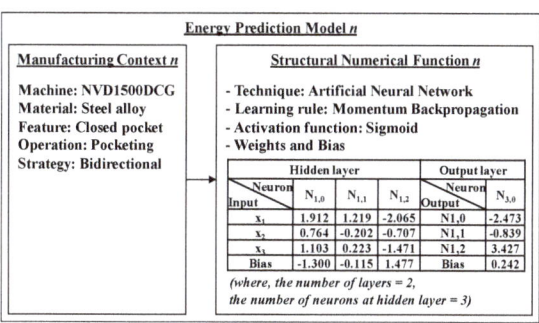

Figure 6. Structure of an energy prediction model.

(5) Model validation involves the quantification of model significance and reliability to validate model conformance. This process checks whether model performance satisfactorily meets a threshold by measuring learning error (the deviation between training data and a numerical function) and prediction error (the deviation between the numerical function and real values to be predicted). Root mean square error (RMSE) is widely used as a performance metric for measuring the learning error [54]. Cross validation is useful for measuring the prediction error. It splits the full dataset into training and test data folds, measures the trained model's performance using the test data fold, and then repeats this procedure by changing the roles of the data folds [54].

(6) Model storage and retrieval involves the storage of validated models in a model repository with their structural forms, and retrieval of the models when requested. Such numerical functions expressed in Equation (1) are quite hard to store in and retrieve from the database. The tabular model representation illustrated in Figure 6 makes this model storage and retrieval efficient. As common relational database systems store and retrieve data records in tabular form, the attributes of ANN functions can be identified as columns and their instances can be recorded as rows in tables. A metadata-based search is also useful as the metadata act as model navigators, as explained in Section 4.1.1 (1). The metadata of a model also need to consider accessibility (as with the metadata of the data) because accessibility is the common sense of storage and retrieval in a database. However, the metadata of the model need to be designed in accordance with specificity because manufacturing contexts depend on and vary with characters of manufacturing systems (e.g., types and complexity of production). The models requested need to be accurately retrieved, and thus the metadata of the model should be able to represent the manufacturing context in a straightforward manner. The metadata of a model for machining processes can be identified as follows.

Metadata of model = {UUID, Group ID, Creator, Source, Means, Creation, Machine, Material, Feature, Operation, Strategy}.

- Machine: an identifier for the machine tool that creates a model.
- Material: an identifier for a workpiece material.
- Feature: an identifier for a machining feature.
- Operation: an identifier related to a machining operation.
- Strategy: an identifier related to a machining strategy that identifies the tool path pattern.

4.1.2. Transfer Learning Method

Transfer learning enables indirect model acquisition through knowledge transfer from existing models when data do not exist, as addressed in Section 3.2. Our method is inductive transfer learning in that the source and target domain (machining process) is the same, but the target task (manufacturing context) is different from the source tasks. On the assumption that the similarity analysis has been investigated, the transfer-learning method consists of: (1) model substitution, (2) model validation, and (3) model storage and retrieval. We skip (3) because it is the same as in Section 4.1.1 (6). We will further discuss the assumption in Section 5.3 (3). It is worth mentioning that traditional machine learning needs to be prior to transfer learning because the former builds on real data, whereas the latter is based on transferred knowledge. When training datasets are available and can be learned, transfer-learned models need to be replaced by machine-learned models to ensure data-driven predictive capability.

(1) Model substitution involves the generation of an alternative energy model by selecting the model that is most similar to the target manufacturing context. Figure 7 shows two methods of model creation: cloning and competing. When there is only one model with a high-level of similarity, the cloning just copies and pastes the original model to a new model, as shown in Figure 7a. When the number of such models is greater than one, the competing is required to choose the best model based on the criteria including default, preference and likeness, as illustrated in Figure 7b. For example, if a new model for a contour strategy is requested and two bidirectional and spiral strategies indicate a high-level of similarity with the former strategy, one of the latter models needs to be chosen. The spiral

strategy model can be chosen if it turns out to be more like a contour strategy model with regard to their machining power distributions.

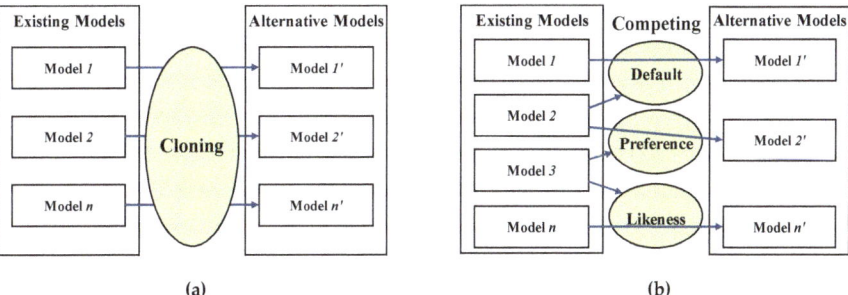

Figure 7. Two methods of transfer learning; (**a**) cloning, (**b**) competing.

(2) Model validation involves quantification of model significance and reliability to validate model conformance. However, validating transfer-learned models is harder because the conformance of a transfer-learned model may not be assured in the target task, although the original model proved to be significant and reliable in the source task. The most obvious method of validation is to measure prediction error by gathering real test data in the target manufacturing context. Reverse validation is recommended when a few of datasets exist for the target manufacturing context [55]. It approximates the difference between the estimated and true conditional distributions in the context of data limitation, although it still requires a minimum dataset at the target task. In reverse validation, a transfer-learned model is re-learned by combining $\{X_s, Y_{s,pred}\}$ (output dataset of the original model) and $\{X_t, Y_t\}$ (real dataset gathered in the target task). In turn, the difference between $Y_{t,pred}$ (output of the new model) and Y_t (real output) is measured to quantify the model approximation for the true conditional distribution (*s*: dataset in the source tasks, *t*: dataset in the target task, *pred*: predicted value).

4.2. System Architecture

Section 4.1 explained the modeling methods and described how to create and use models. It is necessary to identify objects and their functions to allocate such methods from a software architecture perspective. We designed a holonic-based system architecture, as shown in Figure 8. The PROSA architecture underlies this architectural design and can be used to pursue goal-oriented systemization through virtualizing object agents that have autonomous and collaborative capabilities.

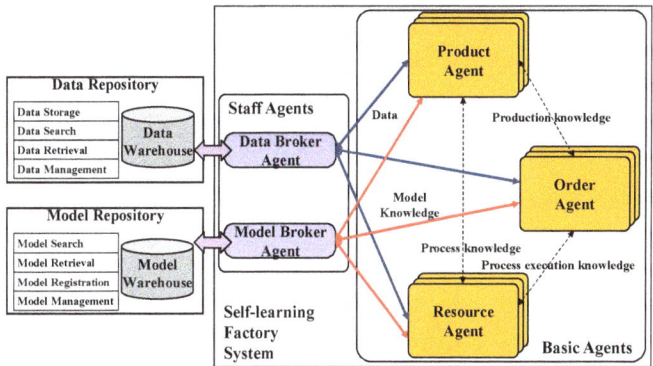

Figure 8. A system architecture for a self-learning factory.

In the PROSA architecture, basic agents consisting of product, order and resource agents mutually exchange production knowledge, process knowledge and process execution knowledge. A product agent performs the functions of request, allocation, confirmation, and supervision of tasks. An order agent is an orchestrator that takes charge of Calls-For-Proposals (CFP), bid evaluation and selection, task allocation, and task progress supervision. Meanwhile, a resource agent receives proposals, checks availability, creates bids, accepts allocations and executes tasks.

In our architecture, data and models should be interconnected within the HMS architecture for integrating the modeling method described in Section 4.1. The proposed architecture thereby gains the capabilities of predictive model-based bid submission for resource agents and predictive value-based bid evaluation for order agents. For this purpose, a data broker agent and a model broker agent are added as staff agents.

The data broker agent acts as a mediator connecting basic agents and a data repository. The data broker thus helps basic agents acquire manufacturing context information, training data and task details. It receives the metadata of data from basic agents when these agents need to gather data to create models or check availability. It returns the resulting data instances to the basic agents through the metadata-based search in the data repository. Meanwhile, the model broker agent is a mediator to connect basic agents and a model repository. It stores models in the model repository once basic agents create models using the acquired data. It searches and returns the models requested by the basic agents when the latter need to use the former. Likewise, the metadata of model is applied to enable the metadata-based search in model requests, searches, and returns.

4.3. Operational Procedure

This sub-section describes the operational procedure to specify agents' activities and interactions in a sequential order based on the architecture proposed above. Figure 9 shows the operational procedure represented by a sequence diagram in Unified Modeling Language. This figure is focused on model creation and usage of resource agents.

Figure 9a illustrates the procedure for model creation, substitution and registration to prepare the self-learning ability. If the target model already exists in the model repository (5), this procedure is terminated (7). If not (8.1), the procedure is invoked and starts with a training data request (10). If training datasets are available (13.1), models are created using the machine-learning method (16). If training datasets are not available (13.2), models are substituted from existing models using the transfer-learning method (24). The models created by the two different methods are requested to register (26) and are then registered in the model repository (28).

Figure 9b shows the procedure for model usage to apply the self-learning ability. This procedure builds upon CNP but extends to accommodate the activities and interactions associated with model-based bidding and evaluation. An order agent requests the task taken for fabricating a product (1), and a product agent provides task metadata to the order agent (3). The order agent issues CFP to resource agents (4). The resource agents check their availability with respect to their capability (whether they can fabricate or not) and idleness (whether they are occupied or not) (5). Available resource agents receive the task details (technical specification of the task) using the task metadata (8), and receive models using the metadata of the model extracted from the task metadata (11). They automatically determine process parameters within their allowable ranges and capacities (12). In turn, they anticipate energy values for the task using the models received (13), and then submit their bids where predictive energy values are recorded (14). The order agent evaluates the resources' bids based on energy values (15), and then chooses and notifies the resource agent who submits the minimum energy value (16). The remainder follows the traditional CNP. While agents communicate, they comply with the Foundation for Intelligent Physical Agents–Agent Communication Language (FIPA-ACL), which defines a set of interaction protocols and their individual communicative acts to coordinate multi-message actions [56].

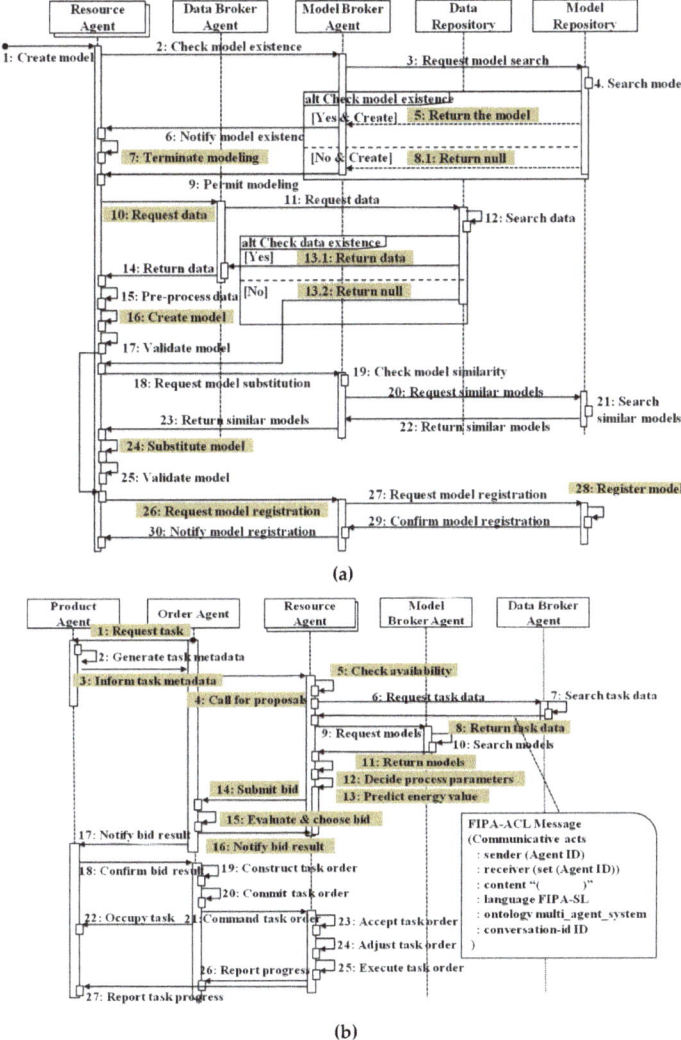

Figure 9. An operational procedure for a self-learning factory (**a**) Model creation, substitution, and registration; (**b**) Contract Net Protocol and model usage. FIPA-SL: Foundation for Intelligent Physical Agents – Semantic Language.

5. Implementation

We implement a prototype system to show the feasibility of the self-learning factory. The prototype demonstrates automatic predictive modeling and autonomous process planning for energy reduction in milling machining. Section 5.1 describes implementation scenarios, and Section 5.2 explains prototype implementation. Section 5.3 discusses implementation results.

5.1. Implementation Scenarios

Figure 10 shows a test part containing 13 machining conditions represented by {machining feature; machining operation; machining strategy}. Here, a machining condition corresponds to a manufacturing context. Implementation scenarios consist of: (1) model creation, substitution, and registration, and (2) CNP and model usage, as explained in Section 4.3.

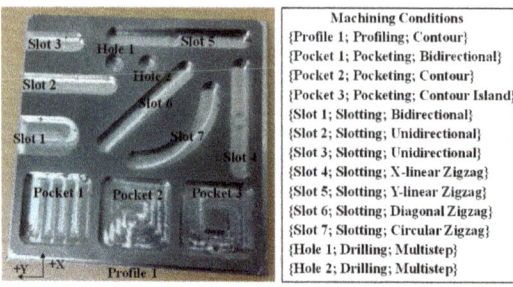

Figure 10. A test part and a set of machining conditions.

Figure 11 presents the two scenarios (the simplification of Figure 9a,b). In Figure 11a, a machine tool generates models using hybrid learning and registers them into the model repository for the next scenario. When the machine asks the model broker for checking the existence of the models associated with the machining conditions, the model broker returns the relevant models if they exist. Otherwise, the machine requests data to the data broker who returns the data requested. If the data exist, the machine creates energy models using machine learning and then requests model registration to the model broker who notifies the registration confirmation to the machine. If the data do not exist, the machine requests similar models to the model broker. The model broker then searches similar models based on model similarity and returns them to the machine. The machine creates alternative energy models using transfer learning and registers the models in the same way. Here, a model for the machining condition {Pocket 1; Pocketing; Spiral}, which is assumedly to be absent in the model repository, is created.

In Figure 11b, a CFP is initiated by an order if a product needs to be machined and informs the relevant tasks. Five machine tools compete for a task. Machines 4 and 5 refuse this task due to their unavailability because Machine 4 is a turning machine and Machine 5 was occupied by another task. The remaining three machines (Machine 1, Machine 2, and Machine 3) decide their process parameters within their allowable ranges or preferences. Process parameters are assumedly determined as follows: Machine 1 (feedrate: 0.0127 mm/tooth, spindle speed: 1750 Revolution per Minute (RPM), cutting depth: 1.5 mm), Machine 2 (feedrate: 0.0178, spindle speed: 2000, cutting depth: 1.0), and Machine 3 (feedrate: 0.0127, spindle speed: 2000, cutting depth: 1.5). They use their energy models to anticipate the energy values consumed during the execution of the given machining conditions. Once their energy values are estimated, the machines send their bids including their predictive energy values to the order. The order evaluates the energy values submitted by the three machines. The order accepts one of the machine tools if its proposal is the minimum energy value.

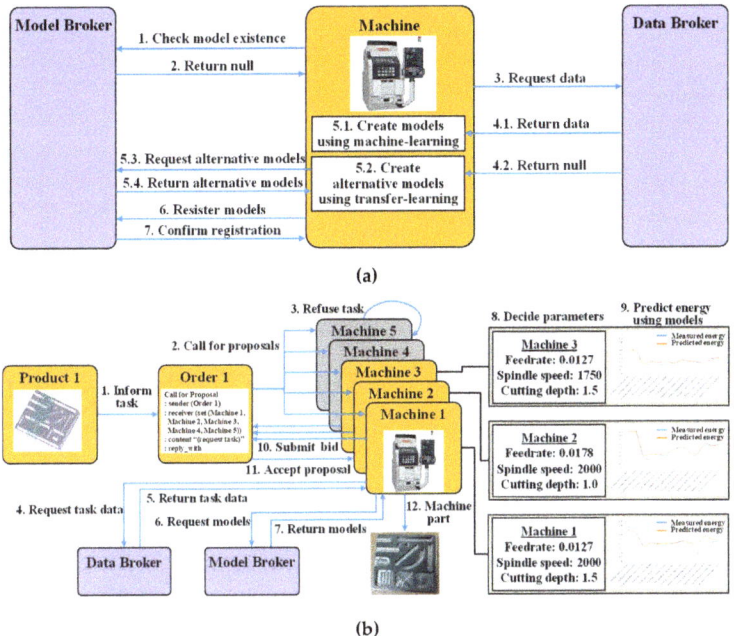

Figure 11. Implementation scenarios for (**a**) Model creation, substitution and registration; (**b**) model usage and machine selection.

5.2. Prototype Implementation

For energy modeling, we fabricated 12 test parts in a milling machine. Table 4 presents a list of process parameters for the 12 individual parts. These parameters were randomly determined within the experimental safety and allowable ranges of the machine and cutting tool used. Figure 12 shows an implementation architecture. We generated ISO14649 programs manually, while NC programs were generated using computer-aided manufacturing software, and MTConnect documents were collected in a physical part to represent the machine-monitoring data heterogeneously sourced from an NC and a power meter. The installations of this experiment included the machine (Mori Seiki NVD 1500 DCG), NC (Fanuc 0i), workpiece (Steel 1018, 10.16 cm × 10.16 cm × 1.27 cm), cutting tool (solid carbide flat-end mill, 8-mm diameter, four flutes), and power meter (high-speed power meter from System insights). Note that only one machine is used due to our experimental limitation. We implemented a prototype system in a cyber part based on the mechanism of the self-learning factory, as explained in Section 4. The installations of this implementation included an integrated development environment (Eclipse Java Oxygen), agent platform (Java Agent Development framework (JADE)), JADE execution and deployment (EJADE), data and model repositories (MySQL), and a Java-based ANN framework (Neuroph).

Table 4. List of process parameters.

Trial	Feedrate (mm/tooth)	Spindle Speed (RPM)	Cutting Depth (mm)
1	0.0127	1500	1.5
2	0.0127	2000	1.5
3	0.0127	1750	1
4	0.0229	1750	1
5	0.0127	1750	2
6	0.0178	1500	1
7	0.0178	2000	1
8	0.0178	2000	2
9	0.0178	1750	1.5
10	0.0076	1750	1.5
11	0.0152	1750	1.5
12	0.0127	1750	1.5

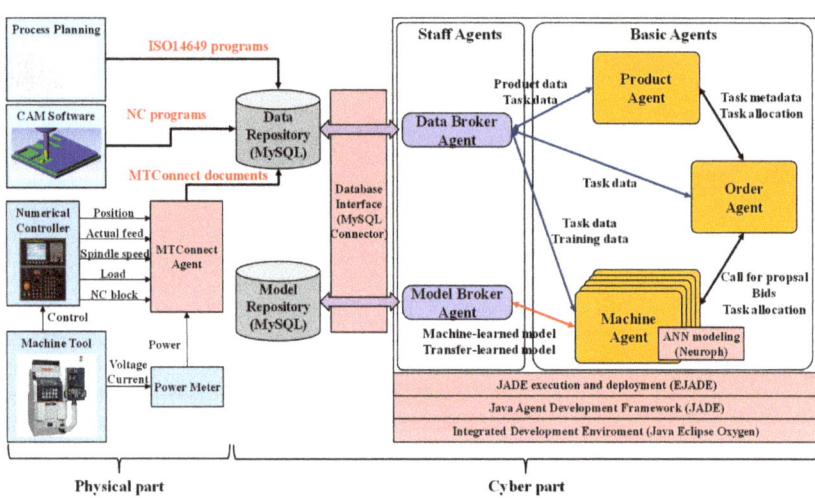

Figure 12. Implementation architecture of the prototype system.

Figure 13 illustrates the screen shots for the implementation of model creation, substitution and registration corresponding to Figure 11a. The screen shots, captured from the JADE sniffer agent, represent FIPA-ACL message exchanges and interactions across individual agents with respect to time, while a computer automatically proceeds (we only click the start button). Note that the arrows only indicate external message exchanges with communicative acts between agents, while internal works inside agents are hidden. In Figure 13a, the machine learning works to create models associated with the 13 machining conditions. The ANN technique is used for this purpose, and the attributes of the ANN-based energy models include (see example in Figure 6) the learning rule (momentum backpropagation), activation function (sigmoid), the number of layers (2), the number of neurons at a hidden layer (3), learning rate (0.3), maximum error (0.01), maximum iteration (1000), and momentum (0.2). As shown in Figure 13b, the transfer learning is activated because the data broker cannot find data in the data repository and then refuses data return. An energy model for {Pocket 1; Pocketing; Spiral}, as described in Section 5.1, is alternatively created through cloning the energy model for {Pocket 1; Pocketing; Bidirectional} due to their high level of similarity. In these ways, energy models are created and registered in the model repository for the next use.

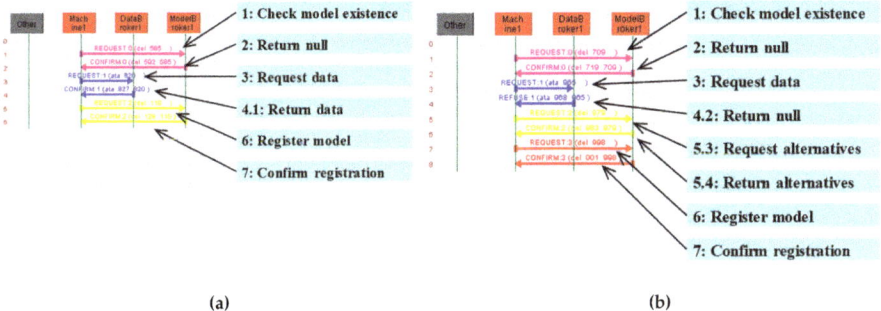

Figure 13. Implementation result: model creation, substitution and registration (**a**) Machine learning-based; (**b**) Transfer learning-based.

Figure 14 shows the screen shot for the implementation of CNP and model usage. The order agent is an initiator in the frame of CNP, which comprises an initiator and participants for requesting a task and performing the task, respectively. The order agent communicates with not only the product agent (a participant) for issuing a task but also the machine agents (participants) for assigning the task. The order agent processes its operations aligning with the scheduling of one-shot, cyclic or conditional behaviors for communicating with the participants using FIPA-ACL messages (see Figure 9b). The order agent sends the messages to the target participants and receives the messages from them based on the behavioral scheduling because it can write or read the FIPA-ACL messages that include a sender, receivers, communicative acts (i.e., a tag for communicative acts; e.g., call for proposal, accept/reject proposal, inform and refuse), contents, conversation ID, and so forth.

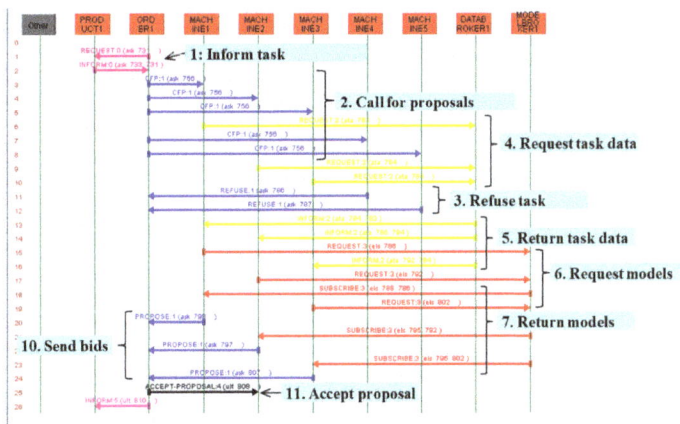

Figure 14. Implementation result: model usage and machine selection.

Order 1 calls for proposals for selecting a machine who can perform the task informed from Product 1. Machines 4 and 5 refuse this task due to their unavailability as explained in Section 5.1. The remaining three machines receive the energy models of the 13 machining conditions and input the determined process parameters. These three machines predict energy values for running the 13 machining conditions and the predictive energy values are, respectively: 11,825 kJ (Machine 1), 11,700 kJ (Machine 2), and 12,957 kJ (Machine 3). These machine agents propose their bids including these predictive energy values to the order agent. The order agent evaluates their energy values to seek the machine agent who submits the minimum energy value in its bid. Because Machine 2 proposes the minimum energy value, the order agent accepts the proposal from Machine 2, transmits the bidding result and instructs Machine 2 to take the task. This selection of the minimum energy-using machine tool (Machine 2) achieves 9.70% energy reduction, compared with the maximum energy-using one (Machine 3).

We measure real energy values from actual machining to check the accuracy of their corresponding predicted values derived from the ANN-based energy models. The real measured energy values were recorded as 11,382 kJ (Machine 1), 11,044 kJ (Machine 2), and 12,580 kJ (Machine 3). These different values come from the application of different process parameters given in Section 5.1. The total relative error, which measures the percentage of predicted energy values − real energy values)/real energy values, were 3.89%, 5.94%, and 3.00%, respectively.

5.3. Discussion

(1) Experimental limitation: our implementation could select the minimum energy-using machine tool among three machines through predicting and competing their energy values. Using three different machine tools is desirable because individual machine tools make different machine-specific energy values due to the differences in their capabilities and performances. However, a single machine had to be used due to our experimental limitations. It will be more realistic to use different machine tools for creating machine-specific energy models through instantiating different values in the attribute 'machine' of the manufacturing context. Alternatively, transfer learning can be applied to create machine-specific models by reflecting the difference of capabilities and performances between the target and source machines (see an example in the below (3)). In addition, virtual simulators can be useful for limited experimental environments. Some machining simulators can generate machine-specific power values affected by machine's capabilities and performances [57,58].

(2) Increase of practicability: our implementation has been made within a single order on a single process. This may be far from the reality in common manufacturing systems where multiple processes deal with various products and orders. Thus, gaining practicability remains critical. In other words, the target application is demanded to extend toward production planning considering multiple products and orders in a production line. It is expected that the adoption of MAS technology makes the practicability achievable because MAS use unique identification and take their autonomous and collaborative actions regardless of the number of product, order and machine agents. The proposed approach needs to be extended to the production planning by adding more product, order, and machine agents, although the difficulty and complexity of implementation increase.

(3) Uncertainty of transfer learning: our implementation shows the feasibility of the acquisition of self-learning ability by machine and transfer learning techniques. Transfer learning creates an energy model for the target manufacturing context {Pocket 1; Pocketing; Spiral}, which was not machined in our experiment, by cloning the energy model for the source manufacturing context {Pocket 1; Pocketing; Bidirectional}. However, we could not quantify significance nor validate conformance of the transferred model because it was not machined just as stated. The model validation of transfer learning remains as a future work. Our transfer-learning approach builds upon the similarity between the two manufacturing contexts and thus the similarity needs to be analyzed and verified in advance. If such similarity is not verified, transfer learning may not work properly and thus need to consider alternative means besides the similarity analysis. Reverse validation needs to be applied if a few of datasets are generated in the target manufacturing context, as described in Section 4.1.2 (3). Otherwise, the properties that characterize a difference between target and source manufacturing contexts can be added as variables of transfer learning. For example, an energy model for a machine tool can be transferred from that for another machine. The former model should be different to the latter one because these machines have different property values in basic power, rotation torque, and motor efficiency, which affect power and energy in machining. These properties can be additional input variables in the

structure of machine-learning models so that transfer learning can derive machine-dependent results through learning the influence of different values of those properties.

(4) Implementation challenge: energy-efficient machining has become a massive trend in some countries; however, it is still far from reality in other countries where many small-and-medium sized manufacturers fabricate products with paying cheap industrial electricity costs. These countries deal with time and quality as critical performances and are less concerned with energy consumption because they regard the energy cost as an endurable expense. Nevertheless, researchers need to keep their efforts on implementing and deploying cost-effective and data-accessible solutions for energy-efficient machining as the metal-cutting industry affects a large portion of the total energy consumption over the world. The use of open sources helps increase cost effectiveness for implementing such solutions. The implementation tools that we used in the prototype system are all open sources, which are publicly accessible without payment (payment may be required for commercial purposes). This implementation strategy can reduce solution development expenses to the reasonable cost level and help the deployment of such solutions toward small-and-medium sized manufacturers. The use of interoperable and open data interfaces comfortably supports the availability of data collection as data are critical for implementing manufacturing intelligence. Recent standardized interfaces including MTConnect and Open Platform Communications–Unified Architecture facilitate data accessibility. These interfaces provide open source tools as well and thus are quite useful for making a data bridge between physical and cyber factories.

6. Conclusions

In the present work, we designed and implemented a holonic-based mechanism for a self-learning factory based on a hybrid-learning approach. The concept of the self-learning factory was proposed to allow manufacturing objects to learn past experience using their real data, to perform predictive analytics and to determine their behaviors and activities for improving a target KPI. The holonic-based mechanism identified a modeling method, system architecture, and operational procedure to implement an autonomous and collaborative prediction environment through the virtual agentization of manufacturing objects under the paradigm of HMS. The hybrid-learning approach was designed to acquire predictive capability independently with the degrees of freedom in the data through the accommodation of machine learning and transfer learning. This hybrid learning can be used to build up a massive knowledge base through the accumulation of models, thereby gaining self-learning ability in manufacturing systems. A prototype demonstrated the feasibility of the proposed mechanism via predictive process planning for energy reduction in milling machining. Autonomous and collaborative activities of manufacturing agents are carried out on a computer to select the minimum energy-using machine tool while minimizing human intervention.

The limitations of the present work are as follows: (1) Our target is limited to process planning for a single product and process and thus cannot ensure the feasibility of the proposed mechanism in a more complex production line, (2) our implementation is restricted by the use of a single machine and thus does not embody more realistic scenarios by multiple machine tools, (3) our experiment does not show the validity of transferred models due to our experimental limitations, and (4) our implementation excludes control and feedback of the cyber and physical parts in a real-time manner as CPPS obviously require mirrored synchronization between the both parts. We plan to overcome these limitations in future work.

Author Contributions: All authors conceived the research idea and the methods of this study. S.-J.S. designed the concept and mechanism and implemented the prototype; Y.-M.K. contributed to specifying the mechanism of the modeling methods; P.M. supported the prototype implementation and analyzed the experiment results.

Funding: This work was supported by the Basic Research Program in Science and Engineering through the Ministry of Education of the Republic of Korea and the National Research Foundation (NRF-2018R1D1A1B07047100).

Conflicts of Interest: The authors declare no conflict of interest.

References

1. Davis, J.; Edgar, T.; Porter, J.; Bernaden, J.; Sarli, M. Smart manufacturing, manufacturing intelligence and demand-dynamic performance. *Comput. Chem. Eng.* **2012**, *47*, 145–156. [CrossRef]
2. Zuehlke, D. SmartFactory—Towards a factory-of-things. *Annu. Rev. Control* **2010**, *34*, 129–138. [CrossRef]
3. Rajkumar, R.; Lee, I.S.; Sha, L.; Stankovic, J. Cyber-physical systems: The next computing revolution. In Proceedings of the 47th ACM/IEEE Design Automation Conference, Anaheim, CA, USA, 13–18 June 2010; pp. 731–736.
4. Monostori, L.; Kadar, B.; Bauernhansl, T.; Kondoh, S.; Kumara, S.R.T.; Reinhart, G.; Sauer, O.; Schuh, G.; Sihn, W.; Ueda, K. Cyber-physical systems in manufacturing. *CIRP Ann.* **2016**, *65*, 621–641. [CrossRef]
5. Schuh, G.; Anderl, R.; Gausemeier, J.; ten Hompel, M.; Wahlster, W. *Industrie 4.0 Maturity Index—Managing the Digital Transformation of Companies*; ACATECH Study; Herbert Utz Verlag: Munich, Germany, 2017.
6. Shen, W.; Maturana, F.; Norrie, D.H. Enhancing the performance of an agent-based manufacturing system through learning and forecasting. *J. Intell. Manuf.* **2000**, *11*, 365–380. [CrossRef]
7. Brussel, H.V.; Wyns, J.; Valckenaers, P.; Bongaerts, L.; Peeters, P. Reference architecture for holonic manufacturing systems: PROSA. *Comput. Ind.* **1998**, *37*, 255–274. [CrossRef]
8. Shen, W.; Hao, Q.; Yoon, H.J.; Norrie, D.H. Applications of agent-based systems in intelligent manufacturing: An updated review. *Adv. Eng. Inform.* **2006**, *20*, 415–431. [CrossRef]
9. Priore, P.; De la Fuente, D.; Puente, J.; Parreno, J. A comparison of machine-learning algorithms for dynamic scheduling of flexible manufacturing systems. *Eng. Appl. Artif. Intell.* **2006**, *19*, 247–255. [CrossRef]
10. Pan, S.J.; Yang, Q. A survey on transfer learning. *IEEE Trans. Knowl. Data Eng.* **2010**, *22*, 1345–1359. [CrossRef]
11. Manufacturing Energy Consumption Survey. Available online: https://www.eia.gov/consumption/manufacturing/data/2014/index.php (accessed on 24 September 2019).
12. Hu, S.; Liu, F.; He, Y.; Hu, T. An on-line approach for energy efficiency monitoring of machine tools. *J. Clean. Prod.* **2012**, *27*, 133–140. [CrossRef]
13. Kara, S.; Li, W. Unit process energy consumption models for material removal processes. *CIRP Ann.* **2011**, *60*, 37–40. [CrossRef]
14. Shin, S.J.; Woo, J.Y.; Rachuri, S.; Meilanitasari, P. Standard data-based predictive modeling for power consumption in turning machining. *Sustainability* **2018**, *10*, 598. [CrossRef]
15. Babiceanu, R.F.; Chen, F.F. Development and applications of Holonic manufacturing systems: A survey. *J. Intell. Manuf.* **2006**, *17*, 111–131. [CrossRef]
16. Monostori, L.; Vancza, J.; Kumara, S.R.T. Agent-based systems for manufacturing. *CIRP Ann.* **2006**, *55*, 697–720. [CrossRef]
17. Leitão, P. Agent-based distributed manufacturing control: A state-of-the art survey. *Eng. Appl. Artif. Intell.* **2009**, *22*, 979–991. [CrossRef]
18. Odell, J.J.; Van Dyke Parunak, H.; Bauer, B. Representing Agent Interaction Protocols in UML. In *Agent-Oriented Software Engineering*; Ciancarini, P., Wooldridge, M.J., Eds.; Springer: Berlin, Germany, 2000; pp. 121–140.
19. Alotaibi, A.; Lohse, N.; Vu, T.M. Dynamic Agent-based Bi-objective Robustness for Tardiness and Energy in a Dynamic Flexible Job Shop. *Procedia CIRP* **2016**, *57*, 728–733. [CrossRef]
20. Marchiori, F.; Belloni, A.; Beninie, M.; Cateni, S.; Colla, V.; Ebeld, A.; Lupinelli, M.; Nastasi, G.; Neuer, M.; Pietrosanti, C.; et al. Integrated dynamic energy management for steel production. *Energy Procedia* **2017**, *105*, 2772–2777. [CrossRef]
21. Giret, A.; Trentesaux, D.; Salido, M.A.; Garcia, E.; Adam, E. A holonic multi-agent methodology to design sustainable intelligent manufacturing control systems. *J. Clean. Prod.* **2017**, *167*, 1370–1386. [CrossRef]
22. Lind, M.; Roulet-Dubonnet, O. Holonic shop-floor application for handling, feeding and transportation of workpieces. *Int. J. Prod. Res.* **2011**, *49*, 1441–1454. [CrossRef]
23. Adam, E.; Berger, T.; Sallez, Y.; Trentesaux, D. Role-based manufacturing control in a holonic multi-agent system. *Int. J. Prod. Res.* **2011**, *49*, 1455–1468. [CrossRef]
24. Barbosa, J.; Leitão, P.; Adam, E.; Trentesaux, D. Dynamic self-organization in holonic multi-agent manufacturing systems: The ADACOR evolution. *Comput. Ind.* **2015**, *66*, 99–111. [CrossRef]
25. Wang, K.; Choi, S.H. A holonic approach to flexible flow shop scheduling under stochastic processing times. *Comput. Oper. Res.* **2014**, *43*, 157–168. [CrossRef]

26. Hsieh, F.S.; Lin, J.-B. A self-adaptation scheme for workflow management in multi-agent systems. *J. Intell. Manuf.* **2016**, *27*, 131–148. [CrossRef]
27. Leitão, P. A holonic disturbance management architecture for flexible manufacturing systems. *Int. J. Prod. Res.* **2011**, *49*, 1269–1284. [CrossRef]
28. Silva, R.; Blos, M.; Junqueira, F.; Filho, D.S.; Miyagi, P. A service-oriented and holonic control architecture to the reconfiguration of dispersed manufacturing systems. In Proceedings of the 5th Doctoral Conference on Computing, Electrical and Industrial Systems (DoCEIS), Costa de Caparica, Portugal, 7–9 April 2014.
29. Jovanovic, M.; Zupan, S.; Prebil, I. Holonic control approach for the "green"-tyre manufacturing system using IEC 61499 standard. *J. Manuf. Syst.* **2016**, *40*, 119–136. [CrossRef]
30. Indriago, C.; Cardin, O.; Rakoto, N.; Castagna, P.; Chacon, E. H2CM: A holonic architecture for flexible hybrid control systems. *Comput. Ind.* **2016**, *77*, 15–28. [CrossRef]
31. Nejad, H.T.N.; Sugimura, N.; Iwamura, K. Agent-based dynamic integrated process planning and scheduling in flexible manufacturing systems. *Int. J. Prod. Res.* **2011**, *49*, 1373–1389. [CrossRef]
32. Quintanilla, F.G.; Cardin, O.; L'Anton, A.; Castagna, P. A Petri net-based methodology to increase flexibility in service-oriented holonic manufacturing systems. *Comput. Ind.* **2016**, *76*, 53–68. [CrossRef]
33. Nouri, H.E.; Driss, O.B.; Ghédira, K. Simultaneous scheduling of machines and transport robots in flexible job shop environment using hybrid metaheuristics based on clustered holonic multiagent model. *Comput. Ind. Eng.* **2016**, *102*, 488–501. [CrossRef]
34. Pascal, C.; Panescu, D. On rescheduling in holonic manufacturing systems. *Comput. Ind.* **2019**, *104*, 34–46. [CrossRef]
35. Monostori, L.; Brussel, H.V.; Westkampfer, E. Machine learning approaches to manufacturing. *CIRP Ann.* **1996**, *45*, 675–712. [CrossRef]
36. Wuest, T.; Weimer, D.; Irgens, C.; Thoben, K.D. Machine learning in manufacturing: Advantages, challenges, and applications. *Prod. Manuf. Res.* **2016**, *4*, 23–45. [CrossRef]
37. Monostori, L. AI and machine learning techniques for managing complexity, changes and uncertainties in manufacturing. *Eng. Appl. Artif. Intell.* **2003**, *16*, 277–291. [CrossRef]
38. Kadar, B.; Monostori, L. *Approaches to Increase the Performance of Agent-Based Production Systems*; Engineering of Intelligent Systems; Springer: Heidelberg, Germany, 2001; pp. 612–621.
39. Liu, Z.; Guo, Y. A hybrid approach to integrate machine learning and process mechanics for the prediction of specific cutting energy. *CIRP Ann.* **2018**, *67*, 57–60. [CrossRef]
40. Garg, A.; Lam, J.S.L.; Gao, L. Energy conservation in manufacturing operations: Modelling the milling process by a new complexity-based evolutionary approach. *J. Clean. Prod.* **2015**, *108*, 34–45. [CrossRef]
41. Bhinge, R.; Park, J.; Law, K.H.; Dornfeld, D.A.; Helu, M.; Rachuri, S. Toward a Generalized Energy Prediction Model for Machine Tools. *J. Manuf. Sci. Eng.* **2017**, *139*, 041013. [CrossRef]
42. Bang, S.H.; Ak, R.; Narayanan, A.; Lee, Y.T.; Cho, H.B. A survey on knowledge transfer for manufacturing data analytics. *Comput. Ind.* **2019**, *104*, 116–130. [CrossRef]
43. Chen, G.; Li, Y.; Liu, X. Pose-dependent tool tip dynamics prediction using transfer learning. *Int. J. Mach. Tools Manuf.* **2019**, *137*, 30–41. [CrossRef]
44. Shin, S.J.; Woo, J.Y.; Rachuri, S. Energy efficiency of milling machining: Component modeling and online optimization of cutting parameters. *J. Clean. Prod.* **2017**, *161*, 12–29. [CrossRef]
45. Lee, J.; Lapira, E.; Bagheri, B.; Kao, H. Recent advances and trends in predictive manufacturing systems in big data environment. *Manuf. Lett.* **2013**, *1*, 38–41. [CrossRef]
46. Li, L.; Liu, F.; Chen, B.; Li, C.B. Multi-objective optimization of cutting parameters in sculptured parts machining based on neural network. *J. Intell. Manuf.* **2015**, *26*, 891–898. [CrossRef]
47. Aramcharoen, A.; Mativenga, P.T. Critical factors in energy demand modelling for CNC milling and impact of toolpath strategy. *J. Clean. Prod.* **2014**, *78*, 63–74. [CrossRef]
48. Han, S.H.; Choi, Y.; Yoo, S.B.; Park, N.K. Collaborative engineering design based on an intelligent STEP database. *Concurr. Eng. Res. Appl.* **2002**, *10*, 239–249. [CrossRef]
49. Xu, X.W.; Wang, H.; Mao, J.; Newman, S.T.; Kramer, T.R.; Proctor, F.M.; Michaloski, J.L. STEP-compliant NC research: The search for intelligent CAD/CAPP/CAM/CNC integration. *Int. J. Prod. Res.* **2005**, *43*, 3703–3743. [CrossRef]
50. Vijayaraghavan, A.; Dornfeld, D. Automated energy monitoring of machine tools. *CIRP Ann.* **2010**, *59*, 21–24. [CrossRef]

51. MIT Critical Data. *Secondary Analysis of Electronic Health Records*; Springer International Publishing: Cham, Switzerland, 2016; pp. 115–141.
52. Nannapaneni, S.; Mahadevan, S.; Rachuri, S. Performance evaluation of a manufacturing process under uncertainty using Bayesian networks. *J. Clean. Prod.* **2016**, *113*, 947–959. [CrossRef]
53. Shin, S.J.; Woo, J.; Rachuri, S.; Seo, W. An energy-efficient process planning system using machine-monitoring data: A data analytics approach. *Comput. Aided Des.* **2019**, *110*, 92–109. [CrossRef]
54. Witten, I.H.; Frank, E. *Data Mining—Practical Machine Learning Tools and Techniques*; Elsevier: San Francisco CA, USA, 2005.
55. Zhong, E.; Fan, W.; Yang, Q.; Verscheure, O.; Ren, J. Cross validation framework to choose amongst models and datasets for transfer learning. In *Machine Learning and Knowledge Discovery in Databases*; Springer: Heidelberg, Germany, 2010; pp. 547–562.
56. Bellifemine, F.; Caire, G.; Greenwood, D. *Developing Multi-Agent Systems with JADE*; John Wiley & Sons, Ltd.: Chichester, UK, 2007.
57. Larek, R.; Brinksmeier, E.; Meyer, D.; Pawletta, T.; Hagendorf, O. A discrete-event simulation approach to predict power consumption in machining processes. *Prod. Eng. Res. Dev.* **2011**, *5*, 575–579. [CrossRef]
58. Shin, S.J.; Woo, J.; Kim, D.B.; Kumaraguru, S.; Rachuri, S. Developing a virtual machining model to generate MTConnect machine-monitoring data from STEP-NC. *Int. J. Prod. Res.* **2016**, *54*, 4487–4505. [CrossRef]

© 2019 by the authors. Licensee MDPI, Basel, Switzerland. This article is an open access article distributed under the terms and conditions of the Creative Commons Attribution (CC BY) license (http://creativecommons.org/licenses/by/4.0/).

Article

Multi-Objective Predictive Control Optimization with Varying Term Objectives: A Wind Farm Case Study

Clara M. Ionescu [1,2,3,*], Constantin F. Caruntu [4], Ricardo Cajo [1,2,5], Mihaela Ghita [1,2], Guillaume Crevecoeur [2,6] and Cosmin Copot [7]

1   Research lab on Dynamical Systems and Control, Ghent University, Tech Lane Science Park 125, 9052 Ghent, Belgium; ricardoalfredo.cajodiaz@ugent.be (R.C.); mihaela.ghita@ugent.be (M.G.)
2   EEDT—Core Lab Decisions and Controls, Flanders Make, Tech Lane Science Park 131, 9052 Ghent, Belgium; guillaume.crevecoeur@ugent.be
3   Department of Automatic Control, Technical University of Cluj-Napoca, Memorandumului 28, 400114 Cluj-Napoca, Romania
4   Department of Automatic Control and Applied Informatics, Gheorghe Asachi Technical University of Iasi, D. Mangeron Blvd 27, 700050 Iasi, Romania; caruntuc@ac.tuiasi.ro
5   Facultad de Ingeniería en Electricidad y Computación, Escuela Superior Politécnica del Litoral, ESPOL, Campus Gustavo Galindo Km 30.5 Vía Perimetral, P.O. Box 09-01-5863, 090150 Guayaquil, Ecuador
6   Electrical Energy Lab, Ghent University, Tech Lane Science Park 131, 9052 Ghent, Belgium
7   Department of Electromechanics, Antwerp University, Op3Mech, Groenenborgerlaan 171, 2020 Antwerp, Belgium; cosmin.copot@uantwerpen.be
*   Correspondence: claramihaela.ionescu@ugent.be; Tel.: +32-9264-5608

Received: 24 September 2019; Accepted: 24 October 2019; Published: 29 October 2019

Abstract: This paper introduces the incentive of an optimization strategy taking into account short-term and long-term cost objectives. The rationale underlying the methodology presented in this work is that the choice of the cost objectives and their time based interval affect the overall efficiency/cost balance of wide area control systems in general. The problem of cost effective optimization of system output is taken into account in a multi-objective predictive control formulation and applied on a windmill park case study. A strategy is proposed to enable selection of optimality criteria as a function of context conditions of system operating conditions. Long-term economic objectives are included and realistic simulations of a windmill park are performed. The results indicate the global optimal criterium is no longer feasible when long-term economic objectives are introduced. Instead, local sub-optimal solutions are likely to enable long-term energy efficiency in terms of balanced production of energy and costs for distribution and maintenance of a windmill park.

Keywords: windmill park; wind speed estimator; multi-objective optimization; sequential optimisation; distributed model predictive control

---

1. Introduction

When it comes to optimization strategies, advanced control methodologies such as model based predictive control (MPC) is of great industrial relevance [1–3]. For large scale systems, its variant as distributed MPC has great added value in terms of numerical optimization and computational efficiency [4,5]. Additionally, it requires a significantly lower amount of information than full multivariable MPC, hence the optimization can be accelerated. The cost function implemented in such MPC schemes is usually tailored upon the specific objectives of the process at hand. Less academic but highly relevant in practice objectives such as performance degradation, failure monitoring and nesting, implementation and training costs are making the MPC an even more appealing strategy for large scale processes. Supply chain optimization is a great example for successful stories of economic and environmental related objective based MPC applications [6,7].

Wide area control systems have been recently undergoing a revision of concepts and relevance going beyond the classical output performance [8,9]. Such examples of revisited control objectives can be found in large scale power systems [10], office buildings [11], traffic control [12], sustainability in company management [13], and distribution networks [9]. A generic feature is to search solutions for optimal operation from the decentralized to distributed control systems and combinations thereof. The features of tomorrow's control systems are based on system-wide control, plug-and-play control, measurement driven, adaptive and reconfigurable architectures.

An application where environmental and economic objectives are core in the long-term cost management and return of investment policies is the area of renewable energy, e.g., wave energy [14]. Example of large scale interacting sub-systems with independent control but global optimization policies is the case of windmill parks, either land- or water-based. Due to their negative impact on aesthetics and noise near populated areas, land based windmill parks lose their popularity as the marine parks gather higher interest [15]. Satellite-based synthetic aperture radar measurements allows a most optimal location along with best geometrical placement of such parks for the function of wind speed and its energy content [16,17]. These issues are now mature and the next at hand objective for optimal operation is derived from evidence-based effects of wind power systems on marine life and related maintenance costs—the latter being obviously quite different from land-based parks [18]. Reports of various agencies and policy makers indicate that long-term and large-scale impacts have cumulative effects and a knowledge gap is identified.

Following the exponential growth of wind capacity since 1996 to now, sufficient data is now available to comprehensively evaluate their efficiency, by taking into account the effective energy production and related costs on the long-term. Recent studies evaluating the distribution of costs for wind power systems indicate that offshore based parks have about double costs with respect to land-based costs [19]. Of the total cost distribution, maintenance and safe operation at lower power production are estimated to be a staggering 30%–40%, and thus the major source of costs [18,20]. The same distribution of costs applies also to a levelised cost of electricity from wind power systems: operation and maintenance remain leaders in the cost indicators [21].

Maintenance costs are related to the operation of the wind parks, in terms of environmental conditions (saline air, saline water, etc) and safety related operational decisions (wind speed and stability of the construction). There are other effects such as variability of the wind speed, affecting the grid where the power is introduced, further enlarging the cost-related issues to the grid itself [22]. Choice of instrumentation is also important to avoid operational problems in the grid [23–25]. Weak grids will operate at a lower power than the capacity for power generation from the wind park itself, and the choice of control strategy (voltage- or power-based criterium) will greatly affect the impact [26]. Limitations in storage capacity will impose limits in peak voltage as a safety margin.

In this paper, we perform a feasibility study of a concept for multi-objective optimization (MOO) scheme integrated with MPC and applied on a relevant study case: a windmill park. The combined scheme addresses optimization criteria on a priority-based algorithm, resulting in a multi-objective optimization priority MPC (MOOP-MPC). Economic costs are taken into account along with safety related limitations [27]. Comparison to a global priority optimization formulation is performed to analyse the best strategy for long-term impact. An existing model for a windmill park and updated cost function is used in Matlab/Simulink. Here, the estimation of wind speed is the basis for control feedback law and a real-time estimation scheme is proposed. The hypothesis tested in this study is whether the choice of the MOOP objectives will have an effect on the global performance output and efficiency of the system.

2. Materials and Methods

Model predictive control is an advanced control strategy well established over the past decades and commonly employed in the so-called money making industry, where economic costs are highly relevant [28–30]. For processes consisting of multiple interacting sub-systems, with highly coupled dynamics, multivariable MPC algorithms are demanding in terms of model data availability and computational costs. Instead, a much lighter version, ignoring interactions (if weak enough) or used with decoupling matrices (if interactions are strong), is that of decentralized MPC. As a trade-off solution, with interaction information communicated among the sub-systems, is the distributed MPC [4,5], also a well established MPC strategy.

In this paper, we use a distributed MPC algorithm presented in [4,5], with a tailored optimization scheme described hereafter. For nonlinear systems, in our version of MPC, linearization of the process model is not necessary, in the condition that the step response of system dynamics matrix is updated at every sampling time and the step input to obtain it has the amplitude in the region of the expected steady state values of the controller output (due to nonlinear dynamics, if large input is used, the information matrix no longer has information upon the specific operation point currently used). This is a special version of the MPC, especially suitable for data driven formulation; see comprehensive details in [31,32].

2.1. Multi Objective Optimization with Priorities

The new approach for wide-area control of systems is to have decentralized and distributed control layers throughout the global system output to be controlled [10,33]. Assuming the inner/lower layer of control works adequately, the upper level can deal with objectives other than absolute output, or combinations thereof with economic and environmental objectives.

At this point, we introduce the prioritized multi-objective optimization (MO) algorithm. This is a simplified approach compared to those proposed in the literature [34–38]. As with any process, the safety constraint is set as a hard constraint, given limit value intervals for all input–output variables. If this condition is not satisfied, a pre-set of (suboptimal) safety values are given to the process operation units. This step is implemented as proposed in [39]. If this condition is fulfilled, then the next priority is to meet the product specifications, i.e., performance tolerance error intervals and/or maintenance costs are evaluated. Finally, if this is also fulfilled, then the optimization minimizes the control effort, i.e., enters energy saving operation of the plant. The performance and other long-term costs are soft constraints, i.e., they are tailored to fit the objective at hand and not to minimize a specific cost goal. This allows a much faster computational convergence while process operation remains active within safety bounds. The sequential (prioritized) flowchart is iterated at every sampling period and the computational time within iteration is recorded.

2.2. Area-Wise Wind Speed Estimation

Bat algorithms and learning machine algorithms are popular estimation tools for predicting wind speed on location of the wind turbine [26,40]. By adjusting the wind turbine speed, the system can operate at optimal rotational speed while achieving the maximal power. However, the operation of the turbine is influenced by wind speed and direction in a complex context given the influence of neighbouring turbines in a windmill park. A good wind estimator is desired to reduce the uncertainty that will directly affect the performance of the proposed controller in a neighbouring area. Figure 1 depicts the position of the wind estimator within the loop.

Drones or in general, unmanned aerial vehicles (UAV) can be used to fly over the park and estimate wind speeds at determined locations in order to update the information for the individual operation of the turbines depending on the time-varying wind conditions. The proposed wind estimation is derived below for the coordinate frames of the UAV.

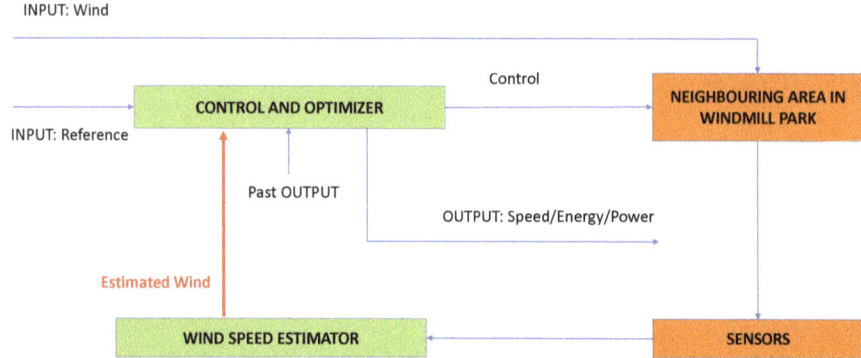

Figure 1. Block scheme with wind estimator and controller on the turbine.

Solutions to estimate the wind using on-board embedded wind sensors, are given in [41,42]. A possible drawback is that the on-board wind sensors use valuable payload that could be otherwise used. From aerodynamics, a nonlinear dependence of the UAV follows, by the wind speeds u_w, v_w, w_w, while the disturbances (external forces F_{Xaero}, F_{Yaero}, F_{Zaero} and moments L_{aero}, M_{aero}, N_{aero}) enter linearly in the drone equations. Hence, the problems of estimation of wind velocities and disturbances can be formulated in two context conditions: piecewise constant or slowly varying. To estimate linear parameters varying models for wind, the literature mainly reports two groups of solutions: with/without airspeed sensor information. For instance, Qu et al. [43] uses airspeed sensor, roll angle, and sideslip angle to estimate the wind speed for small fixed-wing UAV. Rhudy et al. [44] estimated the wind field based on four formulations, combining the data coming from the Pitot -static tube, the global position system (GPS), the inertial measurement unit (IMU) and the angle of attack and sideslip vanes. Otherwise, Xing et al. [45] estimated the shear wind vector at low altitude using IMU and GNSS modules. Pappu et al. [46] used a Kalman filter based gust identification technique for estimating wind gusts. In our work, it is supposed that the estimation algorithm can use IMU (accelerometer, gyroscope) sensors augmented with a motion tracking system and rotor's rotational velocity sensors.

Quadrotor linear velocities (u, v, w) and accelerations are provided by the on-board accelerometer, which measures directly

$$\tilde{\Delta}_a(X) = \Delta_a(X) + \epsilon_a \tag{1}$$

where ϵ_a is the bounded measurement noise and

$$\Delta_a(X) = \begin{bmatrix} \Delta_{au}(X) \\ \Delta_{av}(X) \\ \Delta_{aw}(X) \end{bmatrix} = \begin{bmatrix} \dot{u} \\ \dot{v} \\ \dot{w} \end{bmatrix} + \begin{bmatrix} p \\ q \\ r \end{bmatrix} \times \begin{bmatrix} u \\ v \\ w \end{bmatrix} - \begin{bmatrix} -g \sin \theta \\ g \cos \theta \sin \phi \\ g \cos \theta \cos phi \end{bmatrix}. \tag{2}$$

From the gyroscope, which measures the rotational velocity in body frame with respect to the earth, the other state coordinates are measured

$$\tilde{\Delta}_g(X) = \Delta_g(X) + \epsilon_g \tag{3}$$

where ϵ_g is the bounded measurement noise and

$$\tilde{\Delta}_g(X) = \begin{bmatrix} p & q & r \end{bmatrix}^T. \tag{4}$$

The IMU sensor is augmented with ground based cameras, used to estimate (u, v, w, θ, ϕ) coupled with the gyroscope and accelerometer, enabling drone observability with respect to the inertial frame. The dynamics equation can be summarized as follows:

$$\begin{bmatrix} \Delta_a(X) \\ \Delta_g(X) \end{bmatrix} = f_0(X, U, \omega) + \Omega(\omega) d_w \quad (5)$$

where f_0 is known and Ω needs to be estimated.

Define the predicted acceleration

$$\tilde{\Delta}_a = f_{0a} + \Omega_a \hat{d}_w \quad (6)$$

allowing introduction of the error between the measured and predicted state acceleration as

$$e_a = \tilde{\Delta}_a - \hat{\Delta}_a = \Omega_a(d_w - \hat{d}_w) + \epsilon_a. \quad (7)$$

From Rios et al. [47], we use the finite-time estimation algorithm

$$\hat{d}_w = \gamma_a \Omega_a^T [e_a]^{\alpha_a}, \quad 0 < \alpha_a < 1, \quad \gamma_a \gg 0. \quad (8)$$

The Lyapunov function can be selected

$$V = \frac{1}{2\gamma_a} |d_w - \hat{d}_w|^2 \quad (9)$$

with $||^2$ denoting the norm and for the simplified dynamics of a quadrotor, the matrix Ω_a is invertible and the solution is feasible and stable [47,48].

An adaptive observer equation model can be derived at this point

$$\begin{aligned} \hat{\Omega}_g &= f_{0g} + \Omega_g \hat{d}_w + l_g \text{sign}(\tilde{\Delta}_g - \hat{\Delta}_g) \\ \hat{d}_w &= \gamma_g \Omega_g^T (\tilde{\Delta}_g - \hat{\Delta}_g) \end{aligned} \quad (10)$$

with l_g and $\gamma \gg 0$ tuning parameters. Consider here the Lyapunov function

$$V = \frac{1}{2} \left([\Delta_g - \hat{\Delta}_g]^2 + \frac{1}{\gamma_g} [d_w - \hat{d}_w]^2 \right) \quad (11)$$

which has the derivative in time

$$\dot{V} = -l_g |\Delta_g - \hat{\Delta}_g| \quad (12)$$

and is bounded for all $t > 0$, with the state estimator converging asymptotically to the origin [47]. The wind estimation error converges as well to the origin, due to the persistent excitation in Ω_g.

Some convergence bottlenecks may appear with this algorithm as the computations occur. To increase robustness of convergence, the following algorithm is proposed (recall operations are element-wise):

$$\begin{aligned} \hat{\Delta}_g &= f_{0g} + \Omega_g \hat{d}_w + l'_g (\tilde{\Delta}_g - \hat{\Delta}_g) + \Phi \hat{d}_w \\ \Phi &= -l'_g \Phi + \Omega_g \\ \hat{d}_w &= \gamma'_g \Phi^T |\tilde{\Delta}_g - \hat{\Delta}_g|^{\alpha'_g} \end{aligned} \quad (13)$$

where $l'_g > 0$, $\alpha'_g \in (0, 1)$ and $\gamma'_g \gg 0$ are tuning parameters. The auxiliary matrix in this augmented model has the same dimension of Ω_g and is limited for bounded values of Ω_g and positive values of l'_g.

In the presence of measurement noise and time-varying wind speed conditions, Equation (10) will freeze its estimates, while (8) and (13) will estimate in an aggressive, respectively conservative manner. The fusion of the equations may seem appealing under the form

$$\hat{d}_w^{fusion}(t) = \frac{\sum_{i=1}^{2} e^{-k_i v_i^2(t)} \hat{d}_w^i}{\sum_{i=1}^{2} e^{-k_i v_i^2(t)}} \tag{14}$$

with k_i positive, a tuning parameter for convergence speed, and errors $v_1(t) = e_a(t)$, respectively $v_2(t) = \tilde{\Delta}_g(t) - \hat{\Delta}_g(t)$. This fusion algorithm has the worst estimation error given by the maximum of the two estimation errors for the algorithms (8) and (13).

Important fluctuations in the wind speed such as rapid turbulence will result in voltage fluctuations. The above described algorithm can estimate the basal envelope of the speed, while a faster component can be estimated by filtered white noise. The mean and standard deviation of the wind speed are linearly related with a constant k found experimentally from the park site:

$$\hat{\sigma}_w = k \cdot \bar{v}. \tag{15}$$

Using the shaping filter from Suvire et al. [49]

$$H(j\omega) = \frac{K}{(1 + j\omega T)^{5/6}} \tag{16}$$

we can set the time constant of the filter as

$$T = \frac{L}{\bar{v}} \tag{17}$$

with L the turbulence length (e.g., hundreds of meters). The constant of the filter is set by the condition of coloured noise with unit standard deviation for the wind speed values:

$$K = \sqrt{\frac{2\pi \cdot T}{B(0.5, 0.3) \cdot T_s}} \tag{18}$$

with T_s the sampling period and B denoting the Bessel function.

The global estimation of the wind speed will dictate the decisions in the MOOP-MPC optimization algorithm.

2.3. Windmill Park Simulator

A recent comprehensive literature review of wind farm operation using distributed predictive control illustrates that wind speed is a core variable of information in the optimization of the farm output in terms of generated power [50]. The wake effect is mainly in stream with the geometry of the park and has a decaying effect towards the end area of its direction. Integration of wind mill parks has been discussed comprehensively in a tutorial [10] and models for power control assumed from Ugalde-Loo et al. [51], with Matlab codes available as in Sadamoto et al. [52]. Networked delayed control is relevant for this system, but it is neglected in this study [8].

The simplest manner to model the wind park is to consider it a uniform turbine model. This is useful for global optimization objectives with global wind speed conditions. The other extreme is to consider individual models for each turbine. Wind mill parks can be as large as hundreds of units, hence a significant computational load makes such models restricted to specific analysis such as coherence, correlation, and similar properties of wind effects on energy production. However, area related (i.e., group, or local) effects on the optimization scheme may be investigated if groups of wind turbines are considered bundled into a model, and the wind park consists of several of such lumped models. We use the model presented in Suvire et al. [49] and available in Matlab/Simulink, with values for realistic wind speeds from Degraer et al. [53], but in a simplified version.

The schematic in Figure 2 depicts the concept of area-wise simulation of the windmill park and possible interaction between the areas coming from the wind direction.

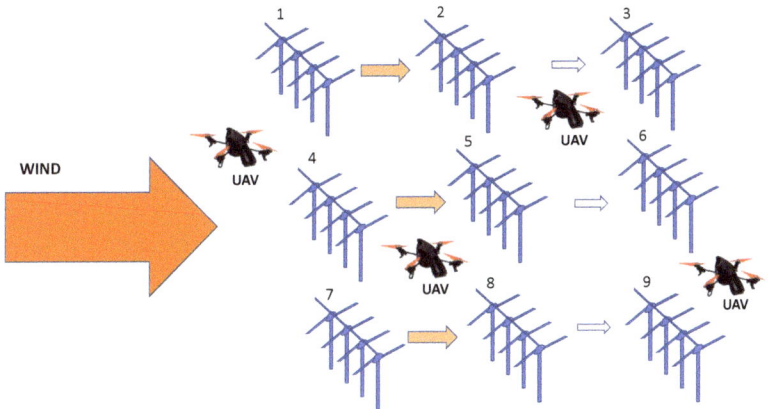

Figure 2. Windmill park conceptualisation as area-wise controlled sub-systems with limited interaction from wind direction and speed. The interaction intensity fades (with colour) as it travels through the system; the direction is limited to the arrow indicators, hence the interaction matrix is not fully coupled, which further motivates the use of distributed control.

The model of each sub-system denoted in Figure 2 by its corresponding number is a simplified one from the one used in Zhang et al. [40] and Suvire et al. [49], which is easily implementable in Matlab/Simulink. The output is the turbine's generated power. The time constant of the low pass filter depends on the average wind speed and can be assumed constant of time-varying. The effect of the wind fluctuations at rated power operation is filtered by a damped second order transfer function. Figure 3 schematically depicts the system to be controlled. This denotes a single grouped area of windmills in the large park. The model has been fitted as a first hand least square optimization algorithm to determine the parameters: $T_b = 11$, $K_p = 2.8$, $T = 12.5$ and $d = 97$, the latter term being related to the distance covered by one sub-system.

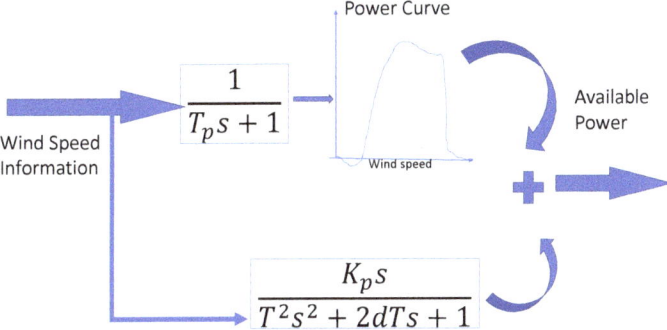

Figure 3. Schematic summary of the over-simplified model used for each sub-system in the windmill park.

3. Results

To verify the result between a global and an area-wise MOOP with distributed MPC strategy, simulations are performed in Matlab/Simulink R2017a. The values of the turbine model are the same as in Zhang et al. [40], and the values for the distributed MPC algorithm are set to 30 samples for the prediction horizon, one sample as the control horizon, and no time delay. The sampling period is 100 milliseconds. The output of the system is the power. The tuning of the MPC follows the rules of thumb given in Ionescu et al. [54].

The MOOP has three parameters sequentially activated as in Figure 4. The safety limits for the input are 0–100 Volts, rotor flux for turbine between 0–1 and angle speed for rotor tracking between 10–40 rad/s. Current is limited to maximum 5 A. These are taken as hard constraints in the global optimization case, with a 100% weight. The maximum power is normalized to 1 pu, reaching maximum output at 10 m/s and safety shut-down resulting in zero power output at 20 m/s. For the simulation purposes, the default values from SimPowerSystems/Simulink Toolbox within Matlab are used, but with wind speed from Degraer et al. [53] with an average varying between 10–20 m/s. The length of the park is assumed to be 300 m. The total output power of the farm is obtained by adding the power from each sub-system area for which model from Figure 3 has been identified from the simulator data.

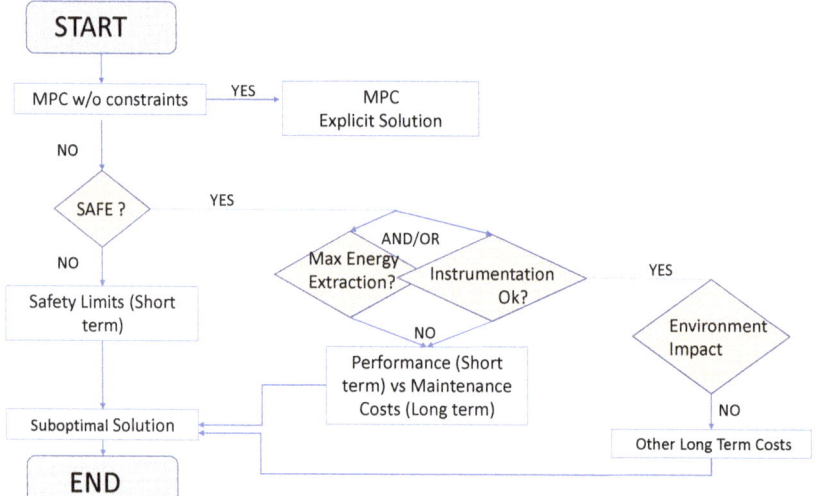

Figure 4. Flowchart of the sequential prioritized optimization scheme. Model based predictive control (MPC).

The performance term is evaluated as quadratic error between the maximal power extracted from the estimated wind speed and actual power from the system, during global optimization at a 70% weight. Maintenance costs are long-term costs and during global optimization they have a weighting factor in the second objective as a 30% of the total 100% weight—the rest is given to the performance as a short-term objective will have more influence on the cost variability. Alternatively, the short-term and long-term costs can be also implemented as a function of the prediction horizon: shorter intervals will give a faster convergence with large fluctuations, whereas longer prediction intervals will provide a more basal variation in the long-term objectives. A ratio of 2:5 is proposed for the short vs. long-term prediction horizon.

When exploring the environmental impact information such as in Degraer et al. [53], this is a cyclic signal, which can be modeled as a slowly moving average signal with increasing offset (i.e., cost). The minimization therefore is in terms of average values bringing them piecewise towards zero. The impact of this term during the global optimization case is 10%.

Figure 5 depicts the various timelines of the optimization parameters and the wind speed profile for the simulation is given in Figure 6.

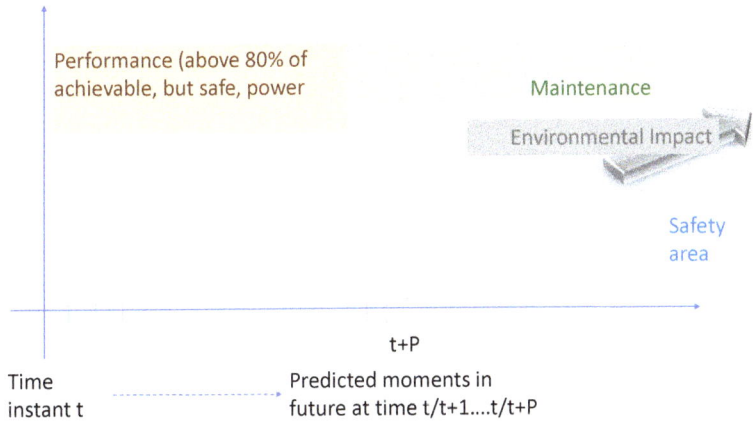

Figure 5. Illustration of the multiple objectives as a function of time and operation range percentage.

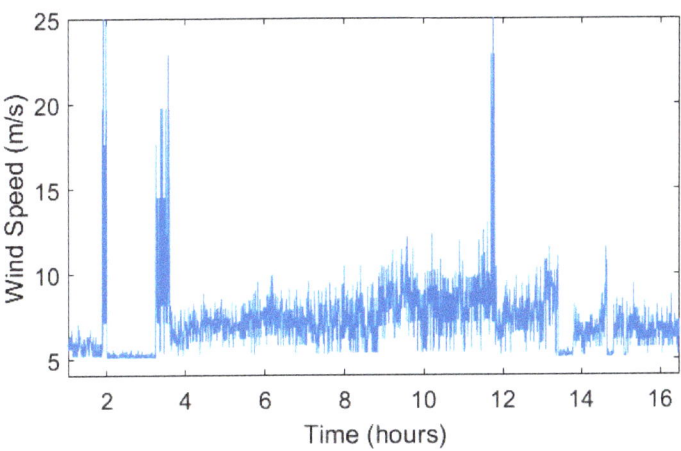

Figure 6. Example of variable wind speed profiles used in simulation.

The safety criterion is tested as being of the foremost importance in the operation of the park. Figure 7 illustrates the simulation of the park with a moment for wind speed above 20 m/s and temporary low windmill operation, resulting in lower output power—once the wind speed is recovered below the maximum allowed, the operation is resumed.

We compare now the two situations in variable speed wind conditions. First, all sub-system areas of the park are globally optimized, with a solidary cost function to take into account all output variables equally. Distributed MPC (dMPC) is still valid in this case, but the objective function is limited to neighbouring areas of each individual group of windmills. Second, the optimization is done individually per group (per sub-system), taking into account interactions from neighbouring areas, but the optimization cost function is a MOOP with weighting factors among the various types of costs. For the same time interval of 20 min, with same wind speed conditions, the power output for the two strategies is depicted in Figure 8.

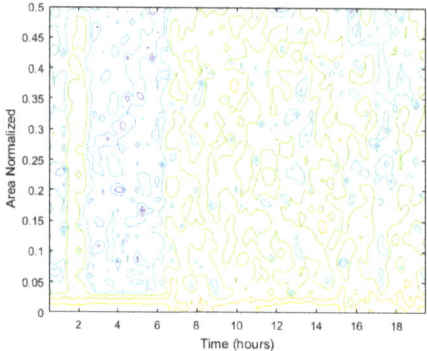

Figure 7. Example of testing the safety limits of operation of the park. Contour plot, blue colour denotes lowest values. The maximum value (red) corresponds to a 90–100% power extraction, while the lowest value (blue) corresponds to 0–10% power extraction.

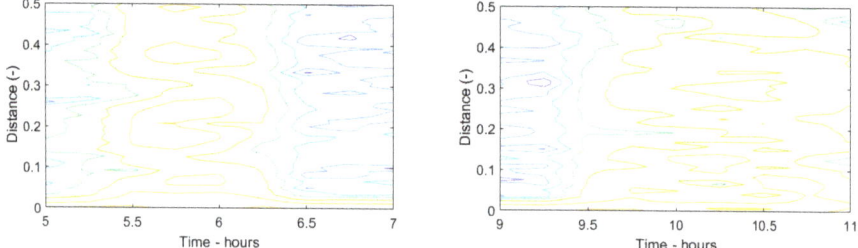

Figure 8. Contour plot of power output for variable wind speed conditions. **Left**: MOOP optimization. **Right**: global optimization. The maximum value (red) corresponds to a 90–100% power extraction, while the lowest value (blue) corresponds to 0–10% power extraction.

In case of constant average speed conditions, the comparison between the two strategies is illustrated in Figure 9.

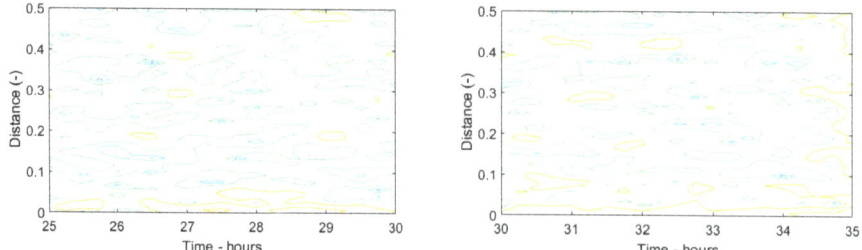

Figure 9. Contour plot of power output for constant wind speed conditions. **Left**: MOOP optimization. **Right**: Global optimization. The maximum value (red) corresponds to a 90–100% power extraction, while the lowest value (blue) corresponds to 0–10% power extraction.

The obtained performance is summarized in Tables 1 and 2 below for the two wind speed situations.

Table 1. Normalized units in percent for variable wind speed conditions. Distributed MPC (dMPC), Global and Multi-objective optimization priority (MOOP) methods

Method	Performance	Total Power Output	Cost
Global dMPC	88	85	91
MOOP dMPC	63	71	68

Table 2. Normalized units in percent for constant wind speed conditions.

Method	Performance	Total Power Output	Cost
Global dMPC	90	95	91
MOOP dMPC	87	91	88

4. Discussion

The idea of using UAVs as a flying sensor is borrowed from precision agriculture applications [55] and emergency medicine [56], but its integration in windmill park utilisation still leaves many open challenges. For instance, the effect of wind gusts and wake on the UAV in flying mode may be destabilizing. Efforts to model such effects are numerous [57,58]. Also, the need for real time measurements of wind speed may not always be justified, but this again fits with the use of UAVs, and this is perhaps less appealing for on-board turbine instrumentation [59]. The advantage of using UAVs for sensing wind speed may be in their versatility to approach different areas of the park at different altitudes and at different moments in time, while still having relatively less costs when compared to on-board instrumentation. The problem of in flight stability has been addressed in manifold applications of UAV and it is considered a rather mature control problem and is beyond the objective of this paper. By contrast, the use of UAVs can be eliminated, at the cost of determining via measurements the wind speed, direction and other useful features for maximizing throughput of the windmill park. In this study we have particularly chosen the use of UAVs, motivated by the manifold applications and versatility of their use in limited environments.

The use of the simplified model for wind turbine and the simplified wind farm scheme have the advantage of allowing the simulations to convey information on the methodology of optimization and control, while keeping to a minimum the effects of difficult dynamics from the system itself. The disadvantage is that the analysis is limited to a concept, an incentive for further development, and does not claim to be a comprehensive feasibility study of the wide area control system at hand.

It is not yet clear whether the increase in output power with maximizing global power extraction from the park is actually more productive than the MOOP optimization scheme. In fact, they have to be analysed in the perspective of their maintenance costs, which as indicated, are higher for the global optimization case than in MOOP. The relatively lower power extraction of the MOOP case may be well justified in long-term economic investigation.

From the simulations we performed, it appears that under constant wind speed conditions, there is no significant difference among the two strategies. This is somewhat expected, as the variability among the sub-systems remains similar for uniform distributions of wind energy in time. As the area-based dynamics are simplified to common dynamics, there is no source of inter-system variability propagated throughout the park. By contrast, variable speed conditions will induce different cost function convergences, as the data used for optimization contains persistent excitation and different weights affect the cost function optimizer. In this case, we observe significant differences among the two strategies.

There are manifold opportunities for control-related items to be further investigated. Moreover, the following questions may be applicable to any wide-area control system.

For instance, how would comparing the global optimization of maximizing output power against a MOOP with a short-term cost function and a long-term cost function? Whereas by long-term cost function, one could use an economic cost of maintenance over several months, and impact on environment cost over several years.

Would it still be interesting/appealing to use the maximizing power strategy?

Or, perhaps, a better solution could be that such a strategy could be used alternatively with the MOOP short–long-term cost objectives?

These are merely a shortlisted enumeration of possible research directions. As mentioned in [10], we are witnessing a transition to a new infrastructure in a manifold of power system networks, creating a shift in energy supply from a centralized to a distributed network of energy supply and demand. Such changes are visible in other than the power sector as well. Climate changes and stringent environmental regulations affect the evaluations of cost-driven investments and this is visible in long-term rather than short-term criteria. The concept and incentive study presented in this paper suggest that operators are well motivated to be inclined to explore new control methods that go far beyond current system management.

5. Conclusions

This paper introduces an incentive for wide-area control systems, with a conceptual study for optimization of output power as a function of short-term and long-term impact objectives. A relevant windmill park case study is presented, in which the initial simulations indicate opportunities and challenges for both control and economic driven studies.

Author Contributions: Conceptualization, C.F.C. and C.M.I.; methodology, C.M.I.; software, C.C. and R.C.; writing—original draft preparation, C.M.I. and M.G.; writing—review and editing, G.C., R.C., M.G., C.C. and C.F.C.; project administration, C.M.I. and C.F.C.; funding acquisition, G.C., C.M.I. and C.F.C.

Funding: C.M.I. gratefully acknowledges the financial support for this work from Ghent University Special Research Fund, project MIMOPREC—Starting grant 2018. C.F.C. gratefully acknowledges the financial support for this work from a research grant of the TUIASI, Romania, project number GI1654/2018. R.C. and M.G. received in part funding from Flanders Make—CONACON project nr HBC.2018.0235. M.G. received in part financial support from research Foundation Flanders, project nr G0D9316N and doctoral grant fundamental research fellowship nr. 1184220N. R.C. gratefully acknowledges the financial support from ESPOL and National Secretariat of Higher Education, Science, Technology and Innovation of Ecuador (SENESCYT).

Conflicts of Interest: The authors declare no conflict of interest.

Abbreviations

The following abbreviations are used in this manuscript:

MPC   Model Predictive Control
dMPC   distributed Model Predicitve Control
MOOP   Multi Objective Optimization Procedure
UAV   Unmanned Aerial Vehicle

References

1. Samad, T. A survey on industry impact and challenges thereof. *IEEE Control Syst. Mag.* **2017**, *37*, 17–18. [CrossRef]
2. Maxim, A.; Copot, D.; Copot, C.; Ionescu, C.M. The 5Ws for control as part of Industry 4.0: Why, what, where, who and when—A PID and MPC control perspective. *Inventions* **2016**, *4*, 10. [CrossRef]
3. Zhao, S.; Cajo, R.; De Keyser, R.; Liu, S.; Ionescu, C.M. Nonlinear predictive control applied to steam/water loop in large scale ships. In Proceedings of the 12th IFAC Symposium on Dynamics and Control of Process Systems, Including Biosystems, Florianopolis, Brazil, 23–26 April 2019; pp. 868–873. [CrossRef]
4. Zhao, S.; Maxim, A.; Liu, S.; De Keyser, R.; Ionescu, C.M. Distributed model predictive control of steam/water loop in large scale ships. *Processes* **2019**, *7*, 442. [CrossRef]

5. Maxim, A.; Copot, D.; De Keyser, R.; Ionescu, C.M. An industrially relevant formulation of a distributed model predictive control algorithm based on minimal process information. *J. Process Control* **2018**, *68*, 240–253. [CrossRef]
6. Fu, D.; Zhang, H.-T.; Yu, Y.; Ionescu, C.M.; Aghezzaf, E.-H.; De Keyser, R. A distributed model predictive control strategy for the bullwhip reducing inventory management policy. *IEEE Trans. Ind. Inf.* **2019**, *15*, 932–941. [CrossRef]
7. Fu, D.; Ionescu, C.M.; Aghezzaf, E.-H.; De Keyser, R. Quantifying and mitigating the bullwhip effect in a benchmark supply chain system by an extended prediction self-adaptive control ordering policy. *Comput. Ind. Eng.* **2015**, *81*, 46–57. [CrossRef]
8. Wang, S.; Meng, X.; Chen, T. Wide-area control of power systems through delayed network communication. *IEEE Trans. Control Syst. Technol.* **2012**, *20*, 495–503. [CrossRef]
9. Allen W. Effects of wide-area control on the protection and operation of distribution networks. In Proceedings of the 2009 Power Systems Conference, Clemson, SC, USA, 10–13 March 2009. [CrossRef]
10. Sadamoto, T.; Chakrabortty, A.; Ishizaki, T.; Imura, J.-C. Dynamic modelling, stability, and control of power systems with distributed energy resources. *IEEE Control Syst. Mag.* **2019**, *39*, 34–65. [CrossRef]
11. MacNaughton, P.; Pegues, J.; Satish U.; Santanam, S.; Spengler, J.; Allen J. Economic, environmental and health implications of enhanced ventilation in office buildings. *Int. J. Environ. Res. Public Health* **2015**, *12*, 14709–14722. [CrossRef]
12. Sutandi, C. Advanced traffic control system impacts on environmental quality in a large city in a developing country. *J. East. Asia Soc. Transp. Stud.* **2007**, *7*, 1169–1179.
13. Caputo, F; Veltri, S.; Venturelli, A. Sustainability strategy and management control systems in family firms. Evidence from a case study. *Sustainability* **2017**, *9*, 977. [CrossRef]
14. De Koker, K.; Crevecoeur, G.; Meersman, B.; Vantorre, M.; Vandevelde, L. A wave emulator for ocean wave energy, a Froude-scaled dry power take-off test setup. *Renew. Energy* **2017**, *105*, 712–721. [CrossRef]
15. Klaeboe, R.; Sundfor, H.B. Windmill noise annoyance, visual aesthetics, and attitudes towards renewable energy sources. *Int. J. Environ. Res. Public Health* **2016**, *13*, 746. [CrossRef] [PubMed]
16. Johannessen, O.M.; Korsbakken, E. Determination of wind energy from SAR images for siting windmill locations. *Earth Obs. Q.* **1998**, *1*. Available online: http://www.esa.int/esapub/eoq/eoq59/JOHANNESSEN.pdf (accessed on 12 September 2019).
17. Stevens, R.J.A.M.; Gayme, D.F.; Meneveau, C. Effects of turbine spacing on the power output of extended wind-farms. *Wind Energy* **2015**, *19*, 359–370. [CrossRef]
18. Bailey, H.; Brookes, K.L; Thompson, P.M. Assessing environmental impacts of offshore wind farms: Lessons learned and recommendations for the future. *Aquat. Biosyst.* **2014**, *10*. [CrossRef]
19. Anaya-Lara, O.; Campos-Gaona, D.; Moreno-Goytia, E.; Adam, G. *Offshore Wind Energy Generation: Control, Protection and Integration to Electrical Systems*; Wiley: Hoboken, NJ, USA, 2014.
20. European Wind Power Association. Operation and Maintenance Costs of Wind Generated Power. Available online: https://www.wind-energy-the-facts.org/operation-and-maintenance-costs-of-wind-generated-power.html (accessed on 12 September 2019).
21. IRENA—International Renewable Energy Agency. Renewable Power Generation Costs in 2017. Abu Dhabi, 2018, 160p. Available online: https://www.irena.org/-/media/Files/IRENA/Agency/Publication/2018 (accessed on 12 September 2019).
22. Muljadi, E.; Butterfield, C.; Parsons, B.; Ellis, A. Effect of variable speed wind turbine generator on stability of a weak grid. *IEEE Trans. Energy Convers.* **2007**, *22*, 29–36 . [CrossRef]
23. Feltes, J.W.; Fernandes, B.S. Wind turbine generator dynamic performance with weak transmission grids. In Proceedings of the IEEE Power & Energy Society General Meeting, San Diego, CA, USA, 22–26 July 2012; 7p. [CrossRef]
24. Kadar, P. Pros and Cons of the renewable energy application. *Acta Polytech. Hung.* **2014**, *11*, 211–224.
25. Tudorache, T.; Popescu, M. FEM optimal design of wind energy based heater. *Acta Polytech. Hung.* **2009**, *6*, 55–70.
26. Bindner, H. Power Control for Wind Turbines in Weak Grids: Concepts Development. Denmark. Forskningscenter Risoe, 1999, Risoe-R, No. 1118 (EN). Available online: https://orbit.dtu.dk/files/7729819/ris_r_1118.pdf (accessed on 12 September 2019).

27. Santo, G.; Peeters, M.; Van Paepegem, W.; Degroote, J. Dynamic load stress analysis of a large horizontal axis wind turbine using full scale fluid-structure interaction simulation. *Renew. Energy* **2019**, *140*, 212–226. [CrossRef]
28. Rossiter, J.A. *A First Course in Predictive Control*, 2nd ed.; Textbook; CRC Press, Taylor and Francis Group: London, UK, 2018; ISBN 9781138099340.
29. Wang, L. *Model Predictive Control System Design and Implementation Using MATLAB*; Advances in Industrial Control Series; Springer: London, UK, 2009; ISBN 9781848823303.
30. Kouvaritakis, B.; Cannon, M. *Model Predictive Control, Advanced Textbooks in Control and Signal Processing*; Springer International Publishing: Cham, Switzerland, 2016; ISBN 9783319248516.
31. De Keyser, R.; Van Cauwenberghe, A. A self-tuning multistep predictor application. *Automatica* **1981**, *17*, 167–174. [CrossRef]
32. De Keyser, R. Model Based Predictive Control for Linear Systems. In *UNESCO Encyclopaedia of Life Support Systems*; Series on Control Systems, Robotics and Automation; Eolss Publishers Co. Ltd.: Oxford, UK, 2003; Volume XI, 30p, Article Contribution 6.43.16.1. Available online: http://www.eolss.net/ebooklib/cart.aspx (accessed on 10 September 2019).
33. Ishizaki, T.; Sadamoto, T.; Imura, J.-C.; Sandberg, H.; Johansson, K.H. Retrofit control: Localization of controller design and implementation. *Automatica* **2018**, *95*, 336–346. [CrossRef]
34. Lesser, K.; Abate, A. Multi-objective optimal control with safety as a priority. In Proceedings of the 2017 International Conference on Cyber-Physical Systems (ICCPS), Pittsburgh, PA, USA, 18–21 April 2017; pp. 25–36.
35. Maree, J.; Imsland, L. On multi-objective economic predictive control for cyclic process operation. *J. Process Control* **2014**, *24*, 1328–1336. [CrossRef]
36. Bemporad, A.; de la Pena, D. Multiobjective model predictive control. *Automatica* **2009**, *45*, 2823–2830. [CrossRef]
37. Yamashita, A.; Zanin, A.; Odloak, D. Tuning of model predictive control with multi-objective optimization. *Braz. J. Chem. Eng.* **2016**, *33*, 333–346. [CrossRef]
38. Tan, K.; Khor, E.; Lee, T.; Sathikannan, R. An evolutionary algorithm with advanced goal and priority specification for multi-objective optimization. *J. Artif. Intell. Res.* **2003**, *18*, 183–215. [CrossRef]
39. Wojsznis, W.; Mehta, A.; Wojsznis, P.; Thiele, D.; Blevins, T. Multiobjective optimization for model predictive control. *ISA Trans.* **2007**, *46*, 351–361. [CrossRef]
40. Zhang, Y.; Zhang, L.; Liu, Y. Implementation of maximum power point tracking based on variable speed forecasting for wind energy systems. *Processes* **2019**, *7*, 158. [CrossRef]
41. Prudden, S.; Fisher, A.; Marino, M.; Mohamed, A.; Watkins, S.; Wild, G. Measuring wind with Small Unmanned Aircraft Systems. *J. Wind Eng. Ind. Aerodyn.* **2018**, *176*, 197–210. [CrossRef]
42. Palomaki, R.T.; Rose, N.T.; van den Bossche M., Sherman, T.J.; De Wekker, S.F.J. Wind Estimation in the Lower Atmosphere Using Multirotor Aircraft. *J. Atmos. Ocean. Technol.* **2017**, *34*, 1183–1191. [CrossRef]
43. Qu, Y.; Duan, J.; Zhang, Y. An algorithm of online wind field estimation for small fixed-wing UAVs. In Proceedings of the 2016 Chinese Control Conference, Chengdu, China, 27–29 July 2016; pp. 10645–10650. [CrossRef]
44. Rhudy, M.B.; Fravolini, M.L.; Gu, Y.; Napolitano, M.R.; Gururajan, S.; Chao, H. Aircraft model-independent airspeed estimation without pitot tube measurements. *IEEE Trans. Aerosp. Electron. Syst.* **2015**, *51*, 1980–1995. [CrossRef]
45. Xing, Z.; Qu, Q.; Zhang, Y. Shear wind estimation with quadrotor UAVs using filtering regressing method. In Proceedings of the 2017 International Conference on Advanced Mechatronic (ICAMechS), Xiamen, China, 6–9 December 2017; pp. 196–201. [CrossRef]
46. Pappu, V.S.R.; Liu, Y.; Horn, J.F.; Cooper, J. Wind gust estimation on a small VTOL UAV. In Proceedings of the 7th AHS Technical Meeting on VTOL Unmanned Aircraft Systems and Autonomy, Mesa, AZ, USA, 24–26 January 2017.
47. Rios, H.; Efimov, D.; Moreno, J.A.; Perruquetti, W.; Rueda-Escobedo, J.G. Time-Varying Parameter Identification Algorithms: Finite and Fixed-Time Convergence. *IEEE Trans. Autom. Control* **2017**, *62*, 3671–3678. [CrossRef]

48. Perozzi, G.; Efimov, D.; Biannic, J.-M.; Planckaert, L. Trajectory tracking for a quadrotor under wind perturbations: Sliding mode control with state-dependent gains. *J. Frankl. Inst.* **2018**, *355*, 4809–4838. [CrossRef]
49. Suvire, G.O.; Mercado, P.E. Wind farm: Dynamic model and impact on a weak power system. In Proceedings of the 2008 IEEE/PES Transmission and Distribution Conference and Exposition: Latin America, Bogota, Colombia, 13–15 August 2008; 8p. [CrossRef]
50. Caruntu, C.F. Distributed predictive control for wind farms efficiency maximization: Challenges and opportunities. In Proceedings of the 6th International Conference on Control, Decision and Information Technologies (CoDIT), Paris, France, 23–26 April 2019; pp. 452–457. [CrossRef]
51. Ugalde-Loo, C.E.; Ekanayake, J.B.; Jenkins, N. State-space modeling of wind turbine generators for power system studies. *IEEE Trans. Ind. Appl.* **2013**, *49*, 223–232. [CrossRef]
52. Sadamoto, T. CSM2018 Matlab Codes. Available online: https://github.com/TSadamoto/CSM2018 (accessed on 10 September 2019).
53. Degraer, S.; Brabant, R.; Rumes, B.; Vigin, L. (Eds.) *Environmental Impacts of Offshore Wind Farms in the Belgian Part of the North Sea*; Scientific Reports Series: Memoirs on the Marine Environment; 2018, 136p. Available online: https://odnature.naturalsciences.be/downloads/mumm/windfarms/winmon_report_2018_final.pdf (accessed on 12 September 2019).
54. Ionescu, C.; Copot, D. Hands-on MPC tuning for industrial applications. *Bull. Pol. Acad. Sci. Tech. Sci.* **2019**, accepted.
55. Hernandez, J.A.; Murcia, H.; Copot, C.; De Keyser, R. Towards the development of a smart flying sensor: Illustration in the field of precision agriculture. *IEEE Sens. J.* **2015**, *15*, 16688–16709. [CrossRef]
56. Van de Voorde, P.; Gautama, S.; Momont, A.; Ionescu C.M.; De Paepe, P.; Fraeyman, N. The drone ambulance [A-UAS]: Golden bullet or just a blank? *Resuscitation* **2017**, *116*, 46–48. [CrossRef]
57. Mian Ab, M.S.; De Kooning, J.; Vandevelde, L.; Crevecoeur, G. Harvesting wind gust energy with small and medium wind turbines using a bidirectional control strategy. *J. Eng.* **2019**, *2019*, 4261–4266.
58. Papadopoulos, V.; Knockaert, J.; Develder, C.; Desmet, J. Investigating the need for real time measurements in industrial wind power systems combined with battery storage. *Appl. Energy* **2019**, *247*, 559–571. [CrossRef]
59. Santo, G.; Peeters, M.; Van Paepegem, W.; Degroote, J. Analysis of the aerodynamic loads on a wind turbine in off-design conditions. In *Recent Advances in CFD for Wind and Tidal Offshore Turbines*; Springer Tracts in Mechanical Engineering; Springer: Cham, Switzerland, 2019; pp. 51–59.

© 2019 by the authors. Licensee MDPI, Basel, Switzerland. This article is an open access article distributed under the terms and conditions of the Creative Commons Attribution (CC BY) license (http://creativecommons.org/licenses/by/4.0/).

Article

The Rotating Components Performance Diagnosis of Gas Turbine Based on the Hybrid Filter

Li Zeng [1],*, Shaojiang Dong [1] and Wei Long [2]

[1] School of Mechatronics & Vehicle Engineering, Chongqing Jiaotong University, Chongqing 400074, China; dongshaojiang100@163.com
[2] School of Mechanical Engineering, Sichuan University, Chengdu 610065, China; scdxlongwei@yeah.net
* Correspondence: zengli@cqjtu.edu.cn or zengli_sichuan@163.com

Received: 12 August 2019; Accepted: 29 October 2019; Published: 5 November 2019

Abstract: Gas turbine converts chemical energy into mechanical energy and provide energy for aircraft, ships, etc. The performance diagnosis of rotating components of gas turbine are essential in terms of the high failure rate of these parts. A problem that the sudden changing of operation state of turbines may lead to the misdiagnosis due to the defect of gas turbine's model. This paper constructs the strong tracking filter based on the unscented Kalman filter to achieve accurate estimation of gas turbine's measured parameters when the state changes suddenly. In the strong tracking filter, a parameter optimization method based on the residual similarity of measured parameters is proposed. Next, adopt the measured parameters filtered by the strong tracking filter to construct the health parameters estimation algorithm based on the particle filter. The particle weight is optimized by the mean adjustment method. Performance diagnosis is realized by checking the changes of health parameters output by particle filter. The results show that the proposed method improves the accuracy of performance diagnosis obviously.

Keywords: Unscented Kalman Filter; particle filter; weight optimization; hybrid filter; gas turbine

1. Introduction

Performance diagnosis is essential to realize the health management of gas turbine, and is absolutely necessary to the concept of on-condition maintenance which is an advanced maintenance idea of gas turbine. There are many ways to achieve the performance diagnosis, the method based on wear particle morphology analysis in lubricating oil, the method based on vibration signal analysis, and the method based on electrostatic signal analysis at the outlet of nozzle [1–10]. Borguet Sebastien and Leonard Olivier combine two diagnostic tools to improve the diagnosis accuracy of gas turbine. One tool is the principal component analysis (PCA) which is used to isolate the components fault, and another one is the Kalman filter in order to realize on-line evaluation of health condition of gas turbine [11]. Lu Feng, Ju Hongfei, and Huang Jinquan propose a nonlinear state estimation algorithm based on the extended Kalman filter. The transformation matrix is used to calculate estimation errors and construct the underdetermined extended Kalman filter [12]. Vanini Sadough put forward multiple dynamic neural networks to learn the different conditions of gas turbine. For each network, residuals between the outputs of network and the measured values are calculated. Furthermore, the thresholds of residuals are obtained, and performance diagnosis can be achieved by comparing the size of residuals and their thresholds [13]. Chen Libo and Song Lanqi propose a hybrid technique which composed of spectrometric oil analysis and auto debris classifier to enhance the diagnosis accuracy of wear fault. The Dempster-Shafer evidence theory is adopted to detect the fault [14]. Huang Qiang, Zhang Guigang, and Zhang Ting optimized the parameters of support vector machine by the genetic algorithm and simulated annealing method. A performance diagnosis approach of aero engine gas path is proposed by the advanced support vector machine [15]. Verma Rajeev, Roy Niranjan, and Ganguli Ranjan developed a fuzzy system based on a linear model to detect the failure of gas turbine [16]. Bachir, A and Hafaifa, A introduce a way to monitor the working condition of gas turbine based on the vibration signal analysis with respect to the principle of principal component analysis [17]. Yang Liu, Ding Shuiting, and Wang Ziyao propose a risk assessment method to evaluate the health status of aeroengine based on probability density evolution, and validate the effectiveness of this method by compare it with the Monte Carlo simulation method [18]. Zeng Li, Long Wei, and Li Yanyan suggest an approach based on the kernel principal component analysis to detect the fault and locate the failure by analyze the influence of fault to gas path components [19]. All the methods mentioned above must obtain the measured values of sensors and it is a consensus that the measurements contain many noise signals. However, these methods do not treat the noise signals before construct the fault detection algorithm, and it is probably lead to misdiagnosis. Furthermore, a defect exists in those methods which based on the physical model of gas turbine is that the modeling errors may lead to the performance diagnosis distortion.

To remove the noise in the measured parameters of airborne sensors and realize the performance diagnosis exactly, Wang Lei, Liu Zhiwen, and Miao Qiang proposed use ensemble local mean decomposition and fast kurtogram decompose the raw signal into the production functions to characterize the fault information. Then the optimal band-pass filter to filter the selected production functions and the impulse signal are obtained. By analyzing the fault characteristic frequencies, fault identification can be realized [20]. Zhang Yongxiang and Randall R.B proposed fast kurtogram and genetic algorithm to diagnose the failure of rolling element bearing. The initial parameters can be given by fast kurtogram and the optimized parameters with minimal constraint can be obtained [21]. Pham Hongthom and Yang Bo-suk adopt the linear ARMA model and nonlinear GARCH model to describe the fault of machine. The hybrid model can predict the future state of machine with high accuracy and give obvious explanation of the state [22].

In this paper, a novelty hybrid filter which composed of strong tracking filter based on unscented Kalman filter and particle filter with weight optimized is proposed to diagnose the performance variance of gas turbine. Firstly, construct a strong tracking filter based on the unscented Kalman filter by constraining the measurement residuals of current and last sampling time to be orthogonal. The calculation process of the scale factor is optimized by the residual similarity of measured parameters. The strong tracking filter is used to filter the noise signals contained in measurements. Next, the particle filter is used to estimate the health parameters of gas turbine. The health parameters consist of the efficiency coefficients and flow coefficients of rotating components. The outputs of strong tracking filter are the input parameters of particle filter. The problem that the distortion of estimated values of health parameter is resolved and the drawback of weight degradation of particle is overcome.

2. Materials and Methods

2.1. Fault Diagnosis Algorithm of Gas Turbine Based on Hybrid Filter

Compare with other component of gas turbine, the performance degradation rate of gas path components is higher. In order to reflect the health status of gas path components accurately, the health parameters include the efficiency coefficients and flow coefficients of turbines can be used to indicate the performance changes of gas turbine [23-25]. Performance diagnosis can be realized by estimating the values of health parameters. The strong tracking filter is constructed to eliminate the noise contained in measured parameters based on unscented Kalman filter. The particle filter is used to estimate the values of health parameters of gas turbine. To solve the problems of weight degradation and degradation of diversity exist in the particle filter, a weight optimization method is proposed. The principle of hybrid filter is shown in Figure 1.

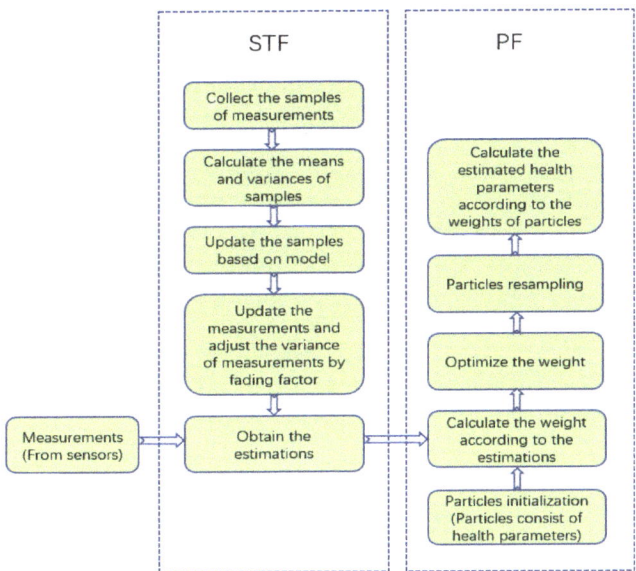

Figure 1. Principle of hybrid filter.

The two-spool turbojet is adopted as the research object. The low-pressure compressor, high-pressure compressor, low-pressure turbine, high-pressure turbine are the principal components of two-spool turbojet. The low-pressure compressor and the low-pressure turbine are connected by the low-pressure rotor. The high-pressure compressor and the high-pressure turbine are connected by the high-pressure rotor. Air enters the engine from Section 0 and compressed by the compressors. Next, compressed air mixed with fuel in the burning room and combusts. The high temperature gas at the exit of the burning room expands and drives the turbines to rotate. Some gas is ejected from the nozzle to generate thrust. Due to the turbines and compressors are connected by rigid rotors, turbines transmit torque to compressors to drive compressors to rotate, and pressurize the air. The structure of the two-spool turbojet is shown in Figure 2.

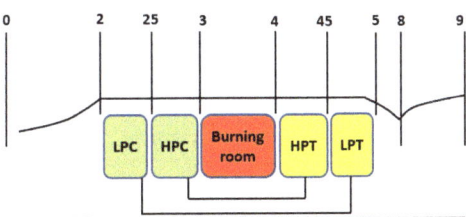

Figure 2. Principal components of two-spool turbo jet.

Measurements include the following contents:

T_{t25} Total temperature at the outlet of LPC
T_{t45} Total temperature at the outlet of LPT
P_{t25} Total pressure at the outlet of LPC
P_{t45} Total pressure at the outlet of LPT

T_{t3} Total temperature at the outlet of HPC
T_{t5} Total temperature at the outlet of HPT
P_{t3} Total pressure at the outlet of HPC
P_{t5} Total pressure at the outlet of HPT

There are:

Low Pressure Compressor (LPC)
Low Pressure Turbine (LPT)

High Pressure Compressor (HPC)
High Pressure Turbine (HPT)

2.2. Problem in the UKF

In order to ensure the good accuracy of the output parameters of unscented Kalman filter, a model that can reflect the real working state of the monitored target must be established. In the condition monitoring of two-spool turbojet based on the unscented Kalman filter, the Component-level Gas Path Model (CGPM) is usually used to predict the values of health parameters [26–32]. The health parameters are the indicators of the health status of turbojet and can be used to illustrate the flowing ability and working efficiency. The CGPM is essentially a series of physical equations based on the principle of aerothermodynamics. By the operation of CGPM, the health parameters (flow coefficients and efficiency coefficient of gas path components) and the measured parameters (total temperature and total pressure of gas path components) can be calculated. The structure of CGPM is shown in Figure 3.

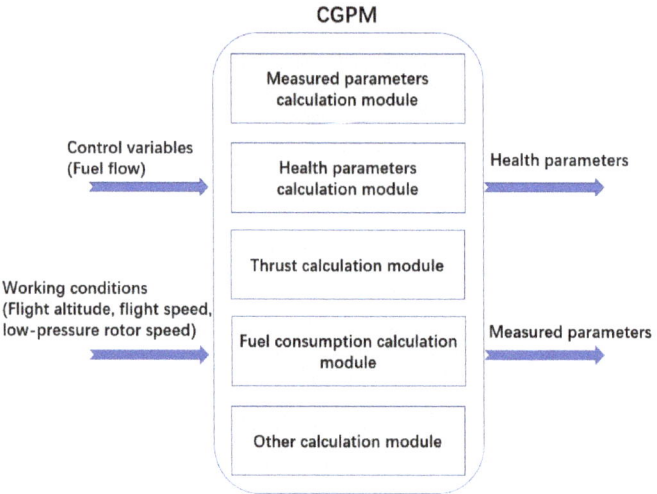

Figure 3. Structure of Component-level Gas Path Model.

The Component-level Gas Path Model of two-spool turbojet can be described as follows [1]:

$$M_{k+1} = f(M_k, C_{k+1}) + v_k \tag{1}$$

$$Y_{k+1} = g(M_{k+1}, C_{k+1}) + w_{k+1} \tag{2}$$

In Formulas (1) and (2), k represents the kth sampling time. M_k is the state parameters vector. Furthermore, all of these parameters are the estimated objects. C_{k+1} is the control variable. There are:

$$M_k = [\eta_{LPC}, F_{LPC}, \eta_{HPC}, F_{HPC}, \eta_{HPT}, F_{HPT}, \eta_{LPT}, F_{LPT},] \qquad C_{k+1} = F_{fuel}$$

η and F are the efficiencies and flow coefficients of different components, respectively, include the low-pressure compressor, high pressure compressor, low pressure turbine, and high pressure turbine. C_{k+1} is the flow of fuel. v_k and w_k are the state transmission noise and measurement transmission noise. f represents the process of predicting the health parameters based on the Component-level Gas Path Model. M_{k+1} is the vector of health parameter. The value of M_{k+1} is determined by C_{k+1} and M_k. Thus, the calculation of f can be realized by the health parameters calculation module, as shown in Figure 3. The content of g is similar with that of f. g represents the process of predicting the measured parameters based on the Component-level Gas Path Model. Y_{k+1} is the vector of measured parameters at the $k + 1$th sampling time. By the operation of measured parameters calculation module, function g can be.

The working state transformation of turbojet is a continuous process, the Component-level Gas Path Model can not accurately reflect all the working state. If the working state of turbojet changes suddenly, the output measured parameters of CGPM deviate greatly from those of the turbojet due to the defect of CGPM. Consequently, the estimations of UKF may be distorted. This circumstance can be illustrated by Figure 4.

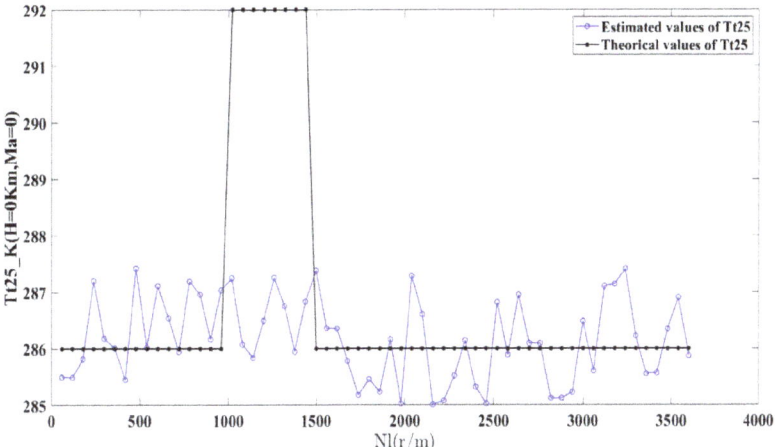

Figure 4. Estimations of T_{t25} by UKF.

In Figure 4, the values of T_{t25} are estimated by the unscented Kalman filter based on the Component-level Gas Path Model. The working state of the turbojet changed abruptly after working for an hour, and the measured parameters include T_{t25} changed widely in short time. However, due to the defect of Component-level Gas Path Model, the estimations of unscented Kalman filter are not consistent with the true values of measurements, as shown in Figure 4.

2.3. Resolution

To overcome above drawback, a strong tracking filter based on the UKF is proposed. The strong tracking filter (STF) satisfies following condition [33–35]:

$$E(\varepsilon_k \cdot \varepsilon_{k+i}^T) \approx 0 \text{ where } \varepsilon_k = Y_k - y_k \tag{3}$$

ε_k and ε_{k+i} are the residuals of measurements and outputs of model at the kth and $(k+i)$th sampling times. Y_k and y_k are the measurements and outputs of model, respectively. Equation (3) means that the residuals of measurements and outputs of model is orthogonal if the UKF is working normally. When the working state of the engine changes abruptly, the residuals are not orthogonal anymore. Paper design STF to adjust the variance ratio of measurements at different times to force the residuals keep orthogonal so that the accuracy of estimated parameters remains high. The steps of STF are as follows:

Samples collection and weight calculation.

$$X_0 = \bar{x}, \qquad W_0 = v/(v+n) \qquad i = 0$$

$$X_i = \bar{x} + \sqrt{(v+n)P_{xx}} \qquad W_i = 1/(2(v+n)), \qquad i = 1, 2, \ldots, n$$

$$X_i = \bar{x} - \sqrt{(v+n)P_{xx}} \qquad W_i = 1/(2(v+n)), \qquad i = n+1, n+2, \ldots, 2n$$

X_i are the estimated objects (state variable) which consist of T_{t25}, T_{t3}, T_{t45}, T_{t5}, P_{t25}, P_{t3}, P_{t45}, and P_{t5}. $\bar{x}$ denotes the mean vector of the estimated objects. W is the weight of estimated object. v is the parameter to reduce prediction error. n is the number of estimated objects and there is $n = 8$. P_{xx} is the covariance matrix of the estimated objects.

State variable calculation based on model.

$$X^i_{k+1} = f(X_k, C_{k+1}) + v_k \tag{4}$$

$$\bar{X}_{k+1} = \sum_{i=1}^{10} W^m_i X^i_{k+1} \tag{5}$$

$$P_{X,k+1} = \sum_{i=1}^{10} W^c_i (X^i_{k+1} - \bar{X}_{k+1})(X^i_{k+1} - \bar{X}_{k+1})^T \tag{6}$$

$$Y^i_{k+1} = g(X_{k+1}, C_{k+1}) + w_{k+1} \tag{7}$$

$$\bar{Y}_{k+1} = \sum_{i=1}^{10} W^m_i Y^i_{k+1} \tag{8}$$

$$P_{Y,k+1} = F_{k+1} * \sum_{i=1}^{10} W^c_i (Y^i_{k+1} - \bar{Y}_{k+1})(Y^i_{k+1} - \bar{Y}_{k+1})^T \tag{9}$$

$$P_{XY,k+1} = \sum_{i=1}^{10} W^c_i (X^i_{k+1} - \bar{X}_{k+1})(Y^i_{k+1} - \bar{Y}_{k+1})^T \tag{10}$$

$$K_{k+1} = P_{XY,k+1} P^{-1}_{Y,k+1} \tag{11}$$

$$X_{k+1} = \bar{X} + K_{k+1}(y_{k+1} - \bar{Y}_{k+1}) \tag{12}$$

$$P_{k+1} = P_{X,k+1} - K_{k+1} P_{Y,k+1} K^{-1}_{k+1} \tag{13}$$

State variable is calculated by Equation (4). C_{k+1} is the value of fuel flow. X_{k+1} and X_k are the estimated variables (state variable) at the $K+1$th and Kth sampling time. $\bar{X}_{k+1}$ is the mean vector of estimated variables and $P_{X,k+1}$ is the covariance matrix. Y^i_{k+1} and $\bar{Y}_{k+1}$ are the estimated values and the mean value of measurements respectively. P_{XY} is the covariance matrix of X and Y. P_Y and P_X are the variances of Y and X, respectively. y_{k+1} is the measurement vector obtained by sensors. K_{k+1} is the Kalman gain. v_k and w_k are the state transmission noise and measurement transmission noise. $v_k \in N(0, 0.002^2)$, $w_k \in N(0, 0.002^2)$.

The need to pay attention is that F_{k+1} is fading factor vector. There is $F_{k+1} = diag(f_1, f_2, \ldots, f_8)$. f_i denotes the fading factor. By regulating the proportion of fading factors, the residual of measurements at the current and last sampling time can be kept orthogonal. The emphasis of STF is to calculate the value of F_{k+1}. For F_{k+1}, set each fading factor as:

$$f_i = a * p_i, \quad i = 1, 2, 3, \ldots, 8 \tag{14}$$

a is the common parameter and p_i is the ratio parameter. The ratio value of fading factors can be determined by experience, there is:

$$f_1 : f_2 : f_3 : \ldots : f_8 = p_1 : p_2 : p_3 : \ldots : p_8 \tag{15}$$

Obviously, f_i can be calculated if a is obtained. Equation (3) can be transformed as:

$$E(\varepsilon_k \varepsilon_{k+i}) \approx P_{XY,k} - K_k C_k = 0 \tag{16}$$

C_k is the residual covariance matrix of measurements. The condition to satisfies Equation (16) is that:
$$I - P_{Y,k}^{-1} C_k = 0 \tag{17}$$

There is:
$$\begin{aligned} C_k &= F_k \sum_{i=1}^{10} W_i^c (Y_k^i - \bar{Y}_k)(Y_k^i - \bar{Y}_k)^T + Q_k \\ &= a * diag(p_1, p_2, \ldots, p_8) * \sum_{i=0}^{16} W_i^c (Y_k^i - \bar{Y}_k)(Y_k^i - \bar{Y}_k)^T + Q_k \end{aligned} \tag{18}$$

Q_k is the noise statistical matrix of measurements. Compute the trace of Equation (18), and the expression of a can be obtained.

$$a = \frac{tr(C_k - Q_k)}{tr(diag(p_1, p_2, \ldots, p_8) \sum_{i=0}^{16} (Y_k^i - \bar{Y}_k)(Y_k^i - \bar{Y}_k)^T)} \tag{19}$$

$$C_k = \begin{cases} \varepsilon_0 \varepsilon_0^T & k = 0 \\ \frac{\sigma C_{k-1} + \varepsilon_k \varepsilon_k^T}{1+\sigma} & k \geq 1 \end{cases} \tag{20}$$

σ named scale factor is used to adjust the ratio of residual covariance matrix at the $k-1$th sampling time. The greater the value of σ, the greater the proportion of C_{k-1}. Otherwise, the greater the proportion of $\varepsilon_k \varepsilon_k^T$. Usually, the value of σ is determined by experience, and there is a drawback that unreasonable value of σ may lead to the distortion of C_k. Paper proposes a method to obtain σ. The steps are as follows:

(1) Construct a variance vector ψ_{k-1} which consist of the diagonal elements of C_{k-1}. Furthermore, obtain the residual vector ζ_k which consist of diagonal elements of $\varepsilon_k \varepsilon_k^T$.
(2) Similarity calculation between ζ_k and ψ_{k-1}.

$$s_k = \frac{\langle \zeta_k, \psi_{k-1} \rangle}{(|\zeta_k| * |\psi_{k-1}|)} \tag{21}$$

s_k is the cosine value between ζ_k and ψ_{k-1}, and $s_k \in [-1, 1]$. Considering the Equation (20), coefficients of C_{k-1} and $\varepsilon_k \varepsilon_k^T$ are $\frac{\sigma}{1+\sigma}$ and $\frac{1}{1+\sigma}$ respectively. Obviously, the sum of $\frac{\sigma}{1+\sigma}$ and $\frac{1}{1+\sigma}$ is 1. Set the angle between ψ_{k-1} and ζ_k as θ, there is:

$$\cos \theta = s_k$$
$$\sin^2 \theta = 1 - s_k^2$$

Replace the original coefficients of C_{k-1} and $\varepsilon_k \varepsilon_k^T$ with s_k and $1 - s_k$. Equation (20) can be transformed as:

$$C_k = \begin{cases} \varepsilon_0 \varepsilon_0^T, & k = 0 \\ s_k^2 C_{k-1} + (1 - s_k^2) * \varepsilon_k \varepsilon_k^T, & k > 0, \quad s_k^2 < 1/2 \\ (1 - s_k^2) C_{k-1} + s_k^2 * \varepsilon_k \varepsilon_k^T, & k > 0, \quad s_k^2 > 1/2 \end{cases} \tag{22}$$

According to the working principle of gas turbine and taking into account that the proportion of current (the kth sampling time) information should be greater than that of previous sampling time. Equation (22) ensures that the coefficient of $\varepsilon_k \varepsilon_k^T$ is greater than that of C_{k-1}. Estimate T_{t25} by above method.

Compare with Figures 4 and 5 accurately reflects the sudden change of measurements. It shows the validity of STF proposed by paper, which compensates the model error and enhances the estimation accuracy.

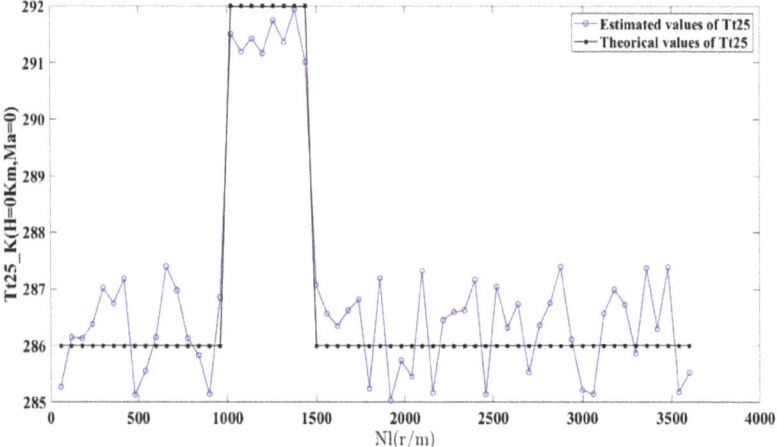

Figure 5. Estimations of T_{t25} by the strong tracking filter (STF).

2.4. Health Parameters Estimation

Paper adopt particle filter to estimate the health parameters. The measurements filtered by the STF are used to determine the posterior probability. Weight degradation that may lead to the accuracy decrease of estimations is a commonly problem exists in the process of particle filter Particle resampling is a traditional way to solve this problem. By increasing the number of larger-weight particles and make all particles have the same weight, the weight degradation has been effectively solved. But the above-mentioned method will lead to another problem, that is, the loss of particle diversity. In order to coordinate these two issues, paper proposes a weight optimization method in the PF. The core idea of this method is to adjust the posterior probability density function of health parameters. By properly increasing the weight of small weight particles and reducing the weight of large weight particles, the diversity of particles can be kept, and the high accuracy of probability density function can be ensured. The steps of health parameters estimation algorithm based on weight optimization PF are as follows [33,34]:

(1) k = 0, particles initialization.

k denotes the sampling time. Set the number of particles is 100. Each particle represents the value of health parameter. Generate particles $\{x_i^0\}_{i=1}^{100}$ according to the importance probability density function $q(x)$. x consists of health parameters which include the efficiency coefficients of LPC, HPC, HPT, LPT, and the flow coefficients of LPC, HPC, HPT, and LPT. $q(x)$ is the uniform distribution function.

(2) k = 1, 2, 3, ... weight update.

Predict the health parameters based on the prior probability distribution function:

$$x_i^k \propto p(x_i^k | x_i^{k-1})$$

Above calculation can be realized based on the component-level model of engine. Weight update:

$$\omega_i^k \propto \omega_i^{k-1} p(y_i^k | x_i^k)$$

y consists of different measurements filtered by STF. There are total temperatures at the outlet of LPC, HPC, HPT, LPT, and total pressures at the outlet of LPC, HPC, HPT, and LPT.

Weight optimization. Calculate the mean of weights, there is:

$$\bar{\omega}_k = \frac{\sum_{i=1}^{100} \omega_i^k}{100}$$

Adjust the weight of each particle:

$$\omega_i^k = \omega_i^k - (\omega_i^k - \bar{\omega}_k)R$$

R is regulator and $R \in (0,1)$. The function of R is to regulate the weight of particles. Normalize the weights:

$$\omega_i^k = \frac{\omega_i^k}{\sum_{i=1}^{100} \omega_i^k}$$

(3) Particles resampling:

$$x_i^k \sim \{x_i^k, \omega_i^k\}, i = 1, 2, 3, \ldots, 100 \qquad \omega_i^k = \frac{1}{100}$$

(4) Optimize the health parameters:

$$\bar{x}_k = \sum_{i=1}^{100} x_i^k * \omega_i^k$$

In order to verify the validity of proposed method, a simulation to detect the failure occurrence of two-spool turbojet is conducted. By suddenly changing the value of health parameters, failure occurrence can be simulated [1,13]. According to the research of previous chapters, by estimate the values of health parameters, failure detect can be realized [13]. The steps to conduct the simulation are as follows:

1. Generate the measured parameters from a software named Gasturb13 (Gasturb 13 is a simulation software for gas turbine performance calculation with high accuracy). Add noise w to these measured parameters. $w \in N(0, 0.002^2)$, N is the normal probability density function.
2. Establish the Component-level Gas Path Model of turbojet. This model is the detailed expression of the Equations (1) and (2).
3. Build the module of strong tracking filter according the method introduced in Section 2.3. The measured parameters including noise are input into the module and output to the particle filter after being processed by the strong tracking filter.
4. Build the module of particle filer with weight optimization according to the method introduced in Section 2.4. This module is used to estimate the health parameters.
5. Input the measured parameters to the particle filter and estimate the health parameters. The way to simulate the failure are listed as follows:

$$F_{LPC} = F_{ini} - \Delta F$$
$$E_{LPC} = E_{ini} - \Delta E$$

F_{LPC} and E_{LPC} are the latest values of low-pressure compressor's flow coefficient and efficiency coefficient after the failure is simulated. F_{ini} and E_{ini} are the initial values of low-pressure compressor's flow coefficients and efficiency coefficient before failure are simulated. $\Delta F = \hat{F}(T - T_{failure})$. ΔF named the failure factor is variation volume of F_{ini}. $\hat{F}$ denotes the degradation value of flow coefficients during every sampling time if failure happen. The meaning of ΔE and $\hat{E}$ are similar to that of ΔF and $\hat{F}$. T and $T_{failure}$ represent current sampling time and failure occurrence time. The design working parameters of engine are as follows:

Efficiency of LPC: $E_{LPC} = 0.868$
Efficiency of HPC: $E_{HPC} = 0.878$
Efficiency of high-pressure rotator: $E_{HPR} = 0.98$
Efficiency of burning room: $E_{BR} = 0.98$
Cooling coefficient of HPT: $C_{HPT} = 0.03$
Cooling coefficient of LPT: $C_{LPT} = 0.01$
Design rotating speed of Low Pressure Rotator: $S_{LPR} = 10^4 \text{r/m}$
Design rotating speed of High Pressure Rotator: $S_{HPR} = 1.6 \times 10^4 \text{r/m}$
Total temperature at the outlet of burning room: $T_{t4} = 1600 \text{ K}$
Heat value of fuel: $FHV = 4.29 \times 10^4$

Pressure ratio of LPC: π_{LPC}
Pressure ratio of HPC: π_{HPC}
Efficiency of low-pressure rotator: $E_{LPR} = 0.98$
Air intake coefficient of cabin: $E_{AI} = 0.01$
Efficiency of HPT: $E_{HPT} = 0.89$
Efficiency of LPT: $E_{LPT} = 0.91$

Due to limitation of space, the estimation of low-pressure compressor's flow coefficient and efficiency coefficient are listed only. The estimation processes of other health parameters of high-pressure compressor, high-pressure turbine, low-pressure turbine are similar with that of low-pressure pressure. Assure that 100 measured parameters are collected. When the engine performance degrades slowly, the efficiency coefficient decreases by 0.6% compared with the initial value, and the flow coefficient decrease by 0.7%. To simulate the failure, at the 11th sampling time, set the flow coefficient and efficiency coefficient decreased by 0.3%.

Figure 6 shows the Estimated health parameters of low-pressure compressor based on the traditional unscented Kalman filter and particle filter. Set there are 100 sampling times. The initial theoretical values of the efficiency coefficient and the flow coefficient are 0.868 and 0.92, respectively. During each sampling time, the variation of efficiency coefficient and flow coefficient are 5.2×10^{-6} and 6.4×10^{-6}, as shown in Table 1. There are:

$$\frac{E_{ini} - E_{end}}{E_{ini}} = 0.6\%$$

$$\frac{F_{ini} - F_{end}}{F_{ini}} = 0.7\%$$

E_{end} and F_{end} are the values of efficiency coefficient and flow coefficient after the performance degrades slowly. It can be seen that the estimations are close to the theoretical values of health parameters basically. However, the estimations curve fluctuates greatly, and the accuracy degree of estimations is not high.

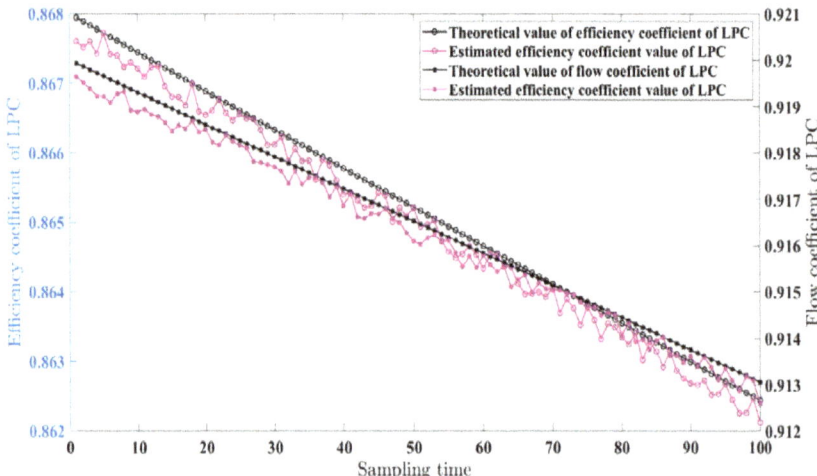

Figure 6. Estimated health parameters of low-pressure compressor by the traditional unscented Kalman filter and particle filter.

Table 1. Error analysis under the condition of slow degradation of performance based on traditional method.

Estimated Parameters	Maximum Error	Mean Value of Error	Variance
Efficiency coefficient	0.162%	0.118%	5.2×10^{-6}
Flow coefficient	0.158%	0.112%	6.4×10^{-6}

Figure 7 shows the estimated values of LPC's flow coefficient and efficiency coefficients when the working state of two-spool turbojet is steady based on the proposed hybrid filter. Under this working condition, the variations range of health parameters (flow coefficient and efficiency coefficient of LPC) are small and slow degradation of performance is happened due to the poor working circumstance of turbojet. The purple curve consists of the estimated values and black curve consists of the theoretical values. Obviously, the method proposed in this paper can accurately characterize the change trend of health parameters. Furthermore, the accuracy of the estimations is also consistent with the theoretical values of health parameters. The estimations variance of efficiency coefficient and flow coefficient are 2.59×10^{-6} and 4.05×10^{-6} respectively, as shown in Table 2.

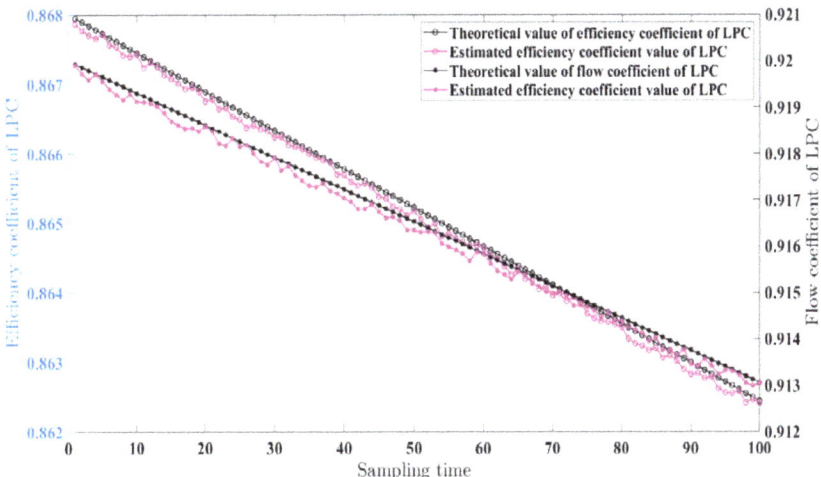

Figure 7. Estimated health parameters of low-pressure compressor (LPC) by the proposed hybrid filter.

Table 2. Error analysis under the condition of slow degradation of performance based on proposed method.

Estimated Parameters	Maximum Error	Mean Value of Error	Variance
Efficiency coefficient	0.094%	0.076%	2.59×10^{-6}
Flow coefficient	0.089%	0.073%	4.05×10^{-6}

Figure 8 shows the estimated values of low-pressure compressor's flow coefficient and efficiency coefficients based on the method proposed by this paper when the working state of two-spool turbojet breaks down. At the 11th sampling time, set the efficiency coefficient and flow coefficient have a sudden change of 0.3%. There are:

$$\frac{E_{ini} - E_{end}}{E_{ini}} = 0.3\%$$

$$\frac{F_{ini} - F_{end}}{F_{ini}} = 0.3\%$$

Paper simulate the failure by changing the health parameters at the tenth sampling time. Due to the occurrence of failure, the measured parameters have a sudden change. By the application of strong tracking filter, the measured parameter can be estimated with high accuracy. According to the introduction of Section 2.1, the output of strong tracking filter is input to the particle filter. Due to the high accuracy tracking ability of the STF to the state mutation and the weight optimization of particle filter, the health parameters are estimated with high accuracy by the particle filter, as shown in Figure 8. The value of efficiency coefficient reduced from 0.8675 to 0.8649, and the value of flow coefficient reduced from 0.9196 to 0.9168. From Figure 8, the mutations in health parameters are accurately reflected on the curve and the occurrence of failure can be detected.

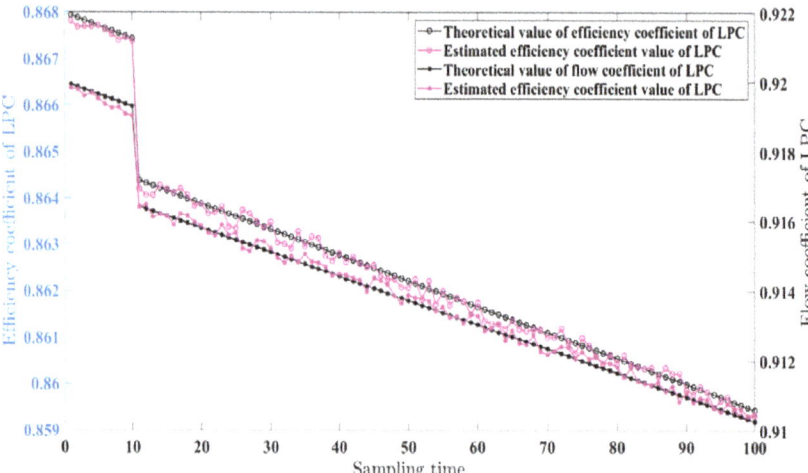

Figure 8. Estimated health parameters of low-pressure compressor in sudden change of working state.

3. Conclusions

In this paper, a developed method based on the unscented Kalman filter and particle filter is proposed. To eliminate the noises contained in measurements which obtained by sensors, UKF is adopted to dispose these noises. Furthermore, in order to enhance the estimation accuracy of measurements when the working state of turbojet changes suddenly, a strong tracking filter is constructed by adjust the variance ratio of measurements at different sampling times based on the UKF. The output (The measurements filtered by STF) of STF is used to determine the weight of each particle. According to the simulations conducted by paper, there are three conclusions can be made.

1. The strong tracking filter is used to eliminate the noise contained in measurements and the accuracy of measured parameters is enhanced when the turbojet performance changes slowly. Besides, the estimation accuracy remains high when the working state of turbojet changes abruptly by adjust the variance ratio of measurements.
2. An optimization method for strong tracking filter is proposed. By calculating the similarity between covariance vectors at different sampling times of measured parameters, the value of scale factor can be obtained. This calculation method replaces the traditional way of relying on experience.
3. In particle filter, to ensure the diversity of particles, paper proposes a weight optimization method to adjust the weights of different particles. The regulation equation is derived according to the regulator R and the mean of all weights. By above method, the high accuracy of probability density function can be ensured.

Author Contributions: Conceptualization, L.Z.; Formal analysis, L.Z.; Methodology, L.Z.; Writing-original draft, L.Z.; Writing-review & editing, L.Z.; Funding acquisition, S.D.; Supervision, W.L.

Funding: This research was funded by the National Natural Science Foundation of China, grant number 51775072. Natural Science Foundation Project of CQ, grant number cstc2017jcyjAX0279.

Conflicts of Interest: The authors declare no conflict of interest.

References

1. Chen., Y. Gas Path Fault Diagnosis for Turbojet Engine Based on Nonlinear Model. Ph.D. Thesis, Nanjing University of Aeronautics and Astronautics, Nanjing, China, 2014.
2. Nada, T. Performance Characterization of Different Configuration of Gas Turbine Engines. *Propuls. Power Res.* **2014**, *3*, 121–132. [CrossRef]
3. Ogaji, S.O.T.; Sampath, S.; Singh, R.; Probert, S.D. Parameter Selection for Diagnosing a Gas-Turbine's Performance-Deterioration. *Appl. Energy* **2002**, *73*, 25–46. [CrossRef]
4. Chowdhary, G.; Jategaonkar, R. Aerodynamic Parameter Estimation from Flight Data Applying Extended and Unscented Kalman Filter. In Proceedings of the AIAA Atmospheric Flight Mechanics Conference and Exhibit, Keystone, CO, USA, 21–24 August 2006.
5. Miller, I.; Gencer, G.; Francis, M. A General Model for Estimating Emissions from Integrated Power Generation and Energy Storage Case Study: Integration of Solar Photovoltaic Power and Wind Power with Batteries. *Processes* **2018**, *6*, 267. [CrossRef]
6. Li, J.; Wang, K.L.; Lian, M.L.; Li, Z.; Du, T.Z. Process Simulation of the Separation of Aqueous Acetonitrile Solution by Pressure Swing Distillation. *Processes* **2019**, *7*, 409. [CrossRef]
7. Simpson, T.W.; Mauery, T.M.; Korte, J.; Mistree, F. Kriging models for global approximation in simulation-based multidisciplinary design optimization. *AIAA J.* **2005**, *43*, 853–863.
8. Sanchez, S.; Escobet, T.; Puig, V.; Odgaard, P.F. Fault Diagnosis of an Advanced Wind Turbine Benchmark Using Interval-Based ARRs and Observers. *IEEE Trans. Ind. Electron.* **2015**, *62*, 3783–3792. [CrossRef]
9. Gao, Z.; Cecati, C.; Ding, S.X. A survey of fault diagnosis and fault-tolerant techniques—Part I: Fault diagnosis with model-based and signal-based approaches. *IEEE Trans. Ind. Electron.* **2015**, *62*, 3757–3767. [CrossRef]
10. Gao, Z.; Cecati, C.; Ding, S.X. A Survey of Fault Diagnosis and Fault-Tolerant Techniques—Part II: Fault diagnosis with knowledge-based and hybrid/active approaches. *IEEE Trans. Ind. Electron.* **2015**, *62*, 3768–3774. [CrossRef]
11. Borguet, S.; Leonard, O. Coupling Principal Component Analysis and Kalman Filtering Algorithms for On-Line Aircraft Engine Diagnostics. *Control Eng. Pract.* **2009**, *17*, 494–502. [CrossRef]
12. Lu, F.; Ju, H.F.; Huang, J.Q. An improved extended Kalman filter with inequality constraints for aero engine health monitoring. *Aerosp. Sci. Technol.* **2016**, *58*, 36–47. [CrossRef]
13. Vanini, Z.S.; Khorasani, K.; Meskin, N. Fault Detection and isolation of a dual spool gas turbine using dynamic neural networks and multiple model approach. *Inf. Sci.* **2014**, *259*, 234–251. [CrossRef]
14. Chen, L.B.; Song, L.Q.; Chen, G. Study on Fusion Diagnosis Techniques of Wear Faults in Synthesized Monitoring of Aero-Engine. *J. Aerosp. Power* **2009**, *24*, 169–175.
15. Huang, Q.; Zhang, G.G.; Zhang, T.; Wang, J. A Kind of Approach for Aero Engine Gas Path Fault Diagnosis. In Proceedings of the IEEE International Conference on Prognostics and Health Management (ICPHM), Dallas, TX, USA, 1 June 2017.
16. Verma, R.; Roy, N.; Ganguli, R. Gas turbine diagnostics using a soft computing approach. *Appl. Math. Comput.* **2006**, *172*, 1342–1363. [CrossRef]
17. Bachir, A.; Hafaifa, A.; Guemana, M.; Hadroug, N. Application of Principal Component Analysis Approach in Gas Turbine Defect Diagnosis. In Proceedings of the International Conference on Applied Smart Systems, Medea, Algeria, 1 November 2018.
18. Yang, L.; Ding, S.T.; Wang, Z.Y.; Li, G. Efficient Probabilistic Risk Assessment for Aeroengine Turbine Disks Using Probability Density Evolution. *AIAA J.* **2017**, *55*, 2755–2761. [CrossRef]
19. Zeng, L.; Long, W.; Li, Y.Y. A Novel Method for Gas Turbine Condition Monitoring Based on KPCA and Analysis of Statistics Tand SPE. *Processes* **2019**, *7*, 124. [CrossRef]

20. Wang, L.; Liu, Z.W.; Miao, Q.; Zhang, X. Time-Frequency analysis based on ensemble local mean decomposition and fast kurtogram for rotating machinery fault diagnosis. *Mech. Syst. Signal Process.* **2018**, *103*, 60–75. [CrossRef]
21. Zhang, Y.; Randall, R.B. Rolling element bearing fault diagnosis based on the combination of genetic algorithm and fast kurtogram. *Mech. Syst. Signal Process.* **2009**, *23*, 1509–1517. [CrossRef]
22. Pham, H.T.; Yang, B.S. Estimation and forecasting of machine health condition using ARMA/GARCH model. *Mech. Syst. Signal Process.* **2010**, *24*, 546–558. [CrossRef]
23. Hu, Y.; Yang, Y.Y.; Zhang, S.Y.; Sun, Z.S.; Zhu, J.T. Turbofan Engine Gas Path Performance Monitoring based on Improved Spherical Simplex Square Root Unscented Kalman Filter. *J. Aerosp. Power* **2014**, *29*, 441–449.
24. Hu, Y.; Zhang, S.Y.; Yang, Y.C.; Zhu, J.T.; Yang, Z.W. Fault Diagnosis of Gas Path Components of Turbofan Engine based on Spherical Square Root Unscented Kalman Filter Algorithm. *J. Aerosp. Power* **2014**, *29*, 689–694.
25. Zhang, P. Aeroengine Fault Diagnosis Based on Kalman Filter. Ph.D. Thesis, Nanjing University of Aeronautics and Astronautics, Nanjing, China, 2014.
26. Kurzke, J.; Riegler, C.; Bauer, M. Some Aspects of Modeling Compressor Behavior in Gas Turbine Performance Simulation. *ASME J. Turbo Mach.* **2001**, *123*, 373–378.
27. Li, R.X.; Prasad, V.; Huang, B. Gaussian Mixture Model-Based Ensemble Kalman Filtering for State and Parameter Estimation for a PMMA Process. *Processes* **2016**, *4*, 9. [CrossRef]
28. Simon, D.; Simon, D.L. Constraint Kalman Filtering Via Density Function Truncation for Turbofan Engine Health Estimation. *Int. J. Syst. Sci.* **2010**, *41*, 159–171. [CrossRef]
29. Simon, D. Kalman Filtering with State Constraints: A Survey of Linear and Non-linear Algorithms. *IET Control Theory Appl.* **2010**, *4*, 1303–1318. [CrossRef]
30. Soken, H.E.; Hajiyev, C. Pico Satellite Attitude Estimation via Robust Unscented Kalman Filter in the Presence of Measurement Faults. *ISA Trans.* **2010**, *49*, 249–256. [CrossRef] [PubMed]
31. Dyke, M.C.; Schwartz, J.L.; Hall, C.D. Unscented Kalman Filtering for Spacecraft Attitude State and Parameter Estimation. In Proceeding of the AAD/AIAA Space Flight Mechanics Conference, Maui, HI, USA, 8–12 February 2004.
32. Liu, H.F.; Yao, Y.; Lu, D.; Ma, J. Study for Outliers based on Kalman Filtering. *Electr. Mach. Control* **2003**, *7*, 40–42.
33. Cheng, F.Z.; Qu, L.Y.; Qiao, W.; Hao, L.W. Enhanced Particle Filtering for Bearing Remaining Useful Life Prediction of Wind Turbine Drivetrain Gearboxes. *IEEE Trans. Ind. Electron.* **2019**, *66*, 4738–4748. [CrossRef]
34. Wang, X.X.; Pan, Q.; Huang, H.; Gao, A. Overview of Deterministic Sampling Filtering Algorithm for Nonlinear System. *Control Decis.* **2012**, *27*, 801–810.
35. Zeng, L.; Long, W.; Li, Y.Y. Performance Degradation Diagnosis of Gas Turbine Based on Improved FUKF. *J. Southwest Jiaotong Univ.* **2018**, *53*, 873–878.

 © 2019 by the authors. Licensee MDPI, Basel, Switzerland. This article is an open access article distributed under the terms and conditions of the Creative Commons Attribution (CC BY) license (http://creativecommons.org/licenses/by/4.0/).

Article

Evolution of High-Viscosity Gas–Liquid Flows as Viewed Through a Detrended Fluctuation Characterization

J. Hernández, D. F. Galaviz, L. Torres *, A. Palacio-Pérez and A. Rodríguez-Valdés and J. E. V. Guzmán

Instituto de Ingeniería, Universidad Nacional Autónoma de México, Mexico City 04510, Mexico; JHernandezGa@iingen.unam.mx (J.H.); DGalavizL@iingen.unam.mx (D.F.G.); APalacioP@iingen.unam.mx (A.P.-P.); ARodriguezV@iingen.unam.mx (A.R.-V.); JGuzmanV@iingen.unam.mx (J.E.V.G.)
* Correspondence: ftorreso@iingen.unam.mx

Received: 13 September 2019; Accepted: 31 October 2019; Published: 6 November 2019

Abstract: We characterize the long-term development of high-viscosity gas–liquid intermittent flows by means of a detrended fluctuation analysis (DFA). To this end, the pressures measured at different locations along an ad hoc experimental flow line are compared. We then analyze the relevant time-series to determine the evolution of the various kinds of intermittent flow patterns associated with the mixtures under consideration. Although no pattern transitions are observed in the presence of high-viscosity mixtures, we show that the dynamical attributes of each kind of intermittence evolves from one point to another within the transport system. The analysis indicates that the loss of a long-range correlation between the pressure responses are due to the discharge processes.

Keywords: high-viscosity; two-phase flow; detrended fluctuation analysis; heavy oils

1. Introduction

Today, the transportation of high-viscosity gas–liquid mixtures constitutes an important flow assurance problem in several oil producing countries. As an example, the production of petroleum in Mexico has declined over the last two decades, while the heavy and extra-heavy crude oil varieties account for nearly 60% of the total reserves. Although diverse technologies are available to increase the productivity (e.g., [1]), their application might not necessarily be possible, or may require to be tailored to suit the specific needs of each particular case. Therefore, understanding how certain flow characteristics evolve as the associated mixtures progress inside the pipelines is crucial to the development of adequate flow enhancement techniques.

Other problems and technical difficulties may arise in the present context. For instance, Matsubara and Naito [2] previously noted that the traditional flow maps were substantially modified with high-viscosity gas–liquid flows. The boundary lines separating intermittent flows from other patterns were drastically changed, such that at low local flow rates (or equivalently, at low local superficial velocities) the pattern belonged exclusively to the intermittent class. Similar observations were made more recently by Hernandez et al. [3] in a longer experimental flow loop.

In addition, after investigating the characteristics of the flow patterns produced with relatively high-viscosity oils and air, Foletti et al. [4] concluded that the agreement with current predictive models was rather poor. This pointed out to the necessity of conducting further work along these lines. As a matter of fact, with lower viscosity oils there was room for improvement, as the results reported by Khaledi et al. [5] clearly indicated.

Experiments carried out with oils (whose viscosities were in the range ≤10 Pa·s) also showed substantial differences between the measured data and the predicted values produced by the existing correlations. In particular, Zhao et al. [6] identified four flow patterns, three of which were intermittent. The need for improved mechanistic models that could handle such viscosities was stressed. Similar patterns were caused at still higher viscosities (~6 Pa·s) in horizontal pipes ([7]).

One relevant quantity affecting the properties of the intermittence is the local liquid holdup. Al-Safran et al. [8] reported a new empirical correlation that is valid in the range of 0.1–0.58 Pa·s. In a preceding article, Farsetti et al. [9] studied the frequencies produced by intermittent oil-air flows with moderate-to-high-viscosities, in addition to other flow parameters. According to the reported results, the frequency appeared to depend on the liquid holdup. Interestingly, only the estimates obtained from the ad hoc correlation proposed by Gokcal et al. [10] compared favourably with the data. The same kind of conclusions were drawn by Okezue [11], who studied high-viscosity oil–gas mixtures (from 1.1 to 4 Pa·s) over a wide range of flow conditions. More recently, Baba et al. [12] conducted additional experiments to further characterize the slug frequencies. It is noted that these flow traits have a definitive influence in the design of field operations, as well as of fluid handling facilities and equipment (e.g., [12]). A statistical analysis of the measured data was previously applied to this kind of flows by Losi et al. [13]. Basic flow features such as slug frequencies, lengths and pressure drops, were discussed in terms of the gas superficial velocities and other parameters. Furthermore, the authors derived improved correlations based on the probability density functions thus obtained.

In view of the forgoing arguments, our present aim is to highlight some of the key traits of high-viscosity gas–liquid flows by means of the well established detrended fluctuation analysis (or DFA). This alternative viewpoint has not been fully exploited in this context, and may provide a robust description of the properties of the intermittent patterns emerging naturally as the flow evolves.

Even though different methods were applied in the past to investigate low viscosity two-phase flows, only the work by Zahi et al. [14] concerns the use of DFA (to the best of our knowledge). As a matter of fact, these authors compared various fractal methods with the DFA. Within the superficial velocities of the low viscosity phases involved, it was found that the numerical value of the scaling exponent was lowest for the bubbly flow pattern, while the highest value was attained with the intermittent slug flow pattern. The former result was attributed to the random dynamics of the flow, and the latter to the periodicity of the slugging. More importantly, the DFA was shown to perform better in comparison with other fractal techniques with varying noise levels. Therefore, Zahi et al. [14] concluded that the DFA based approach can effectively underline the flow pattern characteristics. The ability of the method to provide information of the transitions was also stressed.

The possibilities and implications of the detrended fluctuation analysis were clearly illustrated by Peng et al. [15]. Because the DFA is an approach that relies on a simple scalar parameter (the scaling exponent) to represent the correlation properties of a given signal, it may be applied to extract useful dynamical information from the pressure time-series obtained at various locations. Moreover, the DFA permits the detection of long-range correlations embedded in seemingly non-stationary time-series, while avoiding spurious artifacts related with stationary effects. This technique has been applied to such diverse fields of interest as DNA, heart rate dynamics, neuron spiking, human gait, long-time weather records, cloud structure, economical time-series, solid state physics, and even reservoir characterization (e.g., [16–18]).

2. Experiments

2.1. Test Apparatus

The experiments were conducted in the flow loop shown in Figure 1. This setup was designed to promote the long-term evolution of the mixture, since the pipe's length-to-diameter ratio was of order $L/d \sim 10^2$. The test section was constructed with (schedule 80) steel tubes with an internal diameter of 0.0762 m (3 in) and overall length of 54 m. It should be mentioned that a single 3 m section, made of transparent PVC tube, was fitted 3 m upstream from the discharge point for visualization purposes.

Glycerin was selected as the Newtonian, high-viscosity, liquid phase. Its measured dynamic viscosity was $\eta = 1.1$ Pa·s (1100 cP) at a room temperature of 25 °C. The liquid phase was supplied to the test section by a progressive cavity pump (Seepex Mod. BN35-24). Regardless of the pressure buildup at its discharge plane, the pump was capable of delivering constant mass flow rates in the interval [0.0 kg/s, 6.1 kg/s]. The outlet of the test section was connected to a separator tank with an internal capacity of 1.5 m^3. The pressure in its interior was kept constant.

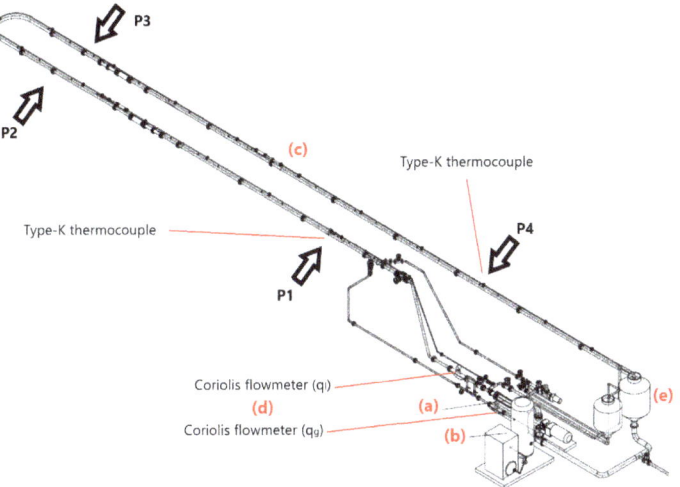

Figure 1. Test apparatus. The main elements of the flow loop are: (a) the pumping subsystem, (b) the pressurized gas subsystem, (c) the flow loop, (d) the inlet flow measuring equipment, and (e) the separator and storage tanks. The flow loop consisted of a 54 m long pipe, along which the pressure transducers, thermocouples, and tomographic systems, were installed.

On the other hand, a twin-scroll Kaeser Aircenter SK.2 compressor supplied a constant mass of dry air (at room temperature) for the pressures within the interval [0.0 Pa, 1.6×10^6 Pa]. The mass flow rate was finely tuned with a regulator and globe valves to produce a choked condition at the injection port. The mixture was produced at the 3-way connection tube shown in Figure 1.

Both mass flow rates were measured at the inlet with Endress-Hauser Coriolis meters. The Promass 83F80 DN80 3" model used with the glycerine had an accuracy of 0.1% across the entire measuring range. Similarly, the Promass 83F50 DN50 2" model used with the air had an accuracy of 0.05% across the entire measuring range. Also, all the pressures were measured with an array of conventional MEAS U5300 transducers with a resolution of 0.1% across their measuring ranges. The two measuring intervals selected for these instruments were [0.0 Pa, 1.03×10^5 Pa], and [0.0 Pa, 3.45×10^5 Pa]. Table 1 summarizes the reference values of the fluid properties.

Table 1. Fluid properties at standard conditions (i.e., @ $P = 1.013 \times 10^5$ Pa, $T = 25$ °C).

Fluid	Viscosity (Pa·s)	Density (kg/m^3)	Interfacial Tension (N/m)
Air	1.8×10^{-5}	1.2	-
Glycerine	1.1	1.2×10^3	6.3×10^{-2}

Impedance computed tomography scans (or ICT scans) provided snapshots of the internal flow configuration. The tomograms were produced with the International Tomography Systems (ITS) ERT and ECT instruments. Each tomogram is constituted by snapshots taken at a sampling rate of 30 frames per second in two separate measuring planes. Then, the computerized system reconstructs the flow structure from the collected images. Although the ECT-ERT systems also yields an accurate measurement of the holdup within the measuring planes, the corresponding times series are similar to those of the pressure. These measurements will be discussed in a future paper.

2.2. Methodology

The experiments were produced in accordance with the inlet mass flow rates indicated in Table 2. In total, 15 different combination pairs (q_g, q_l) were considered, and 10 experiments were conducted for each one of them.

Table 2. Experimental matrix. Mass flow rates of the liquid (q_l) and the gas (q_g) phases injected into the test section.

q_g (kg/s)	q_l (kg/s)
	1.3
0.005	2.5
0.01	3.7
0.015	4.9
	6.1

Every experiment conformed to the following procedure: first, a steady state, single phase flow of glycerin was produced at a given flow rate. Owing to the elevated viscosity, the upper bound on the Reynolds number ($Re_l = (\rho u d/\eta)_l \sim 10^2$) implied that the Poiseuille velocity profile developed fully in just under 20 pipe diameters. The steady state condition was then verified by making sure that $\Delta p/\Delta t \approx 0$ in all pressure transducers. Once this condition was reached, a specific mass flow rate of air was injected (see Table 2). Because $Re_l/Re_g \sim 10^{-2}$–10^{-4}, the head loss was essentially determined by the liquid phase; accordingly, the two-phase (or mixture) Reynolds number was simply $Re_{TP} \sim Re_l$. Even though the two-phase flow developed rather quickly, the system was nevertheless allowed to settle for at least 2 minutes before taking any kind of measurement. At this point the flow properties were measured and collected by the data acquisition system.

In order to remove any external influence that might affect the measurements, only the differential pressures were used to determine the pressure gradients of interest. The measuring ports were located at 0, 18, 22, and 43 m, downstream from the inlet. These ports are labeled with P1, P2, P3 and P4, respectively, in Figure 1.

The experiments were conducted at an average ambient temperature of 25 °C, as measured inside the laboratory. In order to verify that the temperature of the fluids in the flow loop did not vary as the experiments were conducted, two Type K thermocouples were installed at the inlet and outlet sections. It is worth mentioning that no temperature fluctuations were observed within their resolution limit (0.5 °C across the measuring range). Nevertheless, the mixture's viscosity was checked for each test by taking samples (from the previously indicated locations) and by measuring their viscosities with a Brookfield DV2T viscometer.

Following Kline-McClintok's method to determine the experimental uncertainties, it was readily verified that the pressure time-series were measured with an uncertainty of ±10% with a confidence level of 95% ([19,20]). It is worth mentioning that this uncertainty is well within the normally accepted range in most multiphase flow contexts (e.g., [21–23]). Accordingly, at least 3 experiments were conducted for each entry of the experimental matrix (i.e., for every (q_g, q_l) combination). Around 900 samples were gathered for the measured variables per test. The universal set contained nearly 750,000 data points comprising the simultaneous measurement of mass flow rates, pressures and temperatures. In general terms, it was observed that the statistical distribution of the data points conformed to Student's t-distribution with a 95% confidence level.

3. Results and Discussion

The set of electric impedance tomograms shown in Figure 2 portray the internal phase distribution for three different flow regimes (in correspondence with the low, intermediate and high liquid flow rates). Each image constitutes a snapshots of the actual structure by the two phases as they traverse the measuring device.

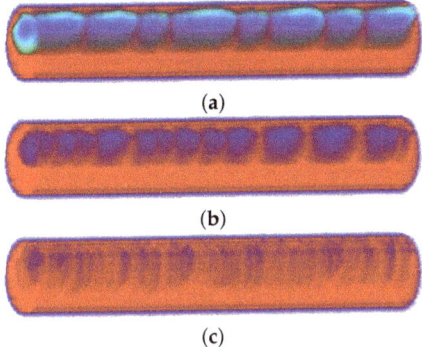

Figure 2. Intermittent flow patterns corresponding to the experimental matrix given in Table 2. From top to bottom, image (**a**) shows the flow structure for $q_g > q_l$, while (**b**) shows it for $q_g \sim q_l$, and (**c**) for $q_g < q_l$.

Respectively, image (a) illustrates the classical structure of an elongated bubble pattern resulting when the relatively low liquid flow rate, $q_l = 1.3$ kg/s, is introduced in the pipe. By increasing q_l to 3.7 kg/s the bubbles become shorter. This leads, in turn, to an augmented frequency. Similarly, by further rising the liquid flow rate to $q_l = 6.1$ kg/s the number of smaller bubbles and the corresponding frequency increase considerably.

3.1. Pressure Response

The present analysis focuses on the pressure signals alone, because the corresponding time-series encode information of every effect related with the flow. This approach proves to be convenient in field applications ([24,25]). For illustration purposes, Figure 3 shows a typical set of pressure measurements obtained at the U-turn section. The curves are deliberately plotted with different scales, in order to highlight their distinctive characteristics for all the flow rate combinations given in Table 2.

Clearly, to every combination (q_g, q_l) corresponds a particular fluctuation pattern. From left to right the fist image shows how, with low liquid flow rates, the resulting fluctuations tend to be relatively large with respect to the average pressure. The converse is true when more liquid is pumped into the system because the mean pressure increases substantially. In general terms, these non-periodic amplitudes become quite important as they account for significant over-pressures inside the pipe. Obvious external manifestations of such dynamical effects may include vibrations, as well as a very noticeable recoil of the pipe during the ejection of the liquid slugs.

In contrast to the preceding behavior, the fluctuation amplitudes seem to be about the same order of magnitude when the gas mass flow rates increase. However, the resulting frequencies appear to increase with q_g. One may also notice that the intermittence tends to become more regular. This is easily explained in view of the fact that: (a) many more bubbles are formed, and (b) these tend to be better distributed throughout the entire length of the pipe. This latter effect is purely conditioned by the elevated viscosity of the liquid phase.

Interestingly, as will be shown, these broad conclusions may not be directly extended to other regions of the flow system. Instead of analyzing the time-series in those regions, the data is subjected to the Detrended Fluctuation Analysis. In preparation for that, Figures 4–6 depict the pressure differences near the inlet, at the mid-section and near the outlet of the pipe. No information has been lost to this process, because the data has not been filtered. These ΔP are calculated at the locations specified previously, while the flow rate combinations are the same in all figures.

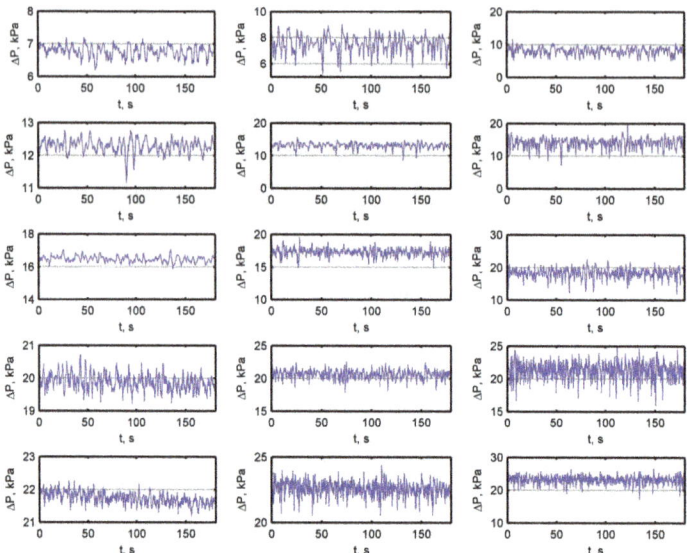

Figure 3. Characteristic fluctuations of the pressure drop across the measuring ports $P_2 - P_3$ (between 18 m and 22 m). These time-series give a clear indication of the dynamical complexity of the flow.

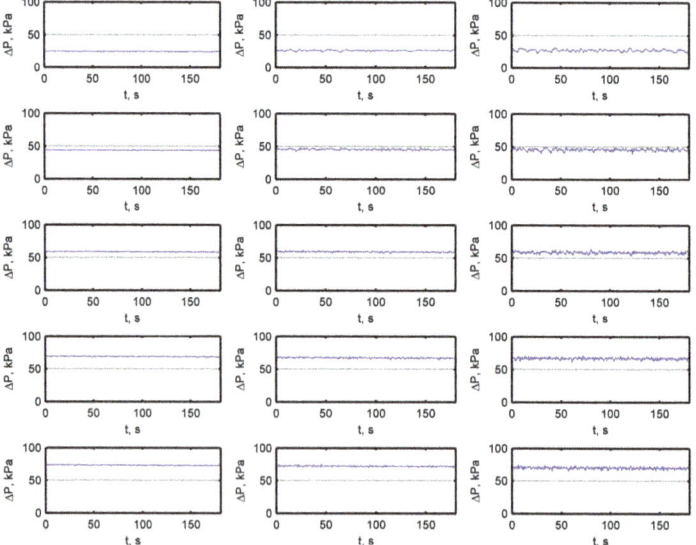

Figure 4. Time series for the pressure difference $P_1 - P_2$. These pressure differences are calculated as part of the FDA analysis and show the relative size of the fluctuations as a function of q_g and q_l.

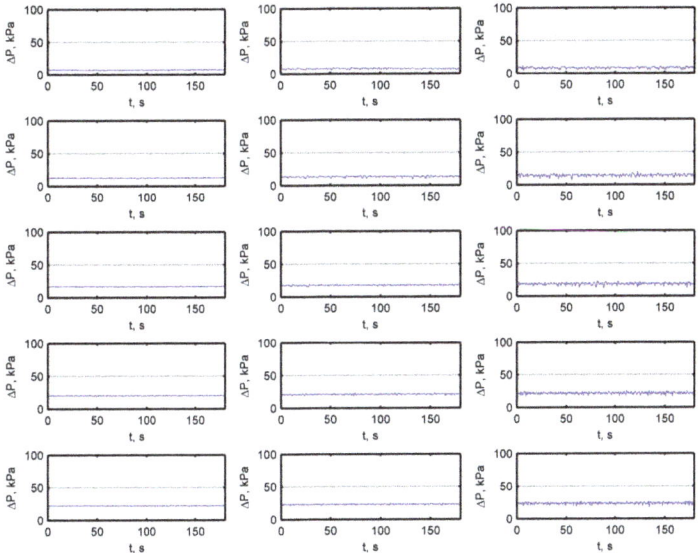

Figure 5. Time series for the pressure difference $P_2 - P_3$. These pressure differences are calculated as part of the FDA analysis and show the relative size of the fluctuations as a function of q_g and q_l.

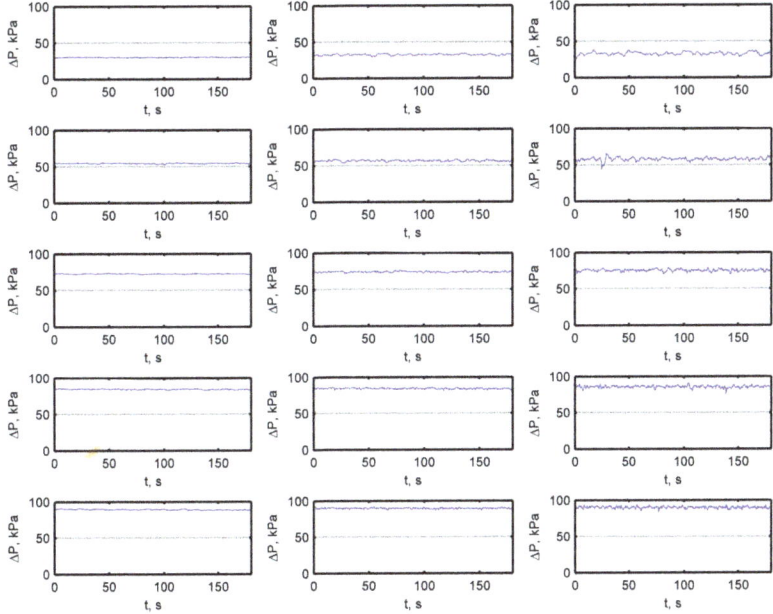

Figure 6. Time series for the pressure difference $P_3 - P_4$. These pressure differences are calculated as part of the FDA analysis and show the relative size of the fluctuations as a function of q_g and q_l.

3.2. Spectral Evolution of the Flow

In order to extract useful information from the pressure time-series, it proves convenient to first analyze the signals in the frequency domain. Since the primary interest of the analysis concerns the pressure fluctuations, the mean pressure is subtracted from the time-series. The set of images in Figures 7–9 show the spectrograms for the flow rates given in the experimental matrix (Table 2). Note that the three sets correspond to the three different locations along the pipe. Again, the columns represent q_l and the rows q_g (their values are given in each snapshot).

These plots provide information about the frequency bandwidths that are excited at specific times for different inlet flow rates (q_g, q_l). Moreover, since the three sets correspond to the three specified locations along the tube, they also convey important information on the evolution of the intermittence. The color scale indicates the amplitude of the excited modes in a given frequency neighbourhood. Thus, for example, in the upper left image of Figure 7 one may observe that modes with frequencies in the interval $(0.02\,\text{Hz} \leq f \leq 0.07\,\text{Hz})$ are excited from 60 s to 120 s. This relatively long time spans most of the experiment. Notice that a strong peak, centered at 0.05 Hz, is produced at approximately $t = 80$ s after the experiment is initiated. In contrast, for $(q_g, q_l) = (0.005\,\text{kg/s}, 3.7\,\text{kg/s})$ and $(q_g, q_l) = 0.005\,\text{kg/s}, 6.1\,\text{kg/s})$ only a very narrow, low-frequency band is excited towards the end of the experiment ($t \approx 140$ s).

One may conclude that, in a broad sense, the periodic fluctuations appear as high intensity (red) spots in the spectrograms. Conversely, noisy flow fields are represented by a rather diffuse, low intensity, distribution of interconnected bands with a color scheme much closer to the blue. Nevertheless, the abundance of spectral realizations induced by most (q_g, q_l) combinations is clearly exemplified by the amount of modes excited in distinct bandwidths, and at different times. For the sake of comparison consider the images corresponding to $(q_g, q_l) = (0.015\,\text{kg/s}, 3.7\,\text{kg/s})$ and $(q_g, q_l) = (0.015\,\text{kg/s}, 3.7\,\text{kg/s})$, i.e., two neighbouring flow rate combinations of the experimental matrix, which clearly illustrate the complexity of the fluctuation patterns.

Perhaps the most interesting feature portrayed by these images is the particular space-wise evolution of the fluctuation patterns. Consider in this case the spectrograms for $(q_g, q_l) = (0.005\,\text{kg/s}, 1.3\,\text{kg/s})$ in Figures 7–9. What initially constitutes a single, narrow frequency band extending in time (Figure 7), eventually unfolds into a couple of short frequency bands excited at the same time (seen as two red spots at $60\,\text{s} \leq t \leq 120\,\text{s}$ in Figure 8), plus a wide range of less energetic modes, in the mid-section of the tube. Finally, on the downstream section of the tube, two similar frequency bands are excited at two different times (i.e., The two red blobs centered at $f \approx 0.05$ Hz now appear at $t \sim 75$ s and $t \sim 125$ s in Figure 9). Notice that the less energetic modes have already decayed at this final stage. Interestingly, other flow combinations seem to be relatively immune to such changes and, therefore, tend to persist in time.

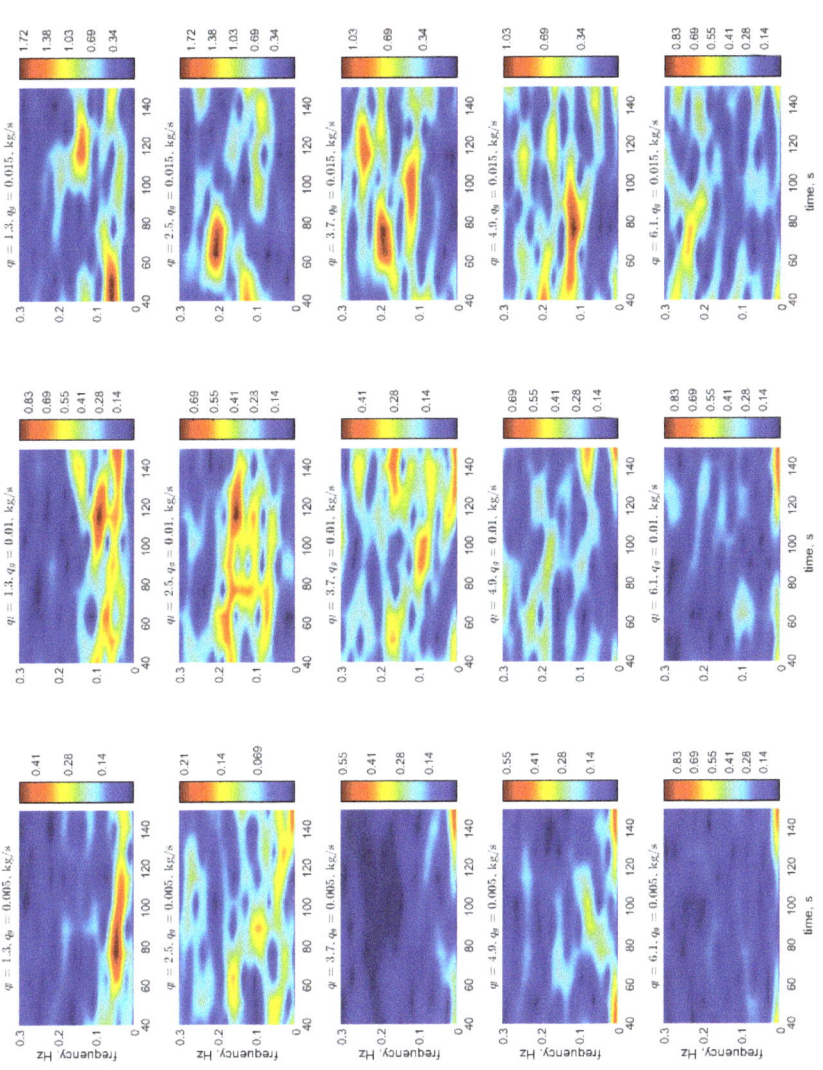

Figure 7. Spectrogram matrix for $\Delta P = P_1 - P_2$. The range of the color bar indicates the pressure level (in kPa). Note that for visualization purposes the respective ranges are not normalized. These spectrograms indicate how different frequency bands are excited at different times in a specific location, while the whole set of images illustrates the difference from one location to another.

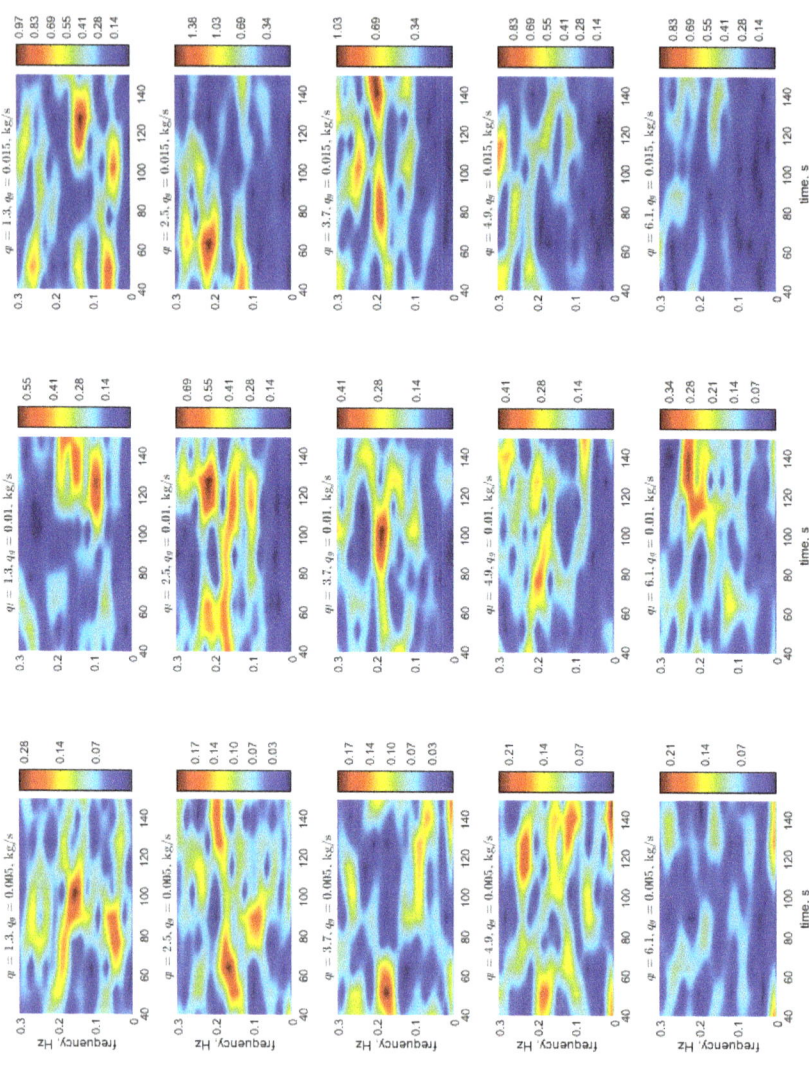

Figure 8. Spectrogram matrix for $\Delta P = P_2 - P_3$. The range of the color bar indicates the pressure level (in kPa). Note that for visualization purposes the respective ranges are not normalized. These spectrograms indicate how different frequency bands are excited at different times in a specific location, while the whole set of images illustrates the difference from one location to another.

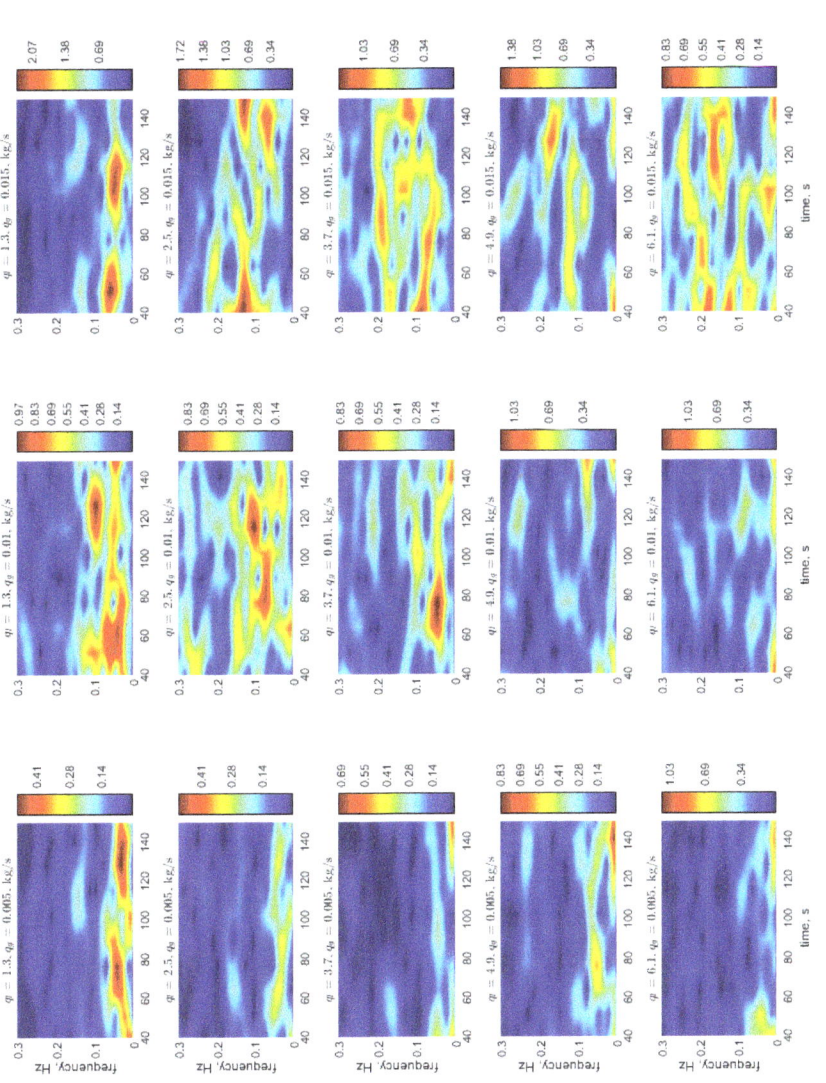

Figure 9. Spectrogram matrix for $\Delta P = P_3 - P_4$. The range of the color bar indicates the pressure level (in kPa). Note that for visualization purposes the respective ranges are not normalized. These spectrograms indicate how different frequency bands are excited at different times in a specific location, while the whole set of images illustrates the difference from one location to another.

3.3. Detrended Fluctuation Analysis (DFA)

The characteristics of the spectral evolution, and consequently of the fow intermittence, may be better understood in terms of the detrended fluctuation analysis (DFA). The reason is that two-phase flows exhibit complex self-similar fluctuations over a broad range of space and time scales, much in the same way as in many physical, biological, physiological and economic systems, where multiple component feedback interactions play a central role.

Because of the nonlinear mechanisms controlling the underlying interactions, the output signals of this two-phase flow are typically non-stationary. Furthermore, they are characterized by embedded trends in heterogeneous segment patches with different (local) statistical properties. In this case, traditional methods, such as the spectral and the auto-correlation analyses, may not be necessarily well suited to study this kind of processes. The DFA methodology, in contrast, enables a proper quantification of long-range correlations with non-stationary, fluctuating signals.

The procedure relies on the profiling of the pressure, $\mathcal{P}$, through an 'integration' of the time-series

$$\mathcal{P}(k) = \sum_{i=1}^{k}[p(i) - \bar{p}], \tag{1}$$

where $\bar{p}$ is the mean value of the signal and $p(i)$ is the value of the pressure at time i. First, the initial data set is divided into equal segments of length n. Then, a linear approximation $\mathcal{P}_n$ is computed with a least squares fit performed over each separate segment. The resulting fit represents the trend in the given section. Next, the average fluctuation $\delta p(n)$ of the signal about the trend is determined with

$$\delta p(n) = \sqrt{\frac{1}{N}\sum_{k=1}^{N}[\mathcal{P}(k) - \mathcal{P}_n(k)]^2}. \tag{2}$$

With these elements one may proceed to directly plot $\log \delta p(n)$ as a function of $\log n$. The scaling exponent α then corresponds to the slope of the least squares, linear regression fit to the data points

$$\delta p(n) \sim n^{\alpha}. \tag{3}$$

Accordingly,

$$\log \delta p(n) \sim \alpha \log n, \tag{4}$$

and the numerical value of the scaling exponent (also known as the fractal scale, autocorrelation exponent, or self-similarity parameter) is given by

$$\alpha = \frac{\sum\left[\left(\log n - \overline{\log n}\right)\left(\log \delta p - \overline{\log \delta p}\right)\right]}{\sum\left(\log n - \overline{\log n}\right)^2}. \tag{5}$$

As in previous cases the overline indicates the mean value of the respective quantity. Power law relationships of the kind represented by Equation (3) suggest the presence of self-similar fluctuations. Furthermore, depending on the actual value of α, the following cases may be considered:

1. $\alpha < 0.5$: The fluctuations are anti-correlated.
2. $\alpha \simeq 0.5$: The fluctuations are uncorrelated and represent $1/f^0$ noise; i.e., white noise.
3. $\alpha > 0.5$: The fluctuations have positive autocorrelation.
4. $\alpha \simeq 1$: The fluctuations represent $1/f^1$ noise; i.e., pink noise.
5. $\alpha > 1$: The fluctuations are non-stationary.
6. $\alpha \simeq 1.5$: The fluctuations represent $1/f^2$ noise; i.e., Brownian noise.

This DFA analysis scheme produced the results summarized in Table 3 for the experimental data sets obtained in the laboratory. It is readily seen that the table is divided in three sections, one for each of the pressure differences $p_1 - p_2$, $p_2 - p_3$ and $p_3 - p_4$ previously discussed. The rows and columns are in full correspondence with the liquid and gas flow rates of the experimental matrix. All flow rate combinations are considered.

Table 3. Numerical values of α. These values provide the DFA footprint of the flows generated with the inlet flow rates given in the experimental matrix (see Table 2).

	$P_1 - P_2$			$P_2 - P_3$			$P_3 - P_4$		
$q_l - q_g$	0.005	0.01	0.015	0.005	0.01	0.015	0.005	0.01	0.015
1.3	0.72	0.43	0.37	0.67	0.27	0.34	0.94	0.52	0.63
2.5	0.84	0.37	0.27	0.79	0.19	0.22	0.99	0.62	0.42
3.7	1.15	0.75	0.45	0.77	0.41	0.24	1.11	0.63	0.33
4.9	1.03	0.88	0.68	0.82	0.46	0.33	1.02	0.97	0.86
6.1	1.18	1.03	0.78	1.04	0.69	0.51	1.08	1.05	0.78

The numerical values acquired by the scaling exponent allows an interpretation of certain aspects stemming from the long range development of the flow. Thus, with high-viscosity gas–liquid mixtures we may draw two general observations:

First, it is noticed that the flow in the exit section of the test section is much noisier than in the inlet section. This effect is caused by the ejection of the irregular liquid slugs at the outlet of the pipe. Accordingly, in the long-term development of the flow the autocorrelated ($\alpha > 0.5$) and anti-correlated ($\alpha < 0.5$) patterns tend to become noisier (i.e., as white, pink or Brownian noises). This means that the slugs' irregularities increase as the flow progresses towards the outlet. In contrast, the U-turn renders the flow less noisy, possibly due to a regularization effect induced by the secondary flows.

Secondly, in relation with the frames corresponding to $(q_g, q_l) = (0.005 \text{ kg/s}, 3.7 \text{ kg/s})$ and $(q_g, q_l) = (0.005 \text{ kg/s}, 6.1 \text{ kg/s})$ in figs. 8 and 10, one may notice two characteristic time scales: one corresponds to the flow time scale, while the other to a low frequency mode with a much larger time scale (easily seen as a small red spot appearing at the bottom of the images at $t = 150$ s). Nonetheless, at higher gas flow rates, the flow in the exit section tends to be more regular, because higher modes are excited (i.e., intense red spots).

For concreteness, however, the flow properties may be described in more detail as follows:

1. The pressure drop exhibits a space-wise evolution. It may be caused by local variations of the phase volume fractions, which take place as the mixture evolves downstream along the pipe. For example, consider the flow produced with $(q_g, q_l) = (0.005 \text{ kg/s}, 1.3 \text{ kg/s})$ (upper most squares of columns 1, 4 and 7 of the table). The pressures appear to be autocorrelated except at the downstream section.

    In other words, the pressure measured at P1 corresponds to the pressures measured at P2 and P3, the three being in-phase because $\alpha > 0.5$ (positive autocorrelation). On the downstream section, however, the pressure measured at P4 does not correspond to any of the measurements at P1, P2 or P3. Hence, three possibilities arise: a) the coherence may be lost to the strong effects induced by the ejection of irregular liquid lumps at the outlet, b) secondary flows at the U-turn may have a disruptive effect on the properties of the pressure waves, and c) a combination of the two. The reason for these possibilities is that the U-turn and the outlet are the only two points in the flow system where the two following effects can take place: secondary flows due to the centripetal acceleration undergone by the flow inside the U-turn section, and the sudden depressurization caused by the ejection of the liquid slugs into the separator tank.

2. Apparently, very few flows exhibit white noise characteristics or non-stationary fluctuations. The former are more related with higher liquid and gas flow rates, while the latter are more related with low q_g and high liquid q_l.

The non-stationary cases represented by $\alpha > 1$ are interesting, because they indicate that the mean values of the pressures are varying with time (or equivalently with position). Even though these variations might be slight, the method is still capable of identifying them. This opens up the prospect of designing techniques for industrial applications. Obvious examples would be the development of methods to detect small leakages in pipelines, or to detect slow corrosion processes.

3. It is worth noticing that pink noise processes are mostly observed at the outlet section of the pipe. In general, the α values of this section appear to show a relative increase with respect to the values of the inlet section. This suggests that the scaling exponent increases in the direction of the flow.

Since pink noise refers to scalability, the reproduction of self-similar patterns and small scale traits of the signal would suggest the existence of an energy distribution process (analogous to the energy cascade in turbulence). However, it is noted that only the cases corresponding to the inlet flow rates $(q_g, q_l) = (0.005 \text{ kg/s}, 4.9 \text{ kg/s})$ and $(q_g, q_l) = (0.01 \text{ kg/s}, 6.1 \text{ kg/s})$ maintain this behavior. On the other hand, only the flow combination $(q_g, q_l) = (0.005 \text{ kg/s}, 6.1 \text{ kg/s})$ seems to correspond to this process in the U-turn. Overall, from the physical point of view, these characteristics would be mostly related with the ejection effects produced at the outlet of the pipe.

4. Interestingly, not a single combination of inlet gas–liquid flow rates produced fully random processes in this kind of flow system. The question still remains whether such Brownian noise patterns would eventually emerge in a longer pipeline, or not. The same question could be asked regarding the flow rates.

Anti-correlated flows for which $\alpha < 0.5$ seem to dominate in the U-turn section of the pipe. This is particularly true with high flow regimes, that is, those produced with elevated inlet mass flow rates. Similar anti-correlations are also observed at the inlet and outlet section of the pipe with high flow rates. These cases indicate that the phases of the pressure waves shift by approximately π radians, as they progress from one pressure port to the next one.

4. Conclusions

Experiments were carried out with high vicosity mixtures of glycerine and air, in a flow loop with a length to diameter ratio of $l/d \sim 10^2$. The liquid phase had a viscosity of 1.1 Pa·s.

A detrended fluctuation analysis, or DFA, was applied to the experimental measurements obtained in this facility. Concretely, the technique was applied to the pressure time-series, under the assumption that relevant information can be extracted from them in a meaningful manner. It is important to note that the mentioned experiments are significantly different to those so far reported in the open literature. Therefore, a longer term development of the flow is possible, and a variety of dynamical effects are reflected in the spectral content of the registered pressure signals.

The DFA showed that each of the 15 mixture combinations formed by different fractions of air and glycerin have a unique head loss pattern, which can be expressed by power law expressions. In principle, this uniqueness can be exploited by pattern recognition algorithms for the identification of flow regimes, which is an important stage to design separation equipment, slug catchers, gas lift operations, wellhead gathering systems, and production management.

It was noticed that the frequencies are relatively low as compared with those measured with low viscosity water-air or nitrogen-water mixtures. The primary factor producing this outcome is the viscosity of the glycerine.

The DFA analysis enabled the proper identification of various energy states of the flow, in accordance with the observed behavior of the flow pattern at different locations. The corresponding states were determined from the variations of the scaling exponent, which was measured at equally spaced measuring ports on the flow loop.

Author Contributions: All authors contributed equally to this work.

Funding: This investigation was partially funded by CONACYT through the Project No. 4730 (Convocatoria de Proyectos de Desarrollo Científico para Atender Problemas Nacionales 2017).

Acknowledgments: J.H. and D.F.G. wish to thank CONACYT for the scholarships provided.

Conflicts of Interest: The authors declare no conflict of interest.

Nomenclature

d	diameter	(m)
f	frequency	(Hz)
l	length	(m)
p	pressure	(kPa)
q	flow rate	(m^3/s)
t	time	(s)
u	velocity	(m/s)
x	length	(m)
$\mathcal{P}$	pressure profile	(kPa)
α	scaling exponent	-
δ	standard deviation	(kPa)
η	dynamic viscosity	(Pa·s)
ρ	density	(kg/m^3)
Re	Reynolds number	-
avg	average	
l	liquid	
g	gas	
i, k, n	dummy indices	
N	number of data points	

References

1. Hart, A. A review of technologies for transporting heavy crude oil and bitumen via pipelines. *J. Pet. Explor. Prod. Technol.* **2013**, *4*, 1–10. [CrossRef]
2. Matsubara, H.; Naito, K. Effect of Liquid Viscosity on Flow Patterns of Gas-Liquid Two-Phase Flow in a Horizontal Pipe. *Int. J. Multiph. Flow* **2011**, *37*, 1277–1281. [CrossRef]
3. Hernandez, J.; Montiel, J.C.; Palacio-Perez, A.; Rodríguez-Valdés, A.; Guzmán, J.E.V. Horizontal Evolution of Intermittent Gas-Liquid Flows With Highly Viscous Phases. *Int. J. Comput. Methods Exp. Meas.* **2018**, *6*, 152–161.
4. Foletti, C.; Farisé, S.; Grassi, B.; Strazza, D.; Lancini, M.; Poesio, P. Experimental investigation on two-phase air/high-viscosity-oil flow in a horizontal pipe. *Chem. Eng. Sci.* **2011**, *66*, 5968–5975. [CrossRef]
5. Khaledi, H.A.; Smith, I.E.; Unander, T.E.; Nossen, J. Investigation of two-phase flow pattern, liquid holdup and pressure drop in viscous oil–gas flow. *Int. J. Multiph. Flow* **2014**, *67*, 37–51. [CrossRef]
6. Zhao, Y.; Yeung, H.; Zorgani, E.E.; Archibong, A.E.; Lao, L. High viscosity effects on characteristics of oil and gas two-phase flow in horizontal pipes. *Chem. Eng. Sci.* **2013**, *95*, 343–352. [CrossRef]
7. Zhao, Y.; Lao, L.; Yeung, H. Investigation and prediction of slug flow characteristics in highly viscous liquid and gas flows in horizontal pipes. *Chem. Eng. Res. Des.* **2015**, *102*, 124–137. [CrossRef]
8. Al-Safran, E.; Kora, C.; Sarica, C. Predictiono of slug liquid holdup in high-viscosity liquid and gas two-phase flow in horizontal pipes. *J. Pet. Sci. Eng.* **2015**, *133*, 566–575. [CrossRef]
9. Farsetti, S.; Farisé, S.; Poesio, P. Experimental investigation of high-viscosity oil-air intermittent flow. *Exp. Therm. Fluid Sci.* **2014**, *57*, 285–292. [CrossRef]
10. Gokcal, B.; Al-sarki, A.M.; Sarica, C.; Al-Safran, E.M. Prediction of slug frequency for high-viscosity oils in horizontal pipes. *SPE Proj. Facil. Constr.* **2009**, *124057*, 447–461.
11. Okezue, C.N. Application of the gamma radiation method in analysing the effect of liquid viscosity and flow variables on slug frequency in high-viscosity oil–gas horizontal flow. *Comput. Methods Multiph. Flow VII* **2007**, *79*, 447–461.

12. Baba, Y.D.; Archibong, A.E.; Aliyu, A.M.; Ameen, A.I. Slug frequency in high-viscosity oil–gas two-phase flow: Experiment and prediction. *Flow Meas. Instrum.* **2017**, *54*, 109–123. [CrossRef]
13. Losi, G.; Arnone, D.; Correra, S.; Poesio, P. Modelling and statistical analysis of high-viscosity oil/air slug flow characteristics in a small diameter horizontal pipe. *Chem. Eng. Sci.* **2016**, *148*, 190–202. [CrossRef]
14. Zahi, L.S.; Jin, N.D.; GAo, Z.K.; Chen, P.; Chi, H. Gas-Liquid Two Phase Flow Pattern Evolution Characteristics Based on Detrended Fluctuation Analysis. *J. Metrol. Soc. India* **2011**, *26*, 255–265.
15. Peng, C.K.; Buldyrev, S.V.; Havlin, S.; Simons, M.; Stanley, H.E.; Goldberger, A.L. Mosaic organization of DNA nucleotides. *Phys. Rev. E* **1994**, *49*, 1685. [CrossRef]
16. Kantelhardt, J.W.; Koscielny-Bunde, E.; Rego, H.H.; Havlin, S.; Bunde, A. Detecting long-range correlations with detrended fluctuation analysis. *Phys. A Stat. Mech. Its Appl.* **2001**, *295*, 441–454. [CrossRef]
17. Subhakar, D.; Chandrasekhar, E. Reservoir characterization using multifractal detrended fluctuation analysis of geophysical well-log data. *Physica A* **2016**, *445*, 57–65. [CrossRef]
18. Ribeiro, R.A.; Mata, M.V.M.; Lucena, L.S.; Corso, G. Spatial analysis of oil reservoirs using detrended fluctuation analysis of geophysical data. *Nonlinear Process. Geophys.* **2014**, *21*, 1043–1049. [CrossRef]
19. Moffat, R.J. Contributions to the Theory of Single-Sample Uncertainty Analysis. *Trans. ASME* **1982**, *104*, 250–258. [CrossRef]
20. Holman, J.P. *Experimental Methods for Engineers*, 6th ed.; McGraw-Hill: New York, NY, USA, 1994.
21. Tompkins, C.; Prasser, H.M.; Corradini, M. Wire-mesh sensors: A review of methods and uncertainty in multiphase flows relative to other measurement techniques. *Nucl. Eng. Des.* **2018**, *337*, 205–220. [CrossRef]
22. Ameran, H.L.M.; Mohamad, E.J.; Muji, S.Z.; RAhim, R.A.; Abdullah, J.; Rashid, W.N.A. Multiphase flow velocity measurement of chemical processes using electrical tomography: A review. In Proceedings of the IEEE 2016 International Conference on Automatic Control and Intelligent Systems (I2CACIS), Bandar Hilir, Malaysia, 16–18 December 2016; pp. 162–167.
23. Arvoh, B.K.; Hoffmann, R.; Halstensen, M. Estimation of volume fractions and flow regime identification in multiphase flow based on gamma measurements and multivariate calibration. *Flow Meas. Instrum.* **2012**, *23*, 56–65. [CrossRef]
24. Hanafizadeh, P.; Eshraghi, J.; Taklifi, A.; Ghanbarzadeh, S. Experimental identification of flow regimes in gas–liquid two phase flow in a vertical pipe. *Meccanica* **2016**, *51*, 1771–1782. [CrossRef]
25. Matsui, G. Identification of flow regimes in vertical gas–liquid two-phase flow using differential pressure fluctuations. *Int. J. Multiph. Flow* **1984**, *10*, 711–719. [CrossRef]

© 2019 by the authors. Licensee MDPI, Basel, Switzerland. This article is an open access article distributed under the terms and conditions of the Creative Commons Attribution (CC BY) license (http://creativecommons.org/licenses/by/4.0/).

Article

Performance Improvement of a Grid-Tied Neutral-Point-Clamped 3-φ Transformerless Inverter Using Model Predictive Control

Hani Albalawi [1,*] and Sherif A. Zaid [2,*]

1. Department of Electrical Engineering, Faculty of Engineering, University of Tabuk, Tabuk 47913, Saudi Arabia
2. Department of Electrical Power, Faculty of Engineering, Cairo University, Cairo 12613, Egypt
* Correspondence: halbala@ut.edu.sa (H.A.); sherifzaid3@yahoo.com (S.A.Z.)

Received: 29 September 2019; Accepted: 5 November 2019; Published: 15 November 2019

Abstract: Grid-connected photovoltaic (PV) systems are now a common part of the modern power network. A recent development in the topology of these systems is the use of transformerless inverters. Although they are compact, cheap, and efficient, transformerless inverters suffer from chronic leakage current. Various researches have been directed toward evolving their performance and diminishing leakage current. This paper introduces the application of a model predictive control (MPC) algorithm to govern and improve the performance of a grid-tied neutral-point-clamped (NPC) 3-φ transformerless inverter powered by a PV panel. The transformerless inverter was linked to the grid via an inductor/capacitor (LC) filter. The filter elements, as well as the internal impedance of the grid, were considered in the system model. The discrete model of the proposed system was determined, and the algorithm of the MPC controller was established. Matlab's simulations for the proposed system, controlled by the MPC and the ordinary proportional–integral (PI) current controller with sinusoidal pulse width modulation (SPWM), were carried out. The simulation results showed that the MPC controller had the best performance for earth leakage current, total harmonic distortion (THD), and the grid current spectrum. Also, the efficiency of the system using the MPC was improved compared to that using a PI current controller with SPW modulation.

Keywords: PV; 3-φ transformerless inverter; NPC; boost converter; model predictive control; maximum power point tracking

1. Introduction

Renewable energy utilization is currently expanding due to global warming awareness and the predicted depletion of fossil fuels. Many governments across the world encourage and motivate people by applying incentive rules to use renewable energies. As a result, grid-connected photovoltaic (PV) systems are now widespread within communities.

Most PV system installations are single-phase installations used for small-scale systems up to 5–6 kW [1]. However, this type of installation has a smooth direct current (DC) input and a pulsating alternating current (AC) output with a large DC capacitor that decreases the system's reliability and lifetime. In contrast, in three-phase systems, the large capacitor is not required, which improves the system's reliability and lifetime as it has a constant AC output [2]. The most critical part of the PV system is the inverter because it works as an interface between the PV system and the utility grid. Usually, the inverter comes with a transformer to isolate the PV panel system from the grid and to match the system voltage to that of the grid [2]. In other words, the transformer helps to boost the PV system's voltage when needed and reduces the harmonic injection into the grid, thereby improving the power quality [3]. The transformer may be integrated into grid-connected PV systems using two configurations. The first involves a high-frequency transformer connected between the power inverter and the PV panel, while the second configuration involves a low-frequency transformer between the power inverter and the grid. These transformers increase the system's weight, size, and cost. Also, they decrease efficiency and introduce more complexity to the system. Recently, new and amazing topologies—called transformerless PV systems—have been introduced [2,3]. Although these topologies reduce the drawbacks of transformer-based systems, they introduce an earth leakage current problem. The source of this problem is the absence of galvanic isolation between the PV system and the grid. Consequently, any potential fluctuations between the PV panels and the ground increase the earth leakage current. This is unfavorable as it generates losses, destroys the system's safety, and distorts the grid current. Research has indicated that fluctuations of the inverter's common-mode voltage (CMV) are the origin of the leakage current [4]. Therefore, to decrease the leakage current issues that appear in transformerless PV systems, it is required to keep the CMV constant.

In the literature, several studies have proposed single-phase transformerless structures to tackle the problem of leakage current [5–8]. In contrast, research on three-phase transformerless topology is still limited. There are two approaches in the literature for manipulating the leakage current problem of three-phase transformerless systems: the inverter modulation technique and the inverter structure or topology. Several modulation methods and conversion structures have been reported recently. Due to the high leakage current of conventional pulse width modulation (PWM), whether discontinuous PWM (DPWM) or space vector PWM (SVPWM), it is not sufficient for three-phase transformerless PV applications. The authors of Reference [2] presented a remote-state PWM (RSPWM) technique for a conventional three-phase transformerless PV system to eliminate leakage current. The main disadvantage of the presented modulation method was that it could only be used for two-level inverters with a 650 V DC link in the case of a 110 V grid phase. In Reference [3], the authors introduced the H8 topology with a modulation technique dependent on conventional sinusoidal PWM (SPWM). However, the high number of power switches increased the system's losses and reduced efficiency. In Reference [9], a Z-source inverter (ZSI) topology was implemented by adding a fast recovery diode to reduce the leakage current using a modified modulation technique. Although the overall efficiency was increased, the system and controller were complex. Furthermore, a new three-phase transformerless inverter topology called H7 [10,11] has been introduced to minimize the leakage current with modified SVPWM. This system provides good results according to the leakage current; however, the efficiency is slightly reduced. Reduced leakage current was also achieved using a new topology and modulation strategy called a zero-voltage state rectifier (ZVR), as presented in Reference [12]. However, the topology suffered from a severe unbalance in the voltages of the capacitors.

Among all topologies of 3-φ transformerless inverters, the most recently used is the multilevel inverter type, especially the neutral-point-clamped (NPC) inverter type [13,14]. This type is vastly employed in industrial applications. NPC features two characteristics; despite the large number of switching devices and diodes, NPC is characterized by a low total harmonic distortion (THD) of output voltage and a low rating of switching devices compared to two-level inverters. The authors in Reference [15] presented two PWM approaches to decrease the common-mode current (CMC) in a three-level NPC inverter. These techniques improved the CMV, but they also increased the voltage ripples and THD. In Reference [16], a multivariable linear quadratic regulator (LQR) was presented for the comprehensive control of a three-level inverter connected to the grid with an inductor/capacitor (LC) filter.

Lately, model predictive control (MPC) methods have attracted many researchers to control power converters [17–26]. MPC is considered a favorable and proper methodology to control power converters as it has a discrete nature. Additionally, there are many advantages of using MPC, such as easy comprehension of the concept, fast dynamic response, and a simple control algorithm. Also, it has the characteristic of dealing with multivariable cases and constraints treatment and dead time compensation. However, implementation of the MPC algorithm requires many calculations which are now solvable due to great developments in digital signal processors (DSPs). The MPC technique is used to control 3-φ transformerless inverters. In Reference [27], a finite control set MPC technique was applied to remove the leakage current of a T-type transformerless three-level inverter. Moreover, in Reference [28], a new MPC control scheme for current control of a three-phase NPC inverter was introduced. Nevertheless, the inverter output filter has not been used in the system model. Usually, tuned LC filters are used to reduce the grid current harmonics injected by transformerless inverters [29].

This paper introduces an MPC algorithm to control and improve the performance of a grid-tied NPC 3-φ transformerless inverter powered by a PV panel. The transformerless inverter was linked to the grid via an LC filter. The objectives of this research were to:

- Apply the MPC controller to the proposed system with appropriate consideration of the LC filter and grid impedance.
- Discuss the effect of the MPC on performance factors, such as the earth leakage current, grid current THD, and efficiency.
- Compare the performance of the system using the MPC controller with the system that used the proportional–integral (PI) controller.

The first step in designing the MPC controller was system modeling. The filter elements, as well as the internal impedance of the grid, were considered in the system model. Then, the discrete-time model was derived, followed by a description of the MPC algorithm for the proposed system. Finally, the Matlab platform was used to simulate the proposed system and a performance comparison between the MPC controller and the ordinary PI with SPWM. The simulation results showed that the MPC controller had the best performance for earth leakage current, THD, and the grid current spectrum. Also, the efficiency of the system that used MPC was improved compared to the system that used the PI current controller with SPW modulation.

The paper is prepared as follows. Section 2 describes the proposed system, while modeling of the system is presented in Section 3. Section 4 discusses the MPC controller design and whole system controllers. Section 5 provides a detailed discussion of the simulation results and Section 6 provides the net conclusions of the paper.

2. System Description

The proposed system is a PV-powered 3-φ transformerless inverter linked to the grid. Shown in Figure 1, the first stage in the system is the PV panel, which is usually linked to a capacitor at its terminals. The capacitor functions to regulate power and improve PV performance [30]. The PV output is coupled to a boost converter, which acts as an adjustable load for the PV panels.

The maximum power point tracking (MPPT) operation of the PV can be reached by regulating the boost converter. Also, the boost converter is capable of stepping up the PV voltage at low insolation levels. Therefore, it supports extracting low power levels from the PV. The terminals of the boost converter output represent the DC link, which is attached to the input of the 3-φ transformerless inverter. The 3-φ transformerless inverter has NPC topology. An LC filter is set at the inverter output terminals which are connected to the utility grid. The filter avoids high-frequency ripples and damps the current dynamics [31]. The model and operation of each part of the system will be explained in the next paragraphs.

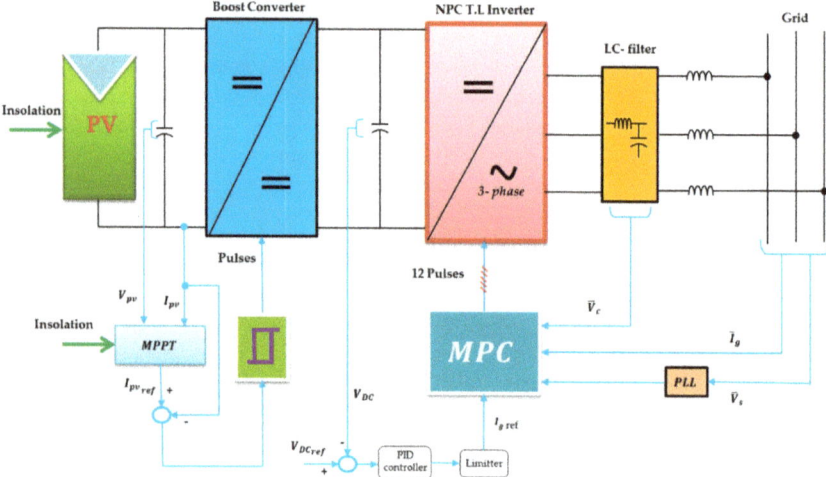

Figure 1. The proposed photovoltaic (PV)-powered 3-level transformerless inverter connected to the grid and its controllers.

3. System Modelling

Predictions of the grid current and filter voltage are essential for MPC controller operation. Modelling the system is the first step in MPC controller design. The following assumptions were used in the mathematical model of the system:

- Boost converter losses are neglected.
- Voltage drops and leakage currents of all the switching devices are neglected.
- Snubber circuits are neglected.
- The grid internal impedance is taken into consideration.

These assumptions were only used for the mathematical model and were not applied in the computer simulation. The dynamic model of all the proposed system parts shown in Figure 1 is explained in the following sections.

3.1. Photovoltaic Panel Model

Figure 2 presents the PV panel model, which consisted of a current source with a parallel diode and series and parallel resistances. If the parallel resistance was large enough, its current was able to be neglected in the cell model. The equations of the model are well known in the literature [32].

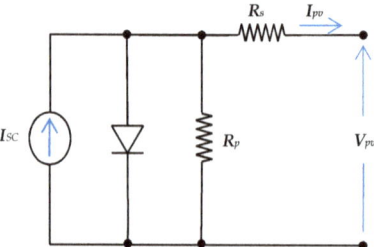

Figure 2. Model of the PV panel, where I_{SC} is the panel short circuit current and (R_p, R_s) is the model parallel and series resistances.

3.2. Boost Converter Dynamic Model

Figure 3 shows the power circuit of the boost converter. Its input is the PV panel output and the output feeds the DC link voltage of the 3-level transformerless inverter. Its function was to regulate the PV power to operate at MPPT conditions. Assuming that the DC link capacitor is large enough, the dynamic model of the converter is specified by the following equations [33]:

$$V_{pv} = L_b \frac{dI_{pv}}{dt}, \quad Q_b \rightarrow on \qquad (1)$$

$$V_{pv} - V_{DC} = L_b \frac{dI_{pv}}{dt}, \quad Q_b \rightarrow off \qquad (2)$$

where L_b is the boost converter input inductance and (V_{DC}) is the DC link voltage.

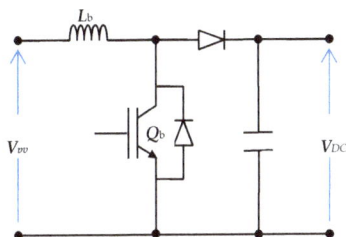

Figure 3. The power circuit of the boost converter.

3.3. Three-Level Inverter and Filter Model

The power stage of the NPC 3-level inverter is shown in Figure 4a. It contains 12 Insolated Gate Bipolar Junction Transistor (IGBTs) and 6 clamp diodes. The DC bus voltage had to be split using two capacitors, as shown in the figure. It is well known that the NPC 3-level inverter has 27 states, as shown in Figure 4b. When these states are represented as space vectors, they produce 19 voltage vectors (V_1, ...V_{19}). As stated in Reference [15], there are only seven states with zero common-mode voltage, named ($V_8, V_{10}, V_{12}, V_{14}, V_{16}, V_{18}, V_0$). Hence, to limit the CMV of the NPC three-level inverter, the previous seven switching states were utilized. Consequently, the earth leakage current can be killed.

As shown in Figure 4a, the transformerless NPC inverter was connected to the grid through an LC filter. The source inductance (l_g) was taken into consideration in the model. The grid was assumed to be an infinite 3-φ bus that had constant frequency and voltage amplitude. All 3-φ voltages and currents are expressed as space vectors using:

$$\underline{U} = 2/3\left(u_a + au_b + a^2 u_c\right) \qquad (3)$$

where (u_a, u_b, and u_c) are the 3-φ quantities, $\underline{u}$ is the equivalent space vector, and $a = e^{j(2\pi/3)}$.

From the circuit's basic laws, the system dynamic behavior can be expressed by:

$$L_f \frac{d\underline{I}_f}{dt} = \underline{V}_i - \underline{V}_c \qquad (4)$$

$$L_g \frac{d\underline{I}_g}{dt} = \underline{V}_c - \underline{V}_s \qquad (5)$$

$$C_f \frac{d\underline{V}_c}{dt} = \underline{I}_f - \underline{I}_g \qquad (6)$$

where (L_f, C_f) is the filter inductance and capacitance. The filter capacitor voltage vector is $\underline{V}_c$, the inverter voltage vector is $\underline{V}_i$, the grid current vector is $\underline{I}_g$, and the filter current vector is $\underline{I}_f$.

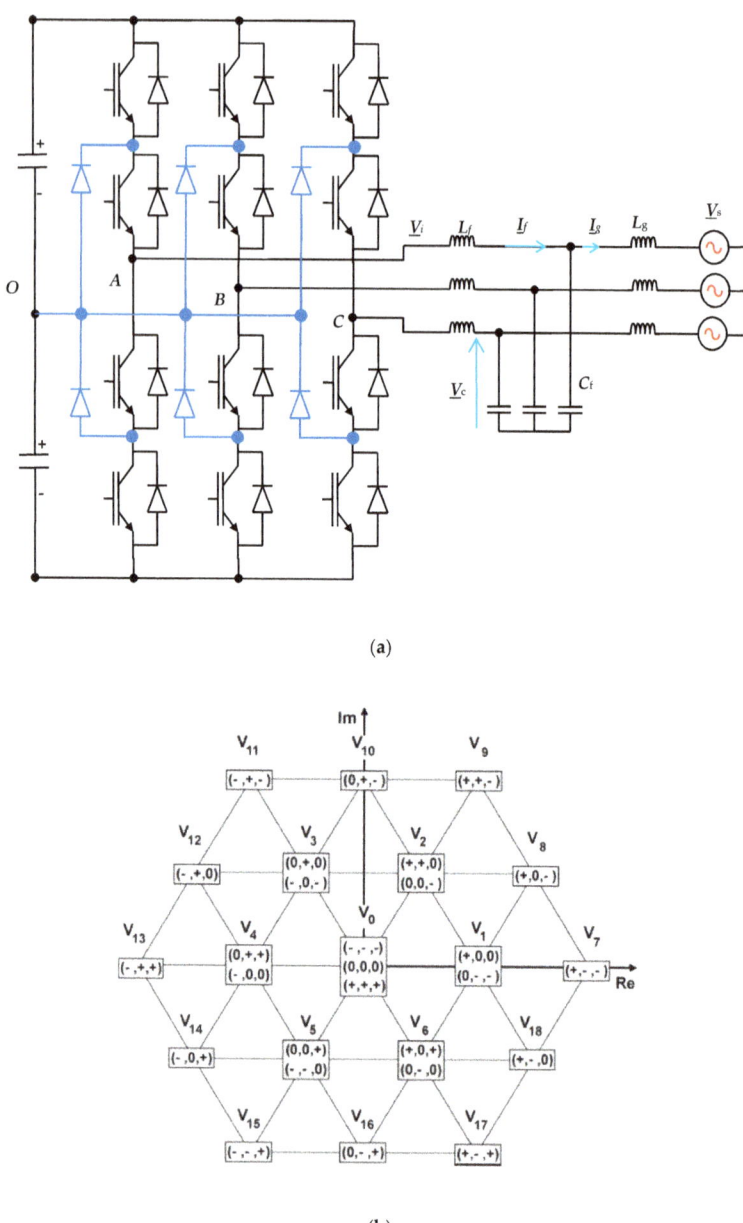

Figure 4. (a) The power circuit diagram of an NPC 3-φ inverter and (b) the inverter space vectors.

3.3.1. State Space Form of the 3-φ Transformerless Inverter Model

Equations (4)–(6) can be rewritten in the state space system matrix form:

$$\underline{X} = \begin{bmatrix} \underline{I_f} \\ \underline{I_g} \\ \underline{V_c} \end{bmatrix} \tag{7}$$

$$\frac{d\underline{X}}{dt} = A\underline{X} + B\underline{V}_i + C\underline{V}_s \tag{8}$$

where

$$A = \begin{bmatrix} 0 & 0 & \frac{-1}{L_f} \\ 0 & 0 & \frac{1}{L_g} \\ \frac{1}{C_f} & \frac{-1}{C_f} & 0 \end{bmatrix}, B = \begin{bmatrix} \frac{1}{L_f} \\ 0 \\ 0 \end{bmatrix}, C = \begin{bmatrix} 0 \\ \frac{-1}{L_g} \\ 0 \end{bmatrix} \tag{9}$$

3.3.2. Discrete-Time Prediction of the 3-φ Transformerless Inverter Model

Implementation of the MPC algorithm requires prediction of the future values of the controlled quantities. The prediction process is accomplished using the system discrete model. Usually, this model is determined by the forward Euler approximation:

$$\frac{d\underline{X}}{dt} = \frac{\underline{X}(k+1) - \underline{X}(k)}{T_s} \tag{10}$$

where T_s is the sampling time. With the help of Equation (10), the state space equation in (8) could be transformed into discrete as:

$$\underline{X}(k+1) = \dot{A}\underline{X}(k) + \dot{B}\underline{V}_i(k) + \dot{C}\underline{V}_s(k) \tag{11}$$

where

$$\text{w; } \dot{A} = e^{AT_s}, \dot{B} = \int_0^{T_s} e^{At} B dt, \dot{C} = \int_0^{T_s} e^{At} C dt \tag{12}$$

From these equations, the system variables for the next sample can be predicted. An optimization process is then adapted to direct the system to the set point. This process will occur using the cost function.

4. System Controllers

Here, three controllers are introduced to represent the control system for the suggested PV grid-tied system. The first controller is the MPPT controller that adjusts the PV operating point to be very close to the MPPT conditions. The MPPT algorithm generates the reference current to a current-regulated boost converter, which in turn maintains the MPPT conditions. The second controller is the DC link voltage controller that regulates V_{DC} at a specified value. The third controller is used to regulate the grid current of the NPC transformerless inverter controller.

4.1. MPPT Controller

For better utilization of the PV systems, extracting maximum power is very important. Consequently, the operation at the MPPT condition is essential in these systems. Many techniques have been prepared for MPPT [34,35]. In this work, the incremental conductance MPPT approach was applied to utilize the maximum permissible PV power. The approach is based on tracking the slope of the PV power and voltage curve (dPpv/dVpv) until reaching zero, according to [30]:

$$\begin{array}{ll} \frac{\Delta P}{\Delta V} < 0 & \text{on the right of the MPPT condition at the curve} \\ \frac{\Delta P}{\Delta V} = 0 & \text{at the MPPT condition} \\ \frac{\Delta P}{\Delta V} > 0 & \text{on the lift of the MPPT condition at the curve} \end{array} \tag{13}$$

The incremental conductance approach produces the reference current to a current-regulated boost converter, as shown in Figure 5. The reference current is compared to the PV current generating an error signal that derives the hysteresis controller. In turn, the hysteresis controller produces the required duty cycle signals for the boost converter transistor.

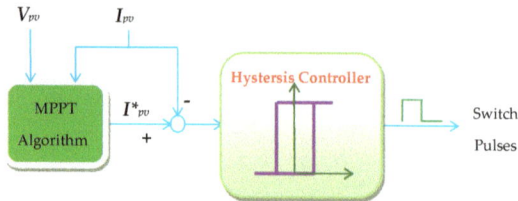

Figure 5. Maximum power point tracking (MPPT) controller block diagram.

4.2. DC Link Voltage Controller

This controller regulates the DC link voltage which has an important role in the power transfer and stability of the whole system. It generates the reference grid's current value. For stability issues, the response of this controller must be slower than the inverter controller. Fortunately, the huge capacitor value at the DC link terminals decelerates the response of the system. As the set value is constant, the PI controller is adequate. The PI controller parameters are tuned by the Nichehols–Ziegler procedure.

4.3. NPC Inverter Controller Implemented Using the MPC Algorithm

The MPC scheme was based on predicting the future manipulated variables of the model to improve system performance. MPC schemes with power electronic systems are different since power electronic systems always use power converters. These converters usually have a limited number of feasible switching states. In those cases, the procedure depends on selecting the switching state which makes the system output as close as possible to its respective reference for each sampling period. For each sampling state, the behavior of the variables can be predicted by using the system model. Then, an optimization is adapted and applied to ensure selection of the appropriate and optimal switching state. This optimization is defined as a cost function that will be assessed for every promising switching state. Then, the optimal and suitable switching state is selected based on the minimization of the cost function obtained. The control structure of the proposed system is illustrated in Figure 6. The goals of this controller were to control the grid current vector ($\underline{I_g}$) to track its sinusoidal reference and achieve unity power factor operation for the power supplied to the grid.

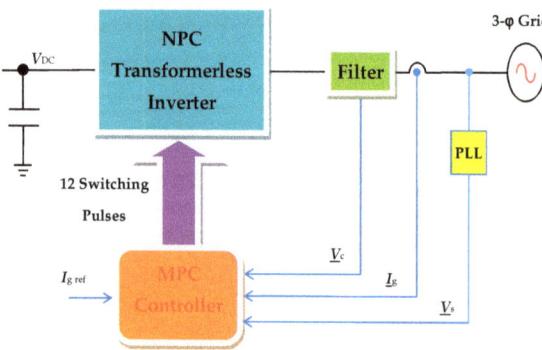

Figure 6. Model predictive control (MPC) block diagram.

The prediction process relied on the measured variables, which were ($\underline{I_g}(k)$, $\underline{V_s}(k)$, $\underline{V_c}(k)$). Next, the prediction of $\underline{I_g}(k + 1)$ for each effective switching state was obtained using the system model and measurements. In turn, the prediction assessed the cost function to obtain the control goals. Afterward, the valid switching state—which provides the minimum cost function—was designated for the next sampling period. Figure 7 shows a flow chart for the MPC controller of the NPC transformerless inverter.

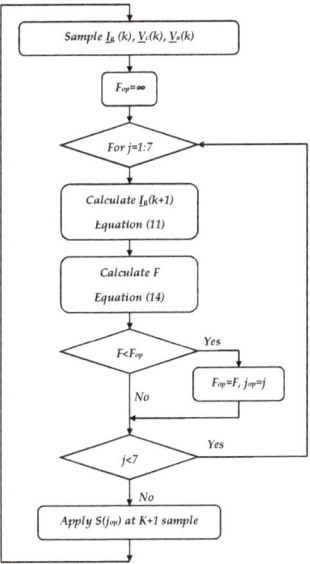

Figure 7. MPC algorithm flowchart.

Cost Function (F)

The cost function is the core of the MPC optimization process. To achieve the MPC goals, F is chosen to minimize the grid current error. F is defined as the square of grid current error, as given by:

$$F = (i_{g\alpha} - i^*_{g\alpha})^2 + (i_{g\beta} - i^*_{g\beta})^2 \tag{14}$$

where $(i_{g\alpha}, i_{g\beta})$ are the real and imaginary components of I_g and $(i^*_{g\alpha}, i^*_{g\beta})$ are the real and imaginary components of the grid current reference.

5. Simulation Results

The proposed system shown in Figure 1 is simulated using the Matlab/Simulink platform. The system parameters are listed in Table 1. Typically, the power ratings of three-phase systems are 10–15 kW in the case of rooftop applications. This research used a 10 Kw power rating. Assuming that the 3-φ grid was (230 V, 50 Hz), the typical DC bus voltage V_{DC} for the transformerless inverter was 650 V [1]. To achieve that value of V_{DC} and output power of 10 Kw, the PV panel structure was 960 series cells × 6 parallel strings. The leakage capacitance (C_{earth}) between the cells and the grounded frame was modeled with a simple capacitance. It can have values up to 50–150 nF [36], depending on the atmospheric conditions and the structure of the panels. However, the value of (C_{earth}) in simulation was selected to be 100 nF. The sampling time (T_s) was selected based on the actual time for completing one control algorithm process. The remaining parameters were selected based on the fact they are commonly used, in practice, for 3-φ inverters. Figure 8 is a comparison of the results of the proposed NPC transformerless inverter controlled by the MPC controller (Figure 8a) and the PI current controller with SPW modulation. The 3-φ grid currents, with the two controllers, are sinusoidal and in phase with the grid voltage (unity power factor). The grid current THD with the MPC controller is 1.22% and with the PI controller is 2.23%. The inverter output line voltages have different waveforms as the controller action in each case is different. The PV currents for the two controllers are the same, as the same MPPT controller is used for each case. For earth leakage current, it is very clear that the MPC controller case is much smaller than the PI controller.

Table 1. System parameters.

Parameter	Value
PV SC current	24.53 A
PV OC voltage	633 V
C_{earth}	100 nF
V_{DC}	650 V
C_f	2 μF
L_f	3 mH
Utility voltage	230 V
Utility frequency	50 Hz
PWM carrier frequency	10 KHz
DC link capacitor	1000 μF
T_s	35 μs

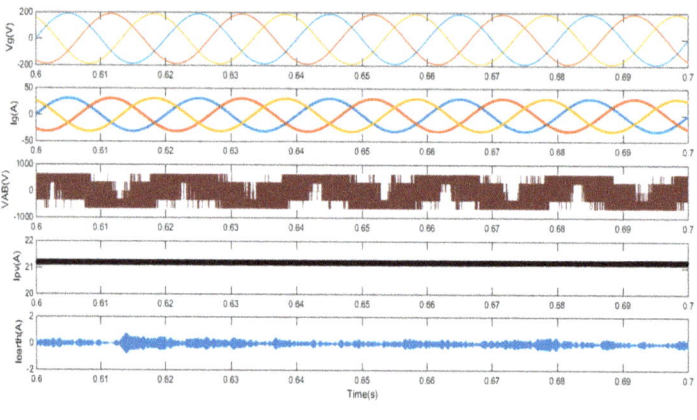

(a)

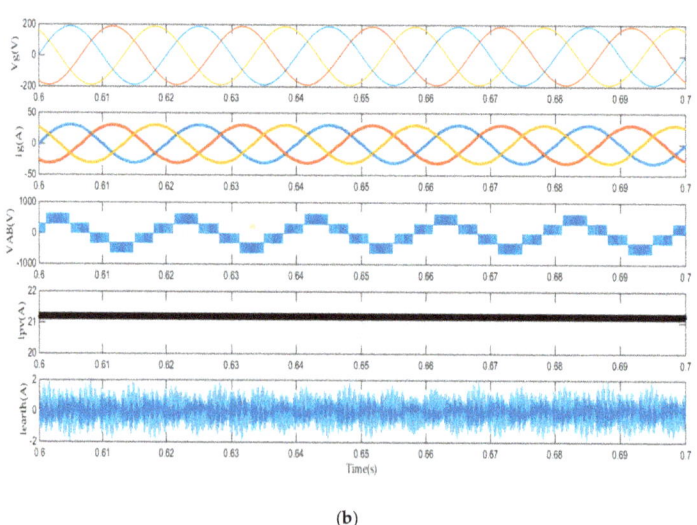

(b)

Figure 8. Simulation results of the proposed neutral-point-clamped (NPC) transformerless inverter in terms of grid voltage, grid current, PV current, and earth leakage current with (**a**) the MPC controller and (**b**) the proportional–integral (PI) current controller with sinusoidal pulse width modulation (SPWM).

Figure 9 shows the variation of the Root Mean Square (RMS) value of the leakage current with the insolation level. The leakage current with the MPC case is less than one-third the value of the PI controller case. In MPC, not all the voltage vectors are used, only those that minimize leakage currents. The PI controller with SPW modulation utilizes all the voltage vectors. Hence, the leakage current is smaller in the MPC case. Furthermore, the leakage current drops for the MPC controller, while remaining nearly constant with the PI controller. This phenomenon is explained in the following paragraphs.

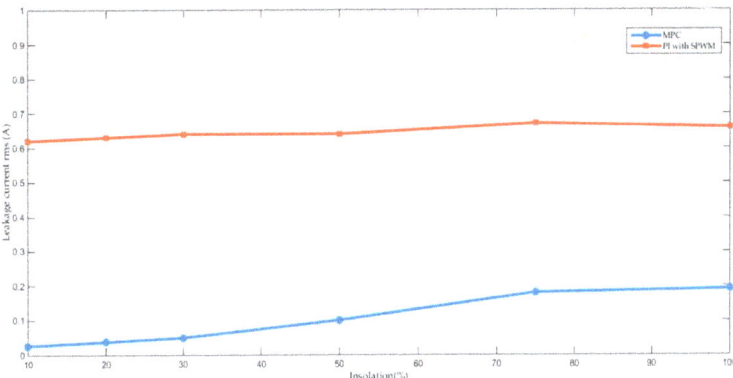

Figure 9. Variation of the leakage current with the insolation level for the MPC controller and PI current controller with SPW modulation.

It is well known in the literature that CMV fluctuations are the main cause of leakage current. By checking the CMV of the two controllers, it was observed that with PI there were small CMV variations with the insolation variation. The MPC case had moderate CMV variations.

Figure 10 shows the variation of the grid current THD with the insolation level. Comparing the THD for the two cases shows that THD with MPC produces less than 50% of the value than THD with the PI controller. This result can be explained by the current-error-minimization process that occurs when the MPC controller is used.

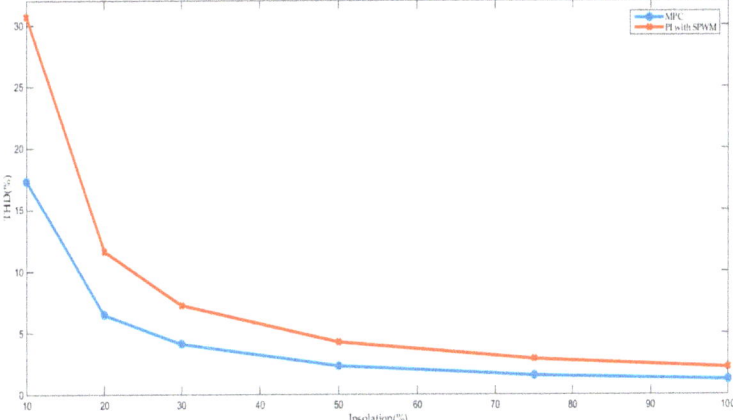

Figure 10. Variation of the grid current total harmonic distortion (THD) with the insolation level for the MPC controller and PI current controller with SPW modulation.

Figure 11 shows the variation of the frequency spectrum of the grid current with the insolation level. The figure shows that lower order harmonics in the MPC case are less than the lower order harmonics in the PI controller case. Generally, the value of the harmonics is lowest for MPC cases. The optimization mechanism in the MPC minimizes THD in the grid current since the cost function focuses on the error present in the grid current. The PI controller with the SPW modulation does not possess this optimization. For this reason, the THD is smaller in the MPC case than the other case.

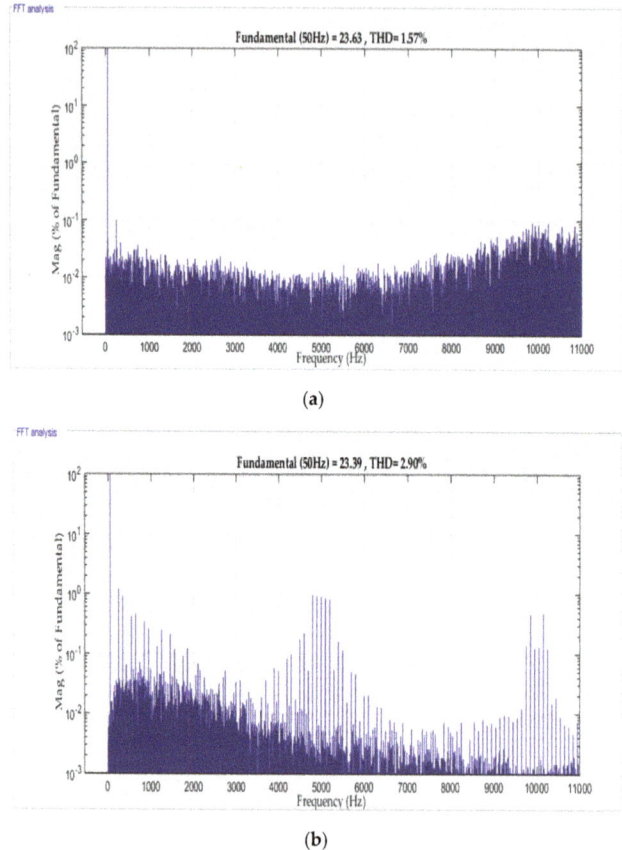

Figure 11. The spectrum of the grid current for the (**a**) MPC controller and (**b**) PI current controller with SPW modulation (@75%insolation).

Figure 12 presents the response of the output power and maximum power point (MPP) power in the case of MPC controller (Figure 12a) and the PI current controller with SPW modulation (Figure 12b). The output power of the two cases tracks the MPP power with a forced steady-state error representing system losses. The losses in the MPC controller cases are smaller than in the PI controller cases. The reason for this difference is because a small THD produces harmonic losses. Losses increase with increasing power levels, which is considered a normal issue since current values increase as power levels increase.

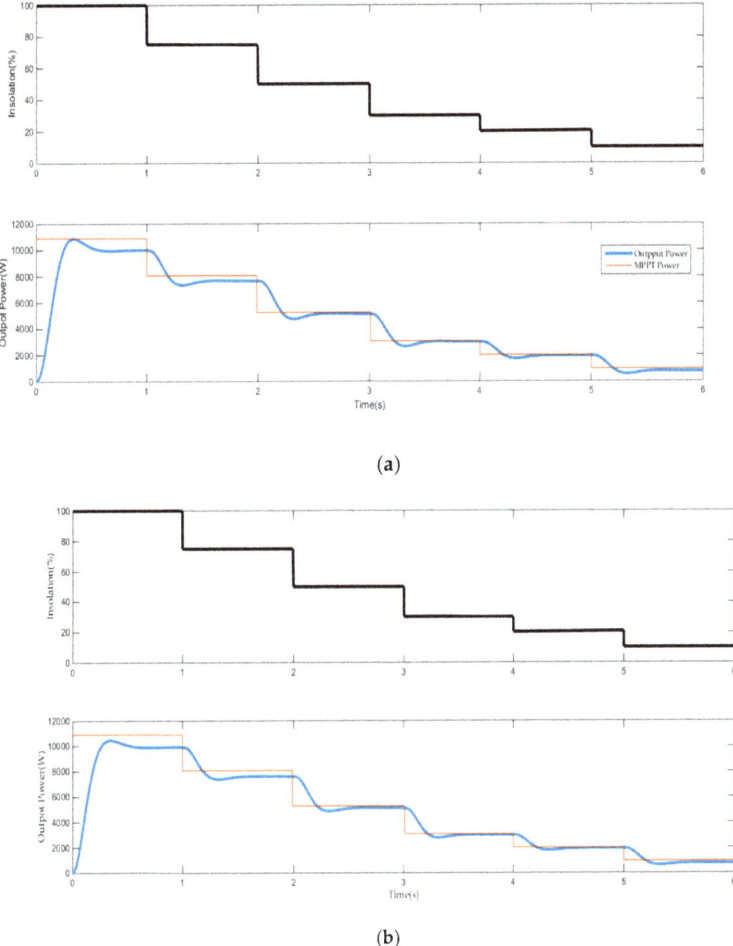

Figure 12. The maximum power point (MPP) power and output power at step insolation variations for the (**a**) MPC controller and (**b**) PI current controller with SPWM.

Figure 13 shows the variation of system efficiency with insolation level for the MPC controller and PI current controller with SPW modulation. Under all insolation levels, the efficiency of the MPC controller is higher than that for the PI controller. The proposed system Californian efficiency ($\eta\ \eta_c$) has been determined for the two controllers using the following equation [37]:

$$\eta_C = 0.05\eta_{100\%} + 0.21\eta_{75\%} + 0.53\eta_{50\%} + 0.12\eta_{30\%} + 0.05\eta_{20\%} + 0.04\eta_{10\%} \quad (15)$$

The efficiency is 95.62% for the MPC controller case and 94.99% for the PI current controller case. As the two compared cases use the same hardware, but have a different controller, the inverter switching pattern and harmonics are different. These differences are because the MPC controller provides better switching patterns and lower harmonics than other controllers. The higher efficiency is thought to come from the small switching losses and low harmonic losses from the MPC controller. Figure 14 provides a line diagram of the major performance factors, including THD, efficiency, and leakage current. The figure shows great improvement in leakage current reduction.

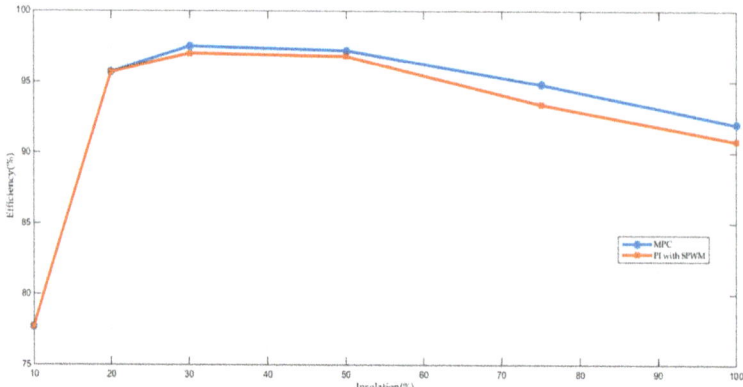

Figure 13. Variation of the system efficiency with the insolation level for the MPC controller and PI current controller with SPW modulation.

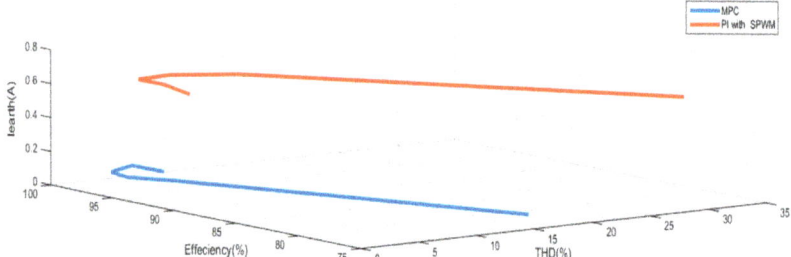

Figure 14. Line diagram of the THD, efficiency, and leakage current for the MPC controller and PI current controller with SPW modulation.

6. Conclusions

This article proposed application of an MPC controller with a PV-powered grid-tied NPC transformerless inverter. The transformerless inverter was linked with the grid through an LC filter. The filter elements, as well as the internal impedance of the grid, were taken into consideration in the system model. The discrete model of the proposed system was determined and the algorithm of the MPC controller was established. Matlab simulations of the proposed system (controlled by MPC) and a system that used an ordinary PI current controller with SPW modulation were carried out. The simulation results showed the following:

(1) The MPC controller had the best performance for all factors of comparison.
(2) The 3-φ grid currents, with the two controllers, were sinusoidal and in phase with the grid voltage (i.e., unity power factor).
(3) The grid current THD with the MPC controller was 1.22% and 2.23% for the PI controller.
(4) The leakage current in the MPC case was less than one-third of the value of the PI controller case.
(5) The efficiency of the system that used the MPC was improved compared to the system that used the PI current controller with SPW modulation.

Author Contributions: S.A.Z. conceived and designed the system model. S.A.Z. and H.A. discussed the results. H.A. reviewed the paper.

Funding: This research was funded by University of Tabuk, grant number S-1440-0067 at https://www.ut.edu.sa/web/deanship-of-scientific-research/home.

Conflicts of Interest: The authors declare no conflict of interest.

Nomenclatures

CMV	Common-mode voltage
CMC	Common-mode current
C_{earth}	The leakage capacitance
DPWM	Discontinuous PWM
DSP	Digital signal processor
F	The cost function
$\underline{I_g}$	The grid current vector
$\underline{I_f}$	The filter current vector
I_{earth}	The earth leakage current
I_{SC}	The panel short circuit current
$(i_{g\alpha}, i_{g\beta})$	The real and imaginary components of $\underline{I_g}$
$(i^*_{g\alpha}, i^*_{g\beta})$	The real and imaginary components of the grid current reference
l_g	The source inductance
L_b	The boost converter input inductance
(L_f, C_f)	The filter inductance and capacitance.
MPP	The maximum power point
MPC	Model predictive control.
MPPT	Maximum power point tracking
NPC	Neutral-point-clamped.
RSPWM	Remote-state PWM
(R_p, R_s)	The model parallel and series resistances
SPWM	Sinusoidal pulse width modulation
SVPWM	Space vector pulse width modulation
THD	Total harmonic distortion
T_s	The sampling time
$(u_a, u_b, \text{and } u_c)$	The 3-φ quantities
$\underline{u}$	The equivalent space vector
(V_{pv}, I_{pv})	The PV terminal voltage and current
V_{DC}	The DC link voltage
$\underline{V_c}$	The filter capacitor voltage vector
$\underline{V_i}$	The inverter voltage vector
ZSI	Z-Source Inverter
ZVR	Zero-voltage state rectifier
η_χ	Californian efficiency

References

1. Kerekes, T.; Teodorescu, R.; Liserre, M.; Klumpner, C.; Sumner, M. Evaluation of three-phase transformerless photovoltaic inverter topologies. *IEEE Trans. Power Electron.* **2009**, *24*, 2202–2211. [CrossRef]
2. Cavalcanti, M.C.; De Oliveira, K.C.; De Farias, A.M.; Neves, F.A.; Azevedo, G.M.; Camboim, F.C. Modulation techniques to eliminate leakage currents in transformerless three-phase photovoltaic systems. *IEEE Trans. Ind. Electron.* **2009**, *57*, 1360–1368. [CrossRef]
3. Gupta, A.K.; Agrawal, H.; Agarwal, V. A Novel Three-Phase Transformerless H-8 Topology With Reduced Leakage Current for Grid-Tied Solar PV Applications. *IEEE Trans. Ind. Appl.* **2018**, *55*, 1765–1774. [CrossRef]
4. Cavalcanti, M.C.; Farias, A.M.; Oliveira, K.C.; Neves, F.A.; Afonso, J.L. Eliminating leakage currents in neutral point clamped inverters for photovoltaic systems. *IEEE Trans. Ind. Electron.* **2011**, *59*, 435–443. [CrossRef]
5. González, R.; Lopez, J.; Sanchis, P.; Marroyo, L. Transformerless inverter for single-phase photovoltaic systems. *IEEE Trans. Power Electron.* **2007**, *22*, 693–697. [CrossRef]
6. Xiao, H.; Xie, S.; Chen, Y.; Huang, R. An optimized transformerless photovoltaic grid-connected inverter. *IEEE Trans. Ind. Electron.* **2010**, *58*, 1887–1895. [CrossRef]
7. Kerekes, T.; Teodorescu, R.; Rodríguez, P.; Vázquez, G.; Aldabas, E. A new high-efficiency single-phase transformerless PV inverter topology. *IEEE Trans. Ind. Electron.* **2009**, *58*, 184–191. [CrossRef]

8. Freddy, T.K.S.; Rahim, N.A.; Hew, W.P.; Che, H.S. Comparison and analysis of single-phase transformerless grid-connected PV inverters. *IEEE Trans. Power Electron.* **2013**, *29*, 5358–5369. [CrossRef]
9. Bradaschia, F.; Cavalcanti, M.C.; Ferraz, P.E.; Neves, F.A.; dos Santos, E.C.; da Silva, J.H. Modulation for three-phase transformerless Z-source inverter to reduce leakage currents in photovoltaic systems. *IEEE Trans. Ind. Electron.* **2011**, *58*, 5385–5395. [CrossRef]
10. Guo, X. Three-phase CH7 inverter with a new space vector modulation to reduce leakage current for transformerless photovoltaic systems. *IEEE J. Emerg. Sel. Top. Power Electron.* **2017**, *5*, 708–712. [CrossRef]
11. Freddy, T.K.S.; Rahim, N.A.; Hew, W.P.; Che, H.S. Modulation techniques to reduce leakage current in three-phase transformerless H7 photovoltaic inverter. *IEEE Trans. Ind. Electron.* **2014**, *62*, 322–331. [CrossRef]
12. Guo, X.; Zhang, X.; Guan, H.; Kerekes, T.; Blaabjerg, F. Three-phase ZVR topology and modulation strategy for transformerless PV system. *IEEE Trans. Power Electron.* **2018**, *34*, 1017–1021. [CrossRef]
13. Krishna, R.A.; Suresh, L.P. A brief review on multilevel inverter topologies. In Proceedings of the International Conference on Circuit, Power and Computing Technologies (ICCPCT), Nagercoil, India, 18–19 March 2016; pp. 1–6. [CrossRef]
14. Akbari, A.; Poloei, F.; Bakhshai, A. A Brief Review on State-of-the-art Grid-connected Inverters for Photovoltaic Applications. In Proceedings of the IEEE 28th International Symposium on Industrial Electronics (ISIE), Vancouver, BC, Canada, 12–14 June 2019; pp. 1023–1028. [CrossRef]
15. Zhang, H.; Jouanne, A.V.; Dai, S.; Wallace, A.K.; Wang, F. Multilevel inverter modulation schemes to eliminate common-mode voltages. *IEEE Trans. Ind. Appl.* **2000**, *36*, 1645–1653. [CrossRef]
16. Alepuz, S.; Busquets-Monge, S.; Bordonau, J.; Gago, J.; Gonzalez, D.; Balcells, J. Interfacing Renewable Energy Sources to the Utility Grid Using a Three-Level Inverter. *IEEE Trans. Ind. Electron.* **2006**, *53*, 1504–1511. [CrossRef]
17. Correa, P.; Rodriguez, J.; Rivera, M.; Espinoza, J.R.; Kolar, J.W. Predictive Control of an Indirect Matrix Converter. *IEEE Trans. Ind. Electron.* **2009**, *56*, 1847–1853. [CrossRef]
18. Mohamed, I.S.; Zaid, S.A.; Elsayed, H.M.; Abu-Elyazeed, M.F. Implementation of model predictive control for three-phase inverter with output LC filter on eZdsp F28335 Kit using HIL simulation. *Int. J. Model. Identif. Control* **2016**, *25*, 301–312. [CrossRef]
19. Gulbudak, O.; Santi, E. A predictive control scheme for a dual output indirect matrix converter. In Proceedings of the IEEE Applied Power Electronics Conference and Exposition (APEC), Charlotte, NC, USA, 15–19 March 2015; pp. 2828–2834. [CrossRef]
20. Vazquez, S.; Leon, J.; Franquelo, L.; Rodriguez, J.; Young, H.A.; Marquez, A.; Zanchetta, P. Model Predictive Control: A Review of Its Applications in Power Electronics. *IEEE Ind. Electron. Mag.* **2014**, *8*, 16–31. [CrossRef]
21. Corts, P.; Kazmierkowski, M.P.; Kennel, R.M.; Quevedo, D.E.; Rodrguez, J. Predictive control in power electronics and drives. *IEEE Trans. Ind. Electron.* **2008**, *55*, 4312–4324. [CrossRef]
22. Corts, P.; Ortiz, G.; Yuz, J.I.; Rodrguez, J.; Vazquez, S.; Franquelo, L.G. Model predictive control of an inverter with output LC filter for UPS applications. *IEEE Trans. Ind. Electron.* **2009**, *56*, 1875–1883. [CrossRef]
23. Mohamed, I.S.; Zaid, S.A.; Elsayed, H.M.; Abu-Elyazeed, M.F. Improved model predictive control for three-phase inverter with output LC filter. *Int. J. Model. Identif. Control* **2015**, *23*, 371–379. [CrossRef]
24. Wang, J. Model Predictive of Power Electronic Converter. Master's Thesis, Norwegian University of Science and Technology, Norway, Trondheim, 2012.
25. Vaccari, M.; Pannocchia, G. A Modifier-Adaptation Strategy towards Offset-Free Economic MPC. *Processes* **2017**, *5*, 2. [CrossRef]
26. Li, W.; Kong, D.; Xu, Q.; Wang, X.; Zhao, X.; Li, Y.; Han, H.; Wang, W.; Chen, Z. A Wind Farm Active Power Dispatch Strategy Considering the Wind Turbine Power-Tracking Characteristic via Model Predictive Control. *Processes* **2019**, *7*, 530. [CrossRef]
27. Xiaodong, W.; Zou, J.; Ma, L.; Zhao, J.; Xie, C.; Li, K.; Meng, L.; Guerrero, J.M. Model predictive control methods of leakage current elimination for a three-level T-type transformerless PV inverter. *IET Power Electron.* **2018**, *11*, 1492–1498. [CrossRef]
28. Vargas, R.; Cortes, P.; Ammann, U.; Rodriguez, J.; Pontt, J. Predictive Control of a Three-Phase Neutral-Point-Clamped Inverter. *IEEE Trans. Ind. Electron.* **2007**, *54*, 2697–2705. [CrossRef]
29. Rockhill, A.A.; Liserre, M.; Teodorescu, R.; Rodriguez, P. Grid-Filter Design for a Multimegawatt Medium-Voltage Voltage-Source Inverter. *IEEE Trans. Ind. Electron.* **2011**, *58*, 1205–1217. [CrossRef]

30. Zaid, S.A.; Kassem, A.M. Review, analysis and improving the utilization factor of a PV-grid connected system via HERIC transformerless approach. *Renew. Sustain. Energy Rev.* **2017**, *73*, 1061–1069. [CrossRef]
31. Liserre, M.; Teodorescu, R.; Blaabjerg, F. Stability of photovoltaic and wind turbine grid-connected inverters for a large set of grid impedance values. *IEEE Trans. Power Electron.* **2006**, *21*, 263–272. [CrossRef]
32. Atia, Y. Photovoltaic Maximum Power Point Tracking Using SEPIC Converter. *Eng. Res. J.* **2009**, *36*, 33–40.
33. Rashid, M. *Power Electronics Handbook*, 2nd ed.; Elsevier Academic Press: Amsterdam, The Netherlands, 2011.
34. Husain, M.A.; Tariq, A.; Hameed, S.; Arif, M.S.B.; Jain, A. Comparative assessment of maximum power point tracking procedures for photovoltaic systems. *Green Energy Environ.* **2017**, *2*, 5–17. [CrossRef]
35. Das, P. Maximum power tracking based open-circuit voltage method for PV system. *Energy Procedia* **2016**, *90*, 2–13. [CrossRef]
36. Elbalawi, H.; Zaid, S.A. H5 Transformerless Inverter for Grid Connected PV System with Improved Utilization Factor and Simple Maximum Power Point Algorithm. *Energies* **2018**, *11*, 2912. [CrossRef]
37. Schmidt, H.; Burger, B.; Siedle, C. Gefährdungspotenzial transformatorloser Wechselrichter—Fakten und Gerüchte. In Proceedings of the 18th Symposium Photovoltaische Sonnenenergie, Staffelstein, Germany, 12–14 March 2003; pp. 89–98.

© 2019 by the authors. Licensee MDPI, Basel, Switzerland. This article is an open access article distributed under the terms and conditions of the Creative Commons Attribution (CC BY) license (http://creativecommons.org/licenses/by/4.0/).

Article

Generalized Proportional Model of Relay Protection Based on Adaptive Homotopy Algorithm Transient Stability

Feng Zheng [1,*], Jiahao Lin [1], Jie Huang [1,*] and Yanzhen Lin [2]

[1] The School of Electrical Engineering and Automation, Fuzhou University, Fuzhou 350116, China; jh_fd5678@163.com
[2] The Fuzhou Power Supply Company, Fuzhou 350116, China; lin_yanzhen@163.com
* Correspondence: zf_whu@163.com (F.Z.); jie.huang@fzu.edu.cn (J.H.)

Received: 1 November 2019; Accepted: 26 November 2019; Published: 2 December 2019

Abstract: Relay protection equipment is important to ensure the safe and stable operation of power systems. The risks should be evaluated, which are caused by the failure of relay protection. At present, the fault data and the fault status monitoring information are used to evaluate the failure risks of relay protection. However, there is a lack of attention to the information value of monitoring information in the normal operation condition. In order to comprehensively improve monitoring information accuracy and reduce, a generalized proportional hazard model (GPHM) is established to fully exploit the whole monitoring condition information during the whole operation process, not just the monitoring fault condition data, with the maximum likelihood estimation (MLE) used to estimate the parameters of the GPHM. For solving the nonlinear equation in the process of parameter estimations, the adaptive homotopy algorithm is adopted, which could ensure the reversibility of the Jacobi matrix. Three testing cases have been reviewed, to demonstrate that the adaptive homotopy algorithm is better than traditional algorithms, such as the Newton homotopy algorithm, regarding the calculation speed and convergence. Therefore, GPHM could not only reflect the real time state of the equipment, but also provide a sound theoretical basis for the selection of equipment maintenance types.

Keywords: relay protection equipment; whole monitoring data; generalized proportional hazard model (GPHM); adaptive homotopy algorithm; jacobi matrix

---

1. Introduction

Because a relay protection device can curb the deterioration of a power grid by its fast and correct action [1], it is always seen as the first line of defense to ensure the safe and stable operation of power systems. In recent years, there have been frequent blackouts around the world, and most of them are related to the incorrect action of relay protection. Therefore, evaluation of the reliability of relay protection is the focus of many scholars, and the relay protection failure rate is one of the indexes to estimate its reliability [2,3].

At present, the research on the failure rate model is generally based on the time-failure rate model and equipment state model [4]. The commonly used fault distribution forms for the time-failure rate model are gamma distribution, Weibull distribution, and exponential distribution [5]. The failure rate of exponential distribution is constant and it only represents the accidental failure period. However, the failure rates of most electrical equipment follow the typical curve, namely the bathtub curve, and the bathtub curve includes the early failure period, the accidental failure period, and the three stages of loss failure period. Therefore, the exponential distribution generally is not adopted. By contrast, the Weibull distribution can match the bathtub curve well, so that it has been widely used.

Nevertheless, the Weibull distribution only pays attention to the effect of equipment running enlistment age on the failure rate, which ignores the effect of some external factors such as the equipment maintenance on the equipment failure rate [6]. Accordingly, Weibull distribution has some limitations. The equipment state model is established based on the current state of the equipment [7], so that it does not take the effects of man-made maintenance and historical condition on the failure rate into account, and so it is difficult to predict the future failure rate. Furthermore, there are differences between countries/power companies. The British EA company and Canada Kinectrics Company focus on the health state [8,9]. In China, the failure rate model is an exponential function, which considers the health state as the independent variable [10–12], and from the model, the failure rate may increase exponentially when the equipment state worsens, and the failure rate is beyond 1, so that it doesn't conform to the actual situation. In other words, it is difficult to predict the future failure rate. Aiming at the shortcoming of the two kinds of failure rate model, a Weibull proportional hazards model (PHM), which considered the fault diagnosis value of failure time and used maximum likelihood estimation (MLE) to estimate the model parameters, has been proposed in [13]. In [14], the proportional covariate model (PCM) was put forward in order to solve the lack of data in the fault interval. For a repairable system, the proportional intensity model (PIM) were first proposed by Kumar [15]. Kumar said that the fault rate in a repairable system was affected by many factors, such as operating environment, equipment materials, history operation, design features, and so on. However, the PIM always assumed that the covariates were changed only when failure/maintenance occurred, and maintained constant during the interval of failure/maintenance [16,17]. This assumption ignored the influence of the concomitant variables on the failure rate during the failure/maintenance interval. In [18], the PHM was used to estimate the reliability of thin oxide dielectrics, and used the partial maximum likelihood method to estimate the parameters. In [19], the scholars pointed out that the environmental factors influencing relay reliability mainly included temperature, humidity, vibration, and so on. Additionally, the application of Cox-proportional hazards modeling with respect to the effect of ambient temperature on electromagnetic relays was discussed.

For the above model, ignoring the whole monitoring condition values is the common point. Therefore, based on the above model, this paper analyzes the influence of variables during the failure/maintenance interval, and takes the time varying covariates in the failure/maintenance interval into account, not only the monitoring state value at failure time; then, the generalized proportional hazards model (GPHM) Weibull is built. In order to get the expressions of fault rate, the parameters of Weibull distribution are needed to estimate, which involves the solution for nonlinear equations. At present, there are many the solutions for nonlinear equations, such as the Newton method, the least square method [20], the Marquardt method [21], and so on. There are also many achievements on the solution of nonlinear equations, whereas they still have a fatal defect, namely, local convergence [22]. Because the initial value must be close to the exact one, the requirement of the initial value is very harsh. In the meantime, the calculation is a large amount which brings certain challenges to the running time and space. Actually, for many nonlinear equations, the initial value is not easy to set, which brings inconvenience to the solution of nonlinear equations. Fortunately, the homotopy algorithm has a large convergence range and its requirement on the initial value is not strict, so that it brings a breakthrough to solve nonlinear equations [23]. Howver, the Jacobi matrix of the homotopy algorithm must be reversible, otherwise the homotopy algorithm loses its significance [24]. For solving this problem, the adjusting factor is introduced to construct an adaptive homotopy algorithm to ensure the non-singularity of the Jacobi matrix in this paper. Then, the nonlinear equation can be solved. In summary, in order to fully consider the influence of time–varying covariates in the failure/maintenance interval on the failure rate, the GPHM-Weibull is proposed, which can reflect the real-time state. MLE is used to estimate parameters of GPHM, and the adaptive homotopy algorithm is used to solve nonlinear equation, where the piecewise function expression of fault rate is solved. According to the failure rate, the operation personnel can make the differential operation strategy and realize the economic, stable operation of the power system.

Section 2 presents the generalized proportional hazard model. Section 3 describes the adaptive homotopy algorithm. Section 4 discusses the estimation of Weibull distribution parameters, the solution of nonlinear equations and the calculation of the initial value. Section 5 presents a summary of the proposed method and draws relevant conclusions.

2. Generalized Proportional Hazard Model

With reference to survival function model in medical science [25], based on GPHM, the failure rate model is constructed, and its mathematical expression is as follows:

$$\lambda(t|Z) = \lambda_0(t)\psi(Z(t)) \tag{1}$$

Here, $\lambda_0(t)$ is the basic failure rate. $\Psi(Z(t))$ is the link function, representing the impact of different states $Z(t)$ on failure rate. $Z(t)$ is a vector of covariates, which is composed of n time-varying covariates. Each covariate can represent a particular state. The expression is $Z(t) = [Z_1(t), Z_2(t) \ldots Z_n(t)]$. In practice, the covariate could be an internal variable which can reflect the state of the device, such as the detection information of device. It can also be an external variable which can affect the operation of the device, such as the environmental conditions. In general, the link function can be expressed as follows:

$$\psi(Z(t)) = exp(\gamma_1 Z_1(t) + \gamma_2 Z_2(t) + \cdots + \gamma_n Z_n(t)) \tag{2}$$

Here, $\gamma = (\gamma_1, \gamma_2 \ldots \gamma_n)$ represents the corresponding regression coefficient of each covariate. The assumptions of GPHM are as follows:

(1) The basic failure rate $\lambda_0(t)$ subjects to Weibull distribution, and its expression is

$$\lambda_0(t) = \frac{\beta}{\eta}(\frac{t}{\eta})^{\beta-1}. \tag{3}$$

Here, β is the shape parameter, and η is the scale parameter.
(2) The fault interval is longer than maintenance time, so that the maintenance time can be neglected.
(3) The effect of covariates on the failure rate maintain constant and it cannot be changed with time.

Choosing the best covariates is the key to establish the GPHM. Age, operating environment, maintenance times, health index, and manufacturer are selected as covariates, as shown in Figure 1 [25].

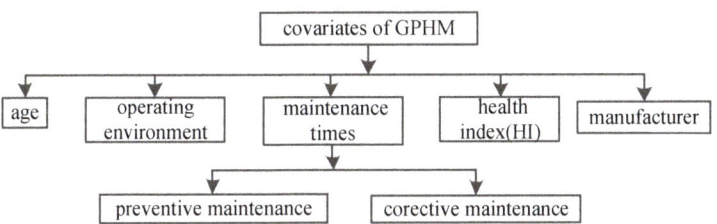

Figure 1. Selection of covariates.

Health index (HI) reflects the overall health level of the relay protection equipment, which is closely related to the equipment failure rate. In order to facilitate quantitative comparison, HI is divided into five levels (normal, attention, serious, emergency, and fault), and their corresponding values are listed in Table 1.

Table 1. Values of Health Index (HI).

Grade	Normal	Attention	Emergency	Serious	Fault
values of HI	0	0.1	0.3	0.6	1

In [26], research's results indicated that when the equipment is in the loss period, the failure rate meets Weibull distribution, and Weibull distribution is a function of time. It means that the failure rate is related to age. In [27], the scholars pointed out that the failure rate function directly is multiplied by the age reduction factor, and failure rate increase factor after repairmen. Because the age reduction factor and failure rate increase factor are related to the maintenance type and maintenance times, the failure rate $\lambda_{k+1}(t)$ after kth maintenance is defined recursively as:

$$\lambda_{k+1}(t) = \beta_k \lambda_k(t + \alpha_k T_k) \quad (4)$$

Here, k is the kth maintenance. T_k represents the interval for the kth maintenance. β_k is age reduction factor, which can simulate the equipment damage caused by each maintenance. α_k is failure rate increase factor, which can describe the degree of improvement in equipment failure rate after maintenance.

According to [28], the failure rate can be affected the operating environment. Therefore, the business district and the industrial area differ considerably. Different manufacturers may have different familial defects which may affect the history data. With reference to [27], the values of the operating environment and manufacturers are given as Tables 2 and 3 shown.

Table 2. Values of operating environment.

Operating Environment	Business District	Industrial Area
value	1	2

Table 3. Of the manufacturer.

Manufacturer	NR Electronic	Beijing Sifang	Changyuan Shenrui	Guodian Nanzi	Xuji Dianqi
value	1	2	3	4	5

Therefore, the expression of the failure rate $\lambda(t|Z)$ which takes the influence of covariates into consideration can be expressed as

$$\lambda(t|Z) = \frac{\beta}{\eta}(\frac{t}{\eta})^{\beta-1} \exp(\gamma_{pm} Z_{pm}(t) + \gamma_{cm} Z_{cm}(t) + \gamma_{age} Z_{age}(t) \\ + \gamma_{env} Z_{env}(t) + \gamma_{HI} Z_{HI}(t) + \gamma_{mau} Z_{mau}(t)) \quad (5)$$

Here, covariant Z_{pm}, Z_{cm}, Z_{age}, Z_{env}, Z_{HI}, and Z_{mau} respectively represent the times of preventive maintenance and corrective maintenance, age, the operating environment, the HI of equipment and manufacturer; the corresponding coefficient of the covariates are λ_{pm}, λ_{cm}, λ_{age}, λ_{env}, λ_{HI}, and λ_{mau}. Equation (5) illustrates that if you want to get to the expression of failure rate, $\beta/\eta/\gamma_i$ is needed in order to estimate. The maximum likelihood function (MLE) and the adaptive homotopy algorithm are used to solve the nonlinear equations.

3. Adaptive Homotopy Algorithm

3.1. The Basicprinciple of Homotopy Algorithm

Considering the following nonlinear Equation (6):

$$\begin{cases} f_1(x_1, x_2, x_3, x_4 \ldots \ldots, x_n) = 0 \\ f_2(x_1, x_2, x_3, x_4 \ldots \ldots, x_n) = 0 \\ \vdots \\ f_n(x_1, x_2, x_3, x_4 \ldots \ldots, x_n) = 0 \end{cases} \quad (6)$$

Here, $x = (x_1, x_2, x_3, x_4 \ldots x_n) \in R^n$ can be obtained. $f_i(x_1, x_2, x_3, x_4 \ldots x_n)$ is a real function defined on a regional D, $i = (1, 2, 3 \ldots n)$. Its vector notation is:

$$\vec{F}(x) = \begin{bmatrix} f_1(x) \\ f_2(x) \\ \vdots \\ f_n(x) \end{bmatrix} \quad \vec{x} = \begin{bmatrix} x_1 \\ x_2 \\ \vdots \\ x_n \end{bmatrix} \in R^n \quad \vec{0} = \begin{bmatrix} 0 \\ 0 \\ \vdots \\ 0 \end{bmatrix} \tag{7}$$

Then, Equation (7) can be converted to:

$$\vec{F}(x) = \vec{0} \tag{8}$$

In Equation (8), parameter t is introduced to construct a set of homotopy mapping $H(x, t)$ which subjects to:

$$\begin{cases} H(x,0) = G(x_0) = 0 \\ H(x,1) = F(x) + (t-1)G(x) \end{cases} \tag{9}$$

From Equation (9), the equations can be obtained, which are $t = 0$, $H(x, 0) = G(x_0)$ and $t = 1$, $H(x, 1) = F(x)$. Then, the solution of equation $F(x) = 0$ is transformed into the solution of equation $x = x(t)$ which subjects to equation $H(x, 1) = 0$. Equation (9) indicates that due to different $G(x)$, there are different homotopy equations.

A Fixed Point Homotopy Algorithm

If $G(x) = x - x_0$, then a fixed point homotopy algorithm is formed as:

$$H(x,t) = F(x) + (t-1)(x - x_0) \tag{10}$$

Newton Homotopy Algorithm

If $G(x) = F(x) - F(x_0)$, then a Newton homotopy algorithm is formed as:

$$H(x,t) = tF(x) + (1-t)(F(x) - F(x_0)) \tag{11}$$

The derivative of parameter t in $H(x, t) = 0$ is:

$$\frac{\partial H}{\partial x}\frac{dx}{dt} + \frac{\partial H}{\partial t} = 0 \tag{12}$$

If the inverse matrix $\left(\frac{\partial H}{\partial x}\right)^{-1}$ exists, then:

$$\frac{dx}{dt} = -\left(\frac{\partial H}{\partial x}\right)^{-1} \cdot \frac{\partial H}{\partial t} \tag{13}$$

Adaptive Homotopy Equation

However, when inverse matrix $(\partial H/\partial x)^{-1}$ doesn't exist, the homotopy algorithm will lose its significance. Because the diagonal factor $G(x) = diag[e^{g_i(x)}]$ in the exponential homotopy method is multiplied by F to construct a new homotopy algorithm, it is only feasible in theory, and the calculation is complicated and not suitable for the large-scale nonlinear equation. However, in reference to the idea of exponential homotopy algorithm, the equation for adaptive homotopy algorithm can be obtained as:

$$H(x,t) = F(x) - (1-t)[F(x_0) - a(1+t)(x - x_0)] \tag{14}$$

$$\left(\frac{\partial H}{\partial x}\right)^{-1} = [F'(x) + a(1-t^2)I]^{-1} \tag{15}$$

Because $a(1-t^2)I$ is a nonsingular matrix and when $(\partial H/\partial x)^{-1}$ is singular, through adjusting the parameter a, it can account for the diagonal dominance, as long as a is large enough. In the actual calculation, the initial value of parameter a is set as 0. When the Jacobi matrix $F'(x)$ becomes singular after some calculation steps, a automatic increase Δa. Thus, the solution of the Equation (7) can be obtained by finding the solution of the homotopy Equation (9).

3.2. Numerical Calculation of the Adaptive Homotopy Algorithm

Equation (13) presents that the calculation of nonlinear equations can be converted into the calculation of IVP (initial value problem), which can be expressed as:

$$\begin{cases} \frac{dx}{dt} = -(\frac{\partial H}{\partial x})^{-1} \cdot \frac{\partial H}{\partial t} = [F'(x) + a(1-t^2)I]^{-1} \cdot \frac{\partial H}{\partial t} \\ x_0 = x(0) \end{cases} \quad (16)$$

In order to solve the Equation (16), the Euler method is used to estimate, and the Runge Kutta method is used to correct.

Euler method

We begin to track the path from the starting point (t_0, x_0) of the homotopy path, and the Euler method is adopted to estimate the next approximate point $(t_1, \tilde{x}_1)$, so that the expression is:

$$\tilde{x}_1 = x_0 + \frac{dx}{dt}\Delta t = x_0 - (\frac{\partial H}{\partial x_0})^{-1} \frac{\partial H}{\partial t_0}\Delta t \quad (17)$$

With the iterative equation as:

$$\begin{cases} \tilde{x}_n = x_{n-1} - (\frac{\partial H}{\partial x_{n-1}})^{-1} \frac{\partial H}{\partial t_{n-1}}\Delta t \\ t_{n-1} = t_1 + (n-1) \times h \\ h = \frac{1}{N} \end{cases} \quad (18)$$

Here, h is the step size, and n is the number of iterations.

The Fouth Runge Kutta method

Through using the Runge Kutta method, the local truncation error of the fourth Runge Kutta method is about $o(h5)$ [13], and its calculation speed is fast. Therefore, the fouth Runge Kutta method is adopted to calculate the initial parameters.

Assuming $\frac{dx}{dt} = -J(x)^{-1}F(x_0) = y(x_n, t_n)$, according to Equation (18), the point (t_1, x_1) can be obtained. Then, the point (t_1, x_1) is set as the starting point. For obtaining the next round of prediction-correction, the equations are used as follows:

$$\begin{cases} x_n = \tilde{x}_n - (\frac{\partial H}{\partial x})^{-1} H(t_n, \tilde{x}_n) \\ t_n = t_0 + h \times n \end{cases} \quad (19)$$

$$\begin{cases} \tilde{x}_{n+1} = x_n + h(\frac{k_1 + 2k_2 + 2k_3 + k_4}{6}) \\ k_1 = y(x_n, t_n) \\ k_2 = y(x_n + \frac{1}{2}h, t_n + \frac{1}{2}hk_1) \\ k_3 = y(x_n + \frac{1}{2}h, t_n + \frac{1}{2}hk_1) \\ k_3 = y(x_n + h, t_n + hk_3) \end{cases} \quad (20)$$

And the prediction-correction process is stopped until $t = 1$. After several iterations, x_{n+1} may not be the exact solution x^*, according to mathematical convergence theory. If $|x_{n+1} - x_n| < \varepsilon$ (ε is the set of coefficients of accuracy), it is considered that the exact solution is found, otherwise the above steps are repeated to perform the predictive-correction process. In order to elucidate the adaptive homotopy algorithm further, Figure 2 shows a flow diagram of this algorithm, which was implemented in MATLAB.

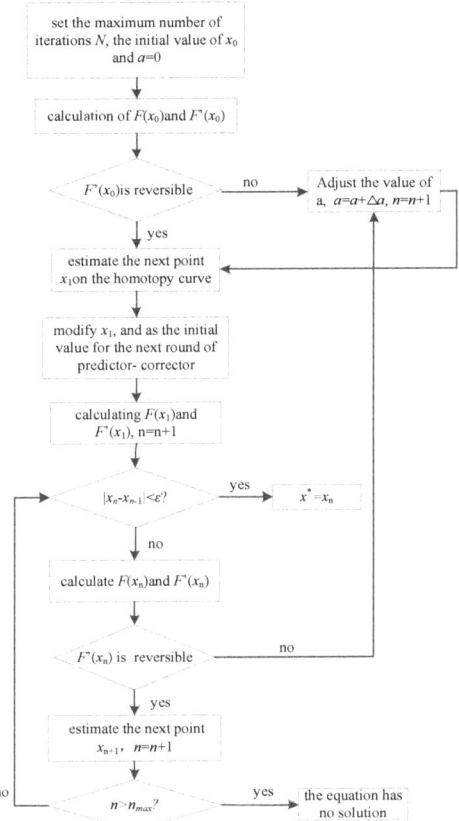

Figure 2. Calculation flow chart of adaptive homotopy algorithm.

Step 1: Set the maximum number of iterations n_{max}, the initial value x_0 and $a = 0$;
Step 2: According to Equation (9), calculate $F(x_0)$ and $F'(x_0)$;
Step 3: Judge the reversibility of $F'(x_0)$, if YES, then turn to step 4, if not, then turn to step 9;
Step 4: Based on Equation (18), estimate the next point x_1;
Step 5: According to Equation (20), modify x_1 and choose the modified x_1 as the initial value for the next round. Meanwhile calculate $F(x_1)$ and $F'(x_1)$, n set as $n + 1$;
Step 6: Assume that nth step has been carried out and obtain the solution x_n. judge whether the inequality $|x_n - x_{n-1}| < \varepsilon$ is true or not. If YES, output the exact solution $x^* = x_n$, otherwise turn step 7;
Step 7: Calculate $F(x_n)$ and $F'(x_n)$, and judge whether $F'(x_n)$ is reversible or not. If YES, based on Equation (18), estimate the next point x_{n+1}, and n automatically add 1, namely, $n = n + 1$. If not, turn step 9;
Step 8: Judge whether the inequality $n > n_{max}$ is true or not. If YES, the equation without solution. If not, then turn step 6;
Step 9: Adjust the value of a, and define $a = a + \Delta a$. n automatically add 1, namely, $n = n + 1$, and turn to step 3.

4. Parameter Estimation

4.1. Weibull Distribution Parameters Estimation

There are two reasons which may lead to the relay protection equipment withdrawal from the power system. One is its failure, and the other one is the maintenance. The former possesses fault data and belongs to the corrective maintenance (CM). The latter possesses the truncated data and belongs to the preventive maintenance (PM). If the relay protection equipment is still running at the end of the observation, the data can also be censored. Considering the censored data, based on maximum likelihood, the parameters' estimation can be described as follows:

The failure time at $0 < t_i < \ldots < t_n$ (T) is observed in the time interval (0 T]. MLE is used to calculate the parameters of GPHM. It is supposed that (t_i, z_i, δ_i) and $(i = 1, 2, \ldots, n)$ are the records of failure and maintenance, respectively; n is the total number of events, including all the CM times and all the PM times, so that n can be expressed as $n = N_{cm}(t) + N_{pm}(t)$. t_i is the failure time of PM and CM; z_i is the state information of the equipment at t_i, δ_i is corresponding censoring indicator variables, and the equation $\delta_i = 0$ represents no failure at t_i. The equation $\delta_i = 1$ represents failure at t_i.

The corresponding likelihood function is given as:

$$L(\beta, \eta) = \prod_{i=1}^{l} f(t_i)^{\delta_i} R(t_i)^{1-\delta_i} \tag{21}$$

Defining $f(t) = R(t)\lambda(t)$, then

$$R(t) = \exp[-\int_0^t \lambda_0(t) \exp(\gamma Z(t))] \tag{22}$$

Therefore, Equation (21) can be converted into

$$\begin{aligned} L(\beta, \eta) &= \prod_{i=1}^{l} f(t_i)^{\delta_i} R(t_i)^{1-\delta_i} \\ &= \prod_{i=1}^{r} \lambda(t_i) \prod_{i=1}^{l} R(t_i) \\ &= \prod_{i=1}^{r} \frac{\beta}{\eta}(\frac{t_i}{\eta})^{\beta-1} \exp(\sum_{j=1}^{6} \gamma_j Z_j(t_i)) * \\ &\quad \prod_{i=1}^{l} \exp(-\int_0^T \frac{\beta}{\eta}(\frac{t_i}{\eta})^{\beta-1}) \exp(\sum_{j=1}^{6} \gamma_j Z_j(t_i))) \end{aligned} \tag{23}$$

Here, l represents the total number of the relay protection equipment and r represents the number of faulty relay protection equipment. Then, the corresponding log likelihood function is:

$$\begin{aligned} \ln L = &\ r \ln \frac{\beta}{\eta} + \sum_{i=1}^{r} [(\beta-1) \ln(\frac{t_i}{\eta}) + \sum_{j=1}^{6} \gamma_j Z_j(t_i)] \\ &- \sum_{i=1}^{n} ((\frac{t_i}{\eta})^{\beta} \cdot \exp \sum_{j=1}^{6} \gamma_j Z_j(t_i)) \end{aligned} \tag{24}$$

Taking the partial derivatives of β and η separately:

$$\frac{\partial \ln L}{\partial \beta} = \frac{r}{\beta} \sum_{i=1}^{r} \ln(\frac{t_i}{\eta}) - \sum_{i=1}^{n} (\frac{t_i}{\eta})^{\beta} \cdot \ln(\frac{t_i}{\eta}) \cdot \exp(\sum_{j=1}^{6} \gamma_j Z_j(t_i)) \tag{25}$$

$$\frac{\partial \ln L}{\partial \eta} = -\frac{\beta r}{\eta} + \sum_{i=1}^{n} +\frac{\beta}{\eta}(\frac{t_i}{\eta})^{\beta} \exp \sum_{j=1}^{6} \gamma_j Z_j(t_i) \tag{26}$$

The maximum likelihood functions are:

$$\frac{\partial \ln L}{\partial \beta} = 0 \tag{27}$$

$$\frac{\partial \ln L}{\partial \eta} = 0 \tag{28}$$

Taking the second derivative operations, then:

$$\frac{\partial^2 \ln L}{\partial \beta^2} = -\frac{r}{\beta^2} - \sum_{i=1}^{n} \left(\frac{t_i}{\eta}\right)^\beta [\ln(\frac{t_i}{\eta})]^2 \exp(\sum_{j=1}^{6} \gamma_j Z_j(t_i)) \tag{29}$$

$$\frac{\partial^2 \ln L}{\partial \eta^2} = -\frac{r\beta}{\eta^2} - \sum_{i=1}^{n} \left(\frac{t_i}{\eta}\right)^\beta \cdot \frac{\beta^2 + \beta}{\eta^2} \exp(\sum_{j=1}^{6} \gamma_j Z_j(t_i)) \tag{30}$$

$$\frac{\partial^2 \ln L}{\partial \beta \partial \eta} = -\frac{r}{\eta} + \sum_{i=1}^{n} \frac{1}{\eta}\left(\frac{t_i}{\eta}\right)^\beta [1 + \beta \ln t_i] \exp(\sum_{j=1}^{6} \gamma_j Z_j(t_i)) \tag{31}$$

$$\frac{\partial^2 \ln L}{\partial \eta \partial \beta} = -\frac{r}{\eta} + \sum_{i=1}^{n} \frac{1}{\eta}\left(\frac{t_i}{\eta}\right)^\beta [1 + \beta \ln t_i] \exp(\sum_{j=1}^{6} \gamma_j Z_j(t_i)) \tag{32}$$

Then, a second order derivative matrix can be obtained, namely Jacobi matrix:

$$J = \begin{bmatrix} \frac{\partial^2 \ln L}{\partial \beta^2} & \frac{\partial^2 \ln L}{\partial \beta \partial \eta} \\ \frac{\partial^2 \ln L}{\partial \eta \partial \beta} & \frac{\partial^2 \ln L}{\partial \eta^2} \end{bmatrix} \tag{33}$$

4.2. The Solution of Nonlinear Equations Based on Adaptive Homotopy Algorithm

According to the adaptive homotopy algorithm mentioned, the iteration equation is:

$$\begin{pmatrix} \beta \\ \eta \end{pmatrix}_{k+1} = \begin{pmatrix} \beta \\ \eta \end{pmatrix}_k - (J_k + a_k(1-t^2)I)^{-1} \begin{pmatrix} \frac{\partial \ln L}{\partial \beta} \\ \frac{\partial \ln L}{\partial \eta} \end{pmatrix} \tag{34}$$

Here, β_0 and η_0 are the initial values of corresponding parameters when $k = 0$. After the initial values are selected, two parameters can be calculated according to the Equation (34).

4.3. Calculation of the Initial Value

Selecting two data points, namely, (t_d, λ_d) and (t_g, λ_g) in the Weibull distribution, the initial values can be solved by the following equation.

$$\begin{cases} \frac{\beta_0}{\eta_0}\left(\frac{t_d}{\eta_0}\right)^{\beta-1} = \lambda_d \\ \frac{\beta_0}{\eta_0}\left(\frac{t_g}{\eta_0}\right)^{\beta-1} = \lambda_g \end{cases} \tag{35}$$

According to the above analysis, MLE can be used to estimate the parameter of GPHM. And the adaptive homotopy algorithm can be used to solve the nonlinear equation in the parameter estimation process. Then, the fault rate model is established.

5. Case Analysis

Case 1: The machine account and defect information of the relay protection equipment are collected, which are running in the similar environment or have the same type, and its failure rate is shown in Table 4. According the above failure rate, the failure rate curve can be drawn as Figure 3 shown. Based on the adaptive homotopy algorithm and the Newton homotopy algorithm, the failure rate parameters are computed iteratively. Then, failure rate parameters of the Weibull distribution can be obtained and shown in Tables 5 and 6, so that their fitting curve can also be seen in Figure 3.

Table 4. Statistics of failure rate of relay protection.

Running Time/Year	Failure Rate/(Times/Device. Year)	Running Time/Year	Failure Rate (Times/Device. Year)
1	0.0251	7	0.0272
1.5	0.0202	7.5	0.0342
2.5	0.0226	8	0.0311
3	0.025	8.5	0.0496
3.5	0.0177	9	0.0364
4	0.0268	9.5	0.0774
4.5	0.0232	10	0.133
5	0.0261	10.5	0.189
5.5	0.0283	11	0.232
6	0.0253	12	0.374
6.5	0.0296	13	0.593

Table 5. Estimation of parameters based on adaptive homotopy algorithm.

Stage of Fault Distribution	Random Failure Period		Loss Failure Period	
	β	η	β	η
parameter values	1.2903	29.1759	7.818	12.526
relative error	0.0971		0.1711	
iteration number		5		

Table 6. Estimation of parameters based on Newton homotopy algorithm.

Stage of Fault Distribution	Random Failure Period		Loss Failure Period	
	β	η	β	η
parameter values	1.177	31.936	7.248	12.435
relative error	0.1532		0.2518	
iteration number		11		

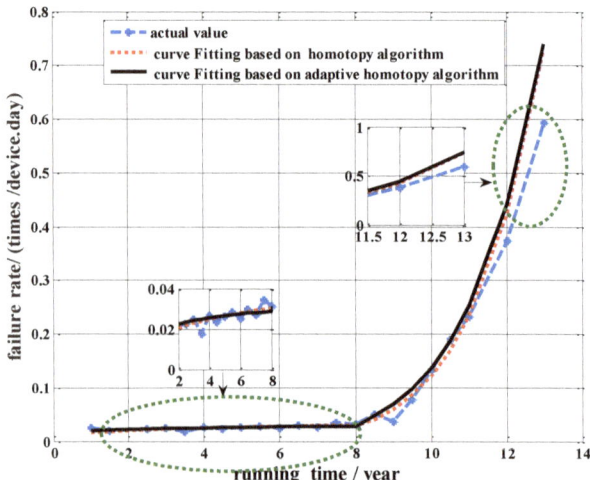

Figure 3. Curve fitting of failure rate.

Through analyzing the data of Tables 5 and 6 and the curves in Figure 3, the iterations number of the adaptive homotopy algorithm is significantly smaller the iterations number of other algorithms. When the equipment is in the random failure period, the relative error of the adaptive homotopy algorithm is 0.0971, and the Newton homotopy algorithm is 0.1532. When the equipment is in the loss failure period, the relative error of the adaptive homotopy method is 0.1711 and the Newton homotopy algorithm is 0.2518, whose error is large. This is caused by the fact that when the equipment is running, all parts of the equipment occur material fatigue, aging or rust and other undesirable conditions. In order to ensure the normal operation of the equipment, the appropriate maintenance should be done, namely, PM, or CM. However, these two types of maintenance inevitably affect the equipment failure rate. Therefore, the model considering the run-time regardless of the current state will be not correct. The result has big difference with the actual operation.

Case 2: Similar to case 1, the operation data of relay protection equipment is shown in Table 7. Based on Table 7, the failure rate curve of relay protection equipment can be obtained in Figure 4.

Table 7. Operation data of relay protection.

Running Time/Year	Failure Rate (Times/Device. Year)	Running Time/Year	Failure Rate (Times/Device. Year)
1.5	0.0224	7.0	0.0268
2	0.0202	7.5	0.0342
2.5	0.0226	8.0	0.0327
3	0.024	8.5	0.0453
3.5	0.0175	9.0	0.0411
4	0.0252	9.5	0.0726
4.5	0.0267	10.0	0.0693
5	0.0223	10.5	0.116
5.5	0.0283	11	0.132
6	0.0253	11.5	0.187
6.5	0.0296	12	0.213

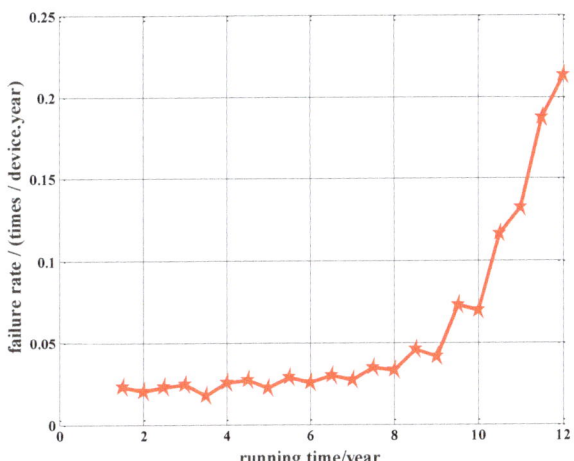

Figure 4. The failure rate curve of equipment.

Based on the data from Table 7, the convergence characteristics of different algorithms can be obtained. From Table 8, it can be found that the Newton homotopy algorithm is non-converging. However, the adaptive homotopy algorithm can guarantee the singularity of the equation by controlling the parameter a. According to Equation (15), a_n affects singularity of nonlinear equations, so that the optimal parameter a is selected which can make the iteration number need less. According to Table 8, although there will be differences between the initial values/parameter values/iterations number, the difference of the final parameter estimates' results are in the allowed error range. Therefore, it can be proved that the homotopy algorithm is independent of the setting of the initial value, and the different initial values are the different optimal values of a. For example, when the initial value is (1, 1), the iteration number is 10, and then the optimal value of a is 3. While the initial value is (11, 1), the iteration number is 35, and then the optimal value of a is −1. Figure 5 shows the curves corresponding to different initial values. From Figure 5, the trend of the curve can fully verify the above analysis results.

Case 3: It is necessary to do the corresponding maintenance for equipment after the device is put into operation for some time. According to Equation (15), it can be found that the choice of maintenance can affect the failure rate. Table 9 gives the operating data of the equipment maintenance. In order to solve the parameters of GPHM, the life data of relay protection equipment is analyzed, firstly. Then, the regression coefficient vector γ is estimated by Statistical Analysis Software (SPSS). On the basis of γ, the adaptive homotopy algorithm is used to estimate the other parameters.

Table 8. Comparisons of two algorithms.

Initial Value	Newton Homotopy Algorithm		Adaptive Homotopy Algorithm			
	convergence	(β, η)	convergence	iteration number	a	(β, η)
(1, 1)	non-converging	-	convergence	10	3	6.201, 13.790
(11, 1)	non-converging	-	convergence	35	−1	6.201, 12.971

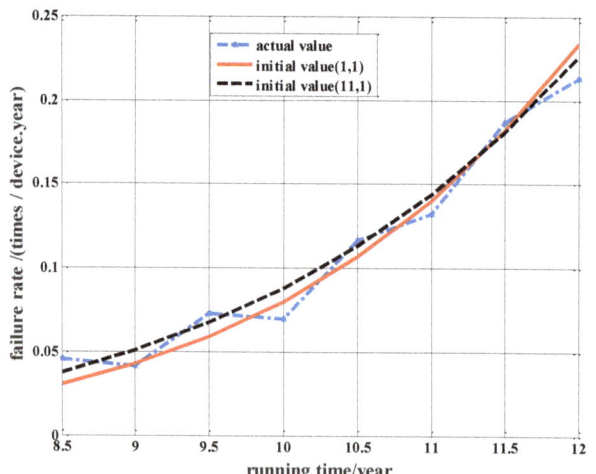

Figure 5. The failure rate curve of equipment.

Table 9. Operation data of relay protection equipment.

Running Time/Day	Failure Rate/(Times/Device. Day)	Running Time/Day	Failure Rate/(Times/Device. Day)
215	0.0130	1300	0.0186
310	0.0141	1557	0.0212
471	0.0182	1754	0.0215
680	0.024	1891	0.0200
763	0.0221	2058	0.0211
841	0.0202	2177	0.0200
1008	0.0168	2482	0.066168
1193	0.019		

Model I: selecting the times of maintenance as a covariate; *Models II/III*: on the basis of model I, the condition monitoring data of the corrective or preventive maintenance moment are fitted, such as age, the operating environment, the HI of equipment, and the manufacturer; *Model IV*: on the basis of *Models II/III*, the condition monitoring data of the whole period (such as age, the operating environment, the HI of equipment, and the manufacturer) are fitted, not just the state of corrective or preventive maintenance moment.

From Table 10, by comparing the estimated log likelihood, the fourth methods are optimal and its log likelihood is −53.930. Obviously, based on the monitoring data of the entire running time, GPHM has better fitting characteristics. According to parameter estimation results, it can be found that the failure maintenances frequency, γ_{cm}, is negative in the above three approaches, so that CM can effectively reduce the failure rate. The preventive maintenances frequency, γ_{pm}, is positive, so that PM can't effectively reduce the failure rate. This conclusion can provide a theoretical basis for maintenance personnel to choose the appropriate maintenance mode, and improve maintenance efficiency in case of the blind maintenance. The results also show that when the environment and the internal state of the covariates deviate from the rated or normal state, the failure rate will be higher. The quantitative analysis is consistent with the experience. Based on GPHM model, the curve of equipment failure rate can be obtained as Figure 5 shown. From Figure 6, we can find that the curve based on GPHM is closed to the actual value, while the Weibull model is away from the actual value, especially in the loss period. That is because in the loss period the equipment has to implement the maintenance. It is inevitable to affect the fault rate.

Table 10. Parameter estimation of models with different factor combinations.

Model	Estimated Log Likelihood	Parameter Estimation							
		β	η	γ_{pm}	γ_{cm}	γ_{age}	γ_{env}	γ_{HI}	γ_{mau}
I	−55.409	1.424	29.58	0.013	−0.001	0	0	0	0
II	−57.940	1.25	33.49	0	−0.001	−0.001	0.911	−0.079	−0.288
III	−54.404	1.434	25.673	0.0137	0	0.0004	0.728	0.053	0.482
IV	−53.930	1.424	29.581	0.0138	−0.001	0.0003	0.773	−0.047	0.516

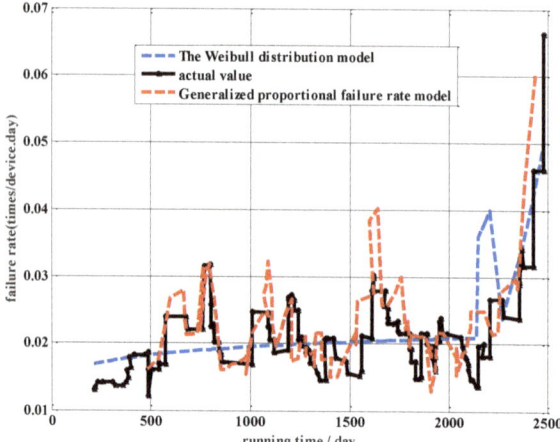

Figure 6. Curve fitting based on the generalized proportional hazard model (GPHM).

6. Conclusions

In order to fully analyze the impact of the whole process of the monitoring state on the failure rate, this paper presents GPHM-Weibull, whose covariates include the times of PM and CM, age, the operating environment, HI, and manufacturer. The baseline function obeys Weibull distribution, and the adaptive homotopy algorithm is adopted to estimate the parameters of Weibull distribution. The regression coefficient vector γ is estimated by SPSS. Finally, three cases have been presented to demonstrate the following conclusions.

(1) The curve drawn by the GPHM is very close to the actual value. However, due to GPHM taking the whole running state of the equipment into consideration, the curve drawn by the Weibull model is away from the actual value, especially in the loss period.

(2) The adaptive homotopy algorithm ensures the singularity of the equation by adjusting the parameter a, and the result of the parameter estimation is less affected by the initial value.

Author Contributions: F.Z. and J.L. conceived and designed the study; J.H. performed the simulation; Y.L. provided the simulation case; F.Z. and J.L. wrote the paper; J.H. and Y.L. reviewed and edited the manuscript. All authors read and approved the manuscript.

Funding: This research was funded by the Natural Science Foundation of Fujian Province, China through grant number 2019J01249.

Conflicts of Interest: The authors declare no conflict of interest.

Nomenclature

λ_0	basic failure rate
$Z(t)$	failure rate
β/η	shape/scale parameter
α_k	failure rate increase factor
Z_{pm}/Z_{cm}	times of preventive/corrective maintenance
Z_{age}/Z_{env}	times of age/operating environment
Z_{HI}/Z_{mau}	times of equipment HI/manufacturer
δ_i	censoring indicator variables
r	faulty relay protection equipment number
$\lambda_{pm}/\lambda_{cm}/\lambda_{age}$ $\lambda_{env}/\lambda_{HI}/\lambda_{mau}$	coefficient of covariates
T_k	kth maintenance interval
k	kth maintenance
β_k	age reduction factor
h	step size
n	iterations number
γ	corresponding regression coefficient
l	relay protection equipment total number

References

1. Lin, X.; Yu, K.; Tong, N.; Li, Z.; Khalid, M.S.; Huang, J.; Li, Z.; Zhang, R. Countermeasures on preventing backup protection mal-operation during load flow transferring. *Int. J. Electr. Power Energy Syst.* **2016**, *79*, 27–33. [CrossRef]
2. Alam, M.N.; Das, B.; Pant, V. An interior point method based protection coordination scheme for directional overcurrent relays in meshed networks. *Int. J. Electr. Power Energy Syst.* **2016**, *81*, 153–164. [CrossRef]
3. Dai, Z.H.; Wang, Z.P. Overview of research on protection reliability. *Power Syst. Prot. Control* **2010**, *38*, 161–167.
4. Zhou, X. *New Technology of Power System Reliability*; China Electric Power Press: Beijing, China, 2014.
5. Moeini, A.; Jenab, K.; Mohammadi, M.; Foumani, M. Fitting the three-parameter Weibull distribution with Cross Entropy. *Appl. Math. Model.* **2013**, *37*, 6354–6363. [CrossRef]
6. Dombi, J.; Jónás, T.; Tóth, Z.E. Clustering Empirical Failure Rate Curves for Reliability Prediction Purposes in the Case of Consumer Electronic Products. *Qual. Reliab. Eng. Int.* **2015**, *32*, 1071–1083. [CrossRef]
7. Wang, H.; Yang, H.; He, B.; Wang, Y. Improvement of state failure rate model for Power transmission and transforming equipment. *Autom. Electr. Power Syst.* **2011**, *15*, 27–31.
8. Jahromi, A.; Piercy, R.; Cress, S.; Service, J.; Fan, W. An approach to power transformer asset management using health index. *IEEE Electr. Insul. Mag.* **2009**, *2*, 20–34. [CrossRef]
9. Hughes, D.; Dennis, G.; Walker, J.; Williamson, C. *Condition Based Risk Management (CBRM)—Enabling Asset Condition Information to be Central to Corporate Decision Making*; Engineering Asset Management; Springer: London, UK, 2006; pp. 1212–1217.
10. Earp, G. Condition based risk assessment of electricity towers using high resolution images from a helicopter Electricity Distribution. In Proceedings of the 18th International Conference and Exhibition on Electricity Distribution (CIRED 2005), Turin, Italy, 6–9 June 2005; IET (Britain): London, UK, 2005; pp. 1–5.
11. Chan, J.K.; Shaw, L. Modeling repairable systems with failure rates that depend on age and maintenance. *IEEE Trans. Reliab.* **1993**, *42*, 566–571. [CrossRef]
12. Dui, H.; Si, S.; Zuo, M.J.; Sun, S. Semi-Markov process-based integrated importance measure for multi-state systems. *IEEE Trans. Reliab.* **2015**, *64*, 754–765. [CrossRef]
13. Jardine, A.; Anderson, P.; Mann, D. Application of the Weibull proportional hazards model to aircraft and marine engine failure data. *Qual. Reliab. Eng. Int.* **1987**, *3*, 77–82. [CrossRef]
14. Sun, Y.; Ma, L.; Mathew, J.; Wang, W.; Zhang, S. Mechanical systems hazard estimation using condition monitoring. *Mech. Syst. Signal Process.* **2006**, *20*, 1189–1201. [CrossRef]

15. Kumar, D. Proportional hazards modelling of repairable systems. *Qual. Reliab. Eng. Int.* **1995**, *11*, 361–369. [CrossRef]
16. Syamsundar, A.; Naikan, V.N.A. Mathematical modelling of maintained systems using point processes. *IMA J. Manag. Math.* **2009**, *20*, 275–301. [CrossRef]
17. Lugtigheid, D. Modelling repairable system reliability with explanatory variables and repair and maintenance actions. *IMA J. Manag. Math.* **2004**, *15*, 89–93. [CrossRef]
18. Elsayed, E.; Chan, C. Estimation of thin-oxide reliability using proportional hazards models. *IEEE Trans. Reliab.* **2002**, *39*, 329–335. [CrossRef]
19. Li, L.; Ma, D.; Li, Z. Cox-Proportional Hazards Modeling in Reliability Analysis—A Study of Electromagnetic Relays Data. *IEEE Trans. Compon. Packag. Manuf. Technol.* **2015**, *5*, 1582–1589.
20. Zhang, R.; Cai, W.; Ni, L.; Lebby, G.L. Power system load forecasting using partial least square method. In Proceedings of the 40th Southeastern Symposium on System Theory (SSST), New Orleans, LA, USA, 16–18 March 2008; pp. 169–173.
21. Bouaricha, A.; Schnabel, R.B. Tensor methods for large sparse systems of nonlinear equations. *Math. Program.* **1998**, *82*, 377–400. [CrossRef]
22. Reif, K.; Weinzierl, K.; Zell, A.; Unbehauen, R. A homotopy approach for nonlinear control synthesis. *IEEE Trans. Autom. Control* **1998**, *43*, 1311–1318. [CrossRef]
23. Li, T.Y.; Zeng, Z.; Cong, L. Solving eigenvalue problems of real nonsymmetric matrices with real homotopies. *SIAM J. Numer. Anal.* **1992**, *29*, 229–248. [CrossRef]
24. Bayat, M.; Pakar, I.; Bayat, M. Approximate analytical solution of nonlinear systems using homotopy perturbation method. *J. Process Mech. Eng.* **2016**, *230*, 10–17. [CrossRef]
25. Kumar, D.; Westberg, U. Proportional hazards modeling of time-dependent covariates using linear regression: A case study: Mine power cable reliability. *IEEE Trans. Reliab.* **1996**, *45*, 386–392. [CrossRef]
26. Wang, H.; Zhao, W.; Du, Z. Economic Life Prediction of Power Transformers Based on the Lifetime Data. *Power Syst. Technol.* **2015**, *39*, 810–816.
27. Percy, D.F.; Kobbacy, K.A.H.; Ascher, H.E. Using proportional-intensities models to schedule preventive–maintenance intervals. *IMA J. Math. Appl. Bus. Ind.* **1998**, *9*, 289–302. [CrossRef]
28. Liu, X.; Shahidehpour, M.; Cao, Y.; Li, Z.; Tian, W. Risk Assessment in Extreme Events Considering the Reliability of Protection Systems. *IEEE Trans. Smart Grid* **2015**, *6*, 1073–1081. [CrossRef]

 © 2019 by the authors. Licensee MDPI, Basel, Switzerland. This article is an open access article distributed under the terms and conditions of the Creative Commons Attribution (CC BY) license (http://creativecommons.org/licenses/by/4.0/).

Article

Evolutionary Observer Ensemble for Leak Diagnosis in Water Pipelines

A. Navarro [1,*,†], J. A. Delgado-Aguiñaga [2,†], J. D. Sánchez-Torres [3,†] and O. Begovich [4,†] and G. Besançon [5,†]

1. Escuela de Ingeniería y Ciencias, Tecnológico de Monterrey, Av. General Ramón Corona 2514, Zapopan C.P. 45138, Jalisco, Mexico
2. Centro de Investigación, Innovación y Desarrollo Tecnológico CIIDETEC-UVM, Universidad del Valle de México, Periférico Sur Manuel Gómez Morín 8077, Tlaquepaque C.P. 45601, Jalisco, Mexico; jorge.delgado@uvmnet.edu
3. OPTIMA Lab, Departamento de Matemáticas y Física, ITESO, Periférico Sur Manuel Gómez Morín 8585, Tlaquepaque C.P. 45604, Jalisco, Mexico; dsanchez@iteso.mx
4. CINVESTAV Guadalajara, Av. del Bosque 1145, Col. El Bajío, Zapopan C.P. 45019, Jalisco, Mexico; obegovi@gdl.cinvestav.mx
5. Université Grenoble Alpes, CNRS, Grenoble INP, Institute of Engineering Université Grenoble Alpes, GIPSA-lab, 38000 Grenoble, France; gildas.besancon@gipsa-lab.grenoble-inp.fr

* Correspondence: adrian.navarro@tec.mx
† These authors contributed equally to this work.

Received: 30 September 2019; Accepted: 22 November 2019; Published: 3 December 2019

Abstract: This work deals with the Leak Detection and Isolation (LDI) problem in water pipelines based on some heuristic method and assuming only flow rate and pressure head measurements at both ends of the duct. By considering the single leak case at an interior node of the pipeline, it has been shown that observability is indeed satisfied in this case, which allows designing an observer for the unmeasurable state variables, i.e., the pressure head at leak position. Relying on the fact that the origin of the observation error is exponentially stable if all parameters (including the leak coefficients) are known and uniformly ultimately bounded otherwise, the authors propose a bank of observers as follows: taking into account that the physical pipeline parameters are well-known, and there is only uncertainty about leak coefficients (position and magnitude), a pair of such coefficients is taken from a search space and is assigned to an observer. Then, a Genetic Algorithm (GA) is exploited to minimize the integration of the square observation error. The minimum integral observation error will be reached in the observer where the estimated leak parameters match the real ones. Finally, some results are presented by using real-noisy databases coming from a test bed plant built at Cinvestav-Guadalajara, aiming to show the potentiality of this method.

Keywords: leak isolation; nonlinear observer; genetic algorithm; fault diagnosis

---

1. Introduction

Fluid transport is a significant issue in the world today. Currently, cities are continually demanding utilities, including drinking water, the distribution of oil products, the treatment of wastewater, etc., and pipelines are predominantly used to do this. The pipeline networks have increased the growth and comfort of society. Nevertheless, there is also a constant risk (in particular, for fuel pipelines) that accidents, environmental pollution or economic losses may occur if the fluid spreads through leaks. In this context, several critical incidents have recently occurred within Mexico, such as San Martín Texmelucan, Puebla in 2010, and more recently in the Tuxpan-Tula poly-duct in the municipality of Tlahuelilpan, Hidalgo in 2019, where many people died as a result of an explosion caused by illegal fuel extraction. On the other hand, according to the National Water Committee (CONAGUA) [1], about 40% of drinking water is lost due to leakage. Although there are entirely different explanations for each problem, both can be solved by using similar techniques.

The scientific community has paid attention to that problem and has proposed several methodologies for monitoring and supervision purposes in order to avoid losses and accidents (see, e.g, [2–11]). In particular, in Begovich et al. [2], a LDI algorithm based on Billman and Isermann [3] has been implemented and tested with accurate results and based on steady-state conditions, which increase the convergence time in the leak parameter estimation process. The proposal in Verde et al. [5] deals with the location of multiple leaks in a pipeline. The key to the leak detector, which should operate in quasi-real-time, is a family of parameterized transient models for all scenarios in the pipeline. In this case, the equivalence in the steady-state of a leak at a position with two leaks allows obtaining the family of dynamic models. Then, to estimate the specific parameter of the leak, an off-line identification process is performed.

Likewise, a multi-leak diagnostic scheme has been suggested in Delgado-Aguiñaga et al. [4] based on Kalman observers. In general, it considers a model-based approach for detecting and isolating several non-concurrent leaks. The method modifies the nonlinear model for each new leakage event. Thus, it is an extension of the single-leak isolation problem. Although this scheme shows acceptable results, the complexity of computation increases as an additional leak occurs. In Rubio Scola et al. [12], the authors presented the development of a nonlinear state observer to locate a blockage in a pipeline. The technique uses a mathematical model derived from the equations of the water hammer together with the method of finite differences for its solution, providing a suitable location for the blockage. Besides, concerning the implementation problem, a recent algorithm based on the extended Kalman filter Delgado-Aguiñaga and Begovich [10] has successfully identified a leak in an aqueduct in Guadalajara, Mexico. A posterior study estimated that approximately 130 million liters of drinking water had been lost in this incident.

There are also other methods with successful application. For example, the approach presented in Ostapkowicz [6] uses a pressure wave method, and Liu et al. [11] presented a system based on acoustic waves. A hybrid approach based on a real-time transient simulation system, and a negative pressure wave method is proposed in Zhang et al. [7]. In the last reference, the authors argued that the most likely future development in pipeline leak detection and location tends to be the use of two or more different methods. Finally, Tian et al. [8] proposed an algorithm to locate leaks based on the pressure difference profile along the pipeline. It considers the effect of the static pressure increases at the leakage point.

On the other hand, analytical redundancy methods (the technique of several model-based methods) have demonstrated to be useful to improve the precision, reliability, and performance of a system. Notably, in the field of fault detection and isolation, the attention on this class of methods has increased lately in several topics, such as robotics Lyu et al. [13], control theory Chouchane et al. [14], diagnosis system Lunze [15], and the application of evolutionary algorithms and neural networks to fault diagnosis Witczak [16]. In particular, several works dealing the leak diagnosis problem in Water Distribution Networks (WDN) have also been proposed on the basis of genetic algorithms. In Vitkovsky et al. [17], a technique in conjunction with the inverse transient method is used to detect leaks and friction factors. Additionally, in [18], a model calibration process is formulated as an nonlinear optimization problem that is solved by using a genetic algorithm. Case studies are presented to demonstrate how the integrated approach is applied to water leak detection.

The framework previously stated encourages researchers to propose new model-based approaches that can be used in combination with other methods and thus contribute to the development of a robust leak diagnostic tool for single pipelines on the basis of analytical redundancy model.

By relying on an observability property, fulfilled for the single leak case, our approach considers building an observer ensemble together with a genetic algorithm to minimize the observation error and, in this way, estimate the leak parameters, i.e., position and magnitude. The extended Luenberger observer has been chosen to estimate the internal state variable (the pressure at the leak point). Such an observer is exponentially stable only if the parameters of the model are known. Otherwise, the observation error is, at last, uniformly ultimately bounded. Then, if only leak position and magnitude are the unknown parameters (the rest of the pipeline mathematical model parameters are well-known), it is possible to design a bunch of observers, each with different values of leak position and magnitude (the search space). Thus, the best estimation of the leak parameters provided by the observer ensemble is the one that gives the minimum residual. Now, the potential of the genetic algorithm could be exploited to find the best estimation of such parameters.

The paper is organized as follows. Section 2 provides the mathematical model. Section 3 describes the Leak Detection and Isolation scheme. Section 4 presents some successful experimental results. Finally, in Section 5, some conclusions and future work are discussed.

2. Pipeline Mathematical Model

The pipeline model is classically derived under the following assumptions: the pipeline is considered to be straight without any fitting and without slope; the fluid is slightly compressible; the duct wall is slightly deformable; and the convective velocity changes are negligible. Likewise, the pipeline cross-section area and fluid density are constant. Then, the Partial Differential Equations (PDE) governing the fluid transient response, can be written as Roberson et al. [19]:

Momentum Equation

$$\frac{\partial Q(z,t)}{\partial t} + gA_r \frac{\partial H(z,t)}{\partial z} + \mu Q(z,t)|Q(z,t)| = 0 \tag{1}$$

Continuity Equation

$$\frac{\partial H(z,t)}{\partial t} + \frac{b^2}{gA_r}\frac{\partial Q(z,t)}{\partial z} = 0 \tag{2}$$

where Q is the flow rate $[m^3/s]$; H is the pressure head $[m]$; z is the length coordinate $[m]$; t is the time coordinate $[s]$; g is the gravity acceleration $[m/s^2]$; A_r is the cross-section area $[m^2]$; b is the pressure wave speed in the fluid $[m/s]$; $\mu = f(Q)/2\phi A_r$, where ϕ is the inner diameter $[m]$ and f is the friction factor; and the rest of physical parameters are computed as in Delgado-Aguiñaga et al. [20] considering a constant water temperature of 20 °C. The dynamics in Equations (1) and (2) is fully defined by related pairs of initial and boundary conditions.

Leak model: Furthermore, one leak arbitrarily located at point $z \in (0, L)$ (where L is the total length of the pipeline), can be modeled as follows Roberson et al. [19] (see Figure 1):

$$Q_l = \lambda \sqrt{H_l} \tag{3}$$

where the constant λ is function of the orifice area and the discharge coefficient (for simplicity, the λ coefficient is referred as "leak magnitude" from now on); Q_l is the flow through the leak; and H_l is the head pressure at the leak point Navarro et al. [21].

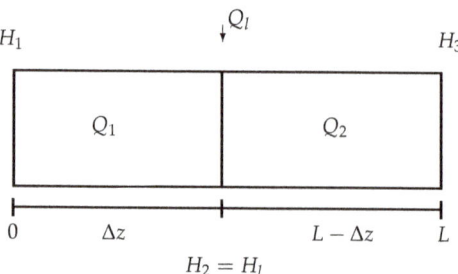

Figure 1. Discretization of the pipeline with a leak Q_l.

This leak produces a discontinuity in the system. Furthermore, due to the law of mass conservation, Q_l must satisfy the next relation:

$$Q_2 = Q_1 + Q_l \tag{4}$$

where Q_1 and Q_2 are the flows before and after of the leak point, respectively.

Friction model: In modern pipes (pipes with a relative roughness usually less than 1×10^{-3}), it is difficult to reach a complete turbulence zone (i.e., the zone where friction factor is almost constant, see Figure 2).

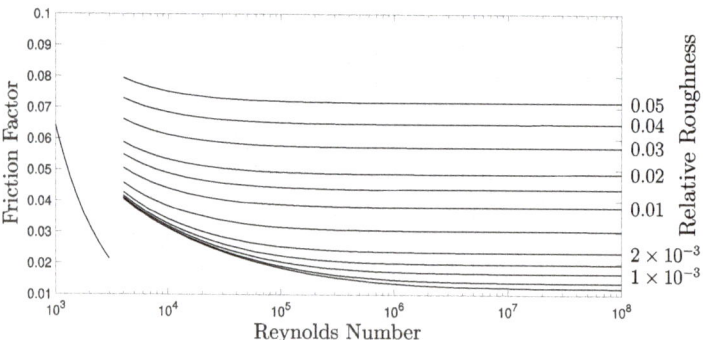

Figure 2. Moody chart.

Therefore, a friction factor deemed as a constant value could yield a poor mathematical model. For this reason, in the present work, the friction factor is calculated by using the well-known Swamee–Jain equation Brkić [22], Swamee and Jain [23]:

$$f(Q) = \frac{0.25}{\left[\log_{10}\left(\frac{\epsilon}{3.7D} + \frac{5.4}{Re^{0.9}}\right)\right]^2} \tag{5}$$

which is suitable for flow regime in the transition zone (as occurs in plastic pipelines) and where $\epsilon \in [0.000001, 0.05]$ [m] is the roughness height, $Re \in [5000, 10^8]$ is the Reynolds number given by

$$Re = \frac{QD}{vA}$$

and v is the kinematic viscosity [m^2/s].

Spatial Discretization of the Modeling Equations

In order to obtain a finite dimensional model from (1) and (2), such PDE's are discretized with respect to the spatial variable z, as in Verde [24], Besançon et al. [25], by using the following relationships:

$$\frac{\partial H(z_i, t)}{\partial z} \simeq \frac{H_{i+1} - H_i}{z_i} \quad \forall i = 1, \cdots, n \tag{6}$$

$$\frac{\partial Q(z_{i-1}, t)}{\partial z} \simeq \frac{Q_i - Q_{i-1}}{z_{i-1}} \quad \forall i = 2, \cdots, n \tag{7}$$

where H_i, Q_i stand for $H(z_i, t)$, $Q(z_i, t)$, and n the number of pipeline sections. Assuming only one partition in the pipeline, as shown in Figure 1, z_i ($i = 1, 2$) becomes the distance from upstream to the point of the leak and from the point of the leak to downstream, respectively. Notice that $\Delta z = z_1$ and $z_2 = L - \Delta z$. The leak position is assumed to be $\Delta z \in (0, L)$ in this description. Applying the approximations in (6) and (7) to equations (1) and (2) together with (3), we get:

$$\begin{bmatrix} \dot{x}_1 \\ \dot{x}_2 \\ \dot{x}_3 \end{bmatrix} = \begin{bmatrix} -\frac{A_r g}{\Delta z}(x_2 - u_1) - \mu_1 x_1 |x_1| \\ -\frac{b^2}{A_r g \Delta z}(x_3 - x_1 + \lambda \sqrt{x_2}) \\ -\frac{A_r g}{L - \Delta z}(u_2 - x_2) - \mu_2 x_3 |x_3| \end{bmatrix} \tag{8}$$

Here, the state vector is defined as $x = [x_1 \ x_2 \ x_3]^T = [Q_1 \ H_2 \ Q_2]^T$, the input vector is $u = [u_1 \ u_2]^T = [H_1 \ H_3]^T$, and the pressure output vector is $y = [x_1 \ x_3]^T = [Q_1 \ Q_2]^T$, $\mu_i = f(Q_i)/2\phi A_r$ with $i = 1, 2$. It is worth noting that H_2 represents the pressure head at the leak point. This value is impossible to measure since the leak position is not known "a priori", but this value could be observed because, for the system (8), the observability property is fulfilled, as seen in the next section.

Notice that the mathematical model in (8) assumes a straight pipe without loss of generality, as, even if the pipe is not straight, it is possible to obtain an Equivalent Straight Length (ESL) of the pipe. This is done by considering losses due to each "non-straight element" (i.e., fitting). The equivalent straight pipe L_e can be calculated as Mott [26]:

$$L_e = L_r + \frac{D \sum_{j=i}^{n} K_j}{f} \tag{9}$$

where L_r stands for the pipeline physical length [m] measured between the sensors placed at the ends of the pipeline, K_j is the fitting loss coefficient for the jth fitting, and n the total number of the pipeline fittings.

3. LDI Scheme Approach

The leak diagnosis process (the task of determining the magnitude λ and location Δz of the leak) proposed in this work is carried out by the design of a bank of observers together with a genetic algorithm method whose selection rule is to minimize the integration error of each observer.

Since the observability property of the system (8) is fulfilled, it is possible to design an extended Luenberger observer to estimate H_2. Such an observer is exponentially stable only if the parameters of the model (A_r, g, b, L, μ_1, μ_2, λ, and Δz, in (8)) are known; otherwise, the observation error will be uniformly ultimately bounded. Then, if only leak position (Δz) and magnitude (λ) are the unknown parameters (the rest of the pipeline mathematical model parameters are known), it is possible to design a bunch of observers each with different values of λ and Δz, i.e., search space. Thus, the best leak position and leak magnitude estimation of the ensemble is the one that gives the minimum residual. Now, the potential of the genetic algorithm could be exploited to find the best estimation of such parameters.

The minimum integral observation error will be reached when the leak position and magnitude match the real ones.

3.1. Extended Luenberger Observer for MIMO Systems

First, let us consider that the space-state representation given by (8) can be rewritten in compact form:

$$\dot{x} = f(x, u)$$
$$y = h(x) \tag{10}$$

with the state $x \in \mathbb{R}^3$, the input $u(t) \in \mathbb{R}^2$, and the output $h(x) \in \mathbb{R}^2$ (with two components, h_1 and h_2). Then, the observability is guaranteed by invertibility of the following map (where L_f denotes the Lie derivative):

$$Y = \begin{bmatrix} y_1 \\ \dot{y}_1 \\ y_2 \end{bmatrix} = \begin{bmatrix} h_1 \\ L_f h_1 \\ h_2 \end{bmatrix} = \begin{bmatrix} x_1 \\ -\frac{A_{rg}}{\Delta z}(x_2 - u_1) - \mu_1 x_1 |x_1| \\ x_3 \end{bmatrix} \tag{11}$$

which is in fact uniform in u. If one considers $x_1|x_1| = x_1^2$ (for unidirectional flow), such a map induces the following rank observability condition:

$$\text{rank}\left(\frac{\partial Y(x)}{\partial x}\right) = \begin{bmatrix} 1 & 0 & 0 \\ -2\mu_1 x_1 & -\frac{A_{rg}}{\Delta z} & 0 \\ 0 & 0 & 1 \end{bmatrix} = 3 \tag{12}$$

such that the system in (10) is locally observable and satisfies the condition for the extended Luenberger observer design for MIMO systems Birk and Zeitz [27]. Then, the system (10) can be rewritten to obtain its additive output nonlinearity form:

$$\dot{x} = Ax + \varphi(x) + \phi(u) + \xi(y)$$
$$y = Cx \tag{13}$$

where matrices A, $\varphi(x)$, $\phi(u)$, $\xi(y)$, and C are given by:

$$A = \begin{bmatrix} 0 & -\frac{a_1}{\Delta z} & 0 \\ \frac{a_2}{\Delta z} & 0 & -\frac{a_2}{\Delta z} \\ 0 & \frac{a_1}{L-\Delta z} & 0 \end{bmatrix} \quad \varphi(x) = \begin{bmatrix} 0 \\ -a_2 \frac{\lambda}{\Delta z}\sqrt{x_2} \\ 0 \end{bmatrix}$$

$$\phi(u) = \begin{bmatrix} \frac{a_1}{\Delta z} u_1 \\ 0 \\ -\frac{a_1}{L-\Delta z} u_2 \end{bmatrix} \quad \xi(y) = \begin{bmatrix} -\mu y_1 |y_1| \\ 0 \\ -\mu y_2 |y_2| \end{bmatrix}$$

$$C = \begin{bmatrix} 1 & 0 & 0 \\ 0 & 0 & 1 \end{bmatrix}$$

where $a_1 \doteq A_{rg}$ and $a_2 \doteq \frac{b^2}{A_{rg}}$. Here, the additive output nonlinearity can be built from direct measurements and thus compensated in the observer design (as it was originally proposed by the authors in Krener and Isidori [28], J. Krener and Respondek [29], for instance). The representation (13) admits an observer of the form:

$$\dot{\hat{x}} = A\hat{x} + \varphi(\hat{x}) + \phi(u) + \xi(y) + K(y - C\hat{x})$$
$$\hat{y} = C\hat{x} \tag{14}$$

By defining the estimation error as $e = x - \hat{x}$, the dynamic error model is:

$$\dot{e} = (A - KC)e + \varphi'(x, \hat{x}) \tag{15}$$

where $\varphi'(x,\hat{x}) = \begin{bmatrix} 0 & -a_2\frac{\lambda}{\Delta z}(\sqrt{x_2} - \sqrt{\hat{x}_2}) & 0 \end{bmatrix}^T$.

In this equation, $e = 0$ is clearly an equilibrium. In addition, K can be chosen so that $(A - KC)$ is Hurwitz (since (A,C) is observable), that is for any $Q = Q^T > 0$, there exists $P = P^T$ satisfying the Lyapunov equation $P(A - KC) + (A - KC)^T P = -Q$.

Notice then that $\varphi(x,\hat{x})$ satisfies a linear growth bound $\|\varphi(x,\hat{x})\| \leq \gamma\|e\|$ on the region of operation, and thus if $\gamma < \zeta_{min}(Q)/2\zeta_{max}(P)$, where $\zeta_{min}(\cdot)$ and $\zeta_{max}(\cdot)$ denote the minimum and maximum eigenvalue of a matrix, one can conclude that the origin of the error system in (13) is exponentially stable (see Khalil [30] for more details).

3.2. Genetic Algorithm

In computer science, the genetic algorithm is an algorithm inspired in the biological evolution that offers a suitable solution to optimization and search problems. The GA is a recursive algorithm where the aptest individuals of a population are discovered, emphasized, and recombined (reproduction) in order to produce descendants of the next generation. Six phases are considered in a genetic algorithm:

1. *Initial population.* The first step of the process is to obtain a set of individuals randomly generated (initial population) in which each such individual is a candidate solution to a problem.

    As in the natural selection process, an individual is characterized by a set of parameters called genes. The solutions, known as chromosomes, are genes joined into a string.

    In a genetic algorithm, the chromosome is represented using a string in terms of an alphabet. Binary encoding (a string of ones and zeros) is the most common procedure to encode the genes in a chromosome.

2. *Fitness function.* The fitness function defines how close an individual fits a solution and, in this way, determines which will reproduce and survive into the next generation. The fitness function provides a "fitness score" to each individual. Such "fitness score" settles the probability that an individual will be selected for reproduction.

3. *Selection.* In this phase, the chromosomes in the population that more closely match the fitness function are selected. The solution (chromosome) that fits better during iteration is more likely to be selected to reproduce.

4. *Crossover.* After the selection process, a recombination of the chromosomes is carried out in order to generate a new population for the next iteration. Crossover is applied to randomly pair strings and exchanges the sub-sequences before and after to create two offspring.

5. *Mutation.* To preserve diversity within the population and prevent premature convergence, a mutation process is done. The mutation operation is applied after the crossover process is achieved. For each bit in a subset of the new offspring, some of their genes can be mutated with a low probability. This is done by flipping some bits in the chromosome bit string.

6. *Termination.* If the algorithm does not produce new populations that are sufficiently different from the previous generation, the algorithm has converged. Then, the genetic algorithm has found a set of solutions to the problem, and it is terminated. Such a criteria is predefined by the designer according to specific constraints. In particular, for the proposed scheme, the algorithm is kept in operation during the entire experiment since a permanent pipeline monitoring is assumed no matter if a leak is occurring or not.

Some final remarks. The GA discussed thus far uses a binary string to encode the genes in a chromosome. Nevertheless, for many engineering problems, it is nearly impossible to represent the solution with a binary encoding (as in the case of the leak diagnosis). Thus, it is necessary to make a mapping between binary and real numbers before the process (crossover and mutation) is started. Such a mapping is built in two stages: First, a function $m = \psi(r)$, which assigns a real number r of a given search interval $r \in [0, R_{max}]$ to a closed set of integer number, $m \in [0, 2^n]$, is defined:

$$\psi(r) = round\left((2^n - 1)\frac{r}{R_{max}}\right) \tag{16}$$

where *round* function rounds each element to the nearest integer, R_{max} stands for the maximum real number of the interval, and n is a natural number. Naturally, the longer n is, the more accurate the mapping will be. Then, once m is obtained, the process to convert m into a binary number follows immediately.

To return from binary to a real number and, in this way, apply the fitness function and selection processes, the inverse mapping ψ^{-1} is applied.

Figure 3 depicts the flowchart of GA (for more information, see Schmitt [31], Mitchell [32], Whitley [33]).

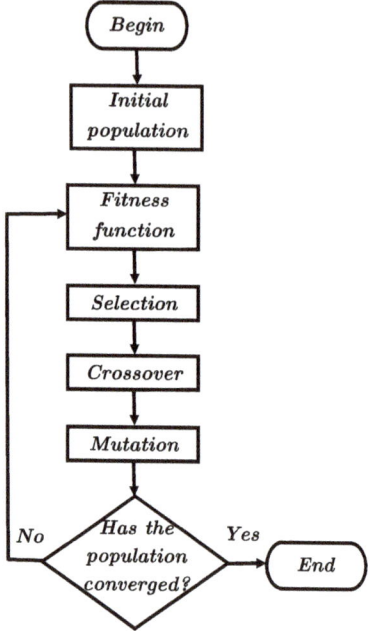

Figure 3. Genetic algorithm block diagram.

3.3. Evolutionary Ensemble of Observers

At this point, let us analyze an observer with the structure of (14), in the presence of parameter errors. First, one can consider that the leak parameters (λ and Δz) match with real values. Then, if a parametric error appears (i.e., there is a deviation between the current λ or Δz and the real ones), the observer structure given by (14) changes in the following form:

$$\dot{\hat{x}} = (A + \delta(A))\hat{x} - a_2 \begin{bmatrix} 0 \\ \left(\frac{\lambda}{\Delta z} + \delta\left(\frac{\lambda}{\Delta z}\right)\right) \\ 0 \end{bmatrix} \sqrt{\hat{x}_2}$$

$$+ \begin{bmatrix} \left(\frac{a_1}{\Delta z} + \delta\left(\frac{a_1}{\Delta z}\right)\right) & 0 \\ 0 & 0 \\ 0 & -\left(\frac{a_1}{L-\Delta z} + \delta\left(\frac{a_1}{L-\Delta z}\right)\right) \end{bmatrix} u \quad (17)$$

$$+ \xi(y) + K(y - \hat{y})$$

$$\hat{y} = C\hat{x}$$

where the symbol $\delta(\bullet)$ denotes a parametric deviation from the real value. This means that $\delta(\lambda)$ is the difference of λ as a sum error (i.e., the wrong value λ_w could be separated as follows: $\lambda_w = \lambda + \delta(\lambda)$, where λ is the real value). Then, (17) yields an error model in the following form:

$$\dot{e} = (A - KC)e + \varphi'(x, \hat{x}) + \delta(A)\hat{x} + \begin{bmatrix} 0 \\ a_2\delta(\frac{\lambda}{\Delta z}) \\ 0 \end{bmatrix} \sqrt{\hat{x}_2} \quad (18)$$

From (18), we have that

$$\delta(A)\hat{x} + \begin{bmatrix} 0 \\ a_2\delta(\frac{\lambda}{\Delta z}) \\ 0 \end{bmatrix} \sqrt{\hat{x}_2}$$

changes the equilibrium point of (15) away from 0. Thus, a residual $r(t)$ is induced in the output error $y - C\hat{x}$ when an error presented in λ or Δz and this residual $r(t)$ is zero only when the λ and Δz match the real values. The present work exploits this system property (as long as the rest of the parameters are properly tuned). It is interesting to see that the residuals do not depend on the input signal $u(t)$.

Hence, it is possible to design an ensemble of observers, each with different values of λ and Δz, such that the residuals of individual observers (namely, $r_1, r_2, ..., r_n$) go away from zero as long as these values do not match with the real ones. Figure 4 depicts this idea.

If the residual is minimized somehow, then it is possible to estimate the correct values of the leak parameters. The present work proposes a GA that searches the correct values of λ and Δz by minimizing the integral squared residual of the ensemble of observers. In this GA, the population is built with the combination (cartesian product) of the possibles values of λ and Δz. The following optimization problem is considered. Find $(\Delta z, \lambda)$ such that:

$$\text{minimize} \int_{t_0}^{t_0+T} (r(t))^2 \, dt \quad (19)$$

where the residual vector is defined as:

$$\mathbf{r}(t) = \begin{bmatrix} r_1 \\ r_2 \\ \vdots \\ r_n \end{bmatrix} \quad (20)$$

Here, $r_i(t) = y(t) - \hat{y}_i(t)$ is the residual of the ith observer and n is the cross product between the number of the position and magnitude that we are looking for, i.e., $n = l \times m$. Here, l and m

are the number of position and magnitude, respectively, arbitrarily proposed by the designer. This work suggests to set $l = m$ such that each value of the variables Δz_i and λ_i belongs to a set of equally separated values, i.e., $\Delta z_i \in \{\Delta z_{min}, 2\Delta z_{min}, ..., m\Delta z_{min} = \Delta z_{max}\}$ and $\lambda_i \in \{\lambda_{min}, 2\lambda_{min}, ..., l\lambda_{min} = \lambda_{max}\}$. Initial time t_0 can be deleted, as well as window length T.

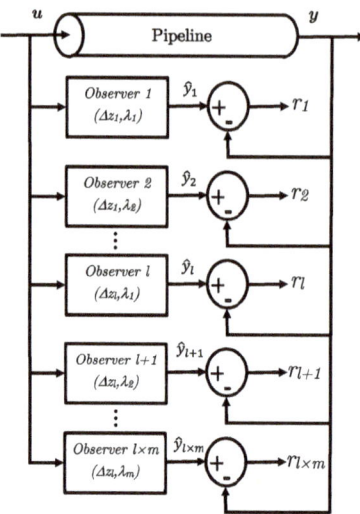

Figure 4. Evolutionary ensemble of observer block diagram.

4. Experimental Results

In this section, the proposed LDI methodology performance is evaluated. To ensure the method's effectiveness, three experiments were carried out by using some database coming from the pilot plant built at Cinvestav-Guadalajara. Three different leaks were emulated by opening three electro-valves located at diverse positions along the pilot pipe.

The section continues as follows: First, a brief description of the pilot pipe is given, and then the experimental setup is stated. Finally, each experiment is described in detail.

4.1. Experimental Setup

The pilot plant referred above was manufactured with PP-R (Polypropylene Copolymer Random), and it is equipped with: two flow rate transducers (FT) and two pressure-head transducers (PT) installed at both ends of the pipe. In addition, a 5 HP centrifugal pump was connected to a variable-frequency driver fixed at 50 Hz (to experiment on flow-rate variation effects over the LDI scheme); and three valves were used to emulate the leak effect. Figure 5 depicts a schematic diagram of the pipeline prototype. More information can be found in Begovich et al. [2]. The main parameters of the pipeline system are shown in Table 1. The sampling rate was 300 Hz satisfying the Courant's condition for system (8).

As mentioned above, the mathematical model used to derive the LDI algorithm given by (8) assumes a straight pipeline, and the prototype is not straight. Therefore, it is necessary to find an Equivalent Straight Length (ESL) for this prototype. Expression (9) is useful for this purpose (for more information, see Navarro et al. [34]). Table 2 establishes the ESL between sensors (see Figure 5) and also from upstream to each valve.

Table 1. Pipeline prototype parameters.

Parameter	Symbol	Value
Length between PT sensors	L_r	68.84 m
Internal diameter	ϕ	6.54×10^{-2} m
Friction factor	f	1.635×10^{-2}
Gravity acceleration	g	9.81 m/s^2
Pressure wave speed	b	341 m/s
Fitting loss coefficient sum	$\sum_{j=i}^{n} K_j$	9.09 [-]

Table 2. Distances in ESL.

Estimated in ESL Terms	Symbol	Value
Between PT sensors	L	105.1 m
Upstream to Valve 1	z_{f1}	30.92 m
Upstream to Valve 2	z_{f2}	43.64 m
Upstream to Valve 3	z_{f3}	62.99 m

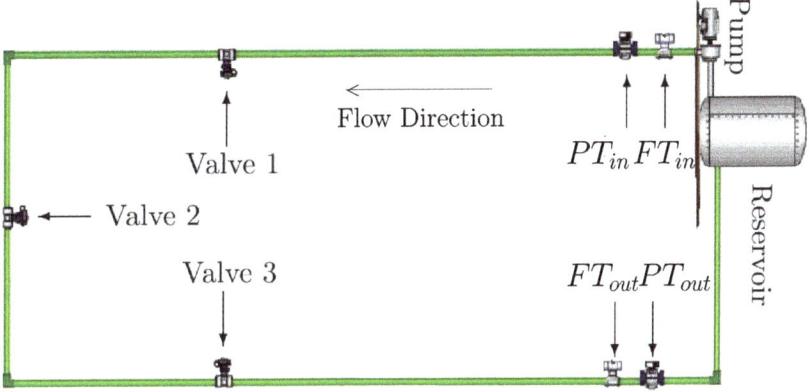

Figure 5. Schematic diagram of the pipeline prototype.

By substituting those pipeline parameters shown in Table 1 into the matrices A, $\varphi(x)$, and $\phi(u)$, $\xi(y)$ in (13), one can see that the observability matrix in (12) has full rank if $z_f \in (0, L_r)$, Torres et al. [35]. The experiments started in a free-leak condition, and at time $t_l \approx 40$ [s], a leak was induced by opening Valves 1–3, respectively. Each leak was detected once the following threshold was triggered:

$$|Q_{in}(t) - Q_{out}(t)| > \delta \tag{21}$$

where $\delta = 1.55 \times 10^{-4}$ [m^3/s] was chosen considering the noise variance of the flow rate measurements in order to avoid false alarms. The ensemble of observers was turned on with the parameters shown in Table 1. The integration error of each residual was computed in a time-window of $T = 22$ [s] in (19), where $T = t_i - t_{i-1}$, with $i \in \mathbb{N}$ and t_0 the time of leak occurrence. This procedure means that, once the leak was detected, executions of the GA were performed with period $T = 22$ [s] (selection, crossover, and mutation). In Figures 7, 8, 10, 11, 13 and 14, this scheduled-process is indicated by: t_1, t_2, t_3, t_4, t_5 and t_6. Therefore, the integral in (19) becomes:

$$\text{minimize} \int_{t_{i-1}}^{t_i} (r(t))^2 \, dt \tag{22}$$

Notice that the selected values of the leak parameters, Δz and λ, were held at time t_{i-1} and reflected up to the next GA-process at time t_i.

In other words, the time-integration-window t_i served for the election of Δz and λ used in next iteration t_{i+1} and so on. In the period t_1, random values of Δz and λ were used.

The initial conditions of the observers were fixed as follows: $\hat{x}_1^0 = \hat{Q}_{in}^0$ and $\hat{x}_3^0 = \hat{Q}_{out}^0$ were equal to the mean values of the input and output flows in steady state in free-leak condition. $\hat{x}_2^0 = \hat{H}_2^0$, the pressure head at the pipeline middle point, was calculated at distance $\hat{\Delta z} = L/2$. Finally, $\hat{\lambda}_i$ (i.e., the leak magnitude of each observer) were fixed as zero, since the pipeline was not leaking. Table 3 summarizes those values.

Table 3. Initial conditions for the observer.

Estimated	Symbol	Value
$\hat{Q}_{in}^0$	$\hat{x}_1^0$	8×10^{-3} [m^3/s]
$\hat{H}_2^0$	$\hat{x}_2^0$	12 [m]
$\hat{Q}_{out}^0$	$\hat{x}_3^0$	8×10^{-3} [m^3/s]

To minimize the integration error in (22), the following considerations were made: the initial population vector of the algorithm was fixed by the cross product of two sets, both formed by uniformly spaced pipe sections $\Delta z_i \in \{\Delta z_{min}, 2\Delta z_{min}, ..., m\Delta z_{min} = L\}$ and $\lambda_i \in \{\lambda_{min}, 2\lambda_{min}, ..., l\lambda_{min} = \lambda_{max}\}$, where $\Delta z_{min} = 3.507$ [m] and $\lambda_{min} = 6.697 \times 10^{-6}$ [m$^{5/2}$/s].

Remark 1. *It is worth noting that λ_{max} was chosen such that the hole induces a flow through a leak 10% the size of the pipeline nominal flow at most. A leak higher than this percent is considered as a failure (a catastrophic breakdown of the system's ability to perform a required function under specified operating condition Isermann [36], and this topic goes beyond the scope of this paper).*

The SNR (signal-to-noise ratio) of each input and output signal is shown in Table 4. The SNR was calculated as the ratio of the signal power to the background noise Papoulis and Pillai [37]:

$$SNR = \frac{E\left[s^2\right]}{\sigma^2}$$

where $E\left[\bullet\right]$ refers to the expected value and σ stands for the standard deviation of the noisy signal.

Table 4. Signal-to-noise ratio of the input and output signals.

Variable	SNR
H_{in}	2.008×10^1
H_{out}	6.689×10^1
Q_{in}	1.338×10^2
H_{out}	1.432×10^2

4.2. Leak Isolation Scheme Results

The proposed scheme was tested by three off-line experiments using some database. First, to ensure the validity of the mathematical model (8) in a free-leak and leak conditions, synthetic data were generated using the ESL parameters (shown in Table 2) and then were compared with their corresponding real data. Some discussions and results are described to demonstrate the Leak Isolation Scheme's effectiveness.

4.2.1. Leak Case in Valve 1

Initially, results in a leak induced in Valve 1 are shown. Figure 6a depicts the measured pressure head at inlet ($u_1 = H_{in}$) and outlet ($u_2 = H_{out}$) of the pipeline (i.e., the observation input). Figure 6b shows the measured flow upstream ($x_1 = Q_{in}$) and downstream ($u_1 = H_{in}$) of the pipe together with their respective synthetic data ($\tilde{Q}_{in}$ and $\tilde{Q}_{out}$), generated by (8). As it can be seen, the mathematical model follows the real data in a proper way despite the measurement noise.

Figure 7a,b shows the evolution of the state observer: upstream and downstream flow rate, respectively. Notice that the inlet and outlet flow rate are well estimated after the first integration time, t_1. This fact shows that the GA chooses the appropriate values of Δz and λ.

The results of the LDI scheme are depicted in Figure 8a,b, where the leak size and its position estimation are shown. As it can be seen, the leak position in all three cases is well estimated despite signal noise.

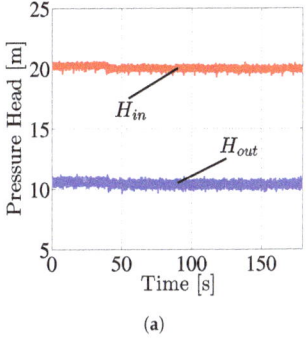 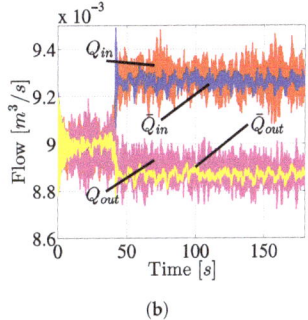

(a) (b)

Figure 6. Model validation for a leak induced in Valve 1: (**a**) pressure head at inlet an outlet of the pipeline $u = [u_1 \; u_2]^T = [H_{in} \; H_{out}]^T$ (input signals); and (**b**) synthetic and real flow rates at inlet ($\tilde{Q}_{in}$ and Q_{in}) and outlet ($\tilde{Q}_{out}$ and Q_{out}) of the pipe.

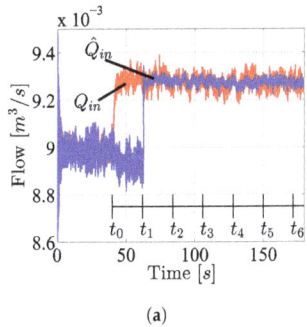 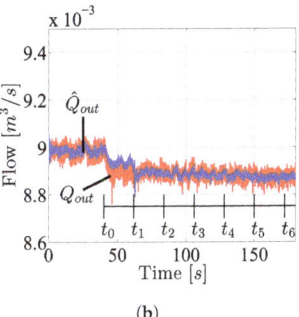

(a) (b)

Figure 7. Flow rate estimations at the ends of the pipeline: (**a**) flow rate at inlet of the pipe (Q_{in}) and its estimation ($\hat{Q}_{in}$); and (**b**) flow rate at outlet of the pipe (Q_{out}) and its estimation ($\hat{Q}_{out}$).

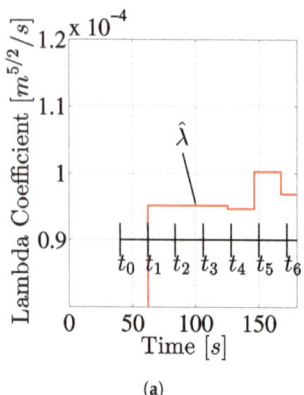
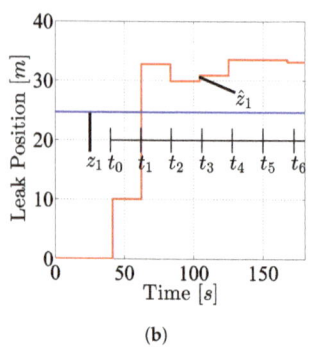

Figure 8. Leak parameters: (**a**) lambda parameter estimation $\hat{\lambda}$ concerning Valve 1 (leak magnitude); and (**b**) leak position estimation Δz concerning Valve 1.

4.2.2. Leak Case in Valve 2

Now, results in a leak induced in Valve 2 are shown. As before, Figure 9a depicts the measured pressure head at inlet ($u_1 = H_{in}$) and outlet ($u_2 = H_{out}$) of the pipeline. Figure 9b shows the measured flow upstream ($x_1 = Q_{in}$) and downstream ($u_1 = H_{in}$) of the pipe together with their respective synthetic data ($\bar{Q}_{in}$ and $\bar{Q}_{out}$). In this second case, the mathematical model follows the real data in a proper way, as well. Figure 10a,b depicts the evolution of the state observer: upstream and downstream flow rate, respectively. Figure 11a,b depicts the leak size and its position estimation.

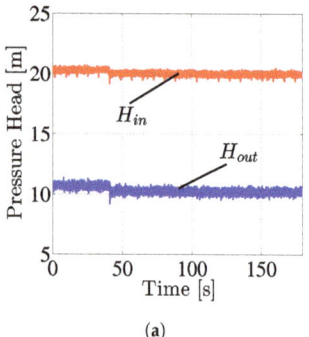
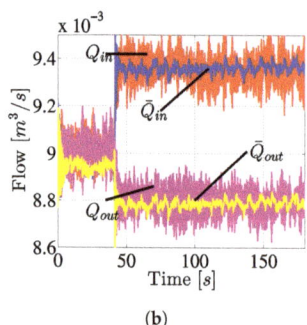

Figure 9. Model validation for a leak induced in Valve 2: (**a**) pressure head at inlet an outlet of the pipeline $u = [u_1 \ u_2]^T = [H_{in} \ H_{out}]^T$ (input signals); and (**b**) synthetic and real flow rates at inlet ($\bar{Q}_{in}$ and Q_{in}) and outlet ($\bar{Q}_{out}$ and Q_{out}) of the pipe.

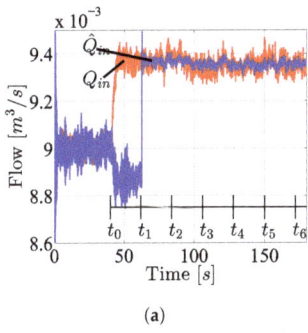
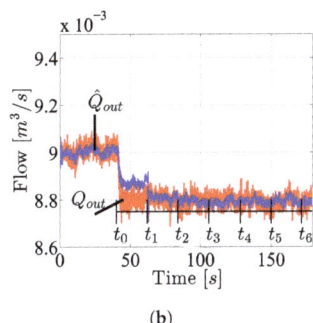

Figure 10. Flow rate estimations at the ends of the pipeline: (**a**) flow rate at inlet of the pipe (Q_{in}) and its estimation ($\hat{Q}_{in}$); and (**b**) flow rate at outlet of the pipe (Q_{out}) and its estimation ($\hat{Q}_{out}$).

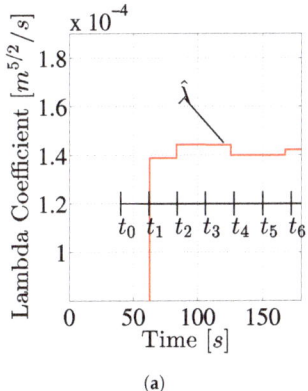
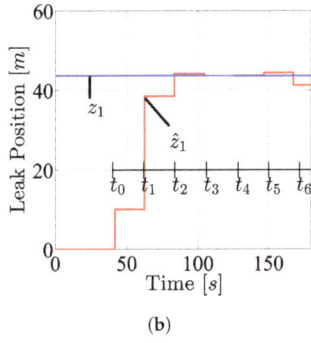

Figure 11. Leak parameters: (**a**) lambda parameter estimation $\hat{\lambda}$ concerning Valve 2 (leak magnitude); and (**b**) leak position estimation Δz concerning Valve 2.

4.2.3. Leak Case in Valve 3

Finally, the results in a leak induced in Valve 3 are shown. Figure 12a depicts the measured pressure head at inlet ($u_1 = H_{in}$) and outlet ($u_2 = H_{out}$) of the pipeline. Figure 12b shows the measured flow upstream ($x_1 = Q_{in}$) and downstream ($u_1 = H_{in}$) of the pipe together with their respective synthetic data ($\tilde{Q}_{in}$ and $\tilde{Q}_{out}$). In the same way as before, the mathematical model follows the real data in a proper way. Figure 13a,b depicts the evolution of the state observer: upstream and downstream flow rate, respectively. Figure 14a,b depicts the leak size and its position estimation.

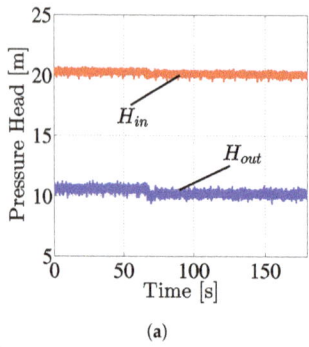 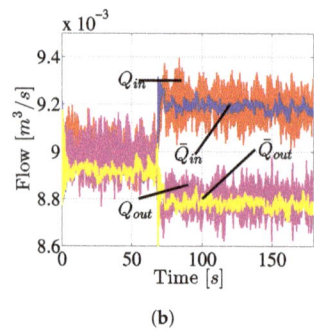

Figure 12. Model validation for a leak induced in Valve 3: (**a**) pressure head at inlet an outlet of the pipeline $u = [u_1\ u_2]^T = [H_{in}\ H_{out}]^T$ (input signals); and (**b**) synthetic and real flow rates at inlet ($\bar{Q}_{in}$ and Q_{in}) and outlet ($\bar{Q}_{out}$ and Q_{out}) of the pipe.

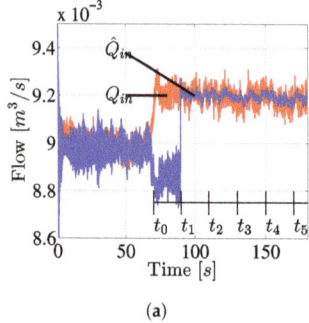 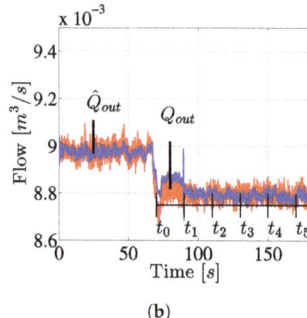

Figure 13. Flow rate estimations at the ends of the pipeline: (**a**) flow rate at inlet of the pipe (Q_{in}) and its estimation ($\hat{Q}_{in}$); and (**b**) flow rate at outlet of the pipe (Q_{out}) and its estimation ($\hat{Q}_{out}$).

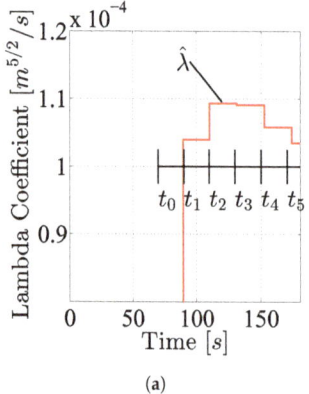 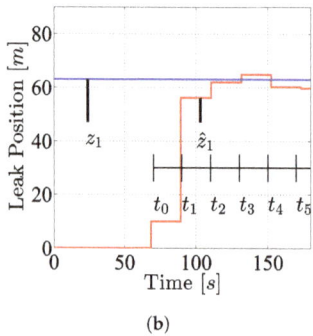

Figure 14. Leak parameters: (**a**) lambda parameter estimation $\hat{\lambda}$ concerning Valve 3 (leak magnitude); and (**b**) leak position estimation Δz concerning Valve 3.

5. Conclusions and Future Work

The present work deals with the leak isolation problem (to estimate the position and magnitude of a leak in a water pipeline) using a heuristic method. The proposed scheme assumes only flow and pressure sensors at the upstream and downstream of the pipeline. Exploiting the fact that the pipeline mathematical model is observable, it is possible to design an observer where the observation dynamic error system is exponentially stable only if the leak size and location parameters are known and, at last, uniform ultimate boundedness in other cases. In this way, the authors propose to design a bank of observers together with a Genetic Algorithm. This scheme allows for minimizing the integral observation error. Then, the minimum integral observation error will be reached when the leak position and magnitude match the real ones.

The approach presented in the paper estimated the leak position and its intensity in a very acceptable way. This is corroborated since both downstream and upstream flow rates were well estimated in the presence of noise. It means that the genetic algorithm chooses the real values of the λ (size of the leak) and Δz (leak location). The use of the integration error as a fitness function helped obtain a good estimation despite the presence of noise.

As future work, this algorithm will be refined to achieve better performance. Moreover, the authors will explore the possibility of extending the present approach to two or more leaks. Finally, the algorithm will be tested to locate leaks in a hydraulic network.

Author Contributions: A.N. proposed the initial idea. A.N., J.A.D.-A. and J.D.S.-T. developed the research, analyzed the results, and wrote the article together. O.B. and G.B. provided critical review.

Funding: This work was funded by the Tecnológico de Monterrey through the School of Engineering and Science.

Acknowledgments: The author would like to thank Tecnologico de Monterrey for the financial support and the facilities granted for fulfillment of this research article. All databases were obtained during the PhD studies of the first author at Cinvestav-Guadalajara.

Conflicts of Interest: The authors declare no conflict of interest.

References

1. CONAGUA. *Sectorización en Redes de Agua Potable*; Technical report; Semarnat: Mexico City, Mexico, 2007.
2. Begovich, O.; Pizano, A.; Besançon, G. Online implementation of a leak isolation algorithm in a plastic pipeline prototype. *Lat. Am. Appl. Res.* **2012**, *57*, 131–140.
3. Billman, L.; Isermann, R. Leak Detection Methods for Pipelines. *Automatica* **1987**, *23*, 381–385. [CrossRef]
4. Delgado-Aguiñaga, J.; Besançon, G.; Begovich, O.; Carvajal, J. Multi-leak diagnosis in pipelines based on Extended Kalman Filter. *Control Eng. Pract.* **2016**, *49*, 139–148. [CrossRef]
5. Verde, C.; Molina, L.; Torres, L. Parameterized transient model of a pipeline for multiple leaks location. *J. Loss Prev. Process. Ind.* **2014**, *29*, 177–185. [CrossRef]
6. Ostapkowicz, P. Leakage detection from liquid transmission pipelines using improved pressure wave technique. *Eksploatacja i Niezawodność Maintenance Reliability* **2014**, *16*, 9–16.
7. Zhang, T.; Tan, Y.; Zhang, X.; Zhao, J. A novel hybrid technique for leak detection and location in straight pipelines. *J. Loss Prev. Process. Ind.* **2015**, *35*, 157–168. [CrossRef]
8. Tian, S.; Du, J.; Shao, S.; Xu, H.; Tian, C. A study on a real-time leak detection method for pressurized liquid refrigerant pipeline based on pressure and flow rate. *Appl. Therm. Eng.* **2016**, *95*, 462–470. [CrossRef]
9. Santos-Ruiz, I.; Bermúdez, J.; López-Estrada, F.; Puig, V.; Torres, L.; Delgado-Aguiñaga, J. Online leak diagnosis in pipelines using an EKF-based and steady-state mixed approach. *Control Eng. Pract.* **2018**, *81*, 55–64. [CrossRef]
10. Delgado-Aguiñaga, J.A.; Begovich, O. Water Leak Diagnosis in Pressurized Pipelines: A Real Case Study. In *Modeling and Monitoring of Pipelines and Networks*; Verde, C., Torres, L., Eds.; Springer International Publishing: Cham, Switzerland, 2017; Volume 7, pp. 235–262. [CrossRef]
11. Liu, C.; Li, Y.; Fang, L.; Xu, M. New leak-localization approaches for gas pipelines using acoustic waves. *Measurement* **2019**, *134*, 54–65. [CrossRef]
12. Rubio Scola, I.; Besancon, G.; Georges, D.; Guillén, M.; Dulhoste, J.F.; Santos, R. On the design of a nonlinear state observer for the location of a blockage in a pipeline. In Proceedings of the XII International Congress on Numerical Methods in Engineering and Applied Sciences, Pampatar, Magarita Island, Venezuela, 24–26 March 2014.
13. Lyu, P.; Liu, S.; Lai, J.; Liu, J. An analytical fault diagnosis method for yaw estimation of quadrotors. *Control Eng. Pract.* **2019**, *86*, 118–128. [CrossRef]

14. Chouchane, A.; Khedher, A.; Nasri, O.; Kamoun, A. Diagnosis of Partially Observed Petri Net Based on Analytical Redundancy Relationships. *Asian J. Control* **2019**. [CrossRef]
15. Lunze, J. A method to get analytical redundancy relations for fault diagnosis. *IFAC-PapersOnLine* **2017**, *50*, 1006–1012.10.1016/j.ifacol.2017.08.208. [CrossRef]
16. Witczak, M. Advances in model-based fault diagnosis with evolutionary algorithms and neural networks. *Int. J. Appl. Math. Comput. Sci.* **2006**, *16*, 85–99.
17. Vitkovsky, J.P.; Simpson, A.R.; Lambert, M.F. Leak Detection and Calibration Using Transients and Genetic Algorithms. *J. Water Resour. Plan. Manag.* **2000**, *126*, 262–265. [CrossRef]
18. Wu, Z.Y.; Sage, P. Water Loss Detection via Genetic Algorithm Optimization-based Model Calibration. In Proceedings of the Water Distribution Systems Analysis Symposium 2006, Cincinnati, OH, USA, 27–30 August 2006; pp. 1–11. [CrossRef]
19. Roberson, J.A.; Cassidy, J.J.; Chaudhry, M.H. *Hydraulic Engineering*; Wiley: Hoboken, NJ, USA, 1998.
20. Delgado-Aguiñaga, J.; Begovich, O.; Besançon, G. Exact-differentiation-based leak detection and isolation in a plastic pipeline under temperature variations. *J. Process. Control* **2016**, *42*, 114–124. [CrossRef]
21. Navarro, A.; Begovich, O.; Sánchez Torres, J.D.; Besancon, G. Real-Time Leak Isolation Based on State Estimation with Fitting Loss Coefficient Calibration in a Plastic Pipeline: Real-Time Leak Isolation based on State Estimation. *Asian J. Control* **2016**. [CrossRef]
22. Brkić, D. Review of explicit approximations to the Colebrook relation for flow friction. *J. Pet. Sci. Eng.* **2011**, *77*, 34–48. [CrossRef]
23. Swamee, P.K.; Jain, A.K. Explicit equations for pipe-flow problems. *J. Hydraul. Div.* **1976**, *102*, 657–664.
24. Verde, C. Accommodation of multi-leak location in a pipeline. *Control Eng. Pract.* **2005**, *13*, 1071–1078. [CrossRef]
25. Besançon, G.; Georges, D.; Begovich, O.; Verde, C.; Aldana, C. Direct observer design for leak detection and estimation in pipelines. In Proceedings of the European Control Conference, ECC'07, Kos, Greece, 2–5 July 2007; pp. 5666–5670.
26. Mott, R.L. *Applied Fluid Mechanics*, 6th ed.; Prentice-Hall: Upper Saddle River, NJ, USA, 2006.
27. Birk, J.; Zeitz, M. Extended Luenberger observer for non-linear multivariable systems. *Int. J. Control* **1988**, *47*, 1823–1836. [CrossRef]
28. Krener, A.J.; Isidori, A. Linearization by output injection and nonlinear observers. *Syst. Control Lett.* **1983**, *3*, 47–52. [CrossRef]
29. J. Krener, A.; Respondek, W. Nonlinear Observer with Linearizable Error Dynamics. *SIAM J. Control Optim.* **1985**, *23*. [CrossRef]
30. Khalil, H.K. *Nonlinear Systems*, 2nd ed.; Prentice-Hall: Upper Saddle River, NJ, USA, 1996.
31. Schmitt, L.M. Theory of Genetic Algorithms. *Theor. Comput. Sci.* **2001**, *259*, 1–61. [CrossRef]
32. Mitchell, M. *An Introduction to Genetic Algorithms*; MIT Press: Cambridge, MA, USA, 1998.
33. Whitley, D. A genetic algorithm tutorial. *Stat. Comput.* **1994**, *4*, 65–85. [CrossRef]
34. Navarro, A.; Begovich, O.; Besançon, G.; Dulhoste, J. Real-time leak isolation based on state estimation in a plastic pipeline. In Proceedings of the IEEE International Conference on Control Applications (CCA), Denver, CO, USA, 28–30 September 2011; pp. 953–957. [CrossRef]
35. Torres, L.; Besançon, G.; Georges, D.; Navarro, A.; Begovich, O. Examples of pipeline monitoring with nonlinear observers and real-data validation. In Proceedings of the 8th International Multi-Conference on Systems, Signals and Devices, Sousse, Tunisia, 22–25 March 2011.
36. Isermann, R. *Fault Diagnosis Systems an Introduction from Fault Detection to Fault Tolerance*; Springer: Heidelberg/Berlin, Germany, 2006.
37. Papoulis, A.; Pillai, S.U. *Probability, Random Variables, and Stochastic Processes*, 4th ed.; McGraw Hill: Boston, MA, USA, 2002.

 © 2019 by the authors. Licensee MDPI, Basel, Switzerland. This article is an open access article distributed under the terms and conditions of the Creative Commons Attribution (CC BY) license (http://creativecommons.org/licenses/by/4.0/).

Article

Fault Diagnosis of the Blocking Diesel Particulate Filter Based on Spectral Analysis

Shuang-xi Liu [1] and Ming Lü [2,*]

[1] National Engineering Laboratory for Mobile Source Emission Control Technology, China Automotive Technology & Research Center, Tianjin 300300, China; liushuangxi1970@126.com
[2] School of Mechanical, Electronic and Control Engineering, Beijing Jiaotong University, Beijing 100044, China
* Correspondence: lvming@bjtu.edu.cn

Received: 23 October 2019; Accepted: 3 December 2019; Published: 10 December 2019

Abstract: Diesel particulate filter is one of the most effective after-treatment techniques to reduce Particulate Matters (PM) emissions from a diesel engine, but the blocking Diesel Particulate Filter (DPF) will seriously affect the engine performance, so it is necessary to study the fault diagnosis of blocking DPF. In this paper, a simulation model of an R425DOHC diesel engine with wall-flow ceramic DPF was established, and then the model was verified with experimental data. On this basis, the fault diagnosis of the blocking DPF was studied by using spectral analysis on instantaneous exhaust pressure. The results showed that both the pre-DPF mean exhaust pressure and the characteristic frequency amplitude of instantaneous exhaust pressure can be used as characteristic parameters of monitoring the blockage fault of DPF, but it is difficult to monitor DPF blockage directly by instantaneous exhaust pressure. In terms of sensitivity, the characteristic frequency amplitude of instantaneous exhaust pressure is more suitable as a characteristic parameter to monitor DPF blockage than mean exhaust pressure. This work can lay an important theoretical foundation for the on-board diagnosis of DPF.

Keywords: DPF; blockage; fault diagnosis; exhaust pressure; spectral analysis

---

1. Introduction

With the development of the economy and the progress of science and technology, the automatic industry has developed rapidly. Diesel engines have been widely used because of their good power, economy, reliability, and emission (lower CO and HC compared with gasoline engines) performances. Not only do diesel engines hold the dominant position in the area of medium- and heavy-duty vehicles, but they are also applied widely in light-duty vehicles in the present situation [1]. While the automobile brings convenience to human life, the related pollution problem is becoming more and more serious. Therefore, many countries have established more and more rigorous regulations to limit engine emissions. In order to reduce the emission pollutants of diesel engines, the researchers have taken many measures, such as improving fuel quality, internal purification technology, and after-treatment technology. For the moment, to satisfy the increasingly stringent emission regulations, we must depend on both internal purification technology and after-treatment technology.

The diesel particulate filter (DPF) is one of the most effective after-treatment techniques to reduce PM emissions from the diesel engine, which has been widely used [2]. At present, the wall flow filter invented by America Corning Company is regarded as the best filter because of its performance and its micro structure [3].

DPFs have been used in diesel engine vehicles for over 10 years [4–9], since the French Peugeot Company invented the DPF system in 2000. During the use of DPF, with the increase in particulate depositions in the DPF, the exhaust resistance of the diesel engine increases, and blocking the DPF will seriously affect the engine performance (particulate depositions increase to a certain extent). To avoid the blockage fault of DPFs and to satisfy the increasingly stringent emission regulations, we need to monitor and diagnose the blockage situation of DPFs so as to clean particulates at the proper time. Therefore, it is necessary to study the fault diagnosis of blocking DPFs.

In recent years, many researchers [10–15] have carried out a lot of research on the fault diagnosis of blocking DPFs. At present, the most common used fault diagnosis method is to monitor the average exhaust pressure pre-DPF [2], but the exhaust pressure is changing constantly, as for on board diagnosis, the method has some disadvantages in sensitivity and aging characteristics. As a result, some researchers want to apply a new method to studying the fault diagnosis of blocking DPFs. Kumar et al. [16] raised a fault diagnosis method based on power spectral density theory, to diagnose the failure status of DPFs by analyzing the power spectral density of upstream and downstream sensor waveforms. Surve et al. [5] conducted the fault diagnosis of DPFs by combining the correlation analysis method with the spectral analysis method—the principle is to diagnose the failure status of DPF by calculating the characteristic value of the transfer function, and the advantages include that the failure status of the DPF can be diagnosed under the transient conditions of the diesel engine and that the slight failure of the DPF can also be found. Gupta et al. [17] raised a new fault diagnosis method based on the adaptive model. This method has great robustness for modeling error, sensor noise, and process variability, and it can be applied to on-board diagnosis (OBD) without any extra sensors.

To our knowledge, although some researchers have conducted a lot of research on the fault diagnosis of blocking DPFs and obtained many research results, there is still no consensus over how to efficiently monitor and diagnose the DPF system (and reach the level of OBD). Also, the methods to monitor and diagnose DPF status are still not comprehensive. Therefore, the spectral analysis method is applied in this paper to study the fault diagnosis of blocking DPF. This work can lay an important theoretical foundation for the on board diagnosis of DPF.

In this paper, a simulation model of a R425DOHC diesel engine with wall-flow ceramic DPF was established, and then the model was verified. On this basis, the effects of different blockage extents on the mean exhaust pressure of pre-DPF and the effects of different blockage extents on instantaneous exhaust pressure and its frequency spectrum were studied, and the sensitivity of mean exhaust pressure and characteristic frequency amplitude of instantaneous exhaust pressure with an increase in particulate depositions in DPF were comparatively studied.

2. Simulation Model and Validation

The research object was an R425DOHC diesel engine with a wall-flow ceramic DPF, and its main technical parameters are shown in Tables 1 and 2. GT-SUITE software was applied in this paper, which can be used for the performance simulation of an engine and after-treatment system. On this basis, a simulation model of the R425DOHC diesel engine with wall-flow ceramic DPF was established, as shown in Figure 1.

Table 1. Main parameters of the R425DOHC diesel engine.

Diesel Parameters	Value	Diesel Parameters	Value
Rated speed (r/min)	4000	Rated power (kW)	105
Max torque (N·m)	340	Displacement (L)	2.499
Bore × Stroke (mm)	92 × 94	Cylinder number	4
Compression ratio	17.5:1	Stroke	4

Table 2. Main parameters of the diesel particulate filter.

DPF Parameters	Value
Filter length (mm)	200
Filter diameter (mm)	190
Cell density (cm^{-2})	16

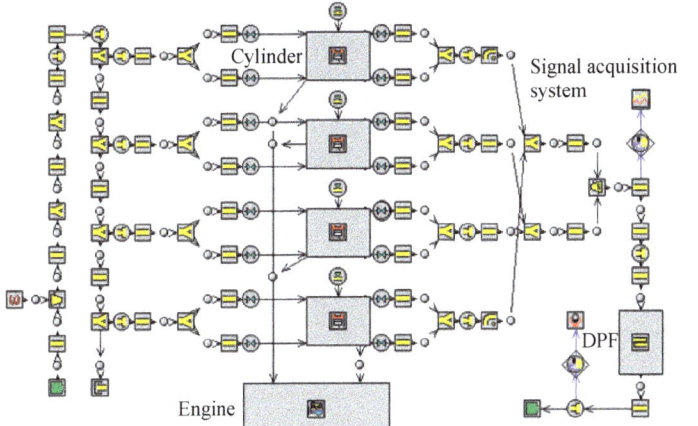

Figure 1. GT-Power model of an R425DOHC diesel engine with a diesel particulate filter (DPF).

To verify the accuracy of simulation model, we compared the simulation results of power performance and fuel economy under external characteristics with experimental data of the diesel engine, and the compared results are shown in Figures 2 and 3 and Table 3. In Figures 2 and 3, we set engine load as full-load, and set engine speed as 1000 r/min, 1500 r/min, 2000 r/min, 2500 r/min, 3000 r/min, 3500 r/min, and 4000 r/min. In Table 3, the max torque point was 100% load, 2000 r/min, and the max power point was 100% load, 4000 r/min. In addition, the experimental data of the diesel engine were supplied by the manufacturer.

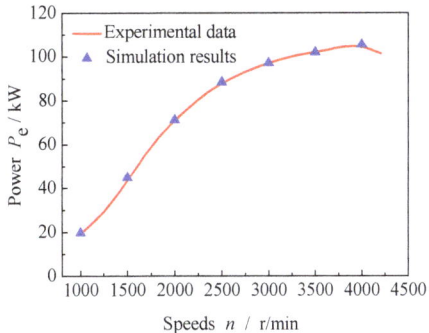

Figure 2. Comparison of experimental engine power with simulation results under full-load conditions.

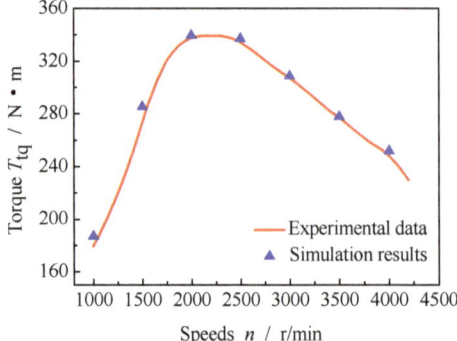

Figure 3. Comparison of experimental engine torque with simulation results under full-load conditions.

Table 3. Comparison of experimental engine specific fuel consumption with simulation results under rated conditions.

Specific Fuel Consumption	Simulation Results	Experimental Data
Max torque point (g/kW·h)	208.9	215
Max power point (g/kW·h)	252.2	256

As shown in Figures 2 and 3 and Table 3, the simulation results of power performance and fuel economy under external characteristics agreed well with the experimental data of the diesel engine, and the calculation errors were less than 5%, which suggests that the simulation model is correct and can be applied to simulating the exhaust characteristics of the diesel engine.

3. Effects of Different Blockage Extents of DPF on Mean Exhaust Pressure

Particulate deposition amounts in a DPF determine the blockage extent of the DPF. Meanwhile, considering the strong pulsation of exhaust pressure, the effects of different blockage extents on the mean exhaust pressure and the instantaneous exhaust pressure will be studied. In this section, the effects of different particulate deposition amounts on mean exhaust pressure will be discussed firstly.

In this section, we set engine speed as 2000 r/min and 4000 r/min—2000 r/min is the maximum torque speed, while 4000 r/min is the maximum power speed (rated speed).

Figure 4 gives the mean exhaust pressure versus particulate deposition amounts in the DPF (0 g, 20 g, 40 g, 60 g,) under different engine conditions.

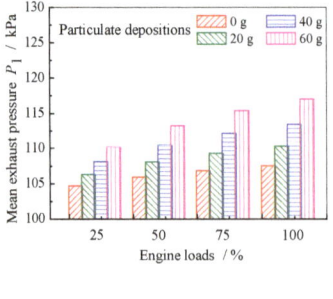

(**a**) 2000 r/min

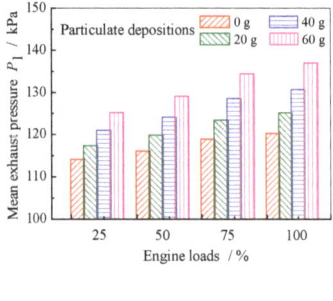

(**b**) 4000 r/min

Figure 4. Effects of particulate depositions on the mean exhaust pressure.

As shown in Figure 4, when the engine speed was 2000 r/min or 4000 r/min, under different engine loads, the mean exhaust pressure increased obviously with the increase in particulate depositions in the DPF. This is because particulate depositions affect the circulation performance of the DPF—the more particulate depositions, the worse the circulation performance of the DPF. The exhaust gas cannot be discharged through the DFP to the external envoriment, which leads to the increase in the mean exhaust pressure pre-DPF.

Also, mean exhaust pressure increased obviously with the increase in diesel engine loads, regardless of the engine speed. In addition, comparing Figure 4a with Figure 4b, under the same engine load, the mean exhaust pressure of pre-DPF when the cngine speed was 4000 r/min was greater than that when the engine speed is 2000 r/min. This is because the flow velocity of exhaust gas at high engine speed is higher than that at low engine speed, while the resistance produced by the DPF at flow velocity is greater than that at low flow velocity.

Based on the above analysis, we can determine that monitoring the mean exhaust pressure pre-DPF is a way to understand the situation of particulate depositions in the DPF so that we can diagnose the blockage extent. This conculation proved again that the commonly used diagnosis method is effective and feasible.

4. Effects of Different Blockage Extents of DPF on the Instantaneous Exhaust Pressure and its Frequency Spectrum

The gas flow in the exhaust system of a diesel engine is an unsteady flow. Exhaust pressure varies not only with engine conditions, but it is also different at different times of one exhaust cycle. So, the blockage extent of a DPF has an effect on the instantaneous exhaust pressure, except for the mean exhaust pressure. In this section, the effects of different blockage extents of DPFs on the instantaneous exhaust pressure will be first studied, then the effects of different blockage extents of DPFs on the frequency spectrum of instantaneous exhaust pressure will be studied, and an attempt will be made to find the eigen value which can be used to diagnose the blockage extents of DPFs.

Similar to the research on mean exhaust pressure, the measurement position of instantaneous exhaust pressure is also at the front of DPFs. Engine load was set as 100%, and engine speeds were set as 1000 r/min, 2000 r/min, 3000 r/min, and 4000 r/min. Particulate deposition amounts in DPF were set as 0 g, 20 g, 40 g, and 60 g, respectively.

Figure 5 shows that the instantaneous exhaust pressure varied with the increase in particulate depositions in DPFs under different engine conditions.

As shown in Figure 5, when engine load was 100%, under different engine speeds, instantaneous exhaust pressure increased with the increase in particulate depositions in the DPF, and the peak and trough values of instantaneous exhaust pressure increased obviously in one exhaust cycle. However, the instantaneous exhaust pressure of a diesel engine varies constantly in the time domain, so it is difficult to diagnose the blockage extents of a DPF by monitoring the instantaneous exhaust pressure in the time domain. Therefore, Fourier transform was applied to process the signal of instantaneous exhaust pressure, and we tried to find the characteristic parameter which can be used to reflect the blockage extents of DPF in frequency domain.

To research whether the frequency amplitude of instantaneous exhaust pressure with different exhaust pulsation frequencies can be used as the characteristic parameter to diagnose the blockage extents of DPF, we need to research the effects of particulate depositions in DPF on the frequency amplitude of instantaneous exhaust pressure with different exhaust pulsation frequencies.

Figure 6 gives the effects of particulate depositions in DPF on the Fourier spectrum of instantaneous exhaust pressure under different engine conditions.

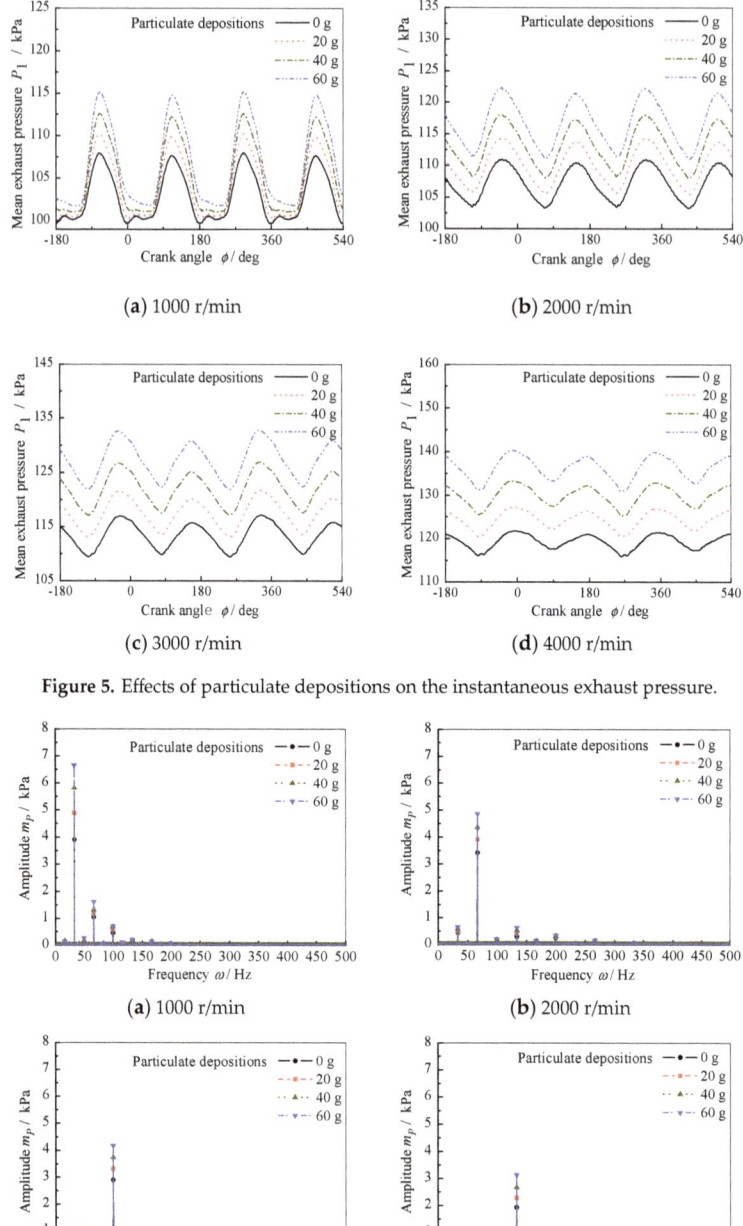

Figure 5. Effects of particulate depositions on the instantaneous exhaust pressure.

Figure 6. Effects of particulate depositions on the Fourier spectrum of instantaneous exhaust pressure.

As shown in Figure 6, in the Fourier spectrum of instantaneous exhaust pressure, when engine load was 100%, under the same engine speed, although particulate depositions in the DPF were different, the maximum frequency amplitude corresponded to the same frequency. The frequency is described as the exhaust pulsation frequency of the diesel engine at this speed. In addition, with the increase in particulate depositions in the DPF, the frequency amplitude of instantaneous exhaust pressure with different frequencies increases, and the variation extent of the frequency amplitude under the exhaust pulsation frequency is the most obvious.

In addition, with the increase in engine speeds, the frequency amplitude under the exhaust pulsation frequency decreased gradually. Also, the increasing extent of the frequency amplitude under the exhaust pulsation frequency decreased gradually with the increase in particulate depositions in the DPF. When the engine speed was 1000 r/min, the exhaust pulsation frequency was 33.33 Hz. The frequency amplitudes under the same exhaust pulsation frequency were 3.9 kPa, 4.9 kPa, 5.8 kPa, and 6.6 kPa when particulate depositions in the DPF were 0 g, 20 g, 40 g, and 60 g, respectively. When the engine speed was 3000 r/min, the exhaust pulsation frequency was 100.00 Hz. The frequency amplitudes under the same exhaust pulsation frequency were 2.9 kPa, 3.3 kPa, 3.7 kPa, and 4.2 kPa when particulate depositions in the DPF were 0 g, 20 g, 40 g, and 60 g, respectively.

Note that the instantaneous exhaust pressure under the constant engine condition is a stationary signal, so Figure 6 gives the calculation results based on the Fourier spectrum. In real engine conditions, the instantaneous exhaust pressure is an unsteady signal. In this paper, we assumed it was quasi-steady state (because the exhaust pulsation frequency is low and constant under the certain engine condition).

To get the sensitivity of the frequency amplitude under the exhaust pulsation frequency versus particulate depositions in the DPF, Figure 7 gives the effects of particulate depositions in the DPF on the frequency amplitude under exhaust pulsation frequency under different engine conditions.

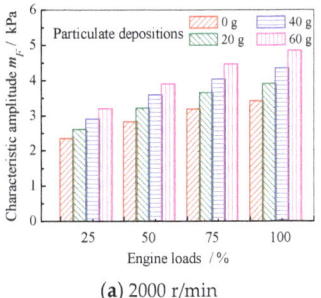
(**a**) 2000 r/min

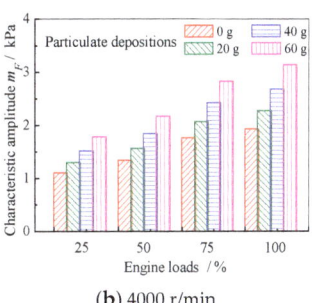
(**b**) 4000 r/min

Figure 7. Effects of particulate depositions in the DPF on the frequency amplitude under exhaust pulsation frequency.

As shown in Figure 7, when the engine speed was 2000 r/min or 4000 r/min, under different engine loads, the frequency amplitude under exhaust pulsation frequency increased obviously with the increase in particulate depositions in the DPF. In addition, the frequency amplitude under exhaust pulsation frequency increased obviously with the increase in diesel engine loads, regardless of the engine speed. Also, compared Figure 7a with Figure 7b, under the same engine load, the frequency amplitude under exhaust pulsation frequency when the engine speed was 4000 r/min was smaller than when the engine speed was 2000 r/min.

Based on the above analysis, we can determine that the frequency amplitude under exhaust pulsation frequency may be used as characteristic parameter to diagnose the blockage extent of DPFs. The frequency amplitude under exhaust pulsation frequency is defined as characteristic frequency amplitude in this paper.

5. Comparison between Characteristic Frequency Amplitude and Mean Exhaust Pressure

According to the analysis of Sections 3 and 4, both mean exhaust pressure and characteristic frequency amplitude of instantaneous exhaust pressure may be used as characteristic parameters to diagnose the blockage extent of DPFs. However, which one is more sentitive to the variation in particulate depositions in DPFs (more qualified to monitor the blockage extents of DPF) requires further comparative research.

Next, this paper will study the sensitivity of mean exhaust pressure change rate and characteristic frequency amplitude change rate under different particulate depositions in DPF. The change rate of mean exhaust pressure is defined by formula (1), while the change rate of characteristic frequency amplitude is defined by formula (2):

$$\theta_{P_1} = \frac{P_{1x} - P_{10}}{P_{10}},\qquad(1)$$

$$\theta_{m_F} = \frac{m_{Fx} - m_{F0}}{m_{F0}},\qquad(2)$$

where P_{1x} represents the mean exhaust pressure when particulate depositions exist in the DPF, P_{10} represents the mean exhaust pressure when particulate deposition is 0 g in the DPF, m_{Fx} represents the characteristic frequency amplitude when particulate depositions exist in the DPF, m_{F0} represents the characteristic frequency amplitude when particulate deposition is 0 g in the DPF.

Figures 8 and 9 provide a comparison of the mean exhaust pressure change rate with characteristic frequency amplitude change rate under different particulate depositions at 2000 r/min and 4000 r/min.

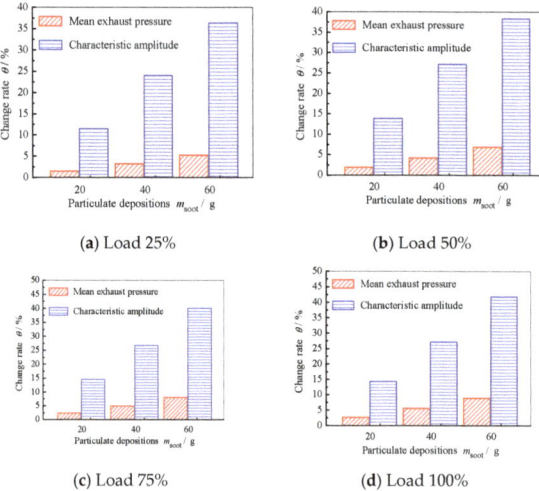

(a) Load 25%   (b) Load 50%

(c) Load 75%   (d) Load 100%

Figure 8. Comparison of mean exhaust pressure change rate with characteristic frequency amplitude change rate under different particulate depositions at 2000 r/min.

As shown in Figures 8 and 9, despite engine conditions and particulate depositions in DPF, the change rate of the characteristic frequency amplitude was always greater than the change rate of mean exhaust pressure. Take Figure 9, for example, when particulate depositions in DPF increased from 0 g to 20 g, the change rates of mean exhaust pressure under different engine loads were less than 5%, while the change rates of characteristic frequency amplitude were greater than 15%; when particulate depositions in the DPF increased from 0 g to 40 g, the change rates of mean exhaust pressure under different engine loads were less than 10%, while the change rates of characteristic frequency amplitude were greater than 25%; when particulate depositions in the DPF increased from 0 g to 60 g, the change rates of mean exhaust pressure under different engine loads were less than 15%, while the change rates of characteristic frequency amplitude were greater than 35%.

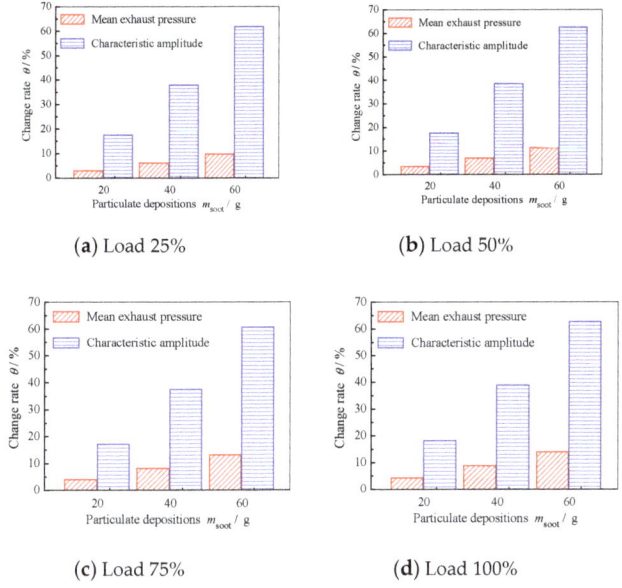

Figure 9. Comparison of mean exhaust pressure change rate with characteristic frequency amplitude change rate under different particulate depositions at 4000 r/min.

The major reason for this is that mean exhaust pressure represents the average value of pressure in a relatively long period, so the value is relatively great and the effect of particulate depositions in the DPF on mean exhaust pressure is small, while characteristic frequency amplitude represents the frequency amplitude component under exhaust pulsation frequency, so the value is relatively small and the effect of particulate depositions in the DPF on characteristic frequency amplitude is great. Therefore, characteristic frequency amplitude is more sentitive than mean exhaust pressure to the variation in particulate depositions in the DPF.

A comparison of Figures 8 and 9 shows that under the same engine speed, the difference in the change rate of characteristic frequency amplitude was not obvious with different engine loads. Also take Figure 9, for example, when particulate depositions in the DPF increased from 0 g to 40 g, the change rates of the characteristic frequency amplitude were 37.8%, 38.4%, 37.4%, and 38.8% with different engine loads (25%, 50%, 75%, 100%, respectively).

In addition, it was also found that when the diesel engine speed was 2000 r/min, the change rate of the characteristic frequency amplitude was less than the value when engine speed was 4000 r/min under the same conditions. When particulate depositions in the DPF increased from 0 g to 60 g, the change rate of the characteristic frequency amplitude was about 40% at 2000 r/min, while the change rate of the characteristic frequency amplitude was about 60% at 4000 r/min.

Therefore, in terms of sensitivity, the characteristic frequency amplitude of instantaneous exhaust pressure is more suitable as a characteristic parameter to monitor DPF blockage than mean exhaust pressure.

Finally, this method is correct under constant engine condition and under the assumption that the instantaneous exhaust pressures in real engine conditions are a quasi-steady state. For more accuracy in further study, wavelet transform and Hilbert Huang Transform (HHT) will be the better research methods.

6. Conclusions

In this paper, a simulation method was used to study the fault diagnosis of blocking DPFs by using spectral analysis on instantaneous exhaust pressure. From the results of this investigation, some conclusions can be drawn:

(1) A simulation model of an R425DOHC diesel engine with wall-flow ceramic DPF was established, and then the correctness of the model was verified with experimental data;

(2) Under different engine conditions, mean exhaust pressure increases obviously with the increase in particulate depositions in DPF. Mean exhaust pressure increases with the increase in diesel engine loads, regardless of the engine speed. Under the same engine load, the mean exhaust pressure of pre-DPF when the engine speed is 4000 r/min is greater than that when the engine speed is 2000 r/min. Monitoring the mean exhaust pressue pre-DPF is a way to understand the situation of particulate depositions in the DPF so that we can diagnose the blockage extents of DPF;

(3) Under different engine conditions, instantaneous exhaust pressure increases with the increase in particulate depositions in the DPF, and the peak and trough values of instantaneous exhaust pressure increase obviously in one exhaust cycle. However, it is difficult to diagnose the blockage extent of a DPF by monitoring instantaneous exhaust pressure in the time domain. Characteristic frequency amplitude decreases gradually an the increase in engine speed, but increases with an increase in engine loads, and the increasing extent of characteristic frequency amplitude decreases gradually with an increase in particulate depositions in the DPF. Characteristic frequency amplitude may be used as a characteristic parameter to diagnose the blockage extent of DPFs;

(4) Despite engine conditions and particulate depositions in DPFs, the change rate of characteristic frequency amplitude is always greater than the change rate of mean exhaust pressure. Under the same engine speed, the difference in the change rate of characteristic frequency amplitude is not obvious with different engine loads. The change rate of characteristic frequency amplitude when engine speed is 2000 r/min is less than the value when engine speed is 4000 r/min under the same conditions;

(5) In terms of sensitivity, the characteristic frequency amplitude of instantaneous exhaust pressure is more suitable as a characteristic parameter to monitor DPF blockage than mean exhaust pressure.

Author Contributions: S.-x.L. and M.L. write the article, collect and analyze data together.

Funding: This research was funded by the National Natural Science Foundation of China, grant number 51606006, 51776016; Beijing Natural Science Foundation, grant number 3182030; and the National Engineering Laboratory for Mobile Source Emission Control Technology, grant number NELMS2017A10.

Acknowledgments: Thanks Zhi Ning from Beijing Jiaotong University for supporting critical revision of this article.

Conflicts of Interest: The authors declare no conflict of interest.

References

1. Johnson, T.V. Diesel Emission Control in Review. *SAE Int. J. Fuels Lubr.* **2009**, *1*, 68–81. [CrossRef]
2. Baek, K.; Choi, H.; Bae, G.; Woo, M. A Study on Regeneration Strategies of Cordierite-DPF Applied to a Diesel Vehicle. In *SAE Paper*; The Automotive Research Association of India: Pune, India, 2011.
3. Bensaid, S.; Marchisio, D.L.; Fino, D.; Saracco, G.; Specchia, V. Modelling of diesel particulate filtration in wall-flow traps. *Chem. Eng. J.* **2009**, *154*, 211–218. [CrossRef]
4. Maricq, M.M. Chemical characterization of particulate emissions from diesel engines: A review. *J. Aerosol Sci.* **2007**, *38*, 1079–1118. [CrossRef]
5. Surve, P.R. Diesel Particulate Filter Diagnostics Using Correlation and Spectral Analysis. ECE Master's Thesis, Purdue University, West Lafayette, IN, USA, 2008.
6. Shim, B.J.; Park, K.S.; Koo, J.M.; Nguyen, M.S.; Jin, S.H. Estimationof soot oxidation rate in DPF under carbon and non-carbon based particulate matter accumulated condition. *Int. J. Automot. Technol.* **2013**, *14*, 207–212. [CrossRef]
7. Shah, C.D. Particulate matter load estimation in diesel particulate filters. Master's Thesis, Purdue University, West Lafayette, IN, USA, 2008.

8. Liu, Z.G.; Berg, D.R.; Swor, T.A.; Schauer, J.J. Comparative analysis on the effects of diesel particulate filter and selective catalytic reduction systems on a wide spectrum of chemical species emissions. *Environ. Sci. Technol.* **2008**, *42*, 6080–6085. [CrossRef] [PubMed]
9. Chen, L.; Ma, Y.; Guo, Y.; Zhang, C.; Liang, Z.; Zhang, X. Quantifying the effects of operational parameters on the counting efficiency of a condensation particle counter using response surface Design of Experiments (DoE). *J. Aerosol Sci.* **2017**, *106*, 11–23. [CrossRef]
10. Bensaid, S.; Marchisio, D.L.; Russo, N.; Fino, D. Experimental investigation of soot deposition in diesel particulate filters. *Catal. Today* **2009**, *147*, S295–S300. [CrossRef]
11. Van Nieuwstadt, M.; Brahma, A. Uncertainty analysis of model based diesel particulate filter diagnostics. *SAE Int. J. Commer. Veh.* **2009**, *1*, 356–362. [CrossRef]
12. Bensaid, S.; Marchisio, D.L.; Fino, D. Numerical simulation of soot filtration and combustion within diesel particulate filters. *Chem. Eng. Sci.* **2010**, *65*, 357–363. [CrossRef]
13. Ochs, T.; Schittenhelm, H.; Genssle, A.; Kamp, B. Particulate matter sensor for on board diagnostics (OBD) of diesel particulate filters (DPF). *SAE Int. J. Fuels Lubr.* **2010**, *3*, 61–69. [CrossRef]
14. Ntziachristos, L.; Fragkiadoulakis, P.; Samaras, Z.; Janka, K.; Tikkanen, J. Exhaust particle sensor for OBD application. In *SAE Technical Paper*; SAE International: Troy, MI, USA, 2011.
15. Cloudt, R. Diagnostics development for cost-effective temperature sensor based particulate matter OBD method. *SAE Int. J. Passeng. Cars-Electron. Electr. Syst.* **2014**, *7*, 573–582. [CrossRef]
16. Kumar, S.V.; Heck, R.M. A new approach to OBD monitoring of catalyst performance using dual oxygen sensors. *SAE Trans.* **2000**, *109*, 586–595.
17. Gupta, A.; Franchek, M.; Grigoriadis, K.; Smith, D.J. Model based failure detection of Diesel Particulate Filter. In Proceedings of the 2011 American Control Conference, San Francisco, CA, USA, 29 June–1 July 2011; pp. 1567–1572.

© 2019 by the authors. Licensee MDPI, Basel, Switzerland. This article is an open access article distributed under the terms and conditions of the Creative Commons Attribution (CC BY) license (http://creativecommons.org/licenses/by/4.0/).

Article

Single Controller-Based Colored Petri Nets for Deadlock Control in Automated Manufacturing Systems

Husam Kaid [1,*], Abdulrahman Al-Ahmari [1,*], Zhiwu Li [2,3] and Reggie Davidrajuh [4]

1. Industrial Engineering Department, College of Engineering, King Saud University, Riyadh 11421, Saudi Arabia
2. Institute of Systems Engineering, Macau University of Science and Technology, Macau 999078, China; systemscontrol@gmail.com
3. School of Electro-Mechanical Engineering, Xidian University, Xi'an 710071, China
4. Faculty of Science and Technology, University of Stavanger, 4036 Stavanger, Norway; reggie.davidrajuh@uis.no
* Correspondence: yemenhussam@yahoo.com (H.K.); alahmari@ksu.edu.sa (A.A.-A.)

Received: 6 November 2019; Accepted: 19 December 2019; Published: 22 December 2019

Abstract: Deadlock control approaches based on Petri nets are usually implemented by adding control places and related arcs to the Petri net model of a system. The main disadvantage of the existing policies is that many control places and associated arcs are added to the initially constructed Petri net model, which significantly increases the complexity of the supervisor of the Petri net model. The objective of this study is to develop a two-step robust deadlock control approach. In the first step, we use a method of deadlock prevention based on strict minimal siphons (SMSs) to create a controlled Petri net model. In the second step, all control places obtained in the first step are merged into a single control place based on the colored Petri net to mark all SMSs. Finally, we compare the proposed method with the existing methods from the literature.

Keywords: Automated manufacturing system; colored Petri net; deadlock prevention; siphon

1. Introduction

An automated manufacturing system (AMS) is a conglomeration of robots, machine tools, fixtures, and buffers. Several types of products enter the manufacturing system at separate points in time; the system can process these parts based on a specified sequence of operations and resource sharing. The sharing of resources leads to the occurrence of deadlock states, in which the local or global system is incapacitated [1–4]. Thus, an efficient deadlock-control algorithm is needed to prevent the deadlocks in an AMS. Petri nets are an excellent mathematical and graphical tool suitable for modeling, analyzing, and controlling deadlocks in AMSs [5,6]. The behavior and characteristics of an AMS (such as synchronization, conflict, and sequences) can be described by using Petri nets. Moreover, Petri nets may be used to provide the liveness and boundedness of a system [7]. To address the deadlock problem in AMSs, several approaches with Petri nets exist. These approaches are categorized into three strategies: (1) deadlock detection and recovery, (2) deadlock prevention, and (3) deadlock avoidance [7,8].

Traditionally, deadlock control approaches for AMS control are evaluated by using three criteria: structural complexity, computational complexity, and behavioral permissiveness [7]. Structural complexity means that a controller can be implemented with fewer monitors and arcs. When the computational complexity of a deadlock control approach is low, it can be applied to a large-scale system [7]. Behavioral permissiveness achieves high resource utilization in a controlled Petri net.

Deadlock control may be implemented in AMS with reliable resources (resources without failures or breakdowns) or unreliable resources (resources with failures or breakdowns). For reliable resources, there are two main techniques to prevent deadlocks in AMSs using a Petri net: reachability graph analysis [9–11] and structural analysis [12,13]. The reachability graph analysis needs listing all or a part of the reachable markings of the Petri net model. There are two parts of the reachability graph: the deadlock zone (DZ) and the live zone (LZ). First-met bad markings (FBMs) are defined in and extracted from the DZ. In this case, the deadlock is eliminated by designing and adding a monitor to prohibit the first-met bad markings from being reached. In this process, all first-met bad markings can be prevented by using iterations [14]. Several policies have been developed to prevent deadlock states, including iterative methods, the theory of region, and siphon control [10,13–19]. The weakness of the reachability graph analysis is that the size of a reachability graph of a Petri net grows quickly and, in the worst case, grows exponentially with respect to the net size and its initial marking, and the net can easily reach an unmanageable level. Structural analysis is often applied to a typical structure of Petri nets, such as siphons. The control steps in this technique are simple: each minimal siphon is prohibited from being non-empty, and each unmarked minimal siphon needs an added monitor to ensure a system to be live. However, the weakness of this technique is that the number of control places will be increased when the size of a Petri net model is increased; hence, this results in high structural complexity [20].

In the literature, deadlock control approaches based on the structural analysis technique (siphons) for AMSs with the Petri nets framework can be implemented by inserting the control places and the associated arcs to the original net, so that its siphons are permanently non-empty. The main disadvantage of the current policies is that many control places and associated arcs are inserted into the original Petri net model, which leads to the increased complexity of the supervisor of the Petri net model, compared with the initial model for the Petri net supervisor. Hence, an efficient approach is needed to minimize the Petri net supervisors' structural complexity for AMS. The objective of this study is to develop a two-step robust deadlock control policy. A technique based on SMSs developed in [21] is used in the first phase to develop a controlled Petri net model. In the second step, all control places obtained in the first step are merged into one control place based on colored Petri nets to make all SMSs marked.

The rest of the paper is organized as follows. Basic concepts of Petri nets are introduced in Section 2. Section 3 describes a deadlock prevention approach based on the SMS and the proposed robust control based on colored Petri nets. Section 4 shows an example from the literature. Finally, Section 5 presents conclusions and future research.

2. Preliminaries

This section introduces the basics of Petri nets and a general Petri net simulator (GPenSIM) tool.

2.1. Basics of Petri Nets

Let $N = (P, T, F, W)$ be a Petri net, where P and T are finite non-empty sets of places and transitions, respectively. Elements in $P \cup T$ are named nodes. Here, $P \cup T \neq \emptyset$ and $P \cap T = \emptyset$; P and T are depicted by circles and bars, respectively. Next, $F \subseteq (P \times T) \cup (T \times P)$ is the set of directed arcs that join the transitions with places (and vice versa), $W: (P \times T) \cup (T \times P) \rightarrow \mathbf{IN}$ is a mapping that assigns an arc's weight, where $\mathbf{IN} = \{0, 1, 2, \ldots\}$.

N is known as an ordinary net if $\forall (p, t) \in F$, $W(p, t) = 1$, where $N = (P, T, F)$. N is named a weighted net if there is an arc between x and y such that $W(x, y) > 1$. Let $N = (P, T, F, W)$ and node $a \in P \cup T$. Then, $\cdot a = \{b \in P \cup T \mid (b, a) \in F\}$ is named the preset of node a, and $a \cdot = \{b \in P \cup T \mid (a, b) \in F\}$ is named the postset of node a.

A marking M of N is a mapping $M: P \rightarrow \mathbf{IN}$. Next, (N, M_0) is a marked Petri net (PN), represented as $PN = (P, T, F, W, M_0)$, where the initial marking of PN is $M_0: P \rightarrow \mathbf{IN}$. A transition $t \in T$ is enabled at marking M if for all $p \in \cdot t$, $M(p) \geq W(p, t)$, which is denoted as $M[t\rangle$. When a transition t fires, it takes

$W(p, t)$ token (s) from each place $p \in \cdot t$, and adds $W(t, p)$ token (s) in each place $p \in t\cdot$. Thus, it reaches a new marking M', denoted as $M[t\rangle M'$, where $M'(p) = M(p) - W(p, t) + W(t, p)$.

We call N self-loop free if for all $a, b \in P \cup T$, $W(a, b) > 0$ implies $W(b, a) = 0$. Let $[N]$ be an incidence matrix of net N, where $[N]$ is an integer matrix that consists of $|T|$ columns and $|P|$ rows with $[N](p, t) = W(t, p) - W(p, t)$. The set of markings that are reachable from M in N is named the set of reachability of the Petri net model (N, M) denoted by $R(N, M)$.

Let (N, M_0) be a marked Petri net. A transition $t \in T$ is live if for all $M \in R(N, M)$, there exists a reachable marking $M' \in R(N, M)$ such that $M'[t\rangle$ holds. A transition is dead at M_0 if there does not exist $t \in T$ such that $M_0[t\rangle$ holds. M' is said to be reachable from M if there exists a finite transition sequence $\delta = t_1 t_2 t_3 \ldots t_n$ that can be fired, and markings $M_1, M_2, M_3, \ldots$, and M_{n-1} such that $M[t_1\rangle M_1[t_2\rangle M_2[t_3\rangle M_2 \ldots M_{n-1}[t_n\rangle M'$, denoted as $M[\delta\rangle M'$, satisfies the state equation $M' = M + [N]\vec{\delta}$, where $\vec{\delta}: T \to \mathbf{IN}$ maps t in T to the number of appearances of t in δ and is called a Parikh vector or a firing count vector.

Let (N, M_0) be a marked Petri net. It is said to be k-bounded if for all $M \in R(N, M_0)$, for all $p \in P$, $M(p) \leq k$ ($k \in \{1, 2, 3, \ldots\}$). A net is safe if all of its places are safe, i.e., in each place p, the number of tokens does not exceed one. In other words, a net is k-safe if it is k-bounded.

P-vectors (place vectors) and T-vectors (transition vectors) are column vectors. A P-vector $I: P \to Z$ cataloged by P is said to be a place invariant or P-invariant if $I \neq \mathbf{0}$ and $I^T \cdot [N] = \mathbf{0}^T$, and a T-vector $J: T \to Z$ cataloged by T is said to be a transition invariant or T-invariant if $J \neq \mathbf{0}$ and $[N] \cdot J = \mathbf{0}$, where Z is the set of integers.

When each element of I is nonnegative, place invariant I is called a place semiflow or P-semiflow. Assume that I is a P-invariant of a net with (N, M_0) and M is a marking reachable from the initial marking M_0. Then, $I^T M = I^T M_0$. Let $\|I\| = \{p \mid I(p) \neq 0\}$ be the support of P-invariant I.

The supports of P-invariant I are classified into three types: (1) $\|I\|^+$ is the positive support of P-invariant I with $\|I\|^+ = \{p \mid I(p) > 0\}$. (2) $\|I\|^-$ is the negative support of P-invariant I with $\|I\|^- = \{p \mid I(p) < 0\}$. (3) I is a minimal P-invariant if $\|I\|$ is not a superset of the support of any other one and its components are mutually prime. Let l_i be the coefficients of P-invariant I if for all $p_i \in P$, $l_i = I(p_i)$.

A colored Petri net (CPN) is described as a nine-tuple $CPN = (P, T, F, SC, C_f, N_f, A_f, G_f, I_f)$, where

1. P and T are finite nonempty sets of places and transitions, respectively, assuming $P \cap T = \emptyset$. F is a set of flows (arcs), from $p_i \in P$ to $t_j \in T$ and from $t_i \in T$ to $p_j \in P$.
2. SC is a color set that contains colors c_i and the operations on the colors.
3. C_f is the color function that maps $p_i \in P$ into colors $c_i \in SC$.
4. N_f is the node function that maps F into $(P \times T) \cup (T \times P)$.
5. A_f is the arc function that maps each flow (arc) $f \in F$ into the term e.
6. G_f is the guard function that maps each transition $t_i \in T$ to a guard expression g that has a Boolean value.
7. I_f is the initialization function that maps each place $p_i \in P$ into an initialization expression.

2.2. GPenSIM Tool

GPenSIM was developed by R. Davidrajuh (the fourth author of our paper) and runs in MATLAB. GPenSIM has been designed to model, control, simulate, and analyze discrete event systems [22]. GPenSIM enables the integration of Petri net models with other toolboxes in MATLAB (e.g., "Control systems" and "Fuzzy logic"). Table 1 shows the advantages and disadvantages of GPenSIM compared to CPN Tools [23]. Compared to the CPN tools, it is simpler to create a colored Petri net in GPenSIM. For instance,

1. Being versatile, CPN allows manipulation of the functions C_f, N_f, A_f, G_f, and I_f independently. However, being simple and crude, these functions (C_f, N_f, A_f, G_f, and I_f) are merged together

in GPenSIM and are coded in the preprocessor files. Hence, GPenSIM allows fewer degrees of freedom when developing a model.
2. In CPN tools, it is possible to impose logical constraints on places, transitions, and arcs. In GPenSIM, logical expressions can only be processed by transitions. Inevitably, this means GPenSIM poses restrictions in modeling compared to CPN. However, this is the price paid for achieving simplicity in GPenSIM (easiness in learning, using, and extending).
3. The arc weights can dynamically alter in CPN tools because of the logic conditions connected to it. GPenSIM does not allow dynamic nets (e.g., dynamic arcs, run-time removal of places and transitions). Once a Petri net is defined in the Petri net definition file (PDF), it cannot be changed.

Table 1. The advantages and disadvantages of GPenSIM compared to CPN Tools.

Tool	Advantages	Disadvantages
GPenSIM	1. Simple, easy to learn, and use. 2. Easy to extend. 3. GPenSIM runs on MATLAB, it is easy to interconnect with other toolboxes.	1. Limited functionality. 2. The user is supposed to extend the primitive functions offered or to develop their own functions.
CPN Tools	1. A large number of functions available. 2. Has been used to model large systems.	1. Quite complicated, as this is a product of several researchers, extending the tool into diverse directions over a period of 20 or more years. 2. Lack of user manual deprives new users.

To model, simulate, analyze, and control the Petri net models in GPenSIM, three files should be coded: Petri net definition file (PDF), main simulation file (MSF), and pre- and postprocessor files.

1. PDF defines the static elements of a Petri net (places, transitions, and arcs).
2. Before the simulation starts, MSF loads a PDF into memory and the workbench, and then the simulation begins. During the simulation runs, MSF will be blocked; the control will be handed back to MSF together with the simulation results when the simulation is finished. Consequently, MSF has no control over what happens during the simulation.
3. Pre-and postprocessors will be called during the simulation before and after firing of the transition. The preprocessor will inspect if the conditions of firing for a certain transition are met, and the postprocessor will execute post-firing activities if needed after a certain transition has been fired. These can be used to control the runtime of the system, as they are called during the simulation.

All tokens are homogeneous inside a place. It does not matter which token is first or last to arrive at the place. Similarly, it does not matter by which transition a token is deposited at the place. However, GPenSIM introduces the token colors. Each token can become identifiable and unique with a unique token identification number (**tokID**). Moreover, we can add some tags ("colors") to each token. The following problems are crucial when using colors in GPenSIM:

1. Only transitions can manipulate colors: the colors of the output tokens can be added, deleted, or changed in the preprocessor.
2. By default, colors are inherited: the system gathers all colored tokens from the input places when a transition fires and then transfers the colored tokens to the output places. However, color inheritance can be avoided by overriding.

3. An enabled transition may choose certain color-based input tokens.
4. An enabled transition may choose certain time-based input tokens (e.g., when the creation time of the tokens is known).
5. A token has the following structure: **tokID**, time of creation, and color setting.

   A. **tokID**: a single token identifier (integer value).
   B. **creation_time**: the transition time (real value) when the token was produced. Importantly, this time may differ when the transition actually deposited the token in an output place.
   C. **t_color** (text string set) is a color setting.

There are several GPenSIM functions used for color manipulation. One of the functions used in this study is **tokenEXColor**, which can be expressed as follows:

[set_of_tokID,nr_token_av] = **tokenEXColor (place, nr_tokens_wanted, t_color)**, where the function requires three input arguments and returns two output values.

1. Input arguments:

   (place, nr_tokens_wanted, t_color):

   - **Place:** from which place the tokens are to be selected.
   - **nr_tokens_wanted:** the number of required tokens with the specified color.
   - **t_color:** a set of colors.

2. Output values:

   [set_of_tokID,nr_token_av]

   - **set_of_tokID:** a set of tokIDs that meet the color specifications. The set length of tokIDs is equal to the input argument of **nr_tokens_wanted**.
   - **nr_token_av:** the number of valid tokIDs available in **set_of_tokID**; the set may have trailing zeros to match the length of **nr_tokens_wanted**.

3. Deadlock Prevention Policy Based on SMSs and Colored Petri Nets

In this section, we use a deadlock-prevention approach based on strict SMSs to design a controlled Petri net model. This approach is adopted from Ezpeleta et al. [1].

Definition 1 [23]. *A PN $N = (P_A \cup \{p^0\}, T, F)$ is said to be a simple sequential process (S^2P), if: (1) N is a strongly connected state machine and (2) each circuit N includes place p^0, where p^0 is named the idle process place and $P_A \neq \emptyset$ is an operation places set.*

Definition 2 [23]. *A PN $N = (\{p^0\} \cup P_A \cup P_R, T, F)$ is said to be a simple sequential process with resources (S^2PR) such that:*

1. The subnet generated by $X = P_A \cup \{p^0\} \cup T$ is an S^2P.
2. $P_R \neq \emptyset$ and $(P_A \cup \{p^0\}) \cap P_R = \emptyset$, where P_R is a resource places set.
3. $\forall p \in P_A, \forall t \in {}^\bullet p, \forall t' \in p^\bullet, \exists r_p \in P_R, {}^\bullet t \cap P_R = t'^\bullet \cap P_R = \{r_p\}$.
4. $\forall r \in P_R, {}^\bullet r \cap P_A = r^\bullet \cap P_A \neq \emptyset$ and ${}^\bullet r \cap r^\bullet \neq \emptyset$.
5. $(p^0)^{\bullet\bullet} \cap P_R = ({}^{\bullet\bullet}p^0) \cap P_R \neq \emptyset$.

Definition 3 [23]. *Let $N = (\{p^0\} \cup P_A \cup P_R, T, F)$ be an S^2PR, and M_o be an initial marking of N. An S^2PR with such a marking is said to be acceptably marked if (1) $M_o(p^0) \geq 1$, $M_o(r) \geq 1$, $\forall r \in P_R$, and (3) $M_o(p) = 0$, $\forall p \in P_A$.*

Definition 4 [23]. *A system of S^2PR, named S^3PR for abbreviation, is defined recursively as follows:*

1. An S^2PR is as well an S^3PR
2. Let N_1 and N_2 be two S^3PRs, where $N_1 = (\{p^0{}_1\} \cup P_{A1} \cup P_{R1}, T_1, F_1)$ and $N_2 = (\{p^0{}_2\} \cup P_{A2} \cup P_{R2}, T_2, F_2)$, such that $(\{p^0{}_1\} \cup P_{A1}) \cap (\{p^0{}_2\} \cup P_{A2}) = \emptyset$, $P_{A1} \cap P_{A2} \neq P_C$, $P_{R1} \cap P_{R2} = P_C$ and $T_1 \cap T_2 \neq \emptyset$. Then, the net $N = (\{p^0\} \cup P_A \cup P_R, T, F)$ is also an S^3PR resulting from the composition of N_1 and N_2 via the set of common P_C and defined as follows: (1) $p^0 = \{p^0{}_1\} \cup \{p^0{}_2\}$, $P_R = P_{R1} \cup P_{R2}$, $P_A = P_{A1} \cup P_{A2}$, $T = T_1 \cup T_2$, $F = F_1 \cup F_2$.

The composition of n S^2PR N_1-N_n via P_C, is denoted by $\bigotimes_{i=1}^{n} N_i$. $\overline{N_i}$ is used to denote the S^2P from which the S^2PR N_i is formed.

Definition 5 [23]. Let $N = (\{p^0\} \cup P_A \cup P_R, T, F)$ be an S^3PR. M_o is an initial marking of N. (N, M_o) is said to be an acceptably marked S^3PR if (1) (N, M_o) is an acceptably marked S^2PR, (2) $N = N_1 \circ N_2$, where (N_i, M_{oi}) is said to be an acceptably marked S^3PR and

- $\forall p \in P_{Ai} \cup \{p^0{}_i\}$, $M_o(p) = M_{oi}(p)$, $\forall i \in \{1,2\}$.
- $\forall r \in P_{Ri} \setminus P_C$, $M_o(r) = M_{oi}(r)$, $\forall i \in \{1,2\}$.
- $\forall r \in P_{Ri}$, $M_o(p) = \max\{M_{o1}(r), M_{o2}(r)\}$, $\forall i \in \{1,2\}$.

Definition 6 [23]. Let N be an S^3PR, A non-empty set $S \subseteq P$ is said to be a siphon in N if $\cdot S \subseteq S\cdot$. When a siphon does not include other siphons, it is said to be a minimal siphon.

Definition 7 [24]. Let S be a minimal siphon in an S^3PR N. A minimal siphon S is said to be strict if $\cdot S \subsetneq S \cdot$. Let $\Pi = \{S_1, S_2, S_3, \ldots, S_k\}$ be a set of SMSs of N. We have $S = S_A \cup S_R$, $S_R = S \cap P_R$, and $S_A = S \setminus S_R$, where S_A denotes the places of operations and S_R denotes the places of resources.

Definition 8 [23]. Let $r \in P_R$ be a reliable resource place in an S^3PR N. The operation places that use r are known as the set of holders of r, indicated by $H(r) = \{p \backslash p \in P_A, p \in \cdot r \cap P_A \neq \emptyset\}$. $[S]$ is said to be the complementary set of S if $[S] = (\cup_{r \in S_R} H(r)) \setminus S_A$.

Definition 9 [24]. Let (N_a, M_a) and (N_b, M_b) be marked Petri nets; $N_i = (P_i, T_i, F_i, W_i)$, where $i = a, b$. We call (N, M) with $N = (P, T, F, W)$ a synchronous net resulting from the integration of (N_a, M_a) and (N_b, M_b) and denote it as $(N_a, M_a) \| (N_b, M_b)$ when the following conditions are satisfied: (1) $P = P_a \cup P_b$, and $P_a \cap P_b = \emptyset$. (2) $T = T_a \cup T_b$. (3) $F = F_a \cup F_b$. (4) $W(e) = W_i(e)$, where $e \in F_i$, $i = a, b$. (5) $M(p) = M_i(p)$, where $p \in P_i$, $i = a, b$.

Definition 10 [25]. Let (N, M_o) be an S^3PR with $N = (P_A \cup \{p^0\} \cup P_R, T, F, M_o)$. The deadlock controller for (N, M_o) developed by Ezpeleta et al. [1] is denoted as $(V, M_{Vo}) = (P_V, T_V, F_V, M_{Vo})$, where (1) $P_V = \{V_S \setminus S \in \Pi\}$ is a set of control places. (2) $T_V = \{t \setminus t \in \cdot V_S \cup V_S \cdot\}$. (3) $F_V \subseteq (P_V \times T_V) \cup (T_V \times P_V)$ is the set of directed arcs that join the control places with transitions (and vice versa). (4) For all $V_S \in P_V$, $M_{Vo}(V_S) = M_{Vo}(S) - 1$, where $M_{Vo}(V_S)$ is called an initial marking of a control place V_S.

We call (N_V, M_{Vo}) a controlled Petri net model resulting from the integration of (V, M_{Vo}) and (N, M_o), denoted as $(V, M_{Vo}) \| (N, M_o)$. A control place or monitor is inserted to each SMS to ensure the liveness of a Petri net, and all SMSs can never be unmarked. The proposed policy is simple and guarantees success. However, it leads to a more complex Petri-net-controlled system than the original Petri net model, because the number of added monitors is equal to that of the SMSs in the target Petri net model, and the number of associated arcs is larger than that of the added monitors. According to the strict minimal siphon concept, the developed deadlock prevention approach proposed by [1] is described by Algorithm 1.

Algorithm 1: Strict minimal siphon-based algorithm

Input: Original S^3PR Petri net model (N, M_0)
Output: Controlled net (N_V, M_{V_0}).
Step 1: Compute the set of SMS Π for N.
Step 2: for each $S \in \Pi$ **do**

- Add a control place V_S. /* By using Definition 10.*/
- Add V_S output arcs; all arc weights are unitary. /* By using Definition 10.*/
- Add V_S input arcs; all arc weights are unitary. /* By using Definition 10.*/
- Compute $M_{V_0}(V_S)$. /* By using Definition 10.*/

end for
Step 3: Output a controlled net (N_V, M_{V_0}).
Step 4: End

Consider the S^3PR Petri net model shown in Figure 1. The Petri net model involves eleven places and eight transitions. The places can be described as the following set partition: $P^0 = \{p_1, p_8\}$, $P_R = \{p_9, p_{10}, p_{11}\}$, and $P_A = \{p_2, p_3, \ldots, p_7\}$. The model has 20 reachable markings and eight minimal siphons, three of which are SMSs. The siphons are $S_1 = \{p_4, p_7, p_9, p_{10}, p_{11}\}$, $S_2 = \{p_4, p_6, p_{10}, p_{11}\}$, and $S_3 = \{p_3, p_7, p_9, p_{10}\}$. According to Definitions 2, 3, and 5

(1) For S_1: $S_A = \{p_4, p_7\}$, $S_R = \{p_9, p_{10}, p_{11}\}$, $H(p_9) = \{p_2, p_7\}$, $H(p_{10}) = \{p_3, p_6\}$, $H(p_{11}) = \{p_4, p_5\}$, $[S_1] = \{p_2, p_3, p_5, p_6\}$, $\cdot V_{S1} = \{t_3, t_7\}$, $V_{S1}\cdot = \{t_1, t_5\}$, and $M_{V_0}(V_{S1}) = 2$.

(2) For S_2: $S_A = \{p_4, p_6\}$, $S_R = \{p_{10}, p_{11}\}$, $H(p_{10}) = \{p_3, p_6\}$, $H(p_{11}) = \{p_4, p_5\}$, $[S_2] = \{p_3, p_5\}$, $\cdot V_{S2} = \{t_3, t_6\}$, $V_{S2}\cdot = \{t_2, t_5\}$, and $M_{V_0}(V_{S2}) = 1$.

(3) For S_3: $S_A = \{p_3, p_7\}$, $S_R = \{p_{10}, p_{11}\}$, $H(p_{10}) = \{p_3, p_6\}$, $H(p_{11}) = \{p_4, p_5\}$, $[S_3] = \{p_2, p_6\}$, $\cdot V_{S3} = \{t_2, t_7\}$, $V_{S3}\cdot = \{t_1, t_5\}$, and $M_{V_0}(V_{S3}) = 1$.

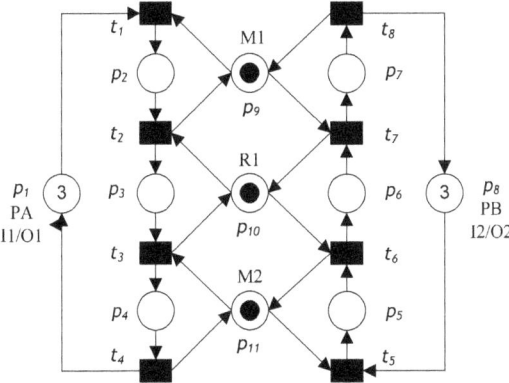

Figure 1. S^3PR Petri net model of an AMS.

After monitors have been added using Algorithm 1, we obtain the controlled Petri net model shown in Figure 2.

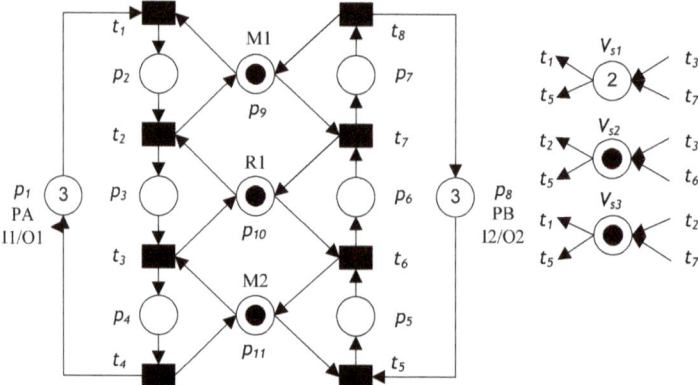

Figure 2. Controlled S³PR Petri net model.

Definition 11. *Let (N, M_0) be an S^3PR with $N = (P_A \cup \{p^0\} \cup P_R, T, F, M_0)$. The deadlock controller for (N, M_0) developed in Ezpeleta et al. [1] is denoted as $(V, M_{V_0}) = (P_V, T_V, F_V, M_{V_0})$. Here, (V, M_{V_0}) can be reduced and replaced by a colored common deadlock control subnet, which is a PN $N_{DC} = (\{p_{combined}\}, \{T_{DCi} \cup T_{DCo}\}, F_{DC}, C_{vsi})$, where $p_{combined}$ is called the merged control place of all monitors $P_V = \{V_S \setminus S \in \Pi\}$. $T_{DCi} = \{t \setminus t \in {}^\bullet V_S\}$. $T_{DCo} = \{t \setminus t \in V_S{}^\bullet\}$. $F_{DC} \subseteq (\{p_{combined}\} \times \{T_{DCi} \cup T_{DCo}\}) \cup (\{T_{DCi} \cup T_{DCo}\} \times \{p_{combined}\})$ is the set of arrows that join the merged control place with transitions (and vice versa). C_{cri} is the color that maps $p_{combined}$ into colors $C_{vsi} \in SC$, where $SC = \cup_{i \in V_S}\{C_{vsi}\}$. (N_{DC}, M_{DCo}) is called a colored common deadlock control subnet. For all $V_S \in P_V$, $M_{DCo}(p_{combined}) = \Sigma M_{V_0}(V_S)$, where $M_{DCo}(p_{combined})$ is an initial token with the colors marking of the merged monitor.*

Figure 3 shows $p_{combined}$, the merged control place of all monitors P_V of the controlled Petri net model from Figure 2, according to Definition 6.

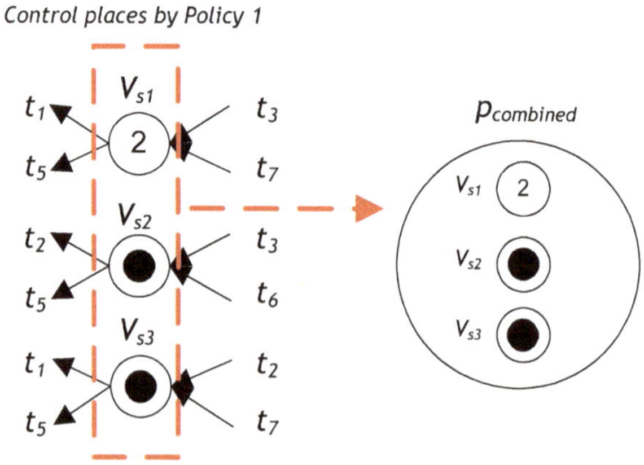

Figure 3. Merged control place for all monitors P_V.

The output arcs of $p_{combined}$ are connected to the source transitions T_{DCo}, which lead to the sink transitions of S. Transitions $V_{si}{}^\bullet$ for all monitors P_V augmented from Algorithm 1 are defined as $V_{S1}\cdot = \{t_1, t_5\}$, $V_{S2}\cdot = \{t_2, t_5\}$, and $V_{S3}\cdot = \{t_1, t_5\}$. Thus, T_{DCo} can be denoted as $T_{DCo} = \cup_{i \in V_S}\{V_{Si}\cdot\}$, so $T_{DCo} = \{2t_1, t_2, 3t_5\}$, as shown in Figure 4.

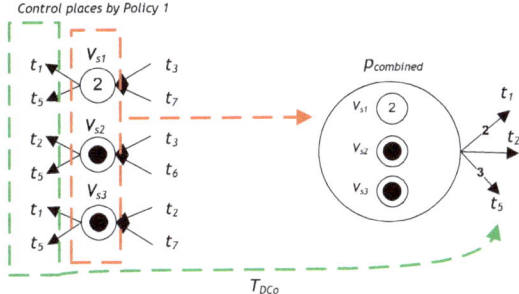

Figure 4. Output arcs of $p_{combined}$ for all monitors P_V.

The input arcs of $p_{combined}$ are connected with the output of S, denoted as T_{DCi}. Transitions $\cdot V_{si}$ for all monitors P_V augmented from Algorithm 1 are defined as $\cdot V_{s1} = \{t_3, t_7\}$, $\cdot V_{s2} = \{t_3, t_6\}$, and $\cdot V_{s3} = \{t_2, t_7\}$. Thus, T_{DCi} can be represented by $T_{DCi} = \cup_{i \in V_S}\{\cdot V_{si}\}$, so $T_{DCi} = \{t_2, 2t_3, t_6, 2t_7\}$, as shown in Figure 5.

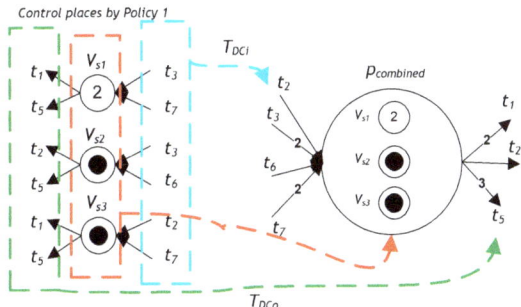

Figure 5. Output arcs of $p_{combined}$ for all monitors P_V.

Moreover, $M_{DCo}(p_{combined}) = \sum M_{V0}(V_S) = M_{V0}(V_{S1}) + M_{V0}(V_{S2}) + M_{V0}(V_{S3}) = 2 + 1 + 1 = 4$. Thus, in the model of the Petri net from Figure 2, we have three color types: $SC = \{C_{vs1}, C_{vs2}, C_{vs3}\}$. Therefore, the total number of colored tokens is 4: we have two tokens of C_{vs1} color, one token of C_{vs2} color, and one token of C_{vs3} color, as shown in Figure 6.

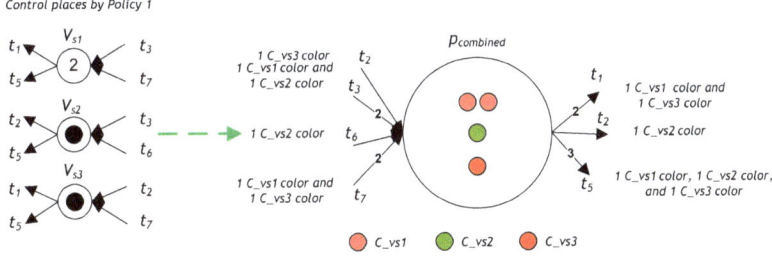

Figure 6. Merged controller for all monitors P_V.

Definition 12. *Let (N, M_0) be an S^3PR with $N = (P_A \cup \{p^0\} \cup P_R, T, F, M_0)$ and (N_{DC}, M_{DC0}) a deadlock controller for (N, M_0) created by Definition 11 with $N_{DC} = \{p_{combined}\}, \{T_{DCi} \cup T_{DCo}\}, F_{DC}, C_{vsi}, M_{DCo})$. We call (N_C, M_{C0}) a controlled marked Petri net, denoted as $(N_C, M_{C0}) = (N, M_0) \| (N_{DC}, M_{DC0})$ and called the composition of (N, M_0) and (N_{DC}, M_{DC0}), where $N_C = (P_A \cup \{p^0\} \cup P_R \cup \{p_{combined}\}, T \cup T_{DCi} \cup T_{DCo}, F \cup F_{DC}, C_R, M_{C0})$ be a colored controlled marked S^3PR, and $R(N_C, M_{C0})$ be its reachable graph.*

Theorem 1. *The colored controlled net (N_C, M_{Co}) is live.*

Proof. We must prove that all transitions T, T_{DCi}, T_{DCo} in (N_C, M_{Co}) are live. No strict minimal siphon is emptied. In addition, no new strict minimal siphon is created, since all $t_1 \in T$ are live. For all $t_2 \in T_{DCi}$; if $\forall p_i \in \bullet\, t_2$, $M_C(p_i) > 0$, then t_2 can fire in any case because it is uncontrollable. Thus, M_C ($p_{combined}$) > 0, for all $t_3 \in T_{DCo}$, if M_C ($p_{combined}$) > 0, then t_3 can fire. Therefore, controlled net (N_C, M_{Co}) is live. □

According to the concepts of SMSs and colored Petri nets, the developed policy is stated in Algorithm 2.

Algorithm 2: Integrated Strict Minimal Siphon And Colored Petri Nets-Based Algorithm

---

Input: Petri net models (N, M_0) and (V, M_{V0}).
Output: Colored controlled S^3PR Petri net (N_C, M_{Co}).
Step 1: Combine all monitors P_V into one monitor ($p_{combined}$), considering the following steps:

1. Add $p_{combined}$ output arcs T_{DCo}. /* By Definition 11.*/
2. Add $p_{combined}$ input arcs T_{DCi}. /* By Definition 11.*/
3. Define colors for monitors P_V /* By Definition 11.*/
4. Compute M_{DCo} ($p_{combined}$) = $\sum M_{V_0}$ (V_S), where M_{DCo} ($p_{combined}$) is an initial token with the colors marking of a merged monitor. /* By Definition 11.*/

Step 2: Insert the combined monitor into the Petri net model (N, M_0). The obtained net is denoted as (N_C, M_{Co}).
Step 3: Output (N_C, M_{Co}).
Step 4: End

Figure 7 shows the proposed single controller for the controlled Petri net model from Figure 2 by using Algorithm 2. In the net shown in Figure 7, when transition t_1 fires, the system selects only one token from input place p_1, one token from resource place p_9, one token of color C_{vs1} from $p_{combined}$, and one token of color C_{vs3} from $p_{combined}$, and it transfers them into p_2. Moreover, when transition t_2 fires, the system selects only one token from operation place p_2, one token from resource place p_{10}, and one token of color C_{vs2} from $p_{combined}$ and transfers them into p_3. When transition t_5 fires, the system selects only one token from input place p_8, one token from resource place p_{11}, one token of color C_{vs2} from $p_{combined}$, one token of color C_{vs2} from $p_{combined}$, and one token of color C_{vs3} from $p_{combined}$ and transfers them into p_5. When transition t_2 fires, it creates a color C_{vs3} on the tokens from p_2 and p_{10} and transfers them into common place $p_{combined}$. In addition, when the transition t_3 fires, it creates two colors C_{vs1} and C_{vs2} on the tokens from p_3 and p_{11} and transfers them into place $p_{combined}$. Moreover, when transition t_6 fires, the system creates a color C_{vs2} on the tokens from p_5 and p_{10} and transfers them into place $p_{combined}$. Finally, when transition t_7 fires, the system creates two colors C_{vs1} and C_{vs3} on the tokens from p_6 and p_9 and transfers them into $p_{combined}$.

By default, colors are inherited: when a transition T_{DCo} fires, the system gathers all colored tokens from the input place $p_{combined}$ and then transfers these colored tokens to the output place p_i. However, color inheritance can be prohibited by overriding.

After implementing a Petri net model shown in Figure 7 by using the GPenSIM code, we usually obtain three files: the Petri net definition file (PDF), common processor file (COMMON_PRE file), and main simulation file (MSF). Figure 8 shows the resulting PDF file and defines the static Petri net model by declaring the sets of places, transitions, and arcs. Figure 9 shows the MSF file used to compute the simulation results. Figure 10 displays the resulting MSF COMMON_PRE file and defines the conditions for the enabled failure and recovery transitions to start firing based on the colored tokens, mean time to failure, and mean time to repair.

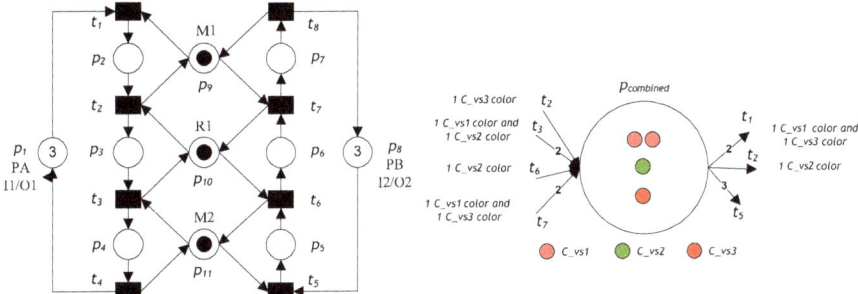

Figure 7. Colored controlled S³PR Petri net model Algorithm 2.

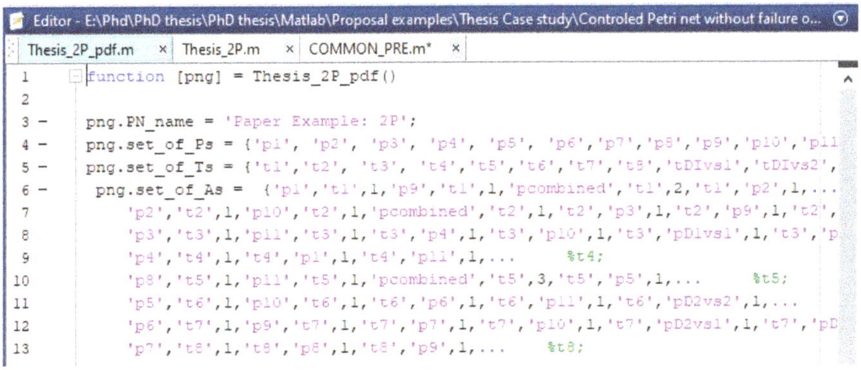

Figure 8. PDF file of the colored controlled S³PR Petri net model from Figure 7.

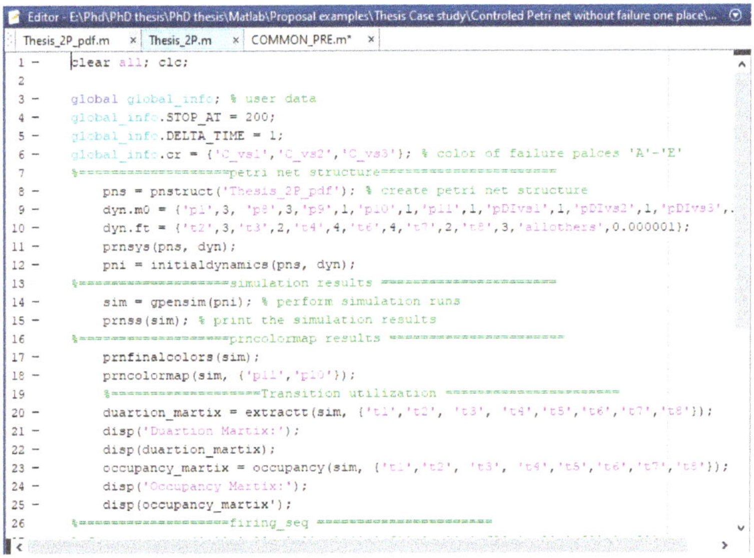

Figure 9. Part of the MSF file of the colored controlled S³PR Petri net model from Figure 7.

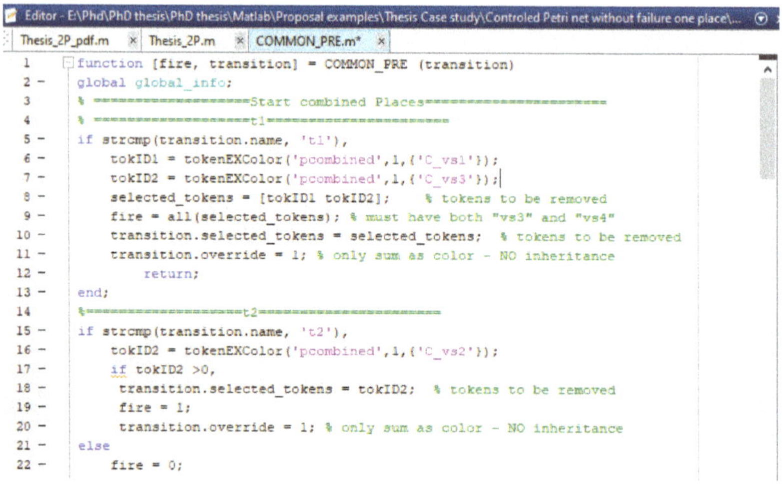
Editor - E:\Phd\PhD thesis\PhD thesis\Matlab\Proposal examples\Thesis Case study\Controled Petri net without failure one place\...
Thesis_2P_pdf.m
Thesis_2P.m
COMMON_PRE.m*
function [fire, transition] = COMMON_PRE (transition)
global global_info;
% ===================Start combined Places=====================
% ===================t1======================
if strcmp(transition.name, 't1'),
tokID1 = tokenEXColor('pcombined',1,{'C_vs1'});
tokID2 = tokenEXColor('pcombined',1,{'C_vs3'});
selected_tokens = [tokID1 tokID2]; % tokens to be removed
fire = all(selected_tokens); % must have both "vs3" and "vs4"
transition.selected_tokens = selected_tokens; % tokens to be removed
transition.override = 1; % only sum as color - NO inheritance
return;
end;
%====================t2======================
if strcmp(transition.name, 't2'),
tokID2 = tokenEXColor('pcombined',1,{'C_vs2'});
if tokID2 >0,
transition.selected_tokens = tokID2; % tokens to be removed
fire = 1;
transition.override = 1; % only sum as color - NO inheritance
else
fire = 0;

Figure 10. Part of the COMMON_PRE file of the colored controlled S^3PR Petri net model from Figure 7.

To make the work more solid and well-positioned, we have developed a Trust-based colored controlled Petri net (TCCPN) [26–29] and comparing the proposed work with relevant TCCPNs.

Definition 13. *Let* (N_{TM}, M_{TMo}) *be a Trust-based colored controlled Petri net (TCCPN)* S^3PR *with* $N_{TM} = (P_A \cup \{p^0\} \cup P_R \cup \{p_{combined}\}, T \cup T_{DCi} \cup T_{DCo}, F \cup F_{DC}, C_R, \eta, \tau, \psi, M_{TMo})$, *and* $R(N_{TM}, M_{TMo})$ *be its reachable graph, where*

1. *η is the arcs weight, which denotes the importance or probability of input arcs into a transition. If there is an arc (p, t), $\eta(p, t) = c$ indicates there is a probability of $\eta(p, t)$ encouraging the token entering t from p. If the token has a capacity h, the new capacity will be $h * c$.*
2. *τ is a time guard for transition $t \in (T \cup T_{DCi} \cup T_{DCo})$, $\tau: t \rightarrow [e, f]$, $\tau(t)$ indicates transition t can only fire during e and f. Particularly, if $e = f$, that indicates the transition can only occur through e.*
3. *ψ is the threshold of token in $p \in (P_A \cup \{p^0\} \cup P_R \cup \{p_{combined}\})$, $\psi: p \rightarrow R$, and R is a real type data. $\Psi(p) = r_1$, indicates when the number of tokens in p is less than or equal to r_1, p can reach a new position.*

To model a TCCPN, there are many types of factors that can have an influence on trust in colored controlled net S^3PR, and a non-negative real number can represent the value of each factor type. An assessment process will consume factors for aggregating a new trust value. We use E^{in} to characterize the input factors consumed and use E^{out} to characterize the aggregation trust value. For an assessment process AP_k, $E^{in}(AP_k)$ is related to the input place and $E^{ou}(AP_k)$ is related to the output place. There are rules for firing transitions in a TCCPN and are stated as below:

Definition 14. *Let* (N_{TM}, M_{TMo}) *be a Trust-based colored controlled Petri net (TCCPN)* S^3PR *with* $N_{TM} = (P_A \cup \{p^0\} \cup P_R \cup \{p_{combined}\}, T \cup T_{DCi} \cup T_{DCo}, F \cup F_{DC}, C_R, \eta, \tau, \psi, M_{TMo})$, *a transition* T_k *under the marking* M *can be fired when the following conditions are satisfied:*

1. $t \in \tau(T_k)$
2. $E^{in}(AP_k) > 0$
3. $E^{out}(AP_k) \geq \Psi(p_i)$
4. *for all* $p \in \cdot t$, $M(p) \geq W(p, t)$

where $\tau(t)$ is the valid firing time in TCCPN, and E^{in} (AP_k) is the current value of the token in the input place. E^{out} (AP_k) is the current value of the token in the output place, and Ψ (p_i) is the threshold for entering p_i. Note that an assessment process AP_k may have more than one input place and output place.

In addition, there are rules for new markings, when a transition t fires, a token value can be changed and it will be kept in a new place because of the threshold, $E^{out}(AP_k) = \sum_{i=1}^{nf} \eta_i * E^{in}{}_i(AP_k)$, where nf is the number of input factors. Then, the new marking will be changed as $M'(p) = M(p) - W(p, t) + W(t, p)$. In order to demonstrate the TCCPN, reconsider Figure 7, it describes a system as follows:

1. $\tau(t)$: $\tau(t_1) = 0$, $\tau(t_2) = 3$, $\tau(t_3) = 2$, $\tau(t_4) = 4$, $\tau(t_5) = 0$, $\tau(t_6) = 4$, $\tau(t_7) = 5$, $\tau(t_8) = 3$;
2. $M_0 = (p_1\ p_2\ p_3\ p_4\ p_5\ p_6\ p_7\ p_8\ p_9\ p_{10}\ p_{11}\ p_{combined}) = (3\ 0\ 0\ 0\ 0\ 0\ 0\ 3\ 1\ 1\ 1\ 4)$;
3. Assessment process is a combined place $Ap_{combined}$
4. The input assessment process is $T_k = \{t_2, 2t_3, t_6, 2t_7\}$
5. The importance on arcs: η $(p_2, t_2) = 0.5$, $\eta(p_{10}, t_2) = 0.5$, $\eta(p_3, t_3) = 0.5$, $\eta(p_{11}, t_3) = 0.5$, $\eta(p_5, t_6) = 0.5$, $\eta(p_{10}, t_5) = 0.5$, $\eta(p_6, t_7) = 0.5$, $\eta(p_9, t_7) = 0.5$
6. The factors' values of E^{in} $(Ap_{combined})$: if t_2 enabled, $E^{in}{}_1(Ap_{combined}) = (1, 1)$, if t_3 enabled, $E^{in}{}_2(Ap_{combined}) = (2, 2)$, if t_6 enabled, $E^{in}{}_3(Ap_{combined}) = (1, 1)$, if t_7 enabled, $E^{in}{}_4(Ap_{combined}) = (2, 2)$.
7. $\Psi(p)$: $\Psi(p_1) = 0$, $\Psi(p_2) = 0$, $\Psi(p_3) = 0$, $\Psi(p_4) = 0$, $\Psi(p_5) = 0$, $\Psi(p_6) = 0$, $\Psi(p_7) = 0$, $\Psi(p_8) = 0$, $\Psi(p_9) = 0$, $\Psi(p_{10}) = 0$, $\Psi(p_{11}) = 0$, and Ψ $(p_{combined}) \leq 4$

Places $p_1, p_2, p_3, p_4, p_5, p_6, p_7, p_8, p_9, p_{10}$ and p_{11} obtain tokens unconditionally, their thresholds are set as zero. t_2, t_3, t_6, t_7 represent the four transitions that can fire. If any of four transitions fires, new tokens with colors are built. For example, when transition t_2 fires, it creates a color C_{vs3} on the tokens from p_2 and p_{10} and transfers them into common place $p_{combined}$. For transition t_2, there are two factors: p_2 and p_{10}

$E^{out}(Ap_{combined}) = \sum_{i=1}^{nf} \eta_i * E^{in}{}_i (Ap_{combined}) = \eta_1 * E^{in}{}_1(Ap_{combined}) + \eta_2 * E^{in}{}_2(Ap_{combined}) = 0.5*1 + 0.5*1 = 1$. Since if the current tokens in $p_{combined}$ place $+ 1 \leq 4$, $p_{combined}$ can be reached.

If transition t_3 fires, it creates colors C_{vs1} and C_{vs2} on the tokens from p_3 and p_{11} and transfers them into common place $p_{combined}$. For transition t_3, there are two factors: p_3 and p_{11}

$E^{out}(Ap_{combined}) = \sum_{i=1}^{nf} \eta_i * E^{in}{}_i (Ap_{combined}) = \eta_1 * E^{in}{}_1(Ap_{combined}) + \eta_2 * E^{in}{}_2(Ap_{combined}) = 0.5*2 + 0.5*2 = 2$. Since if the current tokens in $p_{combined}$ place $+ 2 \leq 4$ $p_{combined}$ can be reached.

When transition t_6 fires, it creates a color C_{vs2} on the tokens from p_5 and p_{10} and transfers them into common place $p_{combined}$. For transition t_6, there are two factors: p_5 and p_{10}

$E^{out}(Ap_{combined}) = \sum_{i=1}^{nf} \eta_i * E^{in}{}_i(Ap_{combined}) = \eta_1 * E^{in}{}_1(Ap_{combined}) + \eta_2 * E^{in}{}_2(Ap_{combined}) = 0.5*1 + 0.5*1 = 1$. Since if the current tokens in $p_{combined}$ place $+ 1 \leq 4$ $p_{combined}$ can be reached.

If transition t_7 fires, it creates colors C_{vs1} and C_{vs3} on the tokens from p_6 and p_9 and transfers them into common place $p_{combined}$. For transition t_7, there are two factors: p_6 and p_9

$E^{out}(Ap_{combined}) = \sum_{i=1}^{nf} \eta_i * E^{in}{}_i(Ap_{combined}) = \eta_1 * E^{in}{}_1(Ap_{combined}) + \eta_2 * E^{in}{}_2(Ap_{combined}) = 0.5*2 + 0.5*2 = 2$. Since if the current tokens in $p_{combined}$ place $+ 2 \leq 4$ $p_{combined}$ can be reached.

Finally, comparing the proposed work with a relevant developed TCCPN is shown in Section 4.

4. Case Study

In this section, we show the results of the experiments with the proposed approach. Specifically, we use an AMS example available in the literature: the AMS Petri net model given in Piroddi et al. [30], Chen et al. [8], Chen and Li [31], Chen et al. [32], and TCCPN [26–29]. The Petri net model is displayed in Figure 11; it includes 14 transitions and 19 places. The places can be described as the following set partition: $P^0 = \{p_1, p_{19}\}$, $P_R = \{p_{13}, \ldots, p_{18}\}$, and $P_A = \{p_2, \ldots, p_{14}\}$. The properties of the developed Petri net models are obtained using the free GPenSIM tool [22]. We find that it has 282 reachable markings, and the system is not live (it has a deadlock).

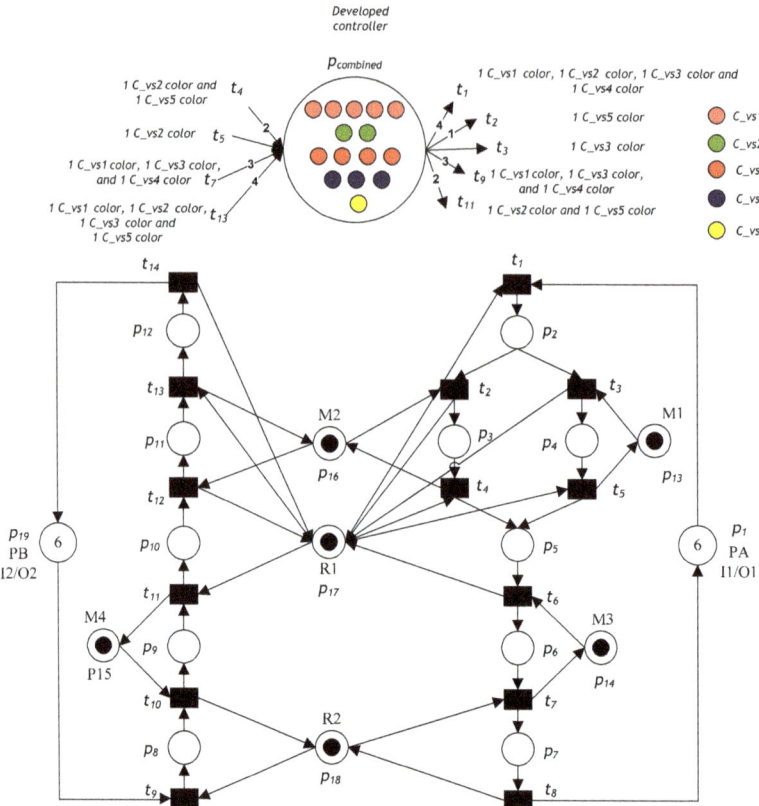

Figure 11. Colored controlled Petri net model of the system.

We apply the proposed deadlock-prevention algorithm to this case study. Without considering recovery subnets, the system model has five SMSs that may be empty: $S_1 = \{p_7, p_{12}, p_{13}, p_{14}, p_{15}, p_{16}, p_{17}, p_{18}\}$, $S_2 = \{p_5, p_{12}, p_{13}, p_{16}, p_{17}\}$, $S_3 = \{p_2, p_7, p_{12}, p_{14}, p_{15}, p_{16}, p_{17}, p_{18}\}$, $S_4 = \{p_2, p_7, p_{10}, p_{12}, p_{14}, p_{15}, p_{17}, p_{18}\}$, and $S_5 = \{p_2, p_5, p_{12}, p_{16}, p_{17}\}$. Based on the suggested deadlock-prevention algorithm (Algorithm 1), five monitors are inserted to protect the five SMSs from being emptied. The required control places using Algorithm 1 are designed as follows:

(1) $^\bullet V_{S1} = \{t_7, t_{13}\}$, $V_{S1}^\bullet = \{t_1, t_9\}$, and $M_{V0}(V_{S1}) = 5$.
(2) $^\bullet V_{S2} = \{t_4, t_5, t_{13}\}$, $V_{S2}^\bullet = \{t_1, t_{11}\}$, and $M_{V0}(V_{S2}) = 2$.
(3) $^\bullet V_{S3} = \{t_7, t_{13}\}$, $V_{S3}^\bullet = \{t_1, t_9\}$, and $M_{V0}(V_{S3}) = 4$.
(4) $^\bullet V_{S4} = \{t_7, t_{11}\}$, $V_{S4}^\bullet = \{t_1, t_9\}$, and $M_{V0}(V_{S4}) = 3$.
(5) $^\bullet V_{S5} = \{t_4, t_{13}\}$, $V_{S5}^\bullet = \{t_2, t_{11}\}$, and $M_{V0}(V_{S5}) = 3$.

By Definition 11, a deadlock control subnet of the Petri net model illustrated in Figure 11 is $N_{DC} = (p_{combined}, \{T_{DCi}, T_{DCo}\}, F_{DC}, C_{vsi})$, where $T_{DCo} = \{4t_1, t_2, t_3, 3t_9, 2t_{11}\}$, and $T_{DCi} = \{2t_4, t_5, 3t_7, 4t_{13}\}$. The initial token with a color marking of a combined monitor is $M_{DCo}(p_{combined}) = \Sigma M_{V0}(V_S) = M_{V0}(V_{S1}) + M_{V0}(V_{S2}) + M_{V0}(V_{S3}) + M_{V0}(V_{S4}) + M_{V0}(V_{S5}) = 5 + 2 + 4 + 3 + 1 = 15$ tokens. Thus, in the Petri net model illustrated in Figure 11, there are five color types, which are $SC = \{C_{vs1}, C_{vs2}, C_{vs3}, C_{vs4}, C_{vs5}\}$. Therefore, the total number of colored tokens is 15: five tokens of color C_{vs1}, two tokens of color C_{vs2}, four tokens of color C_{vs3}, three tokens of color C_{vs4}, and one token of color C_{vs5}, as shown in Figure 11.

In the net displayed in Figure 11, when transition t_1 fires, the system selects only one token from input place p_1, one token from resource place p_{17}. Additionally, the system selects tokens from $p_{combined}$: one token of color C_{vs1}, one token of color C_{vs2}, one token of color C_{vs3}, and one token of color C_{vs4}. If transition t_2 is fired, the system selects only one token from place p_2, one token from resource place p_{16}, and one token of color C_{vs5} from $p_{combined}$. Moreover, when transition t_3 fires, the system selects only one token from input place p_2, one token from resource place p_{13}, and one token of color C_{vs3} from $p_{combined}$.

If transition t_9 fires, the system selects only one token from input place p_{19}, one token from resource place p_{18}, one token of color C_{vs1} from $p_{combined}$, one token of color C_{vs3} from $p_{combined}$, and one token of color C_{vs4} from $p_{combined}$. In addition, when transition t_{11} fires, the system selects only one token from input place p_9, one token from resource place p_{15}, one token from resource place p_{17}, one token of color C_{vs2} from $p_{combined}$, and one token of color C_{vs5} from $p_{combined}$.

When transition t_4 fires, the system creates two colored tokens—one of color C_{vs2} and one of color C_{vs5}—and transfers them into the common place $p_{combined}$. Moreover, when transition t_5 fires, the system adds color C_{vs2} to the tokens and transfers them into the common place $p_{combined}$. If transition t_7 fires, the system creates three colored tokens—one of color C_{vs1}, one of color C_{vs3}, and one of color C_{vs4}—and transfers them into the common place $p_{combined}$. Finally, when transition t_{13} fires, the system creates four colored tokens—one of color C_{vs1}, one of color C_{vs2}, one of color C_{vs3}, and one of color C_{vs5}—and transfers them into the common place $p_{combined}$.

To test and validate the developed GPenSIM code, we compared it with the methods in Piroddi et al. [30], Chen et al. [8], Chen and Li [31], Chen et al. [32], and TCCPN [26–29]. The simulation was undertaken for 480 min. After running and simulating the Petri net model in MATLAB, we obtained the results summarized in Tables 2 and 3. Table 2 shows the results in terms of the number of monitors, number of arcs, liveness, and reachable marking. We observe that the proposed approach provides a supervisor with only a single control place and 9 arcs, both of which are minimal compared with other techniques in Piroddi et al. [30], Chen et al. [8], Chen and Li [31], and Chen et al. [32]. Table 3 displays the results in terms of utilization of the robots and machines, throughput of Part A and Part B, work-in-process (WIP), and total time in system (throughput time). In terms of the resource utilization, all methods obtain approximately the same values, as shown in Figure 12. Moreover, from the viewpoint throughput, the proposed method can provide greater throughput than other techniques as shown in Figure 13. In term of WIP, the proposed method leads to better WIP than the other techniques as shown in Figure 14. With respect to throughput time of Part A and Part B, overall, the proposed method can obtain less throughput time than other techniques as shown in Figures 15 and 16. Therefore, the proposed method is valid, it can give sufficiently accurate results, and it can potentially be applied to other cases.

Table 2. Comparison with the existing policies: number of monitors, number of arcs, liveness, and reachable marking.

Parameters	Chen et al. [8]	Piroddi et al. [30]	Chen and Li [31]	Chen et al. [32]	TCCPN[26–29]	Proposed Method
Monitors	8	5	2	2	1	1
Arcs	37	23	12	12	9	9
Liveness	Live	Live	Live	Live	Live	Live
Reachable marking	205	205	205	205	205	205

Table 3. Comparison with the existing policies: utilization throughput, work-in-process, and throughput time.

Parameter	Chen et al. [8]	Piroddi et al. [30]	Chen and Li [31]	Chen et al. [32]	TCCPN [26–29]	Proposed Method
M 1 utilization%	18.75	17.7083	17.7083	17.7083	17.7083	17.7083
M 2 utilization%	35	33.3333	33.3333	33.3333	34.5833	33.9583
M 3 utilization%	12.5	13.75	14.375	14.375	11.875	12.5
M 4 utilization%	22.5	21.6667	20.8333	20.8333	23.3333	22.5
R 1 utilization%	39.5833	40	40.4167	40.4167	39.1666	39.58333
R 2 utilization%	29.375	30	30	30	30	30
Throughput of Part A (unit)	20	21	20	20	19	24
Throughput of Part B (unit)	26	25	26	26	27	23
Work-In-Process	3.9271	3.93331	3.8480	3.9667	3.4938	3.3854
Throughput time of Part A (min)	23.9500	22.8571	24	23.9235	25.2105	19.9583
Throughput time of Part B (min)	18.4230	19.200	18.3321	18.4615	17.7407	20.8260

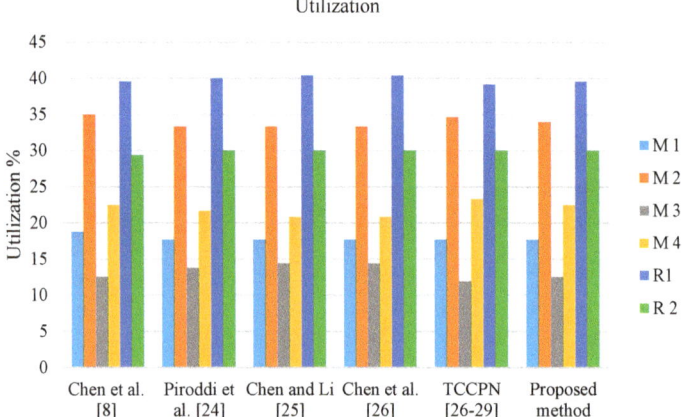

Figure 12. Comparison of utilization for the Petri net model from Figure 11.

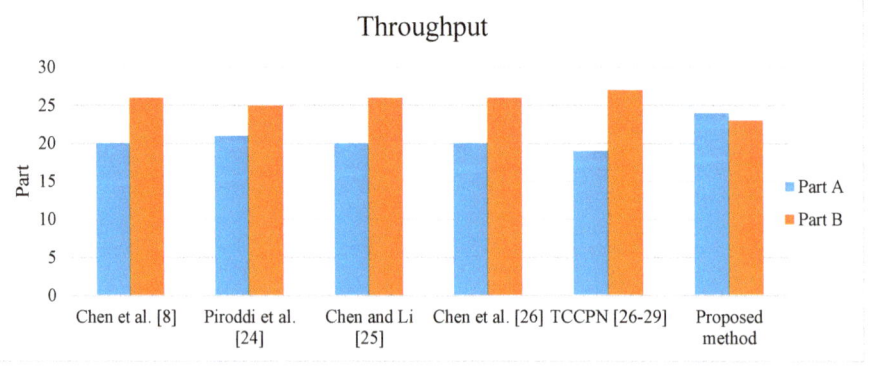

Figure 13. Comparison of throughput for the Petri net model from Figure 11.

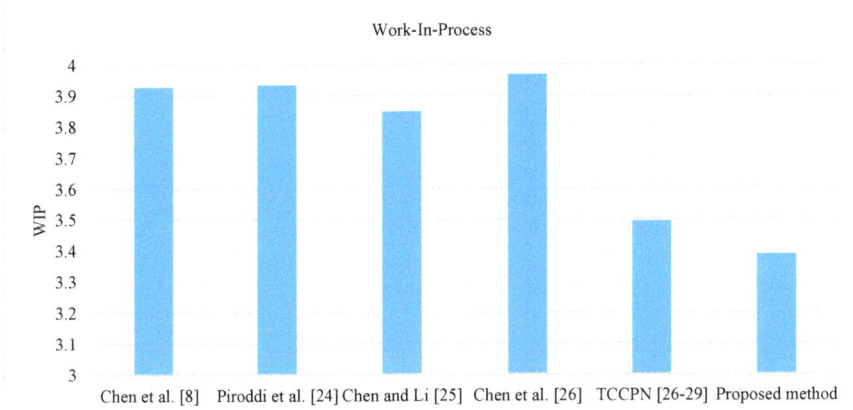

Figure 14. Comparison of work-in-process for the Petri net model from Figure 11.

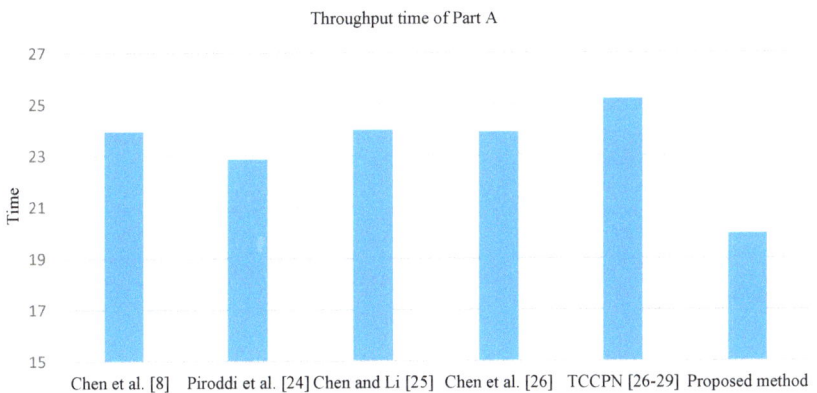

Figure 15. Comparison of throughput time of Part A for the Petri net model from Figure 11.

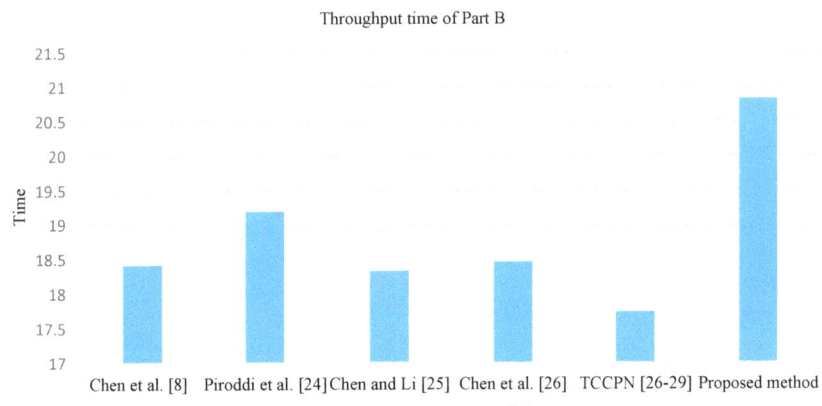

Figure 16. Comparison of throughput time of Part B for the Petri net model from Figure 11.

5. Conclusions

In this paper, we introduce a two-step controlled deadlock policy. In the first step, we create a Petri net controlled model using the deadlock prevention method based on SMSs proposed by [1]. In the second step, all control places obtained after the first step are merged into a single control place based on the colored Petri nets to mark all SMSs. We compare the proposed method with the methods of Piroddi et al. [30], Chen et al. [8], Chen and Li [31], and Chen et al. [32], and TCCPN [26–29]. According to our results, the proposed controller is more powerful, has a simpler structure, and does not need to calculate reachability graphs; therefore, it has low-overhead computation. The most challenging research topic in the future is that the controlled system that developed by previous deadlock control approaches may undergo changes of control requirements and specifications such as:

1. Adding or removing a machine
2. New production ratio
3. Adding new product
4. Changing a resource capacity
5. Resource faults and raw-material processing in a faulty resource
6. The processing routes of the system are changed
7. System contain uncontrollable transitions

When a system has these problems, a system needs to be reconfigurable. Then the deadlock-free system can have deadlocks. Therefore, the proposed robust deadlock control policy needs to be extended to improve efficiency for rapid and valid reconfiguration of Petri net-based supervisory controllers for reconfigurable manufacturing systems.

Author Contributions: Conceptualization, H.K. and A.A.-A.; software, H.K. and R.D.; resources, H.K., A.A.-A. and R.D.; formal analysis, H.K. and A.A.-A.; investigation, H.K., A.A.-A. and Z.L.; validation, H.K., A.A.-A., Z.L. and R.D.; writing—original draft preparation, H.K., A.A.-A. and Z.L.; writing—review and editing, H.K., A.A.-A., Z.L. and R.D.; visualization, H.K., A.A.-A. and Z.L.; supervision, A.A.-A., and Z.L. All authors have read and agreed to the published version of the manuscript.

Funding: This research was funded by King Saud University, grant number (RSP-2019/62), and the APC was funded by King Saud University.

Acknowledgments: The authors would like to thank King Saud University for funding and supporting this research through Researchers Supporting Project Number (RSP-2019/62).

Conflicts of Interest: The authors declare no conflict of interest.

References

1. Li, Z.; Zhou, M.; Wu, N. A survey and comparison of petri net-based deadlock prevention policies for flexible manufacturing systems. *IEEE Trans. Syst. Man Cybern. Part C Appl. Rev.* **2008**, *38*, 173–188.
2. Li, Z.; Wu, N.; Zhou, M. Deadlock control of automated manufacturing systems based on petri nets—A literature review. *IEEE Trans. Syst. Man Cybern. Part C Appl. Rev.* **2012**, *42*, 437–462.
3. Abdulaziz, M.; Nasr, E.A.; Al-Ahmari, A.; Kaid, H.; Li, Z. Evaluation of deadlock control designs in automated manufacturing systems. In Proceedings of the 2015 International Conference on Industrial Engineering and Operations Management, Dubai, UAE, 3–5 March 2015; IEEE: Piscataway, NJ, UAE, 2015.
4. Chen, Y.; Li, Z.; Barkaoui, K.; Giua, A. On the enforcement of a class of nonlinear constraints on petri nets. *Automatica* **2015**, *55*, 116–124. [CrossRef]
5. Nasr, E.A.; El-Tamimi, A.M.; Al-Ahmari, A.; Kaid, H. Comparison and evaluation of deadlock prevention methods for different size automated manufacturing systems. *Math. Probl. Eng.* **2015**, *501*, 1–19. [CrossRef]
6. Kaid, H.; Al-Ahmari, A.; El-Tamimi, A.M.; Abouel Nasr, E.; Li, Z. Design and implementation of deadlock control for automated manufacturing systems. *S. Afr. J. Ind. Eng.* **2019**, *30*, 1–23. [CrossRef]
7. Chen, Y.; Li, Z.; Khalgui, M.; Mosbahi, O. Design of a maximally permissive liveness-enforcing petri net supervisor for flexible manufacturing systems. *IEEE Trans. Autom. Sci. Eng.* **2011**, *8*, 374–393. [CrossRef]

8. Wysk, R.A.; Yang, N.-S.; Joshi, S. Detection of deadlocks in flexible manufacturing cells. *IEEE Trans. Robot. Autom.* **1991**, *7*, 853–859. [CrossRef]
9. Ghaffari, A.; Rezg, N.; Xie, X. Design of a live and maximally permissive petri net controller using the theory of regions. *IEEE Trans. Robot. Autom.* **2003**, *19*, 137–141. [CrossRef]
10. Uzam, M.; Zhou, M. Iterative synthesis of petri net based deadlock prevention policy for flexible manufacturing systems. In Proceedings of the 2004 IEEE International Conference on Systems, Man and Cybernetics, Hague, The Netherlands, 10–13 October 2004; IEEE: Piscataway, NJ, USA, 2004; pp. 4260–4265.
11. Uzam, M. The use of the petri net reduction approach for an optimal deadlock prevention policy for flexible manufacturing systems. *Int. J. Adv. Manuf. Technol.* **2004**, *23*, 204–219. [CrossRef]
12. Chao, D.Y. Direct minimal empty siphon computation using mip. *Int. J. Adv. Manuf. Technol.* **2009**, *45*, 397–405. [CrossRef]
13. Chao, D.Y. Improvement of suboptimal siphon-and fbm-based control model of a well-known. *IEEE Trans. Autom. Sci. Eng.* **2011**, *8*, 404–411. [CrossRef]
14. Uzam, M. An optimal deadlock prevention policy for flexible manufacturing systems using petri net models with resources and the theory of regions. *Int. J. Adv. Manuf. Technol.* **2002**, *19*, 192–208. [CrossRef]
15. Chao, D.Y. Fewer monitors and more efficient controllability for deadlock control in s3pgr2 (systems of simple sequential processes with general resource requirements). *Comput. J.* **2010**, *53*, 1783–1798. [CrossRef]
16. Li, Z.; Zhou, M. Elementary siphons of petri nets and their application to deadlock prevention in flexible manufacturing systems. *IEEE Trans. Syst. Man Cybern. Part A Syst. Hum.* **2004**, *34*, 38–51. [CrossRef]
17. Pan, Y.-L.; Tseng, C.-Y.; Row, T.-C. Design of improved optimal and suboptimal deadlock prevention for flexible manufacturing systems based on place invariant and reachability graph analysis methods. *J. Algorithms Comput. Technol.* **2017**, *11*, 261–270. [CrossRef]
18. Zhao, M.; Uzam, M. A suboptimal deadlock control policy for designing non-blocking supervisors in flexible manufacturing systems. *Inf. Sci.* **2017**, *388*, 135–153. [CrossRef]
19. Cong, X.; Gu, C.; Uzam, M.; Chen, Y.; Al-Ahmari, A.M.; Wu, N.; Zhou, M.; Li, Z. Design of optimal petri net supervisors for flexible manufacturing systems via weighted inhibitor arcs. *Asian J. Control* **2018**, *20*, 511–530. [CrossRef]
20. Lautenbach, K. Linear algebraic calculation of deadlocks and traps. In *Concurrency and Nets*, 1st ed.; Springer: Berlin/Heidelberg, Germany, 1987; pp. 315–336.
21. Ezpeleta, J.; Colom, J.M.; Martinez, J. A petri net based deadlock prevention policy for flexible manufacturing systems. *IEEE Trans. Robot. Autom.* **1995**, *11*, 173–184. [CrossRef]
22. Davidrajuh, R. *Modeling discrete-event systems with gpensim: An introduction*; Springer International Publishing: Gewerbestrasse, Cham, Switzerland, 2018.
23. Ratzer, A.V.; Wells, L.; Lassen, H.M.; Laursen, M.; Qvortrup, J.F.; Stissing, M.S.; Westergaard, M.; Christensen, S.; Jensen, K. Cpn tools for editing, simulating, and analysing coloured petri nets. In Proceedings of the International Conference on Application and Theory of Petri Nets, Eindhoven, The Netherlands, 23–27 June 2003; Springer: Berlin, Heidelberg, Germany, 2003; pp. 450–462.
24. Li, Z.; Zhou, M. *Deadlock Resolution in Automated Manufacturing Systems: A Novel Petri Net Approach*; Springer Science & Business Media: Holborn, London, UK, 2009.
25. Wu, Y.; Xing, K.; Luo, J.; Feng, Y. Robust deadlock control for automated manufacturing systems with an unreliable resource. *Inf. Sci.* **2016**, *346*, 17–28. [CrossRef]
26. Bidgoly, A.J.; Ladani, B.T. Trust modeling and verification using colored petri nets. In Proceedings of the 2011 8th International ISC Conference on Information Security and Cryptology, Mashhad, Iran, 14–15 September 2011; IEEE: Piscataway, NJ, USA, 2011; pp. 1–8.
27. Wahab, O.A.; Bentahar, J.; Otrok, H.; Mourad, A. A survey on trust and reputation models for web services: Single, composite, and communities. *Decis. Support Syst.* **2015**, *74*, 121–134. [CrossRef]
28. Wahab, O.A.; Bentahar, J.; Otrok, H.; Mourad, A. Towards trustworthy multi-cloud services communities: A trust-based hedonic coalitional game. *IEEE Trans. Serv. Comput.* **2016**, *11*, 184–201. [CrossRef]
29. Wang, N.; Wang, J.; Chen, X. A trust-based formal model for fault detection in wireless sensor networks. *Sensors* **2019**, *19*, 1916. [CrossRef] [PubMed]
30. Piroddi, L.; Cordone, R.; Fumagalli, I. Selective siphon control for deadlock prevention in petri nets. *IEEE Trans. Syst. Man Cybern. Part A Syst. Hum.* **2008**, *38*, 1337–1348. [CrossRef]

31. Chen, Y.; Li, Z. Design of a maximally permissive liveness-enforcing supervisor with a compressed supervisory structure for flexible manufacturing systems. *Automatica* **2011**, *47*, 1028–1034. [CrossRef]
32. Chen, Y.; Li, Z.; Zhou, M. Behaviorally optimal and structurally simple liveness-enforcing supervisors of flexible manufacturing systems. *IEEE Trans. Syst. Man Cybern. Part A Syst. Hum.* **2012**, *42*, 615–629. [CrossRef]

© 2019 by the authors. Licensee MDPI, Basel, Switzerland. This article is an open access article distributed under the terms and conditions of the Creative Commons Attribution (CC BY) license (http://creativecommons.org/licenses/by/4.0/).

Article

Robust Fault Protection Technique for Low-Voltage Active Distribution Networks Containing High Penetration of Converter-Interfaced Renewable Energy Resources

Shijie Cui [1,2,3,4], Peng Zeng [1,2,*], Chunhe Song [1,2] and Zhongfeng Wang [1,2]

1. Key Laboratory of Networked Control System, Shenyang Institute of Automation, Chinese Academy of Sciences, Shenyang 110016, China; cuishijie@sia.cn (S.C.); songchunhe@sia.cn (C.S.); wzf@sia.cn (Z.W.)
2. State Key Laboratory of Robotics, Shenyang Institute of Automation, Chinese Academy of Sciences, Shenyang 110016, China
3. Institute for Robotics and Intelligent Manufacturing, Chinese Academy of Sciences, Shenyang 110169, China
4. University of Chinese Academy of Sciences, Beijing 100049, China
* Correspondence: zp@sia.cn; Tel.: +86-186-4203-3625

Received: 3 November 2019; Accepted: 20 December 2019; Published: 30 December 2019

Abstract: With the decentralization of the electricity market and the plea for a carbon-neutral ecosystem, more and more distributed generation (DG) has been incorporated in the power distribution grid, which is then known as active distribution network (ADN). The addition of DGs causes numerous control and protection confronts to the traditional distribution network. For instance, two-way power flow, small fault current, persistent fluctuation of generation and demand, and uncertainty of renewable energy sources (RESs). These problems are more challenging when the distribution network hosts many converter-coupled DGs. Hence, the traditional protection schemes and relaying methods are inadequate to protect ADNs against short-circuit faults and disturbances. We propose a robust communication-assisted fault protection technique for safely operating ADNs with high penetration of converter-coupled DGs. The proposed technique is realizable by employing digital relays available in the recent market and it aims to protect low-voltage (LV) ADNs. It also includes secondary protection that can be enabled when the communication facility or protection equipment fails to operate. In addition, this study provides the detail configuration of the digital relay that enables the devised protection technique. Several enhancements are derived, as alternative technique for the traditional overcurrent protection approach, to detect small fault current and high-impedance fault (HIF). A number of simulations are performed with the complete model of a real ADN, in Shenyang, China, employing the PSCAD software platform. Various cases, fault types and locations are considered for verifying the efficacy of the devised technique and the enabling digital relay. The obtained simulation findings verify the proposed protection technique is effective and reliable in protecting ADNs against various fault types that can occur at different locations.

Keywords: active distribution network; converter; digital relay; DG; fault; protection; power system; renewable energy resource

1. Introduction

Low-voltage active distribution networks (ADNs) comprising distributed generations (DG) such as photovoltaic (PV) solar system, microturbine, wind generation, mini-hydro, and fuel cell have become prominent in the energy sector especially in the existing smart grid setup. This is because the DGs in ADNs are easy accessible, clean, and simplified structures. ADNs are a highly efficient, economic, and reliable form of power grids [1–3].

As a result of the distinct features of ADNs, the conventional power system protection approaches that assume high fault current amplitudes and one-way current flow conventions of radial networks are not adequate to operate ADNs [4].

The main challenge concerning the ADN protection appears where there are high proportion of converter-interfaced renewable energy sources (RESs). In this case, the fault current is relatively low (two times of the peak current) due to the small current rating of the semiconductor apparatuses of the power electronic converters. Consequently, the conventional overcurrent protection scheme cannot sufficiently detect these low fault currents and protect ADNs against severe damages that can be caused by potential network faults [5–8]. Although traditional overcurrent protection schemes can be utilized to protect ADNs when there is a strong main (utility) grid connection, the existing relay configurations must be cautiously attuned since the integration of DGs can challenge the harmonization of the protection plan [4,9–11].

There have been a few research works in area of fault protection for ADNs. Admittance-based ADN protection scheme is devised in [5]. Nevertheless, it could not provide an effective method for determining the precise line admittance for different fault types and places. In addition, the relay coordination was not completely presented in the work.

Network voltage-based fault protection of ADNs and microgrids (MGs) has been proposed by few studies [6,7,12,13]. The method presented in [7], for example, uses Park-transformed (d-q frame) network voltage to detect the occurrence of faults in a MG. Nevertheless, it did not measure the d-q components of the network voltages for all kinds of solid faults. It did not guarantee protection for high impedance faults (HIFs) as well. In addition, the method does not define the configuration of the relay that enables the presented protection scheme. The protection method in [13] applied the d-q components of network voltages for detecting solid faults and wavelet transform-based detection for HIFs. However, the findings of the proposed method are limited to isolated microgrids and its applicability to the ADNs was not considered.

Reference [8] proposed a protection strategy including its enabling relay to protect low voltage power networks. The strategy provides fault protection for both MGs and ADNs. Nonetheless, it might require a comparatively extensive time to sense faults in a medium voltage (MV) power grids because of the definite time grading technique it uses.

Reference [14] devised a fast communication-supported fault protection scheme and a microprocessor-based relay in MV power networks. The scheme delivers speedy and coordinated fault clearance for both ADNs and isolated MGs. Nonetheless, the strategy uses under voltage-based method of fault detection that may lead the relays to command false trip signals to circuit breakers (CBs) in case of temporary occurrence of voltage-sags, which all the time present in the power networks because of dynamic variation of load demands and volatility of RESs. Furthermore, the strategy neither guarantees protection for symmetrical HIFs nor delivers techniques for protecting buses.

This study devices a quick and robust fault protection technique for low voltage (LV) ADNs containing high penetration of converter-interfaced renewable energy resources. It uses microprocessor-based digital relay to enable the proposed protection technique. It explicitly provides the configuration of this digital relay. The digital relays operate in coordination detect and clear faults in the ADN. They exchange information with themselves and the central protection manager (CPM). The CPM also exchanges information with the ADN controllers and demand regulation systems. The devised technique provides primary and secondary protections for all solid fault types and HIFs at various possible fault points in the ADN. Numerous simulations are performed on a complete model of an actual ADN using the PSCAD software platform, for various fault locations and types, to substantiate the success of the devised protection technique and enabling relays.

The remaining sections of the study are prepared as follows. Section 2 offers the configuration of the devised fault protection relay. Section 3 presents the devised protection technique. The case studies and simulation findings are offered in Section 4. The study is summarized at the end in Section 5.

2. Configuration of the Devised Relay

In this study, a communication-aided protection technique is devised for LV active distribution networks. The devised protection technique uses a digital relay to sense the occurrence of fault and segregate the minimum section of the ADN impacted by the fault. The devised protection technique is actuated by the relay that hereafter is said to be the "ADN protection digital relay" (APDR). The focus of Section 2 is towards describing the architecture, operational units and key components of the APDR. If the APDR communicates with other APDRs, the ADN operator and additional components, it is known as a "communication-aided ADN protection digital relay" (CAPDR).

As aforementioned, the integration of DG causes ADNs or traditional distribution grids encounter a number of confronts, concerning control and protection problems. These problems can be summarized as follows:

1. bi-directional power flow
2. limited fault current magnitude
3. dynamic fluctuations of operating conditions
4. uncertainty of power generations

Thus, the traditional fault protection methods and relay algorithms are not enough and hence, the protection scheme and relay configurations must be redesigned and modified to operate ADNs safely and reliably [11]. Particularly, directional components are essential to evade unwanted tripping when faults impact a nearby protection area. The directional component of the neighboring area hastily disable its CB(s), for a specified time, to let the protection (main) components of the fault-impacted area to be activated and remove the fault. If the fault continues, on the other hand, the CB(s) of the neighboring protection area is activated to be opened as secondary protection following the primary protection reverse time-delay.

The proposed protection scheme can be realized using digital relays accessible in the market. Figure 1 illustrates the functional schematic and operational sections of the devised APDR/CAPDR that is the extended form of the relays provided in [13,14]. As depicted in Figure 1, five units present in the APDR/CAPDR: "directional unit", "solid fault detection unit", "HIF detection unit," "the trip unit," and "the auto-reclose unit."

The directional unit decides where the fault current flows using the method that will be discussed in Section 2.3. The solid fault detection unit is responsible for detecting all type of solid faults in the ADN. It uses the Park transformation of the network voltage as a fault detection signal. The detail analysis and derivation of the detection signal will be discussed in Section 2.1. The HIF detection unit is responsible for detection HIFs. It employs wavelet transform-based travelling wave fronts of the network current transients. At the end, the yields from the directional unit, solid fault detection unit, and HIF detection unit are applied to the tripping unit to decide the issuance of a tripping signal. The auto-reclosing unit is responsible for ensuring the seamless recoupling of the isolated section of the ADN to the normal section following the clearance of the fault.

2.1. Solid Fault Detection

As discussed in the previous sections, the proposed solid fault recognition methodology relies on the Park transformation of the ADN system voltage. The measured three-phase voltages at the APDR/CAPDR are first transformed to the direct(d)-axis and quadrature(q)-axis (dq) voltage components [15,16]. Any change in the three-phase voltage is observed by a change in the dq voltage components [13]. This study uses the q-axis voltage as the detection signal and it is expressed as follows [13]:

$$V_q = \frac{2}{3} \cdot \left\{ \cos\theta \cdot V_a + \cos\left(\theta - \frac{2\pi}{3}\right) \cdot V_b + \cos\left(\theta + \frac{2\pi}{3}\right) \cdot V_c \right\} \quad (1)$$

Here, V_a, V_b, and V_c are the 3-phase voltages, and θ is the phase (transformation) angle.

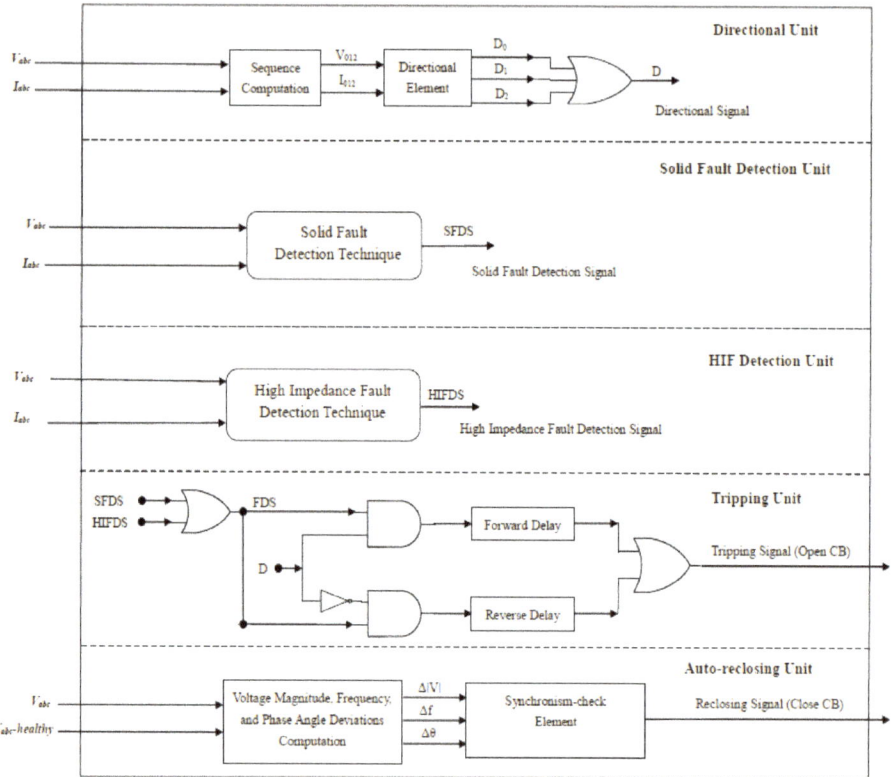

Figure 1. Schematic diagram of the communication-aided active distribution network (ADN) protection digital relay (CAPDR) and its operational units.

The disturbance voltage ($V_{q.dist}$) which is used as the fault detection signal is described as:

$$V_{q.dist} = V_{q.ref} - V_q \quad (2)$$

where, $V_{q.ref}$ is a reference q-axis voltage associated with the normal operation of the ADN before the occurrence of the fault.

Under pre-fault condition, the value of $V_{q.dist}$ is zero. When a fault occurs, $V_{q.dist}$ is a dc signal that changes based on the fault type.

For symmetrical faults, $V_{q.dist}$ is given by:

$$V_{q.dist} = V_{q.ref} - V_m \sin\varphi \quad (3)$$

where, V_m is the peak of phase voltages and φ is the phase angle.

For unsymmetrical faults, $V_{q.dist}$ is expressed as:

$$V_{q.dist} = V_{q.ref} - \{V_{Pm}\sin\varphi_P + V_{Nm}\sin(2\omega t + \varphi_N)\} \tag{4}$$

where, V_{Pm} and V_{Nm} are the maximum +ve and −ve sequence phase voltages correspondingly, ω is frequency, and φ_P and φ_N are phase angles +ve and −ve sequence phase voltage, respectively.

As observed from Equation (3), for symmetrical (three-phase) faults $V_{q.dist}$ is a DC signal. While for unsymmetrical faults, as given in Equation (4), $V_{q.dist}$ is a DC signal plus a sinusoidal component with double frequency (2ω).

Therefore, as per the devised protection technique, the solid fault detection unit of the CAPDR at the end decides the occurrence of a solid fault by contrasting the disturbance voltage ($V_{q.dist}$) with a preset threshold value. When $V_{q.dist}$ is exceeds the preset minimum level, the unit will command a solid fault detection signal (SFDS) to the trip unit of the CAPDR.

2.2. HIF Detection

The traditional overcurrent relays cannot correctly detect HIFs. Although several methodologies have been recommended by prior research works to address the problem (HIF detection) [17–19], there is no comprehensive remedy yet. This study provides a technique for HIF detection based on the observation of travelling wave fronts obtained from current transients measured at fault points (branches) [13,20].

With this technique, the 3-phase currents in the fault-impacted branches are first converted to the modal components (αβ coordinate) by employing the abc-αβ transform. Afterwards, the wave front (discrete wavelet coefficients (DWTCs)) of the modal constituents is obtained utilizing the discrete wavelet transform (DWT). The αβ branch current constituents are mainly used to obtain the propagation modes in the ADN during the fault occurrence. The DWTCs of each modal component is examined and the DWTC having the biggest amplitude is selected to decide the occurrence of the HIF. At the end, the obtained DWTC is contrasted with a preset threshold to decide they HIF occurrence. The technique has the advantage of being deployed into digital relays.

Therefore, according to the devised protection technique, the HIF detection unit of the CAPDR commands a HIF detection signal (HIFDS) to the trip unit if the fault is sensed. The fault detection signal (FDS) in the trip unit is the resultant of the logical OR of SFDS and HIFDS.

2.3. Directional Decision Forming

The devised technique uses directional units that are provided here according to the methods presented in [13,14]. When HIF happens in the ADN, the directional units cannot indicate the precise HIF point. To address this challenge, this study uses zero-sequence directional units; it is similar with the method used in [21]. In addition, negative-sequence directional unit is utilized to ensure reliable protection for unsymmetrical HIFs, for example line-to-line faults [8]. At the end, as depicted in Figure 1, directional commands from zero, −ve, and +ve sequence directional units are merged and used to generate the major directional command D. A unique delay time is employed in either direction of the relays. The relay coordination is performed by regulating the delay times.

3. Proposed Protection Technique

Here, we present the devised communication-aided protection technique, enabled by the devised relay in Section 2. The devised technique offers primary and secondary protections to solve the ADN protection challenges presented above in Sections 1 and 2.

Based on the devised protection technique, a minimal part of the ADN is separated because of the fault from the healthy section of the ADN through the commands transmitted to one or more CAPDRs. The quantity of CAPDRs used in the ADN is determined according to the preferred selectivity and reliability needs. Every CAPDR that is responsible to protect a particular component or area dispatches 2 commands ADN protection manager (APM):

1. Fault detection signal (FDS), which indicates if the CAPDR sensed the fault inside its zone
2. Fault direction signal (D) that specifies the path of the faults from CAPDR perspective.

The calculation of the FDS and D are presented above in Sections 2.1–2.3.

The APM receives the FDS and D commands from every CAPDR and decides using a suitable logical calculation, the fault affected part of the ADN. The logical calculation is alike the "directional decision" protection technique. It is elaborated more in Section 4.

When the incidence of the fault is decided (using the FDS), the APM holds-up for a small preset period to receive another directional command and decide the fault-impacted part of the network. Afterwards, appropriate trip signal(s) are dispatched to the CB(s) linked with the CAPDR(s), to open and segregate the faulty part of the ADN. The trip commands are dispatched following a delay-time (in the range (0.1 s, 0.15 s) [22]) to provide a chance to the adjacent protection elements to operate first. This time setting can guarantee coordination of the CAPDRs with the primary protection adjacent elements.

During a CB malfunctioning, a failure command is commanded to the proximate CBs to isolate the minimum section of the ADN. The CB failure information is sent after a delay-time (in the range (0.3 s, 0.4 s) [22]) if any FDS is still active. The secondary protection is energized following a delay of 0.4 s from the fault incidence and, therefore, provides an opportunity for the aforementioned two signals to be commanded. Accordingly, if the communication malfunctions and the CAPDRs cannot get any information for a preset time, all CAPDRs will be immediately swapped to the secondary protection. The communications are not needed for the secondary protection except that it requires longer activation time than the primary protection.

The segregated (because of the fault occurrence) section of the ADN can be recoupled back and synchronized to the remaining part of the network through the resynchronization setting of its DGs and reclosing ability of its CBs if the fault is short-term and cleared immediately after the segregation.

The devised protection technique can be realized using the communication abilities of smart grids. Wireless communications [23], IEEE-802.11 wireless LAN [24] with Ethernet bridges [25] and IEC 61,850 [26] can be some of the communication channels for the application of the devised protection technique.

4. Results and Discussions

To reveal the success of the devised protection technique, the ADN whose schematic framework illustrated in Figure 2 is used. The ADN in Figure 2 is an actual LV distribution network in Shenyang, China. The network primarily delivers power to industrial park loads with a peak total demand of 5 MVA.

As depicted in Figure 2, the ADN comprises four converter-coupled DGs (CC-DGs) and two synchronous machine DGs (SM-DGs). The CC-DGs are the PV system, wind generation, vanadium redox battery (VRB) and lithium-ion (Li-Ion) battery. The capacity of each of these power sources are indicated in the figure. The SM-DGs are the diesel generator and micro gas turbine generation. Hereafter, the ADN of Figure 2 is called the "case study ADN."

All the DGs in the ADN normally operate in PQ (fixed active and reactive power) control approach as the main utility grid can always supply the reference voltage and frequency. The detail control approaches of the DGs can be referred from [27].

As presented above in Sections 2 and 3, every CAPDR sends 2 signals, FDS and D, to the APM. The APM analyses these signals received from the CAPDRs and determines the precise fault point. Then, the trip signals will be sent to the responsible CB(s) to segregate the fault-impacted part of the ADN following a preset delay-time. Table 1 provides the CAPDRs used for the primary and secondary protections to sense and remove fault from the ADN. During the occurrence faults in the ADN, the D and FDS commands of the corresponding (adjacent) CAPDR(s) are dispatched to the APM. The fault is assumed forward fault if the FDS and D commands have a value of one.

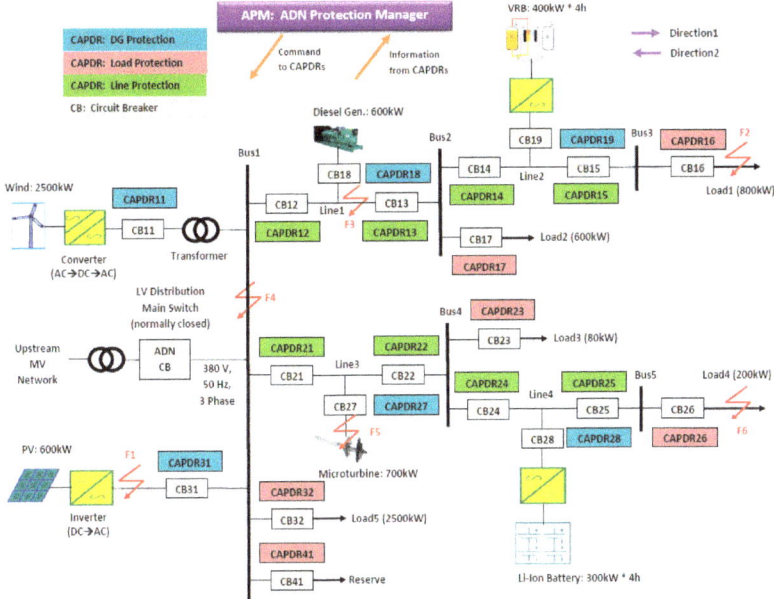

Figure 2. Model of the case study ADN with embedded CAPDRs.

Table 1. Relays and their respective protection responsibility in the case study ADN.

Fault Point		Direction 1		Direction 2	
		Primary	Secondary	Primary	Secondary
DG	Wind		CAPDR 11		
	PV		CAPDR 31		
	Diesel		CAPDR 18		
	Microturbine		CAPDR 27		
	VRB		CAPDR 19		
	Li-Ion		CAPDR 28		
Load	Load 1	CAPDR 16	CAPDR 15	N/A	
	Load 2	CAPDR 17	CAPDR 13	N/A	CAPDR 14
	Load 3	CAPDR 23	CAPDR 22	N/A	CAPDR 24
	Load 4	CAPDR 26	CAPDR 25	N/A	
	Load 5	CAPDR 32	N/A	N/A	CAPDR 12 CAPDR 21
	Reserve	CAPDR 41	N/A	N/A	CAPDR 12 CAPDR 21
Line	Line 1	CAPDR 12	N/A	CAPDR 13	CAPDR 21 CAPDR 14
	Line 2	CAPDR 14	CAPDR 13		N/A
	Line 3	CAPDR 21	N/A	CAPDR 22	CAPDR 12 CAPDR 24
	Line 4	CAPDR 24	CAPDR 22		N/A
Bus	Bus 1	N/A		CAPDR 12 CAPDR 21	CAPDR 13 CAPDR 22
	Bus 2	CAPDR 13	CAPDR 12	CAPDR 14	N/A
	Bus 3	CAPDR 15	CAPDR 14	N/A	
	Bus 4	CAPDR 22	CAPDR 21	CAPDR 24	N/A
	Bus 5	CAPDR 25	CAPDR 24	N/A	

To confirm the efficacy of the devised ADN protection technique and its enabling digital relay, the case study ADN, depicted in Figure 2, is modeled and simulated using the PSCAD simulation platform [28]. For CAPDRs in the major protection, the delay-times are set as 100 ms. This offers sufficient time for the tripping unit not to send false signal for temporary voltage dips in the network. The time-delay CAPDRs/APDRs in the secondary protection is set as 400 ms. The CBs need 20 ms for opening or closing. In addition, the CAPDRs are implemented with double-setting directional units in order to provide both primary and secondary protections when needed.

Likewise, the reverse time-delays of CAPDRs are set according to common practices and the techniques provided in [8] and obviously have dissimilar values from the forward time-delays since there might be distinct DG and load in the reverse direction. The q-axis reference voltage is set as the q-axis rated value. The threshold voltage is taken as 50% of the rated voltage for double line to ground (DLG), 3-line to ground (3LG) and line-to-line (LL) faults, while 20% is used for single line to ground (LG) fault. The Daubechies 8 (Db8) DWT with a sampling frequency of 6 kHz and multiresolution analysis (MRA) of unity resolution is used for the HIF detection.

The simulations consist of faults at various points in the ADN. These faults are F1, F2, F3, F4, F5, and F6 (illustrated in Figure 2) which designate faults at the PV DG terminal, Load1 terminal, distribution line, bus, microturbine DG terminal, and Load4 terminal, correspondingly. All solid faults (LG, DLG, LL, and 3LG) and HIFs are considered in the protection simulation. A resistance value of 60Ω is employed to simulate the HIFs.

For the photovoltaic DG, the primary protection is provided by CAPDR31 while F1 happens near to its terminal. Figure 3 illustrates the 3-phase voltages and q-axis disturbance voltage observed at CB31 while F1 (3LG) occurs. As observed in Figure 3, the disturbance voltage has varied considerably when F1 happened, and surpassed the preset threshold level. As shown, the disturbance voltage is a fixed DC value for a 3LG fault. As Figure 3 shows, F1 happened at 2 s and remained for 0.05 s. The CAPDR sensed F1 using the substantial variation in the q-axis voltage and dispatched the trip command at 2.02 s. CB31 opened at 2.025 s to cut off the DG and segregate it from the remaining part of the ADN. F1 cleared at 2.05 s and the CB reclosed at 2.07 s. It is shown that the ADN voltage has recovered its normal value straightaway following the disappearance of F1 through the reclose function of the devised relay. It demonstrates the rapidity of the devised technique and its quick information exchange capability.

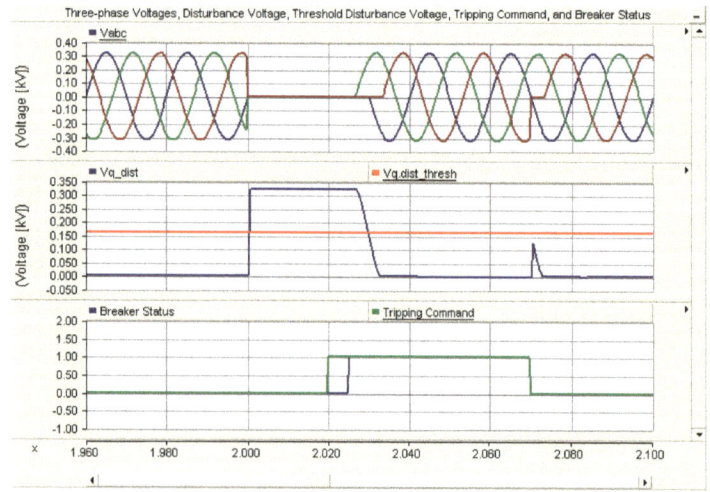

Figure 3. 3-phase voltages, disturbance voltage, and trip signal for fault 1 (F1).

For Load1, primary protection in direction1 is provided by CAPDR16 and secondary protection by CAPDR15 when F2 happens. Figure 4 depicts the 3-phase voltages, q-axis disturbance voltage, and DWTC (wave front) of current-transient observed at CB16, while F2 (DLG HIF) occurs. As shown in Figure 4, the q-axis voltage altered very little when F2 happened. It is smaller than the preset threshold level. Consequently, the q-axis disturbance voltage cannot sufficiently activate the CAPDR to dispatch a tripping command to the responsible CB. Nevertheless, the DWTC has revealed a substantial variation (surpasses the zero threshold level) while the DLG HIF (F2) happened. F2 happened at 2 s. The CAPDR has sensed F2 using the substantial alteration of the DWTC and sent trip signal to CB16 to cut off Load1 and separate it from the healthy part of the ADN.

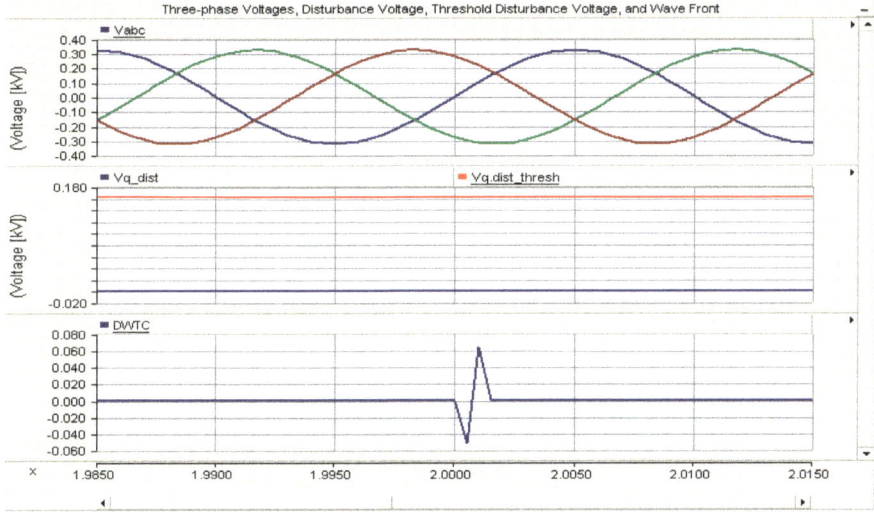

Figure 4. 3-phase voltage, disturbance voltage, and discrete wavelet coefficient (DWTC) of current transient during F2.

CAPDR12 in direction1 and CAPDR13 in direction2 are in charge of the primary protection for Line1 when F3 happens. CAPDR21 for CAPDR12 and CAPDR14 for CAPDR13 are the corresponding secondary protections in direction2. Figure 5 depicts the 3-phase voltages and q-axis disturbance voltage observed at CB13 when F3 (LG) happens. As clearly seen, the q-axis voltage varies considerably while F3 has happened and surpasses the preset threshold. The q-axis voltage in this case is a DC signal plus a ripple element with a double frequency. F3 happened at 2 s and remained for 0.05 s. Both CAPDRs sensed F3 and sent trip commands at 2.02 s. CB12 and CB13 opened at 2.025 s to separate Line1 and separate it from the healthy part of the ADN. F3 cleared at 2.05 s and the CBs reclosed at 2.07 s.

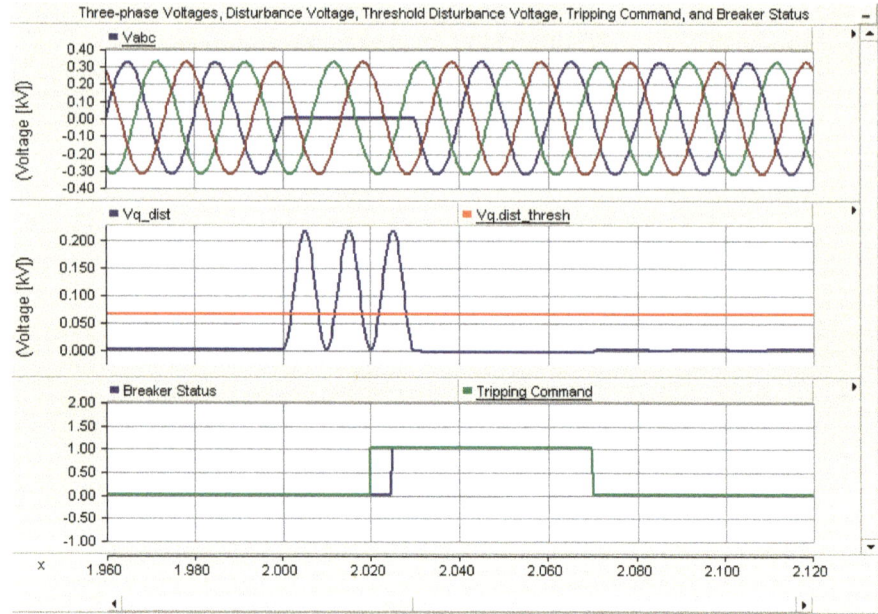

Figure 5. 3-phase voltages, disturbance voltage, and trip signal for F3.

CAPDR12, CAPDR21, CAPDR32, and CAPDR41 are in charge of primary protection of Bus1 when F4 happens. The APM determines bus faults using the direction commands (Ds) obtained from the CAPDRs coupled with the bus. It decides the incidence of the bus fault if all the Ds sent from every relay coupled with the bus are negative one (−1). The trip commands are then dispatched to all responsible CBs. Figure 6 depicts the 3-phase voltages, q-axis disturbance voltage, and current DWTC observed at CB21, while F4 (LL HIF) occurs. Similar outcome can be attained at CB12, CB32, and CB41 as well. As shown, the DWTC surpasses the zero threshold when F4 happened. F4 happened at 2 s. The CAPDRs sensed F4 and sent the trip commands to the responsible CBs to segregate Bus1 from the remaining part of the ADN.

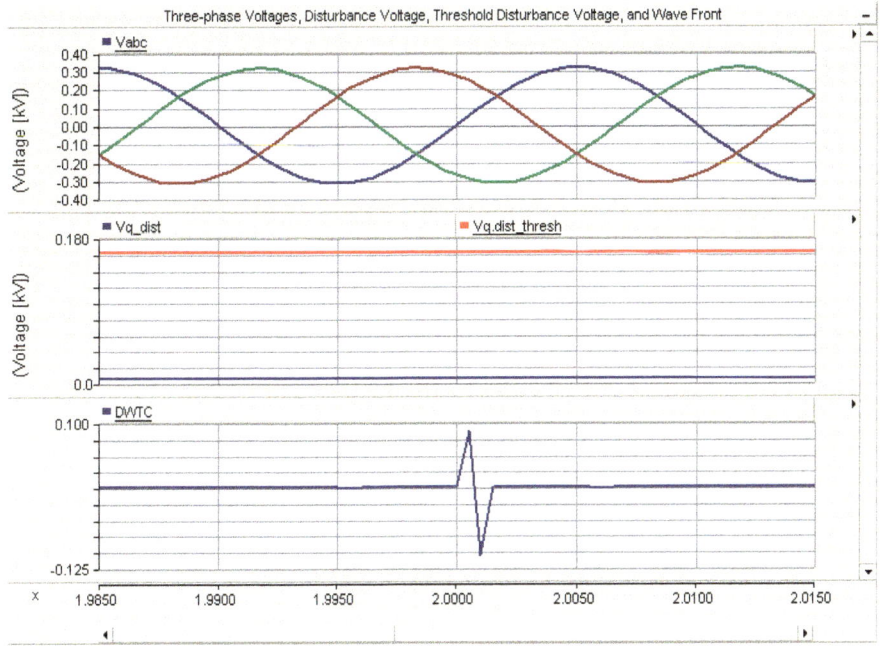

Figure 6. 3-phase voltage, disturbance voltage, and DWTC during F4.

Similarly, Figures 7 and 8 illustrate the 3-phase voltages and q-axis disturbance voltage observed at CB27 when F5 (3LG) occurs and at CB26 when F6 (LG) occurs, respectively.

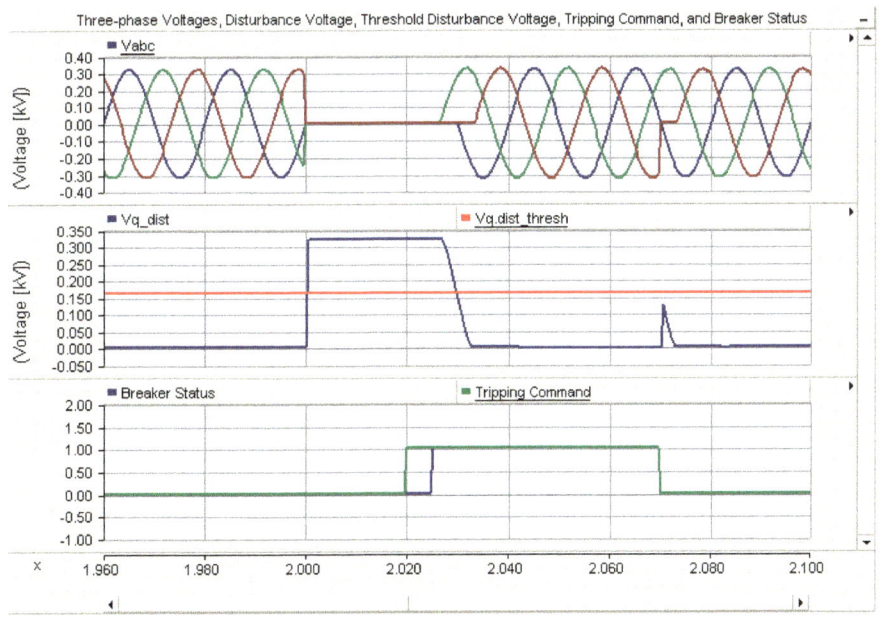

Figure 7. 3-phase voltages, disturbance voltage, and trip signal for F5.

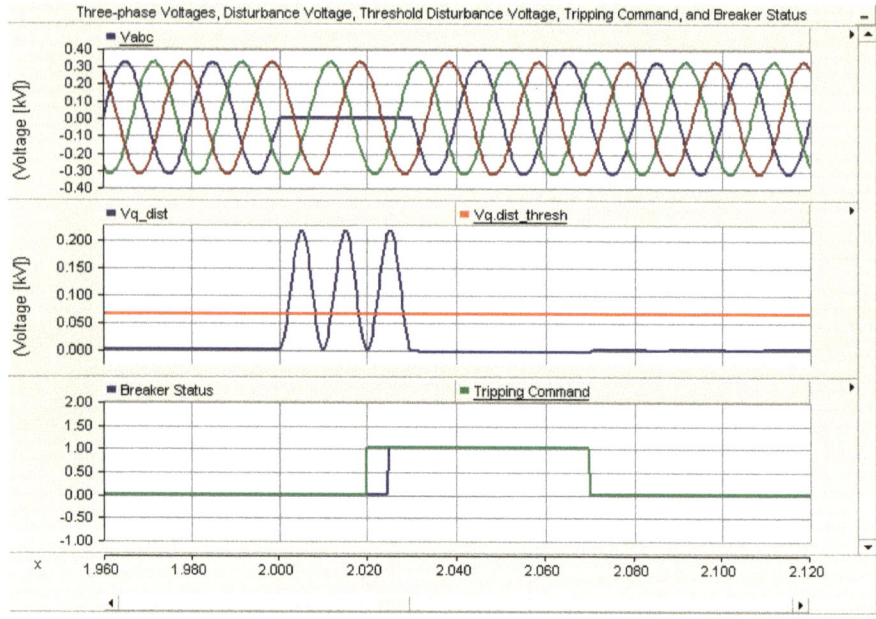

Figure 8. 3-phase voltages, disturbance voltage, and trip signal for F6.

5. Conclusions

We devised a novel communication-aided technique for quick and reliable protection of low voltage converter-dominated and renewable energy-integrated active distribution networks. The technique employs distinct approaches to detect the incidences of solid faults and HIFs. The proposed method solves the challenges brought to ADNs due to small-magnitude fault currents. It also provides both main and backup protections embedded into a proposed digital relay. The devised digital relay has five distinct units with different functions. The relay identifies diverse fault types based on the features of the disturbance voltage. The technique does not require adaptive elements. Plenty of simulations have been executed by employing the PSCAD software platform for various cases and fault locations, to reveal the efficacy of the devised fault protection technique and the actuating digital relay. The devised protection technique is successful irrespective of the position, capacity, and type of the DGs in the ADN. In addition, it is valid for any fault current amplitude, impedance, type, and point.

Author Contributions: S.C. and P.Z. modeled the ADN simulation model, formulated and implemented the protection technique; C.S. and Z.W. accomplished the management and progress evaluations of the work. All authors have checked and agreed the final version. All authors have read and agreed to the published version of the manuscript.

Funding: The research was funded by the National Key R&D Program of China (2017YFB0902900 & 2017YFB0902901), National Natural Science Foundation of China under Grant 61803368, China Postdoctoral Science Foundation under Grant 2019M661156, and Liaoning Provincial Natural Science Foundation of China under Grant 20180540114.

Acknowledgments: The work is supported by the National Key R&D Program of China (2017YFB0902900 & 2017YFB0902901), National Natural Science Foundation of China under Grant 61803368, China Postdoctoral Science Foundation under Grant 2019M661156, and Liaoning Provincial Natural Science Foundation of China under Grant 20180540114.).

Conflicts of Interest: The authors declare no conflict of interest.

Acronyms

ADN	active distribution network
APDR	ADN protection digital relay
APM	ADN protection manager
CC-DG	converter-coupled DG
DG	distributed generation
F	fault
PSCAD	power system computer-aided design
HIF	high impedance fault
dq	d-axis ~ q-axis reference frame
ESS	electrical energy storage
ADNPDR	ADN protection digital relay
CAPDR	communication-aided ADN protection digital relay
abc	three phase reference frame
$\alpha\beta$	alpha-beta reference frame
MG	microgrid
MRA	multi-resolution analysis
MV	medium voltage
DWT	discrete wavelet transform
V_a, V_b and V_c	phase a, b and c voltages, respectively
V_α and V_β	alpha and beta axis voltages, respectively
V_d and V_q	d-axis and q-axis voltages, respectively
$V_{q.dist}$	disturbance voltage signal
$V_{q.ref}$	q-axis reference voltage
V_m	max phase voltage
n	harmonic order
ω	angular frequency of system voltage
f	frequency
φ	initial phase angle of the system voltage
ω_r	angular frequency of the stator voltages
θ	rotor angle of rotation
DC	direct current
V_0	zero-sequence voltage
V_{Pm} and V_{Nm}	peak values of the positive- and negative-sequence fundamental voltages
φ_P and φ_N	initial phase angle values of the positive- and negative-sequence voltages
SFDS	solid fault detection signal
DWTC	discrete wavelet transform coefficient
HIF	high impedance fault
HIFDS	high impedance fault detection signal
FDS	fault detection signal
D	main directional command
D_0, D_2, D_1	zero-, negative-, and positive-sequence directional signals, respectively
DWT	discrete wavelet transform
CB	circuit breaker
CPM	central protection manager
LAN	local area network
LV	low voltage
IEEE	institute of electrical and electronic engineers
IEC	international electro-technical commission
MVA	mega volt ampere
N/A	not applicable
SM-DG	synchronous machine-based DG
VRB	vanadium redox flow battery
Li-Ion	lithium-ion battery
PQ	active-reactive power

U/f	voltage/frequency
PV	photovoltaic
ms	milli second
LG	single-line-to-ground
DLG	double-phase-to-ground
LL	line-to-line
3LG	three-phase-to-ground
Db8	Daubechies 8 wavelet
kHz	killo herz
s	second
RES	renewable energy resource
R&D	research and development
$V_{q.dist\_thresh}$	threshold disturbance voltage signal
V_{abc}	three phase voltages
I_{abc}	three phase currents
V_{012}	zero-, positive-, and negative-sequence voltages
I_{012}	zero-, positive-, and negative-sequence currents
$V_{d.ref}$	d-axis reference voltage
$V_{abc\_healthy}$	three phase voltages of non-fault-impacted or healthy section
$\Delta|V|$	voltage magnitude deviation
Δf	frequency deviation
$\Delta \theta$	phase angle deviation

References

1. Li, H.; Eseye, A.T.; Zhang, J.; Zheng, D. Optimal energy management for industrial microgrids with high-penetration renewables. *Prot. Control Mod. Power Syst.* **2017**, *2*, 12. [CrossRef]
2. Kishore, T.S.; Singal, S.K. Optimal economic planning of power transmission lines: A review. *Renew. Sustain. Energy Rev.* **2014**, *39*, 949–974. [CrossRef]
3. Cicconi, P.; Maneri, S.; Bergantino, N.; Raffaeli, R.; Germani, M. A design approach for overhead lines considering configurations and simulations. *Comput. Aided Des. Appl.* **2019**, *17*, 797–812. [CrossRef]
4. Butler-Purry, K.L.; Funmilayo, H.B. Overcurrent protection issues for radial distribution systems with distributed generators. In Proceedings of the IEEE PES General Meeting, Calgary, AB, Canada, 26–30 July 2009; pp. 1–5.
5. Dewadasa, M.; Ghosh, A.; Ledwitch, G.; Wishart, M. Fault isolation in distributed generation connected distribution networks. *IET J. Gen. Tran. Dist.* **2011**, *5*, 1053–1061. [CrossRef]
6. Tumilty, R.M.; Brucoli, M.; Green, T.C. Approaches to network protection for inverter dominated electrical distribution systems. In Proceedings of the 3rd IET PEMD, Dublin, Ireland, 4–6 April 2006; pp. 622–626.
7. Al-Nasseri, H.; Redfern, M.; Li, F. A voltage based protection for micro-grids containing power electronic converters. In Proceedings of the IEEE PES General Meeting, Montreal, QC, Canada, 18–22 June 2006; pp. 1–7.
8. Zamani, M.A.; Sidhu, T.S.; Yazdani, A. A protection strategy and microprocessor-based relay for low-voltage microgrids. *IEEE Trans. Power Deliv.* **2011**, *26*, 1873–1883. [CrossRef]
9. Hui, W.; Li, K.K.; Wong, K.P. An adaptive multiagent approach to protection relay coordination with distributed generators in industrial power distribution system. *IEEE Trans. Ind. Appl.* **2010**, *46*, 2118–2124.
10. Perera, N.; Rajapakse, A.D.; Buchholzer, T.E. Isolation of faults in distribution networks with distributed generators. *IEEE Trans. Power Deliv.* **2008**, *23*, 2347–2355. [CrossRef]
11. Amin, Z.; Tarlochan, S.; Amir, Y. A strategy for protection coordination in radial distribution networks with distributed generators. In Proceedings of the IEEE PES General Meeting, Providence, RI, USA, 25–29 July 2010; pp. 1–8.
12. Hou, C.; Hu, X. A study of voltage detection based fault judgment method in microgrid with inverter interfaced power source. In Proceedings of the International Conference on Electrical Engineering, Shenyang, China, 5–9 July 2009.

13. Zheng, D.; Eseye, A.; Zhang, J. A communication-supported comprehensive protection strategy for converter-interfaced islanded microgrids. *Sustainability* **2018**, *10*, 1335. [CrossRef]
14. Zamani, M.A.; Yazdani, A.; Sidhu, T.S. A Communication-Assisted Protection Strategy for Inverter-Based Medium-Voltage Microgrids. *IEEE Trans. Smart Grid* **2012**, *3*, 2088–2099. [CrossRef]
15. Edith, C. *Circuit Analysis of AC Power Systems*; Wiley: New York, NY, USA, 1950; Volume 1, p. 81.
16. Park, R.H. Two-reaction theory of synchronous machines: Generalized method of analysis, part I. *Trans. AIEE* **1929**, *48*, 716–727.
17. Christie, R.D.; Zadehgol, H.; Habib, M.M. High impedance fault detection in low voltage networks. *IEEE Trans. Power Deliv.* **1993**, *8*, 1829–1836. [CrossRef]
18. Aucoin, B.M.; Russell, B.D. Distribution high impedance fault detection utilizing high frequency current components. *IEEE Trans. Power Appar. Syst.* **1982**, *PAS-101*, 1596–1606. [CrossRef]
19. Jeerings, D.I.; Linders, J.R. Unique aspects of distribution system harmonics due to high impedance ground faults. *IEEE Trans. Power Deliv.* **1990**, *5*, 1086–1094. [CrossRef]
20. Perera, N.; Rajapakse, A.D.; Muthumuni, D. Wavelet Based Transient Directional Method for Busbar Protection. In Proceedings of the International Conference on Power Systems Transients (IPST2011), Delft, The Netherlands, 14–17 June 2011.
21. Meyer, B. *Directional Ground and Sensitive Ground Fault Settings*; Cooper Power Systems, Cooper Industries: Houston, TX, USA, 2004; pp. 1–17. Available online: https://studylib.net/doc/13718297/directional-ground-and-sensitive-ground-fault-settings--r (accessed on 1 November 2019).
22. IEEE Rec. Practice for Protection & Coordination of Industrial and Commercial Pwr. Sys., IEEE Standard. 242. 1986. Available online: https://standards.ieee.org/standard/242-2001.html (accessed on 1 November 2019).
23. Palak, P.P.; Mitalkumar, G.K.; Tarlochan, S.S. Opportunities and challenges of wireless communication technologies for smart grid applications. In Proceedings of the IEEE PES General Meeting, Providence, RI, USA, 25–29 July 2010; pp. 1–7.
24. IEEE Std. for IT, Telecoms. and Info. Exchange b/n Sys., Local and Metropolitan Area Ntks., IEEE Standard. 802.11. 2007. Available online: https://www.iith.ac.in/~{}tbr/teaching/docs/802.11-2007.pdf (accessed on 1 November 2019).
25. Outdoor Long Range Indus. Wireless Ether. Bridge AFAR Comms. Inc. Available online: http://www.afar.net/wireless/ethernet-bridge/ (accessed on 1 November 2019).
26. IEC Standards and Technical Specifications—IEC 61850-5: Comm. Requirements for Functions and Device Models. 2013. Available online: https://webstore.iec.ch/publication/6012 (accessed on 1 November 2019).
27. Aminr, M.Z.; Amirnaser, Y.; Tarlochan, S.S. A control strategy for enhanced operation of inverter-based microgrids under transient disturbances and network faults. *IEEE Trans. Power Deliv.* **2012**, *27*, 1737–1747. [CrossRef]
28. *PSCAD 4.5*; Manitoba HVDC Research Center: Winnipeg, MB, Canada, 2010; Available online: https://hvdc.ca/uploads/ck/files/reference_material/EMTDC_User_Guide_v4_3_1.pdf (accessed on 1 November 2019).

© 2019 by the authors. Licensee MDPI, Basel, Switzerland. This article is an open access article distributed under the terms and conditions of the Creative Commons Attribution (CC BY) license (http://creativecommons.org/licenses/by/4.0/).

Article

Economic Reliability-Aware MPC-LPV for Operational Management of Flow-Based Water Networks Including Chance-Constraints Programming

Fatemeh Karimi Pour [1], Vicenç Puig [1,*] and Gabriela Cembrano [1,2]

[1] Automatic Control Department, Universitat Politècnica de Catalunya Institut de Robòtica i Informàtica Industrial (CSIC-UPC), C/. Llorens i Artigas 4-6, 08028 Barcelona, Spain; fkarimi@iri.upc.edu (F.K.P.); cembrano@iri.upc.edu (G.C.)
[2] Cetaqua, Water Technology Centre, Ctra. d'Esplugues 75, Cornellà de Llobregat, 08940 Barcelona, Spain
* Correspondence: vpuig@iri.upc.edu; Tel.: +34-934015752

Received: 17 October 2019; Accepted: 24 December 2019; Published: 2 January 2020

Abstract: This paper presents an economic reliability-aware model predictive control (MPC) for the management of drinking water transport networks (DWNs). The proposed controller includes a new goal to increase the system and components reliability based on a finite horizon stochastic optimization problem with joint probabilistic (chance) constraints. The proposed approach is based on a single-layer economic optimization problem with dynamic constraints. The inclusion of components and system reliability in the MPC model using an Linear Parameter Varying (LPV) modeling approach aims to maximize the availability of the system by estimating system reliability. On the other hand, the use of a LPV-MPC control approach allows the controller to consider nonlinearities in the model in a linear like way. Moreover, the resulting MPC optimization problem can be formulated as a Quadratic Programming (QP) problem at each sampling time reducing the computational burden/time compared to solving a nonlinear programming problem. The use of chance-constraint programming allows the computation of an optimal strategy with a pre-established risk acceptability levels to cope with the uncertainty of the demand forecast. Finally, the proposed approach is applied to a part of the water transport network of Barcelona for demonstrating its performance. The obtained results show that the system reliability of the DWN is maximized compared with the other approaches.

Keywords: drinking water networks; model predictive control; reliability; linear parameter varying; operation and management; economic cost

1. Introduction

The real-time control and supervision of drinking water networks (DWNs) is a field of increased interest given the environmental, economic and social impact [1]. DWNs are critical infrastructures in urban environments. These networks provide important services in modern society and maintaining the service availability is an important requirement. Therefore, reliability and resilience are important properties to be guaranteed in DWNs while being subject to constraints and continuously varying conditions of probabilistic nature [2]. DWNs are multivariate dynamic constrained systems that are described by the interconnection of several subsystems (tanks, actuators, sources, nodes and consumer sectors). Moreover, DWN optimal management is a complex challenge for water utilities that can be addressed as a multi-objective optimization problem. This problem can be solved online using a Model Predictive Control (MPC) scheme [3].

Generally, the structure of the MPC approach follows a moving horizon strategy. The control action is obtained solving an optimal control problem that provides a control action sequence in a prediction horizon that minimizes the considered control objectives and satisfies the set of constraints including the system model and physical/operational limitations. Therefore, MPC can provide suitable strategies to achieve the DWN operational control improving their performance, as it allows computing optimal control approaches ahead of time for all the pressure and flow control elements [4]. Revising the literature, different approaches can be found that show the benefits of the optimal DWN management. In [5–7], by optimizing a mathematical function that considers operational goals in a specific time horizon and using a model of the network dynamics and demand forecasts, optimal strategies are computed. These references also assumed predicted disturbances as defined in the model, but involve a soft constraint to penalize evacuation of water volume below a heuristic safety threshold without forcing any target regulation. Regarding optimised control strategies for managing water systems, MPC is not implemented in a classical way, as there is no reference volume to be tracked [8]. The standard MPC forces the system to follow the set point, but it does not guarantee that the system evolution toward the set points is obtained in an economic efficient way. The general aim in the operation of several process industries, as, e.g., DWNs, is the reduction of costs associated to the consumption of energy, which is not the main goal of standard MPC. For this purpose, Economic MPC (EMPC) provides a systematic method for the optimization of economic system operation [9]. The problem of optimization associated to the EMPC strategy aims at obtaining a family of optimal set points considering economic efficiency rather than aiming that the controlled system reach a certain set point [9].

The use of control strategies that take into account the system and component reliability that guarantee the quality of service is necessary. The health monitoring of the actuator and system should be considered for increasing the system reliability, minimising the fault appearance and reducing the operational costs. In the later stages, system reliability in the process of control system has been considered using a Prognosis and Health Management (PHM) framework. This is because reliability is a standard method for evaluating how long the system will achieve its function without malfunctions. Moreover, it can be used to predict future damages in the system according to the health state of its components [10].

In the past few years, the problems of system reliability and actuator lifetime in service has received considerable interest for the researcher community. In [11], to decrease the maintenance cost, the actuator lifetime is regarded as a controlled parameter that is considered as additional goal when using a linear quadratic optimal controller. On the other side, MPC predicts the suitable control actions to obtain optimal performance according to multi-objective cost functions and physical constraints, and therefore it can be considered as a suitable approach for developing health-aware control schemes. An MPC strategy based on distributing the loads among redundant actuators is introduced in [12], while forcing constraints to guarantee that the accumulated actuator degradation will not arrive at the unsafe level at the end of the prediction horizon. In [13], the authors proposed a health-aware MPC controller that incorporates a fatigue-based prognosis into MPC to minimize the component damage. Most of the other methods that consider component health and system reliability management stand within the structure of fault-tolerant control or in the area of preservation scheduling see, e.g., Gallestey et al. [14], Khelassi et al. [15], Salazar et al. [16] and references therein. However, none of these methods consider uncertainty.

The reliability is the system's ability (or component) to carry out its expected functions. The reliability of DWN is influenced by different conditions such as the capacity and the quality of the water accessible at the sources and the pump/pipe failure rates [17,18]. In most of the works, the actuator reliability is assumed that follows an exponential distribution that varies with the control action [19]. The system reliability is characterised according to the interdependence topology based on the combining of each actuator reliability. Subsequently, the system reliability has a demonstrative relationship with the control input that leads to a nonlinear mathematical model.

In several studies, this is achieved by including a damage index in the optimization problem and establishing a trade-off by weight tuning [20] or by imposing constraints with respect to the actuator reliability [17]. However, considering the reliability at the actuator level not at the system level is the main drawback of the previous methods; otherwise, it leads to the use of nonlinear MPC according to nonlinearity of the resulting constraints. Generally, Economic Nonlinear MPC (ENMPC) implies a high computational cost and, the existing gradient-based numerical algorithms do not certify that the obtained solution corresponds to the global one because of the non-convexity of the associated optimization problem. Transforming the nonlinear optimization problem into a quadratic problem through a linearisation method is one way of addressing the non-convexity problem and guaranteeing a unique optimum. In this way, the system is modelled by an incremental model because the model has to be linearised at each iteration. This approach has been improved by means of of the use Linear Parameter Varying (LPV) models that do not require linearisation [21]. The LPV models can describe both nonlinear phenomena and time-varying that can be estimated/measured online.

Another weakness of previous approaches combining reliability analysis and MPC is the conservatism of the resulting control strategies, which affects negatively the efficient DWN operation. Furthermore, in real applications, the assumption of bounded disturbances in real applications is not always satisfied. Thus, constraint violations can not be avoided because of the appearance of faults, unexpected events, etc. A more realistic representation of uncertainty is based on using the stochastic approach that leads to less conservative control methods by incorporating explicit disturbance models in the control design and by converting hard constraints into probabilistic constraints. The stochastic approach is a sophisticated theory in the field of optimization, but a revived consideration has been provided to the stochastic programming methods as powerful tools for the design of controllers, leading to the stochastic MPC, which has a particular alternative called chance-constrained MPC (CC-MPC) [22,23]. The stochastic control approach that represents robustness in terms of probabilistic (chance) constraints, which need that the probability of violation of any operational condition or physical constraint is under a designated value. By placing this value suitably, the user/operator can obtained the desired trade-off between robustness and performance. For related works that proposed the CC-MPC approach in water networks the reader is referred to [24,25]. Some economic-oriented controller that consider the reliability issue has been proposed [20], but without considering reliability at the system level and probabilistic constraints based on the reliability of the system.

The aim of this paper is to include in an EMPC strategy for DWN an additional objective that takes into account PHM information obtained by the online evaluation of the system reliability. The system reliability is incorporated into the control algorithm by using an augmented model that includes both the reliability and DWN models. As the reliability model of the whole DWN is nonlinear, its model is expressed as an LPV model such that at each time instant the varying parameters are updated according to the value of the scheduling variables. This allows to solve the optimization MPC problem associated to the health-aware approach using quadratic programming instead of nonlinear programming. Considering the probabilistic nature of system reliability, it is included in the MPC optimization problem in the form probabilistic constraints as the demands (disturbances) using the chance constraints programming paradigm. The resulting control inputs obtained by the proposed health-aware MPC approach are able to achieve the economic control objectives and simultaneous to increase the lifespan and reliability of the system components.

Chance-constraints programming allow to determine an optimal strategy by establishing the desired level of infeasibility and system reliability. Moreover, it allows considering the system reliability, which is assessed online using an LPV-MPC strategy; representing the main contribution of this paper. The second contribution is to propose an advanced health-aware LPV-MPC approach that formulates a quadratic optimization problem taking into account the functional dependency of scheduling variables and state vector. This approach avoids the use of nonlinear optimization. Moreover, it uses chance constraints programming to manage dynamically designate safety stocks in flow-based networks to satisfy nonstationary flow demands and system reliability.

The structure of the paper is as follows. The control-oriented model considered for DWN when considering the transportation layer is introduced in Section 2. Section 3 presents the chance-constraints programming and the way to use it into the MPC controller. The system reliability modeling and the relationship between reliability and chance constricted are described in Section 4. In Section 5, the economic reliability-aware MPC-LPV including chance-constraints programming is provided. The results of the application of the proposed control strategy to the DWN network using the proposed case study are analyzed and summarized in Section 6. Finally, the conclusions and research future paths are presented in Section 7.

Notation: Throughout this paper $\mathbb{R}, \mathbb{R}_+, \mathbb{R}^n, \mathbb{R}^{m \times n}$ indicate the field of real numbers, the set of non-negative real numbers, the set of column real vectors of length n, and the set of m by n real matrices, respectively. Equivalently, $\mathbb{I}_+$ presents the set of non-negative integer numbers including zero. Define the set $\mathbb{I}_{[a,b]} := \{x \in \mathbb{I}_+ | a \leq x \leq b\}$ for some $a, b \in \mathbb{I}_+$ and $\mathbb{I}_{\geq c} := \{x \in \mathbb{I}_+ | x \geq c\}$ for some $c \in \mathbb{I}_+$. The operator $\oplus$ is direct sum of matrices (block diagonal concatenation). Furthermore, $\|.\|$ denotes the spectral norm for matrices and $\|.\|_2$ is the squared 2-norm symbol. The superscript $\top$ represents the transpose and operators $<, \leq, =, >, \geq$ indicate element-wise relations of vectors.

2. EMPC for Transport Water Networks

2.1. Control-Oriented Model

In the literature, several control-oriented models for DWNs can be found depending if the transportation or distribution layer is considered. (see, e.g., in [26,27]). In this paper, a flow-based control-oriented modeling approach is considered following [6,28] since the transportation layer is considered. A DWN is composed by pipes, water tanks, pumping stations, and valves used for consumer water supply. To derive the control oriented-model, the state vector $x \in \mathbb{R}^{n_x}$ is defined to represent the tank volumes. The vector $u \in \mathbb{R}^{n_u}$ of controlled inputs is associated to the flow rates through the actuators (pumps and valves) of the network, and the vector $d_m \in \mathbb{R}^{n_d}$ of disturbances (demands) as the collection of flow rates required by the consumers at demand nodes. By means fo the flow–mass balance relations in the tanks and nodes, a discrete-time model based on linear differential algebraic equations (DAEs) for all time instant $k \in \mathbb{Z}_{\geq 0}$ can be formulated for a given DWN as follows,

$$x(k+1) = Ax(k) + Bu(k) + B_d d_m(k), \tag{1a}$$
$$0 = E_u u(k) + E_d d_m(k), \tag{1b}$$

where difference Equation (1a) model the dynamics of the storage tanks, whereas the algebraic relations (1b) describe the mass balance at junction nodes. $A \in \mathbb{R}^{n_x \times n_x}, B \in \mathbb{R}^{n_x \times n_u}, B_d \in \mathbb{R}^{n_x \times n_d}$, $E_u \in \mathbb{R}^{n_d \times n_u}, E_d \in \mathbb{R}^{n_d \times n_d}$, and $C \in \mathbb{R}^{n_y \times n_x}$ are time-invariant matrices of that depends on the network topology. The system is subject to physical input and state constraints provided by convex and closed polytopic sets defined as

$$x(k) \in \mathbb{X} := \{x \in \mathbb{R}^{n_x} | Gx \leq g\}, \tag{2a}$$
$$u(k) \in \mathbb{U} := \{u \in \mathbb{R}^{n_u} | Hu \leq h\}, \tag{2b}$$

for all $k \in \mathbb{Z}_{\geq 0}$, where $G \in \mathbb{R}^{m_x \times n_x}$, $g \in \mathbb{R}^{m_x}$, $H \in \mathbb{R}^{m_u \times n_u}$, and $h \in \mathbb{R}^{m_u}$ are vectors/matrices collecting the system constraints, signifying $m_u \in \mathbb{Z}_{\geq 0}$ and $m_x \in \mathbb{Z}_{\geq 0}$, the number of input and state constraints, respectively. Concerning the operation of the considered flow-based networks, the following assumptions are considered in this paper.

Assumption 1. *The demands in $d_m(k)$ and the states in $x(k)$ are observable at each time instant $k \in \mathbb{Z}_{\geq 0}$, also the pair (A, B) is stabilizable.*

Assumption 2. *The demand realizations at the current time instant $k \in \mathbb{Z}_{\geq 0}$ can be represented as*

$$d_m(k) = \bar{d}_m(k) + \tilde{d}_m(k), \tag{3}$$

where $\bar{d}_m(k)$ is the vector of expected disturbances that can be forecast, and $\tilde{d}_m(k)$ is the vector of probabilistic independent forecasting errors with nonstationary uncertainty and a known (or approximated) quasi-concave probability distribution $\mathcal{D}(0, \Sigma(\tilde{d}_{m,(j)}(k)))$. Consequently, each j-th row of $d_m(k)$ is described by an stochastic variable $\bar{d}_{m,(j)}(k)\mathcal{D}(j)(\bar{d}_{m,(j)}(k), \Sigma(\tilde{d}_{m,(j)}(k)))$, where $\bar{d}_{m,(j)}(k)$ represents the mean and $\Sigma(\tilde{d}_{m,(j)}(k))$ the variance.

2.2. EMPC Formulation

Computing the input commands ahead of time, to obtain the optimal performance of the network according to a set of control goals, is the purpose of applying MPC techniques for managing water transportation networks [1]. The control goal is to minimize a convex stage cost function $\ell : \mathbb{Z}_{\geq 0} \times \mathbb{X} \times \mathbb{U} \longrightarrow \mathbb{R}_{\geq 0}$, which might carry any functional relationship with the economics of the system operation. Therefore, the control aim can be expressed for minimization of a convex multi-objective cost function, which involves three functional objects for managing the DWN with different types:

- Economic objective: Minimizing water production and transport costs while providing the demanded volume.
- Safety objective: The safety volumes in the tanks are preserved guaranteeing, up to some level, the water supply under connected variations in the demand.
- Smoothness objective: For avoiding overpressures in pipes and damage in actuators, the actuators are managed based on the smooth control actions.

2.2.1. Economic Cost Minimization

Minimizing the economical costs that include water production and electrical costs related to pumping is the main control objective of the DWN. Transporting drinking water to proper elevation levels by the network involves significant electricity costs due to pumping. Therefore, the cost function related to this objective can be expressed as

$$\ell_e(k) \triangleq \alpha(k)^\top W_e u(k), \tag{4}$$

where $\alpha(k) \triangleq (\alpha_1 + \alpha_2(k)) \in \mathbb{R}^{n_u}$, $\alpha_1 \in \mathbb{R}^{n_u}$ denotes a fixed water production costs that related to the water treatments, and $\alpha_2 \in \mathbb{R}^{n_u}$ corresponds to a time-varying water cost associated to pumping that varies in each time instant k with respect to the dynamic electricity tariff. W_e indicates the weighting term that allows to prioritize the economic control objective in the complete objective function.

2.2.2. Safety Management

To preserve water stocks in spite of unexpected changes in the water demands, an appropriate safety storage level for each tank is required to be guaranteed. This goal can be formulated in the following manner,

$$\ell_s(k) \triangleq \begin{cases} \|x(k) - x_s\|_2, & if\, x(k) \leq x_s \\ 0, & otherwise \end{cases} \tag{5}$$

where x_s indicates the tanks safety levels. This piecewise linear formulation can be avoided by considering that the safety cost function can be expressed through a soft constraint by using a slack variable ξ, which is introduced to retain feasibility of the optimization problem and minimized

$$\ell_s(k) \triangleq \xi^\top(k) W_s \xi(k), \tag{6}$$

and the soft constraint is defined as

$$x(k) \geq x_s - \xi(k), \tag{7}$$

and W_s is diagonal positive definite matrix that allows to prioritize this objective in the complete objective function.

2.2.3. Control Action Smoothness

Pumps and valves are the considered actuators in a DWN. Therefore, the control actions obtained by the MPC controller must be smooth for the purpose of preserving the component lifetime. To achieve the smoothing effect, the variation of the control actions among two consecutive time instants is penalized as follows,

$$\ell_{\Delta u}(k) \triangleq \Delta u(k)^\top W_{\Delta u} \Delta u(k), \tag{8}$$

where $\Delta u(k) \triangleq u(k) - u(k-1)$, and $W_{\Delta u}$ is a weighting matrix that allows prioritizing this objective in the complete objective function.

2.2.4. EMPC Optimization Problem Formulation

The EMPC strategy can be implemented by solving a finite-horizon optimization problem over a prediction horizon N_p, where the multi-objective cost function is minimized subject to the prediction model and a set of system constraints. According to the network model (1), the MPC controller design is based on minimizing the following cost function in the prediction horizon N_p,

$$J = \sum_{l=0}^{N_p} (\ell_e(l|k) + \ell_s(l|k) + \ell_{\Delta u}(l|k)). \tag{9}$$

where at each time instant, the following optimization problem is solved online.

$$\min_{\mathbf{u}(k), \mathbf{x}(k), \xi(k)} J(\mathbf{u}(k), \mathbf{x}(k), \xi(k)), \tag{10a}$$

subject to:

$$x(l+1|k) = Ax(l|k) + Bu(l|k) + B_d d_m(l|k), \ l = 0, \cdots, N_p - 1 \tag{10b}$$
$$0 = E_u u(l|k) + E_d d_m(k), \ l = 0, \cdots, N_p - 1 \tag{10c}$$
$$x(l|k) \geq x_s - \xi(l|k), \ l = 1, \cdots, N_p \tag{10d}$$
$$u(l|k) \in \mathbb{U}, \ l = 0, \cdots, N_p - 1 \tag{10e}$$
$$x(l|k), \in \mathbb{X}, \ l = 1, \cdots, N_p \tag{10f}$$
$$\xi(l|k) \geq 0, \ l = 0, \cdots, N_p \tag{10g}$$
$$x(0|k) = x(k), \tag{10h}$$

The optimal control actions sequence $\mathbf{u}^*(k) = \{u(l|k)\}_{l \in \mathbb{Z}_{[0,N_p-1]}}$, $\mathbf{x}^*(k) = \{x(l|k)\}_{l \in \mathbb{Z}_{[1,N_p]}}$, and $\xi^*(k) = \{xi(l|k)\}_{l \in \mathbb{Z}_{[1,N_p]}}$ are obtained online. Considering the receding horizon philosophy [3], the procedure is based on solving the optimization problem (10a) from the current time instant k to $k + N_p$ by using $x(0|k)$ as the initial condition that is computed from measurements (or state estimation) at time k. Then, by applying the first value $u^*(0|k)$ from the optimal input sequence $\mathbf{u}^*(k)$ to the system, the procedure goes to the next time instant. To calculate $u^*(0|k+1)$ at time $k+1$, the optimization problem (10a) is solved from $k+1$ to $k+1+N_p$, and initial states $x(0|1+k)$ from measurements (or state estimation) are updated at time $k+1$. The same method is iterated for the following time instants.

3. Chance-Constrained Model Predictive Control

If the stochastic nature of disturbances (demands) and reliability of components of the system is not explicitly considered, an optimal solution of (10) satisfying all constraints can not be found in real scenarios. Therefore, to guarantee feasibility of the optimization problem (10), it is appropriate to relax the original constraints that involve stochastic elements with probabilistic statements in the form of chance constraints. In this manner, the constraints are needed to be satisfied with predefined risk levels to manage the uncertainty and component reliability of the system. Chance-constrained programming is a technique of stochastic programming dealing with constraints of the general form as

$$\mathbb{P}[f(v,\zeta) \leq 0] \geq 1 - \delta_\zeta, \tag{11}$$

where $\mathbb{P}$ indicates the probability operator, $v \in \mathbb{R}^{n_v}$ is the decision vector, $\zeta \in \mathbb{R}^{n_\zeta}$ a random variable, and $f: \mathbb{R}^{n_v} \times \mathbb{R}^{n_\zeta} \longrightarrow \mathbb{R}^{n_c}$ a constraint mapping. The level $\delta_\zeta \in (0,1)$ is user given and defines the preference for safety of the decision v. The constraint (11) means that we wish to take a decision v that satisfies the n_c-dimensional random inequality system $f(v,\zeta) \geq 0$ with high enough probability. As demonstrated in [29], if $f(.,.)$ is jointly convex in (v,ζ) and $\Phi \triangleq \mathbb{P}[.]$ is quasi-concave, then the feasible set

$$\Psi(\delta_\zeta) \triangleq \{v | \mathbb{P}[f(v,\zeta) \leq 0] \geq 1 - \delta_\zeta\} \tag{12}$$

is convex for all $\delta_\zeta(0,1)$. All chance-constrained models need prior knowledge of the acceptable risk δ_ζ connected with the constraints. A lower risk acceptability proposes a harder constraint. In general, joint chance constraints lack from analytic expressions because of the involving multivariate probability distribution [30]. In this paper, by following the results in [30,31], a uniform distribution of the joint risk is approximated by upper bounding the joint constraint and assuming a similar distribution of the joint risk among a set of individual chance constraints are transformed inside equivalent deterministic constraints.

Consider the general joint chance constraint (11), and define $f(v,\zeta) \triangleq \zeta - Fv$ with $F \in \mathbb{R}^{\zeta \times n_v}$. Therefore, the additive stochastic element is separable and the following chance constraint is achieved,

$$\mathbb{P}[\zeta \leq Fv] \geq 1 - \delta_\zeta. \tag{13}$$

Then, by rewiring $\omega \triangleq Fv$, for any duple (ζ, ω), it follows that

$$\Phi_\zeta(\omega) = \mathbb{P}[\{\zeta_1 \leq \omega_1, ..., \zeta_{n_c} \leq \omega_{n_c}\}]. \tag{14}$$

Describing the events $C_i \triangleq \{\zeta_i \leq \omega_i\}, \forall i \in \mathbb{Z}_1^{n_c}$ (as e.g., faults in the actuators or unexpected changes in the demand), it follows that

$$\Phi_\zeta(\omega) = \mathbb{P}[C_i \cap ... \cap C_{n_c}]. \tag{15}$$

Indicating the complements of the events C_i by $C_i^c \triangleq \{\zeta_i > \omega_i\}$, and it is obvious from probability theory that

$$C_1 \cap ... \cap C_n = (C_1^c \cup ... \cup C_{n_c}^c)^c, \tag{16}$$

and consequently

$$\Phi_\zeta(\omega) = \mathbb{P}[C_i \cap ... \cap C_{n_c}] \tag{17a}$$
$$= \mathbb{P}[(C_1^c \cup ... \cup C_{n_c}^c)^c] \tag{17b}$$
$$= 1 - \mathbb{P}[(C_1^c \cup ... \cup C_{n_c}^c)^c] \leq 1 - \delta_\zeta. \tag{17c}$$

By using the union bound, the Boole inequality let to bound the result in (17c), declaring that for a countable set of events, the probability that at least one event occurs is not higher than the sum of the individual probabilities [32], such that

$$\mathbb{P}\left[\cup_{i=1}^{n_c} C_i\right] \leq \sum_{i=1}^{n_c} \mathbb{P}[C_i], \tag{18}$$

and, by applying (18) to (17c), it yields to

$$\sum_{i=1}^{n_c} \mathbb{P}[C_i^c] \leq \delta_\zeta \iff \sum_{i=1}^{n_c} (1 - \mathbb{P}[C_i]) \leq \delta_\zeta. \tag{19}$$

Then, a set of constraints rises from previous results as sufficient conditions to enforce the joint chance constraint (13), by allotting the joint risk δ_ζ in n_c separate risks $\delta_{\zeta,i}, i \in \mathbb{Z}_1^{n_c}$. These constraints are described as follows,

$$\mathbb{P}[C_i] \geq 1 - \delta_{\zeta,i}, \quad \forall i \in \mathbb{Z}_1^{n_c} \tag{20a}$$

$$\sum_{i=1}^{n_c} \delta_{\zeta,i} \leq \delta_\zeta, \tag{20b}$$

$$0 \leq \delta_{\zeta,i} \leq 1, \tag{20c}$$

where (20a) produces the set of n_c effective individual chance constraints, which bounds the probability that each inequality of the receding horizon problem could not be satisfied. Moreover, (20b) and (20c) are conditions forced to bound the new single risks in such a way that the joint risk bound is not breached. Each solution that satisfies the aforesaid constraints is guaranteed to provide (13).

According to the satisfaction of each individual constraint is an event $C_i, \forall i \in \mathbb{Z}_1^{n_c}$. A joint chance constraint needs that the connection of all the individual constraints is satisfied with the wanted probability level, such as

$$\mathbb{P}\left[\cap_{i=1}^{n_c} C_i\right] \geq 1 - \delta_\zeta. \tag{21}$$

Considering that each individual constraint is probabilistically dependent, the level of conservatism can be derived by using the inclusion–exclusion principle for the union of finite events, $C_i, \forall i \in \mathbb{Z}_1^{n_c}$, which proves the following equality,

$$\mathbb{P}\left[\cup_{i=1}^{n_c} C_i\right] = \sum_{i=1}^{n_c} \mathbb{P}[C_i] - \sum_{1 \leq i < j \leq n_c} \mathbb{P}\left[C_i \cap C_j\right] \\ + \sum_{1 \leq i < j < k \leq n_c} \mathbb{P}\left[C_i \cap C_j \cap C_k\right] - \ldots + (-1)^{n_c - 1} \mathbb{P}\left[\cap_{i=1}^{n_c} C_i\right]. \tag{22}$$

Note that by considering as an event a fault in an actuator, it can be observed that Equation (22) has a similar as formulation as the one used for evaluating the system reliability based on the component reliability.

In a DWN, the constraints come from models (10b) and (10c) that can be formulated as chance constraints statements taking into account the probabilities associated to the component reliability. Considering only faults in actuators, the reliability of the system is related to the system inputs $u_i(k)$. Therefore, (11) can be formulated in case of the actuators as follows,

$$\mathbb{P}[f(u_i(k), \zeta_i(k)) \leq 0] \geq 1 - \delta_{\zeta_i}, \tag{23}$$

where $\zeta(k) \in \{0,1\}$ is a stochastic variable which considers if the actuator is one of two states $\{Unvailable, Available\}$ (or $\{0,1\}$) defined as follows,

$$\zeta_i(k) = \begin{cases} 1, & R_i(k) > 0 \\ 0, & R_i(k) = 0. \end{cases} \quad (24)$$

where $R_i(k)$ is the actuator reliability. In the case that $\zeta_i(k) = 1$, the input $u_i(k)$ associated to the i-th actuator is bounded by (2b); otherwise, an additional constraint setting $u_i(k) = 0$ should be included. Furthermore, to determine the reliability associated to the system that associates a probability to the system model constraint (1), the joint-chance constraint probability calculation (22) should be used leading to the following probabilistic formulation for the MPC optimization problem (10),

$$\min_{\mathbf{u}(k),\mathbf{x}(k),\boldsymbol{\xi}(k)} J(\mathbf{u}(k),\mathbf{x}(k),\boldsymbol{\xi}(k)), \quad (25a)$$

subject to

$$\mathbb{P}\Big[Ax(l|k) + B(\zeta_i)u(l|k) + B_d d_m(l|k), \quad (25b)$$

$$E_u(\zeta_i)u(l|k) + E_d d_m(k)\Big] \geq 1 - \delta, \quad l = 0, \cdots, N_p - 1 \quad (25c)$$

$$x(l|k) \geq x_s - \xi(l|k), \; l = 1, \cdots, N_p \quad (25d)$$

$$u(l|k) \in \mathbb{U}, \; l = 0, \cdots, N_p - 1 \quad (25e)$$

$$x(l|k), \in \mathbb{X}, \; l = 1, \cdots, N_p \quad (25f)$$

$$\xi(l|k) \geq 0, \; l = 0, \cdots, N_p \quad (25g)$$

$$x(0|k) = x(k). \quad (25h)$$

The main difficulty in solving this stochastic problem using chance constraints is that at each time iteration, the probabilities associated to the system reliability should be updated taking into account the value of the optimal control actions u_i. In the following section, a solution procedure is proposed to solve this problem.

4. Augmenting Network Model with the Reliability Model

As discussed in the introduction, one of the contributions of this work is to integrate the information about system health in the MPC controller by using the reliability approach. In the following, a procedure to derive the reliability model of the DWN is presented, considering that faults can only occur in the actuators.

4.1. Reliability Model

In the literature, different types of distributions have been considered to characterize the evolution of the reliability with time. The most commonly used are exponential, normal, log-normal, and Weibull distributions [33]. Here, the exponential function is considered.

First, define the concept of failure rate which is important to obtain reliability. The general explanation of failure rate, indicated by λ is presented as the fraction of the density of the stochastic lifetime to the remainder function (i.e., conditional probability). Particularly, systems are designed to work under different load values. According to [33], the load firmly affects the component failure rate. Therefore, for presenting system reliability evaluation should be considered the load versus failure rate relationship. A significant amount of literature has been produced to include the impact of the load level in the reliability estimation [33]. In this paper, actuator failure rates under various load levels are considering in function of the applied control input. The following exponential laws is the most widely used relationship to characterize the variation of the actuator fault rates with the load

$$\lambda_i = \lambda_i^0 exp\big(\beta_i u_i(k)\big), \quad i = 1, 2, \ldots, m \tag{26}$$

where λ_i^0 represents the baseline failure rate (nominal failure rate) and $u_i(k)$ is the control action a time k for the i-th actuator. β_i is a constant parameter that depends on the actuator characteristics.

Accordingly to the failure rate definition, reliability of a system or component can be described as follows

Definition 1. *Reliability is determined as the probability that a system (or component) will perform their functioning satisfactorily for a certain period of time subject to operating conditions [34].*

From the mathematical point of view, reliability $R(t)$ is defined as the probability of the successful operation of a system in the intervening period from time 0 to time t:

$$R(t) = \mathbb{P}(T > t), \quad t \geq 0 \tag{27}$$

where T is a stochastic variable that describes time until failure. Furthermore, the unreliability of system (or a component) is represented as follows.

Definition 2. *The unreliability $F(t)$ is determined as the probability that the component or system encounters the first failure or has failed one or more times among the time interval 0 to time t.*

Considering the system (or component) is always in one of the two states introduced in Equation (24), the following relationship is provided,

$$F(t) + R(t) = 1. \tag{28}$$

The reliability of a component $R_0(t)$ in the useful life period can be specified at a certain time t by a starting point. Accordingly, $R_{0,i}(t)$ will denote the i-th actuator reliability determined considering nominal operating conditions

$$R_{0,i}(t) = exp(-\lambda_i^0 t), \quad i = 1, 2, \ldots, m \tag{29}$$

Therefore, the components reliability of a system with the i-th components can be computed by exploiting the exponential function and the baseline reliability level $R_{0,j}$ as follows,

$$R_i(t) = R_{0,i} \, exp\bigg(-\int_0^k \lambda_i(s) \, ds\bigg), \quad i = 1, 2, \ldots, m \tag{30}$$

In discrete-time, Equation (30) can be rewritten as

$$R_i(k+1) = R_{0,i}(k) + exp\bigg(-T_s \sum_{s=0}^{k+1} \lambda_i(s)\bigg), \quad i = 1, 2, \ldots, m \tag{31}$$

where $\lambda_i(s)$ is the failure rate that is acquired from the i-th component under varying load levels u_i and T_s is the sampling time.

4.2. Overall Reliability

The system lifespan can be determined by the reliability of the overall system that is denoted as $R_g(k)$. This reliability is obtained based on the computation of the reliabilities of elementary components (or subsystems). Therefore, $R_g(k)$ is influenced by the configuration of the actuator that can be computed from the combination of parallel and/or series of subsystems (or components) [35]. However, there are several engineering systems that not attending the parallel, series, or connection of parallel and series structures. To manage the more complex situations, a graph model can be used to determine if the component successful path existence can be identified to determine whether the system is working correctly. A path for the graph network is a set of components, in such a way that the system will succeed just when all the components are successful in that set. A minimal path P_s is a set of components that relates to it, but the elimination any one of the components will create the set not to be a successful path [35]. Therefore, the overall system reliability $R_g(k)$ can be counted as

$$R_g(k) = 1 - \prod_{j=1}^{s}\left(1 - \prod_{i \in P_{s,j}} R_i(k)\right), \tag{32}$$

where $j = 1, \ldots, s$ is minimal paths number. As mentioned in previous section, there is an indirect relationship between conservatism of probability and the overall system reliability. In fact, the formula obtained for overall reliability system (Equation (32)) can be obtained from Equation (22).

4.3. System Reliability Modeling

Aiming to include reliability in the MPC model, a transformation is needed allowing to estimate reliability in a LPV framework. The considered transformation relies on applying the logarithm to (32). Then, Equation (32) can be rewritten as follows,

$$\log(Q_g(k)) = \log\left(\prod_{j=1}^{s}\left(1 - \prod_{i \in p_{s,j}} R_i(k)\right)\right), \tag{33}$$

and by introducing an change of variable

$$z_j(k) = 1 - \prod_{i \in p_{s,j}} R_i(k), \tag{34}$$

Equation (33) can be expressed as

$$\log(Q_g(k)) = \sum_{i \in p_{s,j}}^{s} \log(z_j(k)). \tag{35}$$

Considering (34), the $\log(z_j(k))$ can be determined as

$$\log(z_j(k)) = \frac{\log(z_j(k))}{\log(1 - z_j(k))} \sum_{i \in p_{s,j}} \log R_i(k). \tag{36}$$

Afterward, by renaming $\beta_j(k) = \frac{\log(z_j(k))}{\log(1-z_j(k))}$ in (36), (33) can be rewritten as

$$\log(Q_g(k)) = \sum_{i \in p_{s,j}}^{s} \beta_j(k) \sum_{i \in p_{s,j}} \log R_i(k). \tag{37}$$

Finally, the system unreliability can be computed from the unreliability of the baseline system

$$\log(Q_g(k+1)) = \log(Q_g(k)) + \sum_{i \in p_{s,j}}^{s} \beta_j(k) \sum_{i \in p_{s,j}} \log R_i(k). \tag{38}$$

5. Economic Reliability-Aware MPC-LPV Using Chance-Constraints

5.1. Economic Reliability Aware MPC-LPV

In this section, the integration of reliability model in the MPC controller augmenting the DWN model is proposed. As previously discussed, the reliability of the DWN can be determined by employing the control input. Thus, a new objective can be included in the MPC controller that aims to preserve the system reliability additionally to consider the reliability model (38). Figure 1 summarizes the procedure to obtain the augmented control model from the dynamic model of DWN by obtaining the reliability model using Equation (32) or equivalently Equation (22).

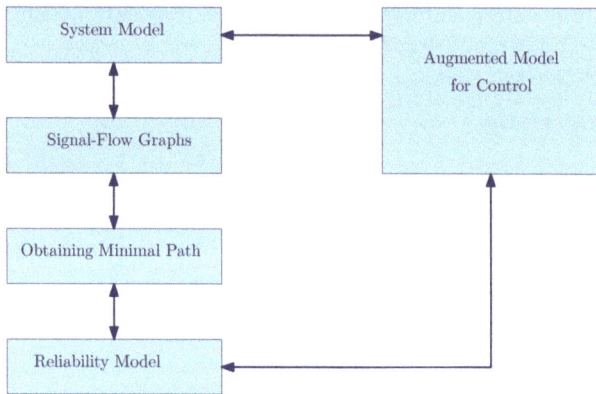

Figure 1. Structure diagram of the proposed approach.

By following this procedure, the augmented MPC model can be formulated as follows,

$$x_g(k+1) = A_g x_r(k) + B_g u(k) + B_{d,g} d_m(k),$$
$$y_g(k) = C_g x(k), \tag{39}$$

where the state and output vector are defined by $x_g = [x, \log(Q_g), \log(R_1), \ldots, \log(R_i)]^T$ and $y_g = [y, \log(Q_g)]^T$, respectively. The augmented matrices are defined as

$$A_g = \begin{bmatrix} A & 0_{n_x \times n_{i+1}} \\ \hline 0_{1 \times n_x} & 1 & \sum_{i \in p_{s,j}}^s \beta_j(k) \\ \hline 0_{n_i \times n_x} & I_{n_i \times n_i} \end{bmatrix}, \quad B_g = \begin{bmatrix} B_{n_u \times n_u} \\ \hline 0 \\ \hline -\lambda_i \times I_{n_i \times n_i} \end{bmatrix},$$

$$B_{d,g} = \begin{bmatrix} B_{d,n_u \times n_u} \\ \hline 0_{n_{i+1} \times n_{B_d}} \end{bmatrix}, \quad C_g = \begin{bmatrix} C & 0 & 0 & \cdots & 0 \\ 0 & 1 & 0 & \cdots & 0 \end{bmatrix}. \tag{40}$$

Considering the control action $u_i(k)$ as the scheduling variable related to each actuator and state in the augmented MPC model, it can be considered (39) as an LPV model. The model (39) cannot be assessed before solving the optimization problem (10), due to the future state sequence is unknown and cannot be determined. In reality, $x(l|k)$ depends on the future control inputs $u(k)$, and also on the future scheduling parameters, thus LPV model can not instantiated offline but instead should be evaluated online at each time instant k. In this way, the MPC optimization problem (10) can be formulated as a QP problem by using an estimation of scheduling variables, $\hat{\theta}$ instead of utilizing θ. That means the scheduling variables in the prediction horizon are estimated using the values from the previous MPC iteration and applied to update the model matrices of the MPC controller. Indeed, the control input sequence is utilized to change the model matrices used in the prediction horizon. Therefore, the predicted parameters and sequence of states are obtained from the optimal control sequence $\mathbf{u}(k)$, such as

$$\tilde{\mathbf{x}}(k) = \begin{bmatrix} x(l+1|k) \\ x(l+2|k) \\ \vdots \\ x(N_p|k) \end{bmatrix} \in \mathbb{R}^{N_p,n_x}, \quad \Theta = \begin{bmatrix} \hat{\theta}(l|k) \\ \hat{\theta}(l+1|k) \\ \vdots \\ \hat{\theta}(N_p-1|k) \end{bmatrix} \in \mathbb{R}^{N_p,n_\theta}. \tag{41}$$

The vector $\Theta(k)$ includes parameters from time k to $k + N_p - 1$ while the state prediction is considered for time $k+1$ to $k + N_p$. Thus, by a small abuse of notation, φ is defined as $\Theta(k) = \varphi([x^T(k) \ \tilde{x}^T(k)], \mathbf{u}(k))$. The vector $\Theta(k)$ consists of parameters from time k to $k + N_p - 1$, whereas the state prediction is performed for time $k+1$ to $k + N_p$.

Therefore, using Equation (41), the predicted states can be directly expressed as follows,

$$\tilde{\mathbf{x}}(k) = \mathcal{A}(\Theta(k))x(k) + \mathcal{B}(\Theta(k))\mathbf{u}(k) + B_{r,d}d_m(k), \tag{42}$$

where $\mathcal{A} \in \mathbb{R}^{n_x \times n_x}$ and $\mathcal{B} \in \mathbb{R}^{n_x \times n_u}$ are provided by Equations (43) and (44).

$$\mathcal{A}(\Theta(k)) = \begin{bmatrix} I \\ A(\hat{\theta}(k)) \\ A(\hat{\theta}(k+1))A(\hat{\theta}(k)) \\ \vdots \\ A(\hat{\theta}(k+N_p-1))A(\hat{\theta}(k+N_p-2))\dots A(\hat{\theta}(k)) \end{bmatrix}, \tag{43}$$

and

$$\mathcal{B}(\Theta(k)) = \begin{bmatrix} 0 & 0 & 0 & \dots & 0 \\ B(\hat{\theta}(k)) & 0 & 0 & \dots & 0 \\ A(\hat{\theta}(k+1))B(\hat{\theta}(k)) & B(\hat{\theta}(k+1)) & 0 & \dots & 0 \\ \vdots & \vdots & \ddots & \ddots & \vdots \\ A(\hat{\theta}_{k+N_p-1})\dots A(\hat{\theta}(k+1))B(\hat{\theta}(k)) & A(\hat{\theta}_{k+N_p-1})\dots A(\hat{\theta}(k+2))B(\hat{\theta}(k+1)) & \dots & B(\hat{\theta}_{k+N_p-1}) & 0 \end{bmatrix}. \tag{44}$$

By exploiting Equation (42), and new weighting matrices $\tilde{w}_1 = diag_{N_p}(w_1)$, and $\tilde{w}_2 = diag_{N_p}(w_2)$, the cost function (9) with the new additional objective that aims to increase the system reliability can be revised in vector form as follows,

$$\min_{\mathbf{u}(k),\xi(k),\mathbf{x}(k)\mathbf{Q}_\mathbf{g}(k)} \sum_{l=0}^{N_p}[\ell_e(l|k) + \ell_s(l|k) + \ell_{\Delta u}(l|k) - \ell_{R_g}(l|k)], \tag{45a}$$

subject to

$$\tilde{x}(k) = \mathcal{A}(\Theta(k))x(k) + \mathcal{B}(\Theta(k))\mathbf{u}(k) + B_{r,d}d_m(k), \tag{45b}$$
$$0 = E_u u(l|k) + E_d d_m(k), \tag{45c}$$
$$x(l+1|k) \geq x_s - \xi(l|k) \tag{45d}$$
$$\log Q_g(l+1|k) = \tilde{x}_{nx+1}(l|k) \tag{45e}$$
$$u(l|k) \in \mathbb{U}, \ l = 0, \cdots, N_p - 1 \tag{45f}$$
$$x(l|k), \in \mathbb{X}, \ l = 1, \cdots, N_p \tag{45g}$$
$$\xi(l|k) \geq 0, \ l = 0, \cdots, N_p \tag{45h}$$
$$x(0|k) = x(k), \tag{45i}$$

where $\ell_{Rg}(k) \triangleq Q_g^\top w_3 Q_g$ is an additional objective including the weight w_3 into the controller cost function to improve the system reliability. The optimization problem is solved as a QP problem according to that the predicted states $\Theta(k)$ in (42) are linear with respect to control inputs $\mathbf{u}(k)$, which is considerably further easier than solving a nonlinear optimization problem. To clarify the proposed approach, Algorithm 1 is presented.

Algorithm 1 LPV-based MPC strategy

1: $k \longleftarrow 0$
2: **repeat**
3: $i \longleftarrow 0$
4: **if** $k = 1$ **then**
5:   To solve the optimization problem (45a), where $\theta(0|k) \simeq \theta(1|k) \simeq \theta(2|k) \simeq \ldots \simeq \theta(N_p - 1|k)$
6:   Calculate $\Theta(k)$ by using $\tilde{x}(k)$ and $\mathbf{u}(k)$
7: **else**
8:   Determine $\Theta(k) = \{\hat{\theta}(i|k)\}_{i=0}^{N_p-1}$ where $\hat{\theta}(i|k) = \psi(x(i|k-1+1), u(i|k-1))$,
9:   Solve the optimization problem (45a)
10: Compute $\tilde{x}(k)$ and $\mathbf{u}(k)$,
11: $i \longleftarrow i + 1$
12: **end if**
13: Apply first element of the input sequence to the plant
14: Define $\Theta_0(k+1) = \psi(\tilde{x}_1(k), \mathbf{u}_0(k))$
15: $k \longleftarrow k + 1$
16: **until** end

5.2. Including Demand Uncertainty Using Chance Constraints

According to the stochastic nature of water demands, the DWN prediction model includes exogenous additive uncertainties. Therefore, the constraint's satisfaction (10) cannot be guaranteed, unless uncertainty it is not explicitly considering in some way. Therefore, the original constraints that include stochastic elements (2a) will formulated by means of probabilistic statements using the chance constraints framework (11). Considering this framework introduced in Section 3, and the form of state constraint set $\mathbb{X}$, the form of a state joint chance constraint is described as

$$\mathbb{P}[G(r)x \leq g(r), \forall r \in \mathbb{Z}_{[1,m_x]}] \geq 1 - \delta_x, \tag{46}$$

where $\delta_x \in (0,1)$ is the risk acceptability level of constraint violation for the states, and $G(r)$ and $g(r)$ indicate the r-th row of G and g, respectively. This entails that all rows r have to be jointly satisfied with the probability $1 - \delta_x$. Also, the form of a state individual chance constraint is described as

$$\mathbb{P}[G(r)x \leq g(r),] \geq 1 - \delta_x, \qquad \forall r \in \mathbb{Z}_{[1,m_x]} \tag{47}$$

which requires that each r-th row of the inequality has to be satisfied individually with the respective probability $1 - \delta_{x,r}$, where $\delta_{x,j} \in (0,1)$. Then, according to Equation (20), the state constraints can be described as follows,

$$\mathbb{P}[G(r)x \leq g(r)] \geq 1 - \delta_{x,r}, \quad \forall r \in \mathbb{Z}_{[1,m_x]} \tag{48a}$$

$$\sum_{r=1}^{m_x} \delta_{x,r} \leq \delta_x, \tag{48b}$$

$$0 \leq \delta_{x,r} \leq 1, \tag{48c}$$

and, as recommended in [31], specifying a constant and equal value of risk to each individual constraint, that is, $\delta_{x,r} = \delta_x/m_x$ for all $r \in \mathbb{Z}_{[1,m_x]}$, then (48b) and (48c) are obtained.

By considering a known (or approximated) quasi-concave probabilistic distribution function for the stochastic disturbance in the dynamic model (1), it yields to

$$\begin{aligned}\mathbb{P}[G(r)x(k+1) \leq g(r)] \geq 1 - \delta_{x,r} &\Leftrightarrow F_{G(r)B_d d_m(k)}(g(r) - G(r)(Ax(k) + Bu(k))) \geq 1 - \delta_{x,r} \\ &\Leftrightarrow G(r)(Ax(k) + Bu(k)) \leq g(r) - F^{-1}_{G(r)B_d d_m(k)}(1 - \delta_{x,r}),\end{aligned} \tag{49}$$

for all $r \in \mathbb{Z}_{[1,m_x]}$, where $F_{G(r)B_d d_m(k)}(.)$ and $F^{-1}_{G(r)B_d d_m(k)}(.)$ are the cumulative distribution and the left-quantile function of $G(r)B_d d_m(k)$, respectively. The use of chance constraints allows to guarantee a safety stock at each storage node of a flow-based network for decreasing the probability of stock-outs due to demand uncertainty. In this way, according to Equation (48a), the safety stocks are optimally assigned and designed by the constraint back-off effect due to the term $F_{G(r)B_d d_m(k)}(1 - \delta_{x,r})$ in Equation (46). Therefore, the original state constraint set $\mathbb{X}$ is adjusted by the effect of the m_x deterministic equivalents in (49) and substituted by the stochastic feasibility set provided by

$$\begin{aligned}\mathbb{X}_s(k) := \{x(k) \in \mathbb{R}^{n_x} | \exists\, u(k) \in \mathbb{U}, \text{such that} \\ G(r)(Ax(k) + Bu(k)) \leq g(r) - F^{-1}_{G(r)B_d d_m(k)}(1 - \delta_{x,r}) \quad \forall r \in \mathbb{Z}_{1,m_x} \\ \text{and} \quad E_u u(k) + E_d \bar{d}(k) = 0\},\end{aligned} \tag{50}$$

where $\bar{d}(k) = \mathbb{E}[d_m]$ is the first moment of d_m for all $k \in \mathbb{Z}_{0 \geq 0}$. The set $\mathbb{X}_s(k)$ is convex when non-empty for all $\delta_{x,r} \in (0,1)$ in most distribution functions, due to the convexity of $G(r)x(k+1) \leq g(r)$ and the log-concavity assumption of the distribution. For some particular distributions, e.g., Gaussian, convexity is preserved for $\delta_{x,r} \in (0, 0.5]$ [30].

5.3. Enhancing System Reliability Using Chance Constraints

According to the Section 5.1, component and system reliability model can be included in the EMPC controller model. Besides, (50) provides a new constraint set according to the deterministic equivalent (49). However, (50) does not consider the states related to the component and system reliability. Therefore, it is necessary to modify the constraint set (50) with probabilistic statements based on the component and system reliability. In this way, the system reliability is formulated in terms of probabilistic constraints as follows,

$$x_{Rg}(k) \in \{x_{Rg} \in \mathbb{R}^{n_R} | \mathbb{P}[G_{Rg} x_{Rg} \geq g_{Rg}] \geq (1 - \delta_{Rg})\} \tag{51}$$

where $x_{Rg}(k) \in \mathbb{R}^{n_{Rg}}$ is system reliability state defined in Equation (39), and $\delta_{Rg} \in (0,1)$ is the corresponding risk acceptability level of constraint violation. According to the above discussion and the effect of stochastic reliability in the model (39), (51) can be rewritten as

$$\begin{aligned}\mathbb{P}[G_{Rg} x_{Rg}(k+1) \geq g_{Rg}] \geq (1 - \delta_{Rg}) &\Leftrightarrow F_{G_{Rg} \eta}(g_{Rg} - G_{Rg} x_{Rg}(k+1)) \geq 1 - \delta_{Rg} \\ &\Leftrightarrow G_{Rg} x_{Rg}(k+1) \geq g_{Rg} + F^{-1}_{G_{Rg}\eta}(1 - \delta_{Rg}),\end{aligned} \tag{52}$$

where η is a random vector whose components follow a normal distribution, and $F_{G_{Rg}\eta}(.)$ and $F_{G_{Rg}\eta}^{-1}(.)$ are the cumulative distribution and the left-quantile functions involved in the state and actuator health deterministic equivalent constraints, respectively. The deficiency of reliability in the system can be caused that the actuator operation compromise the network supply service unless demands result reachable from other redundant flow paths or a fault-tolerant mechanism is activated. Therefore, a preventive strategy can be performed to increase overall system reliability by guaranteeing that the system reliability at each time instant to remain above a safe threshold until a predefined maintenance horizon is reached. Thereupon, the probabilistic constraint (52) can be formulated in the predictive controller as

$$G_{Rg} x_{Rg}(k+N_p|k) \geq g_{Rg}(k) + F_{G_{Rg}\eta}^{-1}(1-\delta_{Rg}), \tag{53a}$$

$$g_{Rg}(k) = x_{Rg,\min}(k) := x_{Rg}(k) + N_p \frac{R_{\text{tresh}} - x_{Rg}(k)}{k_M + N_p + k}, \tag{53b}$$

where $x_{Rg,\min}(k) \in \mathbb{R}^{n_{Rg}}$ is the vector of minimum reliability of the system allowed for time instant k and $R_{\text{tresh}} \in \mathbb{R}^{n_{Rg}}$ is the vector of threshold for the terminal system reliability at a maintenance horizon $k_M \in \mathbb{Z}_{\geq 0}$. The right-hand side of Equation (53b) is an identical restricting of the remaining allowable system reliability $(R_{\text{tresh}} - x_{Rg}(k))$ that is updated at each time step according to the applied control actions and guarantees that $x_{Rg}(k) \geq R_{\text{tresh}}$ for $k = k_M$.

5.4. Chance-Constraints Reliability-Aware EMPC-LPV Reformulation

After the inclusion system reliability in the control low as an additional state of the control model and discussing about how to use the probabilistic statements for demand and reliability constraints and formulating them into deterministic equivalent constraints. Next, the setting of the proposed economic reliability-aware MPC-LPV controller, including deterministic equivalent constraints, is presented. This transformation leads to an optimization problem considering both the dynamic safety stocks and the system reliability theory, in order to improve the flow supply service level in a given network, handling demands uncertainty and equipment damage.

In this way, for a given sequence of demands d, the predicted system reliability R_g, acceptable risk levels δ_x and δ_{Rg}, and the optimization problem associated with the deterministic equivalent for considered transportation DWN at each time step k are expressed as follows,

$$\min_{\mathbf{u}(k), \xi(k), x(k) \mathbf{x_{Rg}}(k)} \sum_{k=0}^{N_p} [\ell_e(k) + \ell_s(k) + \ell_{\Delta u}(k) - \ell_{Rg}(k)], \tag{54a}$$

subject to:

$$\tilde{x}(k) = \mathcal{A}(\Theta(k)) x(k) + \mathcal{B}(\Theta(k)) \mathbf{u}(k) + B_{r,d} d_m(k), \tag{54b}$$

$$0 = E_u u(l|k) + E_d d_m(k), \tag{54c}$$

$$x(k+l+1|k) \leq x_{\max}(r) - \Phi_{k,r}^x(\delta_x), \tag{54d}$$

$$x(k+l+1|k) \geq x_{\min}(r) + \Phi_{k,r}^x(\delta_x), \tag{54e}$$

$$G_{Rg} x_{Rg}(k+N_p|k) \geq x_{Rg,\min}(k) + \Phi_{k,\eta}^{x_{Rg}}(\delta_{Rg}), \tag{54f}$$

$$x(k+l+1|k) \geq x_s - \xi(k+l|k), \tag{54g}$$

$$\xi(k+l|k) \geq 0, \tag{54h}$$

$$x_{Rg}(l+1|k) = R_g(k), \tag{54i}$$

$$u(k), u_{k+1}, \dots, u_{k+N_p-1} \in \mathbb{U}, \tag{54j}$$

$$x(k|k), \bar{d}_m(k|k)) = (x(k), d_m(k)), \tag{54k}$$

for all $l \in \mathbb{Z}_{[0,N_p-1]}$ and all $r \in \mathbb{Z}_{[0,m_r]}$, where the terms $\Phi^x_{k,r}(\delta_x) = F^{-1}_{G(r)}B_d d_m(k)\left(1 - \frac{\delta_x}{n_x N_p}\right)$ and $\Phi^{x_{Rg}}_{k,\eta}(\delta_{Rg}) = F^{-1}_{G_{Rg}\eta}\left(1 - \frac{\delta_x}{N_p}\right)$ are the quantile functions involved in the states and system reliability deterministic equivalent constraints.

6. Application

6.1. Case Study

The system used as a case study is a part of the Barcelona DWN that is presented in [36]. In the considered case study, nine sources were considered, consisting of five underground and four surface sources, which currently provide an inflow of about 2 m^3/s. This part is composed of 17 tanks and 61 actuators (valves and pumps), 12 nodes, and 25 demands. Figure 2 presents the considered network showing the components and the relationships between them.

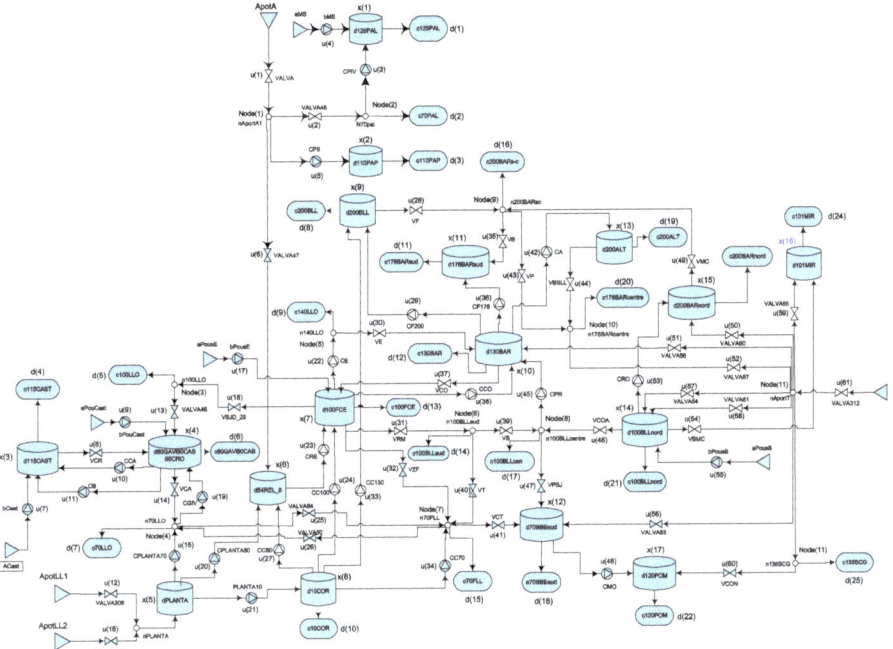

Figure 2. Barcelona drinking water network.

The approach proposed in this paper has been applied to the using the control-oriented model of DWN presented in Figure 2 presented in Section 2. This model can also be represented by means of a graph $G(v, \varepsilon)$, where n_x storage tanks, n_s sources, n_d demands, and n_q intersection nodes are represented by $v \in \nu$ vertices that are connected by $a \in \varepsilon$ links (pipes) (see Figure 3). The graph that shows in figure of the water network was obtained from the state-space representation of the system. This procedure is defined with more detail in [37]. According to the DWN reliability study, demands, sources, pipelines, and tanks are considered completely reliable whereas active elements such as valves and pumps are recognized not completely reliable [38]. The forecast of each demand $\bar{d}_m(k))$ is known as well the distribution of the forecasting error $\tilde{d}_m$ (see Figure 4).

Using the reliability analysis, the states that are structurally controllable can be determined since the path computation analysis gives all possible paths from a source to a consumer node. Furthermore, an approximate operational cost (related to the electricity cost of pump) and a maximal water flow (according to the physical constraints of the actuators) can be obtained for each path.

From the definition of minimal path P_s in Section 4.2, the minimal path sets are determined for Barcelona DWN. A minimal path set is composed by those components which allow a flow path between sources and demands, such as pipes, tanks, and pumps.

Tables 1 and 2 present important number of critical actuators within the network, according to the topology and the way of network elements are linked, as most actuators (pumps or valves) have the unique connection between tanks and demands. Subsequently, if an actuator fails, then the corresponding demand will not be satisfied. Note that the information presented in Tables 1 and 2 is particularly significant for AGBAR because it recognizes the critical elements in the network for monitoring/improvement policies to be performed in the event of element damage [39]. Considering the DWN (Figure 2), Tables 1 and 2, and the study of the success minimal path of the water network, 607 minimal path sets are specified inside the system. Some simplification of success minimal paths from the water network is presented in Table 3.

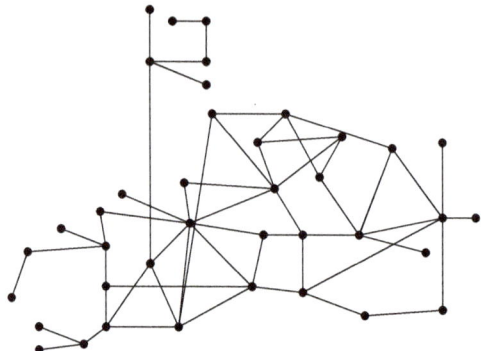

Figure 3. Graph from Barcelona DWN.

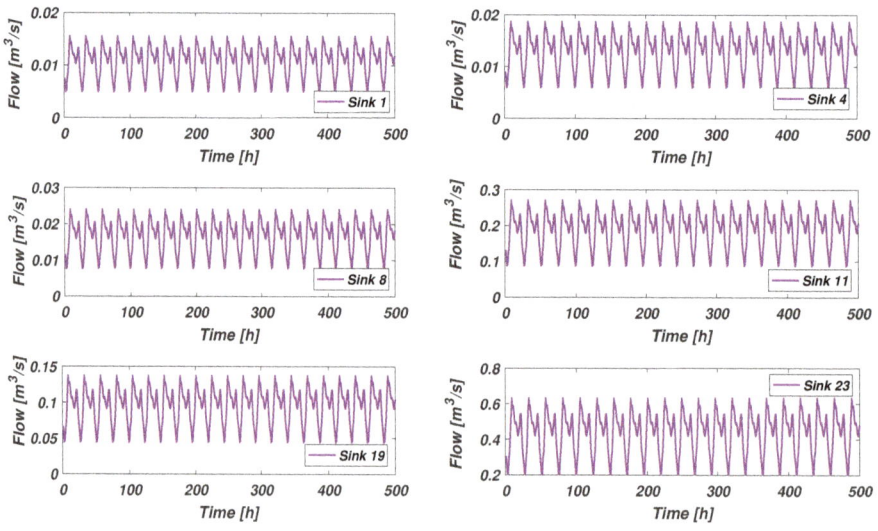

Figure 4. Drinking water demand for several demands.

Table 1. Structural actuators (towards tanks).

No.	Name	No.	Name	No.	Name	No.	Name
u_1	VALVA	u_{16}	VALVA309	u_{33}	CC130	u_{47}	VPSJ
u_3	CPIV	u_{17}	bPousE	u_{34}	CC70	u_{48}	CMO
u_4	bMS	u_{19}	CGIV	u_{35}	VB	u_{49}	VMC
u_5	CPII	u_{20}	CPLANTA50	u_{36}	CF176	u_{50}	VALVA60
u_6	VALVA47	u_{21}	PLANTA10	u_{37}	VCO	u_{51}	VALVA56
u_7	bCast	u_{23}	CRE	u_{38}	CCO	u_{52}	VALVA57
u_8	VCR	u_{24}	CC100	u_{39}	VS	u_{53}	CRO
u_9	bPouCast	u_{25}	VALVA64	u_{40}	V	u_{54}	VBMC
u_{10}	CCA	u_{26}	VALVA50	u_{41}	VCT	u_{55}	bPousB
u_{11}	CB	u_{27}	CC50	u_{42}	CA	u_{56}	VALVA53
u_{12}	VALVA308	u_{28}	VF	u_{43}	VP	u_{57}	VALVA54
u_{13}	VALVA48	u_{29}	CF200	u_{44}	VBSLL	u_{58}	VALVA61
u_{14}	VCA	u_{30}	VE	u_{45}	CPR	u_{59}	VALVA55
u_{15}	CPLANTA70	u_{32}	VZF	u_{46}	VCOA	u_{60}	VCON

Table 2. Structural actuators (towards demands).

No.	Name	No.	Name	No.	Name	No.	Name
u_2	VALVA45	u_{18}	VSJD-29	u_{22}	CE	u_{31}	VRM
u_{61}	VALVA312						

Table 3. Success minimal paths of the Barcelona DWN.

Path	Component Set
P_1	{aMS, bMS, c125PAL}
P_2	{AportA, VALVA, VALVA45, c70PAL}
P_3	{AportA, VALVA, VALVA47, CPIV, c125PAL}
P_4	{AportA, VALVA, CPII, c110PAP}
P_5	{ACast, bCast, c115CAST}
⋮	⋮
P_{607}	{AportT, VALVA312, c135SCG}

The reliability of each minimal path set depends on the reliability of its components; tanks and pipes are supposed perfectly reliable. Although, sources are involved in the minimal path sets only for illustrative purposes of the procedure. The objective of the MPC as has been explained before is to minimize the multi-objective cost function (54). The prediction horizon is 24 h because the demand and also the electrical tariff have periodicity of 1 day. The analysis is carried for a time period of 11 day (264 h) with sampling time of 1 h. The weights of the cost function (54a) are $W_e = 100$, $W_s = 1$, $W_{\Delta_u} = 1$, and $W_{R_g} = 10$. The weighting matrices are founded by iterative tuning until the desired performance is achieved. The tuning of these parameters is arranged based on that the objective with the highest preference is the economic cost, which must be minimized maintaining proper levels of safety volumes and control action smoothness and the same time should maximize the system reliability. The simulation results based on real data are obtained using the Gurobi 6.2 optimization package and Matlab R2015b (64 bits), running in a PC Intel(R) Core(TM)i7-5500 CPU at 2.4 GHz with 12 GB of RAM.

6.2. Results and Discussion

To analyze and assess the benefits of the proposed economic readability-aware MPC-LPV approach, a comparison with respect to baseline control strategies that were earlier proposed for the same case study is considered. In particular, the considered methods are as follows.

- *Reliability-Aware Chance-constrained Economic MPC-LPV (RACCEMPC-LPV):* This is the approach proposed in this paper that is based on solving the optimization problem (54). This approach allows the consideration of nonstationary stochastic demand uncertainty and stochastic whole reliability of the system. Therefore, the base stock constraint, the hard bounds of the states and the terminal constraint of the system reliability are formulated in the framework of chance constraints.
- *Economic MPC-LPV (EMPC-LPV):* This approach is based the optimization problem (45) without including the reliability objective. Moreover, it is not considering the stochastic demand uncertainty, chance constraints, and terminal constraint of the system reliability of the network.
- *Chance-constrained Economic MPC-LPV (CCEMPC-LPV):* This approach is included robustness only for demand uncertainty by replacing the state deterministic constraints with chance-constraints. Moreover, the CCEMPC-LPV controller does not include neither the reliability objective nor the terminal constraint of the system reliability of the network.
- *Reliability-aware economic MPC-LPV (RAEMPC-LPV):* This approach relies on solving problem (45a). In this approach, an additional goal is included to the controller in order to extend the components and system reliability. However, the stochastic demand uncertainty and chance constraints associated to the system reliability are not considered.

Table 4 exhibits the numeric assessment of the above-mentioned controllers through different key performance indicators (KPIs), which are detailed below,

$$KPI_e := \frac{1}{n_s+1} \sum_{k=0}^{n_s} \alpha^\top(k) u_k \Delta_t, \tag{55a}$$

$$KPI_{\Delta_u} := \frac{1}{n_s+1} \sum_{i=1}^{n_u} \sum_{k=0}^{n_s} (\Delta_u(i,k))^2, \tag{55b}$$

$$KPI_s := \sum_{i=1}^{n_x} \sum_{k=0}^{n_s} \max\{0, x_s(i,k) - x(i,k)\}, \tag{55c}$$

$$KPI_R := x_{R_g}(k), \tag{55d}$$

$$KPI_t := t_{opt}(k), \tag{55e}$$

where KPI_e denotes the average economic performance of the water network, KPI_{Δ_u} evaluates the smoothness of the control actions, KPI_s comprises the quantity of water utilized from safety stocks, KPI_R denotes the value of the whole system reliability of the DWN, and KPI_t defines the difficulty to solve the optimization tasks associated with each approach accounting $t_{opt}(k)$ as the average time that gets to solve the corresponding FHOP. In $KPI_e, KPI_{\Delta_u}, KPI_s$, and KPI_t lower values signify better performance results. However, a higher KPI_R value shows better performance in system reliability of the DWN. Furthermore, Table 5 presents the details of the production and operational costs associated with each approach, which are one of the most important objectives for the DWN managers.

Figures 5 and 6 show, respectively, the evolution of the flow actuator commands and the tank volumes for comparison of different considered MPC approaches for the Barcelona DWN. Figure 5 shows that pumps always try to operate at the minimum cost, i.e., when the electrical tariff is cheaper. Figure 6 shows the proper replenishment planning for the tanks that the predictive controller dictates according to the cyclic behavior of demands. Note that the net demand of each tank is properly satisfied along the simulation horizon.

To manage the stochastic demand uncertainty, CCEMPC-LPV and RACCEMPC-LPV controllers incorporated the robustness for demand uncertainty by replacing the state deterministic constraints with chance constraints. Generally, chance constraints create an optimal back-off from real constraints as a risk-averse mechanism to face the nonstationary uncertainty included in the prediction of states. This is reflected in Figure 5, where the behavior in the presented actuators commands in chance constraints approaches (CCEMPC-LPV and RACCEMPC-LPV) are almost the same and larger than in the case of EMPC-LPV and RAEMPC-LPV, that also present a similar behavior. Similarly, this behavior appeared in the volume evolution of the selected storage tanks that are presented in Figure 6. These results are logical since to cope with the uncertainty considered by the chance constraint-based methods additional water is stored in the tanks and this requires more flow to be injected.

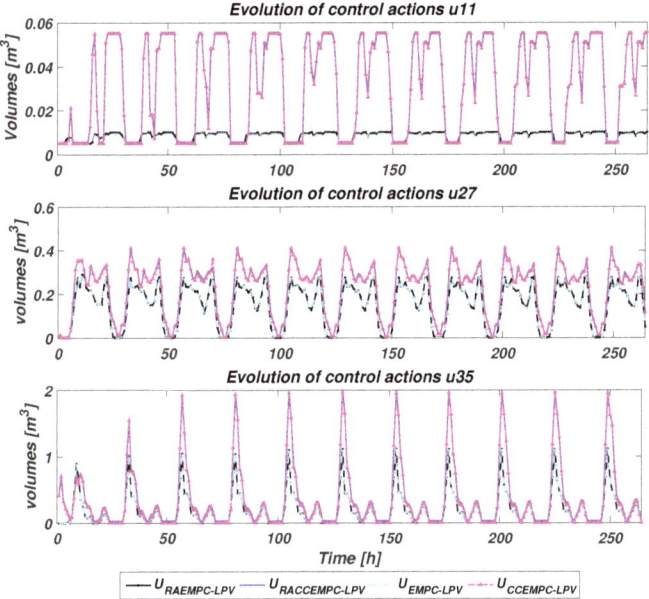

Figure 5. Evaluation of the control actions results.

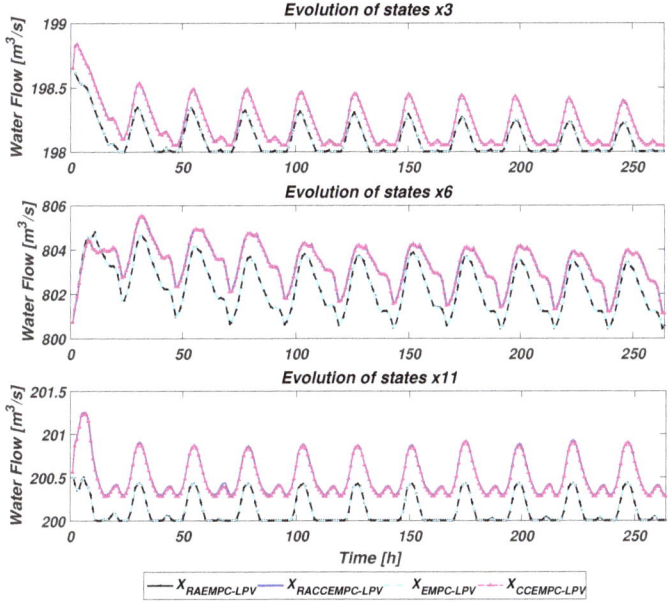

Figure 6. Results of the evolutions of storage tanks.

Table 5 presents the details of water production and electricity cost of each approach. The RACCEMPC-LPV approach has similar costs to those of the baseline CCEMPC-LPV approach, but with the profit of a better handling of constraints and considering the system reliability into control low of the controller. Generally, the proposed RACCEMPC-LPV approach leads to a higher total closed-loop operational cost if considering only the water and electric costs as signs for economic performance. This is the price to pay for increasing the reliability of the system.

On the other hand, Figure 7 shows the comparison of the system reliability predictions and accumulated economic cost of the DWN that obtained from the different MPC approaches. According to this figure and reviewing the results in Tables 4 and 5, it can be observed that the robustness enhancements of the RACCEMPC-LPV approach are larger than the other controllers in terms of reliability. The EMPC-LPV controller has lower values in the economic index KPI_e but, the guarantee of reliability, robustness and feasibility problems are not considered. The main disadvantage of this controller is that control actions are computed based on economic criteria. In this case, the controller overexploits actuators that have lower operational costs, quickening their damage and hazarding the service reliability. The RAEMPC-LPV strategy reached the lowest KPI_e after the EMPC-LPV controller by including the reliability objective in the control low. However, the stochastic demand uncertainty and stochastic uncertainty of the system reliability are not considered.

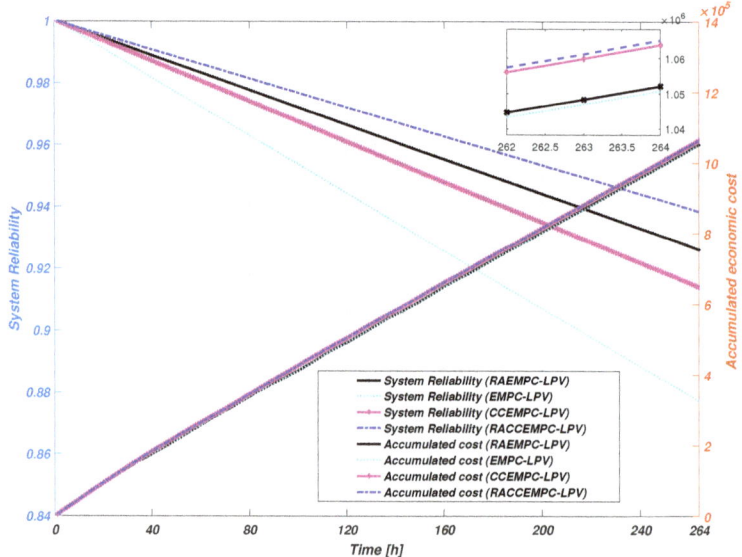

Figure 7. Evaluation of system reliability and accumulated economic cost.

Table 4. Comparison of control performance.

Controller	KPI_e	KPI_{Δ_u}	KPI_s	KPI_{Rg}	KPI_t	Simulation Time
EMPC-LPV	3779.81	0.5271	28951.72	0.8772	1.5628	412.599
CCEMPC-LPV	4029.09	0.4910	28955.69	0.9186	1.9051	502.952
RAEMPC-LPV	3980.07	0.5317	28952.62	0.9263	1.78348	470.841
RACCEMPC-LPV	4029.19	0.4903	28955.90	0.9386	1.9664	519.147

Table 5. Comparison of daily average costs of the MPC approaches.

MPC Approach	Water Average Cost (e.u./day)	Electric Average Cost (e.u./day)	Daily Average Cost (e.u./day)
EMPC-LPV	44162.44	3053.08	47215.53
CCEMPC-LPV	51237.98	3262.43	54500.42
RAEMPC-LPV	44369.90	3121.84	47491.75
RACCEMPC-LPV	51438.13	3262.64	54700.77

7. Conclusions

In this paper, an economic reliability-aware LPV-MPC strategy based on chance constraints for water transport network has been proposed to deal with the management of flow-based networks, considering both demand uncertainty and system reliability in a probabilistic way. The considered control-oriented model of the water transport network is based on a flow modeling approach. By considering chance constraints programming to compute an optimal replenishment policy based on a desired risk acceptability level, the system reliability is introduced as state variables inside the control model, which includes nonlinear term and it is changed in a linear-like form through the LPV structure. Therefore, the LPV model includes both the reliability and DWN models including scheduling parameters are updated with the state vector value at that time. Moreover, nonstationary flow demands and system reliability are satisfied by considering chance-constraint programming. The results obtained show that the system reliability of the DWN network is maximized with the proposed controller while the cost is slightly increased. The level of resultant back-off volume is variable and depend of the forecast demand uncertainty and system reliability at each prediction step based on probabilistic distributions employed to their modeling. As, in practice, disturbances are unbounded, the strategy proposed in this paper is based on a service-level guarantee and a probabilistic feasibility.

Future research will concentrate on the study of predicting the system reliability of water distribution networks by considering the pressure model.

Author Contributions: All the authors have equally contributed. All authors have read and agreed to the published version of the manuscript.

Funding: This work has been partially funded by the Spanish State Research Agency (AEI) and the European Regional Development Fund (ERFD) through the projects SCAV (ref. MINECO DPI2017-88403-R) and also by EU INTERREG POCTEFA (2014-2020) EFA 153/16 SMART.

Conflicts of Interest: The authors declare no conflict of interest.

References

1. Ocampo-Martinez, C.; Puig, V.; Cembrano, G.; Quevedo, J. Application of predictive control strategies to the management of complex networks in the urban water cycle [applications of control]. *IEEE Control Syst.* **2013**, *33*, 15–41.
2. Karimi Pour, F.; Puig, V.; Ocampo-Martinez, C. Health-aware Model Predictive Control of Pasteurization Plant. *J. Phys.* **2017**, *783*, 12030.
3. Maciejowski, J.M. *Predictive Control: With Constraints*; Pearson Education: London, UK, 2002.
4. Cembrano, G.; Quevedo, J.; Puig, V.; Pérez, R.; Figueras, J.; Verdejo, J.; Escaler, I.; Ramón, G.; Barnet, G.; Rodríguez, P.; et al. PLIO: A generic tool for real-time operational predictive optimal control of water networks. *Water Sci. Technol.* **2011**, *64*, 448–459. [CrossRef] [PubMed]
5. Ocampo-Martinez, C.; Ingimundarson, A.; Puig, V.; Quevedo, J. Objective prioritization using lexicographic minimizers for MPC of sewer networks. *IEEE Trans. Control Syst. Technol.* **2008**, *16*, 113–121. [CrossRef]
6. Pascual, J.; Romera, J.; Puig, V.; Cembrano, G.; Creus, R.; Minoves, M. Operational predictive optimal control of Barcelona water transport network. *Control Eng. Pract.* **2013**, *21*, 1020–1034. [CrossRef]

7. Puig, V.; Ocampo-Martinez, C.; De Oca, S.M. Hierarchical temporal multi-layer decentralised MPC strategy for drinking water networks: Application to the barcelona case study. In Proceedings of the 2012 20th Mediterranean Conference on Control & Automation (MED), Barcelona, Spain, 3–6 July 2012; pp. 740–745.
8. Rawlings, J.B.; Mayne, D.Q. *Model Predictive Control: Theory and Design*; Nob Hill Pub.: Santa Barbara, CA, USA, 2009.
9. Karimi Pour, F.; Puig, V.; Ocampo-Martinez, C. Economic Predictive Control of a Pasteurization Plant using a Linear Parameter Varying Model. In *Computer Aided Chemical Engineering*; Elsevier: Amsterdam, The Netherlands, 2017; Volume 40, pp. 1573–1578.
10. Gertsbakh, I. *Theory of Reliability with Applications to Preventive Maintenance*; Springer: Berlin/Heidelberg, Germany, 2000.
11. Gokdere, L.; Chiu, S.L.; Keller, K.J.; Vian, J. Lifetime control of electromechanical actuators. In Proceedings of the 2005 IEEE Aerospace Conference, Big Sky, MT, USA, 5–12 March 2005; pp. 3523–3531.
12. Pereira, E.B.; Galvão, R.K.H.; Yoneyama, T. Model predictive control using prognosis and health monitoring of actuators. In Proceedings of the 2010 IEEE International Symposium on Industrial Electronics (ISIE), Bari, Italy, 4–7 July 2010; pp. 237–243.
13. Karimi Pour, F.; Puig, V.; Ocampo-Martinez, C. Multi-layer health-aware economic predictive control of a pasteurization pilot plant. *Int. J. Appl. Math. Comput. Sci.* **2018**, *28*, 97–110. [CrossRef]
14. Gallestey, E.; Stothert, A.; Antoine, M.; Morton, S. Model predictive control and the optimization of power plant load while considering lifetime consumption. *IEEE Trans. Power Syst.* **2002**, *17*, 186–191. [CrossRef]
15. Khelassi, A.; Theilliol, D.; Weber, P.; Sauter, D. A novel active fault tolerant control design with respect to actuators reliability. In Proceedings of the 2011 50th IEEE Conference on Decision and Control and European Control Conference (CDC-ECC), Orlando, FL, USA, 12–15 December 2011; pp. 2269–2274.
16. Salazar, J.C.; Sarrate, R.; Nejjari, F.; Weber, P.; Theilliol, D. Reliability computation within an MPC health-aware framework. *IFAC-PapersOnLine* **2017**, *50*, 12230–12235. [CrossRef]
17. Salazar, J.C.; Weber, P.; Sarrate, R.; Theilliol, D.; Nejjari, F. MPC design based on a DBN reliability model: Application to drinking water networks. *IFAC-PapersOnLine* **2015**, *48*, 688–693. [CrossRef]
18. Karimi Pour, F.; Puig, V.; Cembrano, G. Health-aware LPV-MPC based on system reliability assessment for drinking water networks. In Proceedings of the 2nd IEEE Conference on Control Technology and Applications, Copenhagen, Denmark, 21–24 August 2018.
19. Chamseddine, A.; Theilliol, D.; Sadeghzadeh, I.; Zhang, Y.; Weber, P. Optimal reliability design for over-actuated systems based on the MIT rule: Application to an octocopter helicopter testbed. *Reliab. Eng. Syst. Saf.* **2014**, *132*, 196–206. [CrossRef]
20. Grosso, J.M.; Ocampo-Martínez, C.; Puig, V. A service reliability model predictive control with dynamic safety stocks and actuators health monitoring for drinking water networks. In Proceedings of the 2012 IEEE 51st Annual Conference on Decision and Control (CDC), Maui, HI, USA, 10–13 December 2012; pp. 4568–4573.
21. Karimi Pour, F.; Puig Cayuela, V.; Ocampo-Martínez, C. Comparative assessment of LPV-based predictive control strategies for a pasteurization plant. In Proceedings of the 4th—2017 International Conference on Control, Decision and Information Technologies, Barcelona, Spain, 5–7 April 2017; pp. 1–6.
22. Calafiore, G.C.; Dabbene, F.; Tempo, R. Research on probabilistic methods for control system design. *Automatica* **2011**, *47*, 1279–1293. [CrossRef]
23. Schwarm, A.T.; Nikolaou, M. Chance-constrained model predictive control. *AIChE J.* **1999**, *45*, 1743–1752. [CrossRef]
24. Geletu, A.; Klöppel, M.; Zhang, H.; Li, P. Advances and applications of chance-constrained approaches to systems optimization under uncertainty. *Int. J. Syst. Sci.* **2013**, *44*, 1209–1232. [CrossRef]
25. Ouarda, T.; Labadie, J. Chance-constrained optimal control for multireservoir system optimization and risk analysis. *Stoch. Environ. Res. Risk Assess.* **2001**, *15*, 185–204. [CrossRef]
26. Mays, L. *Urban Stormwater Management Tools*; McGraw Hill Professional: New York, NY, USA, 2004.
27. Brdys, M.; Ulanicki, B. Operational Control of Water Systems: Structures, Algorithms and Applications. *Automatica* **1996**, *32*, 1619–1620.
28. Cembrano, G.; Quevedo, J.; Salamero, M.; Puig, V.; Figueras, J.; Martı, J. Optimal control of urban drainage systems. A case study. *Control Eng. Pract.* **2004**, *12*, 1–9. [CrossRef]
29. Kall, P.; Mayer, J. *Stochastic Linear Programming*; Springer: Berlin/Heidelberg, Germany, 1976; Volume 7.

30. Grosso, J.M.; Velarde, P.; Ocampo-Martinez, C.; Maestre, J.M.; Puig, V. Stochastic model predictive control approaches applied to drinking water networks. *Optim. Control Appl. Methods* **2017**, *38*, 541–558. [CrossRef]
31. Nemirovski, A.; Shapiro, A. Convex approximations of chance constrained programs. *SIAM J. Optim.* **2006**, *17*, 969–996. [CrossRef]
32. Shapiro, A.; Dentcheva, D.; Ruszczyński, A. *Lectures on Stochastic Programming: Modeling and Theory*; SIAM: Phhiladelphia, PA, USA, 2009.
33. Jiang, R.; Jardine, A.K. Health state evaluation of an item: A general framework and graphical representation. *Reliab. Eng. Syst. Saf.* **2008**, *93*, 89–99. [CrossRef]
34. Gertsbakh, I. *Reliability Theory: With Applications to Preventive Maintenance*; Springer: Berlin/Heidelberg, Germany, 2013.
35. Baecher, G.B.; Christian, J.T. *Reliability and Statistics in Geotechnical Engineering*; John Wiley & Sons: Hoboken, NJ, USA, 2005.
36. Ocampo-Martínez, C.; Puig, V.; Cembrano, G.; Creus, R.; Minoves, M. Improving water management efficiency by using optimization-based control strategies: The Barcelona case study. *Water Sci. Technol. Water Supply* **2009**, *9*, 565–575. [CrossRef]
37. Siljak, D.D. *Decentralized Control of Complex Systems*; Courier Corporation: Chelmsford, MA, USA, 2011.
38. Weber, P.; Simon, C.; Theilliol, D.; Puig, V. Fault-Tolerant Control Design for over-actuated System conditioned by Reliability: A Drinking Water Network Application. *IFAC Proc. Vol.* **2012**, *45*, 558–563. [CrossRef]
39. Robles, D.; Puig, V.; Ocampo-Martinez, C.; Garza-Castañón, L.E. Reliable fault-tolerant model predictive control of drinking water transport networks. *Control Eng. Pract.* **2016**, *55*, 197–211. [CrossRef]

© 2020 by the authors. Licensee MDPI, Basel, Switzerland. This article is an open access article distributed under the terms and conditions of the Creative Commons Attribution (CC BY) license (http://creativecommons.org/licenses/by/4.0/).

Article

Estimation of Actuator and System Faults Via an Unknown Input Interval Observer for Takagi–Sugeno Systems

Citlaly Martínez-García [1], Vicenç Puig [2], Carlos-M. Astorga-Zaragoza [1,*], Guadalupe Madrigal-Espinosa [1,3] and Juan Reyes-Reyes [1]

[1] Department of Electronic Engineering, Tecnológico Nacional de México/CENIDET, Cuernavaca, Morelos 62490, Mexico; mgci@cenidet.edu.mx (C.M.-G.); gmadrigal@cenidet.edu.mx or gme@ineel.mx (G.M.-E.); juanreyesreyes@cenidet.edu.mx (J.R.-R.)
[2] Advanced Control Systems (SAC) Research Group at Institut de Robótica i Informática Industrial (IRI), Universitat Politécnica de Catalunya-BarcelonaTech (UPC), 08028 Barcelona, Spain; vicenc.puig@upc.edu
[3] National Institute of Electricity and Clean Energy, Cuernavaca, Morelos 62490, Mexico
* Correspondence: astorga@cenidet.edu.mx; Tel.: +1-(777)-362-7776 (ext. 2205)

Received: 4 November 2019; Accepted: 24 December 2019; Published: 2 January 2020

Abstract: This paper presents a simultaneous state variables and system and actuator fault estimation, based on an unknown input interval observer design for a discrete-time parametric uncertain Takagi–Sugeno system under actuator fault, with disturbances in the process and measurement noise. The observer design is synthesized by considering unknown but bounded process disturbances, output noise, as well as bounded parametric uncertainties. By taking into account these considerations, the upper and lower bounds of the considered faults are estimated. The gain of the unknown input interval observer is computed through a linear matrix inequalities (LMIs) approach using the robust H_∞ criteria in order to ensure attenuation of process disturbances and output noise. The interval observer scheme is experimentally evaluated by estimating the upper and lower bounds of a torque load perturbation, a friction parameter and a fault in the input voltage of a permanent magnet DC motor.

Keywords: Takagi–Sugeno; fault estimation; unknown input; interval observer; permanent magnet motor

1. Introduction

Typically, an observer is an scheme for state estimation through the system input and output measurements. For instance, in [1] a nonlinear observer is applied to estimate the degree of polymerization in a series of polycondensation reactors. However, an observer can be designed for parameter estimation [2], unknown input estimation [3,4] or fault estimation [5,6] among other important applications where it is important to precisely know the actual value of the states, signals or parameters for multiple purposes.

Sometimes there are many technical difficulties in performing an exact estimation of the state, signals or parameters to be estimated. For instance: (i) Model uncertainties, (ii) simplifying assumptions of physical phenomena for modeling, and (iii) complexity reduction of models or the unmeasured disturbances, represent an important source of mismatch between a real process and a mathematical model. In these cases, an approximation of the estimated values can be performed. These approximations can be very useful in many applications where there is not necessary to know the exact value of a variable.

An alternative to estimate unknown variables in processes with uncertain models, interval observers can be used. These observers provide an interval estimation providing a lower and upper bound of the unknown estimated variables. The actual value of the corresponding unmeasured variable located inside the interval defined by these bounds assuming that the uncertainty bounds are known.

Although it is not possible to estimate the exact value of a variable, the information provided by an interval observer can be very useful for several applications. For instance, the authors in [7] propose an interval observer to estimate the lower and upper bounds of vehicle dynamics regardless of the presence of unknown inputs whose bounded interval is also estimated. The authors in [8] design an interval sliding mode observer for sensor fault detection and applied it to an electrical traction device. Another interesting application of interval observers is given in [9], where a trajectory control based on an interval observer is designed for a quadrotor. The interval observer is synthesized by using an uncertain model where all the uncertainties (parameters, disturbance, noise) are unknown but bounded with known bounds.

The main limitation of recent works regarding interval observers is that in most cases, the interval observer design considers linear systems, or a very particular structure of nonlinear systems which sometimes are transformed into linear ones. For instance, the observer in [7] has been designed for switched systems; therefore, its use is limited. In other cases of interval observer designs such as [9], no faults are considered to be estimated or there is a lack of procedure to detect actuator faults [8].

The objective of this paper is to design an interval observer for a wider variety of nonlinear processes by using the Takagi–Sugeno (T–S) approach. Most of the nonlinear models can be adequately transformed into a T–S model (e.g., [10,11]) by using two different methods [12]:

- The nonlinear sector method, in this case the nonlinear model and its equivalent T–S model have exactly the same behavior. For this reason, this is the method used in this work.
- The linearization method, in which the equivalent T–S model can be dynamically approximated to the original nonlinear model with a certain accuracy, depending on the design requirements.

Besides, many advantageous opportunities arise when interval observers are designed for processes modeled in T–S form: (i) Pole placement via linear matrix inequalities (LMI) regions is considered to compute the observer gains, in contrast with many nonlinear approaches where the observer gains are heuristically tuned; (ii) a standard methodological procedure can be used to compute the observer gains; (iii) many approaches originally conceived for linear systems can be easily extended to T–S systems. For these reasons, the design of interval observers for T–S systems is a recent and interesting research topic. For example in [13], the authors propose an interval observer for the state estimation of systems modeled in T–S form with parametric uncertainty, disturbances, and measurement noise. However, the work is limited to estimate the unmeasured states. The authors in [14] treat the problem of fault diagnosis of proton exchange membrane (PEM) fuel cells. However, this paper deals with only the case of sensor faults by means of a bank of observers. In [15] a robust fault detection procedure for vehicle lateral dynamics using a switched T–S interval observer is presented. The proposed method is conceived to detect but not to estimate faults.

The main contribution of this paper consists in the design of an interval observer that performs a simultaneous estimation of unmeasured states, actuator and system faults for processes modeled in T–S form with uncertainties. The conditions for the existence of such observers are given. Such conditions guarantee the observer stability and they are proved through a Lyapunov analysis combined with a LMI formulation. The interval observer scheme is experimentally evaluated by estimating the upper and lower bounds of a torque load perturbation, a friction parameter and a fault in the input voltage of a permanent magnet direct-current (DC) motor. These cases are typical faults that, if not detected in time, can become catastrophic failures such as short-circuits or machinery damages due to damaged bearings.

2. Problem Formulation and Preliminaries

Consider the following discrete-time T–S system:

$$x(k+1) = \sum_{i=1}^{m} \xi_i(\rho(k))[(A_i + \Delta A_i)x(k) + B_i u(k)] + E_f f(k) + G\theta(k) + E_w w(k), \quad (1)$$
$$y(k) = Cx(k) + E_v v(k),$$

where $x(k) \in \mathbb{R}^{n_x}$, $u(k) \in \mathbb{R}^{n_u}$, $f(k) \in \mathbb{R}^{n_f}$, $\theta(k) \in \mathbb{R}^{n_\theta}$, $w(k) \in \mathbb{R}^{n_w}$ and $v(k) \in \mathbb{R}^{n_v}$ represent the state variable, the input, the actuators fault vector, the unknown parameter, the disturbance and the output noise vector. $A_i, \Delta A_i, B_i, G$ and C are matrices of appropriate dimensions. E_f, E_w and E_v are matrices of the coupling distribution. k denotes the k-th discrete time instant.

The term $\xi_i(\rho(k))$ represents the i-th membership function, which is a weighting of the rule i, where $i = 1, 2, \ldots, m$. The membership functions are normalized, i.e., they satisfy the following conditions [12,16]:

$$\begin{cases} \sum_{i=1}^{m} \xi_i(\rho(k)) = 1 \\ \\ 0 \leq \xi_i(\rho(k)) \leq 1, i = 1, 2, \cdots, m. \end{cases} \quad (2)$$

To obtained a simultaneous estimation of parameters and faults, the system (1) is rewritten as follows

$$x(k+1) = \sum_{i=1}^{m} \xi_i(\rho(k))(A_i + \Delta A_i)x(k) + B_i u(k) + \overline{E} f_{\overline{E}}(k) + E_w w(k) \quad (3)$$
$$y(k) = Cx(k) + E_v v(k),$$

where the vector $f_{\overline{E}}(k)$ is an augmented one, which is defined by the actuator fault vector $f(k)$ and the unknown parameter vector $\theta(k)$; and consequently, the matrix $\overline{E}$ contains the fault coupling distribution matrix E_f and the parameter matrix G, i.e.,:

$$\overline{E} = \begin{bmatrix} E_f & G \end{bmatrix}, \qquad f_{\overline{E}}(k) = \begin{bmatrix} f(k) \\ \theta(k) \end{bmatrix}.$$

The following considerations are taken into account for the T–S system of the Equation (3):

- The augmented fault vector $f_{\overline{E}}(k)$ is defined as:

$$f_{\overline{E}}(k+1) = f_{\overline{E}}(k) + w_{f_{\overline{E}}}(k), \quad (4)$$

where $w_{f_{\overline{E}}}(k)$ is considered as a variation of the actuator fault. Therefore, the estimation of $f_{\overline{E}}(k)$ is equivalent to the estimation of $\hat{f}(k)$ and $\hat{\theta}(k)$.
- The perturbation vector $w(k)$ is considered unknown but bounded as follows:

$$\overline{w}(k) \leq w(k) \leq \underline{w}(k). \quad (5)$$

- The noise vector $v(k)$ is also considered as an unknown but bounded signal, i.e.,:

$$|v(k)| \leq V(k). \quad (6)$$

- The uncertain matrix ΔA_i is considered bounded as follows,

$$\overline{\Delta A_i} \leq \Delta A_i \leq \underline{\Delta A_i}. \quad (7)$$

- Based on previous assumptions, the estimates to be obtained will be as follows

$$\overline{x}(k) \leq x(k) \leq \hat{x}(k), \tag{8}$$

$$\overline{f}_{\overline{E}}(k) \leq f_{\overline{E}}(k) \leq \hat{f}_{\overline{E}}(k). \tag{9}$$

This means that we would get two estimates, i.e., the upper and lower limit of each variable. For that, we consider the following design based on a T-S interval observer.

3. Observer Design

In this section a similar procedure as that in [17] (where no parametric uncertainties nor noise nor disturbances were considered) is presented for the observer design. For this design, first it is considered the output vector at time instant $(k+1)$, i.e.,

$$y(k+1) = Cx(k+1) + E_v v(k+1). \tag{10}$$

Substituting the state equation from system (3), it yields to:

$$y(k+1) = C\left(\sum_{i=1}^{m} \xi_i(\rho(k))[(A_i + \Delta A_i)x(k) + B_i u(k)] + \overline{E}_{f\overline{E}}(k) + E_w w(k)\right) + E_v v(k+1). \tag{11}$$

Next, the following equation can be derived after the pertinent operations

$$C\overline{E}_{f\overline{E}}(k) = y(k+1) - C\sum_{i=1}^{m} \xi_i(\rho(k))(A_i + \Delta A_i)x(k) - C\sum_{i=1}^{m} \xi_i(\rho(k))B_i u(k) - CE_w w(k) - E_v v(k+1), \tag{12}$$

where it is possible to obtain the fault vector $f(k)$ as follows:

$$f(k) = \mathcal{O}\left(y(k+1) - C\left[\sum_{i=1}^{m} \xi_i(\rho(k))[(A_i + \Delta A_i)x(k) - B_i u(k)] - E_w w(k)\right] - E_v v(k+1)\right), \tag{13}$$

such that $\mathcal{O}$ comes from the following condition, which furthermore must be satisfied for the observer to exist [18]:

$$\text{rank}(C\overline{E}_{f\overline{E}}) = \text{rank}(\overline{E}_{f\overline{E}}) = n_\theta + n_f. \tag{14}$$

The decoupling is achieved by computing

$$\mathcal{O} = (C\overline{E}_{f\overline{E}})^+, \tag{15}$$

such that $(C\overline{E}_{f\overline{E}})^+(C\overline{E}_{f\overline{E}}) = I_{n_f}$ is satisfied. Whereas the value of $\mathcal{O}$ is obtained as:

$$\mathcal{O} = \left[(C\overline{E}_{f\overline{E}})^T C\overline{E}_{f\overline{E}}\right]^{-1} (C\overline{E}_{f\overline{E}})^T. \tag{16}$$

Replacing fault vector Equation (13) in system Equation (3), the new T-S discrete-time system is obtained:

$$\begin{aligned} x(k+1) &= \sum_{i=1}^{m} \xi_i(\rho(k))[(\overline{A}_i + \Delta \overline{A}_i)x(k) + B_i u(k)] + \overline{E}_w w(k) + \overline{E}\mathcal{O}y(k+1) - \overline{E}\mathcal{O}E_v v(k+1), \\ y(k) &= Cx(k) + E_v v(k) \end{aligned} \tag{17}$$

with

$$\overline{A}_i = (I - E_f \mathcal{O} C) A_i, \qquad \Delta \overline{A}_i = (I - E_f \mathcal{O} C) \Delta A_i,$$

$$B_i = (I - E_f \mathcal{O} C) B_i, \qquad E_w = (I - E_f \mathcal{O} C) E_w.$$

Now, based on (17), the unknown input T–S interval observer structure can be written as follows [19]:

$$\underline{\hat{x}}(k+1) = \sum_{i=1}^{m} \xi_i(\rho(k))(I - \overline{E}\mathcal{O}C)[(A_i - \underline{L}_i C)\underline{\hat{x}}(k) + B_i u(k) + \underbrace{\Delta A_i x(k)} + \underline{L}_i \underline{y}(k) - |\underline{L}_i| E_v V(k) + E_w \underline{w}(k)$$
$$+ \overline{E}\mathcal{O} y(k+1) - |\underline{L}_i| E_v V(k+1)],$$

$$\overline{\hat{x}}(k+1) = \sum_{i=1}^{m} \xi_i(\rho(k))(I - \overline{E}\mathcal{O}C)[(A_i - \overline{L}_i C)\overline{\hat{x}}(k) + B_i u(k) + \widetilde{\Delta A_i x(k)} + \overline{L}_i y(k) + |\overline{L}_i| E_v V(k) + E_w \overline{w}(k)$$
$$+ \overline{E}\mathcal{O} y(k+1) + |\overline{L}_i| E_v V(k+1)], \qquad (18)$$

$$\underline{\hat{f}}(k) = \mathcal{O}[y(k+1) - C \sum_{i=1}^{m} \xi_i(\rho(k))(A_i \underline{\hat{x}}(k) + B_i u(k) + \underbrace{\Delta A_i x(k)}) - C E_w \underline{w}(k) - E_v V(k+1) - \underline{w}_f(k)],$$

$$\overline{\hat{f}}(k) = \mathcal{O}[y(k+1) - C \sum_{i=1}^{m} \xi_i(\rho(k))(A_i \overline{\hat{x}}(k) + B_i u(k) + \underbrace{\Delta A_i x(k)}) - C E_w \overline{w}(k) + E_v V(k+1) - \overline{w}_f(k)],$$

with

$$\underbrace{\Delta A_i x(k)} = \underline{A}_i^+ \underline{x}^+ - \overline{A}_i^+ \underline{x}^- - \underline{A}_i^- \overline{x}^+ + \overline{A}_i^- \overline{x}^-,$$

$$\widetilde{\Delta A_i x(k)} = \overline{A}_i^+ \overline{x}^+ - \underline{A}_i^+ \overline{x}^- - \overline{A}_i^- \underline{x}^+ + \underline{A}_i^- \underline{x}^-,$$

where $\underline{\hat{x}}(k)$ and $\overline{\hat{x}}(k) \in \mathbb{R}_x^n$ are the interval estimations of $x(k)$, $\underline{\hat{f}}(k)$ and $\overline{\hat{f}}(k) \in \mathbb{R}^s$ are the interval estimations of $f_E(k)$. $\underline{L}_i$ and $\overline{L}_i$ are the observer gains used to compute the upper and lower bounds of the estimated states, faults and parameters, respectively.

The unknown input interval observer can be designer considering (18) in a way that ensures the simultaneous estimation of Equations (8) and (9). The following theorem is introduced to secure the stability analysis and robustness in the presence of unknown entries.

Theorem 1. *Consider the system given by (18) as an interval observer for system (17) for fault and parameter estimation. The observer (18) is stable and robust against the effects of unknown inputs such as bounded disturbances or noise if there exists a symmetric matrix $P = P^T > 0$, a matrix $Q > 0$ and the scalars $\epsilon_1 > 0$, $\gamma > 0$ and $\beta > 0$ such that:*

$$QG_{i,j} - W_i \Gamma > 0, \qquad (19)$$

$$\phi_{i,i} < 0, \qquad (20)$$

$$\phi_{i,j} = \begin{bmatrix} I - P + \gamma \eta^2 I & 0 & 0 & 0 & (QG_{i,j} - W_i \Gamma)^T \\ 0 & \gamma I - \epsilon_1 P & P H_i & P \Phi & 0 \\ 0 & H_i P & -\beta^2 I & 0 & H_i^T Q^T + \Phi^T Q^T \\ 0 & \Phi P & 0 & -\beta^2 I & Q^T \Phi \\ (*) & (*) & (*) & (*) & P - Q - Q^T \end{bmatrix},$$

$$\frac{2}{m-1}\phi_{i,i} + \phi_{i,j} + \phi_{j,i} < 0, \qquad (21)$$

for $i, j = 1, 2, \cdots, m$, $1 \leq i \neq j \leq m$, i.e., for all subsystems. The observer gains are given by

$$\underline{L}_i = Q^{-1} \underline{W}_i, \qquad (22)$$

$$\overline{L}_i = Q^{-1} \overline{W}_i. \qquad (23)$$

Proof. For the stability analysis the following estimation error equations are considered:

$$\underline{e}(k) = x(k) - \underline{\hat{x}}(k), \tag{24}$$

$$\overline{e}(k) = \overline{\hat{x}}(k) - x(k). \tag{25}$$

Substituting the state equation (17) and the estimate state equations (18), (24) and (25) it leads to:

$$\underline{e}(k+1) = \sum_{i=1}^{m} \xi_i(\rho(k))[(A_i + \Delta A_i)x(k) + B_i u(k)] + E_w w(k) + \overline{E}\mathcal{O}y(k+1) - \overline{E}\mathcal{O}E_v v(k+1)$$

$$- \left(\sum_{i=1}^{m} \xi_i(\rho(k))(I - \overline{E}\mathcal{O}C)[(A_i - \underline{L}_i C)\underline{\hat{x}}(k) + B_i u(k) + \underbrace{\Delta A_i x(k)}_{} + \underline{L}_i y(k) - |\underline{L}_i| E_v V(k) \right. \tag{26}$$

$$\left. + E_w \underline{w}(k) + \overline{E}\mathcal{O}y(k+1) - |\underline{L}_i| E_v V(k+1)] \right),$$

$$\overline{e}(k+1) = \sum_{i=1}^{m} \xi_i(\rho(k))(I - \overline{E}\mathcal{O}C)[(A_i - \overline{L}_i C)\underline{\hat{x}}(k) + B_i u(k) + \overbrace{\Delta A_i x(k)}^{} + \overline{L}_i y(k) + |\overline{L}_i| E_v V(k)$$

$$+ E_w \overline{w}(k) + \overline{E}\mathcal{O}y(k+1) + |\overline{L}_i| E_v V(k+1)] - \left(\sum_{i=1}^{m} \xi_i(\rho(k))(A_i + \Delta A_i)x(k) + B_i u(k) \right. \tag{27}$$

$$\left. + E_w w(k) + \overline{E}\mathcal{O}y(k+1) - \overline{E}\mathcal{O}E_v v(k+1) \right),$$

such that the resulting error equations are the following:

$$\underline{e}(k+1) = \sum_{i=1}^{m} \xi_i(\rho(k)) \left[(A_i - \underline{L}_i C)\underline{e}(k) + \Delta A_i x(k) - \underbrace{\Delta A_i x(k)}_{} + E_w(w(k) - \underline{w}(k)) \right.$$

$$\left. + |\underline{L}_i| E_v V(k) - \underline{L}_i E_v v(k) - \overline{E}\mathcal{O}E_v v(k+1) + \underline{L}_i E_v V(k+1) \right], \tag{28}$$

$$\overline{e}(k+1) = \sum_{i=1}^{m} \xi_i(\rho(k)) \left[(A_i - \overline{L}_i C)\overline{e}(k) + \overbrace{\Delta A_i x(k)}^{} - \Delta A_i x(k) + E_w(\overline{w}(k) - w(k)) \right.$$

$$\left. + \overline{L}_i E_v v(k) + |\overline{L}_i| E_v V(k) - \overline{E}\mathcal{O}E_v v(k+1) + \underline{L}_i E_v V(k+1) \right]. \tag{29}$$

By convenience, the estimation error given by equations (28) and (29) are rewritten as follows

$$\varepsilon(k+1) = \sum_{i=1}^{m} \xi_i(\rho(k))[G_i \varepsilon(k) + \Theta_{\Delta A} + H_i \delta(k)] + \Phi \delta(k+1), \tag{30}$$

$$\varepsilon(k) = \begin{bmatrix} \underline{e}(k) \\ \overline{e}(k) \end{bmatrix}, \quad G_i = \begin{bmatrix} A_i - \underline{L}_i C & 0 \\ 0 & A_i - \overline{L}_i C \end{bmatrix}, \quad \Theta_{\Delta A} = \begin{bmatrix} \Delta A_i x(k) - \underbrace{\Delta A_i x(k)}_{} \\ \overbrace{\Delta A_i x(k)}^{} - \Delta A_i x(k) \end{bmatrix},$$

$$H_i = \begin{bmatrix} \begin{bmatrix} E_w & -\underline{L}_i E_v & |\underline{L}_i| E_v \end{bmatrix} & 0 \\ 0 & \begin{bmatrix} E_w & \overline{L}_i E_v & |\overline{L}_i| E_v \end{bmatrix} \end{bmatrix},$$

$$\Phi = \begin{bmatrix} \begin{bmatrix} 0 & -\overline{E}\mathcal{O}E_v & E_v \end{bmatrix} & 0 \\ 0 & \begin{bmatrix} 0 & -\overline{E}\mathcal{O}E_v & E_v \end{bmatrix} \end{bmatrix}, \quad \delta(k) = \begin{bmatrix} \begin{bmatrix} w(k) - \underline{w}(k) \\ v(k) \\ V(k) \end{bmatrix} & \begin{bmatrix} \overline{w}(k) - w(k) \\ v(k) \\ V(k) \end{bmatrix}^T \end{bmatrix}.$$

To show that the observer is stable and robust, the following Lyapunov quadratic function for stability analysis is proposed:

$$V_1(\varepsilon(k)) = \varepsilon(k)^T P \varepsilon(k) > 0 \quad \text{with} \quad P = P^T > 0, \tag{31}$$

whose increment function corresponds to

$$\Delta V_1(\varepsilon(k)) = V_1(\varepsilon(k+1)) - V_1(\varepsilon(k)) \longleftrightarrow \Delta V_1(\varepsilon(k)) = \varepsilon(k+1)^T P \varepsilon(k+1) - \varepsilon(k)^T P \varepsilon(k), \tag{32}$$

Thus, the the stability condition requires $\Delta V_1(\varepsilon(k)) \leq 0$, i.e.,

$$\Delta V_1(\varepsilon(k)) = \sum_{i=1}^{m}\sum_{j=1}^{m} \xi_i(\rho(k))\xi_j(\rho(k))([G_i \varepsilon(k) + \Theta_{\Delta A} + H_i \delta(k)] + \Phi \delta(k+1))^T P,$$
$$([G_i \varepsilon(k) + \Theta_{\Delta A} + H_i \delta(k)] + \Phi \delta(k+1)) - \varepsilon(k)^T P \varepsilon(k) \leq 0. \tag{33}$$

If each function is substituted, Equation (33) can be expressed as:

$$\sum_{i=1}^{m}\sum_{j=1}^{m} \xi_i(\rho(k))\xi_j(\rho(k))\varepsilon(k)^T(G_i^T P G_j - P)\varepsilon(k) + \Theta_{\Delta A}^T P \Theta_{\Delta A} + \delta(k)^T H_i^T P H_j \delta(k)$$
$$+ \delta(k+1)^T \Phi^T P \Phi \delta(k+1) + 2\varepsilon(k)^T (G_{i,j}^T P \Theta_{\Delta A} + G_{i,j}^T P H_j \delta(k) + G_{i,j}^T P \Phi \delta(k+1)) \tag{34}$$
$$+ 2\Theta_{\Delta A}^T (PH_j \delta(k) + P\Phi \delta(k+1)) + 2\delta(k)^T (H_{i,j}^T P \Phi \delta(k+1)) \leq 0.$$

Furthermore, for the unknown input T–S interval observer design, the criterion H_∞ for the robust estimation problem of T-S system is considered to minimize the effects of noise and disturbance signals:

$$\lim_{k \to \infty} \varepsilon(k) = 0 \quad \text{for} \quad \delta(k) = 0 \quad \forall k, \tag{35}$$

$$\| \varepsilon(k) \|_2 < \beta \| \delta(k) \|_2 \quad \text{for} \quad \delta(k) \neq 0, \quad \xi(0) = 0, \tag{36}$$

where $\beta = \begin{bmatrix} \psi \\ \alpha \end{bmatrix}$ correspond to a vector for minimizing the disturbance and noise. The criterion H_∞ corresponds to the following function:

$$\varepsilon(k)^T \varepsilon(k) - \beta^2 \delta(k)^T \delta(k) - \beta^2 \delta(k+1)^T \delta(k+1) \leq 0, \tag{37}$$

such that the increment of the Lyapunov function results in

$$V_1(\varepsilon(k+1)) - V_1(\varepsilon(k)) + \varepsilon(k)^T \varepsilon(k) - \beta^2 \delta(k)^T \delta(k) - \beta^2 \delta(k+1)^T \delta(k+1) \leq 0. \tag{38}$$

In addition to considering the stability analysis and robustness, the next condition is considered for the estimation speed $\Delta V(\varepsilon(k)) \leq (\epsilon_1 P - \gamma)\Delta A_i(k)$ for all trajectory, equivalent to

$$\sum_{i=1}^{m}\sum_{j=1}^{m} \xi_i(\rho(k))\xi_j(\rho(k))\varepsilon(k)^T(G_i^T P G_j - P + I)\varepsilon(k) + \Theta_{\Delta A}^T P \Theta_{\Delta A} + \delta(k)^T H_i^T P H_j \delta(k)$$
$$+ \delta(k+1)^T \Phi^T P \Phi \delta(k+1) + 2\varepsilon(k)^T (G_{i,j}^T P \Theta_{\Delta A} + G_{i,j}^T P H_j \delta(k) + G_{i,j}^T P \Phi \delta(k+1)) \tag{39}$$
$$+ 2\Theta_{\Delta A}^T (PH_j \delta(k) + P\Phi \delta(k+1)) + 2\delta(k)^T (H_{i,j}^T P \Phi \delta(k+1)) - \beta^2 \delta(k)^T \delta(k)$$
$$- \beta^2 \delta(k+1)^T \delta(k+1) + \gamma \Theta_{\Delta A}^T \Theta_{\Delta A} - \epsilon_1 \Theta_{\Delta A}^T P \Theta_{\Delta A} \leq 0,$$

whereas in Equation (39) it can be seen that $\Theta_{\Delta A}^T P \Theta_{\Delta A}$ is a global Lipschitz function such that

$$\underline{f}(\underline{x}, \overline{x}) = (\underline{A}_i^+ - \overline{A}_i^+)\underline{x}^+ - \underline{A}_i^- \overline{x}^+ + \overline{A}_i^- \underline{x}^-, \tag{40}$$

$$\bar{f}(\underline{x},\bar{x}) = (\overline{A_i^+} - \underline{A_i^+})\bar{x}^- - \overline{A_i^-}\underline{x}^+ + \underline{A_i^-}\underline{x}^-, \tag{41}$$

$$|f(\underline{x},\bar{x})| \leq \|\Delta\underline{A_i^+} - \overline{\Delta A_i^+}\|_2 |\underline{x}| + (\|\Delta\underline{A_i^-}\|_2 + \|\overline{\Delta A_i^-}\|_2)|\bar{x}|, \tag{42}$$

$$|\bar{f}(\underline{x},\bar{x})| \leq \|\overline{\Delta A_i^+} - \Delta\underline{A_i^+}\|_2 |\bar{x}| + (\|\overline{\Delta A_i^-}\|_2 + \|\Delta\underline{A_i^-}\|_2)|\underline{x}|, \tag{43}$$

and the resulting functions are given by

$$\eta = 2(\|\Delta\underline{A_i^+} - \overline{\Delta A_i^+}\|_2| + \|\Delta\underline{A_i^-}\|_2 + \|\overline{\Delta A_i^-}\|_2). \tag{44}$$

Consequently, the resulting incremental Lyapunov function can be rewritten as follows

$$\sum_{i=1}^{m}\sum_{j=1}^{m} \xi_i(\rho(k))\xi_j(\rho(k))\varepsilon(k)^T(G_i^T PG_j - P + I + \gamma\eta^2 I)\varepsilon(k) + \delta(k)^T H_i^T PH_j\delta(k) + \delta(k+1)^T \Phi^T P\Phi\delta(k+1) \\ + 2\varepsilon(k)^T(G_{i,j}^T P\Theta_{\Delta A} + G_{i,j}^T PH_j\delta(k) + G_{i,j}^T P\Phi\delta(k+1)) + 2\Theta_{\Delta A}^T(PH_j\delta(k) + P\Phi\delta(k+1)) + 2\delta(k)^T(H_{i,j}^T \\ P\Phi\delta(k+1)) - \beta^2\delta(k)^T\delta(k) + \gamma\Theta_{\Delta A}^T\Theta_{\Delta A} - \epsilon_1\Theta_{\Delta A}^T P\Theta_{\Delta A} - \beta^2\delta(k+1)^T\delta(k+1) \leq 0, \tag{45}$$

and can be expressed in the following form:

$$\begin{bmatrix} \varepsilon(k) \\ \Theta_{\Delta A} \\ \delta(k) \\ \delta(k+1) \end{bmatrix}^T \begin{bmatrix} G_i^T PG_j - P + \eta^2 I + I & G_{i,j}^T P & G_i^T PH_i & G_i^T P\Phi \\ PG_{i,j} & \gamma I - \epsilon_1 P & PH_i & P\Phi \\ H_i PG_i & H_i P & H_i^T PH_i - \beta^2 I & H_i^T P\Phi \\ \Phi PG_i & \Phi P & \Phi PH_i & \Phi P\Phi - \beta^2 I \end{bmatrix} \begin{bmatrix} \varepsilon(k) \\ \Theta_{\Delta A} \\ \delta(k) \\ \delta(k+1) \end{bmatrix} \leq 0 \tag{46}$$

To relax the conservatism of (46), the following theorem is considered.

Theorem 2. *There exists a symmetric matrix $P > 0$ such that [20]*

$$A^T PA - P < 0, \tag{47}$$

and a matrix G such that the following inequality implies (47)

$$\begin{bmatrix} -P & A^T G^T \\ GA & P - G - G^T \end{bmatrix} < 0, \tag{48}$$

Consequently, by applying this theorem, inequality (46) is equivalent to

$$\begin{bmatrix} I - P + \eta^2 I & 0 & 0 & 0 & G_{i,j}^T Q^T \\ 0 & \gamma I - \epsilon_1 P & PH_i & P\Phi & 0 \\ 0 & H_i^T P & -\beta^2 I & 0 & H_i^T Q^T + \Phi^T Q^T \\ 0 & \Phi^T P & 0 & -\beta^2 I & \Phi^T Q^T \\ QG_{i,j} & 0 & QH_i + Q\Phi & Q\Phi & P - Q - Q^T \end{bmatrix} \leq 0, \tag{49}$$

such that denoting the inequality (49) as $\phi_{i,j}$, it follows

$$\sum_{i=1}^{m}\sum_{j=1}^{m} \xi_i(\rho(k))\xi_j(\rho(k)) \begin{bmatrix} \varepsilon(k) \\ \Theta_{\Delta A} \\ \delta(k) \\ \delta(k+1) \end{bmatrix}^T \phi_{i,j} \begin{bmatrix} \varepsilon(k) \\ \Theta_{\Delta A} \\ \delta(k) \\ \delta(k+1) \end{bmatrix} \leq 0. \tag{50}$$

In the inequality, (50) a bilinearity between the GQ matrices appears as can be been in

$$\left(\underbrace{\begin{bmatrix} A_{i,j} + \Delta A_{i,j}^+ & 0 \\ 0 & A_{i,j} + \overline{\Delta A}_{i,j}^+ \end{bmatrix}}_{G_{i,j}} - \underbrace{\begin{bmatrix} \underline{L}_i & 0 \\ 0 & \overline{L}_i \end{bmatrix} \begin{bmatrix} C & 0 \\ 0 & C \end{bmatrix}}_{\Gamma} \right) \overbrace{\begin{bmatrix} Q & 0 \\ 0 & Q \end{bmatrix}}^{Q} \leq 0. \tag{51}$$

To eliminate the bilinearity that there exists with $\underline{L}_i, \overline{L}_i$ and Q matrices, it is possible to use the following change of variables $\underline{W}_i = Q\underline{L}_i$ and $\overline{W}_i = Q\overline{L}_i$. Consequently, the following linear inequality is obtained

$$QG_{i,j} - W_i \Gamma < 0, \tag{52}$$

where W_i correspond to $W_i = \begin{bmatrix} \underline{W}_i & 0 \\ 0 & \overline{W}_i \end{bmatrix}$. Finally, the inequality (21) is the result of using [21], which relaxes the double sum problem. □

4. Simulation Results

4.1. Case Study

A DC motor will be used to illustrate the fault estimation proposed in this paper. The following nonlinear mathematical model represents the dynamics of DC motor [22]:

$$\begin{aligned}
\dot{i}_a(t) &= -\frac{R_a}{L}i_a(t) - \frac{K_e}{L}v_m(t) + \frac{1}{L}u(t), \\
\dot{v}_m(t) &= \frac{K_T}{J_1}i_a(t) - \left(\frac{f_r - f_p v_m(t)}{J_1}\right)v_m(t) - \frac{T_0(t) - T_2(t)}{J_1}.
\end{aligned} \tag{53}$$

where $i_a(t)$ and $v_m(t)$ are the armature current and the rotational speed, $u(t)$ is the input voltage, $T_2(t)$ and $T_0(t)$ correspond to the load and non-load torque. Table 1 summarizes the model parameter values.

Table 1. Parameters of a DC motor.

Parameter	Value
L	850×10^{-3} H
R_a	$1.02\ \Omega$
K_T	0.1 N·m/A
f_p	0.000000075 N/rpm^2
f_r	0.0000035 N/rpm
K_e	0.0134 V/rpm
J_1	0.00668933 N·m·s

L correspond to the inductance, R_a is the armature resistance, K_T is the torque-current coefficient, f_p is the friction coefficient (due to aerodynamics), K_e is the back-emf coefficient, f_r is the friction coefficient (due to the bearing lubrication condition) and J_1 is the normalized inertial moment of the rotor.

The nonlinear model (53) can be transformed first into a continuous T–S representation (3) considering the following assumptions:

Assumption 1. *The torque T_0 and T_2 are considered to be unknown. Therefore, it is necessary to decouple their effect.*

Assumption 2. *The rotational speed v_m is a measurement and is considered as the scheduling parameter.*

Assumption 3. *The armature current i_a is the measured output.*

Consequently, by considering that the rotational speed is scheduling variable $\rho(k) = v_m(k) = x_2(k)$ varying in the interval $\rho(k) \in [\underline{\rho}\ \overline{\rho}]$, being $\underline{\rho} = 100$ and $\overline{\rho} = 300]$ the minimal and maximal rotational speeds. The results T–S representation (3) has the following matrices:

$$A_1 = \begin{bmatrix} -\frac{R_a}{L} & -\frac{K_e}{L} \\ \frac{K_T}{J_1} & -\left(\frac{f_r + f_p \overline{\rho}}{J_1}\right) \end{bmatrix}, A_2 = \begin{bmatrix} -\frac{R_a}{L} & -\frac{K_e}{L} \\ \frac{K_T}{J_1} & -\left(\frac{f_r + f_p \underline{\rho}}{J_1}\right) \end{bmatrix}, B = \begin{bmatrix} \frac{1}{L} \\ 0 \end{bmatrix},$$

$$E = \begin{bmatrix} \frac{1}{L} & 0 & 0 \\ 0 & -\frac{1}{J_1} & -\frac{1}{J_1} \end{bmatrix}, E_w = \begin{bmatrix} \frac{1}{L} \\ 0 \end{bmatrix}, E_v = 0.98,\ C = \begin{bmatrix} 1 & 0 \\ 0 & 1 \end{bmatrix},$$

$$\Delta A_i = 0.01 A_i;\ \text{and}\ f_{\overline{E}} = \begin{bmatrix} f(k) & T_0(k) & T_2(k) \end{bmatrix}^T.$$

The previous continuous-time T–S model can be expressed in discrete time with a sampling time $T_s = 1$. The resulting matrices are

$$A_1 = \begin{bmatrix} 0.2471 & -0.0088 \\ 8.3671 & 0.9178 \end{bmatrix}, A_2 = \begin{bmatrix} 0.2480 & -0.0086 \\ 8.1747 & 0.8820 \end{bmatrix}, B = \begin{bmatrix} 0.6591 \\ 5.9246 \end{bmatrix}, \Delta A_1^+ = \begin{bmatrix} 0.0024 & -0.00008 \\ 0.0836 & 0.0091 \end{bmatrix},$$

$$\Delta A_2^+ = \begin{bmatrix} 0.0024 & -0.00008 \\ 0.0817 & 0.0088 \end{bmatrix}, \Delta A_1^- = \begin{bmatrix} -0.0024 & -0.00008 \\ -0.0836 & -0.0091 \end{bmatrix}, \Delta A_2^- = \begin{bmatrix} -0.0024 & 0.00008 \\ -0.0817 & -0.0088 \end{bmatrix},$$

$$\Delta \underline{A}_1^- = \begin{bmatrix} -0.0024 & 0 \\ -0.0836 & -0.0091 \end{bmatrix}, \Delta \underline{A}_2^- = \begin{bmatrix} -0.0024 & 0 \\ -0.0817 & -0.0088 \end{bmatrix}, \Delta \underline{A}_1^+ = \begin{bmatrix} 0 & 0.8823 \\ 0 & 0 \end{bmatrix} \times 10^{-4},$$

$$\Delta \underline{A}_2^+ = \begin{bmatrix} 0 & 0.8620 \\ 0 & 0 \end{bmatrix} \times 10^{-4}, \Delta \overline{A}_1^- = \begin{bmatrix} 0 & -0.8823 \\ 0 & 0 \end{bmatrix} \times 10^{-4}, \Delta \overline{A}_2^- = \begin{bmatrix} 0 & -0.8620 \\ 0 & 0 \end{bmatrix} \times 10^{-4},$$

$$\Delta \overline{A}_1^+ = \begin{bmatrix} 0.0024 & 0 \\ 0.0836 & 0.0091 \end{bmatrix} \times 10^{-4}, \Delta \overline{A}_2^+ = \begin{bmatrix} 0.0024 & 0 \\ 0.0817 & 0.0088 \end{bmatrix} \times 10^{-4},$$

$$\overline{E} = \begin{bmatrix} 0.0117 & 0 & 0 \\ 0 & -1.4949 & -1.4949 \end{bmatrix} \times 10^2,\ E_w = \begin{bmatrix} 1.1764 \\ 0 \end{bmatrix}, E_v = \begin{bmatrix} 0.08 \\ 0.08 \end{bmatrix}.$$

The solution of LMIs (19)–(21) of Theorem 1 (considering $\gamma = 15$, $\epsilon_1 = 36.46$, $\eta = 0.337$ and $\beta = \begin{bmatrix} 5.6214 \\ 4.1365 \end{bmatrix}$) lead to the following solution

$$P = \begin{bmatrix} 3.6658 & 0.8521 \\ 0.8521 & 6.1022 \end{bmatrix}, Q = \begin{bmatrix} 43.6972 & 14.9802 \\ 14.9802 & 19.6044 \end{bmatrix},$$

$$\underline{L}_1 = \begin{bmatrix} 0.0099 & -0.0111 \\ -0.1453 & 0.2328 \end{bmatrix}, \underline{L}_2 = \begin{bmatrix} -0.0085 & -0.0245 \\ 0.1847 & 0.3049 \end{bmatrix},$$

$$\overline{L}_1 = \begin{bmatrix} -0.1301 & -0.1171 \\ -0.8238 & -0.8665 \end{bmatrix} \times 10^{-13},\ \overline{L}_2 = \begin{bmatrix} 0.07489 & 0.2300 \\ 0.0382 & 0.2643 \end{bmatrix} \times 10^{-13},$$

The initial conditions for the T–S unknown input interval observer are $\underline{\hat{x}}(0) = \begin{bmatrix} 18 & 250 \end{bmatrix}^T$, $\overline{\hat{x}}(0) = \begin{bmatrix} 5 & 150 \end{bmatrix}^T$, $\underline{\hat{f}}(0) = \begin{bmatrix} 0.01 & 0.01 & 0.001 \end{bmatrix}$, $\overline{\hat{f}}(0) = \begin{bmatrix} 0.001 & 0.02 & 0 \end{bmatrix}$. Additionally, the system disturbance system and output noise is considered to be bounded with the following bounds: $-0.98 \le w(k) \le 0.98$ and $|v(k)| \le 0.8$.

4.2. Experimental Tests

Two scenarios are considered for the evaluation of the interval observer. The armature current $i_a(t)$, measurable via an oscilloscope, and the rotational speed $v_m(t)$ of the motor, measurable via an incremental encoder associated with an FPGA myRIO-1900 board of National Instruments is used for implementing the proposed approach.

In the evaluation tests, the laboratory prototype shown in Figure 1 is used. This prototype consists of a DC motor available at the TecNM/CENIDET in Mexico (1) coupled to a bearing train (2), and an incremental encoder (3) through a band, whose mathematical model is presented in Equation (53). The results show the good performance of the interval observer in the event of an actuator fault.

In the first evaluation test, the DC motor is powered with 14 V at time instant 390 s, an abrupt fault, almost instantaneous, is introduced in the motor supply voltage via a programmable testing power source. The fault in the motor input produces a decrease of 3.5 V.

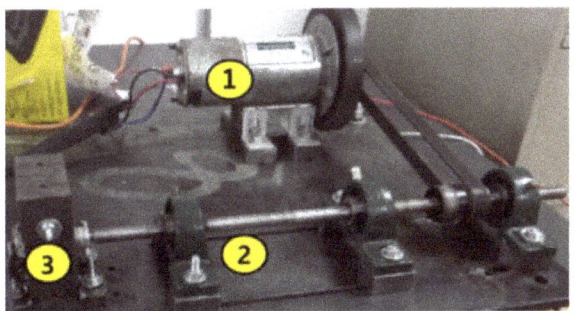

Figure 1. Laboratory prototype.

Figure 2 shows the measurement of the armature current and the limits (upper and lower) estimated by the interval observer. It can be seen in the figure that the current and limits slightly change their value in the presence of the fault.

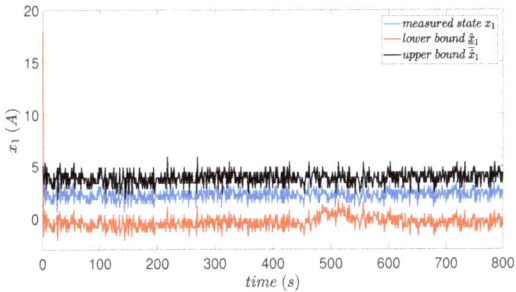

Figure 2. Measurement of armature current $x_1(k)$ and estimation of interval bounds.

Figure 3 shows that the motor speed signal and the limits (upper and lower), estimated by the interval observer (18), present a fairly close dynamic behavior and the speed is always kept within the limits. When the fault disappears, the speed signal recovers its nominal value in approximately 120 s, with the dynamics of the motor coupled to a bearing train.

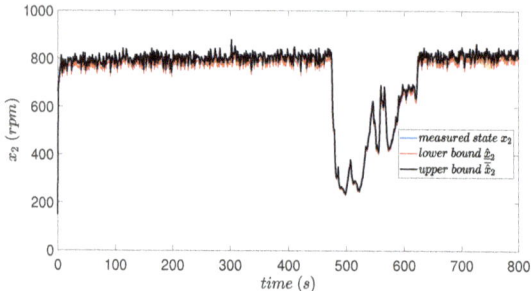

Figure 3. Measurement of rotational speed $x_2(k)$ and estimation of interval bounds.

Figure 4 shows the estimated limits for the input fault, which corresponds to a change in the motor supply voltage. The limits are kept at a value of zero in the absence of failure and change their value when the fault is present.

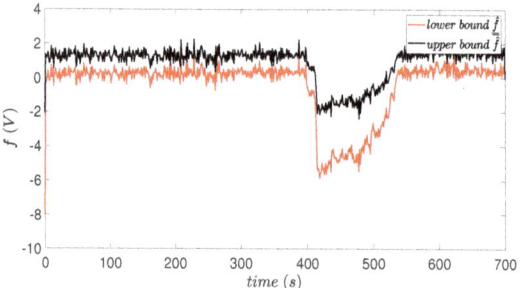

Figure 4. Estimation of the bounds for the input voltage fault.

Figures 5 and 6 show the estimated values of parameters T_0 and T_2, of the parameter vector $\theta(k)$. It can be observed that these parameters remain relatively constant (around 0 and 0.5, respectively) and in the presence of the fault their values are modified. When the fault disappears, they converge again to their initial values.

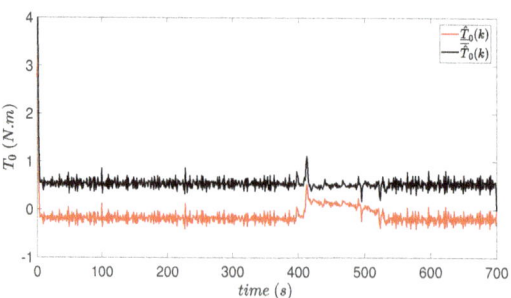

Figure 5. Estimation of $T_0(k)$.

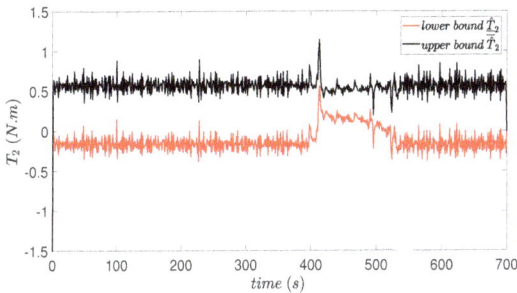

Figure 6. Estimation of $T_2(k)$.

Figure 7 shows the dynamic behavior of the membership functions, which meet the conditions described in Equation (33).

In the second evaluation test, the DC motor is powered with 15 V at time instant 420 s. An intermittent fault occurs in the supply voltage to the DC motor, caused by interruptions in the connection of the power supply.

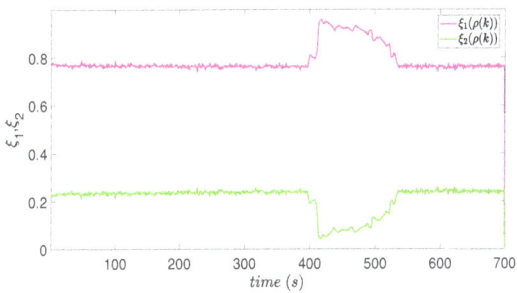

Figure 7. Membership functions.

Figure 8 shows the dynamic behavior of the armature current signal. Figure 9 shows the variations of the motor rotational speed signal. The current signal and the speed signal, both measurable, are maintained within their respective estimated intervals, in the presence of a fault.

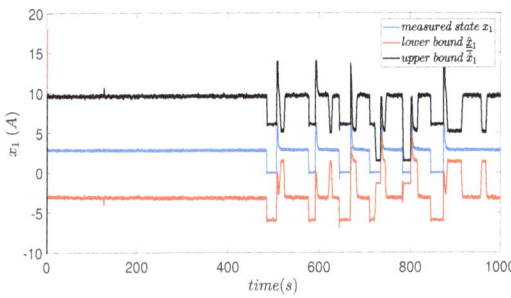

Figure 8. Measurement of armature current $x_1(k)$ and estimation of interval bounds.

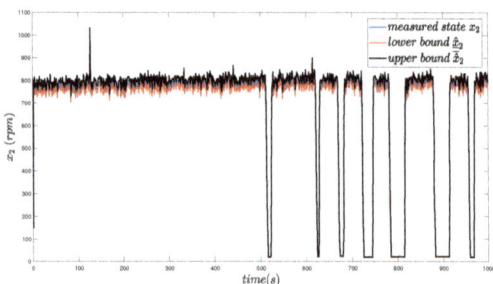

Figure 9. Measurement of rotational speed $x_2(k)$ and estimation of interval bounds.

Figure 10 shows the evolution of the estimated bounds for the input fault, which corresponds to change in the the voltage of the motor power supply. The limits are kept at a value centered around zero in the absence of fault and change their value when the fault is present.

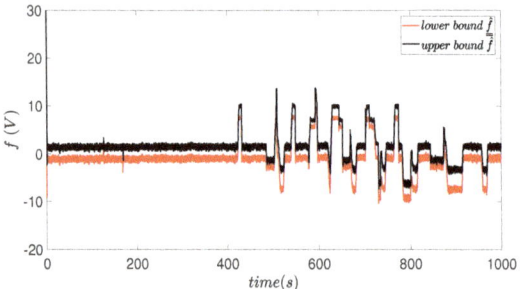

Figure 10. Estimation of the limits of input voltage fault.

Figures 11 and 12 show the estimated values of parameters T_0 and T_2, of the parameter vector $\theta(k)$.

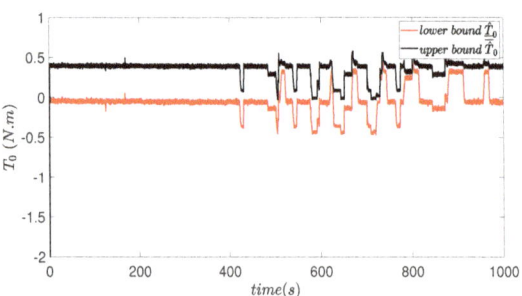

Figure 11. Upper and lower bound estimations of $T_0(k)$.

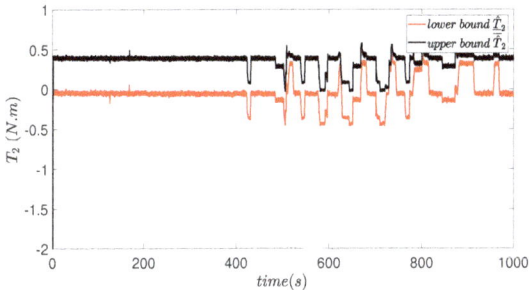

Figure 12. Upper and lower bound estimations of $T_2(k)$.

Figure 13 shows the dynamic behavior of the membership functions, which meet the conditions described in Equation (2).

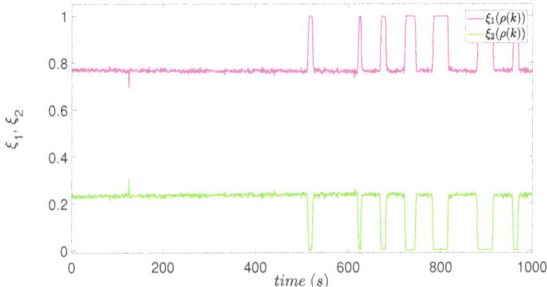

Figure 13. Membership functions.

5. Conclusions

A discrete-time unknown-input interval observer is proposed for a system modeled in T–S form with uncertainties. This observer allows the simultaneous estimation of unmeasured states, actuator and system faults despite disturbances and measurement noise. The structure of the proposed discrete-time T–S model has four additional terms: Three terms in the dynamic structure corresponding to the fault, disturbance and parametric uncertainty, and an additive noise term in the output (measurement noise). The conditions for the existence of the observer are formally given to guarantee the observer stability. Such conditions are derived through a Lyapunov analysis combined with a LMI formulation. The proposed discrete-time interval observer approach is experimentally evaluated by estimating the upper and lower bounds of a torque load perturbation, a friction parameter and a fault in the input voltage, in a permanent magnet DC motor.

The main advantage of the proposed T–S interval observer with respect to Kalman or Luenberger-like observers is that a great amount of nonlinear models can be transformed into the Takagi–Sugeno form, with a consequent benefit of preserving the model dynamics. This feature allows us to use this observer for a great number of nonlinear systems, in contrast with Kalman or Luenberger-like observes which requires linear or linearized systems to be implemented.

Author Contributions: All the authors have equally contributed. All authors have read and agreed to the published version of the manuscript.

Funding: This work has been partially funded by the Spanish State Research Agency (AEI) and the European Regional Development Fund (ERFD) through the projects SCAV (ref. MINECO DPI2017-88403-R) and also by EU INTERREG POCTEFA (2014-2020) EFA 153/16 SMART.

Conflicts of Interest: The authors declare no conflict of interest.

References

1. Ling, C.; Kravaris, C. State observer design for monitoring the degree of polymerization in a series of melt polycondensation reactors. *Processes* **2016**, *4*, 4. [CrossRef]
2. Nagy-Kiss, A.M.; Schutz, G.; Ragot, J. Parameter estimation for uncertain systems based on fault diagnosis using Takagi–Sugeno model. *ISA Trans.* **2015**, *56*, 65–74. [CrossRef] [PubMed]
3. Youssef, T.; Chadli, M.; Karimi, H.R.; Zelmat, M. Design of unknown inputs proportional integral observers for TS fuzzy models. *Neurocomputing* **2014**, *123*, 156–165. [CrossRef]
4. Peng, C.-C. Nonlinear Integral Type Observer Design for State Estimation and Unknown Input Reconstruction. *Appl. Sci.* **2017**, *7*, 67. [CrossRef]
5. Zhang, K.; Jiang, B.; Shi, P.; Xu, J. Fault estimation observer design for discrete-time systems in finite-frequency domain. *Int. J. Robust Nonlinear Control.* **2015**, *25*, 1379–1398. [CrossRef]
6. Van Nguyen, T.; Ha, C. Sensor Fault-Tolerant Control Design for Mini Motion Package Electro-Hydraulic Actuator. *Processes* **2019**, *7*, 89. [CrossRef]
7. Ifqir, S.; Ichalal, D.; Oufroukh, N.A.; Mammar, S. Robust interval observer for switched systems with unknown inputs: Application to vehicle dynamics estimation. *Eur. J. Control* **2018**, *44*, 3–14. [CrossRef]
8. Zhang, K.; Jiang, B.; Yan, X.-G.; Shen, J. Interval Sliding Mode Observer Based Incipient Sensor Fault Detection With Application to a Traction Device in China Railway High-Speed. *IEEE Trans. Veh. Technol.* **2019**, *68*, 2585–2597. [CrossRef]
9. Abadi, A.; El Amraoui, A.; Mekki, H.; Ramdani, N. Guaranteed trajectory tracking control based on interval observer for quadrotors. *Int. J. Control* **2019**, 1–17. [CrossRef]
10. Chang, Y.-C.; Tsai, C.-T.; Lu, Y.-L. Current Control of the Permanent-Magnet Synchronous Generator Using Interval Type-2 TS Fuzzy Systems. *Energies* **2019**, *12*, 2953. [CrossRef]
11. Liu, F.; Li, R.; Dreglea, A. Wind Speed and Power Ultra Short-Term Robust Forecasting Based on Takagi–Sugeno Fuzzy Model. *Energies* **2019**, *12*, 3551. [CrossRef]
12. Lendek, Z.; Guerra, T.M.; Babuska, R.; De Schutter, B. *Stability Analysis and Nonlinear Observer Design Using Takagi-Sugeno Fuzzy Models*; Springer: Berlin/Heidelberg, Germany, 2011.
13. Li, J.; Wang, Z.; Shen, Y.; Wang, Y. Interval Observer Design for Discrete-Time Uncertain Takagi—Sugeno Fuzzy Systems. *IEEE Trans. Fuzzy Syst.* **2019**, *27*, 816–823. [CrossRef]
14. Rotondo, D.; Fernandez-Canti, R.M.; Tornil-Sin, S.; Blesa, J.; Puig, V. Robust fault diagnosis of proton exchange membrane fuel cells using a Takagi-Sugeno interval observer approach. *Int. J. Hydrogen Energy* **2016**, *41*, 2875–2886. [CrossRef]
15. Ifqir, S.; Ichalal, D.; Oufroukh, N.A.; Mammar, S. Adaptive Threshold Generation for Vehicle Fault Detection using Switched TS Interval observers. *IEEE Trans. Ind. Electron.* **2019**. [CrossRef]
16. Ohtake, Hi.; Tanaka, K.; Wang, H.O. Fuzzy modeling via sector nonlinearity concept. *Integr. Comput.-Aided Eng.* **2003**, *10*, 333–341. [CrossRef]
17. Rotondo, D.; Witczak, M.; Puig, V.; Nejjari, F.; Pazera, M. Robust unknown input observer for state and fault estimation in discrete-time Takagi–Sugeno systems. *Int. J. Syst. Sci.* **2016**, *47*, 3409–3424. [CrossRef]
18. Hui, S.; Żak, S. Observer design for systems with unknown inputs. *Int. J. Appl. Math. Comput. Sci.* **2005**, *15*, 431–446.
19. Efimov, D.; Raïssi, T.; Perruquetti, W.; Zolghadri, A. Estimation and control of discrete-time LPV systems using interval observers. In Proceedings of the 52nd IEEE Conference on Decision and Control, Florence, Italy, 10–13 December 2013; pp. 5036–5041.
20. De Oliveira, M.C.; Bernussou, J.; Geromel, J.C. A new discrete-time robust stability condition. *Syst. Control Lett.* **1999**, *37*, 261–265. [CrossRef]
21. Tuan, H.D.; Apkarian, P.; Narikiyo, T.; Yamamoto, Y. Parameterized linear matrix inequality techniques in fuzzy control system design. *IEEE Trans. Fuzzy Syst.* **2001**, *9*, 324–332. [CrossRef]
22. Liu, X.-Q.; Zhang, H.-Y.; Liu, J.; Yang, J. Fault detection and diagnosis of permanent-magnet DC motor based on parameter estimation and neural network. *IEEE Trans. Ind. Electron.* **2000**, *47*, 1021–1030.

© 2020 by the authors. Licensee MDPI, Basel, Switzerland. This article is an open access article distributed under the terms and conditions of the Creative Commons Attribution (CC BY) license (http://creativecommons.org/licenses/by/4.0/).

Article

Improved Genetic Algorithm Tuning Controller Design for Autonomous Hovercraft

Huu Khoa Tran [1,2], Hoang Hai Son [3], Phan Van Duc [4], Tran Thanh Trang [5] and Hoang-Nam Nguyen [6,*]

1. Industry 4.0 Center, National Taiwan University of Science and Technology, Taipei 10607, Taiwan; khoa.tran@mail.ntust.edu.tw
2. Center for Cyber-Physical System Innovation, National Taiwan University of Science and Technology, Taipei 10607, Taiwan
3. Faculty of Mechanical, Electrical, Electronic and Automotive Engineering, Nguyen Tat Thanh University, Ho Chi Minh City 700000, Vietnam; hhson@ntt.edu.vn
4. Faculty of Automobile Technology, Van Lang University, Ho Chi Minh city 700000, Vietnam; phanvanduc@vanlanguni.edu.vn
5. Faculty of Engineering and Technology, Van Hien University, 665-667-669 Dien Bien Phu, Ho Chi Minh city 700000, Vietnam; trangtt@vhu.edu.vn
6. Modeling Evolutionary Algorithms Simulation and Artificial Intelligence, Faculty of Electrical & Electronics Engineering, Ton Duc Thang University, Ho Chi Minh City 700000, Vietnam
* Correspondence: nguyenhoangnam@tdtu.edu.vn

Received: 27 November 2019; Accepted: 26 December 2019; Published: 3 January 2020

Abstract: By mimicking the biological evolution process, genetic algorithm (GA) methodology has the advantages of creating and updating new elite parameters for optimization processes, especially in controller design technique. In this paper, a GA improvement that can speed up convergence and save operation time by neglecting chromosome decoding step is proposed to find the optimized fuzzy-proportional-integral-derivative (fuzzy-PID) control parameters. Due to minimizing tracking error of the controller design criterion, the fitness function integral of square error (ISE) was employed to utilize the advantages of the modified GA. The proposed method was then applied to a novel autonomous hovercraft motion model to display the superiority to the standard GA.

Keywords: modified GA; fuzzy-PID control; autonomous hovercraft; ISE criterion

1. Introduction

John Henry Holland, by imitating Darwin's biological evolution process, proposed the powerful stochastic global search method genetic algorithm (GA) first in 1975 [1,2]. Both the two reproduction mechanisms of genetic algorithm, including crossover and mutation, are used to find the convergence of optimal solutions. These values show an important effect on both behavior and performance. Therefore, GA instructions for choosing appropriate values are introduced by many researchers. In 1986, Grefenstette et al. proposed a method for optimizing the control gains for the genetic algorithm [3]. Then, in 1994, Srinivas and Patnaik demonstrated the adaptive probabilities of crossover and mutation [4]. In 1997, the bounded difficulty problems were considered by Harik [5]. Later in 2004, Zlochin et al. suggested model-based search for implementing combinatorial optimization [6]. In 2007, Zhang et al. adaptively adjusted crossover and mutation probabilities by utilizing a clustering-based technique [7]. Preceding this paper, GA and its innovations have been successfully deployed in a wide range of non-trivial complicated real-world issues, from optimization of flight control laws [8] to aerodynamic optimization problems [9]; from small wind turbine design [10] to path planning of a space-based of a manipulator system [11]; and from modeling collinear data [12] to ship navigation in collision situations [13].

The genetic algorithm optimization offers a powerful methodology for solving single to multivariable problems. Each GA represents a problem solution encoded by the form of binary strings or chromosomes. Its fitness is employed to measure how good of one solution by increasing the best bit patterns. Hence, the maxima/minima of fitness function value are then optimized during the GA course.

Nevertheless, the standard GA (denoted shortly as sGA) still has some drawbacks; for example, low convergence speed and premature convergence because of requiring hundreds of updated generations. Hence, we proposed a modified genetic algorithm (denoted shortly as mGA) to improve the optimal process. This modified algorithm, just in short operating generations [1,2,14], can improve the global search efficiency and increase the convergence speed of the optimal control design. Next, the proposed control design was a fuzzy-proportional-integral-derivative (F-PID) control [15–25] that comprises the fuzzy logic control (FLC) and the common PID controller. The first, fuzzy term is employed to increase the stability and the robustness of the controller design by tuning the membership functions and by selecting suitable methods of fuzzification and defuzzification [15–19,24,25]. The second, PID control term is separated in two small sub-terms: the PD is employed to maintain the system stability while the term I is utilized to eliminating the steady-state error of the controlled system response [20–23]. Based on the calculation of criterion error of the control system, the "integral of square error" (ISE) [20–22] was chosen as a fitness function to show the controller performance index.

A hovercraft, also known as an air cushion vehicle (ACV), moves smoothly on any surface [26–31] from the ground-land to the mud, water, sand, and even on ice. Because a hovercraft is very active and agile, its models was chosen for the control implementation.

To summarize, the modified GA is proposed to take advantages of the fuzzy-PID controller design in shorter operation period. Thus, the dynamic of Hovercraft models could be mobility in high stability with fast response and less error.

2. Hovercraft Prototype Model

A hovercraft, which is known as an underactuated system and named an air cushion vehicle (ACV), has rotors and a cushion, where inside air pressure enables floating and smooth movement on any surface [26–31], such as land, mud, water, sand, and even on ice. The hovercraft is very active and agile; hence, it is applied widely in the coastguard, army, rescue operations, civil engineering, etc. The hovercraft is mounted with a single tilt servomotor on the fin tail. As shown in Figure 1, the rotor duct fan of the hovercraft is settled along the y-axis, while the propeller is attached along the z-axis. Firstly, the lift propeller provides the internal cushion pressure to lift up in a long operation period. Next, the forward moving is created by the rear rotor duct-fan. Finally, the turning typically is operated by directing the thrust airflow through rotor duct fan, which is steered by a tilt servomotor placed at the rear. The subsequently generated momentum is used to maneuver the craft. Although many modern technologies are utilized, the hovercraft still requires an advanced maneuvering system to achieve optimized performance.

The hovercraft dynamic models were derived in [28–31] using right-hand convention coordinate systems. The positive x-axis covers the lateral factors, namely, sway motion or surge position, while the y-axis is the direction along the hovercraft body, covering surge motion or sway position; and the positive z-axis defines the downwards direction. The hovercraft's kinematics can be expressed as Equation (1):

$$\begin{cases} \dot{x} = u\cos\psi - v\sin\psi \\ \dot{y} = v\cos\psi + u\sin\psi \\ \dot{\psi} = r \end{cases} \quad (1)$$

where $u \in R$ and $v \in R$ represent linear velocities in surge direction and sway direction, respectively. The angular velocity is represented as $r \in R$. Using Equation (1), we can derive the kinetic energy T

and potential energy V to define the Lagrange $L = T - V$ by applying Euler–Lagrange formulation as in following equation:

$$M(q)\ddot{q} + C(q,\dot{q})q = \begin{bmatrix} F \\ \tau \\ 0 \end{bmatrix} \quad (2)$$

where τ is the torque in yaw rotation and F is the force acting along surge direction.

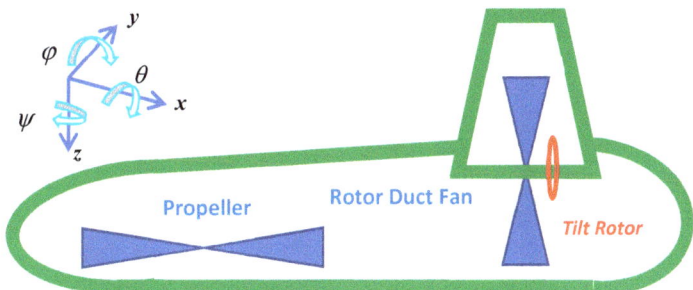

Figure 1. The hovercraft configuration model.

3. Modified GA Optimal Controller Gains

3.1. The Controller Design

Fuzzy logic concepts are extremely modest but powerful and effective for applications in the control of various machines. The fuzzy control takes advantages in stability and robustness since its aptitude is deal with the nonlinear and uncertain systems. Fuzzy control rules are designed based on the center of area method (COA) for defuzzification and Mandani's MIN–MAX inference engine type. It includes the tracking error $e(t)$ and the differential tracking error $de(t)$ as the inputs. The fuzzy control output to PID control is defined as $e_{Fuzzy}(t)$. Seven partitions of the fuzzy rule-table control, which exploit the triangular membership functions (as shown in Figure 2), are noted in Table 1, which include negative small (NS), negative medium (NM), negative big (NB), positive big (PB), positive medium (PM), positive small (PS), and zero (Z).

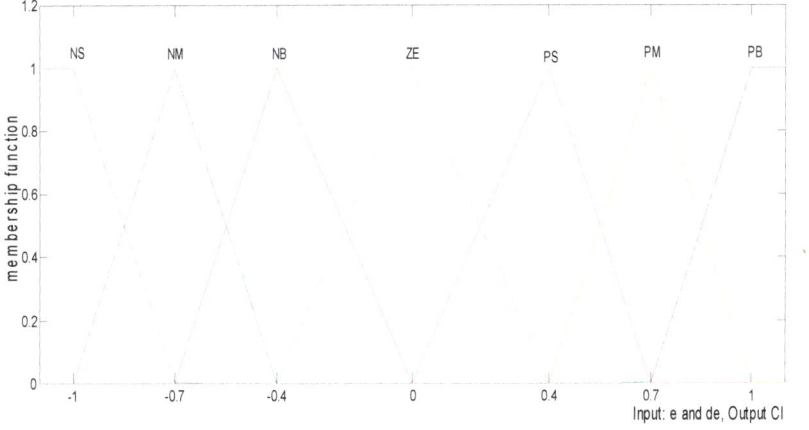

Figure 2. The fuzzification membership function.

Table 1. Fuzzy rule-table.

	$e_{Fuzzy}(t)$	NB	NM	NS	Z	PS	PM	PB
de(t)	PB	Z	PS	PM	PB	PB	PB	PB
	PM	NS	Z	PS	PM	PB	PB	PB
	PS	NM	NS	Z	PS	PM	PB	PB
	Z	NB	NM	NS	Z	PS	PM	PB
	NS	NB	NB	NM	NS	Z	PS	PM
	NM	NB	NB	NB	NM	NS	Z	PS
	NB	NB	NB	NB	NB	NM	NS	Z

PID controllers are designed to satisfy dynamic response, and reduce and/or eliminate error of physical empirical systems. Thus, the fuzzy-PID controller [24,25] output u to the system is proposed and expressed on Equation (3) as:

$$u = u_{Fuzzy} + u_{PID} = u_{Fuzzy} + k_P \times e + k_I \int edt + k_D \times \frac{d}{dt}e(t) \quad (3)$$

where u_{Fuzzy} and u_{PID} are the Fuzzy and PID controller output signals. k_P, k_I, k_D are the PID controller proportional, integral and derivative constants.

The fuzzy-PID controls is highly effective, as it is simple, has less overshoot, and is able to eliminate the steady state error, and especially smooth control signal. Its block diagram is demonstrated in Figure 3.

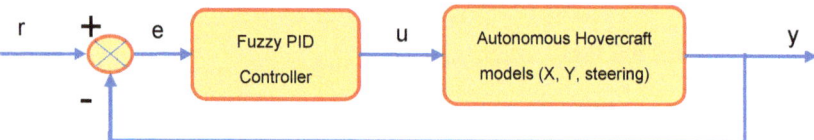

Figure 3. The controller block diagram.

3.2. Improved GA in the Optimization Process

The standard GA's characteristics [3], which includes three main parameters—mutation rate (*Rm*), crossover rate (*Rc*), and population size (*N*)—were employed to verify the optimized gains of the proposed control system. This optimal program usually takes hundreds of generations to update and find out convergence gains. Therefore, the GA is powerful, but it still has a critical drawback, which has been investigated for a solution by many researchers. By checking the GA process, it was found out that chromosome decode is not compulsory due to the evaluation of the cost function [1,2,14]. Hence, we proposed the novel modified GA method where the chromosome decoding step is totally neglected, as shown in Figure 4. This improvement makes modified GA more effective than the conventional GA in the aspects of storage (less) and convergence speed (naturally higher) [1,2,14]. The optimal process of the modified GA to Fuzzy-PID control parameters using the ISE (integral of square errror) fitness function is formulated as $ISE = \int_0^\infty e^2(t)dt$.

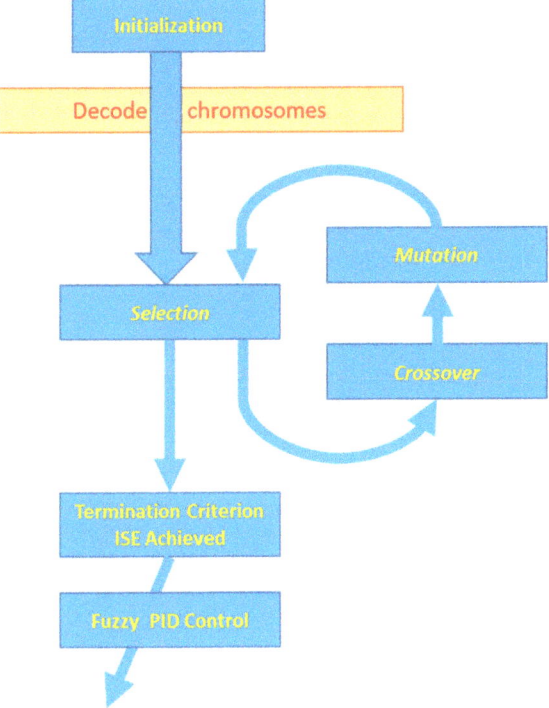

Figure 4. The proposed fuzzy-PID-mGA (fuzzy-proportional-integral-derivative modified genetic algorithm; the decode chromosome step is neglected).

4. Numerical Simulation Results

The autonomous hovercraft simulation parameters were derived from [28–31] and denoted with the mass m = 2.1 kg and the inertia moment I = 0.000257 kg.m². The modified GA had the mutation rate (Rm) of 0.08, the crossover rate (Rc) of 0.95, and one hundred individuals (n = 100). We proposed to tune the fuzzy-PID controller gains in only 20 population-generations of the simulation process. The PID gains were chosen arbitrary initialized in range (0,50) while e and de of fuzzy control changed slightly around 1; the control parameters after optimization by the mGA are in Table 2. In comparison with the standard GA, the modified GA operated in shorter generations (just 20 generations) and rapidly updated the convergence speed of fitness function.

The hovercraft was tested by moving forward (x-direction); stability was tested when it was attacked on the side, as it would be by a wave or the wind (y-direction); and steering was by pilot control (z-direction). Hence, the subsystems of the hovercraft are separated in each channel for the strategy of design the controller. Performance of the proposed controller of the autonomous Hovercraft motions are shown on surge position on x-axis in Figure 5, sway position on y-axis in Figure 6, and yaw angle on z-axis in Figure 7, respectively. The numerical simulation result, achieved in just after 20 generations, clearly proves that the proposed methodology gives a fast response, less error, and zero overshoot. Moreover, the ISE fitness function, as shown in part b of each of Figures 5–7, gives the better process of the minimum error. In all three of Figures 5–7, the modified GA error from start to finish is faster than the standard GA, as shown on Table 3. The 1/Fitness error being larger means that the speed of the convergence from start toward finish is faster. All the simulation performance work was done on the MATLAB/Simulink platform (R2018a, MathWorks, Natick, MA, USA).

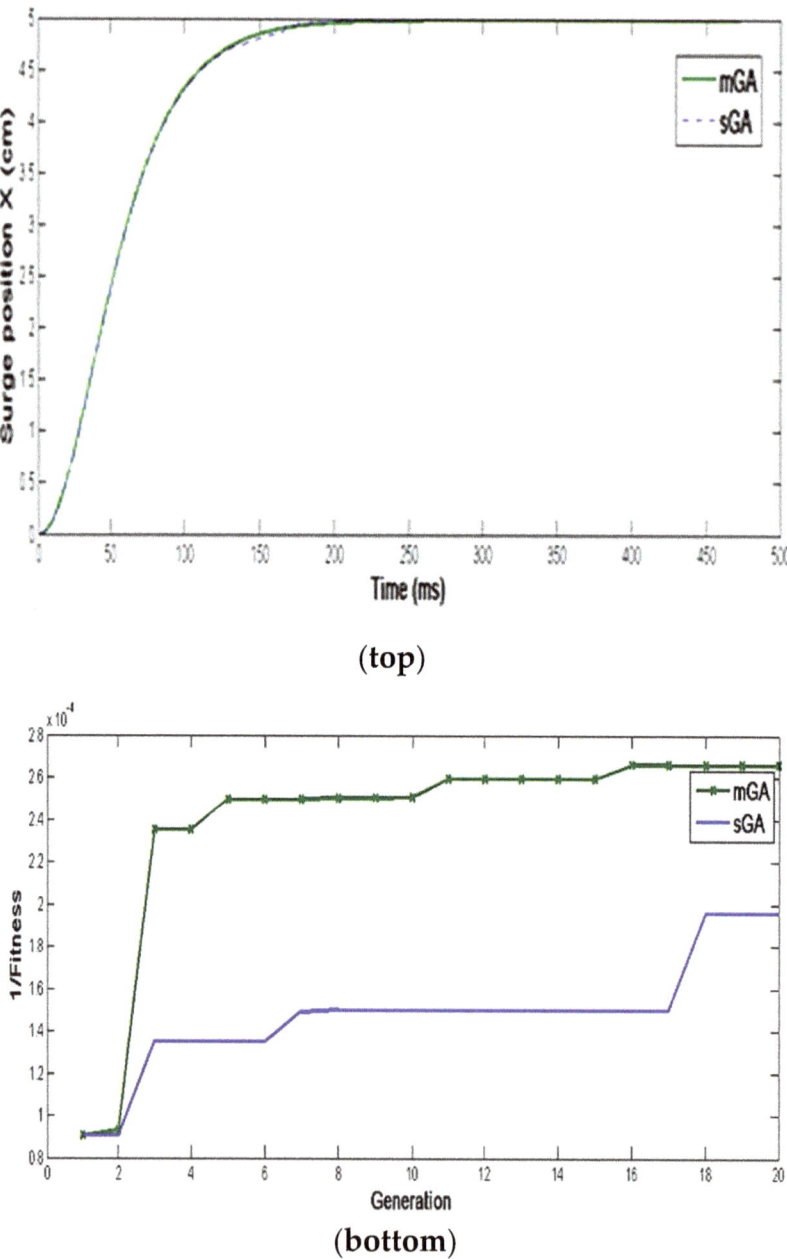

Figure 5. (**Top**) Surge—position control; (**bottom**) improved GA (mGA) versus standard GA (sGA) fitness functions.

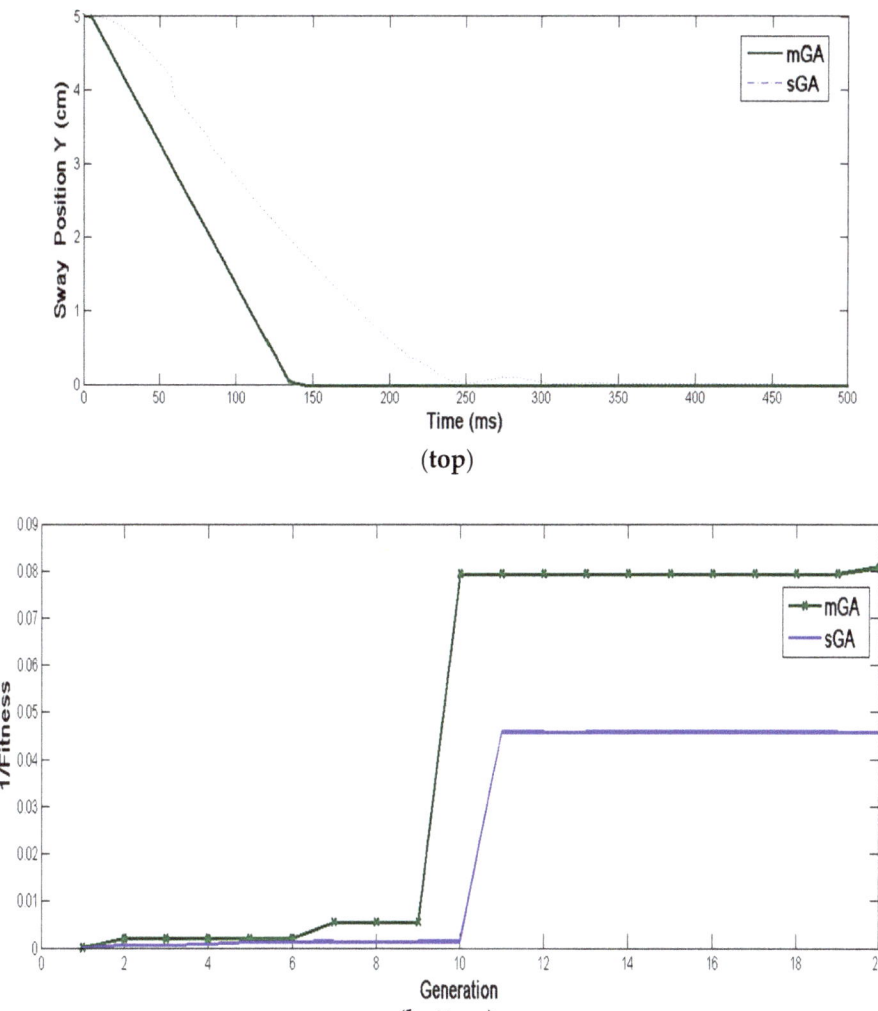

Figure 6. (**Top**) Sway—position control; (**bottom**) improved GA (mGA) versus standard GA (sGA) fitness functions.

Table 2. The fuzzy-PID-mGA tuning results.

	Generation	e	de	K_I	K_P	K_D
Surge position x (5 cm)	20	1	1	41.61	8.75	1.42
Sway position y (5 cm)	20	0.98	1.05	29.3	3.32	0.66
Steering-Yaw angle (5 degree)	20	1.1	1.03	32.8	5.91	3.26

Table 3. The error fitness function by generations.

1/Fitness Error	i Gene-Ration	sGA	mGA	i Gene-Ration	sGA	mGA
Surge position x (5 cm)	3	0.00137	0.00236	20	0.0018	0.00263
Sway position y (5 cm)	10	0.045	0.08	20	0.045	0.081
Steering-Yaw angle (5 degree)	10	0.00035	0.00067	20	0.00036	0.00095

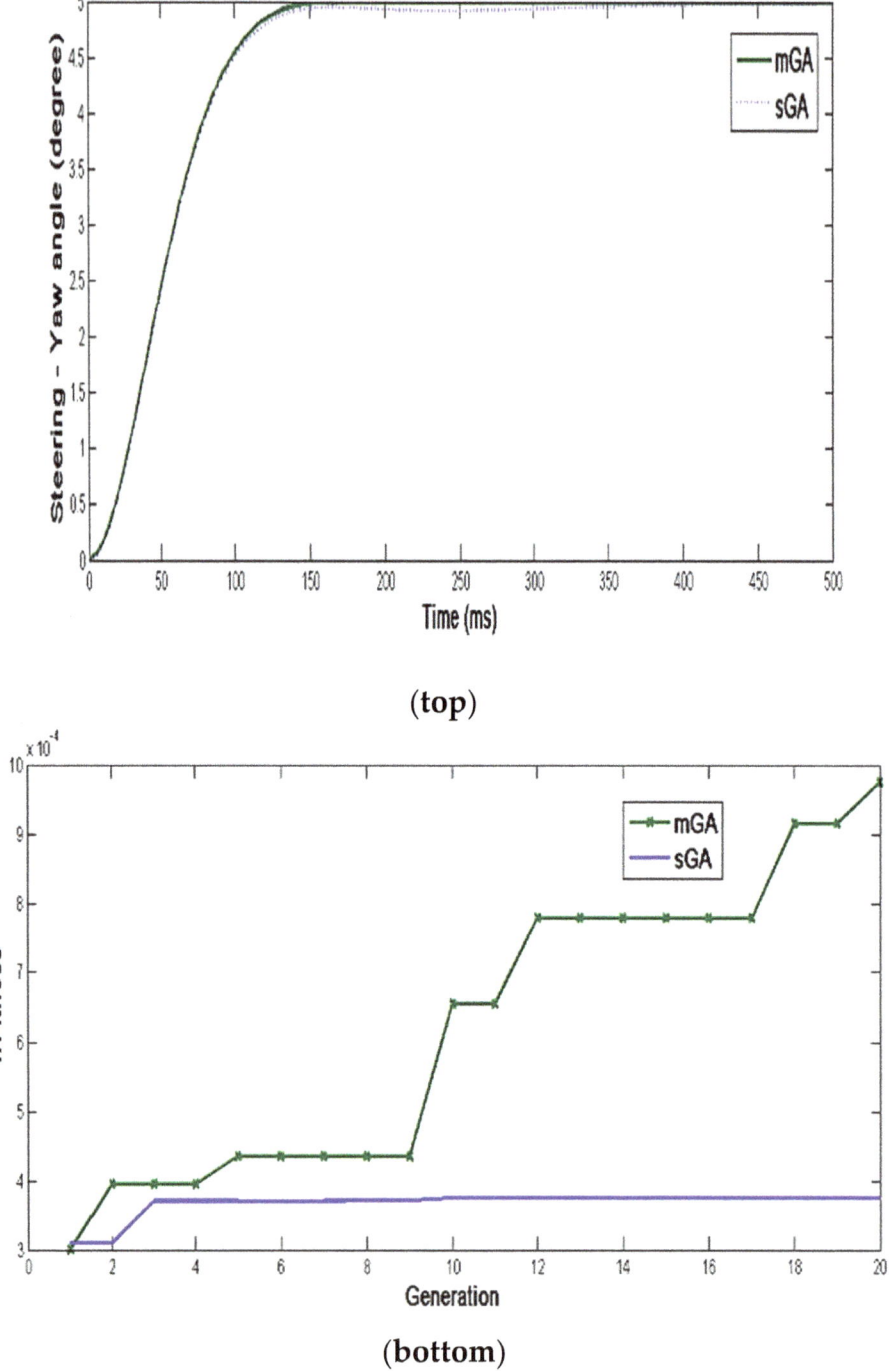

Figure 7. (**Top**) Steering—Yaw control; (**bottom**) improved GA (mGA) versus standard GA (sGA) fitness functions.

5. Conclusions

In this study, genetic algorithms, during a short operation period, were utilized to optimize the parameters of an autonomous hovercraft controller. The proposed method achieved good performances in terms of response (fast), stability (high), error (low), and overshoot (none at all). In addition, the improved GA methodology, which was implemented by make some simple changes inside the standard GA's process that neglects/eliminates the chromosome decode step, displayed better performances in convergence speed. Especially, the modified GA can update the error fitness function in a smaller number of generations. It is undeniable that the improved GA is valuable in the optimization processes, particularly in optimizing the controller parameters. In further research, the authors would like to enhance the efficiency of the modified GA by using another error criteria, such as: ITSE (integral of time weighted square error) and MSE (mean square error). The disturbances attack to system models will be also considered to verify the efficiency of the process of optimizing control parameters.

Author Contributions: Methodology, H.K.T.; Supervision, H.K.T.; Visualization, H.K.T.; Resources, P.V.D., H.H.S.; Validation, T.T.T.; Writing—Review & Editing, H.-N.N. All authors have read and agreed to the published version of the manuscript.

Funding: This work was financially supported by the "Center for Cyber-physical System Innovation" from The Featured Areas Research Center Program within the framework of the Higher Education Sprout Project by the Ministry of Education (MOE) in Taiwan.

Conflicts of Interest: The authors declare that there is no conflict of interests regarding the publication of this paper.

References

1. Yang, X.-S. *Engineering Optimization: An Introduction with Metaheuristic Applications*; John Wiley & Sons, Inc.: Hoboken, NJ, USA, 2010.
2. Haupt, R.L.; Haupt, S.E. *Practical Genetic Algorithms*, 2nd ed.; John Wiley & Sons, Inc.: Hoboken, NJ, USA, 2004.
3. Grefenstette, J.J. Optimization of Control Parameters for Genetic Algorithms. *IEEE Trans. Syst. Man Cybern.* **1986**, *16*, 122–128. [CrossRef]
4. Srinivas, M.; Patnaik, L. Adaptive Probabilities of Crossover and Mutation in Genetic Algorithms. *IEEE Trans. Syst. Man Cybern.* **1994**, *24*, 656–667. [CrossRef]
5. Harik, G. Learning Linkage to Efficiently Solve Problems of Bounded Difficulty Using Genetic Algorithms. Ph.D. Thesis, Dept. Computer Science, University of Michigan, Ann Arbor, MI, USA, 1997.
6. Zlochin, M.; Birattari, M.; Meuleau, N.; Dorigo, M. Model-Based Search for Combinatorial Optimization: A Critical Survey. *Ann. Oper. Res.* **2004**, *131*, 373–395. [CrossRef]
7. Zhang, J.; Chung, H.; Lo, W.L. Clustering-Based Adaptive Crossover and Mutation Probabilities for Genetic Algorithms. *IEEE Trans. Evol. Comput.* **2007**, *11*, 326–335. [CrossRef]
8. Al-Asasfeh, A.; Hamdan, N.; Abo-Hammour, Z. Flight Control Laws Verification Using Continuous Genetic Algorithms. *ISRN Robot.* **2013**, *2013*. [CrossRef]
9. Antunes, A.P.; Azevedo, J.L.F. Studies in Aerodynamic Optimization Based on Genetic Algorithms. *J. Aircr.* **2014**, *51*, 1002–1012. [CrossRef]
10. Hamdan, M.; Abderrazzaq, M.H. Optimization of Small Wind Turbines Using Genetic Algorithms. *Int. J. Appl. Metaheuristic Comput.* **2016**, *7*, 50–65. [CrossRef]
11. Chen, Z.; Zhou, W. Path Planning for a Space-Based Manipulator System Based on Quantum Genetic Algorithm. *J. Robot.* **2017**. [CrossRef]
12. Tang, J.; Zhang, J.; Wu, Z.; Liu, Z.; Chai, T.; Yu, W. Modeling Collinear Data Using Double-Layer GA-Based Selective Ensemble Kernel Partial Least Squares Algorithm. *Neurocomputing* **2017**, *219*, 248–262. [CrossRef]
13. Ni, S.; Liu, Z.; Cai, Y. Ship Manoeuvrability-Based Simulation for Ship Navigation in Collision Situations. *J. Mar. Sci. Eng.* **2019**, *7*, 90. [CrossRef]
14. Tran, H.K. Modified GA Tuning IPD Control for a Single Tilt Tri-Rotors UAV. *Int. Rev. Aerosp. Eng. (IREASE)* **2018**, *11*, 1–5. [CrossRef]
15. Passino, K.M.; Yurkovich, S. *Fuzzy Control*; Addison-Wesley Reading: Boston, MA, USA, 1998.
16. Precup, R.-E.; Preitl, S. PI-Fuzzy Controllers for Integral Plants to Ensure Robust Stability. *Inf. Sci.* **2007**, *177*, 4410–4429. [CrossRef]

17. Sanchez, E.N.; Becerra, H.M.; Velez, C.M. Combining Fuzzy, PID and Regulation Control for an Autonomous Mini-Helicopter. *Inf. Sci.* **2007**, *177*, 1999–2022. [CrossRef]
18. Feng, J.Z.; Li, J.; Yu, F. GA-Based PID and Fuzzy Logic Control for Active Vehicle Suspension System. *Int. J. Automot. Technol.* **2003**, *4*, 181–191.
19. Juang, Y.T.; Chang, Y.T.; Huang, C.P. Design of Fuzzy PID Controllers Using Modified Triangular Membership Functions. *Inf. Sci.* **2008**, *178*, 1325–1333. [CrossRef]
20. Martins, F.G. Tuning PID Controllers Using the ITAE Criterion. *Int. J. Eng. Educ.* **2005**, *21*, 867–873.
21. Skogestad, S. Simple Analytic Rules for Model Reduction and PID Controller Tuning. *J. Process. Control.* **2003**, *13*, 291–309. [CrossRef]
22. Tan, W.; Liu, J.; Chen, T.; Marquez, H.J. Comparison of Some Well-Known PID Tuning Formulas. *Comput. Chem. Eng.* **2006**, *30*, 1416–1423. [CrossRef]
23. Wu, Y.; Zhao, X.; Li, K.; Zheng, M.; Li, S. Energy Saving—Another Perspective for Parameter Optimization of P and PI Controllers. *Neurocomputing* **2016**, *174*, 500–513. [CrossRef]
24. Tóthová, M.; Piteľ, J. Simulation of Fuzzy Adaptive Position Controllers for Pneumatic Muscle Actuator. In Proceedings of the IEEE 13th International Symposium on Intelligent Systems and Informatics (SISY), Subotica, Serbia, 17–19 September 2015; pp. 55–59.
25. Phu, D.X.; Choi, S.-B. A New Adaptive Fuzzy PID Controller Based on Riccati-Like Equation with Application to Vibration Control of Vehicle Seat Suspension. *Appl. Sci.* **2019**, *9*, 4540. [CrossRef]
26. Sira-Ramirez, H.; Ibanez, C.A. On the Control of the Hovercraft. *Dyn. Control.* **2000**, *10*, 151–163. [CrossRef]
27. Balemi, S.; Bucher, R.; Guggiari, P.; Furlan, I.; Kottmann, M.; Chapuis, J. Rapid Control of Prototyping of a Hovercraft. In Proceedings of the MSy'02, Embedded Systems Conference, Shanghai, China, 11–16 October 2002.
28. Rashid, M.Z.A.; Aras, M.S.M.; Kassim, M.A.; Ibrahim, Z.; Jamali, A. Dynamic Mathematical Modeling and Simulation Study of Small Scale Autonomous Hovercraft. *Int. J. Adv. Sci. Technol.* **2012**, *46*, 95–114.
29. Chaos, D.; Moreno-Salinas, D.; Muñoz-Mansilla, R.; Aranda, J. Nonlinear Control for Trajectory Tracking of a Nonholonomic RC-Hovercraft with Discrete Inputs. *Math. Probl. Eng.* **2013**, *2013*, 589267. [CrossRef]
30. Garcia, D.I.; White, W.N. Control Design of an Unmanned Hovercraft for Agricultural Applications. *Int. J. Agric. Biol. Eng.* **2015**, *8*, 72–79.
31. Cabecinhas, D.; Batista, P.; Olivera, P.; Silvestre, C. Hovercraft Control with Dynamic Parameters Identification. *IEEE Trans. Control. Syst. Technol.* **2018**, *26*, 785–796. [CrossRef]

© 2020 by the authors. Licensee MDPI, Basel, Switzerland. This article is an open access article distributed under the terms and conditions of the Creative Commons Attribution (CC BY) license (http://creativecommons.org/licenses/by/4.0/).

Article

Model-Based Safety Analysis for the Fly-by-Wire System by Using Monte Carlo Simulation

Zhong Lu [1,*], Lu Zhuang [1], Li Dong [1] and Xihui Liang [2]

[1] College of Civil Aviation, Nanjing University of Aeronautics and Astronautics, Nanjing 211106, China; zhuanglu@nuaa.edu.cn (L.Z.); dong_li@nuaa.edu.cn (L.D.)
[2] Department of Mechanical Engineering, University of Manitoba, Winnipeg, MB R3T5V6, Canada; Xihui.Liang@umanitoba.ca
* Correspondence: luzhong@nuaa.edu.cn

Received: 27 November 2019; Accepted: 7 January 2020; Published: 9 January 2020

Abstract: Safety analysis is one of the important means to show compliance with airworthiness requirements. The traditional safety analysis methods are significantly dependent on analysts' skills and experiences. A model-based safety analysis approach is proposed for typical fly-by-wire (FBW) systems based on the system development model built via Simulink, by which the response of system performances can be simulated. The safety requirements of the FBW system are defined by presenting the thresholds of system performance metrics, and the effects of failure conditions on aircraft safety are determined according to the system response simulation by injecting failures or failure combinations into the Simulink model. The Monte Carlo simulation method is used to calculate the probability of unsafe conditions, whose effects are determined by the system response simulation with fault injections. Finally, a case study is used to illustrate the effectiveness and advantages of our proposed approach.

Keywords: system safety assessment; fly-by-wire system; fault injection; Monte Carlo simulation; dynamic behavior mode

1. Introduction

Safety is the most important characteristic of aviation products. The flight control system is a typical safety-critical system of modern aircraft, whose failures or malfunctions will lead to an unsafe flight path or structural failure preventing continued safe flight and landing. In the modern transport category of airplanes, fly-by-wire (FBW) systems have been widely used to replace hydro-mechanical ones. By utilizing the FBW system, pilots' commands are converted to electronic signals transmitted by wires to flight control computers, and control commands are calculated by flight control computers based on control laws to determine the movements of the actuators at each control surface. Therefore, the mechanical circuit consisting of rods, cables and pulleys is not required anymore, and the weight of the airplane can be reduced.

In aircraft or system development, the safety assessment process is an integral process that is used to show compliance with airworthiness requirements such as 14CFR/CS 23.1309, 14CFR/CS 23.1309, 14CFR 33.75, CS-E 510 and so on. At present, the safety assessment for civil airborne systems and equipment is usually conducted according to the standard ARP4761 issued by the Society of Automotive Engineer (SAE) [1,2]. In this document, it is recommended that traditional safety analysis techniques including Dependence Diagram Analysis (DDA), Fault Tree Analysis (FTA), Markov Analysis (MA), Failure Mode and Effect Analysis (FMEA) are applied in the safety assessment process [2]. These techniques are based on information synthesized from several sources including informal design models and requirements documents, and they are usually performed manually by safety engineers whose experiences and skills will affect the analysis significantly. Therefore, the results of traditional safety analysis techniques are incomplete, inconsistent and highly subjective [3].

To overcome the deficiencies of the traditional techniques, model-based safety analysis (MBSA) has been proposed. MBSA focuses on modeling the system in a formal specification (model), which can be extended by injecting failure modes of the physical system or conducting safety analysis automatically. In this way, the completeness, consistency and correctness of the safety analysis results can be ensured, and the dependency on the engineers' skills and experiences can be avoided [4].

Over the past decade, different kinds of modeling tools have been applied in the modeling of the formal specification for MBSA. These tools can be classified into three categories [5], which are graphical modeling tools, system modeling languages and failure logic modeling techniques. Graphical modeling tools include Matlab-Simulink [6–8], Modelica [9,10], Petri Net [11–13] and SCADE [3,14]; system modeling languages include SysML [15,16], AADL [17–19], AltaRica [20–22], and NuSMV [23,24]; and failure logic modeling techniques include HiP-HOPS [25,26] and failure propagation approaches [27].

Although all the above-mentioned tools can be applied in the modeling of a formal specification, some of them are used to build models for system development, such as Matlab-Simulink, Modelica, and AADL; others are used to build models for failure analysis specifically, such as AltaRica, HiP-HOPS and FPTN. In the case that the models for system development are applied, the fault can be injected to the model directly and the consistency between system development and safety analysis can be maximized. The model of a flight control system used in development is usually expressed in system dynamics (control laws), which are given in the form of differential equations, transfer functions or state equations, and Matlab-Simulink is the most widely used tool in the development of control systems to build system dynamics models. Meanwhile, other MBSA tools such as AADL, AltaRica and HiP-HOPS are suitable for describing the failure propagations of avionics systems.

In this study, an MBSA method is proposed for typical FBW systems based on the system development model built via Simulink, by which the response of system performances can be simulated. The safety requirements of the FBW system are defined by presenting the thresholds of system performance metrics, and the effects of failure conditions on aircraft safety are determined according to the system response simulation by injecting failures or failure combinations in the Simulink model. The Monte Carlo simulation method is used to calculate the probability of failure conditions (unsafe conditions), whose effects are determined by the system response simulation with fault injections.

The system safety process of an aircraft is usually composed of four parts, which are Functional Hazard Assessment (FHA), Preliminarily Aircraft/System Safety Assessment (PASA/PSSA), Aircraft/System Safety Assessment (ASA/SSA) and Common Cause Analysis (CCA). This study focuses on the ASA/SSA process, the response of the Simulink model with fault injection is used to determine the failure effects of failure modes and their combinations from FMEA, and the Monte Carlo simulation method is applied to calculate the probability of top failure conditions instead of FTA, DDA and MA. The rest of this paper is structured as follows. In Section 2, the nominal (failure-free) model of the FBW system is presented by using Simulink, and the safety requirements are defined by giving the thresholds of system responses for performance metrics. In Section 3, the typical failure modes of the FBW system are modeled by Simulink, and the fault injecting approach is proposed to extend the nominal model. In Section 4, the probability calculation of unsafe conditions based on the Monte Carlo simulation is presented as a step-by-step procedure. In Section 5, a lateral-directional flight control system is used as a case study to show the accuracy and advantages of our proposed Monte Carlo simulation-based method. In Section 6, concluding remarks are presented.

2. Nominal Model of a Typical FBW System

The model of a typical FBW system under normal operating conditions, which is called the nominal or failure-free model, is built with Simulink in this section. Then, the system performance responses under the failed configurations can be simulated by injecting the corresponding failure mode into the nominal model.

2.1. System Modeling via Simulink

In aircraft design, the development model of the FBW system as well as the aircraft dynamics are expressed by mathematical models such as state-space functions, transfer functions and differential equations, which are usually modeled with Simulink. In this section, a lateral-directional flight control system [6] is applied as an example to illustrate how to build the nominal model with Matlab-Simulink (2017a, MathWorks, Natick, MA, USA, 2017).

The lateral-directional flight control system is composed of a flight control computer subsystem, an actuation subsystem, a sensor subsystem and control surfaces. The flight control computer subsystem has a dual redundant architecture that is composed of two identical primary flight computers (PFCs). The actuation subsystem includes the left aileron actuation (LAA), the right aileron actuation (RAA) and the rudder actuation (RA). Each of them is composed of two redundant actuators that are connected with a combiner. The sensor subsystem includes the right aileron position sensors (RAPS), the left aileron position sensors (LAPS), the rudder position sensors (RPS) and the inertial measurement units (IMU). All of them have a triple modular redundant architecture. The control surfaces for lateral control include the left aileron, the right aileron and the rudder. Pilots' commands from pedals and control sticks are converted to electronic signals transmitted by wires to the two PFCs, and control commands are calculated in terms of control laws in the PFCs to determine the movements of the actuators for all control surfaces. The architecture of the lateral-directional flight control system is shown in Figure 1.

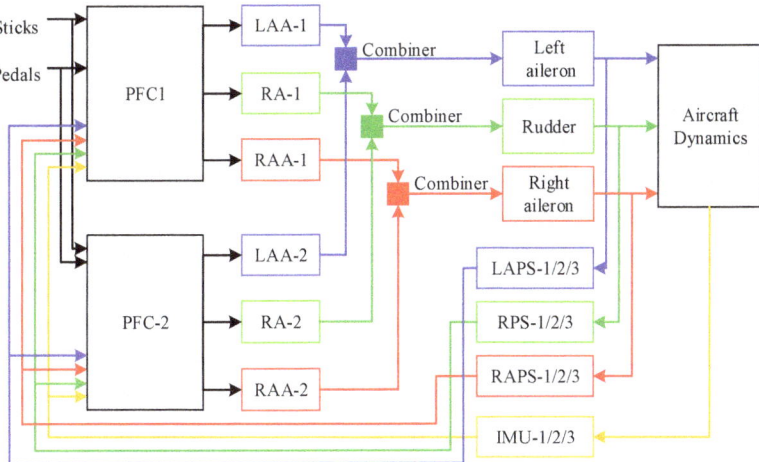

Figure 1. Architecture of the lateral-directional flight control system.

According to the dynamic models of the lateral-directional flight control system given in [6,28], the corresponding Simulink model can be built. Taking the roll control law of the PFC as an example, the mathematical model is expressed as

$$\begin{aligned} R_r(s) &= K_{r_1}\phi_c(s) + K_{r_2}R_b(s) + K_{r_3}\frac{s+z_r}{s+p_r}P_b(s) \\ \delta_{a_r^*}^{l(r)}(s) &= \left(P_r + \frac{I_r}{s} + D_r s\right)\left(R_r(s) + K_r\delta_a^{l(r)}(s)\right) \end{aligned} \quad (1)$$

where $\phi_c(\cdot)$ is the roll command, $R_b(\cdot)$ is the yaw rate, $P_b(\cdot)$ is the roll rate, $\delta_a^{l(r)}(\cdot)$ is the angle of the left (right) aileron, and $\delta_{a_r^*}^{l(r)}(\cdot)$ is the output response of the roll control law. The values of the coefficients are given as $K_{r_1} = 0.66$, $K_{r_2} = -0.145$ s, $K_{r_3} = 2.16$ s, $z_r = 11.1$ s^{-1}, $p_r = 25$ s^{-1}, $P_r = 0.45$ A, $I_r = 6$ A/s, $D_r = 0.01$ As, and $K_r = -1.33$ [6,28].

The corresponding nominal Simulink model is shown in Figure 2.

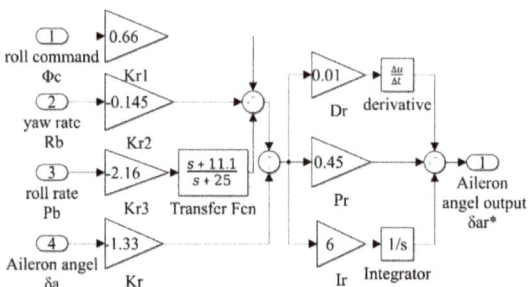

Figure 2. Simulink model of the roll control law of the primary flight computer (PFC).

By combining the Simulink models of all components, we can build the Simulink model of the lateral-directional flight control system, which is shown in Figure 3. In both Figures 2 and 3, the symbol "*" is used to note the output variables of the control laws.

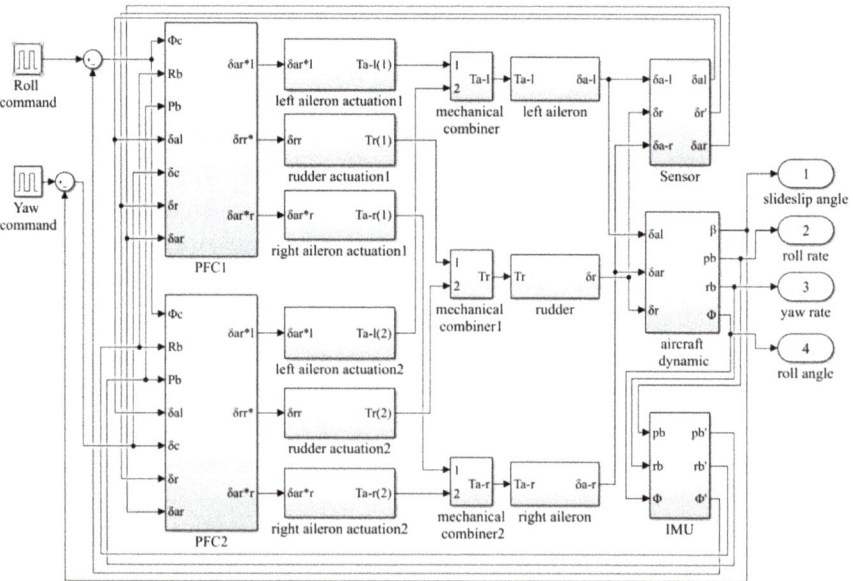

Figure 3. Nominal Simulink model of the lateral-directional flight control system.

2.2. The Definition of the FBW System Safety Requirement

When the aircraft is in a safe condition, the output response of each performance metric should be restricted to within an acceptable region, in which the requirements of the performance metrics can be satisfied. For the lateral-directional flight control system, the performance metrics include the sideslip angle $\beta(t)$, the roll rate $p_b(t)$, the yaw rate $r_b(t)$ and the roll angle $\phi(t)$, thus the safety requirement of the system can be defined as

$$\begin{cases} |\beta(t) - \beta_r(t)| \leq r_\beta \\ |p_b(t) - p_{b_r}(t)| \leq r_{p_b} \\ |r_b(t) - r_{b_r}(t)| \leq r_{r_b} \\ |\phi(t) - \phi_r(t)| \leq r_\phi \end{cases} \quad (2)$$

where $\beta_r(t)$, $p_{b_r}(t)$, $r_{b_r}(t)$ and $\phi_r(t)$ are the reference values of the sideslip angle, the roll rate, the yaw rate and the roll angle, respectively, which are the responses of these parameters in the failure-free configuration; r_β, r_{p_b}, r_{r_b} and r_ϕ are the thresholds of the sideslip angle, the roll rate, the yaw rate and the roll angle, respectively. Here, we take $r_\beta = 0.15$ rad, $r_{p_b} = 0.45$ rad/s, $r_{r_b} = 0.45$ rad/s and $r_\phi = 0.15$ rad from [28].

3. Extension of the Nominal Model

The objective of extending the nominal model is to inject different kinds of failure modes into the failure-free model. In this way, the performance responses of the FBW system under failed configurations can be obtained, and the unsafe conditions can be determined by comparing these responses with the performance thresholds given in Equation (2). In this section, the failure modes as well as their mathematical model are given, the Simulink tool is also used to build the models of different kinds of failure modes, and the fault injecting method is proposed.

3.1. Failure Modes and Their Mathematical Model

The failure modes of the components of the FBW system, as well as their failure rates, are given in Table 1.

Table 1. Component failure modes of the fly-by-wire (FBW) system [6].

Component	Failure Mode	Failure Mode Description	Failure Rate (1/h)
PFCs	Omission	Null output	2×10^{-7}
	Random	Random output between −5 and 5	1×10^{-7}
	Stuck	Output stuck at the last correct value	1×10^{-7}
	Delayed	Output delayed by 0.2 s	1×10^{-7}
Actuators	Omission	Null output	1×10^{-6}
	Stuck	Output stuck at the last correct value	1×10^{-6}
Control Surfaces	Stuck	Output stuck at the last correct value	1×10^{-8}
	Trailing	Output decided by the aero-dynamics	1×10^{-8}
Inertial Measurement Units (IMU)	Omission	Null output	4×10^{-7}
	Gain change	Output scaled by a factor of 1.5	3×10^{-7}
	Biased	Output biased by a factor of 0.3 deg	3×10^{-7}
Position Sensors	Omission	Null output	4×10^{-7}
	Gain change	Output scaled by a factor of 1.5	3×10^{-7}
	Biased	Output biased by a factor of 0.3 deg	3×10^{-7}

Reproduced with permission from (Dominguez-Garcia A. D., Kassakian J. G., Schindall J. E., et al.), (Reliability Engineering & System Safety); published by (Elsevier), 2008.

The dynamic behavior of each component can be expressed as the state-space function:

$$\begin{cases} \dot{x}(t) = Ax(t) + Bu(t) \\ y(t) = Cx(t) + Du(t) \end{cases} \quad (3)$$

where $x(t)$ is the vector of state variables, $u(t)$ is the vector of input variables, $y(t)$ is the vector of output variables, A is the system matrix, B is the control matrix, C is the output matrix and D is the feedforward matrix.

We make the assumption that the output of the failure-free configuration is $y(t)$, and the output of different failure modes can be expressed as

$$\bar{y}(t) = \begin{cases} 0 & \text{ommision} \\ rand & \text{random} \\ y(\tau_s) & \text{stuck} \\ y(t-\tau_d)1(t-\tau_d) & \text{delayed} \\ \vartheta \text{ or } \psi & \text{trailing} \\ Gy(t) & \text{gain change} \\ y(t)+b & \text{biased} \end{cases} \quad (4)$$

where rand is a random value expressing the random output, τ_s is the time point when "stuck" occurs, τ_d is the delayed time, ϑ is the pitching angle (for the ailerons), ψ is the heading angle (for the rudder), G is the gain change factor and b is the biased factor.

3.2. Failure Mode Modeling via Simulink

The Simulink tool is also used to build the models of the seven failure modes expressed via Equation (4), which are given in Figure 4.

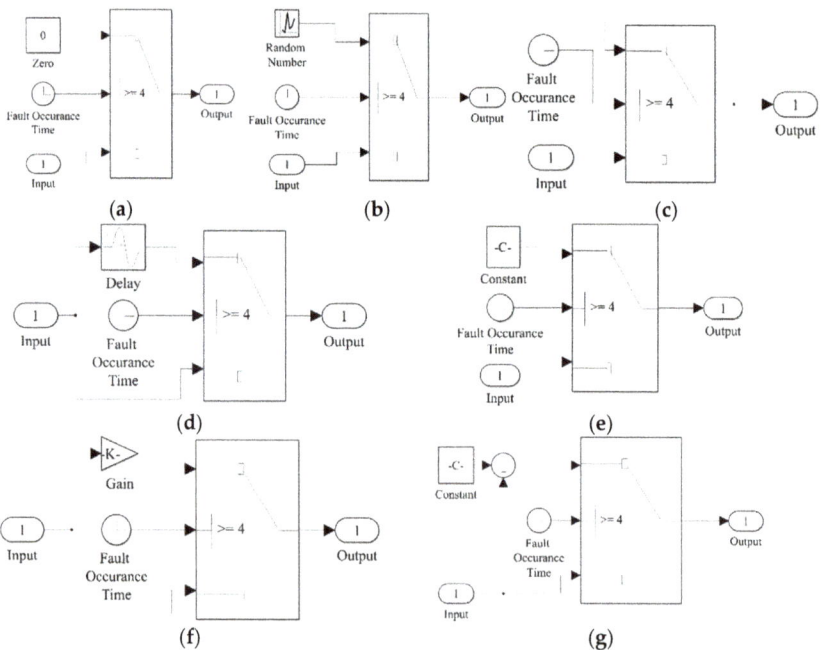

Figure 4. Simulink models of the seven failure modes: (**a**) "omission", (**b**) "random", (**c**) "stuck", (**d**) "delayed", (**e**) "trailing", (**f**) "gain change", and (**g**) "biased".

In all the models, the time point of fault injection is set as the 4th second, and the input is the failure-free response of the related component. Therefore, the output of each model before the 4th second will be the failure-free response of the related component as well. Figure 4a shows the Simulink model of "omission"; the "Output" block will connect to the "Zero" block 4 s later, thus the output after the 4th second will be null. Figure 4b shows the Simulink model of "random"; the "Output" block will connect to the "Random Number" block 4 s later, then the output after the 4th second will be a random value. Figure 4c shows the Simulink model of "stuck"; the "Output" block will connect to the output value of "4 s" after the 4th second, then the output will be stuck at its value of "4 s". Figure 4d shows the Simulink model of "delay"; the "Output" block will connect to the "Delay" block 4 s later, then the output after the 4th second will be delayed. Figure 4e shows the Simulink model of "trailing"; the "Output" block will connect to the "Constant" block 4 s later, then the output after the 4th second will be a constant value. The constant value C in the "Constant" block will be the heading angle ψ for the rudder and the pitching angle ϑ for the ailerons. Figure 4f shows the Simulink model of "gain change"; the "Output" block will connect to the "Gain" block 4 s later, then the output will be scaled by the gain change factor, which is shown as K in the "Gain" block. Figure 4g shows the Simulink model of "biased"; the "Output" block will connect to the "Sum" block 4 s later, then the output will be the summation of the initial output and the biased factor, which is shown as C in the "Constant" block.

3.3. The Fault Injecting Method

The fault injection is conducted by a "fault injector", which is a "Variant Subsystem" block placed between each component block and its output. Figure 5 shows the fault injector and its inner structure. The fault injector includes the blocks of the seven failure modes given in Figure 4a–g as well as an additional "normal" block. The "normal" block denotes the failure-free configuration; its inner "Input" connects "Output" directly.

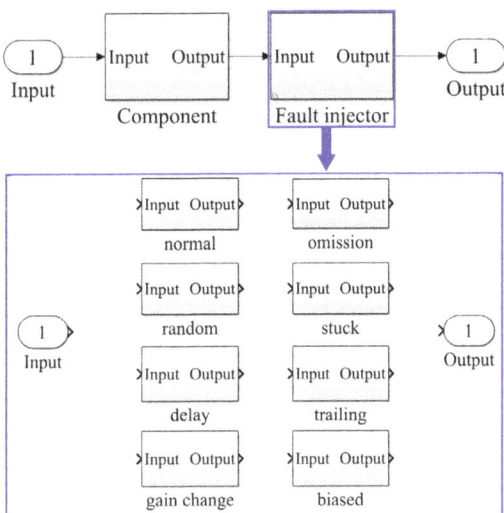

Figure 5. Simulink model of the fault injector.

The block parameters of the "fault injector" are used as the control variables for fault injections. By selecting different control variables, the responses of different kinds of configurations can be obtained.

Figure 6 shows the performance responses of the FBW system. Figure 6a shows the performance responses of the FBW system in the failure-free configuration when the roll command is a 0.2 rad, 0.1 Hz square wave. Figure 6b shows the system performance responses of the same input command when the "random" failure mode of one PFC has been injected. We can see that the difference between the roll angle response and its reference value has exceeded the threshold (0.15 rad). Thus, there is an unsafe condition. The results shown in these figures are similar to those in [6]. Figures 6a and 6b here correspond to Figures 2b and 6b in [6], respectively.

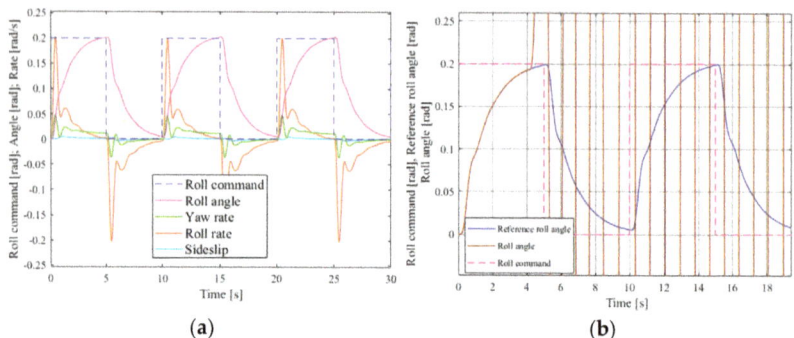

Figure 6. Performance responses of the FBW system: (**a**) the failure-free configuration; (**b**) the configuration of the "PFC random" failure mode.

4. Probability Calculation of Unsafe Conditions Based on Monte Carlo Simulation

Random numbers are used to denote the failure time for each component failure mode. These numbers will be the fault occurrence time of each corresponding failure mode. We order these numbers from smallest to largest, and inject the failure modes at their occurrence time one by one according to the fault injecting method. When the responses do not satisfy the safety requirement given in Equation (2), one simulation will terminate and a sample of time to the unsafe condition can be obtained. In terms of several time samples, the probability distributions of the time to unsafe conditions can be obtained. In this way, the probability of unsafe conditions can be calculated at different time points.

We make the assumption that the system is composed of n components and the ith ($i = 1, 2, \ldots, n$) component has m_i failure modes. N is used to denote the ordinal of the current simulation. T_N is the time to unsafe conditions obtained from the Nth simulation.

The step-by-step procedure of the Monte Carlo simulation-based method is as follows. The flowchart of the step-by-step procedure is shown in Figure 7.

Step 1   Initialization

Before the simulation starts, the ordinal of the current simulation is zero, namely $N = 0$; the corresponding time to unsafe conditions is also set as zero, namely $T_0 = 0$.

Step 2   Start a new simulation

Let $N = N + 1$, and the $(N + 1)$th simulation will start.

Step 3   Generate random numbers

Generate random numbers as the failure time for all failure modes. A common situation is that the failure time of each failure mode follows the exponential distribution, and the random number can be expressed as

$$t_{ij} = -\frac{1}{\lambda_{ij}} \ln(1 - r_{01}) \ (i = 1, 2, \cdots, n; \ j = 1, 2, \cdots, m_i) \tag{5}$$

where λ_{ij} is the constant failure rate of the ith component's jth failure mode, r_{01} is a random number generated from the uniform distribution defined over the interval (0, 1), and t_{ij} is the failure time of the ith component's jth failure mode. For other distributions, we can also obtain a random number by their cumulative density function.

Step 4   Obtain the failure time for each component

As each component has several failure modes and the component will fail when one of the failure modes occurs, the failure time of each component will be the minimum value of all its failure modes' occurrence time. We have

$$t_{ij_i} = \min_{j=1,2,\cdots,m_i} (t_{ij}) \tag{6}$$

where t_{ij_i} is the failure time of the ith component, and j_i is the ordinal of the ith component's failure mode that has the minimum failure time.

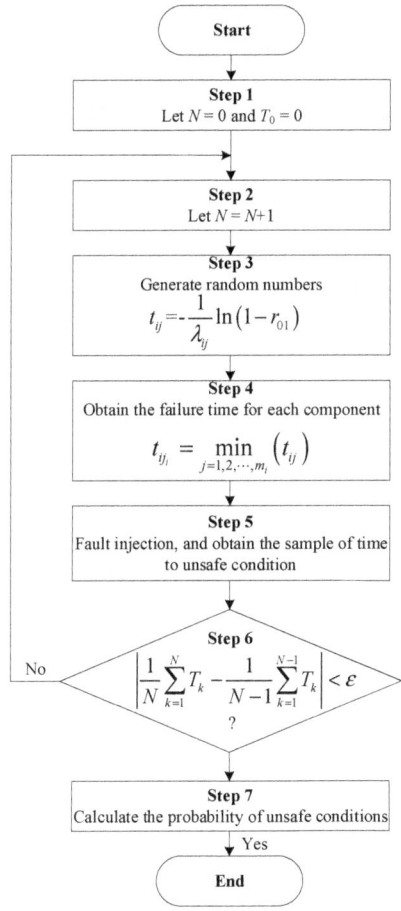

Figure 7. Flowchart of the Monte Carlo simulation.

Step 5   Inject the fault of each component

The failure time for each component is ordered from smallest to largest, and the ith component's j_ith failure mode will be injected into the nominal model at t_{ij_i} one by one according to the fault injection method proposed in Section 3. When the unsafe condition occurs, the corresponding t_{ij_i} will be the Nth sample of the time to unsafe conditions obtained from the Nth simulations (T_N).

Step 6   Decide whether the simulations will end or not

The simulations will stop when the mean time to unsafe conditions converges, thus the simulation ending criteria can be expressed as

$$\left|\frac{1}{N}\sum_{k=1}^{N}T_k - \frac{1}{N-1}\sum_{k=1}^{N-1}T_k\right| < \varepsilon \qquad (7)$$

where ε is an arbitrarily small positive real, and we usually let $\varepsilon = 0.1$.

If Equation (7) can be satisfied, the simulation procedure will go to Step 7; otherwise, it will go back to Step 2.

Step 7 Calculate the probability of unsafe conditions

According to the N samples of time to unsafe conditions, we perform distribution selections, parameter estimations and goodness-of-fit tests. Then, we can obtain the probability distribution of time to unsafe conditions, and the probability of the time to unsafe conditions can be calculated.

5. Case Study and Discussion

The lateral-directional flight control system given in Figure 1 is applied as the case study here. The failure rates of all components' failure modes are already given in Table 1.

By conducting the Monte Carlo simulation procedure given in Section 4, we obtain nearly 2000 samples of time to unsafe conditions, whose histogram is shown in Figure 8. According to the shape of the histogram, it can be estimated that these samples may follow Weibull distribution or Lognormal distribution. For the kth sample T_k, the estimate of the related cumulative distribution function (CDF) can be expressed as

$$\hat{F}(T_k) = \frac{k}{N} \qquad (8)$$

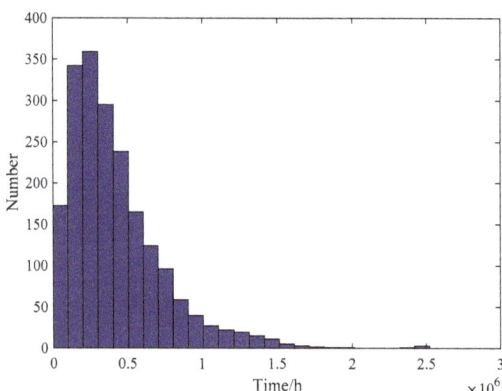

Figure 8. Histogram of time samples.

The CDF of Weibull distribution is expressed as

$$F(t) = 1 - \exp\left[-\left(\frac{t}{\alpha}\right)^\beta\right] \qquad (9)$$

We let

$$\begin{cases} x_k = \ln T_k \\ y_k = \ln\left\{\ln\left[\frac{1}{1-\hat{F}(T_k)}\right]\right\} \end{cases} \qquad (10)$$

Thus, if the samples follow a Weibull distribution, the plotting of x_k versus y_k expressed by (10) should look like a straight line. This plotting is shown in Figure 9a.

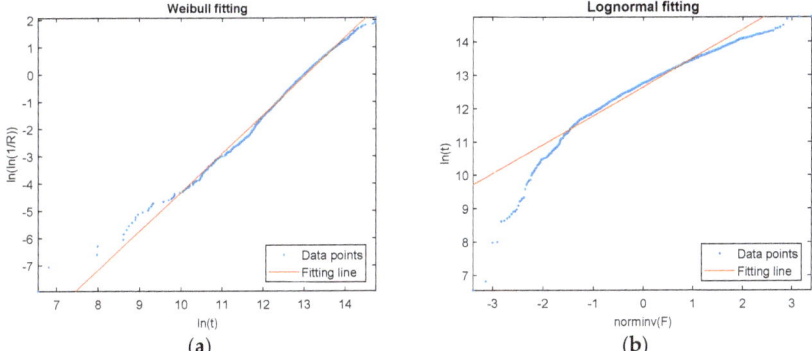

Figure 9. (a) The plotting of Weibull fitting; (b) The plotting of lognormal fitting.

Likewise, the CDF of a lognormal distribution is expressed as

$$F(t) = \Phi\left(\frac{\ln t - \mu}{\sigma}\right) \tag{11}$$

where $\Phi(\cdot)$ is the CDF of the standard normal distribution. Let

$$\begin{cases} x_k = \Phi^{-1}\left[\hat{F}(T_k)\right] \\ y_k = \ln T_k \end{cases} \tag{12}$$

Thus, if the samples follow a lognormal distribution, the plotting of x_k versus y_k expressed by (12) should look like a straight line. This plotting is shown in Figure 9b.

It is shown that the plotting of x_k versus y_k in Figure 9a is much more like a straight line compared with Figure 9b. Therefore, the Weibull distribution is preferred for these samples.

By using the maximum likelihood estimation, we can obtain $\alpha = 2.6062 \times 10^5$ h and $\beta = 1.211$. In addition, the Kolmogorov–Smirnov test shows that we should not reject the hypothesis that the samples of time to unsafe conditions are following the Weibull distribution with $\alpha = 2.6062 \times 10^5$ and $\beta = 1.211$ [29]. Hence, the probability of the unsafe condition at time t can be expressed as

$$P(t) = 1 - \exp\left[-\left(\frac{t}{2.6062 \times 10^5}\right)^{1.211}\right] \tag{13}$$

For the FBW system, the scheduled inspection interval is set as 500 flight hours, which means the function of the FBW system will be thoroughly checked every 500 flight hours and the FBW system will be restored to the perfect condition if it is degraded. Thus, we can obtain the probability of the unsafe condition in a scheduled inspection interval as 5.1234×10^{-4}. Therefore, the average probability of the unsafe condition per flight hour is 1.0247×10^{-6}.

In [6], the Markov process is applied to calculate both the upper and lower bounds of the probability of the unsafe condition in the scheduled inspection interval, which are 5.1178×10^{-4} and 5.8211×10^{-4}, respectively, for this case. We have also used the state enumerating method to obtain the minimal cut sets that cause unsafe conditions. In the state enumerating method, we simulate the model of the FBW system with all component failure modes and their combinations, and the minimal cut sets can be obtained in terms of the system response. Moreover, the probability of the unsafe condition or its interval can be calculated via the probability additive formula. The upper and lower bounds of the probability of the unsafe condition calculated by the minimal cut sets are 5.1088×10^{-4} and 5.2084×10^{-4}, respectively. The results of the three methods are given in Table 2.

Table 2. Probability of the unsafe condition calculated by different methods.

Monte Carlo Simulation	Markov Process	State Enumerating
5.1234×10^{-4}	(5.1178×10^{-4}, 5.8211×10^{-4})	(5.1088×10^{-4}, 5.2084×10^{-4})

Table 2 shows that the result of the Monte Carlo simulation method is located just between the upper and lower limits obtained from both the Markov process and the state enumerating, which illustrates the accuracy of our Monte Carlo simulation method. The advantages and disadvantages of the above-mentioned three methods are discussed in Table 3.

Table 3. Comparison of the three methods.

Methods	Advantages	Disadvantages
Monte Carlo simulation	• The cumbersome work of building a Markov model can be avoided. • The simulation algorithm does not have to be modified when the system is modified.	• Time-consuming for large systems. • A stochastic method, and it does not achieve exact results.
Markov process	• The exact value or interval of the probability can be calculated. • More efficient than Monte Carlo simulation after the Markov model has been built.	• Faced with the state explosion problem. • A new Markov model is needed when the system is modified.
State enumerating	• The probability additive formula is used to calculate the exact value or interval of the probability for an unsafe condition.	• Impossible to enumerate all the states for large systems. • A new formula is required when the system is modified.

6. Conclusions

In this study, an MBSA approach is proposed based on the system development model built by Simulink for the FBW system, and Monte Carlo simulation is used to obtain the probability of unsafe conditions. Our proposed approach has the following advantages:

(1) The performance responses of the system with fault injection are used to determine the effect of component failures or failure combinations on system safety. Compared with the traditional safety analysis methods, the determination of failure effects is no longer dependent on analysts' specific knowledge about the aircraft system.

(2) By using the Monte Carlo simulation method, the cumbersome work of building a Markov model can be avoided, and the state explosion problem of the Markov process can be resolved to some extent. Additionally, when the system is modified or changed, the Markov model should be rebuilt; however, our Monte Carlo simulation algorithm should not be updated.

(3) By using the system development model built by Simulink, the safety assessment can be carried out in the early stage of system development. Moreover, it is easy to update the safety assessment results with the design improvement of the FBW system.

Author Contributions: Conceptualization, Z.L.; methodology, Z.L. and L.Z.; software, Z.L. and X.L.; formal analysis, Z.L., L.Z. and X.L.; data curation, L.Z. and L.D.; writing—original draft preparation, Z.L. and L.D.; project administration, Z.L.; funding acquisition, Z.L. All authors have read and agreed to the published version of the manuscript.

Funding: This research was supported in part by the National Natural Science Foundation of China under Grant U1733124, in part by the Aeronautical Science Foundation of China under Grant 20180252002, and in part by the Fundamental Research Funds for the Central Universities of NUAA under Grant 3082018NT2018019.

Conflicts of Interest: The authors declare no conflict of interest.

References

1. SAE International S-18 Committee. *ARP4754A: Guidelines for Development of Civil Aircraft and Systems*; Society of Automotive Engineers: Warrendale, PA, USA, 2010; pp. 31–37.
2. SAE International S-18 Committee. *ARP4761: Guidelines and Methods for Conducting the Safety Assessment Process on Civil Airborne System and Equipment*; Society of Automotive Engineers: Warrendale, WA, USA, 1996; pp. 22–28.
3. Joshi, A.; Heimdahl, M.P.E. Model-Based Safety Analysis of Simulink Models Using SCADE Design Verifier. In Proceedings of the 24th International Conference on Computer Safety, Reliability and Security, Fredrikstad, Norway, 28–30 September 2005.
4. Joshi, A.; Heimdahl, M.P.E. *NASA/CR-2006-213953: Model-Based Safety Analysis*; NASA: Washington, DC, USA, 2006; pp. 1–22.
5. Chen, L.; Jiao, J.; Zhao, T. Review for model-based safety analysis of complex safety-critical system. *J. Syst. Eng. Electron.* **2017**, *39*, 1287–1291.
6. Dominguez-Garcia, A.D.; Kassakian, J.G.; Schindall, J.E.; Zinchuk, J.J. An integrated methodology for the dynamic performance and reliability evaluation of fault-tolerant systems. *Reliab. Eng. Syst. Saf.* **2008**, *93*, 1628–1649. [CrossRef]
7. Dominguez-Garcia, A.D.; Kassakian, J.G.; Schindall, J.E.; Zinchuk, J.J. On the use of behavioral models for the integrated performance and reliability evaluation of fault-tolerant avionics systems. In Proceedings of the IEEE/AIAA 25th Digital Avionics Systems Conference, Portland, OR, USA, 15–18 October 2006.
8. Chiacchio, F.; D'Urso, D.; Compagno, L.; Pennisi, M.; Pappalardo, F.; Manno, G. SHyFTA, a Stochastic Hybrid Fault Tree Automaton for the modelling and simulation of dynamic reliability problems. *Expert Syst. Appl.* **2016**, *47*, 42–57. [CrossRef]
9. Ping, M.; Zhang, X.; Gao, Z.; Li, W. Simulation model development of three-stage synchronous generator for aircraft power systems based on modelica. In Proceedings of the 19th International Conference on Electrical Machines and Systems, Chiba, Japan, 13–16 November 2016.
10. Hu, W.; Wei, X.; Xie, X. Modeling and Simulation of Micro Hydraulic Transducer Based on Component Library. *Adv. Mater. Res.* **2012**, *505*, 106–111. [CrossRef]
11. Goncalves, P.; Sobral, J.; Ferreira, L.A. Unmanned aerial vehicle safety assessment modelling through Petri Nets. *Reliab. Eng. Syst. Saf.* **2017**, *167*, 383–393. [CrossRef]
12. Flammini, F.; Marrone, S.; Mazzocca, N.; Vittorini, V. A new modeling approach to the safety evaluation of N-modular redundant computer systems in presence of imperfect maintenance. *Reliab. Eng. Syst. Saf.* **2009**, *94*, 1422–1432. [CrossRef]
13. Kleyner, A.; Volovoi, V. Application of Petri nets to reliability prediction of occupant safety systems with partial detection and repair. *Reliab. Eng. Syst. Saf.* **2010**, *95*, 606–613. [CrossRef]
14. Wang, H.; Ning, B.; Chen, T. Route safety verification of train control system by FTA modeling in SCADE. In Proceedings of the 2018 IEEE International Conference on Intelligent Transportation Systems, Maui, HI, USA, 4–7 November 2018.
15. Wang, H.; Zhong, D.; Zhao, T. Integrating Model Checking with SysML in Complex System Safety Analysis. *IEEE Access* **2019**, *7*, 16561–16571. [CrossRef]
16. Clegg, K.; Mole, L.; Stamp, D. A SysML Profile for Fault Trees-Linking Safety Models to System Design. In Proceedings of the 38th International Conference on Computer Safety, Reliability, and Security, Turku, Finland, 11–13 September 2019; pp. 85–93.
17. Bozzano, M.; Cimatti, A.; Katoen, J.P. Safety, Dependability and Performance Analysis of Extended AADL Models. *Comput. J.* **2011**, *54*, 754–775. [CrossRef]

18. Correa, T.; Becker, L.B.; Farines, J.M.; Bodeveix, J.P.; Filali, M.; Vernadat, F. Supporting the Design of Safety Critical Systems Using AADL. In Proceedings of the 15th IEEE International Conference on Engineering of Complex Computer Systems, Oxford, UK, 22–26 March 2010.
19. Gu, B.; Dong, Y.; Wei, X. A Qualitative Safety Analysis Method for AADL Model. In Proceedings of the 2014 IEEE 8th International Conference on Software Security and Reliability-Companion, San Francisco, CA, USA, 30 June–2 July 2014.
20. Batteux, M.; Prosvirnova, T.; Rauzy, A.; Kloul, L. The AltaRica 3.0 Project for Model-Based Safety Assessment. In Proceedings of the 11th IEEE International Conference on Industrial Informatics, Bochum, Germany, 29–31 July 2013.
21. Bozzano, M.; Cimatti, A.; Lisagor, O.; Mattarei, C.; Mover, S.; Roveri, M.; Tonetta, S. Safety assessment of AltaRica models via symbolic model checking. *Sci. Comput. Program.* **2015**, *98*, 464–483. [CrossRef]
22. Ding, Y.; Li, W.; Zhong, D.; Huang, H.; Zhao, Y.; Xu, Z. System States Transition Safety Analysis Method Based on FSM and NuSMV. In Proceedings of the 2nd International Conference on Management Engineering, Software Engineering and Service Sciences, Wuhan, China, 13–15 January 2018.
23. Bozzano, M.; Villafiorita, A. The FSAP/NuSMV-SA safety analysis platform. *Int. J. Softw. Tools Technol. Transf.* **2007**, *9*, 5–24. [CrossRef]
24. Lipaczewski, M.; Ortmeier, F.; Prosvirnova, T. Comparison of modeling formalisms for Safety Analyses: SAML and AltaRica. *Reliab. Eng. Syst. Saf.* **2015**, *140*, 191–199. [CrossRef]
25. Septavera, S.; Yiannis, P. Intergrating model checking with Hip-HOPS in model-based safety analysis. *Reliab. Eng. Syst. Saf.* **2015**, *135*, 64–80.
26. Kabir, S.; Walker, M.; Papadopoulos, Y. Dynamic system safety analysis in HiP-HOPS with Petri Nets and Bayesian Networks. *Saf. Sci.* **2018**, *105*, 55–70. [CrossRef]
27. Fukano, Y. Development of safety assessment methodology on fuel element failure propagation in SFR and its application to Monju. *J. Nucl. Sci. Technol.* **2015**, *52*, 178–192. [CrossRef]
28. McRuer, D.; Myers, T.; Thompson, P. Literal singular-value-based flight control system design techniques. *J. Guid. Control Dyn.* **1989**, *12*, 913–919. [CrossRef]
29. Birolini, A. *Reliability Engineering: Theory and Practice*, 7th ed.; Springer: Berlin, Germany, 2014; pp. 533–558.

© 2020 by the authors. Licensee MDPI, Basel, Switzerland. This article is an open access article distributed under the terms and conditions of the Creative Commons Attribution (CC BY) license (http://creativecommons.org/licenses/by/4.0/).

MDPI
St. Alban-Anlage 66
4052 Basel
Switzerland
Tel. +41 61 683 77 34
Fax +41 61 302 89 18
www.mdpi.com

Processes Editorial Office
E-mail: processes@mdpi.com
www.mdpi.com/journal/processes

www.ingramcontent.com/pod-product-compliance
Lightning Source LLC
LaVergne TN
LVHW071933080526
838202LV00064B/6605